Fundamentals of Electronic Engineering

Rajendra Prasad
*Former Professor and Head,
Department of Electrical Engineering,
National Institute of Technology, Patna*

Andover • Melbourne • Mexico City • Stamford, CT • Toronto • Hong Kong • New Delhi • Seoul • Singapore • Tokyo

Fundamentals of Electronic Engineering

Rajendra Prasad

© 2011 Cengage Learning India Pvt. Ltd

ALL RIGHTS RESERVED. No part of this work covered by the copyright herein may be reproduced, transmitted, stored or used in any form or by any means graphic, electronic, or mechanical, including but not limited to photocopying, recording, scanning, digitizing, taping, Web distribution, information networks, or information storage and retrieval systems, without the prior written permission of the publisher.

> For permission to use material from this text or product,
> submit all requests online at
> **www.cengage.com/permissions**
>
> Further permissions questions can be emailed to
> **India.permission@cengage.com**

ISBN 13: 978-81-315-1545-7
ISBN 10: 81-315-1545-1

Cengage Learning India Pvt. Ltd.
418, F.I.E., Patparganj
Delhi 110092

Cengage Learning is a leading provider of customized learning solutions with office locations around the globe, including Andover, Melbourne, Mexico City, Stamford (CT), Toronto, Hong Kong, New Delhi, Seoul, Singapore, and Tokyo. Locate your local office at: **www.cengage.com/global**

Cengage Learning products are represented in Canada by Nelson Education Ltd.

For product information, visit **www.cengage.co.in**

Printed in India
First Impression 2011

To

my wife
Shanti

my beloved granddaughters
Aishwarya and Ananya

and
my charming grandson
Ayaan

PREFACE

Today, electronic engineering is a very important field of engineering. It is a basic course of electronic engineering and is one of the core subjects being taught to engineering students of almost all branches. The book *Fundamentals of Electronic Engineering* has been designed to cover the syllabus of basic electronic engineering of technical institutions and universities in India.

The book has a number of typical solved problems. The theoretical statements have been well illustrated by numerical examples. Each chapter at the end has a number of review questions, many solved problems, and exercises for the students to practice.

The book encompasses 13 chapters.

Chapters 1–6 discuss semiconductor devices. Starting with the atomic structure of the semiconductor materials—silicon and germanium—Chapter 1 deals with conduction in semiconductor materials and junction diodes, which are used in rectifier circuits. This chapter also discusses the properties and parameters of the junction diodes, special types of diodes, and their applications.

Chapter 2 is devoted to transistors and discusses their characteristics. Transistors discussed include bipolar transistors (pnp and npn), field-effect transistor (FET), metal-oxide semiconductor FET (MOSFET), silicon-controlled rectifiers (SCRs), and unijunction transistors.

Chapter 3 looks at different applications of semiconductor junction diodes. Different applications of diodes include different rectifier circuits, clipping circuits, comparators, and sampling gates.

Chapter 4 deals with various types of amplifier circuits using transistors. Amplifiers include the RC-coupled amplifier and different types of feedback amplifiers.

Chapter 5 discusses the operational amplifier and its various uses. Since the differential amplifier is the heart of an operational amplifier, it is discussed in the beginning of this chapter. This chapter also includes power amplifiers, push-pull amplifiers, and class B and class C amplifiers.

Chapter 6 deals with oscillator circuits and functional generators.

Chapter 7 concentrates on attenuators and filters, which are very important components of electronic circuits.

Chapter 8 is devoted to discussing different types of cathode-ray oscilloscope (CRO) and their applications.

Electrical power supply plays an important role in our daily life. However, the power requirements of different users are vastly different. The power supplies required may be variable voltage d.c and a.c. supplies, and for this, converters and inverters, respectively, are used. A popular device which converts a.c. voltage to d.c. voltage is a switched mode power supply (SMPS). We, generally, need an a.c. power supply as a stand-by source when mains power supply is not available. These a.c. power supplies are known as uninterrupted power supply (UPS). These power supplies use SCRs as main components. Chapter 9 discusses all these special power supplies.

The ever-growing importance of digital devices and techniques in engineering design and practice necessitates a discussion on digital devices and techniques in any fundamental book of electronic engineering. The introduction to digital systems is covered in three chapters. Chapter 10 discusses binary algebra in detail. Logic gates such as NOT, AND, NOR, NAND, and XOR are discussed in this chapter. The various number systems and their conversions are also discussed. This chapter includes techniques for simplifying logical functions as well. A rather thorough description of principal components used in digital systems is presented in Chapter 11. Digital devices are divided into two groups, namely, combinational logic circuit and sequential logic circuit. Sequential logic circuits include, in addition to the combination logic circuit, the memory devices such as flip-flops, counters, registers, and random access memories. The last chapter on digital systems is concluded in Chapter 12 by treating at some length the operation of microprocessors, minicomputers, and microcontrollers.

Chapter 13 gives a brief introduction to electrical communication systems. This chapter discusses the analogue, digital, and optical communication systems.

I am thankful to my colleagues of the Department of Electrical Engineering, National Institute of Technology (NIT), Patna, for their suggestions and encouragement. I am thankful to my wife and children to whom I could not give time due to engagement in writing of this book. I mention a special thanks to my son Akhilendra for his suggestions and help in preparing the manuscript.

I hope this textbook will help the readers in acquiring an integrated and in-depth understanding of the subject. Any constructive suggestions for improving the content of the book are warmly appreciated.

Rajendra Prasad

TABLE OF CONTENTS

Preface v

1. SEMICONDUCTOR DEVICES 1

1.1 Semiconductors 2
1.2 Conduction in Semiconductors 4
1.3 p-n Junctions or Diodes 7
1.4 p-n Junction as a Rectifier 10
1.5 *V-I* Characteristic 12
1.6 Diode Parameters 14
1.7 Special Types of Diodes 18
1.8 Review Questions 22
1.9 Solved Problems 23
1.10 Exercises 31

2. TRANSISTORS 33

2.1 Bipolar Junction Transistors 34
2.2 Static Characteristics of Transistors 37
2.3 Low Frequency Analysis of Transistors 41
2.4 Field-Effect Transistor 45
2.5 Metal Oxide Semiconductor FET 51
2.6 Silicon-Controlled Rectifiers 54
2.7 Unijunction Transistors 60
2.8 Review Questions 62
2.9 Solved Problems 63
2.10 Exercises 71

3. DEVICES USING DIODE AS AN ELEMENT   73

3.1 Analysis of Diode Circuit   74

3.2 Piecewise Linear Model of a Diode   76

3.3 Clipping Circuits   78

3.4 Comparators   81

3.5 Sampling Gate   81

3.6 Rectifiers   84

3.7 Other Rectifier Circuits   92

3.8 Additional Diode Circuits   94

3.9 Review Questions   96

3.10 Solved Problems   97

3.11 Exercises   116

4. ELECTRONIC AMPLIFIERS   119

4.1 Basic Transistor Amplifiers   120

4.2 Biasing of Transistor Amplifier   125

4.3 Multistage Amplifiers   132

4.4 RC-Coupled Amplifier   139

4.5 Feedback Amplifiers   141

4.6 Voltage Series Feedback Amplifier   153

4.7 Current Series Feedback Amplifier   159

4.8 Voltage Shunt Feedback Amplifier   164

4.9 Current Shunt Feedback Amplifier   168

4.10 Review Questions   173

4.11 Solved Problems   173

4.12 Exercises   198

5. SPECIAL AMPLIFIERS   201

5.1 Differential Amplifier   202

5.2 Operational Amplifier   209

5.3 Linear Applications of Operational Amplifiers   214

5.4 Power Amplifier   220

5.5 Review Questions   233

5.6 Solved Problems   233

5.7 Exercises   246

6. WAVE GENERATORS   249

6.1 Oscillators   250

6.2 Multivibrators   260

6.3 Blocking Oscillators   265

6.4 Square Wave Generators   270

6.5 Pulse and Pulse Wave Generators   273

6.6 Signal Generators   277

6.7 Function Generators   280

6.8 Sweep-Frequency Generator   281

6.9 Review Questions   283

6.10 Solved Problems   284

6.11 Exercises   291

7. ATTENUATORS AND FILTERS   295

7.1 Attenuators   296

7.2 Symmetrical Attenuators   300

7.3 Padding Sources and Loads   308

7.4 Passive Filters   310

7.5 RC Filters   312

7.6 LC Filters   316

7.7 Active Filters   319

7.8 Active Resonant Bandpass Filter   325

7.9 Review Questions   330

7.10 Solved Problems   330

7.11 Exercises   341

8. CATHODE-RAY OSCILLOSCOPE   343

    8.1   Introduction   344

    8.2   Cathode-Ray Tube   345

    8.3   Focusing Devices   347

    8.4   Deflecting Forces   349

    8.5   Vertical Deflection System   352

    8.6   Horizontal Deflection System   355

    8.7   Synchronization of Sweep   358

    8.8   Methods of Improving Sweep Linearity   360

    8.9   CRO Block Diagram   362

    8.10   Applications of Cathode-Ray Oscilloscopes   364

    8.11   Special Purpose CROs   371

    8.12   Review Questions   377

    8.13   Solved Problems   378

    8.14   Exercises   382

9. ELECTRICAL POWER SUPPLIES   385

    9.1   Introduction   386

    9.2   Controlled Rectifiers   387

    9.3   Inverters   417

    9.4   A.C. Voltage Controllers   433

    9.5   D.C. Voltage Regulators   444

    9.6   Choppers   450

    9.7   Switched Mode Power Supply   457

    9.8   Uninterrupted Power Supply   462

    9.9   Review Questions   464

    9.10   Solved Problems   466

    9.11   Exercises   487

10. DIGITAL SYSTEMS   491

    10.1   Introduction   492

    10.2   Binary Logic and Logic Gates   493

10.3 Number Systems 501

10.4 Boolean Algebra 513

10.5 Simplification of Logical Functions 518

10.6 Review Questions 527

10.7 Solved Problems 528

10.8 Exercises 536

11. COMPONENTS OF DIGITAL SYSTEMS 539

11.1 Introduction 540

11.2 Encoders 543

11.3 Adders 544

11.4 Subtractors 550

11.5 Decoders and Demultiplexers 552

11.6 Data Selectors/Multiplexers 556

11.7 ROM Unit 558

11.8 Flip Flops 561

11.9 Registers 568

11.10 Counters 572

11.11 RAM Unit 581

11.12 D/A Converter 584

11.13 A/D Converter 590

11.14 Review Questions 593

11.15 Solved Problems 595

11.16 Exercises 621

12. DIGITAL COMPUTER 625

12.1 Introduction 626

12.2 Central Processing Unit 628

12.3 Computer Registers 633

12.4 Control Unit 636

12.5 Memory 637

xii Table of Contents

- 12.6 Input-Output Devices 642
- 12.7 Computer Instructions 648
- 12.8 Instruction Cycle 652
- 12.9 Computer Programming 653
- 12.10 Microprocessor 661
- 12.11 Microcomputer 671
- 12.12 Microcontroller 673
- 12.13 Review Questions 677
- 12.14 Solved Problems 678
- 12.15 Exercises 687

13. COMMUNICATION SYSTEMS 689

- 13.1 Introduction 690
- 13.2 Modulation and Demodulation 697
- 13.3 Transmitters 703
- 13.4 Receivers 705
- 13.5 Digital Communication 707
- 13.6 Quantization 711
- 13.7 Digital Modulation 715
- 13.8 Multiplexing 719
- 13.9 Optical Fibre and Communication Systems 722
- 13.10 Review Questions 732
- 13.11 Solved Problems 733
- 13.12 Exercises 738

Answers to Exercises 741

Index 745

1
SEMICONDUCTOR DEVICES

Outline

1.1 Semiconductors 2
1.2 Conduction in Semiconductors 4
1.3 p-n Junctions or Diodes 7
1.4 p-n Junction as a Rectifier 10
1.5 V-I Characteristic 12
1.6 Diode Parameters 14
1.7 Special Types of Diodes 18
1.8 Review Questions 22
1.9 Solved Problems 23
1.10 Exercises 31

1

1.1 SEMICONDUCTORS

Metals having conductivity considerably higher than that of insulators but considerably lower than that of good conductors are called semiconductors. Germanium (Ge) and silicon (Si) are widely used semiconductors. At room temperature, the conductivity of intrinsic or pure germanium is of the order of 2 mho/m. In comparison, copper, a good conductor, has the conductivity of the order of 5×10^5 mho/m, whereas that of glass, a good insulator, is less than 1×10^{-6} mho/m. As a result of the intensive studies on the properties of semiconductors which started in the late forties, semiconductor devices such as semiconductor diodes and transistors were developed. Nowadays, semiconductor devices are used extensively in the electronic industry. A semiconductor diode has properties similar to that of a vacuum diode. A transistor has substituted a grid-controlled vacuum tube in amplifiers. Semiconductor devices have completely replaced vacuum tubes. The following properties made semiconductor devices popular:

1. Because of the absence of a heating element, they do not require heating power and warm-up time.
2. They are relatively very small in size and, thus, are well suited for portable instruments.
3. They consume very less power.
4. They have very long life.

Both silicon and germanium—the most extensively used materials in semiconductor devices—have a valency of 4 and the same crystal structure as diamond. Germanium has atomic number 32, and silicon has atomic number 14. Their atomic weights are 72.60 and 28.08, respectively. There are four valence electrons in the outer subshells of germanium and silicon. Pure germanium and silicon crystals are electrically neutral because each valence electron of an atom within a crystal forms a covalent bond with one valence electron of the neighbouring atom.

At low temperature, the electrons of a semiconductor are strongly bounded to the atom, and almost no conduction takes place. The bonding force is overcome by means of the thermal energy at temperatures greater than 0 K. At 300 K (room temperature), the measured resistivity of pure or intrinsic germanium is 47 Ω-cm and that of intrinsic silicon is several hundred ohm-centimetres. As the temperature of the semiconductor is increased, the total energy of the system as well as the energies of individual electrons increases. The energies of individual electrons in the crystal may be widely different. Only a fraction of the total number of electrons may have enough energy to break the covalent bonds. This fraction is very small at low temperature. As temperature is increased, this fraction increases and more electrons become free. Hence, the conductivity of a semiconductor increases with an increase in temperature.

The conductivity of a semiconductor at room temperature can also be increased by optical illumination. The incidence of photons in the visible range breaks the covalent bonds in the semiconductor.

A very common method used to produce a large number of carriers is to add a small and measured amount of chemical impurity to the intrinsic semiconductor. This impure semiconductor is called the extrinsic semiconductor. The impurities must be of right type and in proper quantity. The amount of impurity required to be added to obtain large conductivity is very small. It may be of the order of 1 atom per million atoms of the intrinsic semiconductor. Since germanium and silicon are tetravalent, the impurity atom required may be either pentavalent or trivalent. The typical pentavalent materials used are arsenic (As) and antimony (Sb), and trivalent materials are gallium (Ga) and boron (B).

The pure semiconductor is electrically neutral, and there is no free electron. A schematic representation of the covalent bonds existing between the valence electrons of the silicon atom is shown in Fig. 1.1. Of the 14 orbital electrons, only four in the outermost orbit are shown. For consistency the nucleus is also shown with the corresponding charge of +4.

When a pentavalent impurity (say arsenic) is added to the silicon crystal, the atom in the crystal lattice structure changes, as shown in Fig. 1.2. Four of the five valence electrons

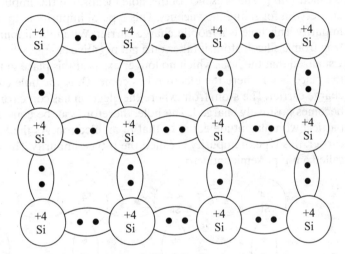

FIGURE 1.1 A neutral silicon lattice structure.

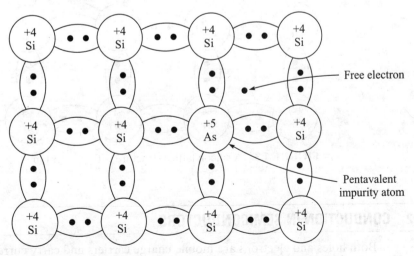

FIGURE 1.2 A silicon lattice structure with pentavalent impurity.

make covalent bonds with valence electrons of neighbouring silicon atoms. The fifth valence electron of arsenic is unloaded and is free to move randomly. As a pentavalent impurity donates one charge carrier in the form of an electron to the semiconductor, this type of impurity is called the donor impurity. As a negative charge carrier is added by this impurity, it is also called the n-type impurity, and the semiconductor formed is called the n-type semiconductor.

A second type of extrinsic semiconductor results when a trivalent impurity (say gallium) is added to pure silicon (or germanium). Since gallium has only three valence electrons in the outer orbit, these valence electrons make covalent bonds with electrons of neighbouring atoms of silicon. The lattice structure of this extrinsic silicon is shown in Fig. 1.3. Note that only three covalent bonds are complete. One more electron is needed to complete the fourth bond. This means that there is a vacancy for an electron. This vacancy is called a *hole*. The presence of this hole means that this impurity atom is ready to accept an electron for filling the vacancy. This type of impurity is, therefore, called the acceptor impurity. This hole is a mobile charge carrier. When the impurity atom receives an electron from a silicon atom, it takes on a net negative charge, which is immobile. So, it can be visualized that the hole, which no longer exists at this place in the material, has moved to the place from where the electron has come. Thus, the hole can be considered a mobile charge carrier. The atom from where the electron has come (or where the hole has gone) becomes positively charged. The hole, therefore, can be considered a positively charged particle with magnitude equal to that of an electron. As the hole is a positively charged carrier, this type of impurity is called the p-type impurity. The extrinsic semiconductor is called a p-type semiconductor.

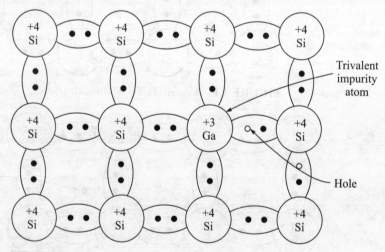

FIGURE 1.3 A silicon lattice structure with trivalent impurity.

1.2 CONDUCTION IN SEMICONDUCTORS

Both holes and electrons are mobile charge carriers and carry current. They are, therefore, referred to as current carriers or simply carriers. Holes and electrons exist together in both intrinsic and extrinsic semiconductors. In intrinsic semiconductors, electrons and holes exist in equal density. In extrinsic semiconductors, the relative concentration of holes and electrons depends on the density of donor or acceptor impurities. In an extrinsic semiconductor, the carriers that are in majority are called majority carriers, whereas the others are called minority carriers. Thus, electrons are majority carriers, whereas holes are minority carriers in n-type semiconductors. In p-type semiconductors, holes are majority carriers, whereas electrons are minority carriers. Majority carriers are produced by impurities and by breaking of covalent bonds by thermal energy, whereas minority carriers are produced only by breaking of covalent bonds by thermal energy.

In a semiconductor, holes and electrons move randomly along straight lines because of their thermal energy. Each carrier moves in a random direction, collides with a nucleus, and gets repelled. The carrier moves in a new direction until it does not collide with another nucleus. Thus, we can say that carriers vibrate. This random motion is defined by an average path length l between two consecutive collisions in an average time t. The average velocity of the movement of carriers is then given by

$$v = \frac{l}{t} \tag{1.1}$$

When the time in consideration is large compared with t in Eq. (1.1), the net movement of a carrier is zero. So, no resultant electric current flows. However, if an electric field is applied across the semiconductor, the carriers get influenced, hence a directed motion is superimposed on the random thermal motion. This results in a net average velocity in the direction of the applied electric field, while the electrons move in the opposite direction. As holes and electrons are opposite charges, both produce current in the same direction. This current is called *drift current*. In extrinsic semiconductors, the current is essentially majority-carrier flow.

In an extrinsic semiconductor, if the concentration of carriers is not uniformly distributed, that is, varies from point to point, then carriers flow away from densely concentrated regions towards the region of low carrier concentration. This flow of carriers is called *diffusion,* and this current is called *diffusion current.*

There are other examples of diffusion phenomena. A gas introduced into a container diffuses to fill it with uniform concentration. An ink drop diffuses when dropped in water and water becomes uniformly coloured, if enough time is given.

Consider a surface S at the interface of region 1 and region 2 with different carrier concentrations n_1 and n_2, respectively, $n_1 > n_2$, as shown in Fig. 1.4. In random thermal motion, as discussed earlier, a carrier in region 1 flows through S into region 2 and a carrier in region 2 flows through S into region 1. This results in a certain percentage of n_1 flowing through S into region 2 in any given period of time, and in the same period of time, the same percentage of n_2 flowing through S in the opposite direction into region 1. As region 1 is more densely concentrated, the rate of flow must be greater in the direction from region 1 to region 2 than in the opposite direction, resulting in a net flow from high- to low-concentration region.

A schematic representation of the n-type semiconductor placed between mounting electrodes is shown in Fig. 1.5. The immobile ionized donor atom is represented by a plus sign enclosed in a circle because after the impurity atom has donated an electron it becomes a positive ion. The n-type semiconductor has a large number of free electrons moving randomly and an equal number of immobile donor atoms. There are also

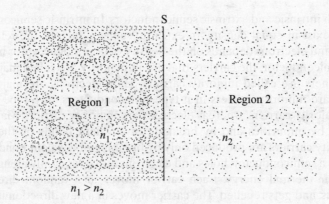

FIGURE 1.4 Interface between two regions with different carrier concentrations.

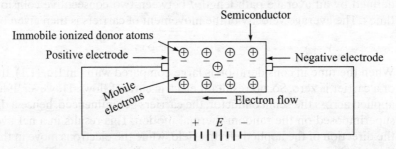

FIGURE 1.5 A schematic representation of the n-type semiconductor.

relatively negligible numbers of holes. Under the influence of electric field directed from the positive electrode to the negative electrode, the free electrons drift towards the positive electrode and enter into it. The uncovered ionized donor atoms near the negative electrode drag electrons from the negative electrode in an exact number to replace the electrons drifted to the metal of the positive electrode. The number of free electrons in the semiconductor remains the same. Thus, under all conditions, the ionized donor atoms are covered. Therefore, the semiconductor remains electrically neutral. The voltage difference applied between the electrodes appears as voltage drop across the semiconductor. Therefore, n-type semiconductors obey Ohm's law.

Figure 1.6 shows a schematic representation of the p-type semiconductor kept between its mounting electrodes. The immobile ionized acceptor atom is represented by

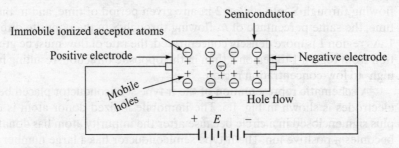

FIGURE 1.6 A schematic representation of the p-type semiconductor.

a minus sign enclosed in a circle. This is because of the reason that after the impurity atom has accepted an electron it becomes a negative ion. The p-type semiconductor has a large number of mobile holes. These holes move randomly. There are an equal number of immobile ionized acceptor atoms and a negligible small number of free electrons. Under the influence of electric field, the mobile holes drift towards the negative electrode. On reaching the negative electrode, the mobile holes pull out free electrons from the negative electrode and combine with them.

As holes drift away from the vicinity of the positive electrode, the immobile ionized acceptor atoms are no longer neutral. An electric field is formed. This electric field detaches the loosely bonded ionizing electrons from the ionized acceptor atoms. Such free electrons then strike the positive electrode and enter the metal of the positive electrode. Each acceptor atom, when loses one electron, steals an electron from a neighbouring ionized acceptor atom. This results in a new hole. The new holes drift towards the negative electrode, whereas the electrons captured by the acceptor atom are readily detached and lost to the positive electrode. The creation of new holes takes place at the same rate to replace the holes that have drifted towards the negative electrode. The number of mobile holes in the semiconductor remains unchanged under all conditions. The semiconductor, therefore, remains neutral. The voltage difference applied between the electrodes appears as voltage drop in the semiconductor. Therefore, p-type semiconductors obey Ohm's law.

1.3 p-n JUNCTIONS OR DIODES

When a junction is formed between a sample of p-type semiconductor and a sample of n-type semiconductor, the device so formed is called the p-n junction. This device has the property of a rectifier. A p-n junction can be formed by making a single crystal of semiconductor material in which the acceptor impurity is made to predominate in one part and the donor impurity to predominate in the other part.

Figure 1.7(a) shows a p-n junction under thermal equilibrium with no external voltage applied. In the beginning, there are only p-type carriers (holes) to the left of the junction and n-type carriers (electrons) to the right of the junction. But, there exists a density gradient across the junction. Hence, holes diffuse to the right and electrons diffuse to the left of the junction.

The holes on diffusing from the p region to the n region combine with the electrons near the junction. With the combination of these electrons, the donor atoms, which were initially neutral, become positively charged. Similarly, electrons diffusing from the n region to the p region combine with the holes near the junction. This leaves the immobile acceptor atom negatively charged. Thus, an electric field is created near the junction directed from the n region to the p region with positive ions in the n region and negative ions in the p region. This electric field opposes the diffusion from both sides. An equilibrium is reached when the electric field becomes large enough to completely stop the process of diffusion. The charge distribution across the junction is shown in Fig. 1.7(b).

As the region in the vicinity of the junction is depleted of mobile charges, this region of junction is known as the depletion region, space charge region, or transition region. The thickness of the region is of the order of 1 Å (1 Å = 10^{-6} m).

8 Chapter 1 / Semiconductor Devices

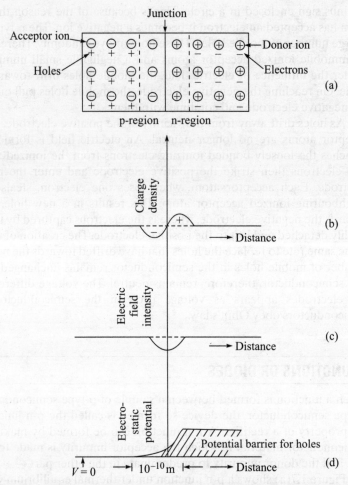

FIGURE 1.7 (a) p-n junction, (b) charge density, (c) electric field intensity, and (d) potential-energy barrier.

From Poisson's equation

$$\frac{d^2v}{dx^2} = -\frac{\rho}{\epsilon} \tag{1.2}$$

where v is the potential, ρ is the charge density, x is the distance from the junction, and ϵ is the permittivity of the medium. The electric field intensity,

$$E = -\frac{dv}{dx} = -\int \frac{\rho}{\epsilon} dx \tag{1.3}$$

thus, is the integral of the charge density ρ. The electric field intensity is shown in Fig. 1.7(c).

The electrostatic potential v is given by

$$v = -\int E\, dx \tag{1.4}$$

The electrostatic voltage v is shown in Fig. 1.7(d). The potential of the p region is taken as reference (or zero) potential. The potential rises from zero to v_{cp} in the depletion region. The potential v_{cp} is a constant potential and constitutes a potential-energy barrier against further diffusion of holes from the p region to the n region and of electrons from the n region to the p region. The contact potential is also known as diffusion potential. The potential-energy barrier magnitude V_o is of the order of a few tenths of a volt.

The concentration of holes in the p region exceeds that in the n region. This results a tendency of a large diffusion current to flow across the junction from the p region to the n region. The electric field developed across the junction is directed from the n region to the p region. This tends to send a drift current across the junction from the n region to the p region. This drift current exactly counter balances the diffusion current. Thus, the net flow of current is zero.

1.3.1 Bias Voltage

The potential barrier developed across the depletion region prevents the flow of carriers across the junction without injecting energy from the external source. An external voltage source is applied between the n region and the p region, as shown in Fig. 1.8(a). The p-n junction circuit symbol of Fig. 1.8(a) is shown in Fig. 1.8(b). The external voltage source provides an additional energy. The applied voltage E is called the bias voltage. The bias voltage decreases or increases the potential barrier and controls the flow of carriers across the junction. When the applied voltage is zero, the potential barrier is unaffected and the circuit of the p-n junction behaves same as the open-circuit p-n junction. No current flows. The positive value of the applied voltage decreases the potential barrier. This increases the number of majority carriers (electrons or holes) crossing the p-n junction. The positive applied voltage is known as forward-bias voltage. When the applied voltage is made negative, it helps the potential barrier and reduces the number of carriers (electrons or holes) diffusing across the junction. The application of negative voltage, as shown in Fig. 1.9(a), is called reverse bias. The p-n junction with reverse bias is shown symbolically in Fig. 1.9(b).

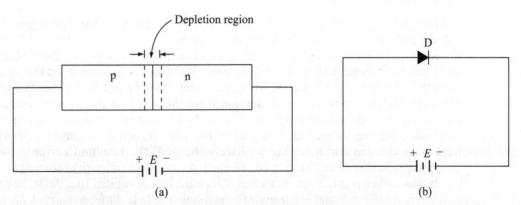

FIGURE 1.8 (a) Forward-biased p-n junction and (b) circuit symbol of p-n junction.

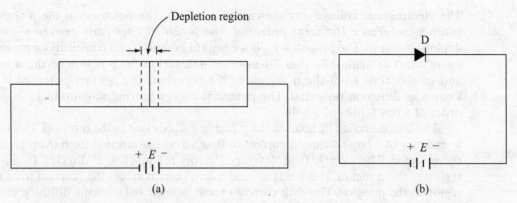

FIGURE 1.9 (a) Reverse-biased p-n junction and (b) circuit symbol of p-n junction.

1.4 p-n JUNCTION AS A RECTIFIER

The essential electric characteristic of a p-n junction is that it possesses the property of a rectifier. This means that a p-n junction permits the easy flow of charge in one direction, but restrain the flow in the opposite direction.

1.4.1 Reverse Bias

The reverse-bias voltage strengthens the potential barrier. This increase in the barrier potential further reduces the flow of majority carriers, holes in the p region and electrons in the n region. However, the minority carriers, holes in the n region and electrons in the p region, are not influenced by the increase of barrier potential. The minority carriers are responsible for the small reverse current. It is also called saturation current. As the densities of majority carriers increase with an increase in temperature, the reverse current also increases with an increase in temperature. This is the reason that the reverse-biased resistance decreases with an increase in temperature.

1.4.2 Forward Bias

A forward bias, as shown in Fig. 1.8, is applied to the p-n junction. The forward-bias voltage reduces the magnitude of the potential barrier. So, the equilibrium between the forces tending to produce the diffusion of majority carriers and the restraining force due to potential barrier at the junction is disturbed. The holes cross the junction from the p region to the n region, where they constitute an injected minority current. Similarly, the electrons cross the junction from the n region to the p region, where they constitute an injected minority current. Holes and electrons travel in opposite directions, but they constitute a current in the same direction because electrons are negatively charged and holes are positively charged. The resultant current crossing the junction is, therefore, the sum of the electron and hole (minority) currents.

It was earlier pointed out that when a forward bias is applied to a diode, holes are diffused into the n side and electrons into the p side. The hole diffusion current, I_{pn}, in the n region decreases exponentially with the distance x into the n region. The electron diffusion current, I_{np}, in the p region, similarly, decreases with distance in the p region. Since doping on the two sides of the junction may not be identical, the plots of I_{pn} and I_{np} are not identical. In this case, it is assumed that the concentration of holes is much greater than the donor density. So, the hole current greatly exceeds the electron current.

Holes crossing the junction at $x = 0$ from left to right constitute a current in the same direction as electrons crossing the junction from right to left. Hence, the total diode current I_t at $x = 0$ is

$$I_t = I_{pn}(0) + I_{np}(0)$$

As the current is the same through a series circuit, I_t is independent of x. However, I_t is dependent on the applied voltage (forward bias) and is given as

$$I_t = I_0 \left(e^{V/\eta V_T} - 1 \right) \tag{1.5}$$

where I_0 is the saturation current and η and V_T are dependent on the parameter and the thermal conditions of the diode, respectively.

In the n region, the total current I_t is constant and the minority (hole) current I_{pn} varies with x. Clearly, there must exist a majority (electron) current I_{nn} in the n region, which must be a function of x, because the total current I_t at any position is the sum of hole and electron currents. Hence,

$$I_{nn}(x) = I_t - I_{pn}(x)$$

I_{pn} is plotted as a solid line and I_{nn} as a dotted line in the n region in Fig. 1.10. The majority hole current I_{pp} in the p region is plotted as a dashed line, whereas the minority electron current I_{np} as a solid line.

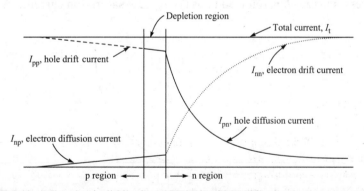

FIGURE 1.10 Minority (solid line) and majority (dotted line) currents versus distance.

Deep into the p region the current is a drift (or conduction) current I_{pp}. As the holes approach the junction, some of them recombine with the electrons that are diffused into the p region from the n region. The current I_{pp}, thus, decreases towards the junction. The reduced current of I_{pp} at the junction enters into the n region and becomes the hole diffusion current I_{pn}. Similarly, drift current I_{nn} in the n region partially combines with the holes diffusing into the n region and the remaining I_{np} diffusing into the p region. It is plotted as solid lines in the p region.

1.5 V-I CHARACTERISTIC

1.5.1 Static Characteristic

As per earlier discussion, the current I in a p-n junction is related to the voltage V applied across the junction by Eq. (1.6)

$$I = I_0 \left(e^{V/\eta V_T} - 1 \right) \tag{1.6}$$

A positive I means that current flows from the p region to the n region. The diode is forward bias when V is positive (as shown in Fig. 1.8). The symbol η is unity for germanium and is approximately 2 for silicon at rated current. The symbol V_T is volt equivalent of temperature and is given by Eq. (1.7) as follows:

$$V_T = \frac{T}{11600} \tag{1.7}$$

where T is the temperature in kelvin. At room temperature ($T = 300$ K), $V_T = 0.026$ V $= 26$ mV.

The volt-ampere characteristic described by Eq. (1.6) is graphically shown in Fig. 1.11. When the voltage V is positive and several times V_T, the unity in the parentheses of Eq. (1.6) is neglected, the current increases exponentially with voltage (except for a small range near the origin). Equation (1.6) reduces to Eq. (1.8).

$$I = I_0 \, e^{V/\eta V_T} \tag{1.8}$$

When the p-n junction is reverse biased and $|V|$ is several times V_T, $I = -I_0$. The reverse current is, therefore, constant and is independent of the applied reverse voltage. The reverse current I_0 is referred to as the reverse saturation current.

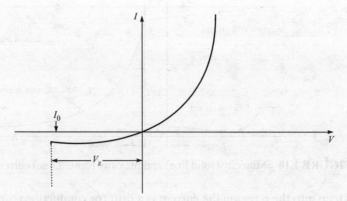

FIGURE 1.11 Volt-ampere characteristic of a diode.

The dashed portion of the curve of Fig. 1.11 indicates that the diode characteristic shows an abrupt and marked departure from Eq. (1.6) at reverse biasing voltage V_Z. At this critical voltage, a large reverse current flows and the diode is said to reach the breakdown condition discussed later.

Example 1.1

A silicon diode is forward biased, with voltage equal to 0.5 V at a temperature of 295 K. The diode current at this biasing is 15 mA. Determine the saturation current for the diode.

Solution

From Eq. (1.7), the volt equivalent of temperature

$$V_T = \frac{T}{11600} = \frac{295}{11600} = 0.0254 \text{ V}$$

From Eq. (1.6), the saturation current,

$$I_0 = \frac{I_T}{e^{V/\eta V_T} - 1} = \frac{15 \times 10^{-3}}{e^{0.5/2 \times 0.0254} - 1} = 0.797 \text{ μA}$$

In the forward-biased volt-ampere characteristic, there is a voltage V_γ below which the current is very small (say about 1% of maximum rated current). This voltage is called cut-in, offset, break point, or threshold voltage. Typical values of cut-in voltage are 0.6 to 0.7 V_γ for silicon device and 0.2 to 0.3 V for germanium device. Beyond the threshold voltage V_γ, the forward-biased current rises very rapidly.

For silicon diodes $\eta = 2$ (for small currents), and the current increases as $e^{V/2V_T}$ for the first several tenths of a volt and increases as e^{V/V_T} only at higher voltages.

The reverse saturation current in a germanium diode is normally larger by a factor of about 1000 than the reverse saturation current in a silicon diode of comparable ratings. For germanium diode, I_0 is in the range of μA (microampere) and for silicon diode, I_0 is in the range of nA (nanoampere) at room temperature.

The *V-I* characteristic of an ideal semiconductor diode is shown in Fig. 1.12.

A large-signal approximation of the *V-I* characteristic of a diode is the piecewise linear representation. This representation leads to a sufficiently accurate engineering solution. The piecewise linear approximation for a diode characteristic is shown in Fig. 1.13. The diode behaves like an open circuit for $V < V_\gamma$ and has a constant incremental resistance $r = dV/dI$ for $V > V_\gamma$. The resistance r is called forward resistance and is represented by R_f.

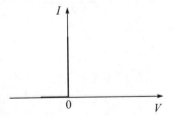

FIGURE 1.12 *V-I* characteristic for an ideal diode.

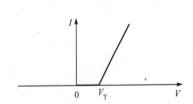

FIGURE 1.13 Piecewise linear characteristic for a diode.

1.5.2 Temperature Dependence of *V-I* Characteristics

In Eq. (1.6), V_T and I_0 are temperature dependent. With the increase in temperature the minority currents (holes in n region and electrons in p region) increase, and so I_0 increases. From experimental data it is found that I_0 increases approximately by 7% per degree centigrade for both silicon and germanium. If $I_0 = I_{01}$ at $T = T_1$, then at temperature $T = T_2$, the saturation current is given by

$$I_{02} = I_{01} \times 2^{(T_2 - T_1)/10} \tag{1.9}$$

As seen from Eq. (1.7), V_T is directly proportional to temperature in kelvin. So, the forward-biased *V-I* characteristic slope increases with the temperature.

1.6 DIODE PARAMETERS

1.6.1 Diode Resistance

The static resistance *R* of a diode is defined as the ratio of operating voltage to operating current. In Fig. 1.14, if *Q* is the operating point, then

$$R = \frac{V_Q}{I_Q} \tag{1.10}$$

that is, it is the reciprocal of the slope of the line joining the operating point Q to the origin (shown by chain line). Thus, the static resistance varies widely with the change of operating point and is not a useful parameter for the diode.

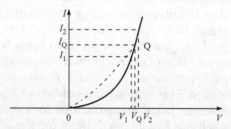

FIGURE 1.14 Forward-biased *V-I* characteristic.

For small-signal operation, the incremental resistance *r* is an important parameter. This resistance is also called dynamic resistance. The dynamic resistance is not a constant, but depends on the operating voltage. So, the dynamic resistance $r = dV/dI$. Differentiating Eq. (1.6), we get

$$\frac{dI}{dV} = \frac{I_0}{\eta V_T} e^{V/\eta V_T}$$

or

$$\frac{dV}{dI} = \frac{\eta V_T}{I_0 e^{V/\eta V_T}} = \frac{\eta V_T}{I + I_0}$$

or

$$r = \frac{\eta V_T}{I + I_0}$$

Since, $I_0 \ll I$,

$$r \approx \frac{\eta V_T}{I} \tag{1.11}$$

For a reverse bias, $dV \gg dI$; hence, the resistance r is very large. At room temperature and $\eta = 1$ (germanium),

$$r = \frac{26}{I} \tag{1.12}$$

where I is in milliamperes and r is in ohms. For a forward current of 26 mA, $r = 1\ \Omega$. The dynamic resistance r varies with the current, but for small-signal model, r is taken as a constant.

Example 1.2

A germanium diode is forward biased at 300 K and the diode current is 15 mA. Determine the dynamic resistance of the diode.

Solution

At 300 K, the volt equivalent of temperature

$$V_T = \frac{T}{11600} = \frac{300}{11600} = 0.026\ V$$

For germanium, $\eta = 1$. So neglecting the saturation current with respect to forward-bias current, the dynamic resistance,

$$r = \frac{26}{15} = 1.733\ \Omega$$

1.6.2 Depletion Capacitance

It has been mentioned earlier that the depletion layer increases with increase in reverse-bias voltage. This means that with the increase of reverse-bias voltage more immobile charges are uncovered. This may be considered a capacitive effect. This incremental capacitance is called *depletion*, *space charge*, or *transition capacitance* and is defined as

$$C_T = \left|\frac{dQ}{dV}\right| \tag{1.13}$$

where dQ is the increase in charge due to an increase dV in the voltage.

Consider a reverse-bias p-n junction shown in Fig. 1.15(a). Let the concentration of acceptor ions on p side be N_A and that of donor ions on n side be N_D. It is not necessary that N_A be equal to N_D. In reality, it is often advantageous to have an unsymmetrical junction. Consider a junction in which there is an abrupt change from acceptor ion on

16 Chapter 1 / Semiconductor Devices

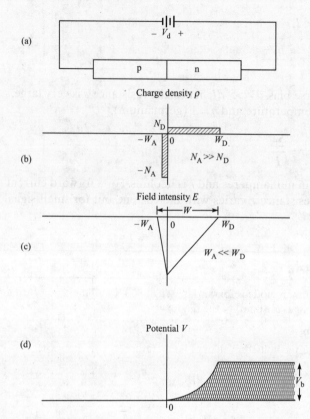

FIGURE 1.15 (a) A reverse-biased p-n junction, (b) the charged density, (c) the field intensity, and (d) the potential variation with x.

one side to donor ion on the other side, as shown in Fig. 1.15(b). Let W_A and W_D be the lengths of depletion on p side and n side, respectively, from this abrupt junction. Since the net charge must be zero,

$$N_A W_A = N_D W_D \tag{1.14}$$

If $N_A \gg N_D$, then $W_D \gg W_A$. From Poisson's equation,

$$\frac{d^2V}{dx^2} = -\frac{\rho}{\epsilon} = -\frac{qN_D}{\epsilon} \tag{1.15}$$

where q is the charge on an electron.

The electric lines of flux start from the positive donor ions on n side and terminate to the negative acceptor ions on p side. Hence, there are no lines of flux to the right of the boundary at $x = W_D$ [Fig. 1.15(c)]. Integrating Eq. (1.15), we obtain

$$\frac{dV}{dx} = -\frac{qN_D}{\epsilon}(x - C) \tag{1.16}$$

where C is a constant.

Since, $\dfrac{dV}{dx} = 0$ at $x = W_D$.

Hence, $C = -W_D$.
Substituting for C in Eq. (1.16), we obtain

$$\dfrac{dV}{dx} = -\dfrac{qN_D}{\epsilon}(x - W_D) = -\epsilon$$

But, $W_D \approx W (= W_A + W_D)$
Hence,

$$\dfrac{dV}{dx} = -\dfrac{qN_D}{\epsilon}(x - W) = -\epsilon \tag{1.17}$$

The small potential drop across W_A may be neglected, and so it may be taken as $V = 0$ at $x = 0$. Then, integrating Eq. (1.17), we get

$$V = -\dfrac{qN_D}{2\epsilon}(x^2 - 2Wx) \tag{1.18}$$

The voltage is plotted in Fig. 1.15(d).
At $x = W$, $V = V_b$ = junction or barrier potential. Hence,

$$V_b = \dfrac{qN_D W^2}{2\epsilon} \tag{1.19}$$

If externally applied voltage is V_d, the net barrier potential,

$$V_b = V_0 - V_d \tag{1.20}$$

where V_d is negative for reverse bias, and V_0 is the barrier potential when there is no externally applied voltage. From Eq. (1.19)

$$W \propto (V_0 - V_d)^{1/2} \tag{1.21}$$

This shows that with applied forward voltage, the depletion layer decreases, whereas it increases with reverse bias.

If A is the area of the junction, the charge at distance W is

$$Q = qN_D W A \tag{1.22}$$

and the depletion capacitance

$$C_T = \left|\dfrac{dQ}{dV_b}\right| = qN_D A \left|\dfrac{dW}{dV_b}\right|$$

From Eq. (1.19),

$$\left|\dfrac{dW}{dV_b}\right| = \dfrac{\epsilon}{qN_D W} \tag{1.23}$$

Hence,

$$C_T = qN_D A \times \frac{\epsilon}{qN_D W} = \frac{\epsilon A}{W} \qquad (1.24)$$

This expression is exactly the same as that for a parallel-plate capacitor of area A, in square meters, and plate separation W, in meters, containing material of permittivity ϵ. In case N_A is not negligible, the results are slightly modified. In Eq. (1.20), $W = W_A + W_D$, and $1/N_D$ is replaced by $1/N_A + 1/N_D$. Equation (1.24) remains unaffected.

1.6.3 Diffusion Capacitance

Figure 1.16 shows the variations of hole density injected into the n region of a p-n junction as a function of distance x from the junction for forward-bias voltages V_1 and V_2. At the junction the variation of the hole density with forward bias is an increasing exponential function as given below

$$\rho_{n1} = \rho_n e^{V_1/V_T} \qquad (1.25)$$

and

$$\rho_{n2} = \rho_n e^{V_2/V_T} \qquad (1.26)$$

where $V_2 > V_1$, ρ_n denotes the hole density at junction (when forward bias is zero) and V_T is given in Eq. (1.7). For a forward bias, however, as one moves away from the junction, the hole density diminishes because of recombination with the electrons.

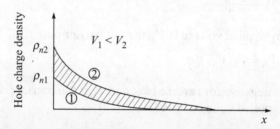

FIGURE 1.16 Variation of hole density as a function of distance.

1.7 SPECIAL TYPES OF DIODES

In this section, special types of junction diodes used for specific purposes are discussed.

1.7.1 Zener Diodes

These types of diodes operate in the breakdown region of the reverse-voltage characteristic. These diodes are designed with adequate power dissipation capabilities to operate in the breakdown region. Zener diodes are used as voltage-reference or constant-voltage devices. They are also known as avalanche or breakdown diodes.

The reverse-voltage characteristic of a Zener diode is shown in Fig. 1.17. The forward-biased characteristic is similar to that of an usual junction diode. A circuit

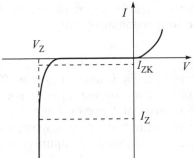
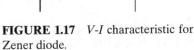
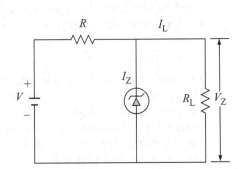

FIGURE 1.17 V-I characteristic for Zener diode.

FIGURE 1.18 A circuit using Zener diode voltage regulation.

utilizing a Zener diode for regulating the voltage across a load is shown in Fig. 1.18. The source voltage V and the series resistance R are selected such that, initially, the diode operates in the breakdown region. The diode voltage and current are V_Z and I_Z, respectively. The diode voltage V_Z is also the voltage across the load R_L. The Zener diode regulates the voltage across R_L against variation in load current I_L and against variation in supply voltage V. Let the load current increase from I_L to $I_L + \Delta I_L$, increasing the drop across R. This decreases the diode voltage, in turn decreasing the diode current I_Z, which decreases the drop across R. Thus, the diode voltage is stabilized to the original diode voltage V_Z. Thus, the load voltage maintains a constant value. Similarly, the variation in the source voltage is also compensated by the Zener diode. The Zener diode regulates satisfactorily for all values of diode current above a minimum value I_{ZK}, shown in Fig. 1.17. The upper limit of the diode current is determined by the power dissipation rating of the diode.

There are two mechanisms of diode breakdown. One is *avalanche breakdown* and other is *Zener breakdown*.

Avalanche breakdown

In reverse-bias condition, a thermally generated carrier comes to the junction barrier and gains energy from the applied voltage. This carrier collides with a crystal ion and imports enough energy to break the covalent bond. Thus, a new electron-hole pair is generated. These carriers may also gain sufficient energy from the applied field and produce still another pair of electron-hole when they collide with another crystal ion. In this chain process, each new carrier may, in turn, produce an additional pair of electron-hole. This chain process results in large reverse currents. The diode is, then, said to be in the region of *avalanche breakdown*.

Zener breakdown

In this mechanism, the existence of the electric field may rapture the covalent bond because of strong force experienced by a bond electron. Thus, a new electron-hole pair is created, increasing the reverse current. This process is called *Zener breakdown* and does not involve collision of carriers with the crystal ions as in the avalanche breakdown.

For a fixed applied voltage, the field intensity increases with the impurity concentration. It is found that the Zener breakdown occurs at a field intensity of about 2×10^7 V/m. For heavily doped diodes, this threshold value is reached at a voltage below 6 V, whereas the breakdown voltage is higher for lightly doped diodes. In lightly doped diodes, avalanche breakdown is, therefore, the predominant effect.

1.7.2 Tunnel Diodes

In normal p-n junction diode, impurity concentration is about 1 part in 10^8. The depletion layer width depends on the amount of impurity concentration. In normal p-n junction, the width of depletion layer is of the order of a micron (10^{-6} m). The potential barrier formed by the depletion layer restrains the flow of the majority carriers across the depletion layer. In a tunnel diode the concentration of impurity atoms is greatly increased, say, to 1 part in 10^3.

The width of the junction barrier varies inversely as the square root of impurity concentration. The depletion layer in a tunnel diode is less than 10^{-4} m. For this thin depletion layer, there is a large possibility that an electron will penetrate the barrier. This behaviour of penetration of an electron through the barrier is called tunneling. Hence, high-impurity p-n junction diodes are called tunnel diodes.

The V-I characteristic of a tunnel diode is shown in Fig. 1.19. It is evident from the V-I characteristic that the tunnel diode is a good conductor in reverse bias. For small forward-bias voltage (up to 50 mV for germanium), the resistance remains small (about 5 Ω). At the peak current I_P, corresponding to voltage V_P, the slope (dI/dV) of the V-I characteristic is zero (i.e., resistance is infinity). If the voltage is increased beyond V_P, the current decreases giving a negative slope. The tunnel diode, thus, exhibits a negative-resistance characteristic between the peak current I_P and the minimum current I_V, called the valley current. The conduction at valley point, where valley voltage and current are V_V and I_V, respectively, is again zero. Beyond this valley point, the resistance characteristic becomes positive and remains positive. At voltage V_F (peak forward voltage), the current again reaches the value I_P. For still larger voltages, the current increases beyond this value. The symbol of the tunnel diode is shown in Fig. 1.20.

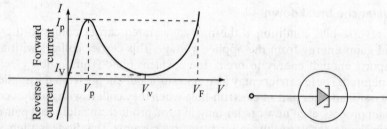

FIGURE 1.19 V-I characteristic of a tunnel diode.

FIGURE 1.20 Circuit symbol of a tunnel diode.

A current between I_P and I_V is obtained for three different values of the voltage. So, the forward-bias curve is triple valued. Because of this multivalued feature, the tunnel diode is used in pulse and digital circuits. Another application of the tunnel diode is as a very high-speed switch. Reasonable switching time is of the order of nanoseconds, although times as low as 50 ps have been obtained. Tunnel diodes are also used as high-frequency (microwave) oscillators.

The commercially available tunnel diodes are mostly made from germanium or gallium arsenide. It is difficult to manufacture a silicon tunnel diode with a high I_P/I_V

ratio. A gallium arsenide tunnel diode has the highest I_p/I_v ratio of 15, whereas those of germanium and silicon diodes are 8 and 3.5, respectively. The peak point (V_p, I_p) is not very sensitive to temperature, whereas valley point (V_v, I_v) is quite temperature sensitive.

The advantages of a tunnel diode are low cost, high speed, simplicity, environmental immunity, and low power. The disadvantages are low output-voltage swing and its being a two-terminal device. Because of being a two-terminal device, it has no isolation between input and output, and this posses serious circuit design difficulties.

1.7.3 Photodiodes

If a reverse-bias p-n junction is exposed to the light, the current varies almost linearly with the light flux. This effect is utilized in a photodiode. This device consists of a p-n junction embedded in a clear plastic. The light is allowed to fall upon one surface across the junction and the rest of the part of the plastic is either painted black or enclosed in a metallic case. This unit has dimensions of the order of tenths of an inch.

V-I characteristic

Photodiodes operate in reverse bias. When no light is falling on any portion of the photodiode, the current, called dark current, is the saturation current I_0 of the diode. As we know, I_0 is due to thermally generated minority carriers. When light falls upon the photodiode surface, additional electron-hole pairs are created. The newly created minority carriers diffuse to the junction, cross it, and form a part of the reverse current.

Typical *V-I* characteristics for a germanium photodiode are shown in Fig. 1.21. The dark current is approximately 10 μA. This current, which is a saturation current, is dependent on temperature. With the fall of light the reverse current increases. It is evident from the characteristics that the curves (except dark curve) do not pass through the origin. Therefore, even for zero reverse voltage, there is a current that depends on the strength of the light falling upon the diode and the distance of the illuminated spot from

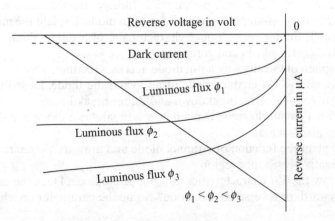

FIGURE 1.21 *V-I* characteristics of a photodiode.

the junction. If the radiation is focused on a spot far away from the junction, the generated minority carriers recombine before diffusing to the junction. Hence, a much smaller current would result than that if the spot to be illuminated is near the junction. Thus, the photocurrent is a function of the distance of the junction from the spot at which the light is focused.

The p-n photodiodes find extensive application in high-speed reading of computer-punched cards and tapes; light-detection systems; reading of film sound track, light-operated switches; production line counting of objects, which interrupt a light beam, etc.

1.7.4 Light-Emitting Diodes

When an electron recombines with a hole, energy is released. In silicon and germanium, the liberated energy goes into the crystal as heat. However, in gallium arsenide, the energy released due to the recombination of electrons with holes appears in the form of radiation. Such a p-n junction is called a *light-emitting diode* (LED). The radiation is mainly in infrared region. The efficiency of the process of light generation increases with the injected current and with a decrease in temperature.

1.8 REVIEW QUESTIONS

1. Describe how charge carriers are made available for conduction in a semiconductor device.
2. What do you understand by donor and acceptor impurities?
3. Describe the diffusion process that takes place at the p-n junction, and explain the presence of depletion region.
4. Sketch the V-I characteristic of the p-n junction diode for forward-bias voltage. Differentiate between static and dynamic resistance of the diode.
5. What are transition and diffusion capacitances of a semiconductor diode?
6. Plot the hole current, the electron current, and the total current as a function of distance on both sides of a p-n junction. Indicate the transition region.
7. Write the volt-ampere equation for a p-n diode. Explain the meaning of each symbol.
8. Sketch the piecewise linear characteristic of a diode. What are approximate cutin voltages for silicon and germanium?
9. Explain physically why a p-n diode acts as a rectifier.
10. Sketch the V-I characteristic of an avalanche diode. Describe the physical mechanism for avalanche breakdown and Zener breakdown.
11. With a circuit diagram, explain how a breakdown diode is used to regulate the voltage across a load.
12. Explain the function of a tunnel diode and draw its V-I characteristic, indicating the negative-resistance region.
13. Draw the V-I characteristics of a p-n photodiode. Does the current correspond to a forward- or reverse-biased diode? Name the parameter on which value each curve is drawn.
14. What is a light-emitting diode. What are its uses?

1.9 SOLVED PROBLEMS

1. A silicon diode at a temperature of 294 K has a forward-biased current of 25 mA. Determine the dynamic resistance of the diode.

   Solution

   The volt equivalent of 294 K

   $$V_T = \frac{294}{11600} = 0.0253 \text{ V} = 25.3 \text{ mA}$$

   From Eq. (1.11), the dynamic resistance

   $$r = \frac{\eta V_T}{I} = \frac{2 \times 25.3}{25} = \mathbf{2.024 \ \Omega}$$

2. A germanium p-n junction diode has a reverse saturation current of 10 µA. The diode is operated in an ambient temperature of 600 K. A forward bias of 0.2 V is applied. Find (a) the diode current, (b) the apparent diode resistance, and (c) the incremental resistance.

   Solution

   The volt equivalent of 600 K

   $$V_T = \frac{600}{11600} = 0.052 \text{ V} = 52 \text{ mA}$$

   (a) The diode current is given by

   $$I = I_0(e^{(V/\eta V_T)} - 1)$$

   For germanium, $\eta = 1$, hence,

   $$I = 10 \times 10^{-6}[e^{(0.2/0.052)} - 1]$$

   $$= 458 \times 10^{-6} \text{ A} = \mathbf{458 \ \mu A}$$

   (b) The apparent resistance

   $$R = \frac{0.2}{458 \times 10^{-6}} = \mathbf{436.6 \ \Omega}$$

   (c) The incremental resistance

   $$r = \frac{\eta V_T}{I} = \frac{1 \times 52}{0.458} = \mathbf{113.5 \ \Omega}$$

3. The v-i characteristic of a p-n junction diode is given by the following equation:

$$I = 10^{-5}(e^{38.5V} - 1)$$

The diode is placed in series with a resistance of 5 Ω and a d.c. source of 1 V. Determine the diode current.

Solution

The circuit is shown in Fig. 1.22.

The loop equation

$$1 = 5I + V$$

or $I = \dfrac{(1 - V)}{5}$ (1)

Also, $I = 10^{-5}(e^{38.5V} - 1)$ (2)

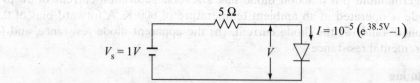

FIGURE 1.22 Circuit for Solved Problem 3.

For different values of V, the values of I from the above two equations are given in the table below:

Voltage drop across diode (V)	0.05	0.1	0.15	0.2	0.25	0.3
$I = (1 - V) \times 10^3/5$ mA	190	180	170	160	150	140
$I = 10^{-5}(e^{38.5V} - 1)$ mA	0.06	0.46	3.21	22.07	151.4	103.7

The two curves of I may be plotted. The approximate value of current is

$$I \approx 150 \text{ mA}$$

4. A junction diode has a reverse saturation current of 5 μA at 300 K. Find the maximum and minimum currents that flow in the diode when a voltage wave of $0.1 \sin \omega t$ is superimposed on a 0.2-V d.c. source and applied to the diode.

Solution

Assume germanium diode. The maximum voltage applied is $0.2 + 0.1 = 0.3$ V and the minimum voltage is $0.2 - 0.1 = 0.1$ V. Hence, the maximum current,

$$I_{max} = 5(e^{0.3/0.026} - 1) \text{ μA} \qquad \left(\because V_T = \dfrac{300}{11600} = 0.026 \text{ V} \right)$$

$$= 513 \text{ mA}$$

The minimum current,

$$I_{min} = 5(e^{0.1/0.026} - 1) \, \mu A$$
$$= 229 \, \mu A$$

5. The p-n junction diode in Solved Problem 3 is placed in series with an unknown resistance and a 1-V d.c. source. The current through the diode is 100 mA. Find the value of the unknown resistance.

Solution

The diode current

$$I = 10^{-5}(e^{38.5V} - 1)$$
or $\quad 0.1 = 10^{-5}(e^{38.5V} - 1)$
or $\quad e^{38.5V} = 10^4 + 1 \approx 10^4$
or $\quad 38.5 V = 4 \log_e 10 = 9.21$
$\therefore \quad V = 0.24 \, V$

Hence, the drop across the series resistance is $1.0 - 0.24 = 0.76$ V. The unknown resistance, therefore, is given by

$$0.76 = 0.1 R$$
$$R = 7.6 \, \Omega$$

6. (a) Find the reverse-bias voltage for which the reverse current in a p-n junction germanium diode reaches 90% of its saturation current at 300 K. (b) Calculate the ratio of the current for a forward bias of 0.05 V to the current for the same magnitude of reverse-bias voltage.

Solution

(a) The diode current is given by

$$I = I_0(e^{(V/\eta V_T)} - 1)$$

For germanium diode, $\eta = 1$ and the volt equivalent of temperature of 300 K (as in Solved Problem 4) is 0.026 V. For reverse bias, V is taken negative. Hence,

$$I_r = I_0(e^{-V/0.026} - 1)$$
or $\quad 0.9 I_0 = I_0(e^{-V/0.026} - 1)$
or $\quad 1.9 = e^{-V/0.026}$
or $\quad V = -0.026 \log_e 1.9 = \mathbf{-0.0167 \, V}$

(b) The forward current for $V = 0.05$ V,

$$I_f = I_0(e^{0.05/0.026} - 1)$$

and the reverse current for $V = -0.05$ V,

$$I_r = I_0(e^{-0.05/0.026} - 1)$$

Hence,

$$\frac{I_f}{|I_r|} = \frac{(e^{0.05/0.026} - 1)}{(1 - e^{-0.05/0.026})} = \frac{5.84}{0.854} = \mathbf{6.84}$$

7. A silicon diode operates at a forward voltage of 0.4 V. Calculate the factor by which the current is to be multiplied when the temperature is increased from 25 to 150°C.

Solution

The two temperatures in kelvin are as follows:

$$273 + 25 = 298 \text{ K}$$
$$273 + 150 = 423 \text{ K}$$

The volt equivalents of temperatures are as follows:

$$V_T^{25} = \frac{298}{11600} = 0.0257 \text{ V}$$

$$V_T^{150} = \frac{423}{11600} = 0.0365 \text{ V}$$

From Eq. 1.9,

$$I_{0150} = I_{025} \times 2^{(150-25)/10}$$
$$= 5794.3 I_{025}$$

The forward currents

$$I_{150} = 5794.3 I_{025}(e^{0.4/2 \times 0.0365} - 1)$$
$$= 5794.3 \times 238.7 \, I_{025}$$

and $\quad I_{25} = I_{025}(e^{0.4/2 \times 0.0257} - 1)$

$$= 2396.3 \, I_{025}$$

Hence, the required factor

$$m = \frac{I_{150}}{I_{25}} = \frac{5794.3 \times 238.7}{2396.3} = \mathbf{577.2}$$

8. A germanium p-n junction diode has a reverse saturation current of 30 μA at a temperature of 125°C. At the same temperature, find the dynamic resistances for 0.2 V bias in (a) the forward direction and (b) the reverse direction.

Solution

The volt equivalent of temperature

$$V_T = \frac{273+125}{11600} = 0.0343 \text{ V}$$

(a) The forward current

$$I_f = 30 \times 10^{-6}(e^{0.2/0.0343} - 1)$$
$$= 1019 \times 10^{-5} \text{ A} = 10.19 \text{ mA}$$

The dynamic resistance is given by

$$r = \frac{\eta V_T}{I} = \frac{1 \times 0.0343}{0.01019} = \mathbf{3.366 \ \Omega}$$

(b) The reverse current

$$I_r = 30 \times 10^{-6}(e^{-0.2/0.0343} - 1)$$
$$= -2.99 \times 10^{-5} \text{ A}$$

The dynamic equation is given by

$$r = \frac{\eta V_T}{I} = \frac{1 \times 0.0343}{2.99 \times 10^{-5}} = \mathbf{1147 \ \Omega}$$

9. Calculate the barrier capacitance of a germanium p-n junction whose area is 1 × 1 mm and space charge thickness is 2 × 10⁻⁴ cm. The relative dielectric constant of germanium is 16.

Solution

The barrier capacitance [Eq. (1.24)],

$$C_b = \frac{\epsilon A}{W} = \frac{16 \times 8.849 \times 10^{-12} \times 1 \times 1 \times 10^{-6}}{2 \times 10^{-6}}$$
$$= \mathbf{70.79 \text{ pF}}$$

10. The Zener diode can be used to prevent overloading of sensitive meter measurements without affecting meter linearity. The circuit shown in Fig. 1.23 consists of a d.c. voltmeter, which reads 20 V at full scale. The meter resistance is 560 Ω and

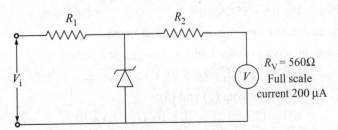

FIGURE 1.23 Circuit for Solved Problem 10.

$R_1 + R_2 = 99.5$ kΩ. If the diode is a 16 V Zener, find the values of R_1 and R_2 so that, when $V_i > 20$ V, the Zener diode conducts, and the overload current is shunted away from the meter.

Solution

For maintaining linearity of the meter, the maximum current through the meter is 200 μA. For 20 V, the Zener does not conduct, so the current through the meter for 20 V

$$I = \frac{20}{99.5 \times 10^3 + 560} \approx 200 \, \mu A$$

but never reaches 200 μA. The maximum voltage across the Zener should be 16 V. Hence, for the maximum current of 200 μA through the meter,

$$R_2 + 560 = \frac{16}{200 \times 10^{-6}} = 80000 \, \Omega$$

Hence, $R_2 = 80000 - 560 = 79440 \, \Omega = 79.44$ kΩ
and $R_1 = 99.5 - 79.44 = $ **20.06 kΩ**

11. Each diode, shown in Fig. 1.24, is described by a linearized v-i characteristic with incremental resistance r and offset voltage V_γ. Diode D_1 is germanium diode with $V_\gamma = 0.2$ V and $r = 20 \, \Omega$, whereas D_2 is silicon diode with $V_\gamma = 0.6$ V and $r = 15 \, \Omega$. Find the diode currents if (a) $R = 10$ kΩ and (b) $R = 1$ kΩ.

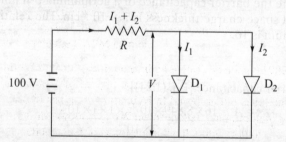

FIGURE 1.24 Circuit for Solved Problem 11.

Solution

The v-i characteristic of the two diodes are given by

$$V = 0.2 + 20I_1 \times 10^{-3} \tag{1}$$
$$V = 0.6 + 15I_2 \times 10^{-3} \tag{2}$$

where I_1 and I_2 are currents in milliamperes.
From the circuit of Fig. 1.24,

$$100 = V + R(I_1 + I_2) \times 10^{-3} \tag{3}$$

(a) $R = 10$ kΩ, so from (1) and (3),

$$100 = 0.2 + 0.02I_1 + 10 \times 10^3 (I_1 + I_2) \times 10^{-3}$$

or $\quad 99.8 = 10.02I_1 + 10I_2 \tag{4}$

and from (2) and (3),

$$100 = 0.6 + 0.015I_1 + 10 \times 10^3 (I_1 + I_2) \times 10^{-3}$$

or $99.4 = 10I_1 + 10.015I_2$ (5)

Solving (4) and (5),

$$I_1 = \frac{\begin{vmatrix} 99.8 & 10 \\ 99.4 & 10.015 \end{vmatrix}}{\begin{vmatrix} 10.02 & 10 \\ 10 & 10.015 \end{vmatrix}} = \frac{5.497}{0.3503} = \mathbf{15.69 \text{ mA}}$$

$$I_2 = \frac{\begin{vmatrix} 10.02 & 99.8 \\ 10 & 99.4 \end{vmatrix}}{\begin{vmatrix} 10.02 & 10 \\ 10 & 10.015 \end{vmatrix}} = \frac{-2.012}{0.3503} = \mathbf{-5.74 \text{ mA}}$$

(b) $R = 1$ kΩ, so from (1) and (3),

$$100 = 0.2 + 0.02I_1 + 1 \times 10^3 (I_1 + I_2) \times 10^{-3}$$

or $99.8 = 1.02I_1 + I_2$ (6)

and from (2) and (3),

$$100 = 0.6 + 0.015I_1 + 1 \times 10^3 (I_1 + I_2) \times 10^{-3}$$

or $99.4 = I_1 + 1.015I_2$ (7)

Solving (6) and (7),

$$I_1 = \frac{\begin{vmatrix} 99.8 & 1.0 \\ 99.4 & 1.015 \end{vmatrix}}{\begin{vmatrix} 1.02 & 1.0 \\ 1.0 & 1.015 \end{vmatrix}} = \frac{1.897}{0.03503} = \mathbf{53.74 \text{ mA}}$$

$$I_2 = \frac{\begin{vmatrix} 1.02 & 99.8 \\ 1.0 & 99.4 \end{vmatrix}}{\begin{vmatrix} 1.02 & 1.0 \\ 1.0 & 1.015 \end{vmatrix}} = \frac{1.588}{0.03503} = \mathbf{44.98 \text{ mA}}$$

12. A symmetrical 5 kHz square wave signal varying between +10 V and −10 V is impressed upon the clipping circuit shown in Fig. 1.25(a). Assume $R_f = 0$, $R_r = 2$ MΩ, and $V_\gamma = 0$. Sketch the steady state output waveform, showing numerical values of the maximum, minimum, and constant portions.

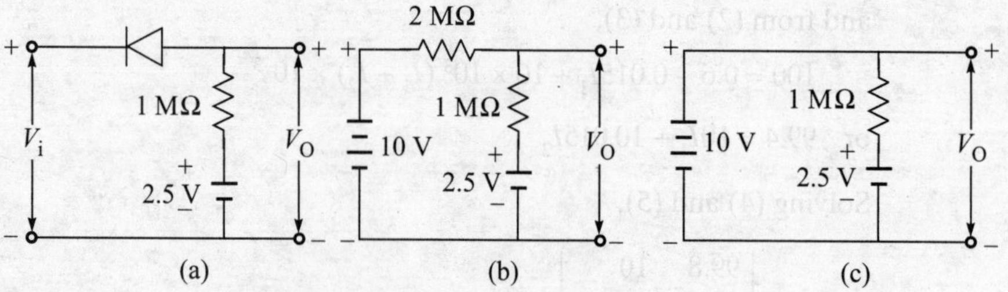

FIGURE 1.25 Circuits for Solved Problem 12.

Solution

For $V_i = +10$ V, the circuit reduces as in Fig. 1.25(b). So, the current,

$$I = \frac{10.0 - 2.5}{2+1} \times 10^{-6} = \frac{7.5}{3} \times 10^{-6} = 2.5\,\mu A$$

$$\therefore \quad v_o = v + 2.5 = 2.5 \times 10^{-6} \times 1 \times 10^6 + 2.5 = 5.0 \text{ V}$$

For $V_i = -10$ V, the circuit reduces as in Fig. 1.25(c). So, the current,

$$I = \frac{10.0 + 2.5}{1} \times 10^{-6} = \frac{12.5}{1} \times 10^{-6} = 12.5\,\mu A$$

$$\therefore \quad v_o = v - 2.5 = 12.5 \times 10^{-6} \times 1 \times 10^6 - 2.5 = 10.0 \text{ V}$$

The output waveform is shown in Fig. 1.26.

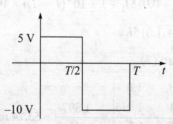

FIGURE 1.26 Circuit for Solved Problem 12.

13. A germanium p-n junction diode has an equilibrium hole density of 2×10^{11} holes/cm^3. The average diffusion length of the holes is 0.1 cm and the length of the n-section is 0.5 cm. The cross-sectional area at the junction is 0.001 cm^2. The diode is used at the room temperature of 300 K. Find the value of diffusion capacitance due to holes associated with a change in forward bias from 0.1 to 0.2 V.

Solution

The hatched area in Fig. 1.16 is given by

$$A = (\rho_{n2} - \rho_{n1}) \int_0^{l_n} e^{-x/l_p} dx$$

where l_n is the length of n-section. Hence,

$$A = (\rho_{n2} - \rho_{n1})[-l_p e^{-x/l_p}]_0^{0.5 \times 10^{-2}}$$
$$= (\rho_{n2} - \rho_{n1})[-0.1 \times 10^{-2}(e^{-0.5/0.1} - 1)]$$
$$= (\rho_{n2} - \rho_{n1}) \times 0.9932 \times 10^{-3}$$

Now,
$$\rho_{n1} = \rho_n e^{0.1/0.026} = 2 \times 10^{11} \times 10^6 \times 46.8 \text{ holes/m}^3$$
and $\quad \rho_{n2} = \rho_n e^{0.2/0.026} = 2 \times 10^{11} \times 10^6 \times 2191.4 \text{ holes/m}^3$

Change of charge per unit area $= A \times$ charge of an electron
$$= 2 \times 10^{17}(2191.4 - 48.6) \times 0.99326 \times 10^{-3} \times 1.602 \times 10^{-19} \text{ C/m}^3$$
$$= 6.82 \times 10^{-2} \text{ C/m}^3$$

Hence $dQ = 6.82 \times 10^{-2} \times 0.001 \times 10^{-4} = 6.82 \times 10^{-9}$ C
and $dV = 0.2 - 0.1 = 0.1$ V

$$\therefore \quad C_D = \frac{dQ}{dV} = \frac{6.82 \times 10^{-9}}{0.1} = 68200 \times 10^{-12} = \mathbf{68200 \text{ pF}}$$

1.10 EXERCISES

1. Find the dynamic resistance of a silicon diode having forward current of 20 mA at room temperature of 25°C.

2. A silicon p-n diode has the reverse saturation current of 20 µA at an ambient temperature of 500 K. A forward bias of 0.4 V is applied. Determine (a) the diode current, (b) the apparent diode resistance, and (c) the dynamic (or incremental) resistance.

3. A germanium p-n diode has a reverse saturation current of 20 µA. The diode is operated in an ambient temperature of 27°C and is placed in series with an unknown resistance and a 1-V d.c. source. The current through the diode is 100 mA. Find the value of unknown resistance.

4. For a germanium p-n diode operating at a temperature of 500 K, calculate the ratio of the current for a forward bias of 0.05 V to the current for the same magnitude of reverse-bias voltage.

5. For the reverse-bias voltage of 0.01 V for a germanium p-n diode, the current is 20 µA at the ambient temperature of 27°C. (a) Determine the saturation current for the diode. (b) At the same ambient temperature, find the incremental resistance for a 0.02 V bias in (i) forward direction and (ii) reverse direction.

6. Two identical silicon p-n diodes with saturation current of 0.1 μA and operating temperature of 18°C are employed in the circuit shown in Fig. 1.27. Find the voltage across each diode and the current in the circuit.

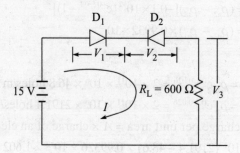

FIGURE 1.27 Circuit for Exercise 6.

7. The diodes in the circuit shown in Fig. 1.28 are ideal. (a) Write the transfer characteristic equation (v_o as a function of v_i). (b) Sketch v_o if $v_i = 40 \sin \omega t$. Indicate all voltage levels.

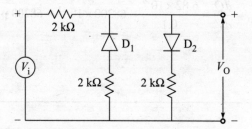

FIGURE 1.28 Circuit for Exercise 7.

2
TRANSISTORS

Outline

2.1 Bipolar Junction Transistors   34
2.2 Static Characteristics of Transistors   37
2.3 Low Frequency Analysis of Transistors   41
2.4 Field-Effect Transistor   45
2.5 Metal Oxide Semiconductor FET   51
2.6 Silicon-Controlled Rectifiers   54
2.7 Unijunction Transistors   60
2.8 Review Questions   62
2.9 Solved Problems   63
2.10 Exercises   71

2.1 BIPOLAR JUNCTION TRANSISTORS

A bipolar junction transistor consists of two back-to-back p-n junctions. As there are two junctions, this device is called bipolar junction transistor. The transistor is formed with a layer of n-type silicon (germanium) sandwiched between two layers of p-type silicon (germanium). This type of transistor is called a pnp transistor. When a layer of p-type is sandwiched between two layers of n-type material, the transistor is called an npn transistor. Figure 2.1 represents the two structures in principle with their circuit symbols. This device is extremely small and is sealed against moisture inside a metal or plastic case.

Figure 2.1 shows the three portions of the transistors known as *emitter*, *base*, and *collector*. The arrow on the emitter lead specifies the direction of current (hole) flow when the emitter-base junction is forward biased. In pnp transistors the arrow is shown going inside, whereas it is coming out in case of npn transistors. In both pnp and npn transistors, however, the emitter, base, and collector currents, I_E, I_B, and I_C, respectively, are taken positive when currents flow into the transistor. The voltages V_{EB}, V_{CB}, and V_{CE} are the emitter-base, collector-base, and collector-emitter voltages, respectively. This means that V_{EB} represents the voltage drop from emitter to base, V_{CB} is the voltage drop from collector to base, and V_{CE} is the voltage drop from collector to emitter.

In pnp transistor the heavily doped p region is called the emitter, the narrow central region is called the base, and the lightly doped p region is called the collector. The doping concentration in each region is assumed to be uniform. Under normal operating conditions, called active region, the emitter-base junction is forward biased and the collector-base junction is reverse biased, as shown in Fig. 2.2(a). The complement structure of the pnp transistor is the npn transistor, which is obtained by interchanging p for n and n for p. The active-mode circuit for the npn transistor is shown in Fig. 2.2(b). The current flows and voltage polarities are reversed. According to Kirchhoff's laws, there

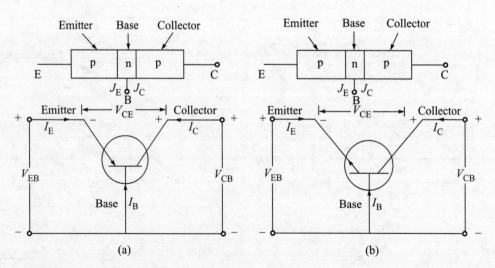

FIGURE 2.1 (a) A pnp transistor with its circuit representation and (b) an npn transistor with its circuit representation.

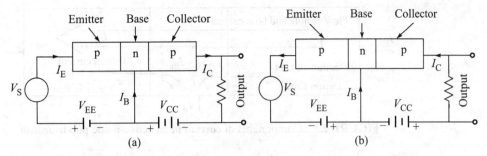

FIGURE 2.2 Active-mode circuits for (a) pnp and (b) npn transistors.

are only two independent currents for a three-terminal device. If two currents are known, the third current is also known.

Figure 2.2(a) and (b) illustrates circuits of transistors connected as amplifiers with common-base configuration. In common-base configuration, base lead is common to the input and output circuits. For further discussion on the operation of transistors, we consider a pnp transistor. The depletion layer width of the emitter-base junction is narrower than that of the collector-base junction. This is because the emitter-base junction is forward biased and collector-base junction is reverse biased. Since emitter-base junction is forward biased, holes are injected (or emitted) from the emitter into the base and electrons are injected from the base into the emitter. Under ideal diode condition, there is no generation-recombination current in the depletion region. The two current components constitute the total emitter current. This means that the total emitter current is the sum of the hole and electron currents, as they are opposite types of charges. The collector-base junction is reverse biased; therefore, a small reverse saturation current flows across the junction. However, if the base width is sufficiently narrow, the holes injected from the emitter can diffuse through the base to reach the collector-base depletion edge and then enter into the collector. If most of the injected holes can reach the collector without recombining with electrons in the base region, then the collector hole current is very close to the emitter hole current. Since the emitter emits holes, it is called the emitter, and the collector collects holes it is called the collector. Hence, when the emitter junction, J_E, is sufficiently close to the collector junction, J_C, the carriers injected from the emitter junction result in a large current flow in the reverse-biased collector junction. This action is the *transistor action*, and it can be realized only when the two junctions are physically close enough. If, on the other hand, the two junctions are so far apart that all the injected holes are recombined in the base before reaching the collector junction, the transistor action is lost. The pnp structure becomes only two diodes connected back-to-back.

2.1.1 Current Gain

Figure 2.3 shows the different current components in an ideal pnp transistor biased in active mode. The holes injected from the emitter form the current I_{Ep}, which is the largest current component in a well-designed transistor. Major portion of I_{Ep} reaches the collector junction and constitutes the current I_{Cp} ($I_{Ep} > I_{Cp}$). There are three components of base current, namely, I_{En}, I_{Bp}, and I_{Cn}. The component I_{En} corresponds to the current due to electrons being injected from the base to the emitter. I_{Bp} is the

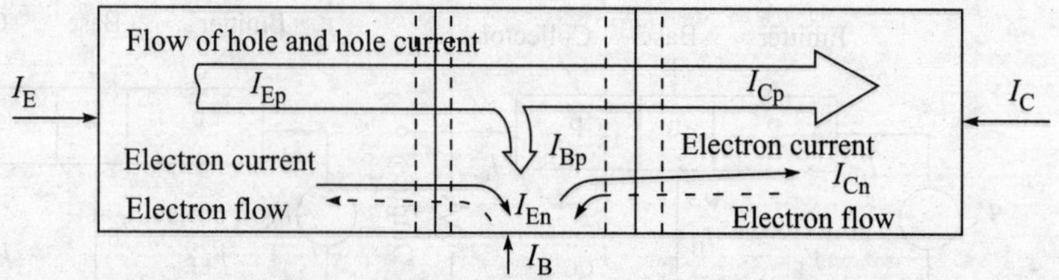

FIGURE 2.3 Components of current in an active-mode pnp transistor.

current that corresponds to electron supplied by the base to recombine with the injected holes. This means that $I_{Bp} = I_{Ep} - I_{Cp}$. I_{Cn} is the thermally generated electrons, which are near the collector junction edge and drift from the collector to the base. As shown in Fig. 2.3, the direction of the electron current flowing from emitter to base (shown by thick arrows) is opposite to the direction of the electron flow from base to emitter (shown by dashed arrows).

The terminal currents I_E, I_C, and I_B are expressed in terms of the various components as follows:

$$I_E = I_{Ep} + I_{En} \tag{2.1}$$

$$I_C = -(I_{Cp} + I_{Cn}) \tag{2.2}$$

$$I_B = -(I_E + I_C) = -(I_{Ep} + I_{En} - I_{Cp} - I_{Cn})$$
$$= -I_{En} - (I_{Ep} - I_{Cp}) + I_{Cn} \tag{2.3}$$

An important parameter of transistors is the *common-base current gain* α_o, which is defined as

$$\alpha_o = \frac{I_{Cp}}{I_E} \tag{2.4}$$

Substituting I_E from Eq. (2.1) and arranging, we obtain

$$\alpha_o = \frac{I_{Cp}}{I_{Ep} + I_{En}} = \frac{I_{Ep}}{I_{Ep} + I_{En}} \times \frac{I_{Cp}}{I_{Ep}} \tag{2.5}$$

The first term on the right-hand side of Eq. (2.5) is called the *emitter efficiency* γ, that is,

$$\gamma = \frac{I_{Ep}}{I_E} = \frac{I_{Ep}}{I_{Ep} + I_{En}} \tag{2.6}$$

The emitter efficiency measures the injected hole current compared to the total emitter current. The component I_{En} of I_E is not desirable and is minimized by using heavier emitter doping. Thus, the emitter efficiency γ is nearly equal to unity.

The second term on the right-hand side of Eq. (2.5) is called the *base transport function*

$$\alpha_T = \frac{I_{Cp}}{I_{Ep}} \tag{2.7}$$

which is the ratio of the hole current reaching the collector to the hole current injected from the emitter to the base. This is also approximately equal to unity. Hence,

$$\alpha_o = \gamma \alpha_T \tag{2.8}$$

As both γ and α_T for a well-designed transistor approach unity, α_o is very close to unity.
Substituting from Eq. (2.4) into Eq. (2.2), we obtain

$$I_C = -\alpha_o I_E - I_{Cn} \tag{2.9}$$

The component current I_{Cn} corresponds to the collector-base current flowing with the emitter open circuited ($I_E = 0$). As I_{Cn} is the current between the collector and the base with emitter open circuited, it is designated as I_{CBO}. Thus, Eq. (2.9) is written as

$$I_C = -\alpha_o I_E - I_{CBO} \tag{2.10}$$

2.2 STATIC CHARACTERISTICS OF TRANSISTORS

There are two static characteristics of transistors. They are input characteristics and output characteristics. The input static characteristics are plots of input voltage versus input current, with output voltage as a parameter. The output static characteristics are plots of output current versus output voltage, with input current as a parameter.

There are three configurations of a transistor, namely, (1) common-base configuration, (2) common-emitter configuration, and (3) common-collector configuration.

2.2.1 Common-Base Characteristics

Common-base circuit for a pnp transistor is shown in Fig. 2.4. This configuration is also known as grounded-base configuration. In this configuration, base is common to the input and the output. For a pnp transistor, the largest current components are due to holes. In Fig. 2.4, all indicated currents and voltages are positive. For an npn transistor, all current and voltage polarities are negative of those for a pnp transistor.

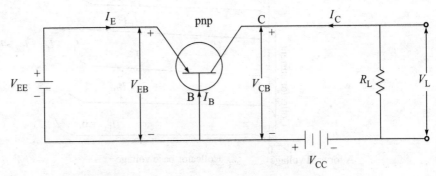

FIGURE 2.4 A common-base circuit for a pnp transistor.

Input static characteristics

If V_{CB} and I_E are chosen as independent variables, the input voltage (emitter-to-base voltage) V_{EB} is completely determined from these two variables. The plots of V_{EB} versus I_E with V_{CB} as a parameter are called input (or emitter) static characteristics. The input characteristics are shown in Fig. 2.5.

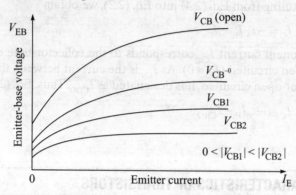

FIGURE 2.5 Input static characteristics.

A transistor, as discussed earlier, consists of two diodes placed in series 'back-to-back'. In the active region, the input diode (emitter to base) is forward biased. The input characteristics of Fig. 2.5 are simply the forward characteristics of the emitter to base diode for various collector-base voltages. Like diode characteristic, there exists a threshold voltage $V\gamma$ below which the emitter current is very small. In general, $V\gamma \approx 0.1$ V for germanium and $V\gamma \approx 0.5$ V for silicon. It can be noted that with the increase of V_{CB}, the emitter current increases for a constant V_{EB}. Thus, the curves shift downwards as V_{CB} increases. The curve with collector open circuited represents the characteristic of the forward-biased emitter-base diode.

Output static characteristics

From Eq. (2.10), we see that the collector (or output) current is completely determined by the emitter current and the saturation current. The saturation current is independent

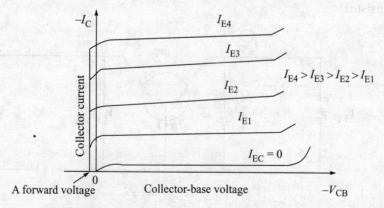

FIGURE 2.6 Output characteristics for pnp transistors.

of the voltage V_{CC} except for very small value of V_{CC}. Also, the saturation current is very small compared to the emitter current. Hence, the output current is approximately equal to the emitter current. The plots of the output current (I_C) versus the collector-base voltage V_{CB}, with emitter current as a parameter, are shown in Fig. 2.6. These characteristics are called output (or collector) static characteristics.

Figure 2.6 shows that the collector current is practically equal to the emitter current and virtually independent of V_{CB}. To reduce the collector current to zero, the base-collector junction voltage is made forward biased to some extent. For a silicon transistor it is approximately 1 V, and for germanium transistor it is 0.25 V.

2.2.2 Common-Emitter Characteristics

Common-emitter configuration is the most-extensively used type of transistor connection, which gives the largest power and current amplification. The common-emitter configuration, which is also called grounded-emitter configuration, is shown in Fig. 2.7.

Neglecting the saturation current, in magnitude, the collector current can be given as

$$I_C = \alpha_o I_E \tag{2.11}$$

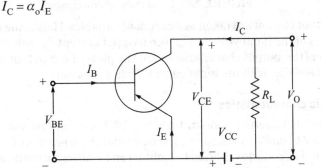

FIGURE 2.7 A common-emitter circuit for a pnp transistor.

Applying Kirchhoff's law, the base current, in magnitude, is given by

$$I_B = I_E - I_C$$
or
$$I_B = (1 - \alpha_o) I_E \tag{2.12}$$

Thus, from Eqs. (2.11) and (2.12),

$$I_C = \left[\frac{\alpha_o}{(1 - \alpha_o)}\right] I_B$$
$$= \beta I_B \tag{2.13}$$

The quantity β is called the d.c. common-emitter short-circuit current gain. For α_o near to unity, β is much larger than unity. For $\alpha_o = 0.99$, the corresponding value of β is 99. Thus, it is evident that a small change in the base current is accompanied by a large change in the collector current.

From above discussion it is clear that the base current has a greater control on the collector current than the emitter current. The common-emitter configuration, shown in Fig. 2.7, for a pnp transistor, is most frequently used in practice.

Like common-base configuration, the input current and the output voltage are taken as independent variables in the common-emitter configuration. The input voltage

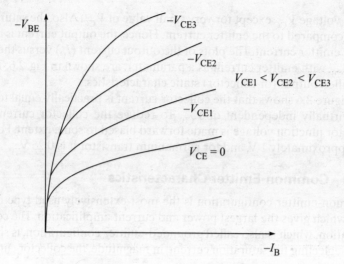

FIGURE 2.8 Input static characteristics.

and the output current are taken as dependent variables. Hence, the input characteristics are the plots of the input voltage V_{BE} versus input current I_B, with output voltage V_{CE} as a parameter. The output characteristics are the plots of the output current I_C versus the output voltage V_{CE}, with the input current I_B as a parameter.

Input static characteristics

Typical input characteristics are shown in Fig. 2.8. For a constant value of V_{CE}, it is a curve of input (base to emitter) voltage V_{BE} versus input (base) current I_B. It can be noted that, with $V_{CE} = 0$ (i.e., collector shorted to emitter) and emitter forward biased, the input characteristic is essentially that of a forward-biased diode. If $V_{BE} = 0$, $I_B = 0$ as both emitter and collector junctions are shorted. In general, with constant V_{BE}, increase in V_{CE} decreases the base current (I_B).

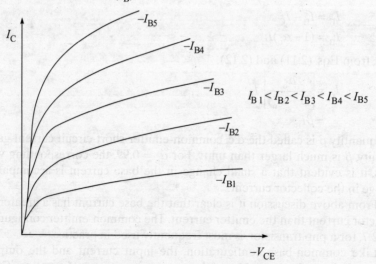

FIGURE 2.9 Output static characteristics.

Output static characteristics

The output characteristic curves for a pnp transistor are shown in Fig. 2.9. The x-axis is the collector-emitter voltage V_{CE}, the y-axis is the collector current I_C, and the curves are given for various values of I_B. For a particular value of I_B, I_C is not very sensitive to V_{CE}. This is because the collector-base junction is reverse bias, and reverse-biased component of collector current is negligible. However, the slopes of the curves in common-emitter case are larger than that in common-base case. Base current is much small than the emitter current.

2.2.3 Common-Collector Characteristics

Common-collector configuration is shown in Fig. 2.10. In this configuration, as in common-emitter configuration, the base current serves as the input current, but the output current is drawn from the emitter. It is the collector grounded configuration. The circuit is basically the same as the circuit of common-emitter configuration. The exception is that the load resistance is in the emitter lead rather than in the collector circuit. When the base current is the saturation current, the emitter current is zero and no current flows through the load. When the base current is increased, the transistor passes through the active mode and eventually reaches saturation. In this condition, all the supply voltage, except for a very small drop across the transistor, appears across the load.

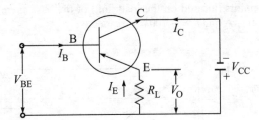

FIGURE 2.10 A common-collector circuit for a pnp transistor.

2.3 LOW FREQUENCY ANALYSIS OF TRANSISTORS

For small signals, the transistor operates with reasonable linearity; therefore, a linear model of the transistor in the active region can be obtained for small signals. The response of the transistor for large signals is obtained graphically. The common-emitter configuration is considered because this configuration is most-commonly used.

2.3.1 Graphical Analysis

The common-emitter configuration for a pnp transistor is shown in Fig. 2.7. The input and output characteristics are shown in Fig. 2.8 and Fig. 2.9, respectively. For graphical analysis, the output characteristic, with a load line for the load resistance R_L, is drawn in Fig. 2.11. The load line is given by Eq. (2.14):

$$V_{CC} + v_C + i_C R_L = 0 \tag{2.14}$$

Before discussing the graphical analysis by an illustrative example, the symbols for different voltages and currents are summarized in Table-2.1.

The following example illustrates the procedure of graphical analysis.

Table-2.1 Symbols of voltages and currents.

	Base (collector) voltages with respect to emitter	Base (collector) current towards electrode from external circuit
Instantaneous total value (d.c. + a.c.)	v_B (v_C)	i_B (i_C)
Quiescent value (d.c.)	V_B (V_C)	I_B (I_C)
Instantaneous value (a.c.)	v_b (v_c)	i_b (i_c)
Effective (r.m.s.) value	V_b (V_c)	I_b (I_c)
Supply voltage (magnitude)	V_{BB} (V_{CC})	

Example 2.1

For a common-emitter circuit of pnp transistor, the output characteristics are shown in Fig. 2.11. The supply voltage V_{CC} = 15 V. The load resistance is 250 Ω. Assume a 200 μA peak sinusoidally varying base current around the quiescent point Q where I_B = − 300 μA.

Solution

The extreme points (located on the load line) of the base current waveform are A and B, where i_B = − 500 μA and i_A = − 100 μA, respectively.

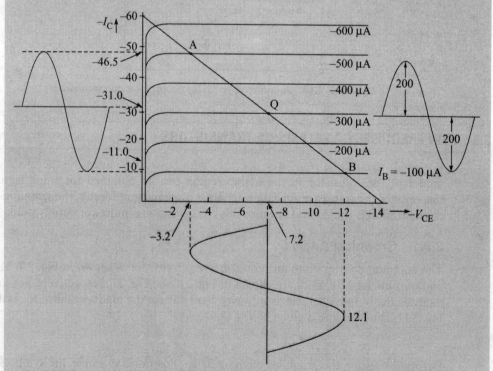

FIGURE 2.11 An experimental plot of output characteristic of common-emitter circuit.

For any value of base current, the corresponding values of i_c and V_{CE} are obtained at the intersection of the load line and the output characteristics. For example, for $i_B = -500$ µA (point A), $i_C = 46.5$ mA and $V_{CE} = -3.2$ V. For $i_B = -100$ µA (point B), $i_C = -11.0$ m and $V_{CE} = 12.1$ V. Hence, the waveforms of i_C and V_{CE} can be obtained as shown in Fig. 2.11.

From i_C and v_{CE} waveforms, it is clear that these waveforms are not purely sinusoidal. Both peaks are not equal. The waveforms are distorted sinusoids. The distortions are because the output characteristics in the neighbourhood of the load line are not parallel lines equally spaced for equal change in i_B. The distortion in waveform is called output non-linear distortion.

The base-to-emitter voltage v_{BE} for any combination of base current and collector-to-emitter voltage can also be obtained from the input characteristics.

2.3.2 Linear Model

As said earlier, the response of a transistor for small signal at low frequency is linear, and it can be analyzed analytically rather than graphically. For very small signals, the analytical technique would offer very accurate results. Therefore, a linear-circuit model for the transistor is obtained. The linear-model network is, then, used to analyze the transistor response.

The transistor is represented by a two-port active device as shown in Fig. 2.12(a). The terminal behaviour of two-port devices is, generally, specified by two voltages and two currents. If the two independent qualities are the input current and the output voltage for the linear model of the two-port active network, such as in a transistor, the two dependent quantities are expressed as follows:

$$v_i = h_i i_i + h_r v_o \tag{2.15}$$

$$i_o = h_f i_i + h_o v_o \tag{2.16}$$

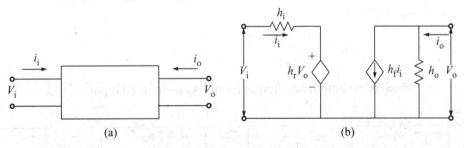

FIGURE 2.12 (a) A two-port active network and (b) its linear model.

These equations represent a hybrid model of a two-port network. The equivalent circuit for the above equations is shown in Fig. 2.12(b).

For common-emitter case, i_b is the input current and v_c is the output voltage. So, the hybrid model equations are expressed as follows:

$$v_b = h_{ie} i_b + h_{re} v_c \tag{2.17}$$

$$i_c = h_{fe} i_b + h_{oe} v_c \tag{2.18}$$

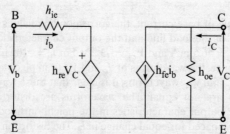

FIGURE 2.13 An equivalent circuit for common-emitter case.

The equivalent circuit for the above equations is shown in Fig. 2.13. The parameter h_{ie} is given by

$$h_{ie} = \left.\frac{v_b}{i_b}\right|_{v_c=0} \tag{2.19}$$

which is the input impedance for a short-circuited output. Therefore, the parameter h_{ie} is obtained by the ratio of the incremental base-emitter voltage to incremental base current, the output terminal being short circuited. The parameter h_{re} is given by

$$h_{re} = \left.\frac{v_b}{v_c}\right|_{i_b=0} \tag{2.20}$$

which is the reverse voltage transfer ratio for an open-circuited input. The parameter h_{fe} given by

$$h_{fe} = \left.\frac{i_c}{i_b}\right|_{v_c=0} \tag{2.21}$$

which is the forward current transfer ratio for a short-circuited output. The parameter h_{oe} is given by

$$h_{oe} = \left.\frac{i_c}{v_c}\right|_{i_b=0} \tag{2.22}$$

which is the output admittance for an open-circuited input.

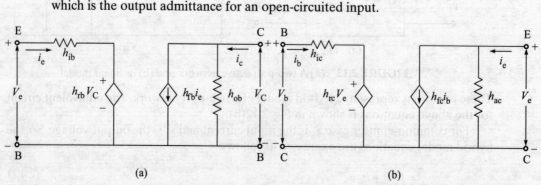

(a) (b)

FIGURE 2.14 An equivalent circuit for (a) common-base and (b) common-collector configurations.

The equivalent circuit for common-base and common-collector configurations are shown in Fig. 2.14(a) and (b), respectively.

2.4 FIELD-EFFECT TRANSISTOR

A field-effect transistor (FET) is a semiconductor device that is used to provide the circuit properties of a controlled source. Usually, this device is referred to as FET. The device is also called junction field-effect transistor (JFET). It is called field-effect transistor, as it depends for its operation on the control of current by an electric field.

The structure of an n-channel FET is shown in Fig. 2.15(a). It is a combination of two p-n junctions. Ohmic contacts are made to the two ends of the n-type semiconductor material. A p-channel FET is shown in Fig. 2.15(b). Current is caused to flow along the channel bar because the voltage supply is connected across its two ends. This current is that of majority carriers, which are electrons in n-channel FET and holes in p-channel FET.

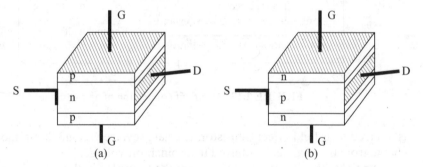

FIGURE 2.15 Structure of FET (a) n channel and (b) p channel.

The terminal from which the majority carriers emit is called the source (marked S) and the terminal marked D collects the majority carriers and is called the drain. The two contacts G are called the upper gate and the lower gate. The region of n-type material between the two gate regions is the channel through which the majority carriers move from source to drain. The p-n junction is reverse biased. For n channel the drain is made positive with respect to the source, whereas for p channel the drain is made negative with respect to the source. The symbol for an n-channel FET is shown in Fig. 2.16(a), whereas that for a p-channel FET is shown in Fig. 2.16(b). The direction of the arrows at the gate indicates the direction in which gate current would flow if the gate junction were forward biased.

The circuit connections for an n channel are shown in Fig. 2.17. There are two depletion regions at two reverse-biased p-n junctions. The movement of the majority carriers through the channel depends on the width of the channel. The effective width of the channel depends on the thickness of the depletion region. The thickness of the depletion region increases from S to D. Here, for a fixed drain-to-source voltage, the drain current is a function of the reverse-biasing voltage across the gate junction. Since the movement of the current is controlled by the field associated with the depletion region, the device

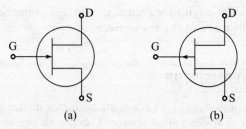

FIGURE 2.16 Symbol for (a) an n-channel FET and (b) a p-channel FET.

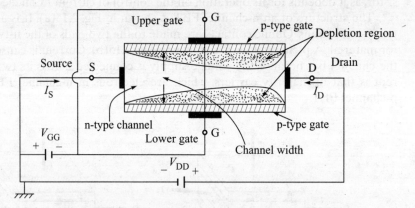

FIGURE 2.17 A circuit connection of an n-channel FET.

is referred to as field-effect transistor. The gate reverse voltage that removes all the free charge from the channel is referred to as pinch-off voltage, V_p.

The field-effect transistor differs from the transistor in the following important characteristics:

1. The operation of an FET depends on the flow of majority carriers only; therefore, it is called a unipolar device, whereas the transistor is called a bipolar device.
2. The fabrication of an FET is simpler, and the space occupied in the integrated form is less as compared to that of a transistor.
3. The input resistance of an FET is high, typically many megaohm, compared to that of a transistor.
4. An FET is less noisy than a bipolar transistor.
5. In FET there is no offset voltage at zero drain current; hence, it makes an excellent signal chopper.

The main disadvantage of the FET is its relatively small gain-bandwidth product.

2.4.1 Static Characteristic

The circuit for common-source configuration for an n-type FET is shown in Fig. 2.18. The drain current I_D and drain-to-source voltage V_{DS} is positive, whereas the gate-to-source voltage V_{GS} is negative. The drain-current characteristic against V_{DS} with V_{GS} as a parameter is shown in Fig. 2.19.

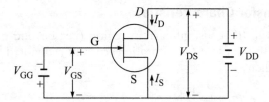

FIGURE 2.18 A circuit for a common-source configuration.

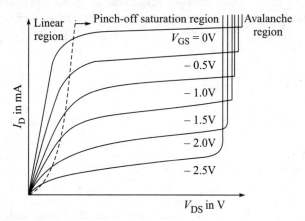

FIGURE 2.19 The drain-current characteristic of an n-channel FET.

Let us consider the characteristic for $V_{GS} = 0$. For $I_D = 0$, the channel between the gate junctions is entirely open. For small applied voltage V_{DS}, the channel acts as a simple semiconductor resistor, and I_D increases linearly with V_{DS}. With increasing I_D, the ohmic voltage drops between the source, and the channel length reverses the junction biasing. So, the width of the channel begins to decrease. The decrease of the width along the length of the channel is not uniform, but the decrease is more at distances farther from the source as shown in Fig. 2.17. There is voltage V_{DS}, equal to V_P, at which the I_D begins to level off and approaches a constant. In principle, at V_{DS} equal to pinch-off voltage V_P, I_D should be zero. But if I_D could be zero, the ohmic drop required to provide reverse bias would itself disappear. Thus, a stable condition is reached at which I_D is constant. It is clear that each characteristic curve has an ohmic region for small values of V_{DS}, where I_D is proportional to V_{DS}. Each characteristic also has a constant-current region for a large value of V_{DS}, where I_D responses very slightly to V_{DS}.

When a reverse V_{GS} is applied, it provides an additional reverse bias, and the pinch-off occurs for smaller values of $|V_{DS}|$. The maximum drain current also reduces. The maximum voltage that can be applied across any two terminals of the FET is the minimum voltage that causes avalanche breakdown across the gate junction. The avalanche breakdown occurs at a lower voltage $|V_{DS}|$ when the gate is reverse biased than when $V_{GS} = 0$. This is because the reverse-biased gate voltage adds to drain voltage. This increases the effective voltage across the gate junction.

2.4.2 Transfer Characteristic

FET is used in the constant-current region (i.e., beyond pinch-off), when used as an amplifier. If the constant (or saturation) current is denoted by I_{DS} and its value for $V_{GS} = 0$ by I_{DSS}, then

$$I_{DS} = I_{DSS}\left(1 - \frac{V_{GS}}{V_P}\right)^2 \tag{2.23}$$

where V_{GS} is the reverse-biased voltage.

The transfer characteristic relates the drain current to reverse-biased voltage between the gate and source. The transfer characteristic of an n-channel FET is shown in Fig. 2.20.

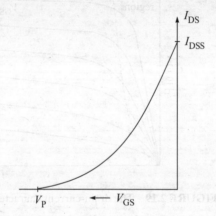

FIGURE 2.20 The transfer characteristic of an n-channel FET.

2.4.3 Signal Transfer Gain

Figure 2.21 depicts an FET circuit containing a time varying signal in the input circuit and a resistive load in the output circuit. As per the earlier discussion, the current i_D in the output circuit can be controlled by varying the voltage applied to input (or gate) circuit.

The output current i_D is composed of a d.c. component I_D and an a.c. component i_d, thereby implementing a signal transfer from the input to the output circuit. The gate circuit contains the reverse-biased gate junction of the FET, and so the gate current is negligible and considered zero. The input signal controls the gate voltage v_G without spending any appreciable amount of power in driving the input. In contrast, the output current i_D may contain an appreciable a.c. component. Thus, a significant amount of signal power is delivered to the load and the transfer signal from the input to the output is accompanied by a power gain.

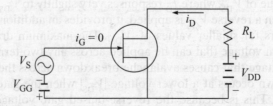

FIGURE 2.21 An FET circuit containing input signal and output load.

Graphical analysis

The following example illustrates the procedure.

Example 2.2

In the circuit of Fig. 2.21 the supply voltage $V_{DD} = 20$ V and $R_L = 5$ kΩ. The drain-current characteristics are shown in Fig. 2.22. Assume a 0.5-V peak sinusoidally varying input signal voltage around the quiescent point Q, where $V_{GG} = +3$ V.

Solution.

The extreme points of the input signal (located on the load line) are A and B, where $V_A = -2.5$ V and $V_B = -3.5$ V, respectively.

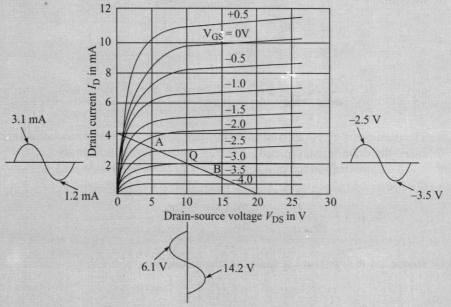

FIGURE 2.22 The drain-current characteristics.

For any value of gate voltage v_{GS}, the corresponding value of drain currents i_D and drain voltage v_{DS} are obtained from the intersection of the drain characteristics and the load line. For example, for $v_{GS} = -2.5$ V (point A), $i_D = 3.1$ mA and $v_{DS} = 6.1$ V. For $v_{GS} = -3.5$V (point B), $i_D = 1.2$ mA and $V_{DS} = 14.2$ V. Similarly, other points of the i_D and v_{DS} waveforms can be obtained as shown in Fig. 2.22.

2.4.4 Low-Frequency Small-Signal Model

As seen earlier, the drain current i_D depends on the gate voltage v_{GS} and drain voltage v_{DS}, and so i_D can be expressed as a function of v_{GS} and v_{DS}, that is,

$$i_D = f(v_{GS}, v_{DS}) \tag{2.24}$$

For small changes in gate and drain voltages, the change in the drain current is approximately expressed by the first two terms of the Taylor's series expansion of Eq. (2.24), that is,

$$\Delta i_D = \left.\frac{\partial f}{\partial v_{GS}}\right|_{v_{DSQ}} \Delta v_{GS} + \left.\frac{\partial f}{\partial v_{DS}}\right|_{v_{GSQ}} \Delta v_{DS}$$

or
$$i_d = g_m v_{gs} + \frac{1}{r_d} v_{ds} \qquad (2.25)$$

as the small-signal notation given in Table-2.1, $\Delta i_D = i_d$, $\Delta v_{GS} = v_{gs}$, and $\Delta v_{DS} = v_{ds}$. Further we define

$$g_m = \frac{\partial f}{\partial v_{GS}} = \frac{\partial i_D}{\partial v_{GS}} = \frac{\Delta i_D}{\Delta v_{GS}} = \frac{i_d}{v_{gs}} \qquad (2.26)$$

and
$$\frac{1}{r_d} = \frac{\partial f}{\partial v_{DS}} = \frac{\partial i_D}{\partial v_{DS}} = \frac{\Delta i_D}{\Delta v_{DS}} = \frac{i_d}{v_{ds}} \qquad (2.27)$$

The parameter g_m is referred to as mutual conductance or transconductance and r_d is called the drain resistance and $g_d = 1/r_d$ is called drain conductance.

An amplification factor μ for an FET can be defined as

$$\mu = -\left.\frac{\partial v_{DS}}{\partial v_{GS}}\right|_{I_{DQ}} = -\left.\frac{\Delta v_{DS}}{\Delta v_{GS}}\right|_{I_{DQ}} = -\left.\frac{v_{ds}}{v_{gs}}\right|_{i_d = 0} \qquad (2.28)$$

Hence, setting $i_d = 0$ in Eq. (2.25), in magnitude,

$$g_m = \frac{1}{r_d} \cdot \frac{v_{ds}}{v_{gs}} = \frac{1}{r_d} \mu$$

or
$$\mu = g_m r_d \qquad (2.29)$$

The linear model for small signal that satisfies Eq. (2.25) is shown in Fig. 2.23.

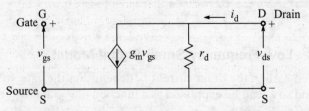

FIGURE 2.23 A low-frequency small-signal FET model.

2.5 METAL OXIDE SEMICONDUCTOR FET

A second type of field-effect transistor is the metal oxide semiconductor field-effect transistor (MOSFET). This is commonly referred to as MOSFET. There are two classes of MOSFET:

1. Enhancement MOSFET
2. Depletion MOSFET

Each of these types can be p-channel MOSFET or n-channel MOSFET. The device consists of four electrodes, namely, source (S), drain (D), gate (G), and substrate (B). In operation, the substrate is almost always directly connected to the source. The p-channel MOSFET and n-channel MOSFET structures are complementary. In a p-channel MOSFET, the substrate is n-type and the source and drain are p-type, whereas in an n-channel MOSFET, the substrate is p-type and the source and drain are n-type. The structures of p-channel and n-channel MOSFETs are shown in Fig. 2.24(a) and (b), respectively. The gate is separated from the semiconductor channel by an insulated layer of silicon dioxide; hence, the device is also called insulated gate field-effect transistor.

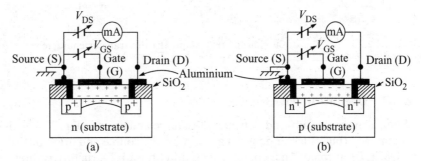

FIGURE 2.24 (a) The p-channel and (b) the n-channel-enhanced MOSFET.

2.5.1 Enhancement MOSFET

The p-channel MOSFET has a lightly doped n-type substrate into which two highly doped p-type regions are diffused as shown in Fig. 2.24(a). These are called p+ regions to emphasize the very large p impurity content. These p+ type regions act as the source and drain. There is a thin layer of insulating silicon dioxide (SiO_2) over the surface of the structure. Two holes are cut into the oxide layer for making metal contacts with the source and drain, respectively. A metal area is overlaid on the oxide layer to form the gate, as shown in Fig. 2.24(a).

If the substrate of the structure in Fig. 2.24(a) is grounded and a negative charge is applied at the gate, an electric field directing perpendicularly through the oxide is produced. Because of the lightly doped n-type substrate and the heavily doped p+ sections of source and drain, a widely spread depletion region is created that completely surrounds the p+ sections. This electric field brings about a polarization of charge within the insulator as shown in Fig. 2.24(a). The negative charges take on a position of alignment along the inner surface of the insulator and these negative charges attract the holes in the depletion region to the area of the insulator. This creates an induced p channel [as shown in Fig. 2.24(a)], which bridges the source and drain sections of the MOSFET. If a negative drain voltage is applied between the D and S terminals, p-channel holes provide a conducting path between the two p+ sections. Thus, an electric current flows in the structure from the source to the drain. The drain current is enhanced by the increase of negative gate voltage and this device is, therefore, called an enhancement MOSFET. The operation of the n-channel enhancement MOSFET of Fig. 2.24(b) is similar to the p-channel operation except that the voltages of opposite polarities are used, as shown in Fig. 2.24(b). The volt-ampere drain characteristics of a p-channel enhancement MOSFET are shown in Fig. 2.25(a) and its transfer characteristics in Fig. 2.25(b).

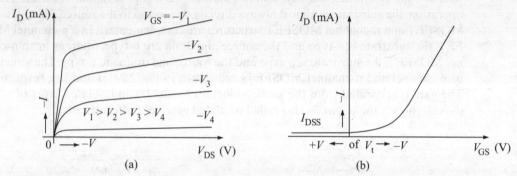

FIGURE 2.25 (a) The drain characteristics and (b) the transfer characteristics of enhancement MOSFET.

2.5.2 Depletion MOSFET

When a channel is diffused between the source and drain in Fig. 2.24, with the same type of impurity (but of lesser density) as used for the source and drain diffusion, a depletion MOSFET is formed. An n-channel depletion MOSFET structure is shown in Fig. 2.26(a). For this device an appreciable drain current flows for $V_{GS} = 0$ compared to that obtained in enhancement MOSFET. When gate voltage is made negative, positive charges are induced in the channel through the insulator of gate capacitor. As the current in the device is due to majority carriers (electrons for an n-type), these induced positive charges make the channel less conductive. Therefore, drain current decreases as V_{GS} is made more negative. The redistribution of positive charges in the channel, as shown in

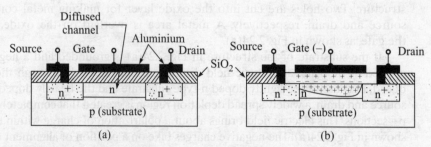

FIGURE 2.26 (a) An n-channel depletion MOSFET and (b) redistribution of positive charges in the channel.

Fig. 2.26(b), causes an effective decrease (or depletion) of majority carriers. Hence, this type of MOSFET is named as depletion MOSFET. Because of the voltage drop due to drain current, the channel region nearest to the drain is more depleted, that is, has more positive charges than the region near to the source, as shown in Fig. 2.26(b). This phenomenon is similar to that of pinch-off occurring in an FET at the drain end of the channel. So, the volt-ampere characteristics of the depletion MOSFET are very similar.

The depletion MOSFET may also be operated with positive gate voltage, that is, $V_{GS} > 0$. When $V_{GS} > 0$, a negative charge is induced in the diffused n channel and thereby conductivity of the channel is increased. This in turn increases the drain current. This is the enhancement mode of operation of the device. Because this device can be operated either in depletion mode or in enhancement mode, it is also called depletion-enhancement MOSFET or simply DE MOSFET. The volt-ampere drain characteristics of the device are shown in Fig. 2.27(a), and the transfer characteristic is shown in Fig. 2.27(b).

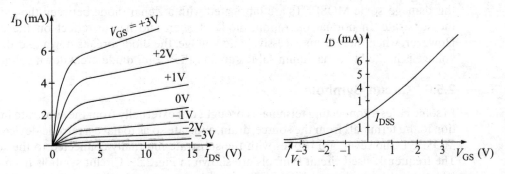

FIGURE 2.27 (a) The drain characteristics and (b) the transfer characteristic of DE MOSFET.

2.5.3 Complementary MOSFET

Both n-channel and p-channel enhancement-type MOSFETs are fabricated together in complementary orientation in the n substrate and the p substrate as shown in Fig. 2.28. The diffusion pattern is such that the drain of the p-channel MOSFET (D_1) is permanently joined to the drain of the n-channel MOSFET (D_2). Such complementary MOSFET (CMOSFETs) are very frequently used in digital system for building inverters and logic gates.

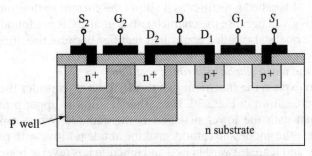

FIGURE 2.28 A complementary MOSFET.

2.5.4 Comparison of n-Channel and p-Channel MOSFET

The p-channel enhancement MOSFET is very popular because it is easier to produce and is cheaper than the n-channel MOSFET. However, p-channel devices have more than twice the area of the n-channel devices to achieve the same resistance. The n-channel MOSFET is three times faster than the equivalent p-channel MOSFET and is often preferred, especially for high-speed applications. However, the fabrication of n-channel devices is expensive, and so n-channel devices are unable to compete economically with p-channel devices.

2.5.5 MOSFET Gate Protection

The silicon dioxide layer of the gate is extremely thin (of the range of 0.2 μm); hence, it may be easily damaged by excessive voltage. The collection of charge on an open-circuited gate may produce a large field that may damage the dielectric. To protect from this damage, some MOSFETs are fabricated with a Zener diode between the gate and the substrate. In normal operation, diode is open and has no effect on the circuit. However, when there is an excessive gate voltage, the diode breaks down and the gate voltage is limited to a maximum value equal to the Zener diode breakdown voltage.

2.5.6 Circuit Symbols

In some cases a connecting terminal is brought out externally from the substrate in addition to the terminals from the source, drain, and gate so as to have a tetrode device. Most MOSFETs, however, are triodes with the substrate internally connected to the source. The frequently used circuit symbols are shown in Fig. 2.29. Circuit symbols from (a) to (f) are used for n-channel MOSFETs and those from (g) to (i) are used for p-channel MOSFETs. Circuit symbols in Fig. 2.29(a), (b), (g), and (h) are used for depletion MOSFETs, whereas those in Fig. 2.29(c), (d), (i), and (j) are used for both depletion and enhancement MOSFETs. Circuit symbols in Fig. 2.29(e), (f), (k), and (i) are used for enhancement MOSFETs.

2.6 SILICON-CONTROLLED RECTIFIERS

Silicon-controlled rectifier (SCR) consists of four alternate layers of p-type and n-type semiconductors. As this device consists of three terminals, it is also known as *thyristor*. The SCR is called a rectifier as it allows the current to flow only in one direction. Silicon name is given because the characteristics of an SCR are found in silicon-type diode only. This device is also called controlled rectifier because the firing of the SCR is controlled by applying a voltage to the third electrode. Therefore, this semiconductor device is given the name *silicon-controlled rectifier*.

To explain the functioning of the SCR, let us consider the four-layer semiconductor structure shown in Fig. 2.30. The terminal with the upper p region is the anode and the terminal with the lower n region is the cathode. When a positive voltage is applied between the anode and cathode making anode positive with respect to the cathode, junctions j_1 and j_2 are forward biased and junction j_2 is reverse biased. Junctions j_1 and j_3 being

Sec. 2.6 / Silicon-Controlled Rectifiers 55

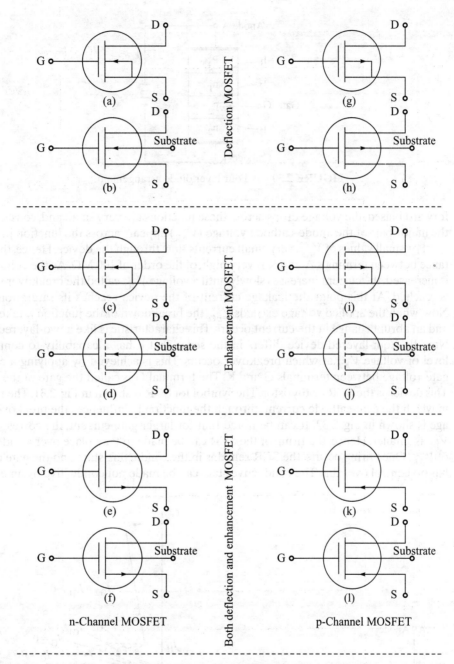

FIGURE 2.29 (a)–(f) Circuit symbols for n-channel MOSFET and (g)–(l) circuit symbols for p-channel MOSFET.

56 Chapter 2 / Transistors

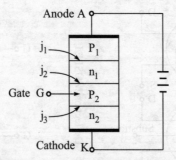

FIGURE 2.30 A Four-layer diode arrangement.

forward biased, the voltage drops across these junctions are very small and, consequently, the major part of the anode-cathode voltage (V_{AK}) appears across the junction j_2.

For small values of V_{AK}, very small currents flow through the device. Hence, the resistance between terminals A and K is very high, of the order of 100 MΩ. As the voltage V_{AK} is increased, the current increases slowly until a voltage V_{BO}, called the breakover voltage, is reached. At this point the leakage current of the device reaches its saturation value. Now, when the applied voltage exceeds V_{BO}, the breakdown of the junction j_2 takes place and an abrupt change in the current occurs. This characteristic is like a two-layered diode. Now, the four-layered device differs in the sense that it has a capability to control the level of voltage V_{AK} at which breakover occurs. This is achieved by applying a cathode gate voltage between terminals G and K. The terminal G is called the gate of the device. This device is the SCR or thyristor. The symbol for SCR is shown in Fig. 2.31. The manner in which the gate-cathode current, through the junction j_3, influences the breakover voltage is shown in Fig. 2.32. It can be noted that for larger gate current, the corresponding V_{BO} is smaller. Hence, the firing of the SCR can be made to take place over a wide range of V_{AK}. Once firing occurs, the SCR remains in the conducting state and the gate current has no control over this. The conductive state can be made nonconductive by an external

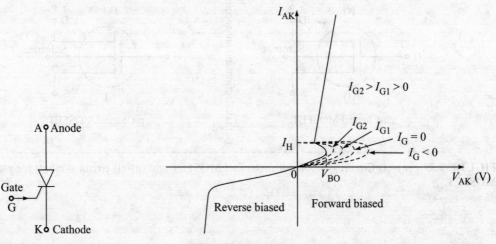

FIGURE 2.31 Symbol for SCR.

FIGURE 2.32 The current-volt characteristic of an SCR.

circuit by reducing the current below I_H, the holding current. In that case, the SCR returns to its OFF state. Thus, the SCR remains either in ON state or in OFF state. When the gate current is very large, the breakover may occur at so a low voltage that the characteristic looks like the characteristic of a simple p-n diode.

Typical ratings of an SCR are 1.5 kA and 10 kV, which corresponds to 15 MW power handling capacity. This power can be controlled by a gate current of the order of 1 A. Turning on the SCR is also known as triggering of the SCR. The various methods of triggering are as follows:

(i) Voltage triggering
(ii) Gate triggering
(iii) Voltage-rate triggering
(iv) High-temperature triggering
(v) Light triggering

Voltage triggering. As said earlier, when the forward voltage is increased to a value equal to or greater than the breakover voltage, the SCR turns on itself. As a high forward voltage may destroy the device, this method is not used in practice.

Gate triggering. A positive gate voltage between the gate and cathode injects a current which, in turn, triggers the SCR when it is in forward blocking state. A higher gate current means lower breakover voltage. The relation between the breakover voltage and gate current is shown in Fig. 2.33.

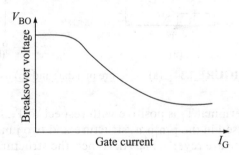

FIGURE 2.33 Breakover voltage versus gate current curve.

Voltage-rate triggering. The reverse-biased junction of the SCR behaves like a capacitor and a high rate of voltage change gives a high charging current through the junction. A high charging current may trigger the SCR. This method is also not used in practice as high charging current may damage the SCR.

High-temperature triggering. The increase in temperature increases the leakage current through the reverse-biased junction, and at a certain temperature, therefore, the junction may breakdown and the SCR starts conducting. High-temperature triggering may cause thermal run away. Therefore, this method is generally avoided.

Light triggering. In this method, light particles are made to strike the reverse-biased junction, which causes an increase in the electron-hole pairs and in turn triggers the SCR. This method is used in light-activated SCRs.

The main field of application of the SCR is in high-power circuits. SCRs are used in rectifiers, inverters, and choppers.

2.6.1 Silicon-Controlled Switch

Silicon-controlled switch (SCS) is similar to an SCR, except it is smaller in size and is designated to operate at lower currents and voltages. These devices are made for low-level applications. These switches have lower leakage and holding currents than SCRs. The silicon-controlled switches require small triggering signals.

2.6.2 Diac (Bidirectional Diode)

A diac is also a four-layered device. This is a diode for a.c. and so it is called diac. The structure of a diac is given in Fig. 2.34(a) and its circuit symbol in Fig. 2.34(b). It has two terminals. It consists of two four layers, $p_1n_1p_2n_2$ and $p_2n_1p_1n'$.

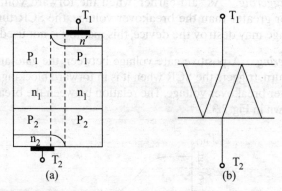

FIGURE 2.34 (a) Structure of a diac and (b) its circuit symbol.

When the terminal T_1 is positive with respect to T_2, the junction between n_1 and p_2 is reverse biased in the $p_1n_1p_2n_2$ structure and two junctions n'-p_1 and n_1-p_2 of the $p_2n_1p_1n'$ structure are reverse biased. Hence, the structure $p_1n_1p_2n_2$ starts conducting when T_1 reaches at a voltage more than V_{BO}. Once the conduction starts, the current through the diac is very large and can be limited by an external circuit resistance. The current-volt characteristic for this structure is given in the first quadrant of Fig. 2.35. Similarly, when T_2 is positive with respect to T_1 (in the next half cycle), the structure $p_2n_1p_1n'$ starts conducting after the voltage reaches the breakover voltage. The current-volt characteristic for this structure is shown in the third quadrant of Fig. 2.35.

In both the cases the currents in blocking region are small leakage currents. Since doping of all the layers is same, the characteristics in both the directions of the currents are identical. In commonly used diac, V_{BO} is about 30 V. In both directions the device exhibits negative-resistance characteristic during the conduction period.

The diac is mostly used in triac circuit to trigger the triac, which requires both positive and negative gate pulses for triggering. Matched diac-triac combinations are manufactured for various control circuits.

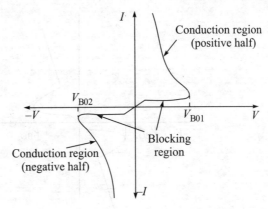

FIGURE 2.35 The i-v characteristic of a diac.

2.6.3 Triac (Bidirectional SCR)

An SCR conducts when the anode is positive with respect to cathode. This means that it is a unidirectional device and can conduct only in one direction. In many a.c. circuits, conductions in both directions are required. This can be achieved by connecting two SCRs in antiparallel as shown in Fig. 2.36.

For low or moderate power level applications, two SCRs are integrated in one device. This device is known as triac. This means triode for a.c. The configuration of a triac is shown in Fig. 2.37(a), and the circuit symbol is shown in Fig. 2.37(b). The device has three terminals, T_1, T_2, and G. T_1 and T_2 are the terminals connected to the supply and G is the gate terminal.

The triac can be triggered with either positive or negative gate voltage when the anode potential is either positive or negative, respectively. Thus, it is an a.c. switch, which can conduct on both half cycle of an a.c. voltage. From the configuration shown in Fig. 2.37(a), the triac is a five-layered ($n_1 p_1 n_2 p_2 n_3$) device. This may be considered to consist of a $p_2 p_1 n_1$ section in antiparallel with an $n_3 p_2 n_2 p_1$ section. An additional n region serves as the gate. The device is, therefore, a double-ended SCR.

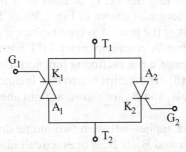

FIGURE 2.36 Two SCRs in antiparallel.

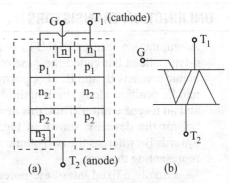

FIGURE 2.37 (a) Configuration of a triac and (b) its circuit symbol.

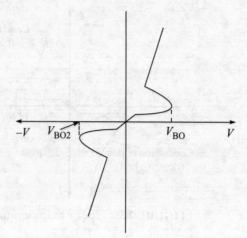

FIGURE 2.38 The v-i characteristics of a triac.

The v-i characteristic of a triac is shown in Fig. 2.38. When anode is positive the triac operates in the first quadrant as an SCR, and when the anode is negative, it operates in the third quadrant. The triac can be triggered by either a positive or a negative gate current in both cycles of operation. However, when the triac operates in the first quadrant (i.e., anode positive), the gate current required for triggering is less if triggering is through positive gate current. Similarly, the gate current required is less if triggering is through the negative gate current when the triac operates in the third quadrant. Hence, preferred modes of operations are as follows:

(i) When anode is positive, the triac is operated through a positive gate current.
(ii) When anode is negative, the triac is operated through a negative gate current.

The triac is widely used for speed control of single-phase induction motor and series motors.

2.7 UNIJUNCTION TRANSISTORS

A unijunction transistor (UJT) is a switching device. It consists of a high-resistivity n-type silicon bar called the *base* of the device as shown in Fig. 2.39(a). There are two ohmic contacts B_1 and B_2 at opposite ends of the base. A p-type emitter is alloyed at the middle position along the length of the bar. A complementary UJT has a p-type base and an n-type emitter. Thus, this device forms a p-n rectifying junction. The circuit symbol for the device is shown in Fig. 2.39(b). The emitter arrow is inclined and points towards B_1, whereas the two ohmic drops B_1 and B_2 are shown at right angles to the line representing the base.

Usually, a fixed inter-base potential is applied between two ohmic drops. The UJT forms an input diode between the emitter and B_1. If B_2 is open circuited, that is, $I_B = 0$, the input V-I relationship is like a usual p-n junction diode and is shown in Fig. 2.40.

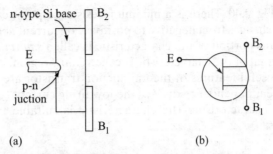

FIGURE 2.39 (a) Constructional details and (b) circuit symbol of a unijunction transistor.

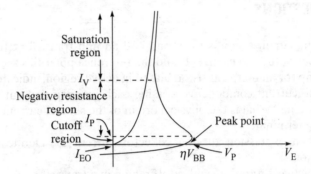

FIGURE 2.40 *V-I* relationship of a unijunction transistor.

As usual, when inter-base voltage V_{BB} is applied and emitter current $I_E = 0$, the n-type silicon bar is an ohmic resistance R_{BB} between B_1 and B_2. Usually, R_{BB} is of order of a few kilohms. Let the resistance between B_1 and the n side of the emitter junction be R_{B1} and that between B_2 and the n side of the emitter junction be R_{B2}. Then, $R_{BB} = R_{B1} + R_{B2}$. Under the condition of open emitter circuit, $V_e = \eta V_{BB}$, where $\eta = R_{B1}/R_{BB}$ is called intrinsic standoff ratio. Usually, the value of η lies between 0.5 and 0.75, and is specified by the manufacturer.

When V_e is less than ηV_{BB}, the input p-n junction is reverse biased and the emitter current I_E is negative as shown in Fig. 2.40. The maximum value of negative current is the reverse saturation current of the p-n junction, I_{EO}. If V_{EE} is increased such that $V_e > \eta V_{BB}$, the input diode becomes forward biased and I_E becomes positive. Until $V_e = \eta V_{BB} + V_\gamma$, that is, the forward bias equals the cut-in voltage V_γ, I_E remains quite small. For silicon junction V_γ is about 0.6 V. This point is peak point and corresponds to the voltage V_p. As $V_e > \eta V_{BB} + V_\gamma$, the emitter current increases very rapidly with a small increase in voltage.

With the increase of I_E, the hole concentration increases in the n region between E and B_1, which in turn increases conductivity of the region (i.e., increases R_{B1}). Hence, V_e decreases with the increase in I_E. Since current increases while voltage decreases, the device has a negative resistance.

When I_E is very large compared to I_{B2}, I_{B2} may be considered zero. Hence, for very large I_E, the input characteristic approaches the curve for $I_{B2} = 0$ asymptotically. As shown in Fig. 2.40. There is a minimum-resistance point called *valley point*, where the resistance changes from negative to positive. For current above I_v, the resistance remains positive. This portion of the characteristic is called *saturation region*. The region left to the peak point of the characteristic is called *cutoff* region.

Most useful features of the unijunction transistor are its stable peak (firing) voltage V_p, linear dependence on V_{BB}, the low firing current (µA) and the stable negative-resistance characteristic. The characteristic is suitable for generating time varying waveforms.

2.8 REVIEW QUESTIONS

1. Draw the circuit symbols for a pnp and an npn transistors. Indicate the reference directions of the three currents and the reference polarities for the three voltages.
2. For a pnp (or an npn) transistor, biased in active region, indicate the various electron and hole current components crossing each junction and entering (or leaving) the base terminal. What is the physical origin of the several current components crossing the base terminal?
3. Obtain an expression for the collector current I_C. Define each symbol in this expression.
4. Define active, saturation, and cutoff region in a transistor.
5. Sketch a family of common-based output and input characteristics for a transistor.
6. Draw the circuit of a transistor in common-emitter configuration. Sketch the output and input characteristics. Indicate the active, saturation, and cutoff regions.
7. Draw the circuit of a transistor in common-collector configuration. Sketch the output and input characteristics. Indicate the active, saturation, and cutoff regions.
8. How does a field-effect transistor differ from a transistor?
9. Describe how the output current responds to changes in the input voltage appearing between the gate and the source in FET.
10. Define the pinch-off voltage of the FET. What is the importance of this voltage?
11. Explain why the input resistance of an FET is very high.
12. What are different parameters of an FET? How are these parameters related to each other?
13. Describe the salient constructional features of a MOSFET and mention how the MOSFET differs from the FET.
14. Describe the kind of operation that takes place in the enhancement-mode MOSFET. How does enhancement-mode type differ from depletion-mode type?
15. How does DE MOSFET differ in construction from the conventional depletion-mode or enhancement-mode MOSFET ?
16. What is an SCR? Explain the function of an SCR. What are the means of firing of the SCR?
17. What is a diac? Describe its construction feature.
18. Describe how the triac differs from the SCR.
19. What is a UJT? Explain its function.

2.9 SOLVED PROBLEMS

1. The common-base configuration of a transistor shown in Fig. 2.4 has the input and output characteristics given in Fig. 2.5 and Fig. 2.6, respectively. Let $V_{CC} = 6V$, $R_L = 200\Omega$ and $I_E = 15$ mA. (a) Find I_C and V_{CB}, (b) find V_{EB} and V_L, and (c) if I_E changes by 10 mA symmetrically around the operating point in (a) and with constant V_{CC}, find the corresponding change in I_C.

Solution

(a) Load line is drawn in Fig. 2.6 by joining points (6,0) and (0,30). From point of intersection of load line and characteristic line for $I_E = 15$ mA,

$I_C = $ **–14.7 mA** and $V_{CB} = $ **–3.06 V**

Alternatively, $V_{CB} = -(6 - 14.7 \times 10^{-3} \times 200)$

$\qquad = $ **–3.06 V**

(b) From Fig. 2.5,

$V_{EB} = 0.145$ V = **145 mV**

The voltage across the load,

$V_L = 14.7 \times 10^{-3} \times 200$

$\qquad = $ **2.94 V**

(c) From Fig. 2.6, the change in I_C

$\Delta I_C \approx $ **–9.8 mA**

2. The common-base configuration of the transistor (Fig. 2.4) has the characteristics given in Fig. 2.5 and Fig. 2.6. Let $I_C = -20$ mA, $V_{CB} = -4$ V, and $R_L = 200 \, \Omega$. (a) Find V_{CC} and I_C and (b) if the supply voltage V_{CC} decreases from its value in (a) by 2 V while I_E retains same as its previous value, find the new value of I_C and V_{CB}.

Solution

(a) The drop in R_L,

$V_L = I_C \times R_L = 20 \times 10^{-3} \times 200$

$\qquad = 4$ V

Hence, $V_{CC} = -V_{CB} - 4 = -4 - 4$

$\qquad = $ **–8 V**

From Fig. 2.6,

$I_E = $ **20.2 mA**

(b) V_{CC} changes to $8 - 2 = 6$ V. Hence,

$V_{CB} \approx $ **–2 V**

From Fig. 2.6,

$I_C \approx $ **19.8 mA**

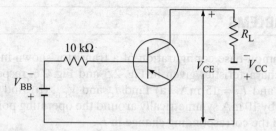

FIGURE 2.41 A common-emitter configuration of a pnp transistor.

3. The common-emitter configuration of the transistor shown in Fig. 2.41 has the characteristics shown in Fig. 2.9. (a) Find V_{BB} if $V_{CC} = 10$ V, $V_{CE} = -1$ V, and $R_L = 250\ \Omega$. (b) If $V_{CC} = 10$ V, find R_L so that $I_C = -20$ mA and $V_{CE} = -4$ V.

Solution

(a) The voltage drop across R_L,

$$V_L = V_{CC} + V_{CE} = 10 - 1 = 9\ \text{V}$$

$$I_C = -\frac{V_L}{R_L} = -\frac{9}{250} = -36\ \text{mA}$$

For $I_C = -36$ mA and $V_{CE} = -1$ V, from output characteristic of Fig. 2.9,

$$I_B = -0.277\ \text{mA}.$$

For $I_B = -0.277$ mA, drop across 10-kΩ resistance is 2.77 V and from input characteristic of Fig. 2.8, for $I_B = -0.277$ mA, $V_{BE} = -0.23$ V,

$$V_{BB} = 2.77 + 0.233 \approx \mathbf{3.0\ V}$$

(b) Drop across R_L,

$$20 \times 10^{-3} R_L = 10 - 4 = 6$$

$$\therefore \quad R_L = \frac{6 \times 10^3}{20} = \mathbf{300\ \Omega}$$

4. (a) Find the silicon transistor currents in Fig. 2.42(a). The transistor has $\beta = 100$. (b) Repeat (a) if a 2-kΩ resistor is added in the emitter circuit, as in Fig. 2.42(b). Neglect saturation current.

Solution

(a) The emitter-base junction is forward biased; hence, it must be either in active region or in saturation. Assume that it operates in active region. Applying KVL to base circuit of Fig. 2.42(a), with I_B in mA,

$$5 - 200 I_B = V_{BE}$$

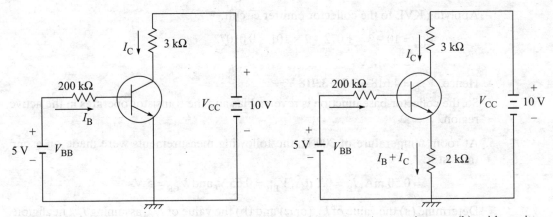

FIGURE 2.42 A common-emitter configuration (a) without emitter resistance (b) with emitter resistance.

The value of V_{BE} may be reasonably taken equal to 0.7 V. Hence,

$$I_B = \frac{5 - 0.7}{200} = 0.0215 \text{ mA}$$

Since $I_{CO} \ll I_B$,

$$I_C = \beta I_B = 100 \times 0.0215 = \textbf{2.15 mA}$$

Now, applying KVL to collector circuit (with I_C in mA),

$$10 - 3I_C = V_{CE}$$

$$\therefore \quad V_{CE} = 10 - 3 \times 2.15 = 3.55 \text{ V}$$

But $\quad V_{CB} = V_{CE} - V_{BE}$

$$= 3.55 - 0.7 = \textbf{2.85 V}$$

So, the collector base junction is reverse-biased, which is essential for the transistor to operate in active region. Thus, our earlier assumption is justified.

(b) The emitter current,

$$I_E = I_B + I_C \approx I_B + \beta I_B = 101 I_B$$

Applying KVL to the base-emitter circuit,

$$5 - 200 \times I_B - 2 \times 101 I_B = V_{BE} = 0.7$$

or $\quad I_B = \dfrac{5 - 0.7}{402} = 0.01072 \text{ mA}$

$$\therefore \quad I_C = 100 \times 0.01072 = 1.072 \text{ mA}$$

Applying KVL to the collector-emitter circuit,

$$V_{CE} = 10 - 3 \times 1.072 - 2 \times 101 \times 0.01072$$
$$= 4.618 \text{ V}$$

Hence, $V_{CB} = 4.618 - 0.7 = \mathbf{3.918 \text{ V}}$

So, the collector-base junction is reverse bias and the transistor operates in the active region.

5. At room temperature of 300 K, the following measurements were made on a pnp transistor:

$$I_C = 0.50 \text{ mA}, I_B = 0.1 \text{ μA}, V_{BE} = 0.65 \text{ V}, \text{ and } V_{CE} = 5 \text{ V}.$$

Determine (a) the value of h_{fe} (or α) and (b) the value of I_{EO}, assuming I_{CO} negligible.

Solution

(a) The emitter current,

$$I_E = I_C + I_B = 0.50 + 0.01 = 0.51 \text{ mA}$$

Hence, $\alpha = \dfrac{I_C}{I_E} = \dfrac{0.50}{0.51} = \mathbf{0.98}$

(b) For room temperature of 300 K,

$$V_T = 0.026 \text{ V}$$

Now, the emitter current is given by

$$I_E = I_{EO}(e^{V/\eta V_T} - 1)$$

$$I_{EO} = \dfrac{I_E}{(e^{V/\eta V_T} - 1)}$$

For $\eta = 1$, $e^{0.65/0.026} - 1 = 7.2 \times 10^{10}$

$$\therefore \quad I_{EO} = \dfrac{0.51 \times 10^{-3}}{7.2 \times 10^{10}} = 70.83 \times 10^{-3} \text{ μA}$$

For $\eta = 2$, $e^{0.65/2 \times 0.026} - 1 = 268336.3$

$$\therefore \quad I_{EO} = \dfrac{0.51 \times 10^{-3}}{268336.3} = 1.9 \times 10^{-3} \text{ μA}$$

6. An n-channel FET with characteristic shown in Fig. 2.43(a) is used in the circuit of Fig. 2.21. The element values are $V_{DD} = 20$ V, $R_L = 2$ kΩ, and $V_{GG} = 2$ V. (a) Find the values of V_{DQ} and I_{DQ} when $v_s = 0$. (b) For $v_s(t) = 1.0 \sin \omega t$, sketch $v_D(t)$ and $i_D(t)$ for one cycle.

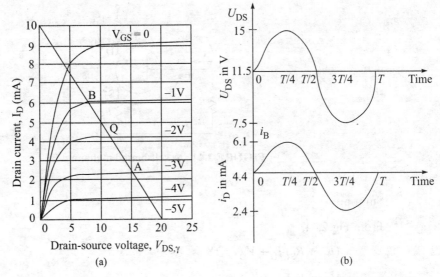

FIGURE 2.43 (a) The drain characteristics of an n-channel FET and (b) Waveforms of v_{DS} and i_{DS}.

Solution

The load line is drawn. The two extreme points of the line are $V_{DS}|_{I_D=0} = 20$ V and $I_D|_{V_{DS}=0} = \dfrac{20}{2 \times 10^3} = 10$ mA. The intersection point of the load line and characteristic for $V_{GS} = -2$ V is point P. So, $V_{DQ} = \mathbf{11.7\ V}$ and $I_{DQ} = \mathbf{4.2\ mA}$.

For $v_s(t) = 1.0 \sin \omega t$, the extreme points of the input signal (located on the load line) are A and B. The points are on the characteristics for $v_{gs} = -3$ V (point A) and $v_{gs} = -1$ V (point B).

The values of i_D and v_{DS} for different values of v_{GS} according to the signal voltage are obtained from the intersections of the load line and the drain characteristics. For the two extreme points the values are given in Table-2.2.

Similarly, other points for v_{Ds} and i_{DS} corresponding to different values of input signal can be obtained. The waveforms of v_{DS} and i_{DS} are plotted in Fig. 2.43(b).

Table-2.2. Extreme point values of v_{DS} and i_{DS}.

	For $v_{GS} = -1$ V	For $v_{GS} = -3$ V
Voltage v_{DS}	7.5 V	15.0 V
Current i_{DS}	6.1 mA	2.4 mA

7. An n-channel FET is biased as in circuit shown in Fig. 2.44. The transfer characteristic is shown in Fig. 2.20. The circuit element values are $R_S = 0.5$ kΩ, $R_D = 2.5$ kΩ, $R_G = 200$ kΩ and $V_{DD} = 18$ V. (a) Find V_{DSQ} and I_{DQ}. (b) Repeat (a) if R_S is changed to 0.6 kΩ.

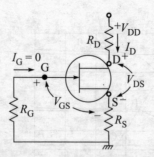

FIGURE 2.44 A self-biased n-channel FET.

Solution

From Fig. 2.44,

$$(R_D + R_S) I_D + V_{DS} = 18 \tag{1}$$

and $\quad V_{GS} + R_S I_D = 0 \tag{2}$

(a) From (2),

$$R_S = -\frac{V_{GS}}{I_D} \tag{3}$$

This gives the line (called biased line) with inclination $R_S = 0.5$ on the transfer characteristic. From intersection of biased line and transfer characteristic,

$V_{GS} = -1.85$ V

and $\quad I_{DQ} = \mathbf{3.7 \text{ mA}}$

Substituting the circuit element values and the value of I_{DQ} in (1),

$(2.5 + 0.5) \times 3.7 + V_{DS} = 18$

or $\quad V_{DS} = 18 - 11.1 = \mathbf{6.9 \text{ V}}$

(b) For $R_S = 0.6$ kΩ, from intersection of biased line and transfer characteristic,

$V_{GS} = -2.1$ V

and $\quad I_{DQ} = \mathbf{3.4 \text{ mA}}$

Substituting the circuit element values and the value of I_{DQ} in (1),

$(2.5 + 0.6) \times 3.4 + V_{DS} = 18$

or $\quad V_{DS} = 18 - 10.54 = \mathbf{7.46 \text{ V}}$

8. A p-channel FET, with $V_{PO} = 4$ V and $I_{DSS} = -5$ mA, is used in the circuit shown in Fig. 2.45. Determine the element values necessary to establish $I_{DQ} = -2$ mA and $V_{DSQ} = -4$ V. Take $V_{DD} = -12$ V.

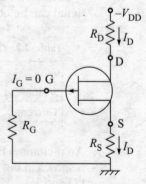

FIGURE 2.45 A self-biased p-channel FET.

Solution

From Eq. (2.23),

$$I_{DQ} = I_{DSS}\left(1 + \frac{V_{GS}}{V_{PO}}\right)^2$$

or

$$-2 = -5\left(1 + \frac{V_{GS}}{4}\right)^2$$

or

$$\sqrt{0.4} = 1 + \frac{|V_{GS}|}{4}$$

$$\therefore \quad V_{GS} = \sqrt{0.4} - 1 = -1.47 \text{ V}$$

But $V_{GS} = I_{DQ} \times R_S$

$$\therefore \quad R_S = \frac{V_{GS}}{I_{DQ}} = \frac{-1.47}{-2} = \mathbf{0.735 \text{ k}\Omega}$$

Also, $(R_D + R_S) I_{DQ} + V_{DSQ} = V_{DD}$

or $(R_D + 0.735) \times 2 + 4 = 12$

or $2R_D = 12 - 5.47$

$$\therefore \quad R_D = 3.265 \text{ k}\Omega$$

9. A MOSFET having drain characteristics as shown in Fig. 2.46(a) is employed in the circuit shown in Fig. 2.46(b). The drain-source voltage is 16 V and gate-source voltage is 2 V. (a) Find the drain current. (b) Find the incremental drain resistance. (c) At a specified operating point determine the transconductance.

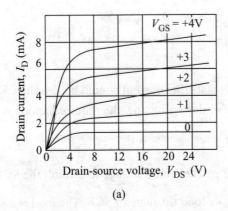
(a)

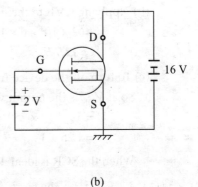
(b)

FIGURE 2.46 (a) The drain characteristics of a MOSFET and (b) circuit using the MOSFET.

Solution

From the drain characteristics,

(a) $I_D = 4$ mA
(b) $r_d = 16/0.9 = $ **17.8 kΩ**
(c) $g_m = 3.4/2 = $ **1.7 mA/V**

10. An SCR circuit is shown in Fig. 2.47. The SCR has a trigger voltage of 0.7 V and a trigger current of 6 mA. The holding current of the SCR is 5 mA. (a) Determine the output voltage in OFF state. (b) Calculate the gate-source voltage necessary to turn on the device. (c) To what value should anode-supply voltage be reduced to turn off the device from ON state?

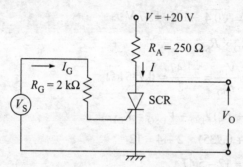

FIGURE 2.47 An SCR circuit.

Solution

(a) When the SCR is in OFF state, there is no current through the SCR. The output voltage
$$V_O = V - IR_A = 20 - 0 \times 250$$
$$= \mathbf{20\ V}$$

(b) Applying KVL to the gate circuit,
$$V_S = 2 \times 10^3 \times 6 \times 10^{-3} + 0.7$$
$$= \mathbf{12.7\ V}$$

(c) To turn off the device from ON state, the current should be less than 5 mA.
So, if V_{ON} is the output voltage while conducting, the required anode voltage
$$V = 5 \times 10^{-3} \times 250 + V_{ON}$$
$$= 1.25 + V_{ON}$$
When the SCR is ideal, $V_{ON} = 0$, so, the anode voltage to turn off the SCR is **1.25 V**.

11. A d.c. supply of 100 V is applied to a load through an SCR. The load is a series combination of 10 Ω resistance and 1 H inductance. Find the minimum gate pulse duration so that the SCR conducts. The latching current is 75 mA.

Solution

The circuit is shown in Fig. 2.48. When the SCR starts conducting, let the voltage drop across the SCR be zero, that is, the SCR be ideal. Hence, the current is given by

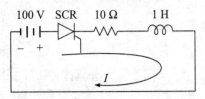

FIGURE 2.48 Circuit for Solved Problem 24.

$$i = \frac{100}{10}\left(1 - e^{-Rt/L}\right)$$
$$= 10\left(1 - e^{-10t}\right)$$

For conduction to be maintained, the conduction current must reached the latching current before the gate voltage is removed. So,

$$75 \times 10^{-3} = 10(1 - e^{-10t})$$

or $e^{-10t} = 1 - 0.0075 = 0.9925$

∴ $t = \mathbf{0.753\ ms}$

2.10 EXERCISES

1. Fig. 2.4 depicts a common-base configuration of a pnp transistor. Its input and output characteristics are shown in Fig. 2.5 and Fig. 2.6, respectively. The supply V_{CC} = 6 V, load resistance R_L = 250 Ω, and emitter current I_E = 20 mA. (a) Determine V_{CB} and I_C and (b) find V_{EB} and V_L. (c) If I_E changes by 5 mA on both sides of the operating point in (a), determine the corresponding change in I_C. Assume V_{CC} as same.

2. Common-emitter configuration of pnp transistor is shown in Fig. 2.41. The input and output characteristics are shown in Fig. 2.8 and Fig. 2.9, respectively. (a) If I_B = −0.1 mA, V_{CC} = 10 V, R_L = 500 Ω, determine V_{BB} and V_{CE}. (b) If R_L = 250 Ω, I_C = −20 mA, and V_{CE} = −4 V, find V_{CC}.

3. A silicon transistor with V_{BE} = 0.7 V, β = 100, V_{CE} = 0.3 V is employed in the circuit shown in Fig. 2.49. Find the value of R_C.

4. An n-channel FET is employed in the circuit shown in Fig. 2.50. The transfer characteristic is shown in Fig. 2.20. The element values are V_{DD} = 18 V, R_D = 2 kΩ and V_{GG} = 2 V. (a) Determine the values of V_{DQ} and I_{DQ} when $v_g(t)$ = 0. (b) For $v_g(t)$ = 1.0 sin ωt, sketch $v_D(t)$ and $i_D(t)$ for one cycle.

5. An n-channel FET is employed in the circuit shown in Fig. 2.44. The transfer characteristic is depicted in Fig. 2.20. The parameter values are V_{GS} = −1 V, V_{DS} = 5 V, and R_G = 500 kΩ. Determine R_S and V_{DD} if (a) R_D = 2 kΩ and (b) R_D = 5 kΩ.

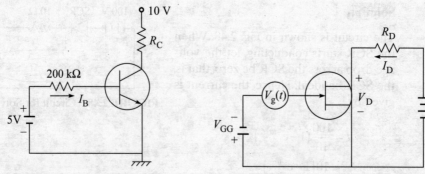

FIGURE 2.49 Circuit for question 3.   **FIGURE 2.50** Circuit for question 4.

6. A thyrister is employed in the circuit shown in Fig. 2.47. It has a trigger voltage of 0.65 V and a trigger current of 5 mA. Its holding current is 4.5 mA. The circuit elements are $V = 18$ V, $R_A = 200$ Ω, and $R_G = 5$ kΩ. (a) Determine v_o while the device is nonconducting. (b) Find V_S to turn on the device. (c) The minimum anode voltage necessary to maintain the conduction, if at the time $v_o = 0.5$ V.

3
DEVICES USING DIODE AS AN ELEMENT

Outline

3.1   Analysis of Diode Circuit   74
3.2   Piecewise Linear Model of a Diode   76
3.3   Clipping Circuits   78
3.4   Comparators   81
3.5   Sampling Gate   81
3.6   Rectifiers   84
3.7   Other Rectifier Circuits   92
3.8   Additional Diode Circuits   94
3.9   Review Questions   96
3.10 Solved Problems   97
3.11 Exercises   116

3.1 ANALYSIS OF DIODE CIRCUIT

A junction diode in series with a resistance is connected to a d.c. source, as shown in Fig. 3.1. The loop equation of the circuit is given by Eq. (3.1).

$$V_{BB} = IR + V_D \tag{3.1}$$

Equation (3.1) represents a straight line. This straight line with V-I characteristic of the diode is shown in Fig. 3.2. The straight line is called *load line*. For a particular value of R, the current I and the voltage V are obtained at the point of intersection Q of the V-I characteristic of the diode and the load line. This point is called *operating point* or *quiescent point*. This is the stable point of the circuit of Fig. 3.1. The stable current and voltage (i.e., current and voltage corresponding to the quiescent point) are I_Q and V_Q, respectively. The voltage V_Q is the voltage drop across the diode. Voltage difference $V_{BB} - V_Q$ is the voltage drop across the resistance. With the change in resistance and same supply voltage V_{BB}, the slope of the load line changes and so does Q, as indicated in Fig. 3.3. In this case, the intercept on the ordinate changes, while the intercept on the abscissa remains the same. If the supply voltage V_{BB} is changed with the load resistance remaining the same, the slope of the load line does not change but the load line moves to a new position parallel to the original line. The load lines for different values of V_{BB} are shown in Fig. 3.4. Again the quiescent point changes with changed intercepts on ordinate and abscissa.

For reverse-biased condition, the diode current is the saturation current I_0, and it is independent of the bias voltage and the load resistance. Hence, the diode voltage can easily be determined from Eq. (3.2).

$$V_{BB} = I_0 R + V_D \tag{3.2}$$

There is no need of the load line.

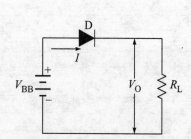

FIGURE 3.1 A forward-biased diode with load resistance.

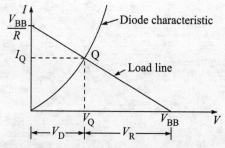

FIGURE 3.2 The diode characteristic and load line for circuit in Fig. 3.1.

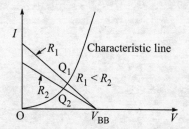

FIGURE 3.3 Load lines for variable resistance.

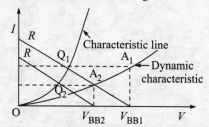

FIGURE 3.4 Load lines for variable supply voltage.

3.1.1 Dynamic Characteristic

The dynamic characteristic is the plot of current versus input voltage. To plot the dynamic characteristic, for a constant load resistance, the input voltage is varied. For each input voltage, a load line is drawn as discussed earlier. In Fig. 3.4, two load lines, parallel to each other, are drawn for the input voltages V_{BB1} and V_{BB2}. For the corresponding two quiescent points, the two currents are I_1 and I_2. The points vertically above the inputs V_{BB1} and V_{BB2} and equal to currents I_1 and I_2 are obtained at A_1 and A_2. A curve starting from the origin and joining the points A_1 and A_2 is drawn. This curve is called the dynamic characteristic. The ordinate in Fig. 3.4 gives the output voltage v_o, if it is multiplied by R_L.

3.1.2 Transfer Characteristic

The transfer characteristic is the curve that relates the output voltage v_o to the input voltage v_i. This is also called the transmission characteristic. Since, in Fig. 3.4, $v_o(t) = i(t) R_L$, for this particular circuit, the transfer characteristic has the same shape as the dynamic characteristic, the ordinate in Fig. 3.4 is, just, multiplied by R_L.

Now, we consider a time-varying voltage, as shown in Fig. 3.5, varying sinusoidally and given by

$$v_s(t) = \sqrt{2}\, V_s \sin \omega t$$

where V_s is the r.m.s. value of the signal and $\omega = 2\pi/T$, T being the time period. It is to be noted that without considering the shape of the static characteristic of the diode, for any waveform of the input signal, the resultant output wave shape can be obtained graphically (at low frequencies) from the transfer curve.

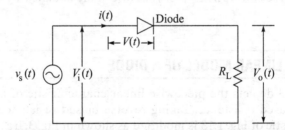

FIGURE 3.5 A diode circuit with time-varying input.

Although we have considered a sinusoidally varying input, it is not necessary to take a sinusoidal input. An input of any shape can be taken. The construction of the output waveform is illustrated in Fig. 3.6. The input signal is drawn with its time axis vertically downwards, so that the input voltage axis is horizontal. Suppose that at time t_1 the input voltage has a value v_{iA} indicated by a point A on the input waveform, the output voltage for this input voltage is obtained by drawing a vertical line through the point A and noting the voltage v_{iA} corresponding to the point of intersection of this vertical line with the transfer curve. This value of output voltage is plotted at an instant of time equal to t_1 along the time axis (horizontal) of the output waveform as point 'a'. Similarly, points b, c, d, ... of the output waveform corresponding to points B, C, D, ... of the input waveform are plotted. A point to be noted is that the output is zero for the input voltage less

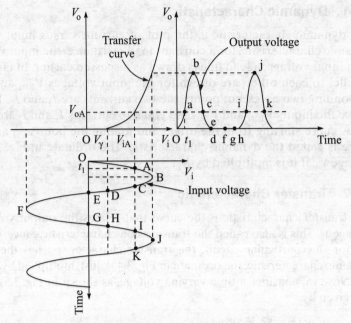

FIGURE 3.6 Construction of output waveform using transfer curve.

than the cut-in voltage V_γ of the diode. Thus, we see that a portion of the input signal does not appear at the output. Therefore, the diode acts as a clipper. It is also to be noted that the output is distorted near $v_i = V_\gamma$ because of the nonlinearity of the transfer curve in this region.

3.2 PIECEWISE LINEAR MODEL OF A DIODE

Figure 1.13 depicts the piecewise linear characteristic of a diode, assuming infinite reverse-biased resistance. Taking reverse-biased resistance R_r, the piecewise linear characteristic of Fig. 1.13 is modified as shown in Fig. 3.7(a). The diode, thus, remains

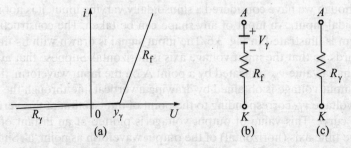

FIGURE 3.7 (a) Piecewise linear v-i characteristic of a diode, (b) ON-state model, and (c) OFF-state model of the diode.

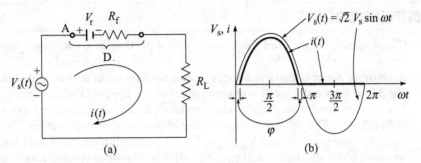

FIGURE 3.8 (a) An equivalent circuit (in ON state) of Fig. 3.5 and (b) the waveforms of $V_i(t)$ and $i(t)$.

ON for $v_i > V_\gamma$ and OFF for $v_i < V_\gamma$. Hence, the diode is a binary device. The forward-biased large-signal (larger than V_γ) model of the diode, for the ON state, is indicated in Fig. 3.7(b). The OFF-state model, for large signal, of the diode is indicated in Fig. 3.7(c).

The ON-state model is represented by a battery of voltage V_γ in series with the low forward resistance R_f (of the order of a few tens of ohm or less). The OFF-state model is represented by a large reverse resistance R_r (of the order of several hundred kilohms or more). Compared to any other diode circuit resistance, usually, R_r is so large that the reverse resistance may be considered infinite.

Let us use the piecewise linear model of Fig. 3.7 (with $R_r = \infty$) for the diode circuit of Fig. 3.5. Then, the equivalent circuit of Fig. 3.5 is as shown in Fig. 3.8(a). The circuit current is given by

$$i(t) = \frac{\sqrt{2}V_s \sin \omega t - V_\gamma}{R_L + R_f} \quad \text{for } v_s(t) \geq V_\gamma$$
$$= 0 \quad \text{for } v_s(t) < V_\gamma \quad (3.3)$$

The waveform of $i(t)$ is plotted in Fig. 3.8(b). The cut-in angle is given by

$$\varphi = \sin^{-1} \frac{V_\gamma}{\sqrt{2}V_s} \quad (3.4)$$

The cut-in angle depends on the peak value of the input. For the peak value equal to twice the cut-in voltage (0.6 V for silicon and 0.2 V for germanium), the cut-in angle is 30°. In contrast, for the peak value of the input greater than ten times the cut-in voltage, the cut-in angle is very small (less than 3.5° for silicon and less than 1.2° for germanium). Hence, the cut-in angle may be neglected in this case.

The piecewise linear characteristic indicates that the change in slope is abrupt at V_γ. But this is not true; the change of state from the OFF state to ON state is not abrupt. However, there is a region over which the slope of the diode characteristic changes gradually from a very small to a very large value. This region is called the *break region*.

3.3 CLIPPING CIRCUITS

Clipping circuits are used for transmitting a portion of a waveform that lies above or below some reference level. Clipping circuits are also referred to as limiter, amplitude selector, or slicer circuits. A clipper circuit with its input and output signals is shown in Fig. 3.9. A sinusoidal input voltage applied to the clipping circuit is shown in Fig. 3.9(a). The clipper circuit using one diode, a resistance, and a reference voltage source is shown in Fig. 3.9(b). The output voltage of the clipper circuit is shown in Fig. 3.9(c). For the input voltage greater than V_B ($V_B = V_R + V_\gamma$), the diode is forward biased and may be assumed to be short circuited (i.e., $R_f = 0$). For the input voltage less than V_B, the diode is reverse biased and may be assumed to be open circuited (i.e., $R_r = \infty$).

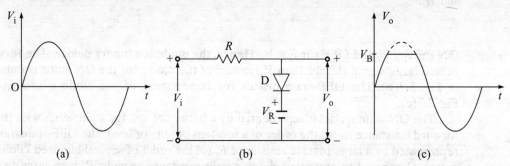

FIGURE 3.9 (a) The input signal, (b) a clipper circuit, and (c) an output of the clipper circuit.

The function of the clipper circuit of Fig. 3.9(b) can be explained as follows. For positive half of the input, until $v_i(t)$ is less than V_B, the diode is reverse biased and the output follows the input. As long as $v_i(t)$ is greater than V_B, the diode remains in forward bias and the output remains constant at V_B. For the negative half of the wave, the diode remains reverse biased and the output again follows the input.

The clipper circuit of Fig. 3.9 is modified in Fig. 3.10 in the sense that the diode is reversed. For the positive half of the input, the diode is reverse biased for $v_i(t)$ greater than V_B ($V_B = V_R - V_\gamma$) and is open circuited. Hence, the output follows the input. As long as $v_i(t)$ is less than V_B, the diode remains in forward bias and is short circuited. The output remains constant at V_B. For the sinusoidal input voltage shown in Fig. 3.10(a), the output is as shown in Fig. 3.10(c).

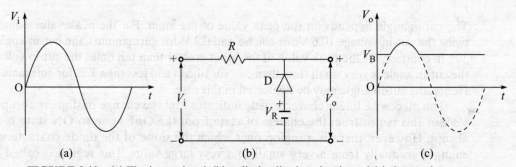

FIGURE 3.10 (a) The input signal, (b) a diode-clipping circuit, and (c) the resultant output.

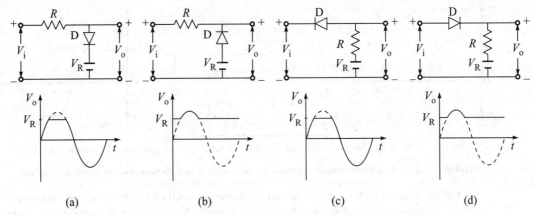

FIGURE 3.11 Four clipping circuits along with their outputs.

In the clipper circuits of Fig. 3.9 and Fig. 3.10, the diode appears as a shunt element. These clipper circuits together with two additional clipping circuits are shown in Fig. 3.11. In the clipping circuits shown in Fig. 3.11(c) and (d), the diodes appear as series elements. Input in each case is a sinusoid. The outputs are shown along with the clipping circuits.

The circuit shown in Fig. 3.9 is a peak clipping circuit, whereas that shown in Fig. 3.10 is a base clipping circuit. The circuit of Fig. 3.9 clips only the positive peak, whereas the circuit of Fig. 3.10 clips only the negative peak. A double-ended clipping circuit using two diodes that clips both positive and negative peaks is shown in Fig. 3.12. While the first diode clips the positive peak, the negative peak is clipped by the second diode. The input sinusoid and the clipped output are shown along with the clipping circuit.

A double-ended clipping circuit using two Zener diodes is shown in Fig. 3.13. When the positive half of the input reaches at the breakdown voltage of the Zener diode D_{z1}, this diode breaks down, making D_{z2} forward biased. The output remains constant at the breakdown voltage. Similarly, when the negative half of the input reaches the breakdown voltage of D_{z2}, it breaks down and D_{z1} becomes forward biased. The negative wave of the output clips at the breakdown voltage.

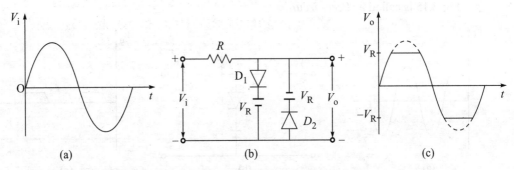

FIGURE 3.12 A double-ended clipping circuit and its input and output.

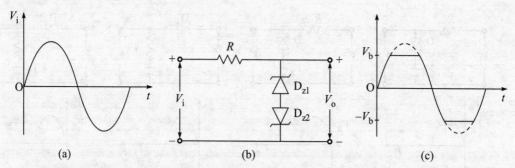

FIGURE 3.13 A double-ended clipping circuit using zener diodes and its input and output.

The circuits described above are all peak clippers. All of them flatten off the positive and/or negative peaks of waves. A clipping circuit that reduces all amplitudes, below some minimum value, to zero is called the base clipper. A base clipping circuit using a diode is shown in Fig. 3.14. The diode is reverse biased to a voltage V_R, which is the desired level of base clipping.

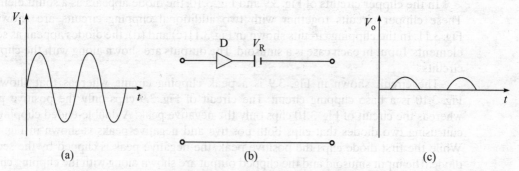

FIGURE 3.14 A base-clipping circuit.

A peak-clipping circuit combined with a base-clipping circuit is shown in Fig. 3.15. The diode D_1, resistance R, and biasing voltage V_R form a peak clipper, whereas the diode D_2 and biasing voltage V_R' function as a base clipper. The combined circuit clips input voltage in excess of V_R, and the input voltage less than V_R' is reduced to zero. Thus, an output voltage having a value less than V_R and greater than V_R' is produced. As the output wave can be regarded as consisting of a slice of the input waveform, the circuit of Fig. 3.15 is called a *slicer circuit*.

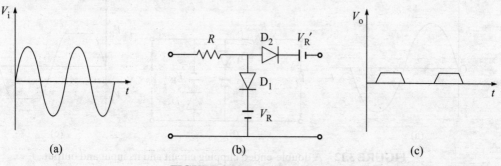

FIGURE 3.15 A slicer circuit.

3.4 COMPARATORS

The clipping circuits used earlier may also be used to perform the operation of comparison. The circuit used for this purpose becomes an element of a comparator system; therefore, the circuit is usually referred to as a comparator. A comparator marks the instant when an arbitrary waveform attains a reference value. Unlike a clipper, a comparator does not reproduce any part of the signal waveform. The output of the comparator is an abrupt change from the quiescent value that occurs at the time the signal reaches the reference value. The output is independent of the signal. Alternatively, the comparator output may be a sharp pulse that occurs at the time when the signal and the reference are equal.

The diode circuit shown in Fig. 3.16, which was used earlier as a clipper in Fig. 3.11(d), is used here as a comparator. Although the input signal may be of any waveform, for the sake of illustrating the operation of a comparator, it is taken as a ramp. The reference voltage is a d.c. voltage of value V_R. Let the input signal v_i be equal to $V_R + V_\gamma$ at the time $t = t_1$. Until $t = t_1$, the output remains quiescent at $v_o = V_R$, and after that v_o rises with the input signal.

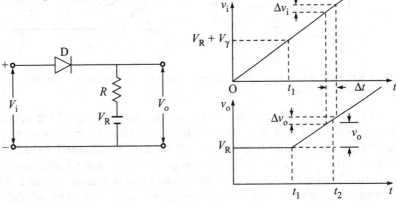

FIGURE 3.16 A diode comparator with a ramp input v_i and the output v_o.

The comparator output is applied to a device that responds when the output voltage rises to a value v_o above V_R. However, as a result of ageing of components, temperature changes, etc., the device does not respond exactly at the voltage v_o above V_R. There is a variation of voltage Δv_o. Hence, there is a variation of the response time Δt and an uncertainty Δv_i in the input voltage corresponding to Δt. The uncertainty range Δv_i can be reduced to $\Delta v_i/A$ by introducing an amplifier (with gain A), preceding the comparator. The amplifier used must be directly coupled and must be highly stable against drift due to ageing of components, change in temperature, etc.

3.5 SAMPLING GATE

An ideal sampling gate is a circuit that transmits an exact input waveform during a selected time interval only. For the remaining time, the output of the sampling gate is zero. The time interval for transmission of the input signal is selected by impressing an external signal. This external signal is known as gating (or control) signal. The gating signal is usually rectangular in shape. The sampling gates are also known as *transmission gates*.

A sampling gate circuit is shown in Fig. 3.17(a). This is in the form of a bridge consisting one diode in each of its four arms. The input signal v_i is applied at node A, and the output v_o is taken across the load R_L at node C. Symmetrical gating voltages $+v_c$ and $-v_c$ are applied at nodes B and D, respectively, through control resistances R_c. Let us consider a rectangular control signal v_c and a sinusoidal input signal v_i; however, the input signal may be of any arbitrary wave shape. The control signal v_c, the input signal v_i, and the sampled output signal v_o are shown in Fig. 3.17(b). The period of v_c need not be the same as the period of v_i, but in most practical systems it is taken to be equal or as an integral multiple of the period of v_i.

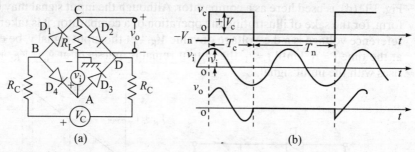

FIGURE 3.17 (a) A four-diode bridge sampling gate circuit and (b) the control, the input, and the output waveforms.

For the sake of simplicity in explaining the working of the system, let us assume ideal diodes with $V_\gamma = 0$, $R_f = 0$, and $R_r = \infty$; during the time interval T_c, $v_c = V_c$. Let us assume that during this time interval all the four diodes are forward biased and so they conduct. The voltages across the diodes are zero and node A and node C are at the same potential. Hence, the output voltage v_o is equal to the input voltage v_i. Thus, the output is an exact copy of the input during the sampling time T_c. During the time interval T_n, when $v_c = -V_n$, let us assume that all the four diodes are reverse biased, and so they are nonconducting. The current through R_L is zero. Hence, the output voltage $v_o = 0$.

The statement made in the previous paragraph is based on whether the diodes are conducting or nonconducting. Biasing of the diodes does not depend only on the control voltage but also on the input voltage.

Consider the time interval T_n during which it is assumed that all the four diodes are nonconducting. In other words, all the four diodes are reverse biased. During this interval, let $v_c = -V_n$ and $v_i = V_m$ (positive peak of input voltage). Under this condition D_4 is reverse biased by $V_n + V_m$, whereas D_1 and D_2 are reverse biased by V_n. The diode D_3 is forward biased by $V_m - V_n$. So, the diodes D_4, D_1, and D_2 are nonconducting for any value of V_n. But the diode D_3 is nonconducting for $V_n \geq V_m$. Hence, the condition for all the four diodes to be nonconducting is given by

$$(V_n)_{min} = V_m \tag{3.5}$$

Hence, there is restriction on the control gate amplitude during the nonconducting interval T_n. The restriction is that the minimum value of V_n must be just equal to the peak value of the input signal.

Now, consider the conducting interval T_c during which all the four diodes must be forward biased. During this interval, $v_c = V_c$ and $v_i = V_m$. Let us assume that all the diodes are conducting. The bridge is balanced, and so there is no current through the load due to control voltage, but there is a current equal to V_m/R_L due to input positive peak voltage. Since the circuit is linear during the conducting interval, the superposition principle applies. Hence, each of the branches has two components of currents; one due to V_c and another due to V_m. The current distributions in the branches are as shown in Fig. 3.18. There are forward currents through D_1 and D_3 for any value of V_c. The net current through the diode D_2 is forward current, when

$$V_c \geq \frac{R_c}{R_L} V_m \tag{3.6}$$

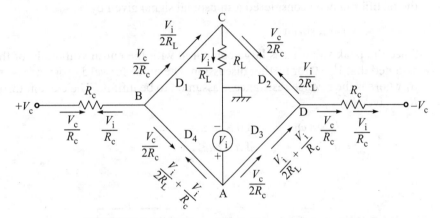

FIGURE 3.18 Current distributions in the branches.

The net current through D_4 is forward current, when

$$V_c \geq V_m \left(2 + \frac{R_c}{R_L}\right) \tag{3.7}$$

Hence, the minimum value of V_c for all the four diodes to be conducting is given by

$$(V_c)_{\min} = V_m \left(2 + \frac{R_c}{R_L}\right) \tag{3.8}$$

In the above discussion, diodes are assumed to be ideal, that is, $V_\gamma = 0$ and $R_f = 0$. In practice this is not true. Also, the four diodes may not have equal values for V_γ and R_f. If it is so, the bridge is not balanced. So, even for no input, that is, $v_i = 0$, there is an output during the conduction period. The output, therefore, is not a replica of the input. This error can be minimized by using matched diodes. Fortunately, integrated circuit techniques have made it possible to fabricate all the four diodes on a tiny silicon chip. It is to be noted that even if matched diodes are used, the above error may exist because of unbalance of control signals. One control signal must be negative of the other.

3.6 RECTIFIERS

A device that converts an a.c. supply into a d.c. supply is called a rectifier. Almost all electronic circuits use a rectifier when they are energized by a power supply. A diode permits easy flow of charge in one direction, while restrains the flow of charge in the opposite direction. Because of this property of the diode, it is used as a main element of a rectifier circuit.

3.6.1 Half-Wave Rectifier

A half-wave rectifier using a p-n junction diode is shown in Fig. 3.19(a). This rectifier converts an alternating signal (say a sinusoidal signal) with zero average value into a unidirectional (though not constant) signal with a non-zero average value. The input to the rectifier may be considered a sinusoidal signal given by

$$v_i = V_m \sin \omega t$$

Since the peak value V_m is large compared with the cut-in voltage V_γ of the diode, it is assumed that $V_\gamma = 0$ for further discussion. The diode forward resistance is assumed to be R_f, whereas the reverse resistance is assumed to be infinite. The current through the load R_L is given by

$$\begin{aligned} i &= I_m \sin \omega t & \text{if } 0 \leq \omega t \leq \pi \\ i &= 0 & \text{if } \pi \leq \omega t \leq 2\pi \end{aligned} \quad (3.9)$$

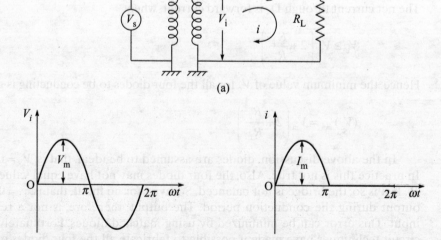

FIGURE 3.19 (a) Basic circuit of the half-wave rectifier, (b) the input signal, and (c) the load current.

The peak value of the current is given by

$$I_m = \frac{V_m}{R_f + R_L} \tag{3.10}$$

The input signal and the load current are shown in Fig. 3.19(b) and (c), respectively.

The load current is unidirectional. When a d.c. ammeter, such as a permanent magnet moving coil ammeter, is connected in the circuit, it measures the average value. When expressed mathematically, the average (d.c.) value is given by

$$I_{d.c.} = \frac{1}{2\pi} \int_0^{2\pi} i \, d\theta \qquad (\theta = \omega t) \tag{3.11}$$

Since $i = 0$ for $\pi \leq \theta \leq 2\pi$, Eq. (3.11) can be written as

$$I_{d.c.} = \frac{1}{2\pi} \int_0^{\pi} I_m \sin\theta \, d\theta = \frac{I_m}{\pi} \tag{3.12}$$

Diode Voltage

The d.c. output voltage measured by a d.c. voltmeter across the load is given by

$$V_{d.c.} = I_{d.c.} R_L = \frac{I_m R_L}{\pi} \tag{3.13}$$

However, a d.c. voltmeter reading, when connected across the diode, is not equal to $I_{dc}R_f$, because the voltage across the diode is not zero during the nonconducting period, rather it is equal to the applied voltage. Of course, it is given by $I_{d.c.}R_f$ during the conducting period. The output waveform across the diode is shown in Fig. 3.20. The voltage across the diode is expressed as

$$\begin{aligned} v &= I_m R_f \sin\theta & \text{if } 0 \leq \theta \leq \pi \\ v &= V_m \sin\theta & \text{if } \pi \leq \theta \leq 2\pi \end{aligned} \tag{3.14}$$

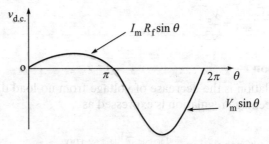

FIGURE 3.20 Voltage across the diode.

The d.c. voltmeter reading across the diode is given by

$$V'_{d.c.} = \frac{1}{2\pi}\left(\int_0^\pi I_m R_f \sin\theta\, d\theta + \int_\pi^{2\pi} V_m \sin\theta\, d\theta\right)$$

$$= \frac{1}{\pi}(I_m R_f - V_m)$$

$$= \frac{1}{\pi}[I_m R_f - I_m(R_f + R_L)]$$

$$= -\frac{I_m R_L}{\pi} \tag{3.15}$$

The d.c. voltage reading across the diode is equal to the negative of the d.c. voltage across the load resistance. The result is correct because the d.c. source voltage is zero and the sum of the d.c. voltages across the loop is also zero.

Peak Inverse Diode Voltage

It is clear from Fig. 3.20 that the peak value of inverse voltage across the diode is V_m. Hence, the diode used in a half-wave rectifier must withstand a voltage V_m.

Effective Values of Current and Voltages

The effective value of a periodic function is the root mean square value of the periodic function. Therefore, this value is also called the r.m.s. value. The effective values of a current and a voltage can be measured by an r.m.s. ammeter and an r.m.s. voltmeter, respectively. For example, a moving iron instrument or a thermocouple-type meter can be used. Mathematically, the r.m.s. value of the rectifier current can be expressed as

$$I_{r.m.s.} = \sqrt{\frac{1}{2\pi}\int_0^{2\pi} i^2 d\theta} = \sqrt{\frac{1}{2\pi}\int_0^\pi I_m^2 \sin^2\theta\, d\theta} = \frac{I_m}{2} \tag{3.16}$$

This is because of the fact that $i = 0$ for $\pi \leq \theta \leq 2\pi$. The r.m.s. value of the output voltage, hence, is given by

$$V_{r.m.s.} = \frac{I_m R_L}{2} \tag{3.17}$$

As the input voltage is a sinusoidal signal, the r.m.s. value of the input voltage is given by

$$V_{irms} = \frac{V_m}{\sqrt{2}} \tag{3.18}$$

Regulation

The regulation is the decrease of voltage from no-load d.c. current to full-load d.c. current. Percentage regulation is expressed as

$$\% \text{ regulation} = \frac{V_{no\,load} - V_{load}}{V_{load}} \times 100 \tag{3.19}$$

where no load means zero current, and load refers to the normal load current. The output voltage is given by

$$V_{d.c.} = I_{d.c.} R_L = \frac{I_m}{\pi} R_L = \frac{V_m R_L}{\pi(R_f + R_L)}$$

$$= \frac{V_m(R_f + R_L - R_f)}{\pi(R_f + R_L)} = \frac{V_m}{\pi} - I_{d.c.} R_f \tag{3.20}$$

The result can be modelled by a circuit shown in Fig. 3.21. So, the rectifier is represented by a constant voltage source $V = V_m/\pi$ in series with an effective internal resistance R_f. In case the input voltage is supplied through a transformer, the internal resistance is represented by $R_t + R_f$, where R_t is the secondary winding resistance of the transformer. The regulation plot of V_{dc} versus I_{dc} is given in Fig. 3.22. The plot is a straight line, and the negative slope of the line gives $R_t + R_f$.

The percentage regulation is, therefore, given by

$$\% \text{ regulation} = \frac{\frac{V_m}{\pi} - I_{d.c.} R_L}{I_{d.c.} R_L} \times 100$$

Subsituting the value of $I_{d.c.} R_L$ from Eq. (3.20), we get

$$\% \text{ regulation} = \frac{\frac{V_m}{\pi} - \left(\frac{V_m}{\pi} - I_{d.c.} R_f\right)}{I_{d.c.} R_L} \times 100$$

$$= \frac{R_f}{R_L} \times 100 \tag{3.21}$$

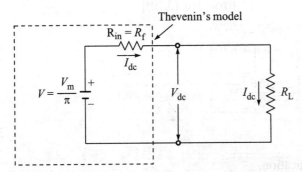
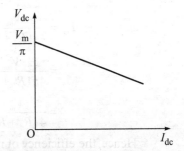

FIGURE 3.21 The circuit representing a half-wave rectifier.   **FIGURE 3.22** The regulation plot.

However, if a transformer is used to supply the rectifier and if its secondary winding resistance is R_t,

$$\% \text{ regulation} = \frac{R_t + R_f}{R_L} \times 100 \tag{3.22}$$

From Thevenin's equivalence of the half-wave rectifier (Fig. 3.21), the d.c. output power is given by

$$P_{d.c.} = I_{d.c.}^2 R_L = \frac{V_m^2 R_L}{\pi^2 (R_f + R_L)^2} \quad (3.23)$$

For maximum d.c. output power

$$\frac{dP_{d.c.}}{dR_L} = \frac{V_m^2 [(R_f + R_L) - 2R_L]}{\pi^2 (R_f + R_L)^2} = 0 \quad (3.24)$$

Hence, $R_L = R_f$. Therefore, for the d.c. output power to be maximum, the load resistance must be equal to the forward resistance of the diode.

Efficiency of Rectification

The efficiency of rectification η_r is defined as

$$\eta_r = \frac{\text{d.c. output power}}{\text{input power}} = \frac{P_{d.c.}}{P_{in}}$$

The d.c. output power of the half-wave single-phase rectifier, as in Eq. (3.23), is

$$P_{d.c.} = I_{d.c.}^2 R_L = \frac{V_m^2 R_L}{\pi^2 (R_f + R_L)^2}$$

The input power,

$$P_{in} = \frac{1}{2\pi} \int_0^{2\pi} vi \, d\theta$$

$$= \frac{1}{2\pi} \int_0^{\pi} V_m \sin\theta \, I_m \sin\theta \, d\theta \quad (\text{as } i = 0 \text{ for } \pi < \theta < 2\pi)$$

$$\therefore \quad I_m = \frac{V_m}{R_f + R_L} \quad \text{[from Eq. (3.10)]}$$

$$\therefore \quad P_{in} = \frac{V_m^2}{4\pi(R_f + R_L)} \int_0^{\pi} (1 - \cos 2\theta) d\theta$$

$$= \frac{V_m^2}{4\pi(R_f + R_L)} \left[\theta - \frac{\sin\theta}{2}\right]_0^{\pi}$$

$$= \frac{V_m^2}{4(R_f + R_L)} \quad (3.25)$$

Hence, the efficiency of rectification,

$$\eta_r = \frac{V_m^2 R_L}{\pi^2 (R_f + R_L)^2} \times \frac{4(R_f + R_L)}{V_m^2}$$

$$= \frac{0.4053}{\left(1 + \frac{R_f}{R_L}\right)} \quad (3.26)$$

Ripple Factor

The ripple factor of a voltage or current wave is defined as the ratio of the r.m.s. (or effective) values of a.c. component to the d.c. (or average) value of the wave. Any voltage or current wave, in general, consists of a d.c. component, a fundamental component, and a number of harmonic components. The r.m.s. value of a voltage wave is given by

$$V_{r.m.s.} = \sqrt{V_{d.c.}^2 + V_1^2 + V_2^2 + V_3^2 + \ldots}$$
$$= \sqrt{V_{d.c.}^2 + V_{a.c.}^2}$$

where $V_{d.c.}$ is the d.c. component and $V_1, V_2, V_3, \ldots$ are the r.m.s. values of fundamental, second, third, ... harmonic components, respectively. The r.m.s. value of a.c. component, $V_{a.c.}$, is given by

$$V_{a.c.} = \sqrt{V_1^2 + V_2^2 + V_3^2 + \ldots}$$

Hence, the ripple factor, as defined above, is given by

$$\gamma = \frac{V_{a.c.}}{V_{d.c.}} = \frac{\sqrt{V_{r.m.s.}^2 - V_{d.c.}^2}}{V_{d.c.}}$$

$$= \sqrt{\left(\frac{V_{r.m.s.}}{V_{d.c.}}\right)^2 - 1}$$

The form factor of a wave, F, is given by

$$F = \frac{V_{r.m.s.}}{V_{d.c.}}$$

Hence, the ripple factor can be given by

$$\gamma = \sqrt{F^2 - 1} \tag{3.27}$$

For a half-wave rectifier output voltage, the form factor

$$F = \frac{V_{r.m.s.}}{V_{d.c.}} = \frac{V_m/2}{V_m/\pi}$$

$$= \frac{\pi}{2} = 1.57$$

Hence, the ripple factor

$$\gamma = \sqrt{F^2 - 1} = \sqrt{(1.57)^2 - 1}$$
$$= 1.21$$

3.6.2 Full-Wave Rectifier

A full-wave rectifier is shown in Fig. 3.23. It uses a central tapped secondary transformer supplying two diodes. During the positive half of cycle of the supply, the diode D_1 is forward biased, whereas the diode D_2 is reverse biased. The diode D_1 conducts through the load from A to B. During the negative half the diode D_2 is forward biased and D_1 is reverse biased. Now, the diode D_2 conducts, again, through the load from A to B. The current through the load, therefore, is in the same direction from A to B. The current during the positive half is shown in Fig. 3.24(a), whereas that during the negative half is shown in Fig. 3.24(b). The current through the load, which is the sum of the two currents, in one cycle, is shown in Fig. 3.24(c).

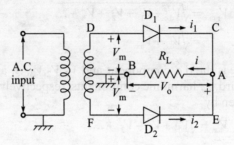

FIGURE 3.23 A full-wave rectifier circuit.

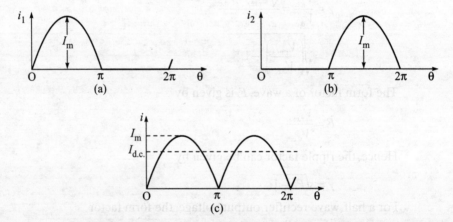

FIGURE 3.24 Individual diode currents and the load current.

The d.c. current, which is clearly double of that for the half-wave rectifier, is given by

$$I_{d.c.} = \frac{2I_m}{\pi} = \frac{2V_m}{\pi(R_f + R_L)} \tag{3.28}$$

The effective or r.m.s. value of current is given by

$$I_{r.m.s.} = \sqrt{2 \times \left(\frac{I_m}{2}\right)^2} = \frac{I_m}{\sqrt{2}} \tag{3.29}$$

and the r.m.s. value of the output voltage,

$$V_{r.m.s.} = \frac{I_m R_L}{\sqrt{2}} \tag{3.30}$$

The reading of an average measuring voltmeter across the load is given by

$$V_{d.c.} = I_{d.c.} R_L = \frac{2 I_m R_L}{\pi} \qquad (3.31)$$

The current I_m is given by Eq. (3.10), and V_m is the peak value of the voltage of the transformer secondary from one end to the central tap. It is clear that the d.c. output voltage for the full-wave rectifier is twice that of the half-wave rectifier. Similar to the half-wave rectifier [Eq. (3.20)], the $V_{d.c.}$ varies with $I_{d.c.}$ as given by

$$V_{d.c.} = \frac{2 V_m}{\pi} - I_{d.c.} R_f \qquad (3.32)$$

Similar to the half-wave rectifier, Thevinin's equivalence for the full-wave rectifier is as shown in Fig. 3.21, with the difference that the voltage source magnitude is double, that is, $2V_m/\pi$. If the secondary winding resistance between one end and central tap is R_t, then $R_{in} = R_t + R_f$.

Peak Inverse Diode Voltage

Let us discuss the situation when the diode D_1 is conducting. Applying KVL around the loop ACDFEA and neglecting the drop across D_1, the voltage across the diode D_2 is $2V_m$. So, the peak inverse voltage across each diode is twice the peak value of the transformer secondary measured from central point to either end.

Regulation

The regulation is the decrease of voltage from no-load d.c. current to full-load d.c. current. Percentage regulation is expressed as

$$\% \text{ regulation} = \frac{V_{no\ load} - V_{load}}{V_{load}} \times 100$$

where no load means zero current and load refers to the normal load current. Using Eq. (3.32), the percentage regulation is, therefore, given by

$$\% \text{ regulation} = \frac{\frac{2V_m}{\pi} - I_{d.c.} R_L}{I_{d.c.} R_L} \times 100 = \frac{\frac{2V_m}{\pi} - \left(\frac{2V_m}{\pi} - I_{d.c.} R_f\right)}{I_{d.c.} R_L} \times 100$$

$$= \frac{R_f}{R_L} \times 100 \qquad (3.33)$$

Efficiency of Rectification

The efficiency of rectification η_r is defined as

$$\eta_r = \frac{\text{d.c. output power}}{\text{input power}} = \frac{P_{d.c.}}{P_{in}}$$

The d.c. output power of the full-wave single-phase rectifier, as from Eq. (3.28), is

$$P_{d.c.} = I_{d.c.}^2 R_L = \frac{4 V_m^2 R_L}{\pi^2 (R_f + R_L)^2}$$

The input power,

$$P_{in} = \frac{1}{2\pi}\int_0^{2\pi} vi\, d\theta$$

$$= \frac{1}{\pi}\int_0^{\pi} V_m \sin\theta\, I_m \sin\theta\, d\theta$$

From Eq. (3.10),

$$I_m = \frac{V_m}{R_f + R_L}$$

$$\therefore\quad P_{in} = \frac{V_m^2}{2\pi(R_f + R_L)}\int_0^{\pi}(1-\cos 2\theta)d\theta$$

$$= \frac{V_m^2}{2\pi(R_f + R_L)}\left[\theta - \frac{\sin\theta}{2}\right]_0^{\pi}$$

$$= \frac{V_m^2}{2(R_f + R_L)} \tag{3.34}$$

Hence, the efficiency of rectification,

$$\eta_r = \frac{4V_m^2 R_L}{\pi^2(R_f + R_L)^2} \times \frac{2(R_f + R_L)}{V_m^2}$$

$$= \frac{0.8106}{\left(1 + \dfrac{R_f}{R_L}\right)} \tag{3.35}$$

Ripple Factor

For a full-wave rectifier output voltage, the form factor

$$F = \frac{V_{r.m.s.}}{V_{d.c.}} = \frac{V_m/\sqrt{2}}{2V_m/\pi} = \frac{\pi}{2\sqrt{2}} = 1.11$$

Hence, the ripple factor, from Eq. (3.27),

$$\gamma = \sqrt{F^2 - 1} = \sqrt{(1.11)^2 - 1} = 0.48$$

3.7 OTHER RECTIFIER CIRCUITS

In this section some of other rectifier circuits are discussed. The circuits discussed here include bridge rectifier and peak rectifier.

3.7.1 Bridge Rectifier

The bridge-rectifier circuit is shown in Fig. 3.25(a). In the circuit, only two diodes conduct at a time. When the node A is positive, as indicated in the circuit, the diodes D_1 and

D_3 are forward biased and the diodes D_2 and D_4 are reverse biased. The diodes D_1 and D_3 conduct and the current flows through the load from B to D. As soon as the supply changes it polarity, the diodes D_1 and D_3 become reverse biased and the diodes D_2 and D_4 become forward biased. Now, the diodes D_2 and D_4 conduct. However, the current through the load flows again from B to D. Thus, the output voltage remains unidirectional. The input and output voltages are shown in Fig. 3.25(b).

The bridge circuit full-wave rectifier has the following features:

1. Both the primary and secondary currents in the transformer are sinusoidal. Hence, a smaller transformer may be used than for the full-wave circuit of Fig. 3.23.
2. A transformer without a central tap is used.
3. The peak reverse voltage of the diode is V_m. Hence, a bridge-circuit rectifier is suitable for high-voltage applications.

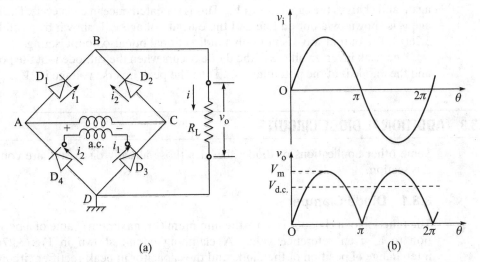

FIGURE 3.25 (a) A bridge-rectifier circuit and (b) the input and output voltages.

3.7.2 Peak Rectifier

The load resistance in the half-wave rectifier is replaced by a capacitor in the circuit of Fig. 3.26(a). To explain the operation of the circuit, assume that the source voltage increases from zero in the positive direction, as shown in Fig. 3.26(b). Assume that the capacitor, initially, is de-energized. As v_i increases, the diode conducts and the capacitor charges and a voltage is built up across the capacitor. At $\theta = \pi/2$, the capacitor is charged to the peak value V_m. As θ increases, the potential of the anode side of the diode decreases from V_m. The capacitor potential remains at V_m; hence, the diode becomes nonconducting. Ideally, the diode remains open for the remaining period. Hence, there is no loss of charge on the capacitor if the leakage current is neglected. The output voltage remains constant at V_m. Practically, the load connected to the peak rectifier is very often resistive. The load resistance across the capacitor provides a path for the capacitor to discharge during that part of the cycle when v_i is less than the capacitor voltage or when it is negative. As a result, the output voltage decreases. The rate of decrease is dependent on the time constant CR_L. Usually, the time constant relative to the period of v_i is very

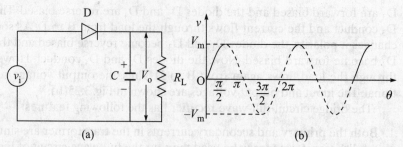

FIGURE 3.26 (a) Peak-rectifier circuit with load resistance and (b) the resulting waveform.

large, so the decrease in output voltage is small. In the next cycle, the diode conducts again and charges the capacitor to V_m. This is repeated once in each cycle. The input voltage v_i is shown by a dotted line and the output voltage v_o is shown by a full line in Fig. 3.26(b). The output wave is not only unidirectional but also nonpulsating.

The peak inverse voltage of the diode occurs when the cathode is at the potential V_m and the anode is at the potential $-V_m$. Thus, the peak inverse voltage is $2V_m$.

3.8 ADDITIONAL DIODE CIRCUITS

Some other applications of diodes, besides those already discussed, are considered in this section.

3.8.1 Diode Clamper

The function of a clamper is to fix the minimum (or maximum) value of a periodic function to a given reference value. A clamping circuit shown in Fig. 3.27(a) is the interchange of position of the diode and the capacitor in peak-rectifier circuit shown in Fig. 3.26(a). As in the peak rectifier, assume that initially the capacitor is de-energized. The input sinusoidal voltage is assumed to start charging the capacitor at the beginning

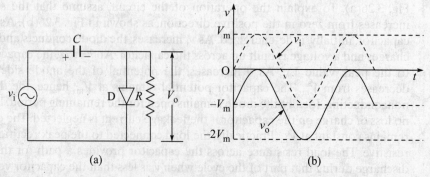

FIGURE 3.27 (a) A clamper circuit with load resistance and (b) the resulting waveform after full charging of the capacitor.

of the positive half cycle. The capacitor is charged to the peak value of input V_m. The capacitor voltage remains constant for all times at V_m, since the time constant RC is large compared to the period of the input signal. After the charging quarter cycle, the diode remains open. The output across the diode, therefore, is given by the following expression:

$$v_o = v_i - V_m \tag{3.36}$$

where v_i is the input voltage. By applying the KVL, the output is given by

$$v_o = -V_m + V_m \sin\theta \tag{3.37}$$

The output waveform is shown in Fig. 3.27(b). The difference between the input and output waveforms is a downward shift of the output relative to the input by a constant voltage V_m. The clamping circuit is said to clamp the positive peak of the input at zero volt at the output terminal.

When the charging process of the capacitor starts at the beginning of the negative half cycle, the capacitor is charged in with opposite polarity, as shown in Fig. 3.28(a). Applying the KVL to the circuit of Fig. 3.28(a), the output is given by

$$v_o = V_m + V_m \sin\theta \tag{3.38}$$

The output waveform is shown in Fig. 3.28(b). The difference between the input and output waveforms is an upward shift of the output relative to the input by a constant voltage V_m. The clamping circuit is said to clamp the negative peak of the input at zero volt at the output terminal.

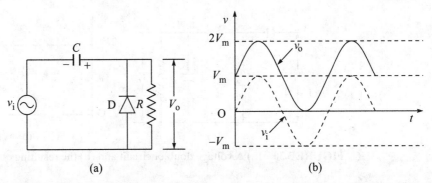

FIGURE 3.28 (a) A clamper circuit with load resistance and (b) the resulting waveform after full charging of the capacitor.

In the above clamping circuit the output voltage is clamped to zero voltage. The circuit shown in Fig. 3.29(a) is used to clamp the output relative to the input at a voltage V_R. Here, the output is clamped to a voltage V_R. The resulting output is shown in Fig. 3.29(b).

It is clear that a clamping circuit takes a purely a.c. input and adds a constant voltage to it. One of the applications of the diode clamper is in television sets. In television sets the clamper is called a *d.c. restorer*.

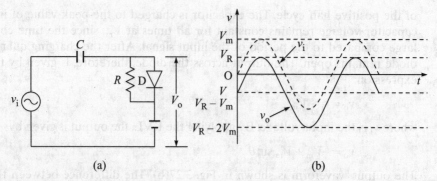

FIGURE 3.29 (a) A clamper circuit that clamps to a voltage V_R and (b) the resulting waveform after full charging of the capacitor.

3.8.2 Voltage Doubler

When the clamping circuit of Fig. 3.28(a) is combined with the peak rectifier (or peak detector) circuit of Fig. 3.26(a), we obtain the voltage doubler circuit shown in Fig. 3.30(a). When the time constant CR is large compared to the period of the sinusoidal input, the clamping portion of the circuit produces a voltage whose negative peak is clamped at zero voltage. Hence, the peak value of the voltage is $2V_m$. Then the peak-detecting circuit produces an output voltage of $v_o = 2V_m$. As the output is twice as large as the input peak, the circuit is called a *voltage doubler*. Since the output is equal to the voltage between the two peaks of the input, the circuit is also called a *peak-to-peak detector*. The resultant output waveform is shown in Fig. 3.30(b).

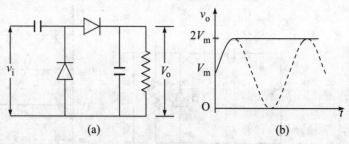

FIGURE 3.30 (a) A voltage doubler circuit and (b) the resulting output waveform.

3.9 REVIEW QUESTIONS

1. Given volt-ampere characteristic of a diode, describe the procedure to obtain the dynamic characteristic.
2. Draw the piecewise linear *V-I* characteristic of a p-n diode considering reverse biased resistance (a) infinite and (b) a very high value. For (b) what is the circuit model for the diode for ON state and for OFF state.
3. For a diode in series with a resistance and a variable voltage source, define (a) static characteristic, (b) dynamic characteristic, and (c) transfer or transmission characteristic. What is the correlation between dynamic and transfer characteristics?

4. Draw the circuit of a peak clipper using an ideal diode that limits the peak output at 10 V. Explain the functioning of the circuit.
5. Draw the circuit of a base clipper using an ideal diode that limits the base output at 10 V. Explain the functioning of the circuit.
6. Draw the clipper circuits in which the diode is used as a series element. Explain the functioning of the circuit.
7. Draw the circuit of a double-ended clipper using ideal diodes that limit the output between ±5 V. Explain the functioning of the circuit.
8. Draw the circuit of a double-ended clipper using Zener diodes. Explain the functioning of the circuit.
9. Describe the working of a comparator circuit. How does a comparator circuit differ from a clipper circuit?
10. Draw the circuit of a four diode sampling gate and explain how it functions.
11. Draw a half-wave rectifier circuit and explain its functioning. Define (a) d.c. current I_{dc}, (b) d.c. voltage V_{dc}, and (c) a.c. current I_{rms}. Also, define the term regulation of a rectifier. What is the peak inverse voltage for a half-wave rectifier?
12. Draw a full-wave rectifier circuit and explain its functioning. Derive the expression for (a) d.c. current, (b) d.c. load voltage, (c) d.c. diode voltage, and (d) r.m.s. current.
13. Derive the equation for regulation of a full-wave rectifier circuit. What is the peak inverse voltage for a full-wave rectifier?
14. Draw the circuit of a bridge rectifier and explain its functioning.
15. Draw the circuit of a peak rectifier and explain its functioning.
16. Draw the circuit of a clamper and explain its functioning.
17. Draw the circuit of a voltage doubler and explain its functioning.

3.10 SOLVED PROBLEMS

1. A p-n germanium junction diode has a reverse saturation current of 10 µA at room temperature of 300 K. The diode has negligible ohmic resistance and a Zener breakdown voltage of 100 V. The diode in series with a 1-kΩ resistance is connected to a 30 V supply. (a) Find the current (i) if the diode is forward biased and (ii) if the supply polarity is reversed. (b) Repeat (a) if the Zener breakdown voltage is 10 V.

Solution

(a) The volt equivalent of temperature at 300 K

$$V_T = \frac{T}{11600} \approx 0.026 \text{ V}$$

(i) The forward current is given by

$$I = I_0(e^{V/0.026} - 1)$$
$$= 10^{-2}(e^{38.48V} - 1) \text{ mA} \qquad (1)$$

The loop equation of the circuit,

Table-3.1. Current versus voltage.

Diode voltage (V)	0.1	0.15	0.2	0.25	0.3
Current in (1) (mA)	0.46	3.2	22.0	150.0	1031.5
Current in (2) (mA)	29.1	29.85	29.8	29.75	29.7

$$30 = 1 \times I + V$$
or $\quad I = 30 - V \quad$ (2)

The currents in Eqs. (1) and (2) for different values of diode voltage V are tabulated in Table-3.1 and plotted in Fig. 3.31.

From the point of intersection of two plots in Fig. 3.31, the forward current,

$$I_f = 29.75 \text{ mA}$$

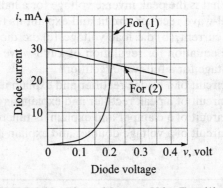

FIGURE 3.31 Plots of I versus V for Eqs. (1) and (2).

(ii) When reversed biased, the current will be equal to the saturation current, that is,

$$I_r = -0.01 \text{ mA}$$

(b) (i) The forward current will be same, that is,

$$I_f = 29.75 \text{ mA}$$

(ii) When reversed biased the diode break downs at 10 V. So, the voltage across the diode will be 10 V. The current is controlled by the resistance R. The voltage drop across R is 20 V; hence, the current,

$$I_r = \frac{20}{1} = -20 \text{ mA}$$

2. Both the diodes shown in Fig. 3.32 are described by linearized V-I characteristics with incremental resistance r and cut-in voltage V_γ. D_1 is a germanium diode with $V_\gamma = 0.2$ and $r = 20 \, \Omega$, whereas D_2 is a silicon diode with $V_\gamma = 0.6$ and $r = 15 \, \Omega$. Find the diode currents if (a) $R = 10 \text{ k}\Omega$ and (b) $R = 1 \text{ k}\Omega$.

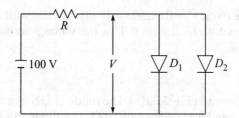

FIGURE 3.32 Circuit for Solved Problem 2.

Solution

There are two possibilities: (i) when $V \geq 0.6$ and (ii) when $V \leq 0.6$. The currents are in mA.
(a) (i) Both diodes are forward biased. Hence,

$$100 = 10(I_1 + I_2) + 0.2 + 0.02I_1$$
or $\quad 99.8 = 10.02I_1 + 10I_2$ \hfill (1)
and $\quad 100 = 10(I_1 + I_2) + 0.6 + 0.015I_2$
or $\quad 99.4 = 10I_1 + 10.015I_2$ \hfill (2)

After solving (1) and (2) for I_1 and I_2,

$$I_1 = 15.7 \text{ mA}$$

and $\quad I_2 = -5.75$ mA.

The result shows that D_2 is not forward biased. So, the assumption is not true.

(ii) The diode D_2 is reversed biased, and so does not conduct. The current through D_1 is the only current flowing through R. Hence,

$$100 = (10 + 0.020)I_1 + 0.2$$

$$I_1 = \frac{99.8}{10.020}$$

$$= 9.96 \text{ mA}$$

and $\quad I_2 = 0$

(b) $R = 1$ kΩ. (i) Let us assume that both the diodes are forward biased. Hence,

$$100 = 1 \times (I_1 + I_2) + 0.2 + 0.02I_1$$
or $\quad 99.8 = 1.02I_1 + I_2$ \hfill (3)
and $\quad 100 = 1 \times (I_1 + I_2) + 0.6 + 0.015I_2$
or $\quad 99.4 = I_1 + 1.015I_2$ \hfill (4)

After solving (3) and (4) for I_1 and I_2,

$$I_1 = 53.74 \text{ mA}$$

and $\quad I_2 = 44.99$ mA.

The assumption made above that both the diodes are forward biased is true.

3. A silicon diode (with $V_\gamma = 0.6$ and $r = 10 \,\Omega$) in series with a load of $100 \,\Omega$ is connected to an input signal $v_i(t) = 2.4 \sin \theta$. Plot the voltage across the diode for one cycle of the input signal.

Solution

The circuit is shown in Fig. 3.33(a). The diode in ON condition is represented by a d.c. source of 0.6 V and a resistance of 10 Ω. The diode in OFF condition is assumed to have infinite resistance.

The current is given by

$$i(t) = \frac{2.4 \sin \theta - 0.6}{10 + 100}$$

$$= 0.0218 \sin \theta - 0.545 \times 10^{-2} \quad \text{for } v_i \geq 0.6 \text{ V}$$

$$= 0 \qquad\qquad\qquad\qquad\qquad \text{for } v_i < 0.6$$

The cut-in angle $\phi = \sin^{-1} \frac{0.6}{2.4} = 14.5°$.

Hence, the current $i(t)$ is zero from $\theta = 0$ to $\theta = 14.5°$ and again from $\theta = 165.5°$ to $\theta = 360°$. When current is zero, the diode is open circuited and the diode voltage follows the input voltage. For the remaining period, the diode voltage is $i(t)R_f$. The input voltage, current, and diode voltage for some values of θ are given in Table-3.2.

The input voltage and the diode voltage are plotted in Fig. 3.33(b).

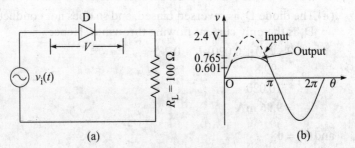

(a) (b)

FIGURE 3.33 (a) The diode circuit and (b) the input and output voltage signals.

Table-3.2. Input voltage, diode current, and diode voltage for different values of θ.

	Angle θ (deg)					
	0	14.5	90	165.5	180	270	360
Input voltage (V)	0	0.601	2.4	0.601	0	−2.4	0
Diode current (mA)	0	0	0.01635	0	0	0	0
Diode voltage (V)	0	0.601	0.765	0.601	0	−2.4	0

4. For the diode-clipping circuit shown in Fig. 3.34(a), $V_R = 3$ V, $v_i(t) = 6 \sin \omega t$, $R_L = 2$ kΩ, $R_f = 100 \,\Omega$, $R_r = \infty$, and $V_\gamma = 0$. Draw the input and output waveforms, if (a) $R = 100 \,\Omega$, (b) $R = 1$ kΩ, and (c) $R = 10$ kΩ.

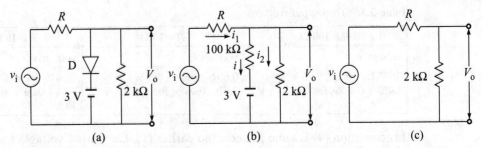

FIGURE 3.34 (a) A diode-clipping circuit, (b) with diode ON, and (c) with diode OFF.

Solution

Let us assume that the diode is forward biased. The diode is ON and the resulting circuit is as shown in Fig. 3.34(b). By Ohm's law,

$$i_1 = \frac{v_i - 3 - 100i}{R} \quad \text{and} \quad i_2 = \frac{3 + 100i}{2000} = 1.5 \times 10^{-3} + 0.05i$$

and by KCL

$$i = i_1 - i_2 = \frac{v_i - 3 - 100i}{R} - 1.5 \times 10^{-3} - 0.05i$$

Since the diode is ON,

$$i = \frac{v_i - 3 - 1.5R \times 10^{-3}}{1.05R + 100} > 0$$

or $v_i > 3 + 1.5R \times 10^{-3}$

We also conclude that the diode is OFF, when

$$v_i \leq 3 + 1.5R \times 10^{-3} \tag{1}$$

and the output voltage,

$$v_o = 3 + 100i = 3 + 100 \times \left(\frac{v_i - 3 - 1.5R \times 10^{-3}}{1.05R + 100}\right) \tag{2}$$

$$= \frac{3R + 100v_i}{1.05R + 100}$$

When the diode is OFF the circuit of Fig. 3.34(a) changes as shown in Fig. 3.34(c). The output voltage is given by

$$v_o = \frac{2000}{R + 2000} v_i \tag{3}$$

The voltage across the diode,

$$v = v_o - 3 = \frac{2000 v_i}{R + 2000} - 3 = \frac{2000(v_i - 3) - 3R}{R + 2000}$$

So, for the diode to be OFF,

$$v = \frac{2000(v_i - 3) - 3R}{R + 2000} \leq 0$$

or $v_i \leq 3 + 1.5R \times 10^{-3}$ \qquad (4)

Table-3.3. The output voltages.

$R = 100\ \Omega$	$R = 1\ k\Omega$	$R = 10\ k\Omega$
voltage v_o	voltage v_o	voltage v_o
$0.9523v_i$ for $v_i \leq 3.15$ V $1.463 + 0.488v_i$ for $v_i > 3.15$ V	$0.667v_i$ for $v_i \leq 4.5$ V $2.61 + 0.087v_i$ for $v_i > 4.5$ V	$0.167v_i$ for $v_i \leq 18$ V $2.83 + 0.009v_i$ for $v_i > 18$ V

This condition (4) is same as deducted earlier (1). The output voltages for the three values of R are given in Table-3.3.

For different values of input voltage, output voltages for three values of R are tabulated in Table-3.4. The inputs and outputs are plotted in Fig. 3.35. For $R = 10$ kΩ, the input voltage never reaches 18 V. Hence, the first condition prevails and the second condition is never satisfied. The output is always $0.167v_i$ as shown in Fig. 3.35(c).

Table-3.4. Input and output voltages for different R.

ωt	0°	45°	90°	135°	180°	225°	270°	315°	360°
v_i	0	4.3	6.0	4.3	0	−4.3	−6.0	−4.3	0
v_o; $R = 100\ \Omega$	0	3.6	4.4	3.6	0	−4.1	−5.7	−4.1	0
v_o; $R = 1\ k\Omega$	0	2.9	3.13	2.9	0	−2.9	−4.0	−2.9	0
v_o; $R = 10\ k\Omega$	0	0.72	1.0	0.72	0	−0.72	−1.0	−0.72	0

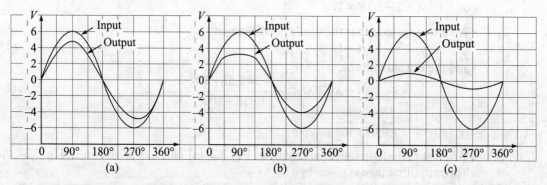

FIGURE 3.35 Input and output voltages for (a) $R = 100\ \Omega$, (b) $R = 1\ k\Omega$, and (c) $R = 10\ k\Omega$.

5. A symmetrical square wave varying between +5 V and −5 V is impressed upon the clipping circuit shown in Fig. 3.36(a). For diode, assume $R_f = 0$, $R_r = 2$ MΩ, and $V_\gamma = 0$. Draw the steady-state output waveform.

Solution

When the diode is forward biased, the circuit reduces as shown in Fig. 3.36(b), and when the diode in reversed biased, the circuit is represented by Fig. 3.36(c).

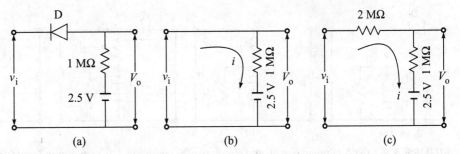

FIGURE 3.36 (a) A clipping circuit, (b) with diode ON, and (c) with diode OFF.

The input has a value either of +5 V or of −5 V. When $v_i = 5$ V, the diode is reverse biased and the current flows as shown in Fig. 3.36(c). The current is given by

$$i = \frac{v_i - 2.5}{(2+1) \times 10^6} = \frac{5 - 2.5}{3} \times 10^{-6}$$

$$= \frac{2.5}{3} \, \mu A$$

The output voltage

$$v_o = \frac{2.5}{3} \times 10^{-6} \times 1 \times 10^6 + 2.5 = \frac{10}{3} V$$

When $v_i = -5$ V, the diode is forward biased and the current flows as shown in Fig. 3.36(b). The current is given by

$$i = \frac{-v_i - 2.5}{1 \times 10^6} = -(5 + 2.5) \times 10^{-6} = -7.5 \, \mu A$$

The output voltage,

$$v_o = -7.5 \times 10^{-6} \times 1 \times 10^6 + 2.5 = -5 \text{ V}$$

The input and output are plotted in Fig. 3.37.

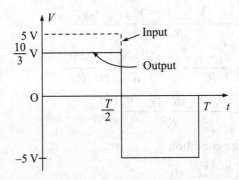

FIGURE 3.37 The input and output waveforms.

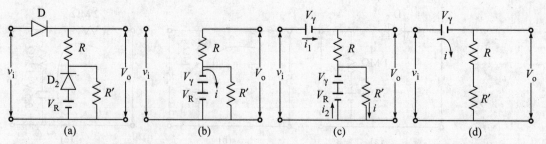

FIGURE 3.38 (a) A clipping circuit, and circuit for (b) condition-1, (c) condition-2 and (d) condition-3.

6. A clipping circuit is shown in Fig. 3.38(a). The forward resistances of diodes are taken as zero, whereas reverse resistances are taken as infinite. The cut-in voltage is V_γ.

   (a) For the current in D_2 being always in forward direction, show that the maximum value of v_i is

   $$v_{i,\max} = V_R + \frac{R}{R'}(V_R - V_\gamma)$$

   (b) Draw the transfer curve v_o versus v_i.

Solution

(a) There are three conditions: (1) when D_1 is reverse biased and D_2 is forward biased, (2) when both D_1 and D_2 are forward biased, and (3) D_1 is forward biased and D_2 is reverse biased.

Condition-1: The circuit is as shown in Fig. 3.38(b). The current,

$$i = \frac{V_R - V_\gamma}{R'} \quad \text{and} \quad v_o = V_R - V_\gamma$$

For D_2 to be forward biased, $i > 0$, that is, $V_R > V_\gamma$ and for D_1 to be reverse biased

$$(v_i - V_\gamma) < V_R - V_\gamma \quad \text{or} \quad v_i < V_R$$

Condition-2: The circuit is as shown in Fig. 3.38(c). The current,

$$i_1 = \frac{(v_i - V_\gamma) - (V_R - V_\gamma)}{R} = \frac{v_i - V_R}{R}$$

and

$$i_2 = \frac{V_R - V_\gamma}{R'} - i_1 = \frac{V_R - V_\gamma}{R'} - \frac{v_i - V_R}{R}$$

$$= \frac{(V_R - V_\gamma)R - (v_i - V_R)R'}{RR'}$$

To satisfy the condition-2,

$$i_1 = \frac{v_i - V_R}{R} > 0$$

Hence, $v_i > V_R$ (1)

$$i_2 = \frac{(V_R - V_\gamma)R - (v_i - V_R)R'}{RR'} > 0$$

or $\quad (v_i - V_R) < \dfrac{R}{R'}(V_R - V_\gamma)$

$\therefore \quad v_i < V_R + \dfrac{R}{R'}(V_R - V_\gamma)$ (2)

To satisfy both (1) and (2),

$$v_{i,\max} = V_R + \frac{R}{R'}(V_R - V_\gamma) \text{ Proved}$$

The output,

$$v_o = v_i - V_\gamma$$

when $v_i = V_R$, $v_o = V_R - V_\gamma$

when $v_i = V_R + \dfrac{R}{R'}(V_R - V_\gamma)$, $v_o = (V_R - V_\gamma) + \dfrac{R}{R'}(V_R - V_\gamma)$

Condition-3: The current, for D_1 being conducting,

$$i = \frac{v_i - V_\gamma}{R + R'} > 0$$

But $iR' > V_R - V_\gamma$ and for D_2 being nonconducting,

$$i > \frac{V_R - V_\gamma}{R'}$$

For D_1 being conducting, while D_2 being nonconducting,

$$\frac{v_i - V_\gamma}{R + R'} > \frac{V_R - V_\gamma}{R'}$$

or $\quad v_i > \dfrac{R + R'}{R'}(V_R - V_\gamma) + V_\gamma$

$\therefore \quad v_i > V_R + \dfrac{R}{R'}(V_R - V_\gamma)$

The output,

$$v_o = v_i - V_\gamma$$

Hence, for D_2 to be always conducting,

$$v_{i,\max} = V_R + \frac{R}{R'}(V_R - V_\gamma)$$

(b) The transfer curve v_o versus v_i is plotted in Fig. 3.39.

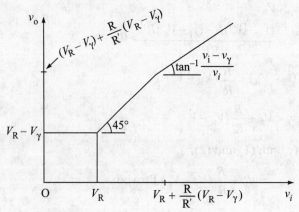

FIGURE 3.39 The transfer curve v_o versus v_i.

7. (a) For the clipping circuit shown in Fig. 3.40(a), the diodes have infinite reverse resistance, a forward resistance of 50 Ω, and $V_\gamma = 0$. Calculate and plot the transfer characteristic v_o versus v_i. Show that the circuit has two breakpoints close together. (b) The circuit is changed by removing the diode D_2 and replacing it by the resistance R. Find the transfer characteristic v_o versus v_i for this changed circuit. (c) If the forward resistance is made very small in comparison with R, then show that the double breakpoints of (a) vanish and only one breakpoint of (b) appears.

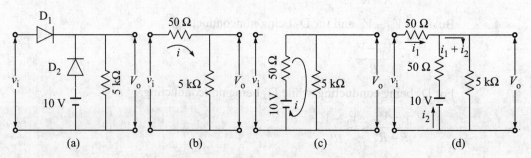

FIGURE 3.40 Circuits for Solved Problem 7.

Solution

(a) When D_1 is forward biased and D_2 is reverse biased, the circuit reduces as shown in Fig. 3.40(b). The current

$$i = \frac{v_i}{5000 + 50} = \frac{v_i}{5.05} \text{ mA}$$

The voltage output,

$$v_o = \frac{5v_i}{5.05} = \frac{v_i}{1.01}$$

For D_1 to conduct and D_2 to open $v_o > 10$ V and $v_i > v_o$; hence,

$$\frac{v_i}{1.01} > 10 \quad \text{or} \quad v_i > 10.1 \text{ V}$$

When $v_i \leq 10.1$ V, D_1 is reverse biased and D_2 is forward biased and the circuit reduces as shown in Fig. 3.40(c). The current,

$$i = \frac{10}{5+0.05} \text{ mA}$$

The voltage output,

$$v_o = \frac{10 \times 5}{5.05} = 9.9 \text{ V}$$

Hence, for D_1 nonconducting and D_2 conducting,

$$v_i \leq 9.9 \text{ V}$$

When both diodes are conducting, the circuit reduces as shown in Fig. 3.40(d). The currents are given by

$$v_i = 0.05i_1 + 5(i_1 + i_2) = 5.05i_1 + 5i_2$$
$$10 = 0.05i_2 + 5(i_1 + i_2) = 5i_1 + 5.05i_2$$

After solving for i_1 and i_2,

$$i_1 = 10.05v_i - 99.5 \text{ mA} \quad \text{and} \quad i_2 = 100.5 - 9.95v_i \text{ mA}$$

For both diodes to be conducting,

$$i_1 = 10.05v_i - 99.5 > 0 \quad \text{or} \quad v_i > 9.9 \text{ V}$$
$$i_2 = 100.5 - 9.95v_i > 0 \quad \text{or} \quad v_i < 10.1$$

The output voltage,

$$v_o = v_i - (10.05 - 99.5) \times 0.05 = 0.5v_i + 4.95$$

The transfer characteristic is plotted in Fig. 3.41. From transfer characteristic, it is clear that there are two breakpoints closed together.

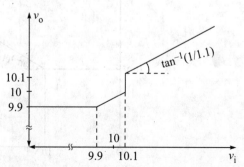

FIGURE 3.41 The transfer characteristic.

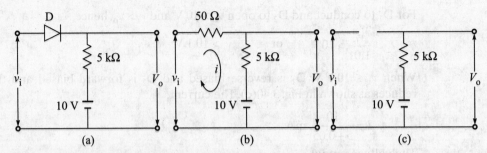

FIGURE 3.42 Circuits for Solved Problem 7.

(b) The changed circuit is shown in Fig. 3.42(a). When diode is conducting, the circuit is as shown in Fig. 3.42(b).

The loop current, in Fig. 3.40(b),

$$i = \frac{v_i - 10}{5 + 0.05} \text{ mA}$$

For D_1 to be conducting, $i > 0$, that is, $v_i > 10$

The output voltage is given by

$$v_o = 10 + 5i$$

$$= 10 + \frac{(v_i - 10)}{1.01} = \frac{0.1 + v_i}{1.01}$$

For $v_i < 10$, the diode is nonconducting and the circuit is as shown in Fig. 3.42(c). From Fig. 3.42(c), the output voltage

$$v_o = 10 \text{ V}$$

The transfer characteristic is shown in Fig. 3.43.

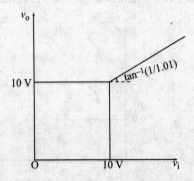

FIGURE 3.43 The transfer characteristic for Fig. 3.42(a).

(c) When $R_f \to 0$, either of the two diodes conducts. When D_1 conducts (for $v_i > 10$ V), the output voltage

$$v_o = v_i$$

For $v_i > 10$ V, D_2 is reverse biased. D_2 conducts when $v_i < 10$ V. Now D_1 does not conduct. The output voltage

$$V_o = 10 \text{ V}$$

So, there is one breakpoint like that of (b).

8. In the circuit shown in Fig. 3.44(a), the diodes used are ideal. (a) Write the transfer characteristic (output v_o as a function of input v_i). (b) Plot the transfer characteristic. (c) Plot v_o and v_i for $v_i = 25 \sin \theta$.

Solution

(a) When D_1 is conducting and D_2 is nonconducting, the circuit is as given in Fig. 3.44(b).

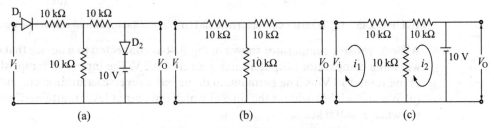

FIGURE 3.44 Circuits for Solved Problem 8.

The current i is given by

$$i = \frac{v_i}{20} = 0.05 v_i \text{ mA}$$

The output

$$v_o = 10 \times 0.05 v_i = 0.5 v_i$$

For D_2 to be nonconducting, $v_o < 10$ V and $v_i < 20$ V.

When both the diodes are conducting, the circuit reduces as shown in Fig. 3.44(c). The loop currents (in mA) are given by

$$20 i_1 - 10 i_2 = v_i$$

$$-10 i_1 + 20 i_2 = -10$$

Solving above equations,

$$i_1 = \frac{v_i - 5}{15} \quad \text{and} \quad i_2 = \frac{v_i - 20}{30}$$

For both D_1 and D_2 to be conducting, $v_i > 20$ V. The output $v_o = 10$ V.

(b) The transfer characteristic is shown in Fig. 3.45(a).

(c) For the given sinusoidal input, input and output are shown in Fig. 3.45(b).

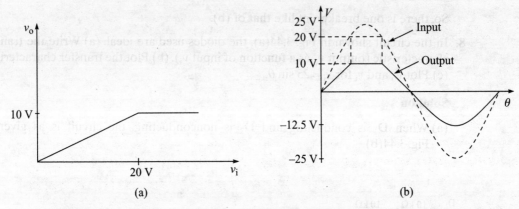

FIGURE 3.45 (a) Transfer characteristic and (b) sinusoidal input and respective output.

9. A diode resistor comparator shown in Fig. 3.46 is connected to a device that responds when the comparator output attains a level of 0.2 V. The input is a ramp, which rises at the rate of 10 V/μs. The germanium diode has a reverse saturation current of 1 μA. (a) When $R = 1$ kΩ, what is the time at which the output level is attained? (b) Repeat (a) when $R = 100$ kΩ.

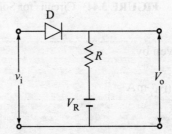

FIGURE 3.46 A diode resistor comparator.

Solution

The output is given by

$$v_o = IR$$

and the current I for germanium at room temperature of 300 K is given by

$$I = I_0(e^{V/V_T} - 1)$$
$$= 1 \times 10^{-6}(e^{38.46V} - 1)$$

(a) When $R = 1$ kΩ for $v_o = 0.2$ V

$$I = \frac{0.2}{1 \times 10^3} = 0.2 \text{ mA}$$

$$\therefore \quad 0.2 \times 10^{-3} = 1 \times 10^{-6} (e^{38.46V} - 1)$$

or $\quad e^{38.46V} = 201$

or $\quad V = \dfrac{\ln 201}{38.46} = 0.138$ V

The input ramp attains this value of voltage in time

$$t_1 = \frac{1 \times 0.138}{10} = \textbf{13.8 ns}$$

(b) When $R = 100$ kΩ for $v_o = 0.2$ V

$$I = \frac{0.2}{100 \times 10^3} = 2 \text{ }\mu\text{A}$$

$$\therefore \quad 2 \times 10^{-6} = 1 \times 10^{-6} (e^{38.46V} - 1)$$

or $\quad e^{38.46V} = 3$

or $\quad V = \dfrac{\ln 3}{38.46} = 0.0286$ V

The input ramp attains this value of voltage in time,

$$t_1 = \frac{1 \times 0.0286}{10} = \textbf{2.86 ns}$$

10. A balancing voltage divider is inserted between diodes D_4 and D_3 of the four diodes sampling gate of Fig. 3.17(a) so as to give zero output for zero input. The divider is assumed to be set at its midpoint. Let the total resistance of the divider be R. If R and forward resistance of the diode R_f are much less than R_c and R_L, show that

$$V_{c, \min} = \left(2 + \frac{R_c}{R_L}\right)\left(1 + \frac{R}{4R_f}\right)$$

Solution

The bridge is balanced for control voltage, so there is no load current due to the control voltage. The load current due to the peak value of input voltage is V_m/R_L. The current distribution in the bridge is as shown in Fig. 3.47. The principle of superposition applies. The currents due to v_c are determined assuming $v_i = 0$. Hence,

$$I_{c1} \times R_f = I_{c2} \times \left(R_f + \frac{R}{2}\right) \qquad (1)$$

$$2V_c = I_c \times 2R_c + I_{c1} \times 2R_f$$

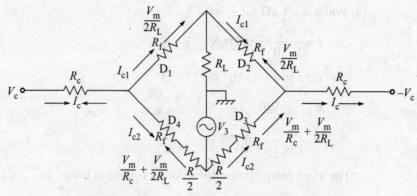

FIGURE 3.47 The current distribution in a bridge.

$$= I_{c1}\left(2R_c + 2R_f + \frac{4R_cR_f}{2R_f + R}\right)$$

$$= I_{c1}\left(\frac{8R_cR_f + 2R_cR + 4R_f^2 + 2R_fR}{2R_f + R}\right)$$

Neglecting small quantities and arranging,

$$I_{c1} = \frac{V_c(2R_f + R)}{4R_cR_f\left(1 + \dfrac{R}{4R_f}\right)} \tag{2}$$

From (1),

$$I_{c2} = \frac{2R_f}{2R_f + R} I_{c1} \tag{3}$$

Substituting for I_{c1} from (2) in (3),

$$I_{c2} = \frac{V_c}{2R_c\left(1 + \dfrac{R}{4R_f}\right)} \tag{4}$$

The current through R_c due to the peak input voltage V_m when $V_c = 0$,

$$I_m = \frac{V_m}{R_c + R_f + R/2} \approx \frac{V_m}{R_c} \tag{5}$$

There are forward currents for D_1 and D_3 for all values of v_c. The net current through D_2 is forward current, when

$$V_c > \frac{2R_cR_f\left(1 + \dfrac{R}{4R_f}\right)}{R_L(2R_f + R)} \tag{6}$$

Sec. 3.10 / Solved Problems   113

The net current through D_4 is forward current, when

$$V_c > \left[2\left(1 + \frac{R}{4R_f}\right) + \frac{R_c}{R_L}\left(1 + \frac{R}{4R_f}\right) \right] V_m$$

or $V_c > \left(2 + \frac{R_c}{R_L}\right)\left(1 + \frac{R}{4R_f}\right) V_m$ (7)

Hence, from (6) and (7),

$$V_{c,\,min} = \left(2 + \frac{R_c}{R_L}\right)\left(1 + \frac{R}{4R_f}\right) V_m$$

11. A diode whose internal resistance is 35 Ω is to supply power to a 1-kΩ load from a 220-V (r.m.s.) supply. Calculate (a) peak load current, (b) d.c. load current, (c) a.c. load current, (d) diode voltage, (e) total input power to the circuit, and (f) percentage regulation from no load to given load.

Solution

The basic circuit of the half-wave rectifier is shown in Fig. 3.19(a).

(a) From Eq. (3.10), the peak load current

$$I_m = \frac{V_m}{R_f + R_L} = \frac{\sqrt{2} \times 220}{35 + 1000}$$

$$= 0.3006 \text{ A}$$

(b) From Eq. (3.12), the d.c. load current,

$$I_{d.c.} = \frac{I_m}{\pi} = \frac{0.3006}{\pi}$$

$$= 0.0957 \text{ A}$$

(c) From Eq. (3.16), the a.c. (r.m.s.) load current,

$$I_{d.c.} = \frac{I_m}{2} = \frac{0.3006}{2}$$

$$= 0.1503 \text{ A}$$

(d) There are two voltages across the diode, one when it is conducting and another when nonconducting. When conducting, the diode voltage is negative of the load voltage and is given by Eq. (3.15),

$$V'_{d.c.} = -\frac{I_m R_L}{\pi} = -\frac{0.3006 \times 1 \times 10^3}{\pi}$$

$$= -95.684 \text{ V}$$

During nonconducting period, the diode voltage is the peak inverse voltage,

$$V_{peak} = V_m = \sqrt{2} \times 220 = 311.13 \text{ V}$$

(e) From Eq. (3.23), the total average power,

$$P_{d.c.} = I_{d.c.}^2 R_L = \frac{V_m^2 R_L}{\pi^2 (R_f + R_L)^2}$$

$$= \frac{2 \times (220)^2 \times 1 \times 10^3}{\pi^2 (35 + 1000)^2} = 9.156 \text{ W}$$

(f) From Eq. (3.21),

$$\% \text{ regulation} = \frac{R_f}{R_L} \times 100 = \frac{35}{1000} \times 100 = 3.5\%$$

12. For the full-wave single-phase rectifier shown in Fig. 3.23, the forward resistance of each diode is 500 Ω and the load resistance is 2 kΩ. The secondary transformer voltage from one terminal to central tap is 230 V. Calculate (a) d.c. load current, (b) d.c. current in each diode, (c) a.c. voltage across each diode, (d) d.c. output power, and (e) percentage regulation.

Solution

(a) From Eq. (3.27), the d.c. load current,

$$I_{d.c.} = \frac{2V_m}{\pi(R_f + R_L)} = \frac{2\sqrt{2} \times 230}{\pi(500 + 2000)}$$

$$= 0.08283 \text{ A} = 82.83 \text{ mA}$$

(b) The current through each diode flows only for half cycle, like the half-wave rectifier. Hence, the d.c. current through the diode,

$$I'_{d.c.} = \frac{I_{d.c.}}{2} = \frac{82.83}{2} = 41.415 \text{ mA}$$

(c) The voltage across each diode while conducting is given by

$$v = I_m R_f \sin\theta = \frac{V_m R_f}{R_f + R_L} \sin\theta$$

and while nonconducting

$$v = 2V_m \sin\theta - \frac{V_m R_f}{R_f + R_L} \sin\theta = \frac{V_m (2R_L + R_f)}{R_f + R_L} \sin\theta$$

Hence, the a.c. (r.m.s.) voltage across each diode is given by

$$V_{a.c.}^2 = \frac{1}{2\pi} \left[\int_0^\pi \frac{V_m^2 R_f^2}{(R_f + R_L)^2} \sin^2\theta \, d\theta + \int_\pi^{2\pi} \frac{V_m^2 (2R_L + R_f)^2}{(R_f + R_L)^2} \sin^2\theta \, d\theta \right]$$

$$\therefore \quad V_{\text{a.c.}} = \frac{V_m \sqrt{R_f^2 + (2R_L + R_f)^2}}{2(R_f + R_L)} = \frac{\sqrt{2} \times 230}{2 \times 2500} \sqrt{(500)^2 + (4500)^2}$$

$$= 294.54 \text{ V}$$

(d) The d.c. output power

$$P_{\text{o, d.c.}} = I_{\text{d.c.}}^2 R_L = (82.83 \times 10^{-3})^2 \times 2 \times 10^3$$

$$= 13.72 \text{ W}$$

(e) From Eq. (3.32),

$$\% \text{ regulation} = \frac{R_f}{R_L} \times 100 = \frac{500}{2000} \times 100$$

$$= 25\%$$

13. A 10-mA d.c. meter having resistance of 10 Ω is calibrated to measure the r.m.s. value of voltage used with a bridge rectifier using ideal diodes. A sinusoidal input voltage is applied in series with a 5-kΩ resistance. What is the full-scale reading of the meter?

Solution

The bridge rectifier with the meter is shown in Fig. 3.48. Let V be the r.m.s. value of the input voltage giving full-scale reading on the meter. Then the peak value of current through the meter

$$I_m = \frac{\sqrt{2} V}{5000 + 10}$$

Hence, the average (d.c.) current on which the deflection of d.c. meter depends is given by

$$I_{\text{d.c.}} = \frac{2 I_m}{\pi} = \frac{2\sqrt{2} V}{5010 \pi}$$

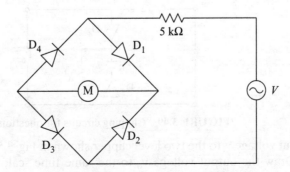

FIGURE 3.48 Circuit for Solved Problem 13.

For full-scale deflection on the meter $I_{d.c.} = 10 \times 10^{-3}$ A. Hence,

$$V = \frac{5010\pi \times 10 \times 10^{-3}}{2\sqrt{2}} = 55.65 \text{ V}$$

3.11 EXERCISES

1. For the diode circuit of Fig. 3.5, use the piecewise linear model with $R_f = 35$ Ω, $R_r = \infty$, $V_\gamma = 0.6$ V, and $R_L = 1$ kΩ. For the input $v_i = 10 \sin \theta$, draw the input voltage and diode current and determine the cut-in angle.

2. For the diode-clipping circuit shown in Fig. 3.34(a), $V_R = 8$ V, $v_i(t) = 15 \sin \omega t$, $R_f = 50$ Ω, $R_r = \infty$, and $V_\gamma = 0$. Draw the input and output waveforms, if (a) $R = 50$ Ω, (b) $R = 0.5$ kΩ, and (c) $R = 5$ kΩ.

3. In the diode-clipping circuit shown in Fig. 3.11(d), $v_i = 15 \sin \theta$, $R = 750$ Ω, and $V_R = 10$ V. The reference voltage is obtained from a 10-kΩ divider connected to a 100-V d.c. source. For the diode $R_f = 45$ Ω, $R_r = \infty$, and $V_\gamma = 0$. Draw the input and output waveforms.
   [**Hint:** Apply Thevenin's theorem to the reference voltage divider network. The reference voltage is a 10-V voltage source in series with a resistance 1 kΩ in parallel with 9 kΩ.]

4. A symmetrical square wave varying between +7.5 V and −7.5 V is impressed upon the clipping circuit shown in Fig. 3.36(a). For diode, assume $R_f = 0$, $R_r = 5$ MΩ, and $V_\gamma = 0$. Draw the steady-state output waveform.

5. In the peak-clipping circuit of Fig. 3.11(a), add another diode D_2 and a resistor R' as shown in Fig. 3.49. (a) Show that the breakpoint of the transmission curve occurs at V_R. Assume $R_r \gg R \gg R_f$. (b) Show that if D_2 is always to remain in conduction it is necessary that

$$v_i < v_{i,\max} = V_R + \frac{R}{R'}(V_R - V_\gamma)$$

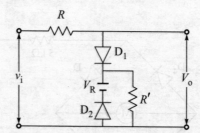

FIGURE 3.49 Clipping circuits for question 5.

6. (a) The input voltage v_i to the two level clipper shown in Fig. 3.50(a) is a ramp from 0 to 120 V. Draw the output voltage v_o to the same time scale as the input voltage. Assume ideal diodes. (b) Repeat (a) for the circuit shown in Fig. 3.50(b).

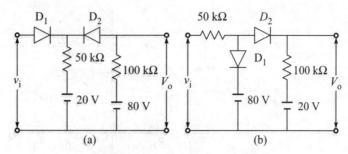

FIGURE 3.50 Circuits for question 6.

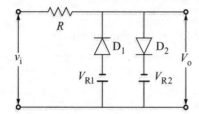

FIGURE 3.51 Circuit for question 7.   **FIGURE 3.52** Circuit for question 8.

7. The circuit shown in Fig. 3.51 is used to square a sinusoidal wave with the peak value of 50 V. It is desired that the output voltage waveform be flat for 90% of the time. Diodes forward resistance is 75 Ω and reverse resistance is 120 kΩ. Find values of reference voltages V_{R1} and V_{R2}. Choose reasonable value for R.

8. The diodes used in circuit of Fig. 3.52 are ideal. (a) Find relation between the output v_o and input v_i. (b) Plot v_o against v_i. (c) Draw v_o if $v_i = 25 \sin \theta$.

9. The diode resistor comparator of Fig. 3.14 is connected to a device that responds at the comparator output of 0.1 V. The input is a ramp with a slope of 5 V/μs. The diode has a saturation current of 10 μA. At what time 0.1 V output level is attained, (a) when $R = 1$ kΩ and (b) when $R = 100$ kΩ?

10. (a) Explain the operation of a sampling gate shown in Fig. 3.53. The supply voltage is a d.c. voltage V. The control voltage is a square voltage, where $v_c = V_c$ for period T_c and $v_c = -V_n$ for period T_n. The diodes are assumed to be ideal. (b) Verify the following relations:

(i) $V_{\min} = \dfrac{R_c}{R_2 + 2R_L \left(1 + \dfrac{R_2}{R_c}\right)}$

(ii) $A = \dfrac{v_o}{v_s} = \dfrac{2R_L}{R_2 + 2R_L \left(1 + \dfrac{R_{2c}}{R_c}\right)}$

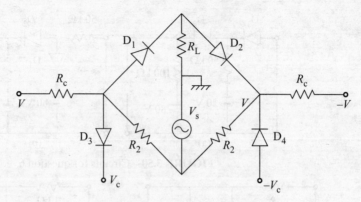

FIGURE 3.53 A sampling gate circuit.

(iii) $V_{n,\min} = AV_s$

(iv) $V_{n,\min} = V_s \dfrac{R_c}{R_c + R_2} - V \dfrac{R_2}{R_c + R_2}$

11. A diode with internal resistance of 15 Ω is used to supply power to a 2-kΩ load from a 230 V (r.m.s.) supply. Calculate (a) peak load current, (b) d.c. load current, (c) a.c. load current, (d) diode voltage, (e) total input power to the circuit, and (f) percentage regulation from no load to given load.

12. For the full-wave single-phase rectifier shown in Fig. 3.23, the forward resistance of each diode is 50 Ω and the load resistance is 1 kΩ. The secondary transformer voltage from one terminal to central tap is 200 V. Calculate (a) d.c. load current, (b) d.c. current in each diode, (c) a.c. voltage across each diode, (d) d.c. output power, and (e) percentage regulation.

13. A 1-mA d.c. meter having resistance of 25 Ω is calibrated to measure r.m.s. voltage used with a bridge rectifier using ideal diodes. The sinusoidal input voltage is applied in series with a resistance. What is the value of series resistance, if 25 V (r.m.s.) voltage gives full-scale reading in the meter?

4

ELECTRONIC AMPLIFIERS

Outline

4.1 Basic Transistor Amplifiers 120
4.2 Biasing of Transistor Amplifier 125
4.3 Multistage Amplifiers 132
4.4 RC-Coupled Amplifier 139
4.5 Feedback Amplifiers 141
4.6 Voltage Series Feedback Amplifier 153
4.7 Current Series Feedback Amplifier 159
4.8 Voltage Shunt Feedback Amplifier 164
4.9 Current Shunt Feedback Amplifier 168
4.10 Review Questions 173
4.11 Solved Problems 173
4.12 Exercises 198

4.1 BASIC TRANSISTOR AMPLIFIERS

In Chapter 2, it has been discussed that the common-emitter mode of operation of the transistor is very commonly used because it offers an appreciable current gain and a reasonable voltage gain. Therefore, a transistor can be used as an amplifying element in an amplifying circuit. In this chapter, therefore, several types of transistor amplifier circuits are discussed.

When a signal source is connected to the input port and a load to the output port of a transistor, it forms a transistor amplifier. The circuit of a basic amplifier represented as a two-port active device is shown in Fig. 4.1. When the two-port active device is a transistor, it is a transistor amplifier.

Figure 4.2 represents a transistor amplifier in which the transistor is represented by a small-signal hybrid model. The circuit is valid for any load. The load may be a pure resistance, an impedance, or another transistor. The voltages and currents are assumed to vary sinusoidally. Therefore, the circuit of Fig. 4.2 is analyzed by using the r.m.s. values of voltages and currents and applying network theories.

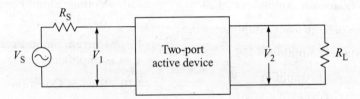

FIGURE 4.1 A basic amplifier circuit.

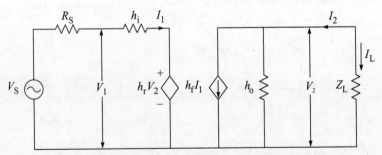

FIGURE 4.2 A transistor amplifier with a voltage source.

4.1.1 Current Gain (or Amplification)

The current gain or amplification denoted by A_I is defined as the ratio of output to input currents. Hence,

$$A_I = \frac{I_L}{I_1} = -\frac{I_2}{I_1} \tag{4.1}$$

But, $\quad I_2 = h_f I_1 + h_o V_2 \tag{4.2}$

and $\quad V_2 = I_L Z_L = -I_2 Z_L \tag{4.3}$

From Eqs. (4.2) and (4.3)

$$I_2 = h_f I_1 - h_o Z_L I_2$$

or

$$I_2 = \frac{h_f}{1 + h_o Z_L} I_1$$

Hence,

$$A_I = -\frac{h_f}{1 + h_o Z_L} \tag{4.4}$$

When the input source is a current source (I_s) with source admittance $1/R_s$, the simplified circuit of Fig. 4.2, replacing V_s by I_s and R_S by $1/R_S$, is shown in Fig. 4.3.

The current gain is defined as

$$A_{IS} = \frac{I_L}{I_S} = -\frac{I_2}{I_S} = -\frac{I_2}{I_1} \times \frac{I_1}{I_S}$$

$$= A_I \frac{I_1}{I_S}$$

But

$$I_1 = \frac{I_S R_S}{Z_i + R_S}$$

Hence,

$$A_{IS} = \frac{A_I R_S}{Z_i + R_S} = -\frac{h_f R_S}{(1 + h_o Z_L)(Z_i + R_S)} \tag{4.5}$$

When $R_S \gg Z_i$, $A_{IS} = A_I$. Hence, A_I is the current gain for an ideal current source.

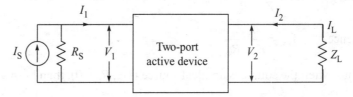

FIGURE 4.3 A transistor amplifier with a current source.

4.1.2 Input Impedance

The input impedance (see Fig. 4.2) denoted by Z_i is given by

$$Z_i = \frac{V_1}{I_1} \tag{4.6}$$

But

$$V_1 = h_i I_1 + h_r V_2$$

$$= h_i I_1 - h_r Z_L I_2$$

Hence,

$$Z_i = h_i - h_r Z_L \frac{I_2}{I_1}$$

Chapter 4 / Electronic Amplifiers

$$= h_i - \frac{h_r Z_L h_f}{1 + h_o Z_L} \qquad (4.7)$$

$$= h_i - \frac{h_r h_f}{Y_L + h_o} \qquad (4.8)$$

From the above equations, it is clear that the input impedance is a function of load impedance. For small values of load, $Z_i \approx h_i$.

4.1.3 Voltage Gain (or Amplification)

The voltage gain or voltage amplification denoted by A_V is defined as the ratio of output voltage to input voltage, that is,

$$A_V = \frac{V_2}{V_1} \qquad (4.9)$$

But $\quad V_2 = -I_2 Z_L$

Hence, $\quad A_V = -\dfrac{I_2 Z_L}{I_1 Z_i} = \dfrac{A_I Z_L}{Z_i} \qquad (4.10)$

The above-defined voltage amplification has ignored the source resistance R_S. The overall amplification A_{VS} is defined by

$$A_{VS} = \frac{V_2}{V_S} = \frac{V_2}{V_1} \cdot \frac{V_1}{V_S} = A_V \frac{V_1}{V_S}$$

But $\quad V_1 = \dfrac{Z_i V_S}{Z_i + R_S}$

Hence, $\quad A_{VS} = \dfrac{A_V Z_i}{Z_i + R_S} \qquad (4.11)$

Thus, when the source is an ideal source (i.e., $R_S = 0$), then both voltage gains are equal, that is,

$$A_{VS} = A_V$$

From Eqs. (4.10) and (4.11),

$$A_{VS} = \frac{A_I Z_L}{Z_i + R_S} \qquad (4.12)$$

From Eqs. (4.5) and (4.12),

$$A_{VS} = A_{IS} \frac{Z_L}{R_S} \qquad (4.13)$$

4.1.4 Output Admittance

Let the source voltage be replaced by its internal resistance R_S (i.e., $V_S = 0$ in Fig. 4.2) and the output terminal be opened by removing the load impedance Z_L (i.e., $Z_L = \infty$).

The transistor amplifier circuit of Fig. 4.2, under this condition, is shown in Fig. 4.4. The output terminals are driven by a voltage source V_2. If the current drawn from the source is I_2, the output admittance is given by

$$Y_o = \frac{I_2}{V_2} \tag{4.14}$$

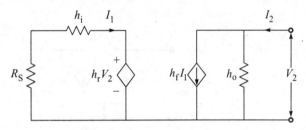

FIGURE 4.4 An equivalent circuit of an amplifier with $V_S = 0$ and $Z_L = \infty$.

From Fig. 4.4, the current I_2 is given by

$$I_2 = h_f I_1 + h_o V_2 \tag{4.15}$$

and the current I_1 is given by

$$I_1 = \frac{-h_r V_2}{R_S + h_i} \tag{4.16}$$

From Eqs. (4.15) and (4.16)

$$I_2 = -\frac{h_f h_r V_2}{R_S + h_i} + h_o V_2$$

Hence, $\quad Y_o = \dfrac{I_2}{V_2} = h_o - \dfrac{h_f h_r}{R_S + h_i} \tag{4.17}$

The different gains, and input impedance and output admittance derived above are summarized in Table-4.1.

Table-4.1 Small-signal parameters of a transistor amplifier.

Current gain $A_I = -\dfrac{h_f}{1 + h_o Z_L}$	Voltage gain $A_V = \dfrac{A_I Z_L}{Z_i}$
Current gain $A_{IS} = \dfrac{A_I R_S}{Z_i + R_S}$	Voltage gain $A_{VS} = \dfrac{A_V Z_i}{Z_i + R_S}$
Input impedance $Z_i = h_i - \dfrac{h_f h_r}{Y_L + h_o}$	Output admittance $Y_o = h_o - \dfrac{h_f h_r}{R_S + h_i}$

4.1.5 Simplified Hybrid Model

In many practical cases, simplified hybrid model of a transistor is justified. For the approximate analysis of low-frequency transistor circuit, only two of the h parameters, h_{ie} and h_{fe}, are sufficient to consider. The only condition for this approximate hybrid model is that the load resistance is small enough to satisfy the condition $h_{oe}R_L < 0.1$. The approximate hybrid model is shown in Fig. 4.5.

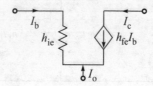

FIGURE 4.5 An approximate hybrid model.

Example-4.1

The h parameters of a transistor connected as a common-emitter amplifier are $h_{ie} = 1.1$ kΩ, $h_{re} = 2.5 \times 10^{-4}$, $h_{fe} = 50$, and $1/h_{oe} = 40$ kΩ. The load resistance $R_L = 1$ kΩ and the source internal resistance $R_S = 1$ kΩ. (a) Determine the various gains, and the input and output impedances. (b) Use simplified model to determine the various gains, and the input and output impedances.

Solution

(a) The current gain with R_S assumed to be zero,

$$A_I = -\frac{h_f}{1 + h_o Z_L} = -\frac{50}{1 + (1/40)} = -48.78$$

The input admittance (resistance in this case),

$$R_i = h_{ie} - \frac{h_{fe}h_{re}}{G_L + h_{oe}} = 1.1 - \frac{50 \times 2.5 \times 10^{-4}}{\frac{1}{1} + \frac{1}{40}} = 1.09 \text{ kΩ}$$

The current gain with $R_S = 1$ kΩ,

$$A_{IS} = \frac{A_I R_S}{R_i + R_S} = -\frac{48.78 \times 1}{1.09 + 1} = -23.34$$

The voltage gain with R_S assumed to be zero,

$$A_V = \frac{A_I R_L}{R_i} = -\frac{-48.78 \times 1}{1.09} = -44.75$$

The voltage gain with $R_S = 1\ \text{k}\Omega$,

$$A_{VS} = \frac{A_V R_i}{R_i + R_S} = \frac{-44.75 \times 1.09}{1.09 + 1} = \mathbf{-23.34}$$

The output admittance (conductance in this case),

$$G_o = h_{oe} - \frac{h_{fe} h_{re}}{R_S + h_{ie}} = \frac{1}{40} - \frac{50 \times 2.5 \times 10^{-4}}{1 + 1.1} = \mathbf{19.05\ \mu A/V}$$

The output impedance (resistance in this case),

$$R_o = \frac{1}{G_o} = \frac{1}{19.05} = 0.0525\ \text{M}\Omega = \mathbf{52.5\ k\Omega}$$

(b) (a) The current gain with R_S assumed to be zero,

$$A_I = -h_{fe} = \mathbf{-50}$$

The input admittance (resistance in this case),

$$R_i = h_{ie} = \mathbf{1.1\ k\Omega}$$

The current gain with $R_S = 1\ \text{k}\Omega$,

$$A_{IS} = -\frac{A_I R_S}{R_i + R_S} = -\frac{50 \times 1}{1.1 + 1} = \mathbf{-23.8}$$

The voltage gain with R_S assumed to be zero,

$$A_V = \frac{A_I R_L}{R_i} = -\frac{-50 \times 1}{1.1} = \mathbf{-45.45}$$

The voltage gain with $R_S = 1\ \text{k}\Omega$,

$$A_{VS} = \frac{A_V R_i}{R_i + R_S} = \frac{-45.45 \times 1.1}{1.1 + 1} = \mathbf{-23.8}$$

The output resistance,

$$R_o = \infty$$

4.2 BIASING OF TRANSISTOR AMPLIFIER

Biasing means establishing the quiescent point of a transistor amplifier in the active region of the transistor characteristics. The biasing of common-emitter configuration is important for two reasons: (1) it is the most-commonly used mode of transistor amplifier and (2) the high-current amplification of this configuration leads to stability problem of the operating point.

There are two techniques normally used for biasing: (1) stabilization techniques and (2) compensation techniques. In stabilization techniques, resistive biasing circuits are used. This allows I_B to vary so as to keep I_C relatively constant, with variations in I_{CO}, β, and V_{BE} due to change in temperature. Compensation techniques use temperature-sensitive devices such as diodes, thermistors, etc., to compensate for the temperature effect.

4.2.1 Resistive Circuit Biasing

A simplest biasing circuit scheme is shown in Fig. 4.6. Applying Kirchhoff's voltage law to the collector-base circuit,

$$V_{CC} - V_{BE} - I_B R_B = 0 \tag{4.18}$$

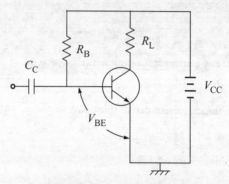

FIGURE 4.6 Fixed biasing.

V_{BE} being very small, the value of base resistance for a specified I_B can be calculated from

$$R_B = \frac{V_{CC}}{I_B} \tag{4.19}$$

As V_{CC} is fixed, therefore for a specified I_B, R_B is fixed, and for this reason this method is called the *fixed bias* method. The capacitor in series with base terminal is inserted for the purpose of blocking any d.c. component in the signal from passing on to the base terminal.

The trouble with this method is that the operating point does not remain stable due to thermal runaway. Thermal runaway is referred to the continuous rising of temperature. The collector current I_C, as we know, is given by

$$I_C = \frac{\alpha}{1-\alpha} I_B + \frac{I_{CO}}{1-\alpha} \tag{4.20}$$

The collector current has two components: one is in response to the base current and second is due to the saturation current I_{CO}. Due to I_C heat is produced at the collector terminal, causing the rise of temperature. As the temperature rises, the saturation current I_{CO} increases rapidly, as it is proportional to the cube of temperature. Because $(1-\alpha)$ is a very small quantity, the increase of I_{CO} causes a greatly magnified increase in I_C. In turn, increased I_C causes a further increase in temperature, which again increases the collector current I_C. This increase continues to that point where I_C is limited by R_L. This means that the operating point moves higher up along the load line to a position almost on the ordinate of the common emitter characteristics. This causes a greatly distorted response of a sinusoidal input.

Example 4.2

In the circuit of Fig. 4.6, $V_{CC} = 12$ V, $R_L = 3$ kΩ, $I_B = 20$ µA, and $V_{BE} = 0.6$ V. Estimate the value of R_B. If $\beta = 99$, determine I_C and V_{CE}, neglecting the saturation current.

Solution

From Eq. (4.18),

$$12 - 0.6 - 20 \times 10^{-6} R_B = 0$$

$$R_B = \frac{11.4}{20 \times 10^{-6}} = 570 \text{ k}\Omega$$

The collector current,

$$I_C = \beta I_B = 99 \times 20 \times 10^{-6} = \mathbf{1.98 \text{ mA}}$$

From collector-emitter circuit,

$$V_{CE} = V_{CC} - I_C R_L$$

$$= 12 - 1.98 \times 3 = \mathbf{6.06 \text{ V}}$$

An improved version of the fixed bias method is the collector to base method depicted in Fig. 4.7. In this arrangement, the operating base current is determined not by V_{CC} but by the collector to base voltage V_{CB}. Any increase in I_C is offset partially by decrease in collector-base voltage, as

$$V_{CB} = V_{CC} - (I_C + I_B) R_L - V_{BE}$$

Example 4.3

For the parameter values in Example-4.2, determine the operating point, that is, I_C and V_{CE} in Fig. 4.7.

Solution

From Example-4.2,

$$R_L = 3 \text{ k}\Omega, V_{BE} = 0.6 \text{ V}, \beta = 99 \quad \text{and} \quad R_B = 570 \text{ k}\Omega$$

But

$$I_B = \frac{V_{CB}}{R_B} = \frac{V_{CC} - (I_C + I_B) R_L - V_{BE}}{R_B}$$

or

$$I_B + \frac{(I_C + I_B) R_L}{R_B} = \frac{12 - 0.6}{570}$$

or $\left(1 + \dfrac{(99+1) \times 3}{570}\right) I_B = \dfrac{11.4}{570}$

or $I_B = \dfrac{11.4}{870} = 13.1\ \mu A$

Hence, $I_C = 99 \times 13.1 = \mathbf{1.3\ mA}$

$V_{CE} = 12 - 1.3 \times 3 = \mathbf{8.1\ V}$

A bias circuit known as self-bias or emitter-bias circuit is shown in Fig. 4.8. This is the most-frequently used bias method. In respect of stability, this method gives very satisfactory results. With this arrangement an increase in I_C due to a temperature rise causes the voltage drop across R_E to increase. Since voltage drop across R_2 is independent of I_C, the voltage drop across V_{BE} decreases with increase in voltage drop across R_E. The decrease in V_{BE} results in decrease in I_B, in turn restoring I_C to its original value.

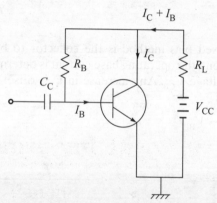

FIGURE 4.7 Improved fixed biasing.

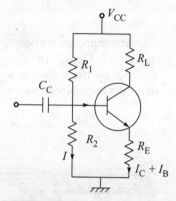

FIGURE 4.8 Self-bias circuit.

Example 4.4

In Fig. 4.8, $R_L = 1\ k\Omega$ and $V_{CC} = 12\ V$. Corresponding to an operating point $I_B = 60\ \mu A$, $V_{CE} = 6\ V$ and $I_C = 4.4\ mA$. Calculate values of $R_E, R_1,$ and R_2 so that I_C and V_{CE} remain unchanged.

Solution

From collector-emitter circuit,

$V_{CC} = I_C R_L + V_{CE} + (I_C + I_B) R_E$

or $12 = 4.4 \times 1 + 6 + (4.4 + 0.06) R_E$

Hence, $R_E = \dfrac{1.6}{4.46} = 0.358\ k\Omega = \mathbf{358\ \Omega}$

Also, $V_{cc} = (I + I_B)R_1 + IR_2$

$V_{BE} = IR_2 - (I_C + I_B)R_E$

The current through R_1 is taken as a fraction of I_C. That fraction may be taken between 0.5 to 0.1. Let $I = 0.5$ mA. Taking $V_{BE} = 0.6$ V,

$$R_2 = \frac{0.6 + 1.6}{0.5} = \frac{2.2}{0.5} = 4.4 \text{ k}\Omega$$

and $$R_1 = \frac{12 - 0.5 \times 4.4}{0.5 + 0.06} = \frac{9.8}{0.56} = 17.5 \text{ k}\Omega$$

The variation of base current due to input signal causes a variation in the emitter resistance drop, V_E. Thus, the biasing changes. This change is avoided by providing a parallel path across R_E through which the varying component of emitter current may bypass with little or no resistance. The parallel path may have a low-impedance capacitor C_E, as shown in Fig. 4.9. The size of C_E is chosen so that its impedance at the input signal frequency is small compared to R_E. However, C_E does not affect the quiescent point because the capacitor is an open circuit to d.c. The capacitor C_C is used to block any d.c. component of the input signal.

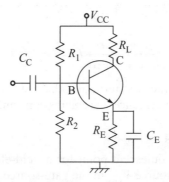

FIGURE 4.9 Biasing with bypass capacitor.

4.2.2 Diode Compensation

A diode-compensation circuit is shown in Fig. 4.10. This circuit offers stabilization against the variation of I_{CO}. For germanium transistors, variations in I_{CO} with temperature play a more important role in I_C stability than for silicon transistors. Hence, this circuit is useful for stabilizing germanium transistors.

The diode and transistor are of same type and material. Let I_O be the reverse saturation current of the diode and I_{CO} be the transistor collector saturation current. The

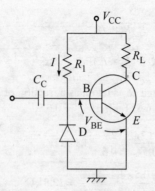

FIGURE 4.10 A diode compensation circuit.

change in I_O with temperature, hence, is same as the change in I_{CO} with temperature. From Fig. 4.9 and taking $V_{BE} = 0.2$ for germanium,

$$I = \frac{V_{CC} - V_{BE}}{R_1} \approx \frac{V_{CC}}{R_1} = \text{constant}$$

The base current

$$I_B = I - I_O$$

But,
$$I_C = \frac{I_{CO}}{1-\alpha} + \frac{\alpha}{1-\alpha} I_B$$

$$= \frac{I_{CO}}{1-\alpha} + \frac{\alpha}{1-\alpha} I - \frac{\alpha}{1-\alpha} I_o \quad (4.21)$$

As $\alpha \approx 1$, change in the first term and the last term due to change in temperature is approximately the same. Hence, I_C remains constant.

4.2.3 FET Biasing

For a given load, the quiescent point for a field-effect transistor (FET) in Fig. 1.48 is determined by a d.c. source V_{GG} in the gate-source circuit. To eliminate this source, the most common procedure is to insert a resistor R_S in the source leg, as shown in Fig. 4.10. The quiescent drain current I_{DQ} flows through R_S establishing a negative gate-source voltage given by

$$V_{GQ} = -I_{DQ} R_S \quad (4.22)$$

A minus sign indicates that the gate terminal G is at a negative potential with respect to source terminal S. As mentioned in transistor biasing, the source bypass capacitor is used to bypass the fluctuating component of the drain current. Thus, the drop across R_S in parallel with C_S remains constant.

Example 4.5

In the circuit of Fig. 4.11, the drain supply voltage, $V_{DD} = 24$ V. Let the quiescent point be located at $I_{DQ} = 3.5$ mA, $V_{DQ} = 15$ V, and $V_{GQ} = 1$ V. Determine the values of R_S and R_L to assure this Q-point.

Solution

From Eq. (4.22),
$$V_{GQ} = -I_{DQ}R_S$$

$$R_S = -\frac{V_{GQ}}{I_{DQ}} = -\frac{-1}{3.5} = 286 \; \Omega$$

Applying Kirchhoff's law to the output circuit at Q point,

$$V_{DD} = I_{DQ}R_L + V_{DQ} + I_{DQ}R_S$$

or
$$24 = 3.5R_L + 15 + 1$$

∴
$$R_L = \frac{8}{3.5} = 2.285 \; k\Omega = 2285 \; \Omega$$

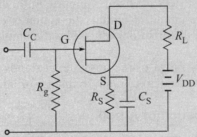

FIGURE 4.11 Biasing for FET.

4.2.4 MOSFET Biasing

Self-bias circuit employed for a transistor amplifier in Fig. 4.8 is used for metal oxide semiconductor field-effect transistor (MOSFET). The circuit is shown in Fig. 4.12. The bias gate-source voltage,

$$V_{GQ} = \frac{R_2 V_{DD}}{R_1 + R_2} - I_{DQ}R_S \tag{4.23}$$

By suitably adjusting the circuit parameters in Eq. (4.23), V_{GQ} may be made to assume either a positive or a negative value to make it consistent with the particular mode of operation.

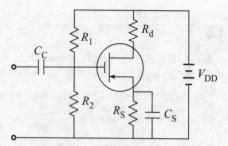

FIGURE 4.12 Biasing for MOSFET.

4.3 MULTISTAGE AMPLIFIERS

An amplifier is used to amplify a signal. The amplification in a single-stage amplifier (i.e., using one active device) may not be sufficient to fulfill our requirement. Therefore, in practice, a number of single-stage amplifiers are used in cascade to get the required amplification of the signal. An important requirement of an amplifier is that it should amplify the signal with a minimum distortion. To fulfill this requirement, the active devices used must operate in the linear region. As discussed earlier, for analysis of the amplifier circuit, the actual circuit is replaced by a linear model. To determine the distortion produced by the amplifier circuit, it is, now, a matter of circuit analysis.

The frequency range of the amplifiers extends from very low (or probably zero) frequency of a few hertz up to radio frequency or ultrahigh frequency.

4.3.1 Classification of Amplifiers

There are many ways of classification of amplifiers. The different ways may be the frequency range, the method of operation, the ultimate use, the type of load, and methods of interstage coupling. The classification on the basis of frequency range includes d.c. amplifiers (from zero frequency), audio amplifiers (from 20 Hz to 20 kHz), video or pulse amplifiers (up to a few MHz), radio amplifiers (from a few kilohertz to hundreds of megahertz), and ultrahigh frequency amplifiers (thousands of megahertz).

Amplifiers may be voltage amplifiers, current amplifiers, transconductance amplifiers, transresistance amplifiers, and power amplifiers. In voltage amplifiers, both the output and input are voltages. The gain is the ratio of the output voltage to the input voltage. Current amplifiers amplify input current and give an output current. The gain is the ratio of the output current to the input current. In transconductance amplifier, input is voltage and output is current whereas in transresistance amplifier, input is current and output is voltage. Power amplifiers amplify the input power.

The load of an amplifier may be a resistance, an impedance, or another transistor. The method of operation depends on the position of the quiescent point and the extent of the characteristic being used. On this basis, amplifiers are classified as class A, class B, class AB, or class C amplifier.

Class A Amplifiers

In class A amplifiers, the quiescent point and the input signal is such that the current in the output circuit flows at all times. This type of amplifiers essentially operates over a linear portion of the device (transistor or FET) characteristics.

Class B Amplifiers

In class B amplifiers, the operating point is at an extreme end of the device characteristics. Under this condition quiescent power is very small. So, either the quiescent voltage or the quiescent current is extremely small. In case the signal is a sinusoidal voltage, amplification takes place for only one half of a cycle.

Class AB Amplifiers

Class AB amplifiers, operate between the conditions of class A and class B operations. Hence, input signals are amplified for more than one half of a cycle.

Class C Amplifiers

The operating point of a class C amplifier is such that the output voltage (or current) is zero for more than one half of a cycle.

4.3.2 Distortions in Amplifiers

In an ideal class A amplifier, the output is a sinusoidal wave for a sinusoidal input signal. However, in general, the output waveform is not an exact replica of the input signal waveform because of various types of distortions that may arise. The distortion may arise either from the internal nonlinearity in the characteristics of the active device (transistor or FET) or from the influence of the associated circuit. Different types of distortions are nonlinear distortion, frequency distortion, and phase shift (or delay) distortion.

Nonlinear Distortion

Nonlinear distortion is also called amplitude distortion and results from the production of new frequencies or harmonics in the output. The new frequencies are not present in the input signal. The harmonics result because of the nonlinear characteristics of the active device. In case of small input signals, the pronounce harmonic is the second harmonic, whereas for large input signals there are other harmonics that influence the output.

Frequency Distortion

Frequency distortion exists when the input signal is composed of sinusoidal waves of different frequencies, and the amplifications of the amplifier for different frequencies are different. Hence, the output is not the exact replica of the input signal. The associated circuit, which includes coupling components or load, produces frequency distortion when the circuit is reactive. Hence, if the frequency response (discussed next in this section) is not a horizontal straight line over the range of frequencies of the input signal, the amplifier exhibits frequency distortion in the output.

Phase Distortion

This distortion results from the unequal phase shifts in different frequency components of the input signal.

The above-mentioned distortions may exist either separately or simultaneously.

4.3.3 Frequency Response of an Amplifier

The input signal to be amplified by an amplifier is not always a pure sinusoidal signal. The signal, in general, may be resolved in a Fourier spectrum. Hence, the signal may be considered to be composed of a number of waves of different frequencies. To obtain the exact replica of the input signal, the amplifier used must amplify the waves of different frequencies with same gain. Therefore, the frequency response of an amplifier is a criterion, which may be used to compare one amplifier with another amplifier. The frequency response of an amplifier is the plot of the gain against frequency. For an amplifier stage, the frequency response may be divided into three regions. The region over which the gain (A_{mb}) is reasonably constant is called the midband frequency region. Below the midband frequencies the gain decreases with the frequency, and gain usually approaches zero at d.c. (frequency = 0). This is the low-frequency region. The behaviour of the amplifier in this region may be considered like a high-pass filter with a gain A_{mb}. Above midband frequencies, the gain decreases with the increase in frequency. The amplifier in the high-frequency region may be considered as a low-pass filter with gain A_{mb}.

4.3.4 Low-Frequency Response

As said above, an amplifier, at low frequencies, behaves like a high-pass filter with a gain equal to the midband gain A_{mb}. Hence, the amplifier at low frequencies can be represented as a filter circuit shown in Fig. 4.13. Using the network theory, the output of the circuit is given by

$$V_o = \frac{A_{mb} V_i R_1}{R_1 + (1/j\omega_L C_1)}$$

$$= \frac{A_{mb} V_i}{1 - j(1/\omega_L C_1 R_1)} \qquad (4.24)$$

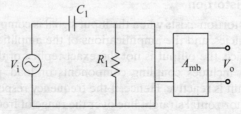

FIGURE 4.13 Amplifier at a low frequency.

The gain at a low frequency,

$$A_L = \frac{V_o}{V_i} = \frac{A_{mb}}{1 - j(1/\omega_L C_1 R_1)} \tag{4.25}$$

Let us define an angular frequency $\omega_{LC} = 1/C_1 R_1$. Now, Eq. (4.25) can be written as

$$A_L = \frac{A_{mb}}{1 - j(\omega_{LC}/\omega_L)} \tag{4.26}$$

The magnitude and the phase angle of the gain at a low frequency are given by Eqs. (4.27) and (4.28), respectively.

$$|A_L| = \frac{A_{mb}}{\sqrt{1 + (\omega_{LC}/\omega_L)^2}} \tag{4.27}$$

$$\theta_L = \tan^{-1}(\omega_{LC}/\omega_L) \tag{4.28}$$

At the angular frequency $\omega_L = \omega_{LC}$, gain,

$$|A_L| = \frac{A_{mb}}{\sqrt{2}} = 0.707 A_{mb} \tag{4.29}$$

From Eq. (4.29), we conclude that at the frequency $f_{LC} = \omega_{LC}/2\pi$, the gain is 0.707 times the midband gain. This frequency is known as the lower cutoff frequency or 3-dB frequency as the gain in dB ($20\log_{10} 0.707$) = -3 dB. In other words, the cutoff frequency is that frequency at which the gain is reduced by 3 dB from midband gain. At this frequency,

$$R_1 = X_{C1} = \frac{1}{2\pi f_{LC} C_1} \tag{4.30}$$

4.3.5 High-Frequency Gain

In high-frequency region, as discussed earlier, an amplifier can be represented by a low-pass filter with a gain A_{mb}, as shown in Fig. 4.14. The output voltage of the circuit of Fig. 4.14 is given by

$$V_o = \frac{A_{mb} V_i (1/j\omega_H C_2)}{R_2 + (1/j\omega_H C_2)}$$

$$= \frac{A_{mb} V_i}{(1 + j\omega_H C_2 R_2)} \tag{4.31}$$

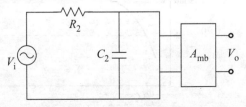

FIGURE 4.14 Amplifier at a high frequency.

The gain at high frequency,

$$A_H = \frac{V_o}{V_i} = \frac{A_{mb}}{1 + j\omega_H C_2 R_2} \quad (4.32)$$

Let us define an angular frequency $\omega_{HC} = 1/C_2 R_2$. Now, Eq. (4.32) can be written as

$$A_H = \frac{A_{mb}}{1 + j(\omega_H/\omega_{HC})} \quad (4.33)$$

The magnitude of the gain and the phase angle at a high frequency are given by Eqs. (4.34) and (4.35), respectively.

$$|A_H| = \frac{A_{mb}}{\sqrt{1 + (\omega_H/\omega_{HC})^2}} \quad (4.34)$$

$$\theta_H = \tan^{-1}\left(\frac{\omega_H}{\omega_{HC}}\right) \quad (4.35)$$

At the angular frequency $\omega_H = \omega_{HC}$, the gain,

$$|A_H| = \frac{A_{mb}}{\sqrt{2}} = 0.707 A_{mb} \quad (4.36)$$

From Eq. (4.36), we conclude that at the frequency $f_{LC} = \omega_{LC}/2\pi$, the gain is 0.707 times the midband gain. This frequency is known as high cutoff frequency or 3-dB frequency as the gain in dB $(20\log_{10} 0.707) = -3$ dB. In other words, the cutoff frequency is that frequency at which the gain is reduced by 3 dB from midband gain. At this frequency,

$$R_2 = X_{C2} = \frac{1}{2\pi f_{HC} C_2} \quad (4.37)$$

The gain magnitude versus frequency plot on a semilog graph is shown in Fig. 4.15. The gain magnitude is taken along the y-axis, which is a normal scale, whereas frequency is taken along the x-axis, which is a log scale.

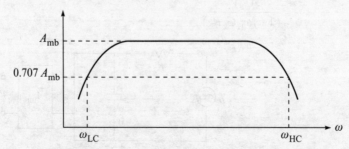

FIGURE 4.15 The gain magnitude versus frequency plot.

4.3.6 Bandpass of Cascaded Stages

To obtain the required gain, a number of amplifiers are cascaded together, as shown in Fig. 4.16. Let us consider that the stages are noninteracting. The cascade of stages is noninteracting if the input impedance of one stage is so high that it does not load the preceding stage. Let f_H^* be the high 3-dB frequency of the above multistage amplifier. This means that f_H^* is equal to the frequency at which the overall voltage gain falls by 3 dB or is equal to $(1/\sqrt{2})$ of the overall midband gain. As the stages of the multistage amplifier are noninteracting, the overall gain is obtained by multiplying gains of individual stages together. If each stage has a high 3-dB frequency, and if the high 3-dB frequency of the ith stage is f_{Hi} ($I = 1, 2, \ldots, n$), then f_H^* can be calculated from Eq. (4.38),

$$\frac{1}{\sqrt{1+(f_H^*/f_{H1})^2}} \cdots \frac{1}{\sqrt{1+(f_H^*/f_{Hi})^2}} \cdots \frac{1}{\sqrt{1+(f_H^*/f_{Hn})^2}} = \frac{1}{\sqrt{2}} \quad (4.38)$$

where $\dfrac{1}{\sqrt{1+(f_H^*/f_{Hi})^2}}$ is the gain of the ith stage at frequency f_H^*, which is high 3-dB frequency (i.e., high cutoff frequency) of the cascaded amplifier. For n stages with identical high 3-dB frequencies, we have $f_{H1} = f_{H2} = \cdots = f_{Hn} = f_H$.

$$\therefore \left[\frac{1}{\sqrt{1+(f_H^*/f_H)^2}}\right]^n = \frac{1}{\sqrt{2}}$$

or $\quad 1+(f_H^*/f_H)^2 = 2^{1/n}$

$$f_H^* = f_H \sqrt{2^{1/n} - 1} \quad (4.39)$$

Hence, the high 3-dB frequency of an n-stage amplifier can be calculated from Eq. (4.39). Thus, we see that the high cutoff frequency decreases with increase in stages. For example, $f_H^* \approx 0.64 f_H$ for a two-stage amplifier. Hence, a two-stage amplifier with individual high cutoff frequency $f_H = 100$ kHz has an overall high cutoff frequency of 64 kHz. Similarly, the overall high cutoff frequency for a three-stage amplifier is 51 kHz.

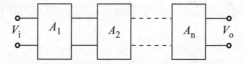

FIGURE 4.16 Cascaded amplifiers.

It can also be shown that if f_L^* is the overall low cutoff frequency and f_L is the individual low cutoff frequency of each stage, the overall low cutoff frequency can be calculated from Eq. (4.40),

$$\left[\frac{1}{\sqrt{1+(f_L/f_L^*)^2}}\right]^n = \frac{1}{\sqrt{2}}$$

or $\quad f_L^* = \dfrac{f_L}{\sqrt{2^{1/n}-1}}$ (4.40)

Hence, the overall low cutoff frequency increases with increase in stages. So, the bandwidth of a multistage amplifier is narrower than that of an individual stage.

Example 4.6

The bandwidth of the individual stage of a multistage amplifier is 20 Hz to 20 kHz. Find the bandwidth of the multistage amplifier when (a) $n = 2$ and (b) $n = 3$.

Solution

The low cutoff frequency f_L^* and the high cutoff frequency f_H^* are given by Eqs. (4.40) and (4.39), respectively.

(a) For $n = 2$,

$$f_L^* = \dfrac{20}{\sqrt{2^{1/2}-1}} = 31 \text{ Hz}$$

and $\quad f_H^* = 20\sqrt{2^{1/2}-1} = 12.87 \text{ kHz}$

Hence, bandwidth = **from 31 Hz to 12.87 kHz.**

(b) For $n = 3$,

$$f_L^* = \dfrac{20}{\sqrt{2^{1/3}-1}} = 39.2 \text{ Hz}$$

and $\quad f_H^* = 20\sqrt{2^{1/3}-1} = 10.2 \text{ kHz}$

Hence, bandwidth = **from 39.2 Hz to 10.2 kHz.**

The amplifiers discussed above have two cutoff frequencies: low cutoff frequency and high cutoff frequency. These amplifiers are grouped under a.c. amplifiers. An a.c. amplifier is also called a bandpass filter. The need often arises for an amplifier whose frequency range extends down to zero frequency (direct current) or very low frequency. The amplifiers used for this purpose are commonly known as direct couple (d.c.) amplifiers. Thus, in case of a d.c. amplifier, there is only one cutoff frequency, which is the high cutoff frequency. Hence, a d.c. amplifier is also a low-pass filter.

4.4 RC-COUPLED AMPLIFIER

A cascaded arrangement of two stages of RC-coupled amplifier is shown in Fig. 4.17. Two pnp transistors are connected in common-emitter configuration. The output of the first stage at the terminal Y_1 is V_{o1}. This output of the first stage is coupled to the input terminal X_2 of the second stage through a blocking capacitor C_b. The blocking capacitor C_b is used to block the d.c. component of V_{o1} from reaching the input terminal X_2. The collector resistor is R_C and the emitter resistor is R_E. The resistors R_1 and R_2 along with R_E are used to establish the bias. The capacitor C_E, as discussed in Section-4.2, is used to bypass any fluctuation in the emitter current. At high frequencies, the junction capacitances of the transistors are also considered. There are also some stray capacitances. For small signals the transistors are assumed to behave linear.

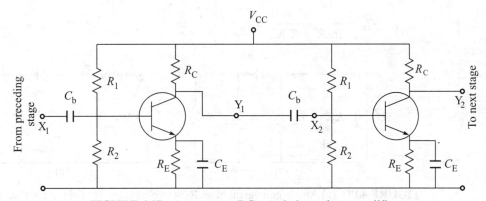

FIGURE 4.17 A two-stage RC-coupled transistor amplifier.

The FET version of the RC-coupled amplifier is shown in Fig. 4.18. The FETs are in common-source configuration. The resistor R_g is connected between the gate and the ground. The source resistor is R_s and R_d is the drain resistor. The source bypass capacitor is C_s.

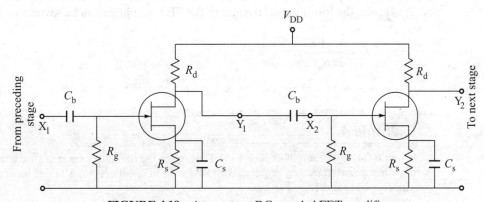

FIGURE 4.18 A two-stage RC-coupled FET amplifier.

4.4.1 Low-Frequency Response of an RC-Coupled Stage

For the time being, let us assume that the bypass capacitors C_E in transistor version and C_S in FET version are arbitrarily large and act as an a.c. short circuit across R_E in transistor and R_S in FET. The a.c. equivalent (Norton equivalent) of a single stage of any of Fig. 4.17 or Fig. 4.18 is shown in Fig. 4.19(a). The preceding stage is represented by a current source $I = h_{fe}I_b$ with its output resistance $R_o = 1/h_{oe}$ (see Fig. 4.2). For FET $R_o = r_d$ (the drain resistance). The resistance R_L represents R_C for transistor and R_d for FET. The resistor R_b is equal to R_1 in parallel with R_2 for transistor and represents R_g for FET. The resistor R_i represents the input resistance of the following stage.

The Thevenin's equivalent of Fig. 4.19(a) is shown in Fig. 4.19(b). The resistor R'_o is equal to R_o in parallel with R_L, and R'_i is equal to R_i in parallel with R_b. Hence, the lower cutoff frequency,

$$f_L = \frac{1}{2\pi(R'_o + R'_i)C_b} \qquad (4.41)$$

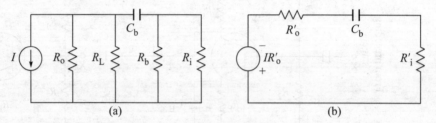

FIGURE 4.19 (a) An a.c. equivalent of an RC-coupled stage. (b) Thevenin's equivalent of (a).

For FET amplifier,

$$R'_i = \frac{R_i R_g}{R_i + R_g} \approx R_g \qquad (\because R_i \gg R_g)$$

As $R_o < R_d$ and R'_o is equal to R_o in parallel with R_d, $R'_o < R_d$. Also, as $R_g \gg R_d$, $R'_i \approx R_g \gg R'_o$. Hence, the lower cutoff frequency for FET amplifier can be given by

$$f_L = \frac{1}{2\pi R_g C_b} \qquad (4.42)$$

Example 4.7

For an RC-coupled amplifier $R_L = 1$ kΩ. Determine the minimum value of coupling capacitor, if low cutoff frequency required is 10 Hz for an FET with $R_g = 1$ MΩ and for transistors with $R_i = 1$ kΩ and $1/h_{oe} = 40$ kΩ.

Solution

(a) $R_d = R_L \ll R_i' \approx R_g$ and $R_o' < R_d$

$\therefore \quad R_o' + R_i' \approx R_g$

Hence, using Eq. (4.42),

$$10 = \frac{1}{2\pi \times 1 \times 10^6 \times C_b}$$

$\therefore \quad C_b = \dfrac{10^{-6}}{20\pi} = \mathbf{0.016\ \mu F}$

(b) $R_o = 1/h_{oe} = 40\ k\Omega$ and $R_L = 1\ k\Omega$,

$\therefore \quad R_o' = \dfrac{40 \times 1}{40+1} \approx 1\ k\Omega$

If it is assumed that $R_b = R_1 \parallel R_2 \gg R_i = 1\ k\Omega$;

$$R_i' = \frac{R_i R_b}{R_i + R_b} \approx R_i = 1\ k\Omega$$

Hence, using Eq. (4.41),

$$10 = \frac{1}{2\pi(1 \times 10^3 + 1 \times 10^3)C_b}$$

$\therefore \quad C_b = \dfrac{1000 \times 10^{-6}}{40\pi} = \mathbf{8.0\ \mu F}$

From the above discussions it is clear that as the input impedance of a transistor is much smaller than that of an FET, a coupling capacitor required for the transistor is much larger than that required for the FET. Fortunately small-size electrolytic capacitors having high value of capacitance at low voltages, at which transistors operate, are available. As for good low-frequency response, the coupling capacitors required are far larger than those obtained in integrated form; cascaded integrated stages are direct coupled.

4.5 FEEDBACK AMPLIFIERS

In a feedback amplifier, a portion of the output signal is combined with the input signal. When the sum of the input signal and a portion of the output signal is fed to the amplifier input, the feedback is positive (or generative) feedback. If the signal fed to the amplifier input is the difference of the input signal and a portion of the output signal, the feedback is negative (or degenerative) feedback. There are many advantages of the negative feedbacks and so negative feedbacks are discussed in this section.

Before discussing about the feedback amplifiers, we introduce four types of amplifiers. The four types of amplifiers are as follows:

1. Voltage amplifier
2. Current amplifier
3. Transconductance amplifier
4. Transresistance amplifier

4.5.1 Voltage Amplifier

In a voltage amplifier, the output voltage is proportional to the input voltage. A Thevenin's equivalent circuit of a voltage amplifier is shown in Fig. 4.20. If $R_s \ll R_i$, then $V_i \approx V_s$. If $R_L \gg R_o$, then $V_o \approx A_v V_i \approx A_v V_s$. The voltage gain V_o/V_i is independent of the source resistance R_s and load resistance R_L. So, for a practical voltage amplifier, the amplifier input resistance is very large compared to the source resistance, and the load resistance is very large compared to the amplifier output resistance.

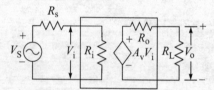

FIGURE 4.20 A voltage amplifier.

4.5.2 Current Amplifier

In a current amplifier, input current is amplified. The ratio of the output current to input current is called the current gain of the amplifier. The input resistance of the amplifier is very small compared to the source resistance and the output resistance is large compared to the load resistance. A Norton's equivalent circuit of the amplifier is shown in Fig. 4.21. As $R_s \gg R_i$, $I_i \approx I_s$, and as $R_L \ll R_o$, $I_o \approx A_I I_i \approx A_I I_s$. The current amplification $A_I = I_o/I_s$ is independent of R_s and R_L.

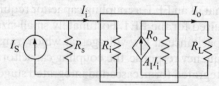

FIGURE 4.21 A current amplifier.

4.5.3 Transconductance Amplifier

Transconductance amplifier is represented by a Thevenin's equivalent in its input circuit and by a Norton's equivalent in its output circuit. This representation is shown in Fig. 4.22. In this amplifier $R_s \ll R_i$ and $R_o \gg R_L$. As $R_s \ll R_i$, $V_i \approx V_s$, and as $R_L \ll R_o$, $I_o \approx G_m V_i \approx G_m V_s$. Therefore, transconductance $G_m = I_o/V_s$ is independent of R_s and R_L.

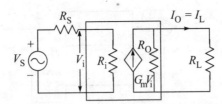

FIGURE 4.22 A transconductance amplifier.

4.5.4 Transresistance Amplifier

In transresistance amplifier, $R_s \gg R_i$ and $R_o \ll R_L$. The amplifier is represented by a Norton's equivalent in its input circuit and by a Thevenin's equivalent in its output circuit. This representation is shown in Fig. 4.23. As $R_s \gg R_i$, $I_i \approx I_s$, and as $R_o \ll R_L$, $V_o \approx R_m I_i \approx R_m I_s$. Hence, transresistance $R_m = V_o/I_s$ is independent of R_s and R_L.

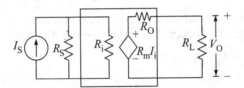

FIGURE 4.23 A transresistance amplifier.

4.5.5 Feedback of Output

In a feedback amplifier, a portion of the output (voltage or current) is mixed with the input signal through a feedback network, as shown in Fig. 4.24. The signal being fed back may be proportional to the voltage output or the current output. First is called voltage feedback, whereas the second is called current feedback. The feedback network is usually a passive two-port network. The network mostly contains resistors, but sometimes capacitors and inductors may be included.

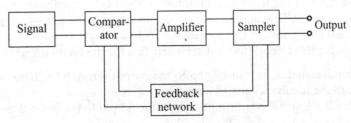

FIGURE 4.24 A feedback amplifier.

Transfer Gain

Let X_s, X_o, and X_f be the input, output, and feedback signals, respectively, of a feedback amplifier. Each signal represents either a voltage or a current. These signals are shown in the schematic diagram of the negative feedback amplifier of Fig. 4.25.

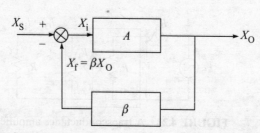

FIGURE 4.25 Schematic diagram of a negative feedback amplifier.

As shown in Fig. 4.25, the feedback signal,

$$X_f = \beta X_o$$

or
$$\beta = \frac{X_f}{X_o} \tag{4.43}$$

The factor β is often a positive or negative real number, but, in general, it is a complex function of the signal frequency. The input signal to the basic amplifier,

$$X_i = X_s - X_f = X_s - \beta X_o$$

The output signal,

$$X_o = AX_i = AX_s - \beta A X_o$$

or
$$(1 + \beta A) X_o = A X_s$$

Hence, the transfer gain of the feedback amplifier,

$$A_f = \frac{X_o}{X_s} = \frac{A}{1 + \beta A} \tag{4.44}$$

In case $\beta A \gg 1$, the transfer gain,

$$A_f = \frac{1}{\beta} \tag{4.45}$$

and so the gain of the feedback amplifier, in this case, depends only on the gain of the feedback network. The factor $-\beta A$ is the gain through the loop starting from X_i to $-X_f$ and is called *loop gain*.

Following three conditions must be satisfied for the feedback network of Fig. 4.25:

1. The input signal X_s is transmitted to the output through the basic amplifier and not through the feedback loop (β network).
2. The feedback signal is transmitted from the output to the input through the feedback loop (β network), and not through the basic amplifier.
3. The reverse transmission factor β of the feedback network is independent of the load resistance R_L and the source resistance R_s.

It is clear from Eq. (4.44) that the overall gain due to negative feedback is reduced. But, there are many advantages of using negative feedback. Hence, even at the cost of reduction in gain, negative feedback amplifiers are used. Now, we discuss some of the advantages of the negative feedback.

Effect of Feedback on Sensitivity

The variations in circuit parameters and transistor or FET characteristics due to ageing and environmental changes cause instability of the transfer gain of the amplifier. The sensitivity of the transfer gain is defined as

$$S = \frac{\partial A_f / A_f}{\partial A / A} = \frac{\text{Percentage change in } A_f}{\text{Percentage change in } A}$$

By using Eq. (4.44),

$$S = \frac{\partial A_f}{\partial A} \times \frac{A}{A_f} = \frac{1}{1+\beta A} \tag{4.46}$$

Hence, sensitivity with the negative feedback is reduced. For example, for $S = 0.1$, the percentage change in gain with negative feedback is one-tenth of the percentage change in gain without negative feedback. The reciprocal of sensitivity is known as *desensitivity*, D, that is,

$$D = (1 + \beta A) \tag{4.47}$$

Effect of Feedback on Stability

The negative feedback improves stability. It can be illustrated as follows: For a unstable gain A, the output is large for a given input. Now, by adding a negative feedback, the gain as well as the output is reduced by a factor D, desensitivity. Thus, the stability improves.

Effect of Feedback on Frequency Distortion

If the feedback network is nonreactive, the overall gain, as from Eq. (4.45), is not a function of frequency. So, a substantial reduction in frequency and phase distortions are achieved.

Effect of Feedback on Bandwidth

To examine the effect of negative feedback on bandwidth, let us consider the midband frequency gain of the basic amplifier to be A_o. The midband frequency gain with feedback,

$$A_{of} = \frac{A_o}{1+\beta A_o} \tag{4.48}$$

The gain of the basic amplifier at a high frequency,

$$A = \frac{A_o}{1+j(f/f_H)} \tag{4.49}$$

where f_H is the high cutoff frequency of the basic amplifier. The gain at high frequency with feedback,

$$A_f = \frac{A}{1+\beta A} = \frac{A_o}{1+\beta A_o + j(f/f_H)} \tag{4.50}$$

Let f_{Hf} be the high cutoff frequency of the basic amplifier with feedback. The gain of feedback amplifier at high frequency, by using Eq. (4.48),

$$A_f = \frac{A_{of}}{1+j(f/f_{Hf})} = \frac{A_o/(1+\beta A_o)}{1+j(f/f_{Hf})}$$

$$= \frac{A_o}{1+\beta A_o + j\dfrac{f(1+\beta A_o)}{f_{Hf}}} \qquad (4.51)$$

Comparing Eqs. (4.50) and (4.51), we get

$$f_{Hf} = (1+\beta A_o)f_H \qquad (4.52)$$

Thus, we see that the high cutoff frequency with feedback is $(1+\beta A_o)$ times higher than the high cutoff frequency without feedback. Similarly, it can be shown that the low cutoff frequency with feedback,

$$f_{Lf} = \frac{f_L}{(1+\beta A_o)} \qquad (4.53)$$

where f_L is the low cutoff frequency of the basic amplifier. Thus, high cutoff frequency increases whereas low cutoff frequency decreases with feedback. So, finally, we can say that the bandwidth increases with the feedback.

Effect of Feedback on Noise

The effect of feedback on noise depends greatly on where the noise signals occur. No general conclusion can be made. But, in many situations, feedback can reduce the effect of noise on the amplifier performance.

4.5.6 Types of Feedbacks

In the feedback amplifier shown in Fig. 4.24, the basic amplifier may be a voltage, current, transconductance, or transresistance amplifier. Depending on which type of basic amplifier is used in the feedback amplifier circuit, there are four types of feedbacks, namely: (1) voltage series feedback, (2) current shunt feedback, (3) current series feedback, and (4) voltage shunt feedback. The four types of feedback amplifiers are shown in Fig. 4.26. The source resistance R_s is considered to be a part of the basic amplifier. Now, we shall discuss these feedback amplifiers.

Voltage Series Feedback

From Figs. 4.20 and 4.26(a),

$$V_s = V_i + \beta V_o = I_i R_i + \beta V_o \qquad (4.54)$$

where R_i is the input resistance of the basic amplifier. The output voltage,

$$V_o = \frac{A_v V_i R_L}{R_o + R_L} = A_V V_i = A_V I_i R_i \qquad (4.55)$$

where

$$A_V = \frac{A_v R_L}{R_o + R_L} \qquad (4.56)$$

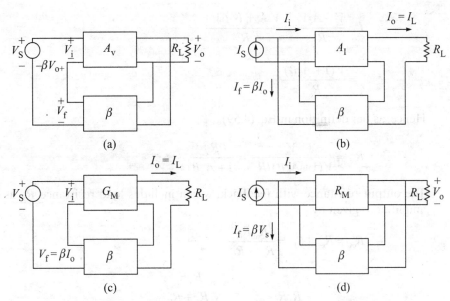

FIGURE 4.26 Feedback amplifier configurations. (a) Voltage amplifier with voltage series feedback. (b) Current amplifier with current shunt feedback. (c) Transconductance amplifier with current series feedback. (d) Transresistance amplifier with voltage shunt feedback.

A_v represents open circuit voltage gain without feedback and A_V is the voltage gain without feedback and taking R_L into account, that is, $A_v = \lim_{R_L \to \infty} A_V$.

From Eqs. (4.54) and (4.55),

$$V_s = \frac{V_o}{A_V} + \beta V_o$$

or $\quad A_V V_s = V_o (1 + \beta A_V)$

Hence, the voltage gain with feedback,

$$A_{Vf} = \frac{V_o}{V_s} = \frac{A_V}{1 + \beta A_V} \tag{4.57}$$

From Eqs. (4.54) and (4.55), the input resistance with feedback is defined as

$$R_{if} = \frac{V_s}{I_i} = (1 + \beta A_V) R_i \tag{4.58}$$

For obtaining the output resistance with feedback, remove external signal and load resistance (i.e., $V_s = 0$ and $R_L = 0$). Apply a voltage V across the output terminals and find the current I. Then, the output resistance with feedback,

$$R_{of} = \frac{V}{I} \tag{4.59}$$

From Figs. 4.20 and 4.26(a),

$$V_i = -\beta V \qquad\qquad (\because V_s = 0)$$

$$\therefore \quad I = \frac{V - A_v V_i}{R_o} = \frac{V - A_v(-\beta V)}{R_o}$$

$$= \frac{V(1 + A_v \beta)}{R_o}$$

Hence, as per definition in Eq. (4.59),

$$R_{of} = \frac{V}{V(1 + A_v \beta)/R_o} = \frac{R_o}{1 + A_v \beta} \qquad (4.60)$$

The output resistance with feedback, which includes load resistance R_L as part of the amplifier, is given by

$$R'_{of} = R_{of} \| R_L = \frac{R_{of} R_L}{R_{of} + R_L}$$

$$= \frac{R_o R_L}{R_o + R_L(1 + A_v \beta)} = \frac{\dfrac{R_o R_L}{R_o + R_L}}{1 + \beta \dfrac{A_v R_L}{R_o + R_L}}$$

$$= \frac{R'_o}{1 + \beta A_V} \qquad (4.61)$$

where $\quad R'_o = R_o \| R_L = \dfrac{R_o R_L}{R_o + R_L}$

and $\quad A_V = \dfrac{A_v R_L}{R_o + R_L}$

Current Shunt Feedback

From Figs. 4.21 and 4.26(b),

$$I_s = I_i + I_f = I_i + \beta I_o \qquad (4.62)$$

and $\quad I_o = \dfrac{A_i R_o I_i}{R_o + R_L} \qquad (4.63)$

where A_i is the short circuit current gain without feedback taking R_s into account. The current gain without feedback, when R_L is taken into account, is given by

$$A_I = \frac{I_o}{I_i} = \frac{A_i R_o}{R_o + R_L} \qquad (4.64)$$

$$\therefore \quad A_i = \lim_{R_L \to 0} A_I$$

Substituting from Eq. (4.64) for I_i into Eq. (4.62)

$$I_s = \frac{I_o}{A_I} + \beta I_o$$

or $\quad (1 + \beta A_I)I_o = A_I I_s$

$\therefore \quad A_{If} = \dfrac{I_o}{I_s} = \dfrac{A_I}{(1+\beta A_I)} \approx \dfrac{1}{\beta} \qquad$ (as $\beta A_I \gg 1$) $\hfill (4.65)$

From Eqs. (4.62) and (4.64),

$$I_s = (1 + \beta A_I)I_i \hfill (4.66)$$

Hence, the input resistance with feedback,

$$R_{if} = \frac{V_i}{I_s} = \frac{V_i}{(1+\beta A_I)I_i} = \frac{R_i}{1+\beta A_I} \hfill (4.67)$$

where R_i is the input resistance without feedback.

As in the case of the voltage series feedback, the external signal and the load resistance are removed and a voltage V is applied across the output terminals. Then the current I supplied by V is given by

$$I = \frac{V}{R_o} - A_i I_i$$

where R_o is the output resistance without feedback and R_L.

Now, $\quad I_i = -I_f = -\beta I_o = \beta I \hfill (4.68)$

$\therefore \quad I = \dfrac{V}{R_o} - A_i \beta I$

or $\quad V = I(1 + \beta A_i)R_o$

$\therefore \quad R_{of} = \dfrac{V}{I} = (1+\beta A_i)R_o \hfill (4.69)$

When R_L is taken into account, the effective output resistance with feedback is given by

$$R'_{of} = R_{of} \| R_L = \frac{R_{of} R_L}{R_{of} + R_L}$$

$$= \frac{R_o(1+A_i\beta)R_L}{R_o(1+A_i\beta) + R_L} = \frac{R_o R_L}{R_o + R_L} \times \frac{(1+A_i\beta)}{1 + \beta \dfrac{A_i R_o}{R_o + R_L}}$$

$$= R'_o \frac{1+\beta A_i}{1+\beta A_I} \hfill (4.70)$$

where $\quad R'_o = R_o \| R_L = \dfrac{R_o R_L}{R_o + R_L}$

and $\quad A_I = \dfrac{A_i R_o}{R_o + R_L}$

Current Series Feedback

From Figs. 4.22 and 4.26(c),

$$V_s = V_i + V_f = I_i R_i + \beta I_o \tag{4.71}$$

and

$$I_o = \frac{G_m V_i R_o}{R_o + R_L} = G_M V_i = G_M I_i R_i \tag{4.72}$$

where G_m is the transconductance without feedback with $R_L = 0$, and G_M is the transconductance with feedback taking R_L into account and is given by

$$G_M = \frac{G_m R_o}{R_o + R_L} \tag{4.73}$$

$$\therefore \quad G_m = \lim_{R_L \to 0} G_M \tag{4.74}$$

From Eqs. (4.71) and (4.72),

$$V_s = \frac{I_o}{G_M} + \beta I_o$$

or

$$G_M V_s = (1 + \beta G_M) I_o$$

Hence, the transconductance with feedback,

$$G_{Mf} = \frac{I_o}{V_s} = \frac{G_M}{1 + \beta G_M} \tag{4.75}$$

From Eqs. (4.71) and (4.72),

$$V_s = R_i (1 + \beta G_M) I_i$$

Hence,

$$R_{if} = \frac{V_s}{I_i} = R_i (1 + \beta G_M) \tag{4.76}$$

As in the earlier cases, the external signal and the load resistance are removed and a voltage V is applied across the output terminals. Then the current I supplied by V is given by

$$I = \frac{V}{R_o} - G_m V_i$$

where R_o is the output resistance without feedback and $R_L = \infty$.

Now,

$$V_i = -V_f = -\beta I_o = \beta I \tag{4.77}$$

$$\therefore \quad I = \frac{V}{R_o} - G_m \beta I$$

or

$$V = I(1 + \beta G_m) R_o$$

$$\therefore \quad R_{of} = \frac{V}{I} = R_o (1 + \beta G_m) \tag{4.78}$$

When R_L is taken into account, the effective output resistance with feedback is given by

$$R'_{of} = R_{of} \| R_L = \frac{R_{of} R_L}{R_{of} + R_L}$$

$$\frac{R_o(1+G_m\beta)R_L}{R_o(1+G_m\beta)+R_L} = \frac{R_o R_L}{R_o + R_L} \times \frac{(1+G_m\beta)}{1+\beta\dfrac{G_m R_o}{R_o + R_L}}$$

$$= R'_o \frac{1+\beta G_m}{1+\beta G_M} \qquad (4.79)$$

where $\quad R'_o = R_o \| R_L = \dfrac{R_o R_L}{R_o + R_L}$

and $\quad G_M = \dfrac{G_m R_o}{R_o + R_L}$

Voltage Shunt Feedback

From Figs. 4.23 and 4.25(d),

$$I_s = I_i + I_f = I_i + \beta V_o \qquad (4.80)$$

and $\quad V_o = \dfrac{R_m I_i R_L}{R_o + R_L} \qquad (4.81)$

where R_m is the open circuit transresistance without feedback taking R_s into account. The transresistance without feedback, when R_L is taken into account, is given by

$$R_M = \frac{V_o}{I_i} = \frac{R_m R_L}{R_o + R_L} \qquad (4.82)$$

$\therefore \qquad R_m = \lim_{R_L \to \infty} R_M$

From Eqs. (4.80) and (4.82),

$$I_s = \frac{V_o}{R_M} + \beta V_o$$

or $\qquad R_M I_s = (1+\beta R_M) V_o$

Hence, the transresistance with feedback,

$$R_{Mf} = \frac{V_o}{I_s} = \frac{R_M}{1+\beta R_M}$$

Substituting from Eq. (4.82) for V_o into Eq. (4.80),

$$I_s = I_i(1+\beta R_M)$$

or $\qquad I_i = \dfrac{I_s}{1+\beta R_M}$

or
$$V_i = I_i R_i = \frac{I_s R_i}{1+\beta R_M}$$

Hence, the input resistance with feedback,

$$R_{if} = \frac{V_i}{I_s} = \frac{R_i}{1+\beta R_M} \quad (4.83)$$

where R_i is the input resistance without feedback.

As in other cases, the external signal and the load resistance are removed and a voltage V is applied across the output terminals. Then the current I supplied by V is given by

$$I = \frac{V - R_m I_i}{R_o} = \frac{V(1+R_m\beta)}{R_o}$$

where R_o is the output resistance without feedback and $R_L = \infty$.

$$\therefore \quad R_{of} = \frac{V}{I} = \frac{R_o}{(1+R_m\beta)} \quad (4.84)$$

When R_L is taken into account, the effective output resistance with feedback is given by

$$R'_{of} = R_{of} \parallel R_L = \frac{R_{of} R_L}{R_{of} + R_L}$$

$$= \frac{R_o R_L}{R_o + R_L + R_m \beta R_L} = \frac{R_o R_L}{R_o + R_L} \times \frac{1}{1+\frac{R_m R_L}{R_o+R_L}\beta}$$

$$= \frac{R'_o}{1+R_M\beta} \quad (4.85)$$

Table-4.2 Input and output resistances of feedback amplifiers.

Types of feedbacks	Input resistances (taking R_s into account)	Output resistances (taking R_L into account)
Voltage series feedback	$R_i(1+A_V\beta)$	$\dfrac{R'_o}{(1+A_V\beta)}$
Current shunt feedback	$\dfrac{R_i}{(1+A_I\beta)}$	$\dfrac{R'_o(1+A_I\beta)}{(1+A_I\beta)}$
Current series feedback	$R_i(1+G_M\beta)$	$\dfrac{R'_o(1+G_m\beta)}{(1+G_M\beta)}$
Voltage shunt feedback	$\dfrac{R_i}{(1+R_M\beta)}$	$\dfrac{R'_o}{(1+R_M\beta)}$

where $\quad R_o' = R_o \parallel R_L = \dfrac{R_o R_L}{R_o + R_L}$

and $\quad R_M = \dfrac{R_m R_L}{R_o + R_L}$

The input and output resistances with feedback in all the four types of feedback amplifiers are given in Table-4.2.

From the above table, we conclude that for series feedbacks, input resistances with feedback are greater than input resistances without feedback (i.e., $R_{if} > R_i$), whereas for shunt feedbacks, $R_{if} < R_i$. The output resistance for voltage feedback with feedback (R_{of}') is less than that of without feedback (R_o') (i.e., $R_{of}' < R_o'$), whereas for current feedback $R_{of}' > R_o'$.

4.6 VOLTAGE SERIES FEEDBACK AMPLIFIER

Two types of amplifiers using voltage series feedback are discussed in this section. The amplifiers are common collector-transistor amplifier (emitter follower) and common-drain FET amplifier (source follower).

4.6.1 Emitter Follower

Emitter follower is a common-collector transistor amplifier, and its circuit diagram is shown in Fig. 4.27(a). The common-collector configuration is called the emitter follower because a change in the base voltage appears as an equal change across the load at the emitter. The voltage gain, therefore, is unity. In other words, the emitter follows the input signal. The common-collector circuit depicted in Fig. 4.27(a) is prone to damage of the transistor due to any accidental short circuit across load resistor R_L or due to any large input voltage swing. To protect the transistor, a resistor is frequently inserted in the collector circuit of an emitter follower. An emitter follower with a collector resistor is shown in Fig. 4.27(b).

The feedback signal is a voltage V_f across R_L and the sampled signal is the voltage V_o across R_L. Hence, this is the case of a voltage series feedback. The circuit of the

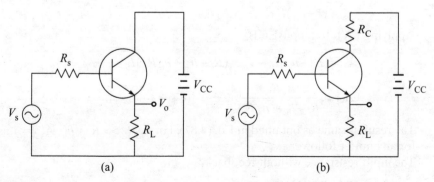

FIGURE 4.27 An emitter follower (a) without collector resistor and (b) with collector resistor.

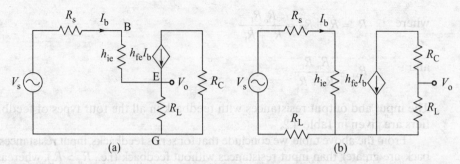

FIGURE 4.28 A low-frequency model of Fig. 4.27(b) (a) with and (b) without feedback.

emitter follower of Fig. 4.27(b) replacing the transistor with its approximate low-frequency model is shown in Fig. 4.28(a). The equivalent circuit without feedback (including the loading of R_L) is shown in Fig. 4.28(b).

The feedback voltage $V_f = V_o$. Hence, $\beta = 1$.

From Fig. 4.28(b),

$$V_s = (R_s + h_{ie} + R_L)I_b$$
$$V_o = h_{fe}I_b R_L$$

Hence, the voltage gain without feedback,

$$A_V = \frac{V_o}{V_s} = \frac{h_{fe}R_L}{R_s + h_{ie} + R_L} \qquad (4.86)$$

The gain with feedback, from Eq. (4.57),

$$A_{Vf} = \frac{A_V}{1+\beta A_V} = \frac{h_{fe}R_L}{R_s + h_{ie} + (1+h_{fe})R_L}$$

$$\approx \frac{h_{fe}R_L}{R_s + h_{ie} + h_{fe}R_L} \qquad (\because \; h_{fe} \gg 1) \qquad (4.87)$$

Alternatively, the voltage gain with feedback can be calculated from Fig. 4.28(a). The output voltage,

$$V_o = (1+h_{fe})I_b R_L \approx h_{fe}R_L I_b \qquad (\because \; h_{fe} \gg 1)$$

Applying KVL to loop BEGB,

$$V_s = (R_s + h_{ie})I_b + V_o = (R_s + h_{ie} + h_{fe}R_L)I_b$$

$$\therefore \quad A_{Vf} = \frac{h_{fe}R_L}{R_s + h_{ie} + h_{fe}R_L}$$

The result is same as obtained in Eq. (4.87). For $h_{fe}R_L \gg R_s + h_{ie}$, $A_{Vf} \approx 1$ as it should be for an emitter follower.

The input resistance without feedback,

$$R_i = R_s + h_{ie}$$

The input resistance with feedback from Eq. (4.58)

$$R_{if} = (1 + A_V)R_i = R_s + h_{ie} + h_{fe}R_L \quad \text{(since } \beta = 1\text{)} \quad (4.88)$$

Alternatively, the input resistance with feedback from Fig. 4.28(a),

$$R_{if} = R_s + h_{ie} + h_{fe}R_L \quad (\because h_{fe} \gg 1)$$

The result is same as in Eq. (4.88).

For output resistance, taking load resistance into account, connect a voltage source of voltage V across the load terminals of Fig. 4.28(a). The circuit reduces to as Fig. 4.29. The current drawn from V,

$$I = I_1 + I_2 + h_{fe}I_2$$

$$= V\left(\frac{1}{R_L} + \frac{(1+h_{fe})}{(R_s + h_{ie})}\right)$$

$$= V\frac{R_s + h_{ie} + (1+h_{fe})R_L}{(R_s + h_{ie})R_L}$$

The output resistance with feedback and load taken into account,

$$R'_{of} = \frac{V}{I} = \frac{(R_s + h_{ie})R_L}{R_s + h_{ie} + (1+h_{fe})R_L} \approx \frac{(R_s + h_{ie})R_L}{R_s + h_{ie} + h_{fe}R_L} \quad (4.89)$$

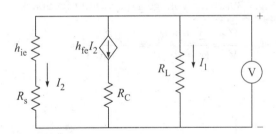

FIGURE 4.29 Circuit for output resistance.

Example 4.8

Calculate A_{Vf}, R_{if}, and R'_{of} for the amplifier of Fig. 4.27(b). Take $R_s = 1$ kΩ, $h_{fe} = 50$, $h_{ie} = 1.1$ kΩ, $h_{re} = h_{oe} = 0$, $R_C = 1$ kΩ, and $R_L = 4$ kΩ.

Solution

The voltage gain without feedback, from Eq. (4.86),

$$A_V = \frac{h_{fe}R_L}{R_s + h_{ie}} = \frac{50 \times 4}{1 + 1.1} = 95.24$$

From Eq. (4.87), the voltage gain with feedback,

$$A_{Vf} = \frac{A_V}{1+\beta A_V} = \frac{95.24}{1+1\times 95.24} = 0.99 \approx 1$$

From Eq. (4.88), the input resistance with feedback,

$$R_{if} = R_s + h_{ie} + h_{fe}R_L = 1 + 1.1 + 50\times 4$$
$$= 202.1 \text{ k}\Omega$$

From Eq. (4.89), the output resistance with feedback and load,

$$R'_{of} = \frac{(R_s + h_{ie})R_L}{R_s + h_{ie} + h_{fe}R_L} = \frac{(1+1.1)\times 4}{1+1.1+50\times 4} = 0.042 \text{ k}\Omega$$

4.6.2 Source Follower

When the transistor in Fig. 4.27(a) is replaced by an FET, the device is called a source follower. A source follower circuit is shown in Fig. 4.30(a). The feedback signal is the voltage V_f across R_L and the sampled signal is the output voltage V_o across R_L. Hence, this is the case of voltage series feedback with $\beta = 1$. Figure 4.30(b) shows the circuit diagram replacing the FET by its low-frequency model. Figure 4.30(c) is an equivalent circuit without feedback (including loading of R_L).

From Fig. 4.30(c), the voltage gain without feedback,

$$A_V = \frac{V_o}{V_s} = \frac{g_m V_s r_d R_L}{(r_d + R_L)V_s} = \frac{\mu R_L}{(r_d + R_L)} \qquad (\because \mu = r_d R_L) \qquad (4.90)$$

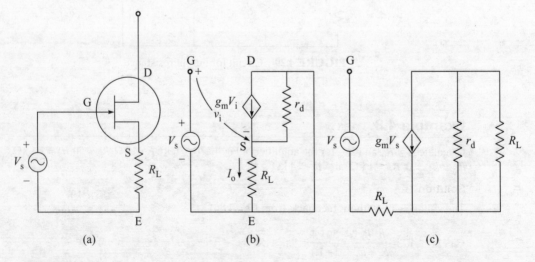

FIGURE 4.30 (a) A source follower, (b) an equivalent circuit of (a), and (c) an equivalent circuit without feedback.

Sec. 4.6 / Voltage Series Feedback Amplifier

From Eq. (4.57), the voltage gain with feedback,

$$A_{Vf} = \frac{A_V}{1 + \beta A_V} = \frac{\mu R_L}{r_d + (1+\mu)R_L} \qquad (4.91)$$

Alternatively, the voltage gain with feedback can be calculated from Fig. 4.30(b). The output voltage,

$$V_o = i_o R_L \qquad (4.92)$$

Applying KVL to loop DSGD,

$$(i_o - g_m V_i) r_d + i_o R_L = 0$$

or

$$i_o r_d - g_m r_d (V_s - i_o R_L) + i_o R_L = 0$$

or

$$i_o [r_d + (1+\mu) R_L] = \mu V_s$$

or

$$V_s = \frac{r_d + (1+\mu) R_L}{\mu} i_o \qquad (4.93)$$

The voltage gain with feedback, from Eqs. (4.92) and (4.93),

$$A_{Vf} = \frac{V_o}{V_s} = \frac{\mu R_L}{r_d + (1+\mu) R_L}$$

The result obtained is same as Eq. (4.91). The input resistance of the FET is infinite (i.e., $R_i = \infty$), and hence the input resistance for voltage series feedback,

$$R_{if} = \frac{R_i}{1 + \beta A_V} = \infty \qquad (4.94)$$

To find the output resistance with feedback, taking load resistance into account, apply a voltage V across the load terminals S and G and determine current I (with $V_s = 0$). The circuit is shown in Fig. 4.31.

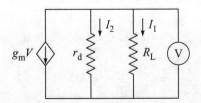

FIGURE 4.31 Circuit for output resistance.

The current drawn,
$$I = I_1 + I_2 + g_m V \tag{4.95}$$

where $\quad I_1 = \dfrac{V}{R_L} \quad$ and $\quad I_2 = \dfrac{V}{r_d}$

$$\therefore \quad I = V\left(\dfrac{1}{R_L} + \dfrac{1}{r_d} + g_m\right)$$

$$= \dfrac{r_d + R_L + g_m r_d R_L}{r_d R_L} V$$

Hence, the output resistance,
$$R'_{of} = \dfrac{V}{I} = \dfrac{r_d R_L}{r_d + (1 + g_m r_d) R_L} = \dfrac{r_d R_L}{r_d + (1 + \mu) R_L} \tag{4.96}$$

Also, $\quad R_{of} = \lim\limits_{R \to \infty} R'_{of} = \dfrac{r_d}{1 + \mu} \tag{4.97}$

Example 4.9

For the source follower of Fig. 4.30(a), calculate A_{Vf}, R_{if}, and R_{of}. Take g_m = 5 mA/V, r_d = 100 kΩ, and R_L = 1 kΩ.

Solution

The voltage gain without feedback, from Eq. (4.90),
$$A_V = \dfrac{g_m r_d R_L}{r_d + R_L} = \dfrac{5 \times 100 \times 1}{100 + 1} = 4.95$$

The voltage gain with feedback, from Eq. (4.91),
$$A_{Vf} = \dfrac{A_V}{1 + \beta A_V} = \dfrac{4.95}{1 + 1 \times 4.95} = 0.832$$

The input resistance,
$$R_{if} = \infty$$

The output resistance with feedback and $R_L = \infty$ (from Eq. 4.97)

$$R_{of} = \dfrac{r_d}{1 + \mu} = \dfrac{100}{1 + 500} \approx \mathbf{0.2 \text{ k}\Omega} \qquad\qquad (\mu = g_m r_d = 5 \times 100)$$

4.7 CURRENT SERIES FEEDBACK AMPLIFIER

Again, two circuits, one using a transistor and another using an FET, are considered in this section. The first is the common-emitter configuration with a resistance in the emitter, whereas in the second an FET is used in common-source configuration with a resistance in the source.

4.7.1 Transistor Configuration

The transistor configuration circuit is given in Fig. 4.32(a), whereas the circuit with low-frequency small-signal approximate model of the transistor is given in Fig. 4.32(b). The feedback signal X_f is a voltage V_f across the emitter resistance R_E, and the sampled signal is, approximately, the load current I_o. Here, the base current is neglected with respect to the load current. Hence, this is an example of current series feedback. As I_o is proportional to V_o, one may put a question that V_f is finally proportional to V_o and so it is the case of a voltage series feedback. If that is the case, then

$$\beta = \frac{V_f}{V_o} = \frac{-I_o R_E}{I_o R_L} = -\frac{R_E}{R_L}$$

But, β should not be a function of the load resistance R_L. Therefore, the example of Fig. 4.32(a) is a case of current series feedback, not a voltage series feedback. So, the output is the current I_o. The equivalent circuit without feedback (including loading of R_E) is shown in Fig. 4.32(c).

From Fig. 4.32(c), the transconductance without feedback,

$$G_M = \frac{I_o}{V_s} = \frac{-h_{fe} I_b}{(R_s + h_{ie} + R_E) I_b} = -\frac{h_{fe}}{(R_s + h_{ie} + R_E)} \tag{4.98}$$

In this case,

$$\beta = \frac{V_f}{I_o} = \frac{-I_o R_E}{I_o} = -R_E \tag{4.99}$$

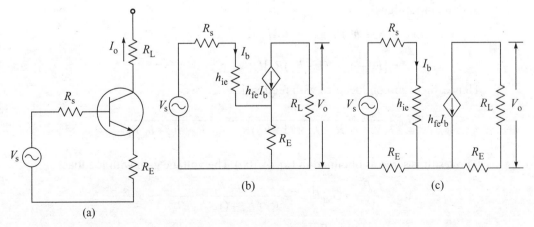

FIGURE 4.32 (a) A common-emitter amplifier, (b) an h-parameter model of (a), and (c) model without feedback.

Hence, from Eq.(4.75), the transconductance with feedback,

$$G_{Mf} = \frac{G_M}{1+\beta G_M} = -\frac{h_{fe}}{R_s + h_{ie} + (1+h_{fe})R_E}$$

$$\approx -\frac{h_{fe}}{R_s + h_{ie} + h_{fe}R_E} \quad (\because h_{fe} \gg 1) \tag{4.100}$$

The voltage gain without feedback,

$$A_V = \frac{I_o R_L}{V_s} = \frac{-h_{fe}I_b R_L}{(R_s + h_{ie} + R_E)I_b} = -\frac{h_{fe}R_L}{R_s + h_{ie} + R_E} \tag{4.101}$$

Hence, the voltage gain with feedback,

$$A_{Vf} = \frac{A_V}{1+\beta G_M} = -\frac{h_{fe}R_L}{R_s + h_{ie} + (1+h_{fe})R_E}$$

$$\approx -\frac{h_{fe}R_L}{R_s + h_{ie} + h_{fe}R_E} = G_{Mf}R_L \tag{4.102}$$

The input resistance without feedback,

$$R_i = R_s + h_{ie} + R_E$$

and the input resistance with feedback,

$$R_{if} = \frac{R_i}{1+\beta G_M} = R_s + h_{ie} + (1+h_{fe})R_E$$

$$\approx R_s + h_{ie} + h_{fe}R_E \tag{4.103}$$

Alternatively, from Fig. 4.32(b), the output current,

$$I_o = -h_{fe}I_b$$

and the input signal,

$$V_s = (R_s + h_{ie})I_b + (1+h_{fe})R_E I_b$$

$$= [R_s + h_{ie} + (1+h_{fe})R_E]I_b$$

Hence, the transconductance with feedback,

$$G_{Mf} = \frac{I_o}{V_s} = -\frac{h_{fe}}{R_s + h_{ie} + (1+h_{fe})R_E} \approx -\frac{h_{fe}}{R_s + h_{ie} + h_{fe}R_E}$$

The result is same as obtained in Eq. (4.100). The voltage gain with feedback,

$$A_{Vf} = \frac{I_o R_L}{V_s} = G_{Mf}R_L = -\frac{h_{fe}R_L}{R_s + h_{ie} + (1+h_{fe})R_E}$$

$$\approx -\frac{h_{fe}R_L}{R_s + h_{ie} + h_{fe}R_E}$$

The result is same as obtained in Eq. (4.102). The input resistance with feedback,

$$R_{if} = \frac{V_s}{I_b} = \frac{[R_s + h_{ie} + (1+h_{fe})R_E]I_b}{I_b} = R_s + h_{ie} + (1+h_{fe})R_E$$

$$\approx R_s + h_{ie} + h_{fe}R_E$$

The result is same as obtained in Eq. (4.103). For $(1+h_{fe})R_E \gg (R_s + h_{ie})$ and $h_{fe} \gg 1$,

$$G_{Mf} = -\frac{1}{R_E} \tag{4.104}$$

$$A_{Vf} = -\frac{R_L}{R_E} \tag{4.105}$$

and $\qquad R_{if} = h_{fe} R_E \tag{4.106}$

From Fig. 4.30(c), for the output resistance without feedback and without load resistance, $R_o = \infty$. Hence,

$$R_{of} = R_o(1 + G_M \beta) = \infty \tag{4.107}$$

The output resistance with feedback and taking R_L into account,

$$R'_{of} = R_{of} \parallel R_L = \frac{R_o(1+G_m\beta)R_L}{R_o(1+G_m\beta) + R_L}$$

As $R_o \to \infty$,

$$R'_{of} = R_L \tag{4.108}$$

Example 4.10

Calculate $G_{Mf}, A_{Vf}, R_{if}, R_{of},$ and R'_{of} for the circuit of Fig. 4.32(a). Take $R_s = 1\ \text{k}\Omega, R_E = 1\ \text{k}\Omega, R_L = 4\ \text{k}\Omega, h_{fe} = 150,$ and $h_{ie} = 1\ \text{k}\Omega$.

Solution

Transconductance without feedback,

$$G_M = -\frac{h_{fe}}{(R_s + h_{ie} + R_E)} = -\frac{150}{1+1+1} = -50$$

Since $\beta = -R_E = -1$

$$G_{Mf} = \frac{G_M}{1 + \beta G_M} = \frac{-50}{1 + (-1) \times (-50)} \approx -1$$

From Eq. (4.102),

$$A_{Vf} \approx -\frac{h_{fe} R_L}{R_s + h_{ie} + h_{fe} R_E} = -\frac{150 \times 4}{1 + 1 + 150 \times 1} \approx -4$$

From Eq. (4.103),

$$R_{if} = R_s + h_{ie} + (1+h_{fe})R_E = 1+1+(1+150)\times 1 = \mathbf{153\ k\Omega}$$

From Eq. (4.107),

$$R_{of} = \infty$$

From Eq. (4.108),

$$R'_{of} = R_L = \mathbf{4\ k\Omega}$$

4.7.2 FET Configuration

The FET common-source configuration with a source resistance R is shown in Fig. 4.33(a) and its low-frequency small-signal model is depicted in Fig. 4.33(b). The model without feedback (including loading of R) is shown in Fig. 4.33(c).

From Fig. 4.33(c), the input signal without feedback is V_s; hence,

$$G_M = \frac{I_o}{V_i} = \frac{I_o}{V_s}$$

where I_o is given by

$$I_o(R_L + R) + (I_o + g_m V_s)r_d = 0$$

or

$$I_o = \frac{-g_m r_d V_s}{R_L + R + r_d}$$

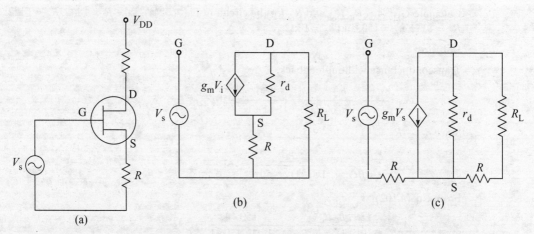

FIGURE 4.33 (a) A common-source amplifier, (b) low-frequency small-signal model of (a), and (c) model without feedback.

Sec. 4.7 / Current Series Feedback-Amplifier

$$\therefore \quad G_M = \frac{-g_m r_d}{R_L + R + r_d} = \frac{-\mu}{R_L + R + r_d} \qquad (4.109)$$

and $\quad \beta = \dfrac{V_f}{I_o} = \dfrac{-I_o R}{I_o} = -R$

Hence $\quad G_{Mf} = \dfrac{G_M}{1+\beta G_M} = \dfrac{\dfrac{-\mu}{R_L+R+r_d}}{1+\dfrac{\mu R}{R_L+R+r_d}} = \dfrac{-\mu}{r_d+R_L+(1+\mu)R} \qquad (4.110)$

The voltage gain with feedback,

$$A_{Vf} = \frac{I_o R_L}{V_s} = G_{Mf} R_L = \frac{-\mu R_L}{r_d + R_L + (1+\mu)R} \qquad (4.111)$$

Since, $R_i = \infty$, hence,

$$R_{if} = R_i(1+\beta G_M) = \infty \qquad (4.112)$$

To find output resistance with feedback, taking load resistance into account, apply a voltage V across the load terminals D and S and determine current I (with $V_s = 0$). The circuit is shown in Fig. 4.34.
The output resistance,

$$R'_{of} = \frac{V}{I}$$

where I is given by

$$I = I_1 + I_2$$

From Fig. 4.34,

$$I_1 = \frac{V}{R_L}$$

and I_2 is given by

$$I_2(1+g_m R)r_d + I_2 R = V$$

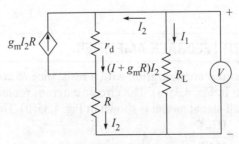

FIGURE 4.34 Circuit for output resistance.

or
$$I_2 = \frac{V}{(1+g_m r_d)R + r_d} = \frac{V}{(1+\mu)R + r_d}$$

hence,
$$I = \frac{V}{R_L} + \frac{V}{(1+\mu)R + r_d} = \frac{R_L + r_d + (1+\mu)R}{R_L[r_d + (1+\mu)R]} V$$

and
$$R'_{of} = \frac{R_L[r_d + (1+\mu)R]}{R_L + r_d + (1+\mu)R} \tag{4.113}$$

The output resistance without taking R_L into account is given by

$$R_{of} = \lim_{R_L \to \infty} R'_{of} = r_d + (1+\mu)R \tag{4.114}$$

Example 4.11

From current series feedback using FET, calculate G_{Mf} and R_{of}. Assume $g_m = 5$ mA/V, $r_d = 100$ kΩ, $R = 1$ kΩ, and $R_L = 4$ kΩ.

Solution

From Eq. (4.109),

$$G_M = \frac{-\mu}{R_L + R + r_d} = \frac{-5 \times 100}{4 + 1 + 100} = -4.76$$

Since, $\beta = -R = -1$, the transconductance with feedback,

$$G_{Mf} = \frac{G_M}{1 + \beta G_M} = \frac{-4.76}{1 + 1 \times 4.76} = -\frac{4.76}{5.76} = -0.826$$

From Eq. (4.114),

$$R_{of} = r_d + (1+\mu)R = 100 + (1 + 5 \times 100) \times 1 = \mathbf{601\ k\Omega}$$

4.8 VOLTAGE SHUNT FEEDBACK AMPLIFIER

A common-emitter configuration with a resistance R' connected from the output to the input is shown in Fig. 4.35(a). The circuit diagram replacing the transistor with its low-frequency small-signal model is shown in Fig. 4.35(b). The feedback current,

$$I_f = \frac{V_i - V_o}{R'}$$

Sec. 4.8 / Voltage Shunt Feedback Amplifier

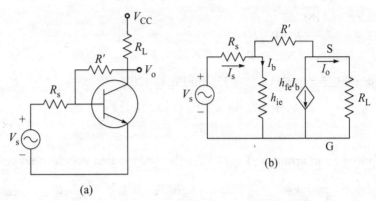

FIGURE 4.35 (a) Voltage shunt feedback amplifier and (b) low-frequency small-signal model of (a).

Also, V_o is much greater than V_i; hence,

$$I_f = -\frac{V_o}{R'} = \beta V_o \tag{4.115}$$

$$\therefore \quad \beta = -\frac{1}{R'} \tag{4.116}$$

Since the feedback current is proportional to the output voltage, this is a case of a voltage shunt feedback.

From Fig. 4.35(b), the base current,

$$I_b = I_s - I_f$$

and $\quad h_{fe}I_b + I_o = I_f$

or $\quad (1 + h_{fe})I_f = h_{fe}I_s + I_o \tag{4.117}$

Substituting for I_f from Eq. (4.115) and $I_o = V_o/R_L$ in Eq. (4.117),

$$\frac{V_o[R' + (1 + h_{fe})R_L]}{R'R_L} = -h_{fe}I_s$$

Hence, the transresistance with feedback,

$$R_{Mf} = \frac{V_o}{I_s} = -\frac{h_{fe}R'R_L}{R' + (1 + h_{fe})R_L} \tag{4.118}$$

Since $(1 + h_{fe})R_L \gg R'$ and $h_{fe} \gg 1$,

$$R_{Mf} = -R' = \frac{1}{\beta} \tag{4.119}$$

From Fig. 4.35(b), the source voltage,

$$V_s = I_s(R_s + h_{ie}) - I_f h_{ie}$$

$$V_s = I_s(R_s + h_{ie}) + \frac{h_{ie}}{R'}V_o \tag{4.120}$$

From Eq. (4.118),

$$I_s = -\frac{R' + (1 + h_{fe})R_L}{h_{fe} R' R_L} V_o \tag{4.121}$$

Substituting for I_s from Eq. (4.121) in Eq. (4.120),

$$V_s = -\frac{h_{ie}(R' + R_L) + R_s[R' + (1 + h_{fe})R_L]}{h_{fe} R' R_L} \times V_o \tag{4.122}$$

Simplifying and arranging Eq. (4.122), the voltage gain with feedback,

$$A_{Vf} = \frac{V_o}{V_s} = -\frac{h_{fe} R' R_L}{h_{ie}(R' + R_L) + R_s[R' + (1 + h_{fe})R_L]} \approx \frac{R'}{R_s} \tag{4.123}$$

The input resistance with feedback, taking R_s into account,

$$R'_{if} = \frac{V_s}{I_s} = \frac{R_{Mf}}{A_{Vf}} = \frac{h_{ie}(R' + R_L) + R_s[R' + (1 + h_{fe})R_L]}{h_{fe} R' R_L} \approx R_s \tag{4.124}$$

To find output resistance with feedback, taking load resistance into account, apply a voltage V across the load terminals S and G and determine current I (with $V_s = 0$). The circuit is shown in Fig. 4.36.

The current drawn,

$$I = I_1 + I_2 + h_{fe} I_b$$

where $\quad I_1 = \dfrac{V}{R_L}$

$$I_2 = \frac{V}{R' + \dfrac{R_s h_{ie}}{R_s + h_{ie}}}$$

$$= \frac{V(R_s + h_{ie})}{R'(R_s + h_{ie}) + R_s h_{ie}}$$

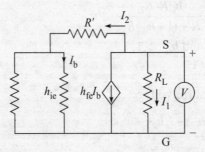

FIGURE 4.36 A circuit for output resistance.

and $I_b = \dfrac{R_s}{R_s + h_{ie}} I_2 = \dfrac{VR_s}{R'(R_s + h_{ie}) + R_s h_{ie}}$

$\therefore \quad I = V\left(\dfrac{1}{R_L} + \dfrac{R_s + h_{ie}}{R'(R_s + h_{ie}) + R_s h_{ie}} + \dfrac{R_s h_{fe}}{R'(R_s + h_{ie}) + R_s h_{ie}}\right)$

$= \dfrac{R_s[R' + (1 + h_{fe})R_L] + h_{ie}(R_s + R' + R_L)}{R_L[R' R_s + h_{ie}(R_s + R')]} V$

Hence, the output resistance with feedback, taking R_L into account,

$$R'_{of} = \dfrac{V}{I} = \dfrac{R_L[R' R_s + h_{ie}(R_s + R')]}{R_s[R' + (1 + h_{fe})R_L] + h_{ie}(R_s + R' + R_L)} \tag{4.125}$$

The output resistance with feedback and $R_L = \infty$,

$$R_{of} = \lim_{R_L \to \infty} R'_{of} = \dfrac{R' R_s + h_{ie}(R_s + R')}{(1 + h_{fe})R_s + h_{ie}} \tag{4.126}$$

Example 4.12

For the voltage shunt feedback amplifier of Fig. 4.35(a), find (a) R_{Mf}, (b) A_{Vf}, (c) R'_{if}, (d) R'_{of} and R_{of}. Assume $R_L = 4$ kΩ, $R' = 40$ kΩ, $R_s = 10$ kΩ, $h_{ie} = 1$ kΩ, $h_{fe} = 100$, and $h_{re} = h_{oe} = 0$.

Solution

(a) From Eq. (4.118), the transresistance with feedback,

$R_{Mf} = -\dfrac{h_{fe} R' R_L}{R' + (1 + h_{fe})R_L}$

$= -\dfrac{100 \times 40 \times 4}{40 + (1 + 100) \times 4} = \mathbf{-36\ k\Omega}$

(b) From Eq. (4.123), the voltage gain with feedback,

$A_{Vf} = -\dfrac{h_{fe} R' R_L}{h_{ie}(R' + R_L) + R_s[R' + (1 + h_{fe})R_L]}$

$= -\dfrac{100 \times 40 \times 4}{1 \times (40 + 4) + 10[40 + (1 + 100) \times 4]} = \mathbf{-3.6}$

(c) From Eq. (4.124), the input resistance with feedback, taking R_L into account,

$R'_{if} = \dfrac{V_s}{I_s} = \dfrac{R_{Mf}}{A_{Vf}} = \dfrac{-36}{-3.6} = \mathbf{10\ k\Omega}$

(d) From Eq. (4.125), the output resistance with feedback, taking R_L into account,

$$R'_{of} = \frac{R_L[R'R_s + h_{ie}(R_s + R')]}{R_s[R' + (1+h_{fe})R_L] + h_{ie}(R_s + R' + R_L)}$$

$$= \frac{4[40 \times 10 + 10 + 40]}{10[40 + (1+100) \times 4] + (10 + 40 + 4)} = 0.4 \text{ k}\Omega$$

(e) From Eq. (4.126), the output resistance with feedback and $R_L = \infty$,

$$R_{of} = \frac{R'R_s + h_{ie}(R_s + R')}{(1+h_{fe})R_s + h_{ie}}$$

$$= \frac{40 \times 10 + 10 + 40}{(1+100) \times 10 + 1} = 0.445 \text{ k}\Omega$$

4.9 CURRENT SHUNT FEEDBACK AMPLIFIER

Two transistor amplifiers in cascade are shown in Fig. 4.37. The feedback is from the second transistor emitter to the first transistor base through the resistor R'. The voltage V_{i2}, being the output of the first amplifier, is much larger than V_{i1}, the input of the first amplifier. Also V_{i2} is 180° out of phase with V_{i1}. Because of emitter follower action, the voltage, V_{e2}, across the emitter resistance R_E of the second transistor, is only slightly smaller than V_{i2}. The voltage V_{e2} and V_{i2} are in phase. Hence, $V_{e2} \gg V_{i1}$, and they are 180° out of phase. If the input signal increase so does I_s. The feedback current I_f also increases and $I_i = I_s - I_f$ is smaller than it would be if there were no feedback. Hence, it is a case of negative feedback.

The feedback current

$$I_f = \frac{V_{i1} - V_{e2}}{R'} \approx \frac{-V_{e2}}{R'} \qquad (\because V_{e2} \gg V_{i1})$$

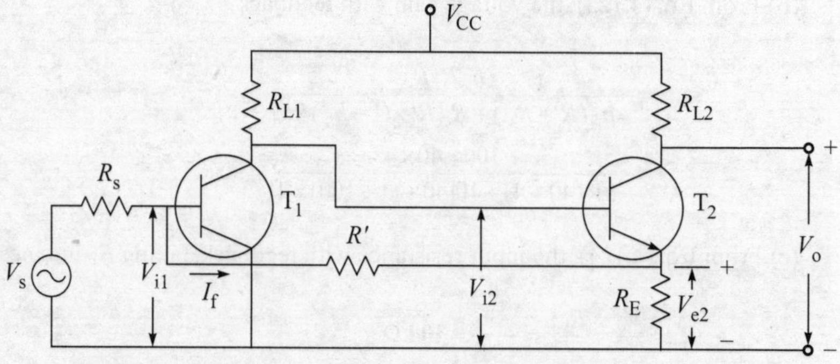

FIGURE 4.37 Two cascade transistor amplifiers with feedback.

Neglecting the base current compared to the collector current,

$$I_f = \frac{(I_o - I_f)R_E}{R'}$$

or

$$I_f\left(1 + \frac{R_E}{R'}\right) = \frac{I_o R_E}{R'}$$

∴

$$I_f = \frac{R_E}{R_E + R'} I_o = \beta I_o \qquad (4.127)$$

where

$$\beta = \frac{R_E}{R_E + R'} \qquad (4.128)$$

Since the feedback current is proportional to the output current, this feedback is an example of current shunt feedback. The current gain without feedback,

$$A_I = \frac{I_o}{I_i}$$

or

$$I_i = \frac{I_o}{A_I}$$

and

$$I_s = I_i + I_f = \frac{I_o}{A_I} + \beta I_o = \frac{1 + \beta A_I}{A_I} I_o$$

Hence, the current gain with feedback,

$$A_{If} = \frac{I_o}{I_s} = \frac{A_I}{1 + \beta A_I} \approx \frac{1}{\beta} = \frac{R_E + R'}{R_E} \qquad (4.129)$$

The source voltage

$$V_s = I_s R_s + (I_s - I_f) h_{ie}$$

Since $(I_s - I_f) \ll I_s$,

$$V_s \approx I_s R_s$$

The voltage gain with feedback,

$$A_{Vf} = \frac{V_o}{V_s} \approx \frac{I_o R_{L2}}{I_s R_s} = A_{If} \frac{R_{L2}}{R_s} = \frac{R_{L2}}{\beta R_s} \qquad (4.130)$$

4.9.1 Amplifier Without Feedback

In the circuit of Fig. 4.37, if the output loop is open at the emitter of T_2, the resistances R' and R_E in series are connected between the base terminal of T_1 and the ground, as shown in Fig. 4.38. When the input terminals of T_1 are shorted, then the output circuit looks as in Fig. 4.38, where resistances R' and R_E are in parallel. Since, the feedback signal is a current, the input source is represented by an equivalent current source (i.e., by Norton's equivalent).

The low-frequency small-signal model of Fig. 4.38 is shown in Fig. 4.39.

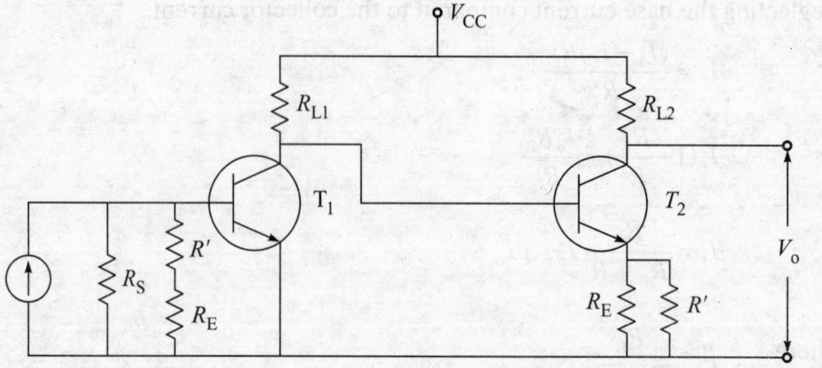

FIGURE 4.38 Amplifier of Fig. 4.37 without feedback but taking loading effect of R' into account.

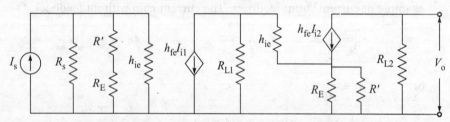

FIGURE 4.39 Approximate equivalent circuit of Fig. 4.38.

From Fig. 4.38, the feedback current I_f is in R' in the output circuit. As $I_o \gg I_{i2}$, the emitter current of T_2 may be taken equal to I_o. Hence,

$$I_f = \frac{R_E}{R_E + R'} I_o$$

and

$$\beta = \frac{I_f}{I_o} = \frac{R_E}{R_E + R'}$$

which is same as obtained in Eq. (4.128).

The equivalent resistance R of $R_s \| (R' + R_E)$,

$$R = \frac{R_s(R'+R_E)}{R_s+R'+R_E} = \frac{R_s R'+R_s R_E}{R_s+R'+R_E}$$

Hence, the input resistance without feedback,

$$R_i = R \| h_{ie} = \frac{R h_{ie}}{R + h_{ie}}$$

The emitter equivalent resistance R'_E of $R' \| R_E$ in T_2,

$$R'_E = \frac{R_E R'}{R_E + R'}$$

The input resistance of stage 2,
$$R_{i2} = h_{ie} + (1+h_{fe})R'_E$$
The output current,
$$I_o = -h_{fe}I_{i2} = h_{fe}\frac{h_{fe}R_{L1}}{R_{L1}+R_{i2}}I_{i1}$$

But $I_{i1} = \dfrac{RI_s}{R+h_{ie}}$, therefore

$$I_o = \frac{h_{fe}^2 R_{L1} R}{[R_{L1}+h_{ie}+(1+h_{fe})R'_E](R+h_{ie})}I_s \qquad (4.131)$$

Hence, the current gain without feedback,

$$A_I = \frac{I_o}{I_s} = \frac{h_{fe}^2 R_{L1} R}{[R_{L1}+h_{ie}+(1+h_{fe})R'_E](R+h_{ie})} \qquad (4.132)$$

From Eq. (4.65), the current gain with feedback,

$$A_{If} = \frac{A_I}{1+\beta A_I} \qquad (4.133)$$

From Fig. (4.38), the voltage gain with feedback,

$$A_{Vf} = \frac{I_o R_{L2}}{I_s R_s} = A_{If}\frac{R_{L2}}{R_s} \qquad (4.134)$$

From Eq. (4.67), the input resistance with feedback,

$$R_{if} = \frac{R_i}{1+\beta A_I} = \frac{R h_{ie}}{(1+\beta A_I)(R+h_{ie})} \qquad (4.135)$$

From Fig. 4.36, the input resistance with feedback, taking R_s into account,

$$R'_{if} = \frac{R_{if} R_s}{R_{if}+R_s} \qquad (4.136)$$

From Fig. 4.35, the input resistance with feedback, taking R_s into account, as seen by the voltage source V_s,

$$R''_{if} = R_s + R_{if} \qquad (4.137)$$

The output resistance with feedback, not taking R_{L2} into account, is given by

$$R_{of} = \frac{V_o}{I_o}$$

where V_o is the open circuit voltage and I_o is the short circuit current. Therefore,

$$R_{of} = \frac{V_o}{I_o} = \frac{V_o}{V_s} \times \frac{V_s}{I_s} \times \frac{I_s}{I_o}$$

$$= \frac{A_{Vf} R_s}{A_{If}} \qquad (4.138)$$

Example 4.13

For the current shunt feedback amplifier of Fig. 4.37, the parameters given are $R_{L1} = 4$ kΩ, $R_{L2} = 0.6$ kΩ, $R_E = 0.05$ kΩ, $R' = 1.2$ kΩ, $R_s = 1.2$ kΩ, $h_{ie} = 1.1$ kΩ, $h_{fe} = 50$, and $h_{re} = h_{oe} = 0$. Calculate (a) A_{Vf}, (b) input resistances (i) R_i, (ii) R_{if}, (iii) R'_{if}, (iv) R''_{if} and, (c) outputs (i) R_{of}, (ii) R'_{of}.

Solution

From Eq. (4.132),

$$A_I = \frac{h_{fe}^2 R_{L1} R}{[R_{L1} + h_{ie} + (1 + h_{fe})R'_E](R + h_{ie})}$$

where

$$R = \frac{R_s(R' + R_E)}{R_s + R' + R_E} = \frac{1.2(1.2 + 0.05)}{1.2 + 1.2 + 0.05} = 0.612 \text{ k}\Omega$$

and

$$R'_E = \frac{R_E R'}{R_E + R'} = \frac{0.05 \times 1.2}{0.05 + 1.2} = 0.048 \text{ k}\Omega$$

$\therefore \quad A_I = \dfrac{(50)^2 \times 3 \times 0.612}{(3 + 1.1 + 51 \times 0.048) \times (0.612 + 1.1)} = 410$

$\therefore \quad \beta = \dfrac{I_f}{I_o} = \dfrac{R_E}{R_E + R'} = \dfrac{0.05}{0.05 + 1.2} = 0.04$

$\therefore \quad D = 1 + \beta A_I = 1 + 0.04 \times 410 = 17.4$

and $\quad A_{If} = \dfrac{A_I}{1 + \beta A_I} = \dfrac{410}{17.4} = 23.56$

(a) $\quad A_{Vf} = A_{If} \dfrac{R_{L2}}{R_s} = 23.56 \times \dfrac{0.6}{1.2} = 11.78$

Also, as given in Eq. (4.130),

$$A_{Vf} \approx \frac{R_{L2}}{\beta R_s} = \frac{0.6}{0.04 \times 1.2} = \mathbf{12.5}$$

(b) (i) $\quad R_i = \dfrac{R h_{ie}}{R + h_{ie}} = \dfrac{0.612 \times 1.1}{0.612 + 1.1} = \mathbf{0.376 \text{ k}\Omega}$

(ii) $\quad R_{if} = \dfrac{R_i}{D} = \dfrac{0.376}{17.4} = \mathbf{20.25 \ \Omega}$

(iii) $\quad R'_{if} = \dfrac{R_{if} R_s}{R_{if} + R_s} = \dfrac{20.25 \times 1200}{20.25 + 1200} = \mathbf{20.2 \ \Omega}$

(iv) $R''_{if} = R_s + R_{if} = 1200 + 20.25 = \mathbf{1.22 \text{ k}\Omega}$

(c) (i) $R_{of} = \dfrac{A_{Vf} R_s}{A_{If}} = \dfrac{11.78 \times 1.2}{23.56} = \mathbf{0.6 \text{ k}\Omega}$

(ii) $R'_{of} = \dfrac{R_{of} R_{L2}}{R_{of} + R_{L2}} = \dfrac{0.6 \times 0.6}{0.6 + 0.6} = \mathbf{0.3 \text{ k}\Omega}$

4.10 REVIEW QUESTIONS

1. Draw the small-signal hybrid model of a transistor amplifier. Derive expressions for current gain, voltage gain, input impedance, and output impedance.
2. Why are biasing circuits used in transistor amplifiers? Describe biasing methods for a transistor.
3. What is the purpose of the shunt capacitor that is placed across the emitter resistor when self-bias method of biasing a transistor amplifier is used?
4. Name different modes of operation of an amplifier and define each of them.
5. Define nonlinear distortion, frequency distortion, and phase distortion.
6. Define the frequency response magnitude characteristic of an amplifier. Sketch a typical response curve. Indicate the high and low 3-dB frequencies. Define bandwidth.
7. Derive the expression for the high 3-dB frequency f_H^* of n identical noninteracting stages in terms of f_H for one stage.
8. Draw two RC-coupled common-emitter transistor stages. Show the low-frequency model for one stage. Derive expression for low cutoff frequency.
9. Draw the equivalent circuits for four types of amplifiers such as (i) voltage amplifier, (ii) current amplifier, (iii) transconductance amplifier, and (iv) transresistance amplifier. What are the values of their input and output resistances?
10. Name the four possible types of feedback amplifiers and draw their configurations indicating the output and feedback signals. Identify the transfer gain for each configuration. Define the feedback factor β.
11. How does negative feedback affect the gain, stability, sensitivity, bandwidth, frequency deviation, and noise of an amplifier?
12. For each of the four configurations of feedback amplifiers, derive the expressions for transfer gain, and input and output resistances.

4.11 SOLVED PROBLEMS

1. The h parameters of a transistor connected as a common-emitter amplifier are $h_{ie} = 1\text{k}\Omega, h_{re} = 2.2 \times 10^{-4}, h_{fe} = 60$, and $1/h_{oe} = 50 \text{ k}\Omega, R_L = 15 \text{ k}\Omega$, and $R_s = 500 \text{ }\Omega$. Determine the various gains, and input and output resistances.

Solution

The current gain when $R_s = 0$

$$A_I = -\frac{h_f}{1+h_o R_L} = -\frac{60}{1+(15/50)} = -46.15$$

The input resistance,

$$R_i = h_i - \frac{h_f h_r}{G_L + h_o} = 1 - \frac{60 \times 2.2 \times 10^{-4}}{\frac{1}{15} + \frac{1}{50}}$$

$$= 0.9086 \text{ k}\Omega = \mathbf{908.6 \ \Omega}$$

The current gain when $R_s = 500 \ \Omega$,

$$A_{Is} = A_I \frac{R_s}{R_i + R_s} = -46.15 \times \frac{500}{908.6 + 500} = \mathbf{-16.38}$$

The voltage gains,

$$A_V = \frac{A_I R_L}{R_i} = -\frac{46.5 \times 15}{0.9086} = \mathbf{-761.5}$$

and

$$A_{VS} = \frac{A_V R_i}{R_i + R_s} = -\frac{761.5 \times 908.6}{908.6 + 500} = \mathbf{-491.27}$$

The output conductance,

$$G_o = h_o - \frac{h_f h_r}{R_s + h_i} = \frac{1}{50} - \frac{60 \times 2.2 \times 10^{-4}}{0.5 + 1} = \mathbf{11.2 \ \mu A/V}$$

The output resistance,

$$R_o = \frac{1}{G_0} = \frac{1 \times 10^3}{11.2} \text{ k}\Omega = \mathbf{89.28 \text{ k}\Omega}$$

2. The h parameters of a transistor connected as a common-emitter amplifier are $h_{ie} = 1.1$ kΩ, $h_{re} = 2.5 \times 10^{-4}$, $h_{fe} = 50$, and $1/h_{oe} = 40$ kΩ. (a) Determine the current gain when $R_L = 0$ and $R_s = \infty$. (b) Calculate the input resistance when (i) $R_L = 0$ and when (ii) $R_L = \infty$. (c) Find voltage gain when $R_L = \infty$ and $R_s = 0$. (d) Determine output resistance (i) when $R_s = 0$ and (ii) when $R_s = \infty$.

Solution

(a) From Eqs. (4.4) and (4.5), the current gain

$$A_{Is} = \frac{h_{fe} R_s}{(1 + h_{oe} R_L)(Z_i + R_s)}$$

When $R_L = 0$ and $R_s = \infty$, the current gain reduces to

$$A_{Is} = -h_{fe} = \mathbf{-50}$$

(b) From Eq. (4.8), the input resistance,

$$R_i = h_{ie} - \frac{h_{fe} h_{re}}{(1/R_L) + h_{oe}}$$

(i) When $R_L = 0$,
$$R_i = h_{ie} - h_{fe}h_{re}R_L = h_{ie} = \mathbf{1.1\ k\Omega}$$

(ii) When $R_L = \infty$,
$$R_i = h_{ie} - \frac{h_{fe}h_{re}}{h_{oe}} = 1.1 - \frac{50 \times 2.5 \times 10^{-4}}{(1/40)} = \mathbf{0.6\ k\Omega}$$

(c) From Eqs. (4.5), (4.8), and (4.13), the voltage gain
$$A_{Vs} = -\frac{h_{fe}R_L}{(1+h_{oe}R_L)\left[\left(h_{ie} - \frac{h_{fe}h_{re}}{G_L + h_{oe}}\right) + R_s\right]}$$

When $R_L = \infty$ and $R_s = 0$
$$A_{Vs} = -\frac{h_{fe}}{h_{ie}h_{oe} - h_{fe}h_{re}} = \frac{50}{1.1 \times (1/40) - 50 \times 2.5 \times 10^{-4}}$$
$$= \mathbf{3333.33}$$

(d) From Eq. (4.17), the output resistance,
$$R_o = \frac{R_s + h_{ie}}{h_{oe}(R_s + h_{ie}) - h_{fe}h_{re}}$$

(i) when $R_s = 0$
$$R_o = \frac{h_{ie}}{h_{oe}h_{ie} - h_{fe}h_{re}} = \frac{1.1}{(1/40) \times 1.1 - 50 \times 2.5 \times 10^{-4}}$$
$$= \mathbf{73.33\ k\Omega}$$

(ii) When $R_s = \infty$,
$$R_o = \frac{1}{h_{oe}} = \frac{1}{2.5 \times 10^{-4}} = \mathbf{4\ k\Omega}$$

3. A transistor having β (h_{fe}) = 99 and V_{BE} = 0.6 V is used in the circuit of Fig. 4.40. The parameter values are V_{CC} = 10 V, R_F = 200 kΩ, and R_C = 2.7 kΩ. (a) Determine the quiescent values of V_{CE} and I_C. (b) If β is changed to 199, determine the new operating point.

Solution

Let currents be in milliamperes.

(a) $\beta = \dfrac{I_C}{I_B}$ or $I_C = 99I_B$

From Fig. 4.40,
$$V_{CB} = R_F I_B = 200 I_B$$

Also, $V_{BE} = V_{CE} - V_{CB} = V_{CC} - (I_C + I_B)R_C - 200I_B$

or $\quad 0.6 = 10 - 2.7(99+1)I_B - 200I_B$

∴ $\quad I_B = \dfrac{9.4}{470} = 0.02 \text{ mA}$

and $\quad I_C = 99 \times 0.02 = \mathbf{1.98 \text{ mA}}$

$V_{CE} = V_{CC} - (I_C + I_B)R_C$

$\quad = 10 - (1.98 + 0.02) \times 2.7$

$\quad = \mathbf{4.6 \text{ V}}$

(b) For $\beta = 199$

$V_{BE} = V_{CE} - V_{CB} = V_{CC} - (I_C + I_B)R_C - 200I_B$

or $\quad 0.6 = 10 - 2.7(199+1)I_B - 200I_B$

∴ $\quad I_B = \dfrac{9.4}{740} = 0.0127 \text{ mA}$

and $\quad I_C = 199 \times 0.0127 = \mathbf{2.527 \text{ mA}}$

$V_{CE} = V_{CC} - (I_C + I_B)R_C$

$\quad = 10 - (2.527 + 0.0127) \times 2.7$

$\quad = \mathbf{3.143 \text{ V}}$

4. In the circuit shown in Fig. 4.40, operating point of the transistor is $V_{CE} = 5$ V, $I_C = 5$ mA, and $V_{CC} = 9$ V. The transistor parameters are $V_{BE} = 0.6$ V and $\beta = 100$. (a) Determine R_F and R_C. (b) Using values obtained in (a), find the new value of I_C and V_{CE} if β changes to 50.

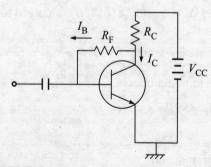

FIGURE 4.40 Circuit for Solved Problem 3.

Solution

From Fig. 4.40

(a) $V_{CE} = V_{CC} - (I_C + I_B)R_C$

or $\quad 5 = 9 - R_C(I_C + 0.01I_C) = 9 - 1.01 \times 5R_C$

or $\quad R_C = \dfrac{4}{5.05} = \textbf{0.792 k}\Omega$

Also, $R_F I_B = V_{CB} = V_{CE} - V_{BE} = 5 - 0.6 = 4.4$

or $\quad R_F = \dfrac{4.4}{5/100} = \textbf{88 k}\Omega$

(b) $V_{CB} = R_F I_B = 88 \times 0.02I_C = 1.76\, I_C$

$V_{CE} = V_{CC} - (I_C + I_B)R_C$

$V_{CB} + V_{BE} = 9 - 0.792 \times 1.02\, I_C = 9 - 0.808\, I_C$ \hfill (1)

$1.76\, I_C + 0.6 = 9 - 0.808\, I_C$

or $\quad I_C = \dfrac{9 - 0.6}{1.76 + 0.808} = \dfrac{8.4}{2.568} = \textbf{3.27 mA}$ \hfill (2)

From (1) and (2)

$V_{CE} = 9 - 0.808 \times 3.27 = \textbf{6.358 V}$

5. Find R_1 in the circuit shown in Fig. 4.41 for the emitter current $I_E = -2$ mA. Take $\alpha = 0.98$, $V_{BE} = 0.7$ V. Neglect the reverse saturation current.

Solution

From Fig. 4.41, the collector current

$I_C = -\alpha I_E = -0.98 \times -2.0 = 1.96$ mA
$I_B = -I_E - I_C = 2.0 - 1.96 = 0.04$ mA

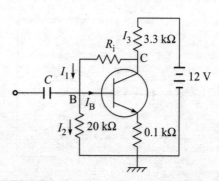

FIGURE 4.41 Circuit for Solved Problem 5.

Applying KVL in base-emitter circuit,

$$0.7 + 0.1 \times 2.0 = 20 I_2$$

$$\therefore \quad I_2 = \frac{0.9}{20} = 0.045 \text{ mA}$$

Applying KCL at node B,

$$I_1 = I_2 + I_B = 0.045 + 0.04$$
$$= 0.085 \text{ mA}$$

Applying KCL at node C,

$$I_3 = I_C + I_1 = 1.96 + 0.085$$
$$= 2.045 \text{ mA}$$

Also, $V_{CE} = 12 - 3.3 \times 2.045 - 0.1 \times 2$

$$= 5.05 \text{ V}$$

Hence, $V_{CB} = 5.05 - 0.7 = 4.35 \text{ V}$

Applying KVL in base-emitter circuit,

$$I_1 R_1 = V_{CB} = 4.05$$

$$\therefore \quad R_1 = \frac{4.05}{0.085} = \mathbf{47.65 \text{ k}\Omega}$$

6. The circuit shown in Fig. 4.42 is to be designed with the transistor with operating point $V_{CE} = 1.0$ V, $I_C = 8$ mA. The parameters of the transistor are $V_{BE} = 0.6$ V, $\beta = 160$, and $V_{CC} = 5$ V. The circuit is to be designed so that the voltage drop across R_E and R_C are equal. (a) Determine R_C, R_B, and R_E. (b) If $\beta = 80$, what are the new values of V_{CE} and I_C?

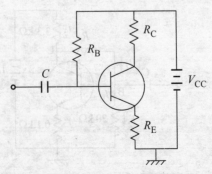

FIGURE 4.42 Circuit for Solved Problem 6.

Solution

(a) From the circuit of Fig. 4.42,

$$V_{CE} = V_{CC} - R_E I_E - R_C I_C$$

$$1.0 = 5.0 - 2 \times 8 R_C$$

$$R_C = \frac{4.0}{16} = \mathbf{0.25 \ k\Omega}$$

$$R_E I_E = R_C I_C = 8 R_C$$

$$R_E = \frac{8 R_C}{I_E} = \frac{8 \times 0.25}{(1 + 1/160) \times 8} \approx \mathbf{0.25 \ k\Omega}$$

From collector-base circuit,

$$R_B I_B = R_C I_C + V_{CB}$$

$$R_B \times \frac{8}{160} = 0.25 \times 8 + (1.0 - 0.6)$$

$$R_B = \frac{2.4 \times 160}{8} = \mathbf{48 \ k\Omega}$$

(b) From the circuit of Fig. 4.42,

$$0.25 \, I_C + V_{CB} = 48 \, I_B$$

$$\left(\frac{48 \times 1}{80} - 0.25 \right) I_C = V_{CE} - V_{CB} = V_{CE} - 0.6 \tag{1}$$

or $\quad 0.35 I_C = V_{CE} - 0.6$

From collector-emitter circuit,

$$R_C I_C + R_E I_E + V_{CE} = V_{CC}$$

or $\quad 2 \times 0.25 \, I_C = 5.0 - V_{CE} \tag{2}$

Solving (1) and (2),

$$I_C = \mathbf{5.18 \ mA}$$

and $\quad V_{CE} = \mathbf{2.41 \ V}$

7. The circuit of Fig. 4.43 employs a transistor having $\beta = 50$ and $V_{BE} = 0.6$ V. The supply voltage is 15 V, and R_E, R_C, and R_B are equal to 1 kΩ, 4 kΩ, and 310 kΩ, respectively. Determine (a) the values of V_{CE} and I_C and (b) the voltage drop from base to ground.

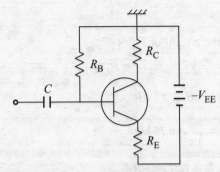

FIGURE 4.43 Circuit for Solved Problem 7.

Solution

(a) Applying KVL to base-emitter circuit,

$$R_B I_B + R_E I_E + 0.6 = 15$$

$$I_B(310 + 51 \times 1) = 14.4$$

$$I_B = \frac{14.4}{361} = 0.04 \text{ mA}$$

∴ $\quad I_C = 0.04 \times 50 = \mathbf{2 \text{ mA}}$

Applying KVL to collector-emitter circuit,

$$R_C I_C + R_E I_E + V_{CE} = 15$$

$$V_{CE} = 15 - (4 \times 50 + 1 \times 51) \times 0.04$$

$$= 4.96 \approx \mathbf{5 \text{ V}}$$

(b) The voltage drop from base to ground,

$$V_B = -R_B I_B = -310 \times 0.04 = \mathbf{-12.4 \text{ V}}$$

Minus sign is because of the fact that the ground is at positive potential.

8. The amplifier circuit shown in Fig. 4.44(a) employs an FET having $I_{DSS} = 16$ mA and $V_p = -3$ V. The circuit parameters are $R_1 = 600$ kΩ, $R_2 = 360$ kΩ, and $V_{DD} = 12$ V. Determine the values of R_S and R_D for which FET is biased in the active region at $I_{DQ} = 4$ mA and $V_{DSQ} = 4$ V.

Solution

Taking Thevenin's equivalent about AB, the circuit reduces as shown in Fig. 4.44(b), where

$$R_G = \frac{R_1 R_2}{R_1 + R_2} = \frac{600 \times 360}{600 + 360} = 225 \text{ kΩ}$$

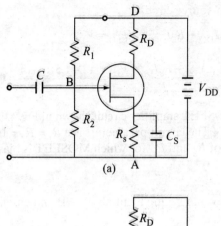

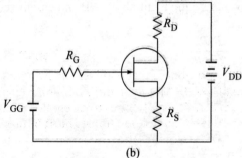

FIGURE 4.44 (a) Circuit for Solved Problem 8 and (b) its equivalent.

and $\quad V_{GG} = \dfrac{V_{DD} R_2}{R_1 + R_2} = \dfrac{12 \times 360}{600 + 360} = 4.5 \text{ V}$

From Eq. (2.23),

$$I_{DQ} = I_{DSS}\left(1 - \dfrac{V_{GS}}{V_p}\right)^2$$

or $\quad 4 = 16\left(1 - \dfrac{V_{GS}}{-3}\right)^2$

∴ $\quad V_{GS} = -1.5 \text{ V}$

From gate-source circuit,

$$R_S I_{DQ} + V_{GS} = V_{GG}$$

$$R_S = \dfrac{-V_{GS} + V_{GG}}{I_{DQ}}$$

$$= \dfrac{-(-1.5) + 4.5}{4} = \dfrac{6}{4}$$

$$= \mathbf{1.5 \text{ k}\Omega}$$

From drain-source circuit,

$$R_D I_{DQ} + R_S I_{DQ} + V_{DSQ} = V_{DD}$$

or $\quad R_D = \dfrac{V_{DD} - R_S I_{DQ} - V_{DS}}{I_{DQ}} = \dfrac{12 - 1.5 \times 4 - 4}{4} = \mathbf{0.5\ k\Omega}$

9. For the MOSFET amplifier circuit shown in Fig. 4.45(a), the MOSFET has $K = 0.2$ mA and $V_t = 2$ V. The circuit parameters are $R_1 = R_2 = 1000$ kΩ and $V_{DD} = 12$ V. Determine the values of R_S and R_D for which MOSFET is biased at $V_{DSQ} = V_{GSQ} = 4$ V.

Solution

Taking Thevenin's equivalent about AB, the circuit reduces as shown in Fig. 4.45(b), where

$$R_G = \dfrac{R_1 R_2}{R_1 + R_2}$$

$$= \dfrac{1 \times 1}{1 + 1} = 0.5\ M\Omega$$

$$= 500\ k\Omega$$

and $\quad V_{GG} = \dfrac{V_{DD} R_2}{R_1 + R_2}$

$$= \dfrac{12 \times 1}{1 + 1}$$

$$= 6.0\ V$$

In Fig. 1.46(b), the voltage V_t is known as threshold voltage, which is defined as the voltage V_{GS} at which the current I_{DS} reaches a small defined value. Taking this value

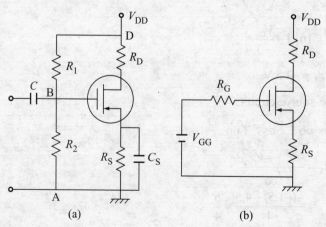

FIGURE 4.45 (a) Circuit for Solved Problem 9 and (b) its equivalent.

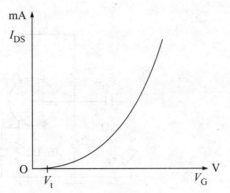

FIGURE 4.46 The transfer characteristic.

approximately zero, the transfer characteristic for n-channel enhancement mode MOSFET is shown in Fig. 4.46 and expressed as

$$I_{DS} = K(V_{GS} - V_t)^2$$

$$I_{DSQ} = 0.2(4-2)^2 = 0.8 \text{ mA}$$

From gate-source circuit,

$$V_{GG} = V_{GSQ} + R_S I_{DSQ}$$

$$R_S = \frac{6-4}{0.8} = \textbf{2.5 k}\Omega$$

From drain-source circuit,

$$V_{DD} = R_D I_{DSQ} + R_S I_{DSQ} + V_{DSQ}$$

$$R_D = \frac{12 - 2.5 \times 0.8 - 4}{0.8} = \textbf{7.5 k}\Omega$$

10. A single-stage transistor amplifier is shown in Fig. 4.47. Circuit parameters are shown in Fig. 4.47. A sinusoidal input with the r.m.s. value of 25 mV and the internal resistance of 1 kΩ is applied to the amplifier. The transistor parameters are $h_i = 1.67$ kΩ, $h_f = 44$, and $1/h_o = 150$ kΩ. Determine the current gain, voltage gain, power gain, input resistance, and output resistance.

Solution

The equivalent circuit for Fig. 4.47 is shown in Fig. 4.48. The resistance

$$R_b = \frac{R_1 R_2}{R_1 + R_2} = \frac{55 \times 4.5}{55 + 4.5} = 4.16 \text{ k}\Omega$$

The source current,

$$I_s = \frac{V_s}{1 + R_p}$$

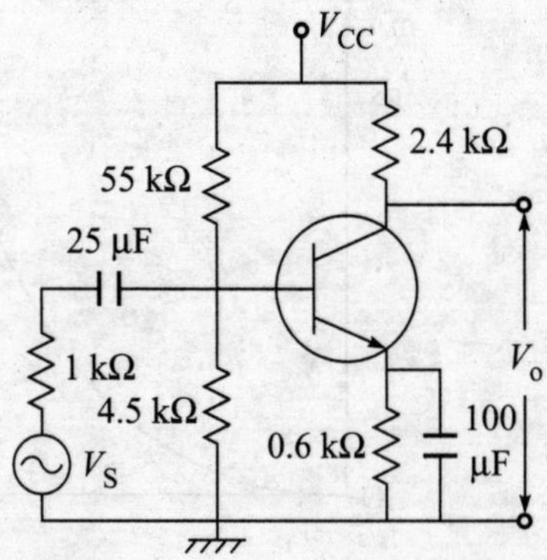

FIGURE 4.47 Circuit for Solved Problem 9.

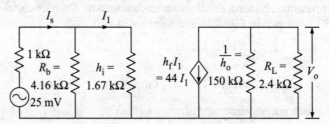

FIGURE 4.48 An equivalent circuit for Fig. 4.47.

where $R_p = \dfrac{h_i R_b}{h_i + R_b} = \dfrac{1.67 \times 4.16}{1.67 + 4.16} = 1.19 \text{ k}\Omega$

Hence, $I_S = \dfrac{0.025}{1 + 1.19} = 11.4 \text{ }\mu\text{A}$

and $I_1 = \dfrac{11.4 \times 4.16}{4.16 + 1.67} = 8.13 \text{ }\mu\text{A}$

The controlled current source value,

$$h_f I_1 = 44 \times 8.13 = 0.358 \text{ mA}$$

Hence, the output current,

$$I_L = -I_2 = -\dfrac{150 \times 0.358}{150 + 2.4} = -0.352 \text{ mA}$$

The transistor current gain,

$$A_I = \dfrac{I_L}{I_1} = \dfrac{-0.352}{8.13 \times 10^{-3}} = \mathbf{-43.3}$$

The amplifier current gain,

$$A_{IS} = \dfrac{I_L}{I_S} = \dfrac{-0.352}{11.4 \times 10^{-3}} = \mathbf{-30.88}$$

The voltage gain,

$$A_{VS} = -\frac{I_L V_L}{V_s} = -\frac{0.352 \times 2.4}{0.025} = \mathbf{-33.79}$$

The power gain,

$$A_p = A_{VS} \times A_{IS} = -33.79 \times -30.88 = \mathbf{1043.4}$$

The input resistance,

$$R_i = R_p = \mathbf{1.19 \; k\Omega}$$

The output resistance,

$$R_O = \frac{(1/h_o)R_L}{(1/h_o) + R_L} = \frac{150 \times 2.4}{150 + 2.4} = \mathbf{2.36 \; k\Omega}$$

11. A transistor amplifier shown in Fig. 4.49 has h parameters given in Solved Problem 2. Calculate $A_I, A_V, A_{Vs}, R_o,$ and R_i.

Solution

The equivalent circuit is shown in Fig. 4.50. All currents are shown in milliamperes. The resistance,

$$R_b = \frac{R_1 R_2}{R_1 + R_2} = \frac{100 \times 10}{100 + 10} = 9.09 \; k\Omega$$

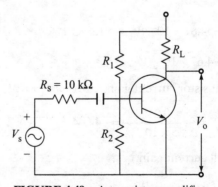

FIGURE 4.49 A transistor amplifier.

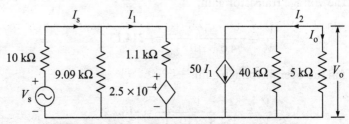

FIGURE 4.50 An equivalent circuit of Fig. 4.49.

The source voltage,

$$V_S = 10I_S + 1.1I_1 + 2.5 \times 10^{-4} V_o \tag{1}$$

$$V_S = 19.09I_S - 9.09I_1 \tag{2}$$

The output current,

$$I_o = -\frac{40 \times 50 I_1}{40 + 5} = \frac{V_o}{5} = 0.2 V_o \tag{3}$$

or $\quad I_1 = -4.5 \times 10^{-3} V_o \tag{4}$

From (1) and (4),

$$V_S = 10I_S - 4.7 \times 10^{-3} V_o \tag{5}$$

From (2) and (4),

$$V_S = 19.09 I_S + 40.905 \times 10^{-3} V_o \tag{6}$$

Solving (5) and (6) for V_S and I_S,

$$V_S = -54.87 \times 10^{-3} V_o \tag{7}$$

and $\quad I_S = -5.02 \times 10^{-3} V_o \tag{8}$

The input voltage to the transistor,

$$V_i = V_S - 10 I_S$$
$$= -54.87 \times 10^{-3} V_o + 50.2 \times 10^{-3} V_o$$
$$= -4.67 \times 10^{-3} V_o \tag{9}$$

Hence, the transistor current gain,

$$A_I = \frac{I_o}{I_1} = -\frac{0.2 V_o}{4.5 \times 10^{-3} V_o} = \mathbf{-44.44}$$

and the overall current gain,

$$A_{IS} = \frac{I_o}{I_S} = -\frac{0.2 V_o}{5.02 \times 10^{-3} V_o} = \mathbf{-39.84}$$

The voltage transistor gain,

$$A_V = \frac{V_o}{V_i} = -\frac{V_o}{4.67 \times 10^{-3} V_o} = \mathbf{-214.13}$$

and the overall voltage gain,

$$A_{VS} = \frac{V_o}{V_S} = -\frac{V_o}{54.87 \times 10^{-3} V_o} = \mathbf{-18.23}$$

The input resistance,
$$R_i = \frac{9.09 \times 1.1}{9.09 + 1.1} = \mathbf{0.99\ k\Omega}$$

The output resistance,
$$R_o = \frac{40 \times 5}{40 + 5} = \mathbf{4.4\ k\Omega}$$

12. The bandwidth of a three-stage amplifier of identical stages is 40 Hz to 10 kHz. Find the bandwidth of individual amplifier.

Solution

The low cutoff frequency of three-stage amplifier is given by Eq. (4.40),
$$f_L^* = \frac{f_L}{\sqrt{2^{1/3} - 1}}$$
$$f_L = f_L^* \times \sqrt{2^{1/3} - 1} = 40 \times \sqrt{2^{1/3} - 1} = \mathbf{20.4\ Hz}$$

The high cutoff frequency of three-stage amplifier is given by Eq. (4.39),
$$f_H^* = f_H \times \sqrt{2^{1/3} - 1}$$
$$f_H = \frac{f_H^*}{\sqrt{2^{1/3} - 1}} = \frac{10}{\sqrt{2^{1/3} - 1}} = \mathbf{19.61\ kHz}$$

Hence, the bandwidth of individual amplifier = **from 20.4 Hz to 19.61 kHz.**

13. For the circuit shown in Fig. 4.51, the transistor parameters are $h_{fe} = 50$, $h_{ie} = 1.1\ k\Omega$, and $h_{re} = h_{oe} = 0$. (a) Find the midband gain and (b) determine the value of coupling capacitance C_b. Take low cutoff frequency as 20 Hz.

Solution

(a) As $R_1 \parallel R_2 \gg h_{ie}$, the approximate mid-frequency equivalent circuit of Fig. 4.51 is shown in Fig. 4.52.

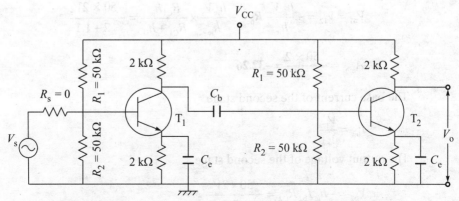

FIGURE 4.51 A two-stage amplifier circuit of Solved Problem-13.

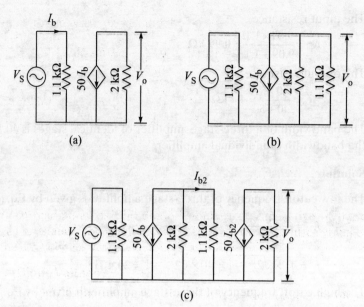

FIGURE 4.52 Approximate mid-frequency equivalent circuit of Fig. 4.51.

The mid-frequency gain of the first stage without the second stage [see Fig. 4.52(a)],

$$A_{mb1} = -\frac{h_{fe}I_b R_C}{V_s} = -\frac{h_{fe}R_C}{h_{ie}} \qquad \left(\because I_b = \frac{V_s}{h_{ie}}\right)$$

$$= -\frac{50 \times 2}{1.1} = -90.91$$

Because of loading effect of the second stage, the mid-frequency gain is reduced. The output voltage of the first stage (the input voltage of the second stage) with the input resistance of the second stage is shown in Fig. 4.52(b). Hence,

$$V_{o1} = V_{i2} = -\frac{h_{fe}V_s}{h_{ie}}R_p = -\frac{h_{fe}V_s}{h_{ie}} \times \frac{R_{C1}h_{ie}}{R_{C1} + h_{ie}} = -\frac{50 \times 2V_s}{2 + 1.1}$$

$$\therefore \quad A_{mb1} = -\frac{50 \times 2}{3.1} = -32.26$$

The base current of the second stage,

$$I_{b2} = \frac{V_{i2}}{1.1}$$

The output voltage of the second stage,

$$V_{o2} = -h_{fe}I_{b2}R_{C2} = -\frac{50 \times 2V_{i2}}{1.1}$$

The midband gain of the second stage [see Fig. 4.52(c)],

$$A_{mb2} = -\frac{100}{1.1} = -90.91$$

The overall gain,

$$A_{mb} = A_{mb1} \times A_{mb2} = -32.26 \times -90.91 = \mathbf{2932.75}$$

The low-frequency equivalent input circuit of the second stage is shown in Fig. 4.53(a) and its Thevenin's equivalent in Fig. 4.53(b).

From Fig. 4.53(b), the low cutoff frequency

$$f_L = \frac{1}{2\pi(R_C + h_{ie})C_b}$$

or $\quad C_b = \dfrac{1}{2\pi(R_C + h_{ie})f_L} = \dfrac{1}{2\pi \times 3.1 \times 10^3 \times 20}$

$\qquad = \mathbf{2.57\ \mu F}$

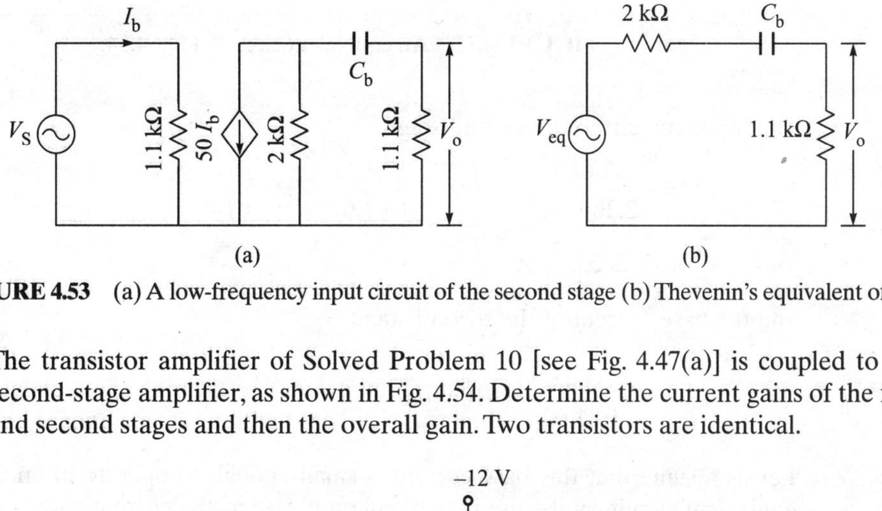

FIGURE 4.53 (a) A low-frequency input circuit of the second stage (b) Thevenin's equivalent of (a).

14. The transistor amplifier of Solved Problem 10 [see Fig. 4.47(a)] is coupled to the second-stage amplifier, as shown in Fig. 4.54. Determine the current gains of the first and second stages and then the overall gain. Two transistors are identical.

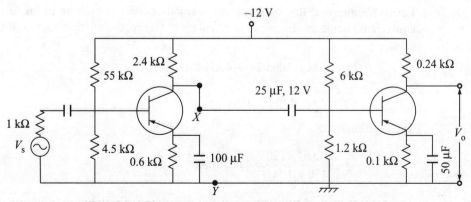

FIGURE 4.54 A two-stage amplifier of Solved Problem 14.

Solution

The input terminals to the second stage are X and Y. As per Solved Problem 10, the first stage may be replaced by an equivalent current source of value $44I_{b1} = 358$ μA with an equivalent shunt resistance equal to the output resistance $(150 \| 2.4)$ of 2.36 kΩ of the first stage. This current source acts as the input signal to the second stage. The equivalent circuit of the second stage is as shown in Fig. 4.55. The equivalent resistance R_{b2} is equal to $6 \| 1.2 = 1$ kΩ. The input current to the second stage is I_{s2} and the base current is I_{b2}.

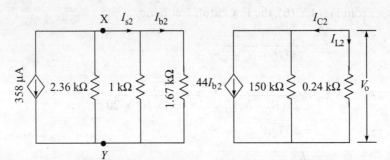

FIGURE 4.55 An equivalent circuit of Fig. 4.54.

The input current to the second stage,

$$I_{s2} = -\frac{358 \times 2.36}{2.36 + 0.625} \quad \left(\because \quad \frac{1 \times 1.67}{1 + 1.67} = 0.625 \right)$$

$$= -283.04 \text{ μA}$$

and the base current of the second stage,

$$I_{b2} = -\frac{283.04 \times 1}{1 + 1.67} = -106.01 \text{ μA}$$

Let us assume that this base current is small enough to operate in linear region. So, equivalent circuit can be used for calculation. Hence, the current source of the second stage,

$$h_{fe}I_{b2} = -44 \times 106.01 = -4664.4 \text{ μA}$$

$$= -4.664 \text{ mA}$$

The load current,

$$I_{L2} = -\frac{-4.664 \times 150}{150 + 0.24} = 4.657 \text{ mA}$$

The current gain of the first stage reduces due to the loading effect of the second stage. The more significant expression of the current gain of the first stage is defined as the ratio of the input current of the second stage to the input current of the first stage, that is,

$$A_{I1} = \frac{I_{s2}}{I_{s1}} = -\frac{283.04}{11.4} = -24.83$$

The current gain of the second stage,

$$A_{I2} = \frac{I_{L2}}{I_{s2}} = -\frac{4.657 \times 10^3}{283.04} = -16.45$$

The overall gain,

$$A_I = A_{I1} \times A_{I2} = -24.83 \times -16.45 = \mathbf{408.45}$$

Alternatively, the overall gain,

$$A_I = \frac{I_{L2}}{I_{s1}} = \frac{4.657 \times 10^3}{11.4} = \mathbf{408.5}$$

15. An emitter follower shown in Fig. 4.56 has $h_{fe} = 100$, $h_{ie} = 1.1$ kΩ, $h_{re} = h_{oe} = 0$, $R_1 = R_2 = 400$ kΩ, $R_E = R_s = 1$ kΩ, $R_L = 9$ kΩ. Find (a) A_I, A_{Is}, (b) A_V, A_{Vs}, and (c) R_{if}.

Solution

The equivalent circuit is shown in Fig. 4.56(b). The resistance,

$$R_p = \frac{400 \times 400}{400 + 400} = 200 \text{ k}\Omega$$

and

$$R_L' = \frac{1 \times 9}{1 + 9} = 0.9 \text{ k}\Omega$$

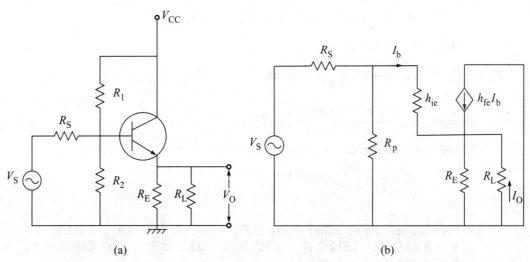

FIGURE 4.56 (a) Circuit for Solved Problem 15 and (b) an equivalent circuit of (a).

The source current,

$$I_s = I_p + I_b = \left[\frac{h_{ie} + (1+h_{fe})R'_L}{R_p} + 1\right]I_b$$

$$= \left[\frac{1.1 + (1+100) \times 0.9}{200} + 1\right]I_b = 1.46 I_b$$

The current through R_L,

$$I_L = \frac{(1+h_{fe}) \times 1}{1+9} I_b = \frac{101 \times 1}{100} I_b = 10.1 I_b$$

Hence, the current gains,

$$A_I = -\frac{I_L}{I_b} = -\mathbf{10.1}$$

$$A_{Is} = \frac{I_L}{I_s} = -\frac{10.1}{1.46} = -\mathbf{6.92}$$

The input voltage,

$$V_i = \left[h_{ie} + (1+h_{fe})R'_L\right]I_b = (1.1 + 101 \times 0.9)I_b = 92 I_b$$

and the output voltage,

$$V_o = I_L R_L = -10.1 \times 9 I_b = -90.9 I_b$$

The source voltage,

$$V_s = I_s R_s + V_i = 1.46 \times 1 I_b + 92 I_b = 93.46 I_b$$

Hence, the voltage gains,

$$A_V = -\frac{V_o}{V_i} = -\frac{90.9}{92} = -\mathbf{0.988}$$

and

$$A_{Vs} = -\frac{V_o}{V_s} = -\frac{90.9}{93.46} = -\mathbf{0.973}$$

The input resistance,

$$R_{if} = \frac{V_i}{I_s} = \frac{92}{1.46} = \mathbf{63.01 \text{ k}\Omega}$$

16. For the transistor amplifier shown in Fig. 4.57(a), $h_{fe} = 100$, $h_{ie} = 1$ kΩ, $h_{re} = h_{oe} = 0$, $R_1 = 300$ kΩ, $R_2 = 150$ kΩ, $R_E = 1$ kΩ, $R_C = 4$ kΩ, and $R_L = 1$ kΩ. Calculate G_{Mf}, A_{Vf}, R_{if}, R_{of}, and R'_{of}.

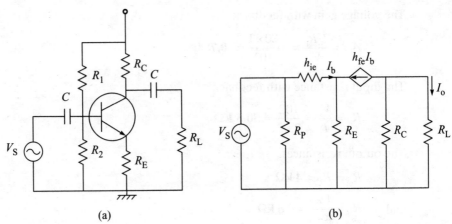

FIGURE 4.57 (a) Circuit for Solved Problem 16 and (b) an equivalent circuit of (a).

Solution

This amplifier is an example of current series feedback. The equivalent circuit is shown in Fig. 4.57(b). The resistance,

$$R_p = \frac{300 \times 150}{300 + 150} = 100 \text{ k}\Omega$$

and

$$R'_L = \frac{1 \times 4}{1 + 4} = 0.8 \text{ k}\Omega$$

From Fig. 4.57(b),

$$I_o = -\frac{h_{fe} I_b R_C}{R_C + R_L} = -\frac{100 \times 4}{4 + 1} I_b = -80 I_b$$

and the input signal,

$$V_s = [h_{ie} + (1 + h_{fe})R_E] I_b$$

$$= (1 + 101 \times 1) I_b = 102 \, I_b$$

$$I_s = I_p + I_b = \left[\frac{h_{ie} + (1 + h_{fe})R_E}{R_p} + 1 \right] I_b$$

$$= \left[\frac{1 + (1 + 100) \times 1}{100} + 1 \right] I_b = 2.02 I_b$$

Hence, transconductance with feedback,

$$G_{Mf} = \frac{I_o}{V_s} = -\frac{80}{102} = \mathbf{-0.784}$$

The voltage gain with feedback,

$$A_{Vf} = \frac{I_o R_L}{V_s} = -\frac{80 \times 1}{102} = -0.784$$

The input resistance with feedback,

$$R_{if} = \frac{V_s}{I_s} = \frac{102}{2.02} = 50.5 \text{ k}\Omega$$

The output resistances,

$$R_{of} = R_C = 4 \text{ k}\Omega$$

and

$$R'_{of} = \frac{4 \times 1}{4+1} = 0.8 \text{ k}\Omega$$

17. For the transistor amplifier shown in Fig. 4.58(a), $h_{fe} = 100$, $h_{ie} = 1 \text{ k}\Omega$, and $R_C = 10 \text{ k}\Omega$. (a) For $R_E = 0$, determine R_{Mf}, A_{Vf}, R_{if}, and R'_{of}. (b) Repeat (a) for $R_E = 1 \text{ k}\Omega$.

Solution

(a) After applying Miller's theorem, the circuit of Fig. 4.58(a) reduces as shown in Fig. 4.58(b), where $A_V = \frac{V_o}{V_i}$. The equivalent circuit is shown in Fig. 4.59. The gain A_V

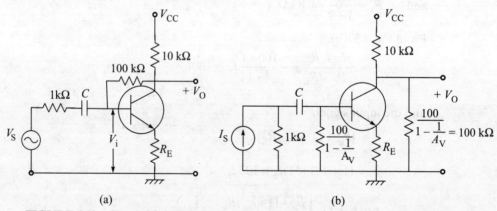

(a)   (b)

FIGURE 4.58 Circuit (a) for Solved Problem 17 and (b) after applying Miller's theorem.

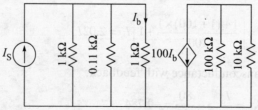

FIGURE 4.59 An equivalent circuit of Fig. 4.58(b) for $R_E = 0$.

is much larger than unity; hence,

$$\frac{100}{1-\frac{1}{A_V}} = 100 \text{ k}\Omega$$

and the effective load resistance,

$$R'_L = \frac{100 \times 10}{100 + 10} = 9.09 \text{ k}\Omega$$

From Fig. 4.59,

$$I_b = \frac{V_i}{h_{ie}} = V_i$$

and $h_{fe}I_b = 100 V_i$

The output voltage,

$$V_o = -100 V_i \times 9.09 = -909 V_i$$

and the voltage gain

$$A_V = \frac{V_o}{V_i} = -909$$

Hence, $\dfrac{100}{1-A_V} = \dfrac{100}{910} = 0.11 \text{ k}\Omega$

The source current,

$$I_s = \left(\frac{1}{1} + \frac{1}{0.11} + \frac{1}{1}\right) V_i = 11.1 V_i$$

The transresistance with feedback,

$$R_{Mf} = \frac{V_o}{I_s} = -\frac{909}{11.1} = \mathbf{-81.89 \text{ V/mA}}$$

The input source voltage,

$$V_s = I_s R_s = 11.1 \times 1 V_i = 11.1 V_i$$

and the voltage gain,

$$A_{Vf} = \frac{V_o}{V_s} = -\frac{909}{11.1} = \mathbf{-81.89}$$

The input resistance,

$$R_{if} = \frac{R_{Mf}}{A_{Vf}} = \mathbf{1 \text{ k}\Omega}$$

The output resistance,

$$R'_{of} = \frac{100 \times 10}{100 + 10} = \mathbf{9.09 \text{ k}\Omega}$$

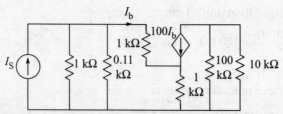

FIGURE 4.60 An equivalent circuit of Fig. 4.58(b) for $R_E = 1$ kΩ.

(b) The equivalent circuit is shown in Fig. 4.60. For the time being, the gain A_V may be assumed to be much larger than unity; hence,

$$\frac{100}{1 - \frac{1}{A_V}} = 100 \text{ k}\Omega$$

and the effective load resistance,

$$R'_L = \frac{100 \times 10}{100 + 10} = 9.09 \text{ k}\Omega$$

From Fig. 4.60,

$$V_i = (1 + 101 \times 1)I_b = 102 I_b$$

and $\quad V_o = -100 \times 9.09 I_b = -909 I_b$

The voltage gain,

$$A_V = \frac{V_o}{V_i} = -\frac{909}{102} \approx -9$$

Now, taking $A_V = -9$

$$\frac{100}{1 - \frac{1}{A_V}} = 90 \text{ k}\Omega$$

and the effective load resistance,

$$R'_L = \frac{90 \times 10}{90 + 10} = 9 \text{ k}\Omega$$

From Fig. 4.60,

$$V_i = (1 + 101 \times 1)I_b = 102 I_b$$

and $\quad V_o = -100 \times 9 I_b = -900 I_b$

The voltage gain

$$A_V = \frac{V_o}{V_i} = -\frac{900}{102} \approx -9$$

Hence, $\dfrac{100}{1-A_V} = \dfrac{100}{10} = 10\ \text{k}\Omega$

The source current,

$$I_s = \left(\dfrac{102}{1} + \dfrac{102}{10} + 1\right)I_b = 113.2 I_b$$

The transresistance with feedback,

$$R_{Mf} = \dfrac{V_o}{I_s} = -\dfrac{900}{113.2} = \textbf{-8.72 V/mA}$$

The input source voltage,

$$V_s = I_s R_s = 113.2 \times 1 I_b = 113.2 I_b$$

and the voltage gain,

$$A_{Vf} = \dfrac{V_o}{V_s} = -\dfrac{900}{113.2} = \textbf{-8.72}$$

The input resistance,

$$R_{if} = \dfrac{R_{Mf}}{A_{Vf}} = \textbf{1 k}\Omega$$

The output resistance,

$$R'_{of} = \dfrac{90 \times 10}{90 + 10} = \textbf{9 k}\Omega$$

18. For a circuit shown in Fig. 4.61, show that

$$A_{Vf} = \dfrac{V_o}{V_s} = -\dfrac{R_f}{R_s} \times \dfrac{1}{1 + \dfrac{R_{in} R_f}{R_m}\left(\dfrac{1}{R_f} + \dfrac{1}{R_{in}} + \dfrac{1}{R_s}\right)}$$

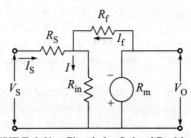

FIGURE 4.61 Circuit for Solved Problem 18.

Solution

From Fig. 4.61,

$$I_s = I + I_f = I + \frac{R_{in}I + R_m I}{R_f}$$

$$= I\left(1 + \frac{R_{in} + R_m}{R_f}\right)$$

and $V_s = V_i + I_s R_s$

$$= \left(R_{in} + \frac{R_f + R_{in} + R_m}{R_f}R_s\right)I$$

The output voltage

$$V_o = -R_m I$$

$$\therefore A_{Vf} = \frac{V_o}{V_s} = -\frac{R_m}{\dfrac{R_s}{R_f}\left(\dfrac{R_{in}R_f}{R_s} + R_f + R_{in} + R_m\right)}$$

$$= -\frac{R_f}{R_s} \times \frac{1}{1 + \dfrac{R_f}{R_m}\left(\dfrac{R_{in}}{R_f} + \dfrac{R_{in}}{R_s} + 1\right)} = -\frac{R_f}{R_s} \times \frac{1}{1 + \dfrac{R_{in}R_f}{R_m}\left(\dfrac{1}{R_f} + \dfrac{1}{R_{in}} + \dfrac{1}{R_s}\right)}$$

4.12 EXERCISES

1. A transistor connected as a common-emitter amplifier has its h parameters as h_{ie} = 1.1 kΩ, h_{re} = 2.5×10^{-5}, h_{fe} = 50, and $1/h_{oe}$ = 40 kΩ, R_L = 12 kΩ, and R_s = 200 Ω. Determine the various gains, and input and output resistances.
2. The h parameters of a transistor connected as a common-emitter amplifier are h_{ie} = 1 kΩ, h_{re} = 2.0 × 10^{-5}, h_{fe} = 40, and $1/h_{oe}$ = 30 kΩ. (a) Determine current gain (i) when R_L = 0 and (ii) when R_s = ∞. (b) Calculate input resistance (i) when R_L = 0 and (ii) when R_L = ∞. (c) Find voltage gain (i) when R_L = ∞ and (ii) when R_s = 0. (d) Determine output resistance (i) when R_s = 0 and (ii) when R_s = ∞.
3. A transistor having β (h_{fe}) = 60 and V_{BE} = 0.7 V is used in the circuit of Fig. 4.40. The parameter values are V_{CC} = 15 V, R_F = 150 kΩ, and R_C = 2.5 kΩ. (a) Determine the quiescent values of V_{CE} and I_C. (b) If β is changed to 99, determine the new operating point.
4. The operating point of the transistor circuit shown in Fig. 4.40 is V_{CE} = 9 V, I_C = 5 mA, and V_{CC} = 15 V. The transistor parameters are V_{BE} = 0.2 V and β = 50. (a) Determine R_F and R_C. (b) Using values obtained in (a), find the new values of I_C and V_{CE} if β changes to 60.

5. Self-biasing circuit of the transistor is shown in Fig. 4.41. For the emitter current $I_E = -1.5$ mA, find the value of R_1. Take $\alpha = 0.98$, $V_{BE} = 0.7$ V. Neglect the reverse saturation current.
6. The operating point of the transistor circuit, shown in Fig. 4.42, is $V_{CEQ} = 5.0$ V, $I_{CQ} = 12$ mA. The parameters of the transistor are $V_{BE} = 0.7$ V, $\beta = 50$, and $V_{CC} = 15$ V. The circuits have to be designed so that the voltage drop across R_E and R_C are equal. (a) Determine R_C, R_B, and R_E. (b) If $\beta = 60$, what are the new values of V_{CEQ} and I_{CQ}?
7. The circuit of Fig. 4.43 employs a transistor having $\beta = 60$ and $V_{BE} = 0.7$ V. The supply voltage is 12 V, and R_E, R_C, and R_B are equal to 2 kΩ, 6 kΩ, and 400 kΩ, respectively. Determine (a) the values of V_{CE} and I_C and (b) the voltage drop from base to ground.
8. An FET amplifier circuit shown in Fig. 4.44(a) has $I_{DSS} = 21$ mA and $V_p = -2.5$ V. The circuit parameters are $R_1 = 300$ kΩ, $R_2 = 200$ kΩ, and $V_{DD} = 10$ V. For biasing the FET in the active region at $I_{DQ} = 4.5$ mA and $V_{DSQ} = 3.5$ V, determine the values of R_S and R_D.
9. The amplifier circuit shown in Fig. 4.45(a) employs a MOSFET having $K = 0.25$ mA and $V_t = 3$ V. The circuit parameters are $R_1 = R_2 = 900$ kΩ and $V_{DD} = 15$ V. Determine the values of R_S and R_D for which it is biased at $V_{DSQ} = V_{GSQ} = 5$ V.
10. A single-stage transistor amplifier is shown in Fig. 4.47. A sinusoidal input with the r.m.s. value of 0.05 V and resistance of 100 Ω is applied to the amplifier. The transistor parameters are $h_i = 1.1$ kΩ, $h_f = 50$, $1/h_o = 250$ kΩ. Determine the current gain, voltage gain, power gain, input resistance, and output resistance.
11. For the circuit shown in Fig. 4.51, the transistor parameters are $h_{fe} = 30$, $h_{ie} = 1$ kΩ, and $h_{re} = h_{oe} = 0$. (a) Find the midband gain and (b) determine the value of coupling capacitance C_b.
12. The bandwidth of a three-stage amplifier of identical stages is 35 Hz to 15 kHz. Find the bandwidth of individual amplifier. If a fourth identical stage is added, determine the bandwidth.
13. The transistor amplifier of question 10 is coupled to the second-stage amplifier, as shown in Fig. 4.50. Determine the current gains of the first and second stages and then the overall gain. Two transistors are identical.
14. An emitter follower circuit of Fig. 4.54 has $h_{fe} = 60$, $h_{ie} = 1$ kΩ, $h_{re} = h_{oe} = 0$, $R_1 = 300$ kΩ, $R_2 = 350$ kΩ, $R_E = R_s = 1$ kΩ, $R_L = 5$ kΩ. Find (a) A_I, A_{Is}, (b) A_V, A_{Vs}, and (c) R_{if}.
15. The amplifier circuit shown in Fig. 4.56(a) employs a transistor having $h_{fe} = 50$, $h_{ie} = 1.1$ kΩ, $h_{re} = h_{oe} = 0$, $R_1 = 600$ kΩ, $R_2 = 300$ kΩ, $R_E = 1.1$ kΩ, $R_C = 5$ kΩ, and $R_L = 1.5$ kΩ. Calculate G_{Mf}, A_{Vf}, R_{if}, R_{of}, and R'_{of}.
16. The transistor of the amplifier circuit of Fig. 4.57(a) has $h_{fe} = 60$, $h_{ie} = 1$ kΩ. (a) For $R_E = 1$ kΩ, determine R_{Mf}, A_{Vf}, R_{if}, and R'_{of}. (b) Repeat (a) for $R_E = 5$ kΩ.

5

SPECIAL AMPLIFIERS

Outline

5.1 Differential Amplifier 202
5.2 Operational Amplifier 209
5.3 Linear Applications of Operational Amplifiers 214
5.4 Power Amplifier 220
5.5 Review Questions 233
5.6 Solved Problems 233
5.7 Exercises 246

5.1 DIFFERENTIAL AMPLIFIER

In a differential amplifier, the output is taken as a difference of outputs of two transistors, as shown in Fig. 5.1. In this amplifier, drift in a direct-coupled amplifier is more efficiently reduced. The main advantage of this amplifier is that it rejects the unwanted common signals present in the inputs of the two transistors while amplifying the difference between the two inputs. The two transistors in the differential amplifier are identical. Thus, the amplifier circuit is balanced.

A few terms that are used in connection with the differential amplifier are defined here.

Common-mode signal: The signal which is common to both inputs of the differential amplifier is known as common-mode signal.

Differential-mode signal: The difference between two input signals is known as differential-mode signal. Two arbitrary signals can be separated into common-mode and differential-mode signals. For the two input signals of Fig. 5.1, the differential-mode signal v_d is given by

$$v_d = v_{i1} - v_{i2} \tag{5.1}$$

The differential-mode signal, v_d, is equally divided between input terminals of the two transistors, as shown in Fig. 5.1. From Fig. 5.1, the two inputs are given by

$$v_{i1} = v_c + \frac{v_d}{2}$$

$$v_{i2} = v_c - \frac{v_d}{2}$$

Hence, $$v_c = \frac{v_{i1} + v_{i2}}{2} \tag{5.2}$$

If $v_{i1} = v_{i2}$, the differential-mode signal,

$$v_d = 0 \tag{5.3}$$

and $$v_c = v_{i1} = v_{i2} \tag{5.4}$$

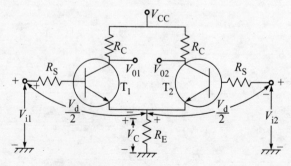

FIGURE 5.1 A differential amplifier.

Single-ended output: The output from one of the collectors to ground is known as a single-ended output. Therefore, in Fig. 5.1, both outputs v_{o1} and v_{o2} are single-ended outputs.

Differential output: The difference between two single-ended outputs is known as differential output. Hence, differential output is given by

$$v_{od} = v_{o1} - v_{o2} \tag{5.5}$$

Common-mode rejection ratio (CMRR): In an ideal differential amplifier, the gains of the two transistors are exactly equal. But, in practice, it is not true. It may be approximately equal. Let A_1 and A_2 be the gains of two transistors, then the outputs are given by

$$v_{o1} = A_1 \left(\frac{v_d}{2} + v_c \right)$$

and

$$v_{o2} = A_2 \left(-\frac{v_d}{2} + v_c \right)$$

Hence, the differential output,

$$v_{od} = v_{o1} - v_{o2}$$

$$= \frac{A_1 + A_2}{2} v_d + (A_1 - A_2) v_c$$

$$= A_d v_d + A_c v_c \tag{5.6}$$

$$= A_d v_d \left(1 + \frac{A_c}{A_d} \times \frac{v_c}{v_d} \right) \tag{5.7}$$

where A_d and A_c are differential-mode gain and common-mode gain, respectively.

Since it is desirable that a differential amplifier amplifies the differential-mode input and suppresses the common-mode input, the ratio A_d/A_c must be very high. This ratio is called CMRR and is given by

$$\text{CMRR} = \frac{A_d}{A_c} \tag{5.8}$$

For high CMRR the differential-mode gain, A_d, must be high and the common-mode gain, A_c, must be low.

Example 5.1

Two outputs of a differential amplifier are 10 mV and 8 mV, when a common-mode input of 1 V is applied. When a differential input of 1 mV is applied to the amplifier, the resulting differential output is 75 mV. Determine CMRR of the amplifier.

Solution

When only common mode is applied, from Eq. (5.6) and $v_d = 0$, the common-mode gain with differential output,

$$A_c = \frac{v_o}{v_c} = \frac{(10-8) \times 10^{-3}}{1} = 2 \times 10^{-3}$$

When only differential input is applied, from Eq. (5.6) and $v_c = 0$, the differential-mode gain with differential output,

$$A_d = \frac{v_o}{v_d} = \frac{75}{1} = 75$$

$$\therefore \quad \text{CMRR} = \frac{A_d}{A_c} = \frac{75}{2 \times 10^{-3}} = 37500$$

A practical differential amplifier is considered to be a balanced circuit, as a small amount of unbalance, which may exist, does not affect its gain appreciably. When one half of the differential input, v_d, is applied across one transistor input and second half across the second transistor input, then the two single-ended outputs are given by

$$v_{o1} = A_1 \frac{v_d}{2}$$

and

$$v_{o2} = A_2 \left(-\frac{v_d}{2} \right) = -A_2 \frac{v_d}{2}$$

where A_1 and A_2 are gains of the two transistors. As the amplifier circuit is balanced, $A_1 = A_2$ and the differential output,

$$v_{od} = v_{o1} - v_{o2} = A_1 v_d$$

If A_{dd} is the gain with differential input and differential output, then

$$v_{od} = A_{dd} v_d$$

or

$$A_{dd} = \frac{v_{od}}{v_d}$$

and

$$|A_{dd}| = |A_1| = |A_2| \tag{5.9}$$

From Eq. (5.9), it is clear that the gain of the differential amplifier with differential output and differential input is same as the gain of one side of the amplifier circuit with the earthed emitter, as shown in Fig. 5.2(a).

If single-ended output, v_{o1}, is defined as v_{os}, then

$$v_{os} = A_1 \frac{v_d}{2} = \frac{A_1}{2} v_d$$

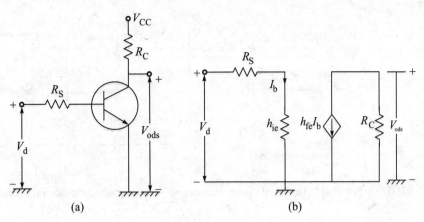

FIGURE 5.2 (a) One side of differential amplifier with differential input and earthed emitter and (b) its equivalent circuit.

or
$$A_{ds} = \frac{v_{os}}{v_d} = \frac{A_1}{2} = \frac{A_{dd}}{2} \tag{5.10}$$

where A_{ds} is the gain with differential input and single-ended output. The equivalent circuit of Fig. 5.2(a) is shown in Fig. 5.2(b). From Eq. (5.10) and Fig. 5.2(b), the single-ended gain of the amplifier with differential input,

$$A_{ds} = \frac{A_1}{2} = -\frac{h_{fe} R_C}{2(R_s + h_{ie})} \tag{5.11}$$

For discussing the gain with common-mode input, Fig. 5.1 can be redrawn as Fig. 5.3. In this case, the same input is applied to both sides. The circuit of Fig. 5.3 is symmetrical about XX' and so this circuit can be bisected about XX' without changing the voltage and current in any part of the circuit. For a perfectly balanced circuit, the differential output with common-mode input,

$$v_{oc} = v_{o1} - v_{o2} = 0$$

which means that the CMRR is infinite.

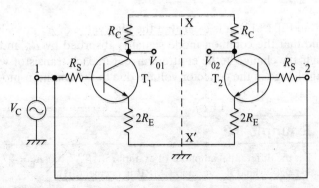

FIGURE 5.3 A differential amplifier with common-mode input.

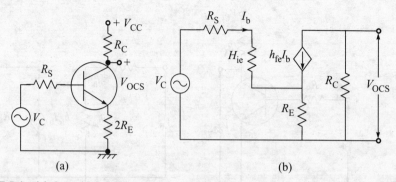

FIGURE 5.4 (a) One side of differential amplifier with common-mode input and (b) its equivalent circuit.

In practice, however, there is some amount of unbalance between the two sections of the amplifier; therefore, CMRR is finite. As the nature of unbalance is unknown, the gain with differential output and differential input, A_{dd}, cannot be calculated, and hence CMRR cannot be calculated with differential output. Therefore, CMRR for a single-ended output is calculated. One section of Fig. 5.3 is shown in Fig. 5.4(a). The equivalent circuit is shown in Fig. 5.4(b). If A_{cs} is the gain of the amplifier of Fig. 5.4(b) with single-ended output and common-mode input, then

$$A_{cs} = \frac{v_{cs}}{v_c} = \frac{-h_{fe}R_C}{R_s + h_{ie} + 2(1+h_{fe})R_E}$$

$$\approx \frac{-h_{fe}R_C}{2(1+h_{fe})R_E} \qquad (5.12)$$

The CMRR, therefore, can be given by

$$\text{CMRR} = \frac{A_{ds}}{A_{cs}} = \frac{-h_{fe}R_C}{2(R_s + h_{ie})} \times \frac{2(1+h_{fe})R_E}{-h_{fe}R_C}$$

$$\approx \frac{(1+h_{fe})R_E}{R_s + h_{ie}} \qquad (5.13)$$

From Eq. (5.13), it is clear that the high value of R_E gives high CMRR. Hence, it can be said that the common-mode signal is absorbed by R_E and it appears across it. If R_E is high, the change in the emitter current in each transistor would be small and the resulting change in the collector voltage due to the common-mode signal would be small.

Example-5.2

For the differential amplifier of example 5.1, $h_{fe} = 100$, $R_E = 4.7$ kΩ, $R_C = R_s = h_{ie} = 1$ kΩ, and $V_{cc} = 20$ V. Find A_{ds}, A_{cs}, and CMRR in decibel (dB).

Solution

From Eq. (5.11),

$$A_{ds} = \frac{-h_{fe}R_C}{2(R_s + h_{ie})} = \frac{-100 \times 1}{2(1+1)} = -25$$

From Eq. (5.12),

$$A_{cs} = \frac{-h_{fe}R_C}{2(1+h_{fe}) \times R_E} = \frac{-100 \times 1}{2(1+100) \times 4.7} = \mathbf{-0.105}$$

$$\therefore \quad \text{CMRR} = \frac{A_{ds}}{A_{cs}} = \frac{25}{0.105} = \mathbf{238}$$

and CMRR = $20 \log_{10}(238)$ = **47.5 dB**

Example-5.3

For the differential amplifier, the first set of inputs is $v_1 = 50$ mV and $v_2 = -50$ mV and the second set is $v_1 = 1000$ mV and $v_2 = 900$ mV. (a) For CMRR = 100, calculate the percentage difference in output voltages obtained for the two sets of inputs. (b) Repeat (a) for CMRR = 1000.

Solution

(a) For the first set, $v_d = v_1 - v_2 = 100$ mV and $v_c = \frac{v_1 + v_2}{2} = 0$. So from Eq. (5.7), the output for CMRR = 100,

$$v_o^1 = 100 A_d \text{ mV}$$

For the second set, $v_d = v_1 - v_2 = 100$ mV and $v_c = \frac{v_1 + v_2}{2} = 950$ mV. So from Eq. (5.7), the output for CMRR = 100,

$$v_o^2 = 100 A_d \left(1 + \frac{1}{100} \times \frac{950}{100}\right) = 100 A_d \left(1 + \frac{9.5}{100}\right) \text{mV} = 109.5 A_d \text{ mV}$$

Hence, the difference of two outputs = $\frac{(109.5 - 100) A_d}{100 A_d} \times 100$ = **9.5%**.

(b) For the first set of inputs, the output for CMRR = 1000,

$$v_o^1 = 100 A_d \text{ mV}$$

For the second set of inputs, the output for CMRR = 1000,

$$v_o^2 = 100 A_d \left(1 + \frac{1}{1000} \times \frac{950}{100}\right) = 100 A_d \left(1 + \frac{0.95}{100}\right) \text{mV} = 100.95 A_d \text{ mV}$$

Hence, the difference of two outputs = $\frac{(100.95 - 100) A_d}{100 A_d} \times 100$ = **0.95%**.

In a practical differential amplifier, A_{cs} must be very small, which requires very high value of R_E. For this purpose R_E is replaced by a current source with a high parallel resistance. A differential amplifier in which R_E is replaced by a current source is shown in Fig. 5.5. A transistor circuit employing transistors T_3, T_4, T_5 acts approximately as a constant current source and gives high resistance.

Transistors T_4 and T_5 are acting as diode for temperature compensation. Assuming base current of T_3 very small, for the base circuit of T_3,

$$I_o \approx I_3 = \frac{1}{R_3}\left[\frac{V_{EE}R_2}{R_1+R_2} + \frac{V_D R_1}{R_1+R_2} - V_{BE3}\right]$$

Since, T_4 and T_5 are same as T_3, $V_D = 2V_{BE3}$ and for $R_1 = R_2$, the emitter current,

$$I_o = \frac{V_{EE}}{2R_3} \tag{5.14}$$

which is a constant.

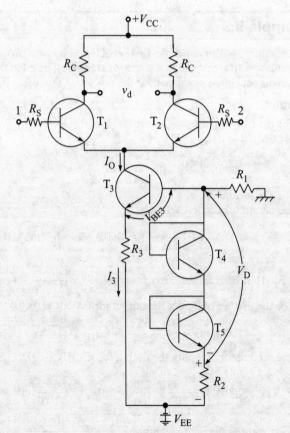

FIGURE 5.5 A differential amplifier with a current source.

5.2 OPERATIONAL AMPLIFIER

Operational amplifiers are direct-coupled high-gain amplifiers and are abbreviated as **OP AMPs**. A feedback is added to OP AMPs to control their response characteristic. They are used to perform mathematical operations such as summation, subtraction, integration, etc. The building block in the OP AMP is the differential amplifier, already discussed in Section-5.1. OP AMPs were first employed in analogue computers for problem simulation and were often referred to as analogue integrated circuits. Now, with its availability in the integrated circuit form at an economical price, OP AMPs find wide applications in wave operation and shaping, analogue-to-digital conversion, digital-to-analogue conversion, comparators, multivibrators, and a variety of other switching operations. It offers all the advantages of integrated circuits such as small size, high reliability, reduced cost, temperature tracking, and low offset voltage and current. By simply changing the feedback impedance in the OP AMP, it can be used for variety of applications.

An ideal OP AMP has the features of very high open-loop d.c. gain (tending to infinity); infinite input impedance; zero output impedance; infinite CMRR; and infinite bandwidth, extending right from d.c. to infinite frequency. An ideal OP AMP is perfectly balanced, that is, $V_o = 0$ when $V_i = V_n$, where V_i is inverting and V_n is non-inverting inputs.

An integrated circuit OP AMP consists of four blocks, as shown in Fig. 5.6(a). The first block consists of a differential amplifier. Since the input resistance of a simple differential amplifier is not necessarily high, the modified version of the differential amplifiers is used or field-effect transistors are used in the place of transistors. In a particular model, this block is made of a differential amplifier with a double-ended output followed by a second differential amplifier with a single-ended output. A typical value of input resistance is 1 MΩ or more. The second block is a high-gain amplifier giving additional gain. The typical overall gain of an OP AMP is 100 000 (or 100 dB) or more. The third block is a buffer and level shifter. Buffer, usually, is an emitter follower and the level shifter is used to keep the output voltage zero when the input voltage is zero. The last block, the driver, is a power amplifier, usually, a push-pull class B amplifier to provide power gain and low output resistance.

Figure 5.6(b) is the schematic diagram of the OP AMP showing different terminals. Terminals 2 and 3 are input terminals, where terminal 2, with (−) sign, is the inverting

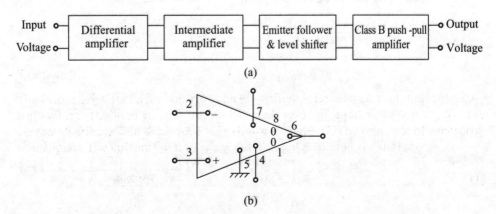

FIGURE 5.6 (a) A block diagram and (b) terminals of an OP AMP.

terminal and terminal 3, with (+) sign, is the noninverting terminal. Terminals 7 and 4, indicating $+V_{CC}$ and $-V_{CC}$, respectively, are positive and negative terminals for connections of positive and negative terminals of power supply. Terminal 5 is to be earthed. Terminals 1 and 8 are for balancing the offset voltage.

5.2.1 Basic Operational Amplifier

For simplicity, the schematic diagram of an OP AMP is shown in Fig. 5.7(a). The equivalent circuit of Fig. 5.7(a) is shown in Fig. 5.7(b). Voltages V_i and V_n are applied to inverting terminal 2 and noninverting terminal 3, respectively. Almost all OP AMPs have only one output terminal. In a single-input amplifier, one of the input terminals is grounded. The output voltage,

$$v_o = -A_v v_{in}$$

where $v_{in} = v_i - v_n$

As discussed earlier, for an ideal OP AMP, $R_i = \infty$ and $R_o = 0$. Hence, an OP AMP with noninverting terminal grounded and with a feedback impedance between inverting and output terminals is shown in Fig. 5.8(a). There is also a series impedance with the inverting terminal. Since the input impedance is considered to be very high, the total input current I_i flows through the feedback impedance Z_f. Hence, the input current is given by

$$I_i = \frac{v_{in} - v_o}{Z_f} = \frac{v_{in} + A_v v_{in}}{Z_f}$$

or

$$v_{in} = \frac{Z_f}{1 + A_v} I_i \tag{5.15}$$

Hence, Fig. 5.8(a) can be represented by Fig. 5.8(b). The input current is also given by

$$I_i = \frac{v_s - v_{in}}{Z_s}$$

Substituting this value in Eq. (5.15),

$$v_{in} = \frac{Z_f (v_s - v_{in})}{(1 + A_v) Z_s} \tag{5.16}$$

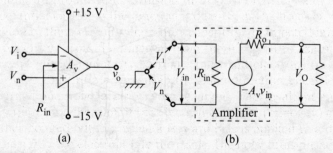

FIGURE 5.7 (a) A schematic diagram of OP AMP and (b) its equivalent circuit.

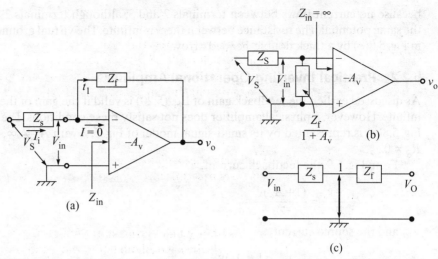

FIGURE 5.8 (a) An OP AMP with feedback, (b) an equivalent circuit of (a), and (c) virtual ground.

$$\left[1 + \frac{Z_f}{(1+A_v)Z_s}\right] v_{in} = \left[\frac{Z_f}{(1+A_v)Z_s}\right] v_s$$

$$v_{in} = \frac{Z_f v_s}{(1+A_v)Z_s + Z_f} \approx \frac{Z_f v_s}{A_v Z_s + Z_f} \qquad (\because A_v \gg 1) \tag{5.17}$$

But, the output voltage,

$$v_o = -A_v v_{in}$$

or

$$v_o \approx -\frac{A_v Z_f v_s}{A_v Z_s + Z_f} = -\frac{Z_f v_s}{Z_s + (Z_f/A_v)} \tag{5.18}$$

As A_v is very high, $Z_s \gg Z_f/A_v$, and hence Z_f/A_v can be neglected. The output can, therefore, be given by

$$v_o = -\frac{Z_f}{Z_s} v_s \tag{5.19}$$

or

$$A_{vf} = \frac{v_o}{v_s} = -\frac{Z_f}{Z_s} \tag{5.20}$$

Thus, we see that the gain is independent of the gain of the OP AMP and depends only on the ratio of two passive components Z_f and Z_s.

As the gain A_v is very high, the input impedance $Z_{in} \gg \dfrac{Z_f}{1+A_v}$ [see Fig. 5.8(b)]. For v_o $(-A_v v_i)$ to be finite, for $A_v \to \infty$, v_{in} should tend to zero, that is, $v_{in} \to 0$. So, the inverting terminal, in Fig. 4.48(b), is called *virtual ground or short circuit*. This is called virtual

5.2.2 Practical Inverting Operational Amplifier

As discussed earlier, the feedback gain of Eq. (5.20) is valid if the gain of the OP AMP is infinite. However, a physical amplifier does not satisfy these conditions. The amplifier in Fig. 5.8(a) is represented by its small-signal model of Fig. 5.9, with $A_v \neq \infty$, $R_i \neq \infty$, and $R_o \neq 0$.

From Fig. 5.9, the feedback current,

$$I = \frac{v_i - v_o}{Z_f} = \frac{(1+A_v)v_i}{Z_f}$$

and the source current,

$$I_s = I_i + I = \frac{v_i}{R_i} + \frac{(1+A_v)v_i}{Z_f} \tag{5.21}$$

Hence, $v_s = v_i + I_s Z_s$

$$= v_i \left[1 + \frac{Z_s}{R_i} + \frac{Z_s(1+A_v)}{Z_f}\right]$$

$$= \frac{R_i Z_f + Z_s Z_f + Z_s R_i (1+A_v)}{R_i Z_f} v_i \tag{5.22}$$

The output voltage,

$$v_o = -A_v v_i + I R_o$$

$$= -\left(A_v + \frac{R_o(1+A_v)}{Z_f}\right) v_i$$

$$= -\frac{A_v Z_f - R_o(1+A_v)}{Z_f} v_i \tag{4.162}$$

because no current flows between terminals 2 and 3, although terminals 2 and 3 are at the same potential. The resistance between them is infinite. The virtual ground is depicted in Fig. 5.8(c) by a thick double-headed arrow.

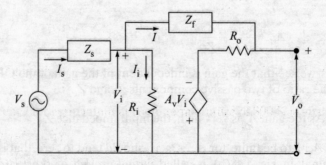

FIGURE 5.9 A small-signal model of Fig. 5.8(a).

Hence, the gain with feedback,

$$A_{vf} = \frac{v_o}{v_s} = -\frac{A_v Z_f R_i - R_i R_o (1+A_v)}{R_i Z_f + Z_s Z_f + Z_s R_i (1+A_v)} \quad (5.23)$$

As $A_v \to \infty$, $R_i \to \infty$ and R_o is small,

$$A_{vf} = \frac{v_o}{v_s} = -\frac{Z_f}{Z_s} \quad (5.24)$$

The result is same as obtained in Eq. (5.20).

5.2.3 Noninverting Operational Amplifier

An OP AMP with input in noninverting terminal is shown in Fig. 5.10.

The voltage across Z_1,

$$v_1 = \frac{v_o Z_1}{Z_1 + Z_f} \quad (5.25)$$

and $\quad v_1 - v_2 = v_i = -\dfrac{v_o}{A_v}$

or $\quad v_2 = v_1 + \dfrac{v_o}{A_v} = \left[\dfrac{Z_1}{Z_1 + Z_f} + \dfrac{1}{A_v}\right] v_o$

$$= \left[\frac{A_v Z_1 + Z_1 + Z_f}{(Z_1 + Z_f) A_v}\right] v_o \quad (5.26)$$

As A_v is very high, $Z_1 A_v \gg (Z_1 + Z_2)$; hence,

$$v_2 = \frac{v_o Z_1}{Z_1 + Z_f} = v_1 \quad (5.27)$$

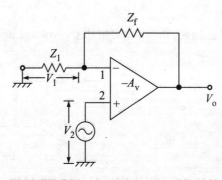

FIGURE 5.10 A noninverting OP AMP.

Hence, the gain with feedback,

$$A_{vf} = \frac{v_o}{v_2} = \frac{Z_1 + Z_f}{Z_1} \qquad (5.28)$$

Hence, the closed-loop gain is always greater than unity. If Z_1 is infinite or Z_f is zero, $A_{vf} = +1$ and the amplifier acts as a voltage follower.

5.3 LINEAR APPLICATIONS OF OPERATIONAL AMPLIFIERS

OP AMPs have important applications in analogue computers. They are used to perform the following mathematical operations:

1. Sign change
2. Scale change
3. Phase shift
4. Summation
5. Integration
6. Differentiation

In addition, an operation amplifier is used as a voltage-to-current converter and a current-to-voltage converter. There are many other applications of OP AMPs and these applications are discussed in sequel.

An OP AMP with a series and feedback impedances is shown in Fig. 5.11. The gain, as shown earlier, is given by

$$A_{vf} = \frac{v_o}{v_i} = -\frac{Z_f}{Z_s} \qquad (5.29)$$

Different mathematical operations performed by the OP AMP are discussed below.

5.3.1 OP AMP as Sign Changer

If the feedback impedance Z_f is made equal to the series impedance Z_s, then

$$v_o = -v_i \qquad (5.30)$$

Thus, the OP AMP changes only the sign of the voltage applied to it.

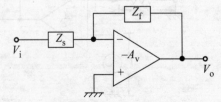

FIGURE 5.11 An OP AMP.

5.3.2 OP AMP as Scale Changer

If the ratio $Z_f/Z_s = K$, a real constant, then the output voltage,

$$v_o = -Kv_i \tag{5.31}$$

In this case, both Z_f and Z_s are pure resistances. The OP AMP, therefore, works as a scale changer.

5.3.3 OP AMP as Phase Shifter

If the magnitudes of Z_f and Z_s are equal, while their phases differ, then

$$\frac{v_o}{v_i} = -\frac{|Z_f|e^{j\theta_f}}{|Z_s|e^{j\theta_s}} = -e^{j(\theta_f-\theta_s)} = e^{j(\pi+\theta_f-\theta_s)} \tag{5.32}$$

Hence, the phase angle can be changed from 0° to 360° (or −180° to 180°).

5.3.4 OP AMP as Summer

For summing operation, a number of resistances (which are equal to the number of voltages to be summed) are connected to the inverting input terminal, as shown in Fig. 5.12. As terminals 1 and 2 are at the same (ground) potential, the output voltage

$$V_o = -IR$$

where $I = \sum_{i=1}^{n} I_i = \sum_{i=1}^{n} \frac{V_i}{R_i}$

$$\text{Hence, } V_o = -\sum_{i=1}^{n} \frac{V_i}{R_i} R \tag{5.33}$$

If $R_1 = R_2 = \ldots = R_n = R$, then

$$V_o = -\sum_{i=1}^{n} V_i = -(V_1 + V_2 + \ldots + V_n) \tag{5.34}$$

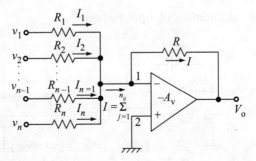

FIGURE 5.12 An operational amplifier as a summer.

There may be other methods to be used for summing voltages. But, this method has the advantage that it may be extended to a very large number of inputs requiring only one additional resistor for each additional input.

5.3.5 OP AMP as Integrator

An integrator circuit using an OP AMP is shown in Fig. 5.13. The feedback impedance is a pure capacitance, whereas the series impedance is a pure resistance. As terminals 1 and 2 are at the same potential, the current I is given by

$$I = \frac{V_i}{R}$$

So, the output voltage,

$$V_o = -\frac{1}{C}\int I\,dt$$

$$= -\frac{1}{CR}\int V_i\,dt$$

If $\frac{1}{CR} = 1$, the output voltage,

$$V_o = -\int V_i\,dt \tag{5.35}$$

$$= -\text{integral of input voltage}$$

5.3.6 OP AMP as Differentiator

A differentiator circuit using an OP AMP is shown in Fig. 5.14. The resistance R and capacitance C in the integrator circuit of Fig. 5.13 is interchanged. The current I is given by

$$I = C\frac{dV_i}{dt}$$

and $$V_o = -IR = -CR\frac{dV_i}{dt}$$

If $CR = 1$, the output,

$$V_o = -\frac{dV_i}{dt} \tag{5.36}$$

$$= -\text{differential of input voltage}$$

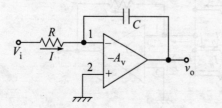
FIGURE 5.13 An integrator circuit.

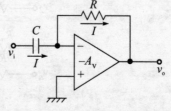
FIGURE 5.14 A differentiator circuit.

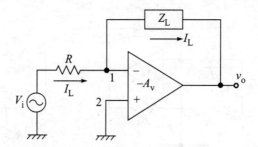

FIGURE 5.15 A voltage-to-current converter.

5.3.7 Voltage-to-Current Converter

It is often desired to convert a voltage signal into a current signal. For example, a voltage-to-current converter is required for driving a deflection coil of a television tube. Consider a circuit shown in Fig. 5.15.

As usual, the load current,

$$I_L = \frac{V_i}{R} \tag{5.37}$$

The current I_L is proportional to the input voltage. One point to be noted here is that, as the same current flows through the signal source and the load, the source should be capable of providing this load current. Consider another circuit shown in Fig. 5.16, where the source is connected to noninverting terminal of the OP AMP. The potential of terminal 1 is given by

$$V_1 = I_L R$$

or $$I_L = \frac{V_1}{R}$$

As the potentials of terminals 1 and 2 are same,

$$I_L = \frac{V_i}{R} \tag{5.38}$$

Here, the load current is again proportional to the input voltage, but the current through the input source is very small because of very high input resistance.

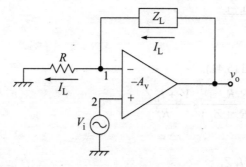

FIGURE 5.16 Another form of voltage-to-current converter.

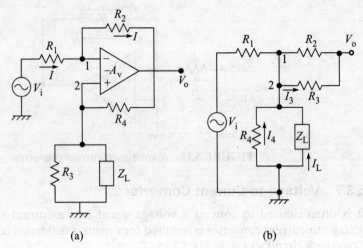

FIGURE 5.17 (a) A voltage-to-current converter and (b) an equivalent circuit of (a).

Now, let us consider a circuit shown in Fig. 5.17(a) where the load is grounded. An equivalent circuit is shown in Fig. 5.17(b).

From Fig. 5.17(b),

$$IR_2 = I_3 R_3$$

or

$$I_3 = \frac{R_2}{R_3} I \qquad (5.39)$$

The load current,

$$I_L = \frac{I_3 R_4}{R_4 + Z_L} = \frac{R_2 R_4}{R_3 (R_4 + Z_L)} I \qquad (5.40)$$

But

$$I = \frac{V_i - V_1}{R_1} = \frac{V_i - V_2}{R_1} \qquad (\because V_1 = V_2)$$

and

$$V_2 = -I_L Z_L$$

$$\therefore \quad I = \frac{V_i + I_L Z_L}{R_1} \qquad (5.41)$$

From Eqs. (5.40) and (5.41),

$$I_L = \frac{R_2 R_4}{R_3 (R_4 + Z_L)} \times \frac{V_i + I_L Z_L}{R_1}$$

If $R_2 R_4 = R_1 R_3$,

$$I_L = \frac{V_i + I_L Z_L}{R_4 + Z_L}$$

or
$$I_L \left[1 - \frac{Z_L}{R_4 + Z_L}\right] = \frac{V_i}{R_4 + Z_L}$$

or
$$\frac{R_4}{R_4 + Z_L} I_L = \frac{V_i}{R_4 + Z_L}$$

or
$$I_L = \frac{V_i}{R_4} \tag{5.42}$$

Hence, the current I_L is proportional to the input voltage V_i.

5.3.8 Current-to-Voltage Converter

A current-to-voltage converter is shown in Fig. 5.18. The current through resistance R_s is zero because of the virtual ground at the input. So, the source current, I_s, flows through the feedback resistance R. Hence, the output voltage,

$$V_o = -I_s R \tag{5.43}$$

which is proportional to the source current.

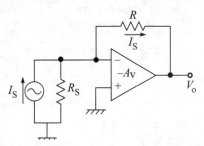

FIGURE 5.18 A current-to-voltage converter.

Example-5.4

For the circuit shown in Fig. 5.19, determine Y_{of}. Assume $A_v = -1000$, $R = 10\,k\Omega$, $R' = 40\,k\Omega$, and $R_o = 50\,\Omega$.

Solution

The equivalent circuit, after short circuiting the source, is shown in Fig. 5.20. The voltage V is applied to the output terminals.

From Fig. 5.19,

$$I_1 = \frac{V}{R + R'}$$

and
$$V_i = I_1 R = \frac{VR}{R + R'}$$

Also,
$$I_2 = \frac{V - A_v V_i}{R_o} = \frac{V(R + R' - A_v R)}{R_o (R + R')}$$

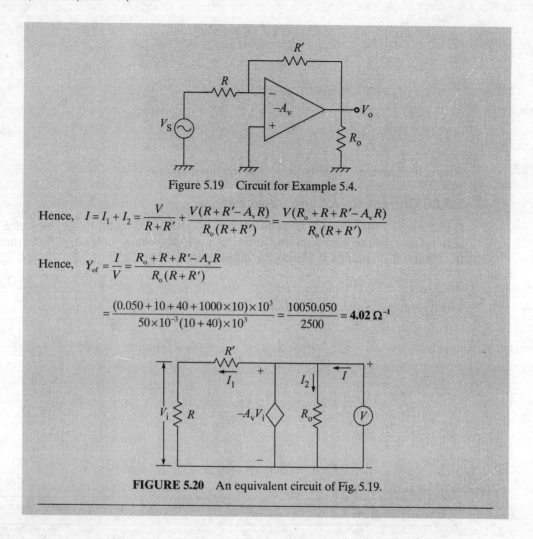

Figure 5.19   Circuit for Example 5.4.

Hence, $I = I_1 + I_2 = \dfrac{V}{R+R'} + \dfrac{V(R+R'-A_v R)}{R_o(R+R')} = \dfrac{V(R_o + R + R' - A_v R)}{R_o(R+R')}$

Hence, $Y_{of} = \dfrac{I}{V} = \dfrac{R_o + R + R' - A_v R}{R_o(R+R')}$

$= \dfrac{(0.050 + 10 + 40 + 1000 \times 10) \times 10^3}{50 \times 10^{-3}(10+40) \times 10^3} = \dfrac{10050.050}{2500} = 4.02\ \Omega^{-1}$

FIGURE 5.20   An equivalent circuit of Fig. 5.19.

5.4   POWER AMPLIFIER

To obtain the required amplification, usually several stages of amplifiers in cascade are used. The input amplifier and intermediate amplifiers are of small-signal class A amplifiers. The function of these amplifiers is to amplify the small input excitation to a large value. Thus, the final-stage amplifiers are excited by large signals. The final-stage amplifiers, called output-stage amplifiers, give outputs of values large enough to drive the final devices. The final devices are transducers such as a loudspeaker, a servomotor, a cathode ray tube, etc. Hence, an output amplifier must be capable of delivering a large voltage or current swing or an appreciable amount of power. These amplifiers are, therefore, called power amplifiers. Thus, power amplifiers have the capacity to dissipate large amounts of power. Hence, a special class of transistors known as power transistors are used. As power transistors deal with large a.c. signals, it is required that their characteristics be more linear than that it were for small-signal transistors.

Until now, we have analyzed small-signal amplifiers, and for analysis the transistors are replaced by their linear models. The transistor response is obtained by the use of the linear-circuit analysis methods. For large-signal excitation, the output voltage or current swings are large and the transistors, therefore, cannot be represented by their linear models. Large swings of the output voltage or current introduce harmonic components in the output. Due to large excitation, the transistors operate over nonlinear region of the dynamic transfer characteristics causing the output voltage and current waveforms to differ from the input voltage and current waveforms. This type of distortion is called the nonlinear distortion or amplitude distortion.

For the input wave considered to be cosine function of time, the distorted output current may be represented by

$$i_C = I_C + i_c' = I_C + B_0 + B_1 \cos \omega t + B_2 \cos 2\omega t + B_3 \cos 3\omega t + \ldots \ldots \tag{5.44}$$

The ith harmonic distortion is defined as

$$D_i = \frac{|B_i|}{|B_1|} \quad (i = 2, 3, 4, \ldots) \tag{5.45}$$

The power output delivered at the fundamental frequency,

$$P_1 = \left(\frac{B_1}{\sqrt{2}}\right)^2 R_C = \frac{B_1^2 R_C}{2} \tag{5.46}$$

Similarly, the power delivered at the ith harmonic,

$$P_i = \frac{B_i^2 R_C}{2} \tag{5.47}$$

Hence, the total power, excluding d.c. power, delivered is given by

$$P = \left(B_1^2 + B_2^2 + B_3^2 + \ldots\right)\frac{R_C}{2}$$

$$= \left(1 + D_2^2 + D_3^2 + \ldots\right)\frac{B_1^2 R_C}{2}$$

$$= \left(1 + D_2^2 + D_3^2 + \ldots\right) P_1 \tag{5.48}$$

$$= \left(1 + D^2\right) P_1 \tag{5.49}$$

where D is the *total distortion* or *distortion factor* and is defined as

$$D = \sqrt{D_2^2 + D_3^2 + D_4^2 + \ldots} \tag{5.50}$$

If the total harmonic distortion is within 10%, the output power,

$$P = [1 + (0.1)^2] P_1 = 1.01 P_1 \tag{5.51}$$

Thus, for $D = 10\%$, if we take P_1 as the output power, the error is only 1%. Hence, we may use fundamental power P_1 in place of the total power P without introducing much error.

The different types of power amplifiers named earlier, that is, class A, class B, class AB, and class C amplifiers, are discussed next.

5.4.1 Class A Power Amplifier

In class A operation, the transistor used in the amplifier operates entirely in the active or linear region. In earlier discussions, we considered small amplifier in which transistors always operated in the active region. The a.c. signals considered were small enough to have class A operation.

For class A operation, the transistor must not go into saturation or cutoff. As the transistor operates along the load line, for the largest possible a.c. signal, the quiescent point should be at the centre of the load line. The a.c. load line for a class A amplifier is shown in Fig. 5.21. The a.c. power to the load is due to varying voltage and current around the quiescent point. The collector current varies around its quiescent current I_{CQ}. The a.c. power delivered to the load (total resistance to the output circuit), R_T, is given by

$$P_o = \left(\frac{I_{C\,peak}}{\sqrt{2}}\right)^2 R_T = \frac{1}{2} I_{C\,peak}^2 R_T$$

The output voltage swing,

$$V_{C\,peak} = I_{C\,peak} R_T \tag{5.52}$$

Hence, $$P_o = \frac{1}{2} V_{C\,peak} I_{C\,peak} \tag{5.53}$$

The quiescent power dissipation P_{DQ} of the transistor is defined as

$$P_{DQ} = V_{CEQ} I_{CQ} \tag{5.54}$$

The quiescent power dissipation is the power dissipated by the transistor when no a.c. signal is present. P_{DQ} is maximum when the Q point is centered on the load line.

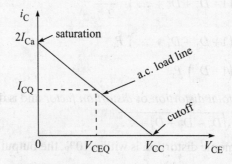

FIGURE 5.21 Load line for class A amplifier.

However, a centered Q point is required for maximum a.c. output power. Hence, the maximum average power that can be safely dissipated by a transistor (P_{Dmax}) is given by the manufacturer. Hence, for a transistor to be selected

$$P_{Dmax} \geq P_{DQ} = V_{CEQ} I_{CQ} \tag{5.55}$$

Since I_C is much greater than I_B, the average power delivered by the supply V_{CC} is given by

$$P_{d.c.} = V_{CC} I_{CQ} \tag{5.56}$$

Even if there is no input signal, the d.c. current drawn from the supply V_{CC} is the collector bias current and the power drawn from the supply is the same as given by Eq. (5.56).

Power Efficiency

The percentage efficiency, also known as *conversion efficiency* of the amplifier, is defined as

$$\eta = \frac{\text{Output power of transistor}}{\text{Power delivered by supply}} \times 100 = \frac{P_o}{P_{d.c.}} \times 100 \tag{5.57}$$

When the Q point is at the centre of the load line, the largest possible swing (the amplitude) of sinusoidal collector current (I_C) is I_{CQ}. Corresponding to this current swing the output voltage swing (amplitude),

$$V_{CEQ} = I_{CQ} R_T = \frac{V_{CC}}{2}$$

Hence, from Eq. (5.53), the maximum possible a.c. power output,

$$P_{o\,max} = \frac{1}{2} V_{CEQ} I_{CQ} = \frac{V_{CC} I_{CQ}}{4} \tag{5.58}$$

The maximum possible efficiency for class A amplifier is

$$\eta_{max} = \frac{P_{o\,max}}{P_{d.c.}} \times 100$$

$$= \frac{V_{CC} I_{CQ}}{4 V_{CC} I_{CQ}} \times 100 = 25\% \tag{5.59}$$

5.4.2 Transformer-Coupled Audio Power Amplifier

The quiescent current passes through the load resistance when it is connected directly in the output circuit of the power amplifier stage. This current produces a considerable amount of power wastage, as it does not contribute to the a.c. (useful) component of power. Furthermore, it is not advisable to pass the d.c. component of current through the output device, such as the voice coil of a loudspeaker. For these reasons, an output transformer, as shown in Fig. 5.22, is usually employed in power amplifier. The input circuit

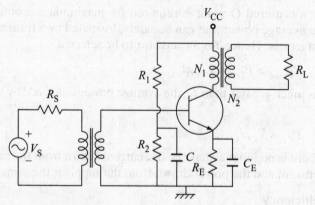

FIGURE 5.22 A transformer-coupled amplifier.

also contains a transformer, but it is also possible to excite the power stage through an RC coupling.

Impedance Matching

The output transformer also serves the purpose of impedance matching. Impedance matching is essential to transfer a significant amount of power to the output device, such as a loudspeaker. The voice coil impedance of a loudspeaker may vary from 5 to 15 Ω. The resistance of the amplifier may be very much higher than that of the loudspeaker. So major portion of power generated would be lost in the active device.

To discuss the impedance matching properties of the transformer, let us consider an ideal transformer. For an ideal transformer,

$$\frac{V_1}{V_2} = \frac{N_1}{N_2} \tag{5.60}$$

and

$$\frac{I_1}{I_2} = \frac{N_2}{N_1} \tag{5.61}$$

where

V_1, V_2 = Primary and secondary voltages, respectively.

I_1, I_2 = Primary and secondary currents, respectively.

N_1, N_2 = Numbers of primary and secondary turns, respectively.

The transformer is a step-down transformer; hence, $N_2 < N_1$. It reduces the voltage in proportion to the turns ratio, $n = N_1/N_2$, and increases the current in same proportion. The load resistance in the secondary side,

$$R_L = \frac{V_2}{I_2} = \frac{V_1/n}{n I_1} = \frac{1}{n^2} \frac{V_1}{I_1}$$

$$\therefore R'_L = \frac{V_1}{I_1} = n^2 R_L \tag{5.62}$$

where R'_L is the load resistance reflected on the input side. Now, a practical problem is to find the turns ratio, n, of the transformer for a given value of load resistance, R_L, so that the power output be maximum for a small allowable distortion. This depends on the characteristics of the transistor.

Power Efficiency

For the d.c. biasing, the transformer winding resistance is small and so may be taken zero. Hence, the biasing collector voltage

$$V_{CEQ} = V_{CC}$$

Hence, from Eq. (5.53), the maximum possible a.c. power output,

$$P_{o\,max} = \frac{1}{2} V_{CEQ} I_{CQ} = \frac{V_{CC} I_{CQ}}{2} \tag{5.63}$$

The maximum possible efficiency for class A amplifier is

$$\eta_{max} = \frac{P_{o\,max}}{P_{d.c.}} \times 100$$

$$= \frac{V_{CC} I_{CQ}}{2 V_{CC} I_{CQ}} \times 100 = 50\% \tag{5.64}$$

5.4.3 Push-Pull Amplifiers

We have seen that a large-signal amplifier using single transistor introduces appreciable distortion due to nonlinearity of transfer characteristic. Even harmonics result in a wave that has a positive half different from the negative half, whereas with odd harmonics both halves of the wave are identical. The distortion in the output is considerably reduced by the circuit, as shown in Fig. 5.23. This arrangement is known as *push-pull configuration*. There are two central taped transformers, input transformer and output transformer.

Two transistors are excited through the input transformer. The inputs to two transistors T_1 and T_2 are in push-pull, that is, in phase opposition. If the input to T_1 is given by

$$v_1 = V_m \cos \omega t$$

then the input to T_2 is given by

$$v_2 = -v_1 = V_m \cos(\omega t + \pi)$$

The output current of T_1 is, therefore, given by Eq. (5.44),

$$i_1 = I_c + B_0 + B_1 \cos \omega t + B_2 \cos 2\omega t + B_3 \cos 3\omega t + \ldots \tag{5.65}$$

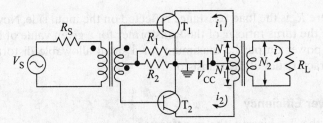

FIGURE 5.23 Push-pull arrangement of two transistors.

and the output current of T_2 is given by

$$i_2 = I_c + B_0 + B_1 \cos(\omega t + \pi) + B_2 \cos 2(\omega t + \pi) + \ldots$$
$$= I_c + B_0 - B_1 \cos \omega t + B_2 \cos 2\omega t - B_3 \cos 3\omega t + \ldots \quad (5.66)$$

The current i_1 and i_2 are in the opposite directions through the output transformer primary winding. The resultant output current is then proportional to the difference between the collector currents of the two transistors. Hence, the output current,

$$i = K(i_1 - i_2) = K(B_1 \cos \omega t + B_3 \cos 3\omega t + \ldots\ldots\ldots) \quad (5.67)$$

The above expression shows that a push-pull circuit balances out all even harmonics in the output. Thus, the remaining distortion is due to odd harmonics only. Since the amplitudes of fifth and higher odd harmonics are very small; hence, the principal cause of the source of distortion is the third harmonic. The above conclusion was derived on the assumption that the two transistors are identical. If their characteristics differ, the even harmonics appear in the output.

Since no even harmonic is present in the output of the push-pull amplifier, the output has identical positive and negative halves. Hence, the output current satisfies the condition,

$$i(\omega t) = -i(\omega t + \pi) \quad (5.68)$$

which can be shown by substituting $(\omega t + \pi)$ for (ωt) in Eq. (5.67).

Advantages of a Push-Pull Amplifier

Followings are the advantages of a push-pull amplifier:

1. As even harmonics are absent from the output, a push-pull amplifier gives more output power per transistor for a given amount of distortion.
2. As the d.c. component of the collector currents of the two transistors oppose each other magnetically, the output transformer core remains unsaturated. Hence, the possibility of distortion in the output due to saturation of the transformer is completely eliminated.
3. Ripple voltages may be present in the power supply due to inadequate filtering. The currents produced due to these ripple voltages flow in the opposite direction in the output transformer winding. Hence, the effect of the ripple voltages on the output is eliminated.

For the effective load R'_L, the r.m.s. value of the load current

$$I_L = 2\frac{I_m}{\sqrt{2}} = \sqrt{2}\, I_m$$

Hence, the power output,

$$P_o = (\sqrt{2} I_m)^2 R'_L = 2 I_m^2 R'_L = 2 I_{CQ}^2 R'_L$$

For maximum power output,

$$I_{CQ} R'_L = V_{CEQ} = \frac{V_{CC}}{2}$$

$$\therefore \quad P_o = 2 V_{CEQ} I_{CQ} = V_{CC} I_{CQ}$$

The d.c. power supplied by the source,

$$P_{d.c.} = 2 V_{CC} I_{CQ}$$

Hence, efficiency

$$\eta = \frac{V_{CC} I_{CQ}}{2 V_{CC} I_{CQ}} \times 100 = \mathbf{50\%}$$

Thus, the maximum efficiency of class A push-pull amplifier is 50%.

5.4.4 Class B Power Amplifier

Although class A amplifiers are simple and stable, the maximum power rating (P_{Dmax}) of a transistor must be equal to or greater than the maximum a.c. power output. Even when no a.c. signal is present, there is an appreciable power dissipation. Also, centered Q point means that the current drain from the supply (for no a.c. signal) is half of the saturation current. For low-power applications these drawbacks may not affect seriously, but in some cases class A amplifiers give poor efficiency. A class B amplifiers are an efficient alternative in those cases.

In class B operation, the transistor is biased at cutoff. The transistor remains cutoff for half the cycle for an a.c. signal and operates in the active region for the other half of the cycle. Since the transistor is biased at cutoff, the quiescent current $I_{CQ} = 0$ for class B operation. With the assumption that the output characteristics are equally spaced and that the dynamic characteristic is a straight line with minimum current zero, the class B operation is shown graphically in Fig. 5.24.

For the total resistance R_T in the output circuit, the average power dissipated in R_T is given by

$$P_o = I_o^2 R_T$$

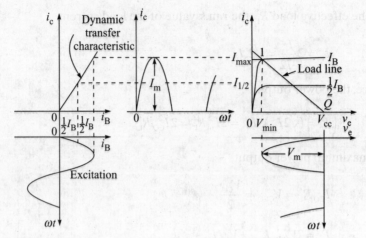

FIGURE 5.24 The output waveform of a single-stage class B amplifier.

Assuming the sinusoidal output current (flowing for one half cycle only), $I_o = \dfrac{I_m}{2}$.

$$P_o = \frac{I_m^2 R_T}{4} = \frac{V_{CEQ} I_m}{4} = \frac{V_{CC} I_m}{4} \qquad (\because \ I_m R_T = V_{CEQ} = V_{CC})$$

The d.c. collector current under load is equal to the average value of half sine wave, that is,

$$I_{d.c.} = \frac{I_m}{\pi}$$

Hence, the d.c. power input,

$$P_{d.c.} = \frac{V_{CC} I_m}{\pi}$$

Hence, efficiency

$$\eta = \frac{P_o}{P_{d.c.}} = \frac{V_{CC} I_m \pi}{4 V_{CC} I_m} \times 100 = \frac{\pi}{4} \times 100 = \mathbf{78.5\%}$$

5.4.5 Class B Push-Pull Amplifier

A class B push-pull amplifier is same as class A push-pull amplifier as far as circuit is concerned. But in the class B amplifier, transistors are biased approximately at cutoff. The class B operation has the following advantages compared to class A operation:

1. Greater power output.
2. Higher efficiency.
3. Negligible power loss at no signal.

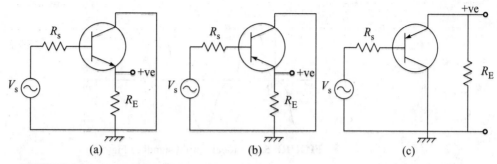

FIGURE 5.25 An emitter follower (a) with npn transistor, and (b) and (c) with pnp transistor.

For these reasons a class B push-pull amplifier is used where power supply is limited, such as those operating from solar cells or batteries. However, there are following disadvantages of the class B amplifier:

1. Harmonic distortion is higher.
2. Self-bias cannot be used.
3. Supply voltage must have good regulation.

To understand the class B push-pull amplifier, let us consider that the emitter follower of Fig. 4.27(a) is biased at cutoff. Let us also assume that the transistor is ideal, that is, $V_{BE} = 0$ and cut-in voltage $V_\gamma = 0$. The a.c. equivalent circuit of Fig. 4.27(a) is shown in Fig. 5.25(a). Thus, when the a.c. signal voltage $v_s > 0$, the base-emitter junction is forward biased; hence, the output voltage $v_e = v_s$. But, when $v_s \leq 0$, the base-emitter junction is reversed biased and $v_e = 0$.

When the npn transistor of Fig. 5.25(a) is replaced with a pnp transistor, the a.c. equivalent circuit is as shown in Fig. 5.25(b). The transistor is again biased at cutoff. Again considering an ideal transistor as above, the output voltage $v_e = v_s$ for $v_s < 0$. For $v_s \geq 0$, the output voltage $v_e = 0$. Since $(1 + h_{fe})i_b \approx h_{fe}i_b$, for $h_{fe} \gg 1$, the circuit of Fig. 5.25(b) can be redrawn as in Fig. 5.25(c).

Now, let us combine the emitter followers, using npn and pnp transistors, shown in Fig. 5.25(a) and 5.25(c), as shown in Fig. 5.26. When $v_s > 0$, for an ideal condition,

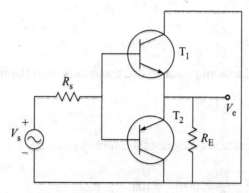

FIGURE 5.26 A push-pull emitter follower.

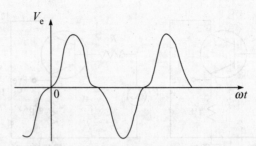

FIGURE 5.27 Ideal output signal of Fig. 5.26.

transistor T_1 conducts and T_2 remains cutoff. The output voltage $v_o = v_s$. Similarly, when $v_s < 0$, transistor T_2 conducts and T_1 remains cutoff. Again the output voltage $v_o = v_s$. The circuit of Fig. 5.26 is a push-pull circuit for two emitter followers. The ideal output signal is shown in Fig. 5.27. It is a replica of the input sinusoidal voltage v_s. Although the output voltage v_o is equal to the input voltage v_s, the transistor has a significant current gain of h_{fe}. Thus, the power gain of the circuit is significant.

The circuit of Fig. 5.23 for class A push-pull amplifiers can be used for class B push-pull amplifiers. But in class B amplifiers, transistors are biased approximately at cutoff. A transistor operates as class B if $V_{BE} = 0$, that is, the base is shorted to the emitter. This condition can be obtained in Fig. 5.23 if $R_2 = 0$.

Since the excitations of the transistors T_1 and T_2 in class B push-pull amplifiers is out of phase by 180°, the outputs of T_1 and T_2 are half sine wave with the phase displacement of 180°. The output of the amplifier is as shown in Fig. 5.27. The effective load, as given in Eq. (5.62), is $R'_L = (N_1/N_2)^2$. The load current, which is proportional to the difference of two collector currents, is a perfect sine wave for the ideal conditions as assumed for Fig. 5.26. The power output is

$$P_o = \frac{V_m I_m}{2} = \frac{I_m V_{CC}}{2} \qquad (5.69)$$

The d.c. collector current of each transistor is equal to an average value of half sine wave, that is,

$$I_{d.c.} = \frac{I_m}{\pi}$$

Hence, the d.c. input power to two transistors from the supply

$$P_i = 2\frac{I_m V_{CC}}{\pi}$$

and the percentage efficiency, therefore, is given by

$$\eta = \frac{P_o}{P_i} \times 100 = \frac{\pi}{4} \times 100 \qquad (5.70)$$

From the above expression, it is clear that the maximum possible conversion efficiency is $25\pi = 78.5\%$ for a class B amplifier compared to 50% for class A amplifier. There is no current in class B amplifier when it is not excited. But, in class A amplifier, there is a flow of current from the power supply even with no excitation. At the quiescent state, dissipation at the collector in class B amplifier is zero and it increases with excitation. In case of class A amplifier, the dissipation at the collector is maximum at zero input and it decreases as the input signal increases. Since the direct current increases with the input signal in class B amplifier, the power supply must have good regulation.

Distortion in Push-Pull Amplifiers

As already discussed, the even harmonics are absent from a push-pull amplifier, the main contribution to distortion is from the third harmonic. Hence, the distortion is given by

$$D_3 = \frac{|B_3|}{|B_1|} \tag{5.71}$$

The power output, taking distortion into account, is given by

$$P_o = \left(1 + D_3^2\right) \frac{B_1^2 R_L}{2} \tag{5.72}$$

Class B Amplifier without Output Transformer

A class B push-pull amplifier circuit without using an output transformer is shown in Fig. 5.28. The input transformer does not have central-tapped secondary power supply; rather it has two identical secondary windings. However, base-to-emitter voltages of two transistors are again in push-pull. The collector currents flow in the opposite direction through the common load. Hence, the output voltage developed across the load is proportional to the difference of two collector currents.

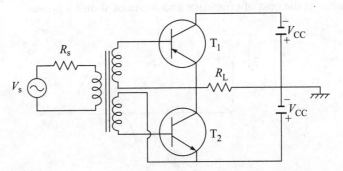

FIGURE 5.28 A class B push-pull amplifier circuit without output transformer.

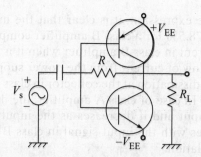

FIGURE 5.29 A push-pull amplifier using complementary transistors.

Class B Amplifier without Transformer

A push-pull circuit that uses neither an output nor an input transformer is shown in Fig. 5.29. The arrangement uses two complementary transistors (one pnp and another npn). The difficulty with this circuit is that of obtaining a matched pair of transistors. If the characteristics of the two transistors differ, then a considerable amount of distortion is introduced. So, even harmonics are no longer absent. Negative feedback is very often used in power amplifiers to reduce distortion.

Crossover Distortion Until now, we considered the distortion due to nonlinearity of the output characteristics and that introduced by unmatched transistors. In addition to above causes of distortion, there is one more cause of distortion. This is the nonlinearity of the input characteristics. As we know, no appreciable base current flows until the emitter is forward biased by the cut-in voltage V_γ. For germanium $V_\gamma = 0.1$ V and for silicon $V_\gamma = 0.5$ V. Hence, even a pure sinusoidal base voltage does not result in a sinusoidal output current. The distortion caused by nonlinearity in the input characteristic of a transistor is shown in Fig. 5.30. For $v_B < V_\gamma$, the output is much smaller than what it would be if the response curve were linear. This effect is called *crossover distortion*.

For minimizing crossover distortion, the transistors are operated on class AB mode. In class AB, a small standby current flows at zero excitation. If the voltage drop across R_2 in Fig. 5.23 is adjusted approximately equal to V_γ, then the transistors T_1 and T_2 operate in class AB mode. Class AB operation results in less distortion than class B operation, but at the cost of efficiency and waste of standby power.

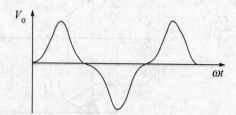

FIGURE 5.30 Crossover distortion in a class B amplifier.

5.5 REVIEW QUESTIONS

1. In differential amplifier define the terms (i) differential-mode signal, (ii) common-mode signal, and (iii) CMRR. How can CMRR be determined in practice?
2. What are advantages of an OP AMP? Explain how an OP AMP can be used as summer, integrator, and differentiator.
3. With neat circuit diagram explain how an OP AMP is used to construct a voltage-to-current converter and a current-to-voltage converter.
4. Define the conversion efficiency η of a power amplifier. Derive expression for conversion efficiency. Compare the maximum efficiency of a series-fed and transformer-coupled class A single transistor power amplifier.
5. Why are even harmonics not present in a push-pull amplifier? Give two additional advantages of the push-pull circuit amplifier over that of a single-transistor amplifier.
6. Derive an expression for the output power of an idealized class B push-pull amplifier. Show that the maximum efficiency of this amplifier is 78.5%.
7. Draw a class B push-pull amplifier using two complementary silicon transistors and without using an output transformer.
8. Explain the origin of crossover distortion. Suggest a method for minimizing this distortion.

5.6 SOLVED PROBLEMS

1. (a) In Fig. 5.1, the two inputs are $v_1 = 0.1$ mV and $v_2 = -0.1$ mV. If CMRR is 1000 and the differential gain is 100, then determine the output of the amplifier. (b) If two inputs are $v_1 = 1.15$ mV and $v_2 = 0.95$ mV and the differential output is 1.1 V, then for CMRR = 100, calculate the differential gain.

Solution

From Eq. (5.7)

$$v_o = A_d v_d \left(1 + \frac{1}{\text{CMRR}} \times \frac{v_c}{v_d}\right)$$

(a) The differential input

$$v_d = 0.1 - (-0.1) = 0.2 \text{ mV}$$

and the common-mode input

$$v_c = \frac{0.1 - 0.1}{2} = 0$$

$$\therefore v_o = A_d v_d = 100 \times 0.2 = \mathbf{20 \text{ mV}}$$

(b) The differential input

$$v_d = 1.15 - 0.95 = 0.2 \text{ mV}$$

and the common-mode input

$$v_c = \frac{1.15 + 0.95}{2} = 1.05 \text{ mV}$$

$$\therefore \quad 1.1 = A_d \times 0.2 \times 10^{-3} \left(1 + \frac{1}{100} \times \frac{1.05}{0.2}\right)$$

$$A_d = \frac{1.1 \times 10^3}{0.2 \times 1.0525} = \mathbf{5225.65}$$

2. For the differential amplifier shown in Fig. 5.1, $h_{fe} = 100$, $h_{ie} = 1 \text{ k}\Omega$, $R_C = R_E = 4.7 \text{ k}\Omega$, $R_s = 0$. Find CMRR in decibel.

Solution

From Eq. (5.13),

$$\text{CMRR} = \frac{(1 + h_{fe})R_E}{R_s + h_{ie}} = \frac{101 \times 4.7}{1} = 474.7$$

and   CMRR in dB = $20 \log_{10}(474.7)$ = **53.53 dB**

3. For an OP AMP circuit shown in Fig. 5.31, $R = 1 \text{ k}\Omega$, $R' = 10 \text{ k}\Omega$, $A = 10^5$, and $V_s = 1$ V. (a) Find V_o, V_1, I_1, and I_2. (b) Find the power absorbed by each resistor, the input source, and the amplifier.

Solution

(a) As the input resistance of the amplifier is infinite, $I_{in} = 0$ and the current $I_1 = I_2$; hence,

$$\frac{V_s - V_1}{R} = \frac{V_1 - V_o}{R'}$$

or $$V_s = \left(1 + \frac{R}{R'}\right)V_1 - \frac{R}{R'}V_o$$

$$= 1.1 V_1 - 0.1 V_o$$

But $V_o = -A V_1 = -10^5 V_1$

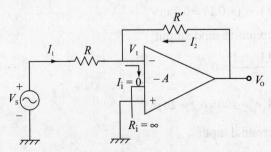

FIGURE 5.31 Circuit for Solved Problem 3.

Hence, $V_s = -(1.1 \times 10^{-5} + 0.1)V_o \approx -0.1 V_o$

$$V_o = -\frac{V_s}{0.1} = -\frac{1}{0.1} = -10 \text{ V}$$

and $V_s = (1.1 + 10000)V_1 \approx 10000 V_1$

$$V_1 = \frac{V_s}{10000} = \frac{1}{10000} = 0.1 \text{ mV}$$

The currents

$$I_1 = \frac{V_s - V_1}{R} = \frac{1 - 0.1 \times 10^{-3}}{1} \approx 1 \text{ mA}$$

and $I_2 = -I_1 = -1 \text{ mA}$

(b) Power absorbed in R

$$P_R = (1 \times 10^{-3})^2 \times 1 \times 10^3 = 1 \text{ mW}$$

Power absorbed in R'

$$P_{R'} = (1 \times 10^{-3})^2 \times 10 \times 10^3 = 10 \text{ mW}$$

Power absorbed by the source

$$P_s = V_s \times I_1 = -1 \times 1 = -1 \text{ mW}$$

Negative sign indicates that the power is supplied by the source.

Power absorbed by the amplifier

$$P_a = -V_o \times I_2 = -10 \times -1 = 10 \text{ mW}$$

4. For a noninverting OP AMP circuit shown in Fig. 5.32, $R_1 = 1 \text{ k}\Omega$, $R_2 = 10 \text{ k}\Omega$, and $V_s = 1$ V. (a) Find V_o, I_1, and I_2. (b) Find the power absorbed by each resistor, the input source, and the amplifier.

Solution

(a) As the input resistance of the amplifier is infinite, the voltages of inverting and noninverting terminals are equal. Hence, $V_n = V_i = V_s$.

$$\therefore \quad I_1 = \frac{V_n}{R_1} = \frac{V_s}{R_1}$$

and $$I_2 = \frac{V_s - V_o}{R_2}$$

But by KCL

$$I_1 = -I_2$$

or $$\frac{V_s}{R_1} = -\frac{V_s - V_o}{R_2}$$

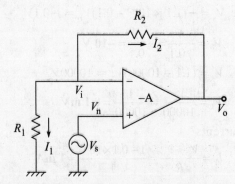

FIGURE 5.32 Circuit for Solved Problem 4.

or $\quad V_o = \dfrac{R_1 + R_2}{R_1} V_s = \dfrac{1+10}{1} \times 1 = \mathbf{11\ V}$

The currents

$$I_1 = \dfrac{V_s}{R_1} = \dfrac{1}{1} = \mathbf{1\ mA}$$

and $\quad I_2 = -I_1 = \mathbf{-1\ mA}$

(b) Power absorbed in R_1

$$P_{R1} = (1 \times 10^{-3})^2 \times 1 \times 10^3 = \mathbf{1\ mW}$$

Power absorbed in R_2

$$P_{R2} = (1 \times 10^{-3})^2 \times 10 \times 10^3 = \mathbf{10\ mW}$$

Power absorbed by the source

$$P_s = V_s \times 0 = \mathbf{0\ W}$$

Power absorbed by the amplifier

$$P_a = -V_o \times I_2 = -11 \times 1 = \mathbf{-11\ mW}$$

Negative sign indicates that the power is supplied.

5. In the circuit of Fig. 5.33, $R_1 = 1\ k\Omega$, $R_2 = 9\ k\Omega$, $V_{s1} = 1\ V$, and $V_{s2} = 2\ V$. Find the output voltage.

Solution

Because of infinite input resistance,

$$I_1 = \dfrac{V_{s1} - V_o}{R_1 + R_2}$$

$$I_2 = \dfrac{V_{s2}}{R_1 + R_2}$$

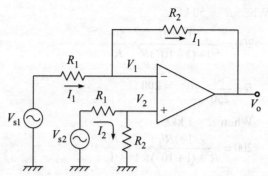

FIGURE 5.33 Circuit for Solved Problem-5.

and $V_1 = V_2$

Also, $V_{s1} - I_1 R_1 = V_{s2} - I_2 R_1$

or $I_1 = \dfrac{V_{s1} - V_{s2}}{R_1} + I_2$

Substituting for I_1 and I_2,

$$\dfrac{V_{s1} - V_o}{R_1 + R_2} = \dfrac{V_{s1} - V_{s2}}{R_1} + \dfrac{V_{s2}}{R_1 + R_2}$$

or $V_o = \left(1 + \dfrac{R_2}{R_1}\right)(V_{s2} - V_{s1}) - (V_{s2} - V_{s1})$

$\qquad = \dfrac{R_2}{R_1}(V_{s2} - V_{s1}) = \dfrac{9}{1}(2-1) = \mathbf{9\ V}$

6. For the OP AMP circuit shown in Fig. 5.31, $R_{in} = \infty$, $R_o = 0$, and $A = 10^5$. The gain with feedback, $A_f = 200$. (a) Determine R when $R' = 50$ kΩ. (b) Determine R' when $R = 1$ kΩ.

Solution

As $R_{in} = \infty$,

$$\dfrac{V_s - V_1}{R} = \dfrac{V_1 - V_o}{R'}$$

$$V_s = \left(\dfrac{R + R'}{R'}\right)V_1 - \dfrac{R}{R'}V_o = \left(\dfrac{R + R'}{R'}\right) \times \dfrac{-V_o}{A} - \dfrac{R}{R'}V_o$$

$$= -\left(\dfrac{R' + (1+A)R}{R'}\right) \times \dfrac{V_o}{A}$$

$\therefore\qquad \dfrac{V_o}{V_s} = -A_f = -\dfrac{AR'}{R' + (1+A)R}$

(a) When $R' = 50 \text{ k}\Omega$,

$$200 = \frac{10^6 \times 50}{50 + (1 + 10^6)R} \approx \frac{50}{R}$$

$\therefore \quad R = \frac{50}{250} \times 10^3 = \mathbf{200 \ \Omega}$

(b) When $R = 1 \text{ k}\Omega$,

$$200 = \frac{10^6 R'}{R' + (1 + 10^6) \times 1} \approx \frac{R'}{1}$$

$\therefore \quad R' = \mathbf{200 \ k\Omega}$

7. An OP AMP, shown in Fig. 5.34, has $R_{in} = \infty$, $R_o = 0$, and finite gain A. Find $A_f = i_o/v_{in}$.

Solution

Since $R_{in} = \infty$,

$$v_1 = i_o R = \frac{v_o}{R + R'} R$$

and $\quad v_1 - v_{in} = -\dfrac{v_o}{A}$

or $\quad v_{in} = \dfrac{v_o R}{R + R'} + \dfrac{v_o}{A}$

$\quad \quad \quad = v_o \dfrac{R' + (AR + R)}{(R + R')A}$

$\therefore \quad \dfrac{v_o}{v_{in}} = \dfrac{A(R + R')}{R' + R(1 + A)}$

or $\quad \dfrac{i_o}{v_{in}} = \dfrac{A}{R' + (R(1 + A))}$

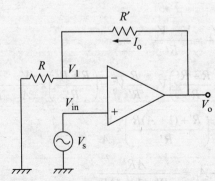

FIGURE 5.34 Circuit for Solved Problem 7.

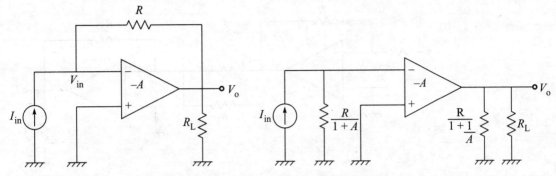

FIGURE 5.35 (a) Circuit for Solved Problem-8 and (b) circuit after applying Miller's theorem.

8. The OP AMP shown in Fig. 5.35(a) has $R_{in} = \infty$, $R_o = 0$, and finite gain A. Find $A_f = v_o/i_{in}$ and $R_{if} = v_{in}/i_{in}$.

Solution

Applying Miller's theorem, Fig. 5.35(a) reduces as shown in Fig. 5.35(b).
From Fig. 5.35(b) the input voltage,

$$v_{in} = \frac{i_{in} R}{1+A}$$

The amplifier gain

$$-A = \frac{v_o}{v_{in}} = \frac{v_o(1+A)}{i_{in} R}$$

$$\therefore \quad A_f = \frac{v_o}{i_{in}} = -\frac{AR}{(1+A)}$$

and $\quad R_{if} = \dfrac{v_{in}}{i_{in}} = \dfrac{R}{(1+A)}$

9. Using an OP AMP, design a circuit to implement
$$V_o = 2v_1 - 5v_2$$

Solution

The circuit to implement the above relation, a summer circuit, as shown in Fig. 5.36, can be used. To change the sign of first input, an OP AMP with unity gain is used. The resistances R may be taken equal to 40 kΩ. The current,

$$I = -\frac{v_1}{R_1} + \frac{v_2}{R_2}$$

and the output voltage,

$$v_o = -IR_3 = \frac{v_1 R_3}{R_1} - \frac{v_2 R_3}{R_2}$$

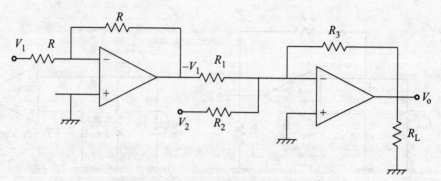

FIGURE 5.36 Circuit for Solved Problem 9.

As required $\dfrac{R_3}{R_1} = 2$ and $\dfrac{R_3}{R_2} = 5$.

Let $R_2 = 30\ \text{k}\Omega$, then $R_3 = 150\ \text{k}\Omega$, and $R_1 = 75\ \text{k}\Omega$.

The typical value of R_L may be taken as $10\ \text{k}\Omega$.

10. The dynamic characteristic of a transistor is represented by a quadratic equation

$$i_c = K_1 i_b + K_2 i_b^2$$

If the input signal is given by $i_b = I_1 \cos \omega_1 t + I_2 \cos \omega_2 t$, show that the output contains a d.c. term and the sinusoidal terms of frequencies $\omega_1, \omega_2, 2\omega_1, 2\omega_2, \omega_1 + \omega_2$, and $\omega_1 - \omega_2$.

Solution

The output is given by

$$i_c = K_1(I_1 \cos \omega_1 t + I_2 \cos \omega_2 t) + K_2(I_1 \cos \omega_1 t + I_2 \cos \omega_2 t)^2$$

$$= K_1 I_1 \cos \omega_1 t + K_1 I_2 \cos \omega_2 t + K_2 I_1^2 \cos^2 \omega_1 t + K_2 I_2^2 \cos^2 \omega_2 t$$
$$\quad + 2K_2 I_1 I_2 \cos \omega_1 t \cos \omega_2 t$$

$$= K_1 I_1 \cos \omega_1 t + K_1 I_2 \cos \omega_2 t + \dfrac{K_2 I_1^2}{2}(1 + \cos 2\omega_1 t) + \dfrac{K_2 I_2^2}{2}(1 + \cos 2\omega_2 t)$$
$$\quad + K_2 I_1 I_2 \cos(\omega_1 + \omega_2)t + K_2 I_1 I_2 \cos(\omega_1 - \omega_2)t$$

$$= I_{\text{d.c.}} + B_1 \cos \omega_1 t + B_2 \cos \omega_2 t + B_3 \cos 2\omega_1 t + B_4 \cos 2\omega_2 t$$
$$\quad + B_5 \cos(\omega_1 + \omega_2)t + B_6 \cos(\omega_1 - \omega_2)t$$

where $I_{\text{d.c.}} = \dfrac{K_2 I_1^2 + K_2 I_2^2}{2}$

$$B_1 = K_1 I_1 \quad \text{and} \quad B_2 = K_1 I_2$$

$$B_3 = \frac{K_2 I_1^2}{2} \quad \text{and} \quad B_4 = \frac{K_2 I_2^2}{2}$$

$$B_5 = B_6 = K_2 I_1 I_2$$

Thus, it shows that the output contains a d.c. term and sinusoidal terms of frequencies $\omega_1, \omega_2, 2\omega_1, 2\omega_2, \omega_1 + \omega_2$, and $\omega_1 - \omega_2$.

11. A transistor supplies an a.c. power of 1.0 W to a load of 5 kΩ. The zero signal d.c. collector current is 37 mA and the d.c. collector current with a signal is 40 mA. Determine the percentage second harmonic distortion.

Solution

Let the dynamic curve of the transistor be given by

$$I_o = K_1 I + K_2 I^2$$

Hence, for $I_{in} = I \cos \omega t$, the output current

$$I_o = K_1 I \cos \omega t + K_2 I^2 \cos^2 \omega t$$

$$= K_1 I \cos \omega t + \frac{K_2 I^2}{2} + \frac{K_2 I^2}{2} \cos 2\omega t$$

$$= I_{av} + I_1 \cos \omega t + I_2 \cos 2\omega t$$

where $I_2 = I_{av}$

This output is only due to the a.c. input signal. The total d.c. component of the actual output is equal to the d.c. collector current when there is no signal plus I_{av}. Hence,

$$I_{av} = 40 - 37 = 3 \text{ mA}$$

Hence, the second harmonic amplitude

$$I_2 = I_{av} = 3 \text{ mA}$$

The total a.c. power is given by

$$P = (I_1^2 + I_2^2) \frac{R_L}{2}$$

or

$$1.0 = (I_1^2 + 3^2) \frac{5 \times 10^{-3}}{2}$$

$$\therefore \quad I_1 = 19.77 \text{ mA}$$

Hence, the percentage second harmonic distortion,

$$D = \left|\frac{I_2}{I_1}\right| \times 100 = \frac{3}{19.77} \times 100 = \mathbf{15.17\%}$$

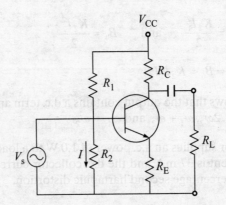

FIGURE 5.37 Circuit for Solved Problem 12.

12. For the amplifier shown in Fig. 5.37, suppose that $h_{fe} = 30$, $R_1 = 600\ \Omega$, $R_2 = 300\ \Omega$, $R_C = R_E = R_L = 100\ \Omega$, and $V_{CC} = 30$ V. Determine the Q point. Hence, find the efficiency of the amplifier.

Solution

For obtaining the Q point, let us consider the base circuit; hence,

$$V_{BE} = IR_2 - (I_{CQ} + I_{BQ})R_E$$

or $0.7 = 300I - 31 \times 100 I_{BQ}$

or $I = \dfrac{0.7 + 3100 I_{BQ}}{300}$

Also, $V_{CC} = (I + I_{BQ})R_1 + IR_2$

or $V_{CC} = I(R_1 + R_2) + I_{BQ} R_1$

or $30 = 900I + 600 I_{BQ}$

Substituting for I,

$30 = 2.1 + (9300 + 600) I_{BQ}$

or $I_{BQ} = \dfrac{30 - 2.1}{9.9} \times 10^{-3} = 2.82$ mA

Hence, the collector current,

$I_{CQ} = 30 \times 2.82 = \mathbf{84.6\ mA}$

From the collector circuit,

$$V_{CC} = I_{CQ} R_C + V_{CEQ} + (I_{CQ} + I_{BQ})R_E$$

or $30 = 84.6 \times 0.1 + V_{CEQ} + 31 \times 2.82 \times 0.1$

or $V_{CEQ} = 30 - 17.2 = \mathbf{12.8\ V}$

∴ $\dfrac{V_{CEQ}}{I_{CQ}} = \dfrac{12.8}{84.6 \times 10^{-3}} = 151$

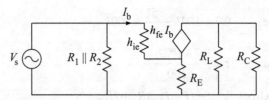

FIGURE 5.38 An equivalent circuit of Fig. 5.37.

From the a.c. equivalent circuit of Fig. 5.37 shown in Fig. 5.38,

$$R_T = R_C \| R_L + R_E$$
$$= \frac{100 \times 100}{100 + 100} + 100 = 150$$

which is approximately equal to V_{CEQ}/I_{CQ}. This shows that the Q point is approximately centered on the load line. The output power,

$$P_o = \frac{1}{2} V_{CEQ} I_{CQ} = 0.5 \times 12.8 \times 84.6 \times 10^{-3} = 0.541 \text{ W}$$

The d.c. power input,

$$P_{d.c.} = V_{CC} I_{CQ} = 30 \times 84.6 \times 10^{-3} = 2.54 \text{ W}$$

$$\therefore \quad \eta = \frac{0.541}{2.54} \times 100 = 21.3\%$$

13. For the transformer-coupled amplifier shown in Fig. 5.39, suppose that $h_{fe} = 50$, $R_1 = 200 \ \Omega$, $R_2 = 100 \ \Omega$, $R_L = 10 \ \Omega$, $R_E = 100 \ \Omega$, and $V_{CC} = 15$ V. Determine the Q point. Hence, find the efficiency of the amplifier.

Solution

From the base circuit,

$$V_{BE} = IR_2 - (I_{CQ} + I_{BQ})R_E$$

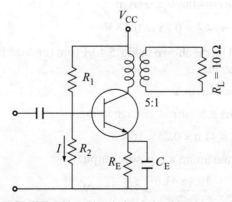

FIGURE 5.39 Circuit for Solved Problem 13.

or $\quad 0.7 = 100I - \dfrac{51}{50} \times 100 I_{CQ}$

or $\quad I = \dfrac{0.7 + 102 I_{CQ}}{100}$

Also, $V_{CC} = (I + I_{BQ})R_1 + IR_2$

or $\quad V_{CC} = I(R_1 + R_2) + I_{BQ} R_1$

or $\quad 15 = 300 I + 4 I_{CQ}$

Substituting for I,

$\quad 15 = 2.1 + (9300 + 600) I_{BQ}$

or $\quad I_{CQ} = 3.225 - 76.5 I_{CQ}$

$\quad I_{CQ} = \dfrac{3.225}{77.5} \times 10^3 = \mathbf{41.6\ mA}$

From collector circuit,

$\quad V_{CC} = V_{CEQ} + (I_{CQ} + I_{BQ}) R_E$

or $\quad 15 = V_{CEQ} + (51/50) \times 41.6 \times 0.1$

or $\quad V_{CEQ} = 15 - 4.243 = \mathbf{10.76\ V}$

Effective load resistance

$\quad R'_L = (5)^2 \times 10 = 250$

One point of the a.c. load line is the Q point and the other point of the a.c. load line is on the I_C axis and is given by

$$I_C = I_{CQ} + \dfrac{V_{CEQ}}{R'_L}$$

$\quad = 41.6 + \dfrac{10.76}{0.25}$

$\quad = 84.6\ mA$

The load line cuts the V_{CE} axis at

$\quad V_{CE} = 84.6 \times 0.25 = 21.15\ V$

The a.c. load line is shown in Fig. 5.40. From the load line it is clear that the maximum a.c. current swing,

$\quad I_{CQ} = 41.6\ mA$

and hence the a.c. output voltage swing,

$\quad V_{CQ} = 41.6 \times 0.25 = 10.4\ V$

Hence, the maximum a.c. power output

$$P_{o\ max} = \dfrac{10.4 \times 41.6}{2} = 216.32\ mW$$

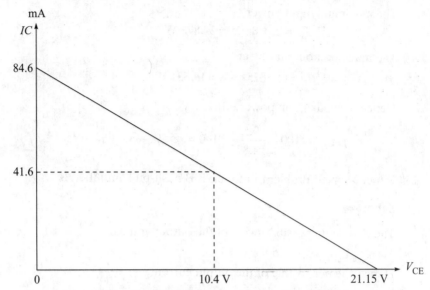

FIGURE 5.40 The a.c. load line for Figure 5.39.

The d.c. power drawn from the supply
$$P_{d.c.} = V_{CC}I_{CQ} = 15 \times 41.6 = 624 \text{ mW}$$
The percentage efficiency,

$$\eta = \frac{216.32}{624} \times 100 = \mathbf{34.67\%}$$

14. A class B amplifier with supply voltage of 25 V drives a load of 8 Ω. Determine the maximum power input, maximum power output, and maximum efficiency of the amplifier.

Solution

The maximum possible swing of the output voltage,
$$V_m = V_{CC} = 25 \text{ V}$$
Hence, the peak value of the output current
$$I_m = \frac{25}{8} = 3.125 \text{ A}$$

The average value of the current, as it is like half-wave rectified current,
$$I_{av} = \frac{I_m}{\pi} = \frac{3.125}{\pi} = 0.995 \text{ A}$$

and the r.m.s. value of current,
$$I_o = \frac{I_m}{2} = \frac{3.125}{2} = 1.5625 \text{ A}$$

The maximum input power,
$$P_{in} = V_{cc}I_{av} = 25 \times 0.995 = \mathbf{24.875 \text{ W}}$$

The maximum output power,
$$P_o = I_o^2 R_L = (1.5625)^2 \times 8 = \mathbf{19.531 \text{ W}}$$

Hence, the maximum power efficiency,
$$\eta = \frac{P_o}{P_{d.c.}} \times 100 = \frac{19.531}{24.875} \times 100 = 78.5\%$$

15. Repeat Solved Problem 14 for class B push-pull amplifier.

Solution

The maximum possible swing of the output voltage,
$$V_m = V_{CC} = 25 \text{ V}$$

Hence, the peak value of the output current,
$$I_m = \frac{25}{8} = 3.125 \text{ A}$$

The average value of the current, as it is like full-wave rectified current,
$$I_{av} = \frac{2I_m}{\pi} = \frac{2 \times 3.125}{\pi} = 1.99 \text{ A}$$

and the r.m.s. value of current,
$$I_o = \frac{I_m}{\sqrt{2}} = \frac{3.125}{\sqrt{2}} = 2.21 \text{ A}$$

The maximum input power,
$$P_{in} = V_{cc}I_{av} = 25 \times 1.99 = \mathbf{49.75 \text{ W}}$$

The maximum output power,
$$P_o = I_o^2 R_L = (2.21)^2 \times 8 = \mathbf{39.07 \text{ W}}$$

Hence, the maximum power efficiency,
$$\eta = \frac{P_o}{P_{d.c.}} \times 100 = \frac{39.07}{49.75} \times 100 = \mathbf{78.5\%}$$

5.7 EXERCISES

1. For a common-mode input of 1 V, the two outputs of a differential amplifier are 8 mV and 12 mV. The differential output of the amplifier for a differential input of 1 mV is 75 mV. Calculate the CMRR of the differential amplifier.

2. For the amplifier of Fig. 5.1, the two sets of inputs are:

   First set: $v_1 = 75$ mV and $v_2 = -25$ mV

   Second set: $v_1 = 50$ mV and $v_2 = -50$ mV

   (a) For CMRR = 1000, calculate the percentage difference in output voltages obtained for the two sets of inputs. (b) Repeat (a) for CMRR = 10000.

3. For differential amplifier in Fig. 5.1, $h_{fe} = 50$, $h_{ie} = 1.1$ kΩ, $R_C = R_E = 10$ kΩ, and $R_s = 1$ kΩ. Determine CMRR in decibel.

4. For an OP AMP shown in Fig. 4.31, $R = 1$ kΩ, $R' = 15$ kΩ, $A = 10^5$, and $V_s = 1$ V. (a) Find $v_o, v_1, I_1,$ and I_2. (b) Find the power absorbed by each resistor, the source, and the amplifier.

5. The OP AMP shown in Fig. 5.32 has $R_{in} = \infty$, $R_o = 0$, and finite gain A. Find $A_f = i_o/v_{in}$.

6. The OP AMP shown in Fig. 5.34 has finite gain A, finite input resistance R_{in}, and $R_o = 0$. Show that when $R_{in} >> R_1$, then

$$R_{if} = \frac{v_{in}}{i_{in}} = \left(1 + \frac{AR_1}{R_1 + R_2}\right)$$

7. In the circuit of Fig. 5.33, $R_1 = 1.2$ kΩ, $R_2 = 11$ kΩ, $V_{s1} = 1.5$ V, and $V_{s2} = 3$ V. Find the output voltage.

8. For the OP AMP circuit shown in Fig. 5.31, $R_{in} = \infty$, $R_o = 0$, and $A = 10^6$. The gain with feedback, $A_f = 250$. (a) Determine R when $R' = 30$ kΩ. (b) Determine R' when $R = 1.2$ kΩ.

9. The OP AMP shown in Fig. 5.35(a) has $R_{in} = \infty$, $R_o = 0$, and finite gain A. Find $R_{of} = v_o/i_o$ and $A_f = i_o/v_{in}$.

10. Using an OP AMP, design a circuit to implement

$$V_o = 3v_1 + 4v_2$$

11. An amplifier supplies an a.c. power of 1.2 W to a load of 8 kΩ. The zero-signal d.c. collector current is 35 mA and the d.c. collector current with a signal is 45 mA. Determine the percentage second harmonic distortion. Let the dynamic curve of the transistor be given by a quadratic equation.

12. For the transformer-coupled amplifier shown in Fig. 5.41, suppose that $h_{fe} = 20$, $R_B = 1$ kΩ, $R_C = 25$ Ω, and $V_{CC} = 20$ V. Determine the Q point. Hence, find the efficiency of the amplifier.

13. For the amplifier shown in Fig. 5.39, suppose that $h_{fe} = 50$, $R_1 = 200$ Ω, $R_2 = 100$ Ω, $R_L = 8$ Ω, $R_E = 100$ Ω, and $V_{CC} = 12$V. Determine the Q point. Hence, find the efficiency of the amplifier.

14. A class A amplifier drives an 8-Ω speaker through a 3:1 transformer. Using a power supply of 30 V, the amplifier delivers 2 W to the load. Calculate (a) a.c. power across

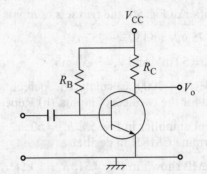

FIGURE 5.41 Circuit for question 12.

transformer primary, (b) a.c. voltage drop across the load, (c) a.c. voltage across transformer primary, (d) the r.m.s. value of load and primary currents, and (e) the efficiency of the circuit.

15. A class B push-pull amplifier provides a 20-V peak signal to a 12-Ω load. Using power supply of 25 V, determine the input power, output power, and circuit efficiency.

6

WAVE GENERATORS

Outline

6.1   Oscillators   250
6.2   Multivibrators   260
6.3   Blocking Oscillators   265
6.4   Square Wave Generators   270
6.5   Pulse and Pulse Wave Generators   273
6.6   Signal Generators   277
6.7   Function Generators   280
6.8   Sweep-Frequency Generator   281
6.9   Review Questions   283
6.10   Solved Problems   284
6.11   Exercises   291

6.1 OSCILLATORS

Consider the feedback system shown in Fig. 6.1. The feedback gain is given by

$$A_f = \frac{A}{1+\beta A} \tag{6.1}$$

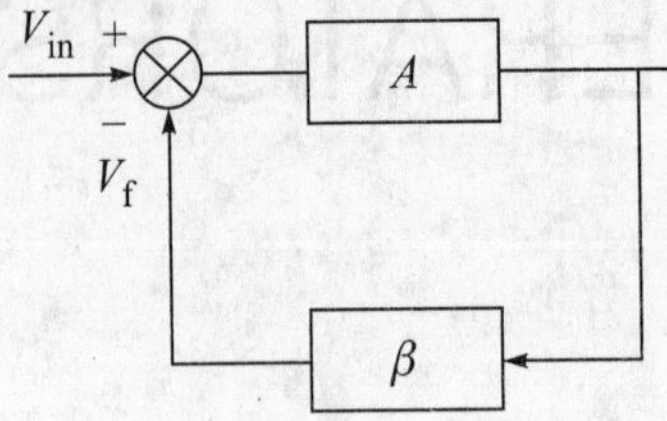

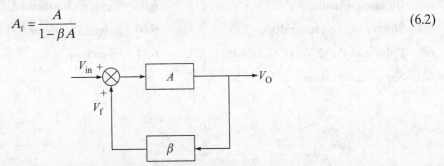

FIGURE 6.1 A feedback system.

where A is the forward path gain and β is the feedback gain. For $\beta A > 0$, the feedback gain $A_f < A$. As discussed earlier, this feedback is negative (or degenerative) feedback. A number of advantages such as stability in voltage gain, higher input resistance, and lower output resistance were listed for this type of feedback in Chapter 4. In contrast, positive feedback, as shown in Fig. 6.2, has feedback gain $A_f > A$ and is given by

$$A_f = \frac{A}{1-\beta A} \tag{6.2}$$

FIGURE 6.2 A positive feedback system.

The feedback loop gain βA can be adjusted to unity at a particular frequency as A and β are frequency dependent. At this frequency, therefore, the gain of the feedback system, A_f, is infinite. Infinity gain theoretically means that there is an output even with zero input at a particular frequency. Under this condition the system is called an *oscillator*. Condition for a positive-feedback system to work as an oscillator can be stated as follows:

The frequency of oscillation of an oscillator is the frequency for which the loop gain is unity and the phase shift is zero.

The condition mentioned above, however, is difficult to maintain in practice. Even if the condition is satisfied initially, the loop gain $|\beta A|$ becomes either less or greater than unity with lapse of time. The reasons for deviation may be variations in characteristics of the circuit components, more importantly of transistors, with time, temperature, voltage, etc. For $|\beta A| < 1$, the oscillation simply stops, and for $|\beta A| > 1$, the amplitude of oscillation is limited by the onset of nonlinearity. Therefore, in a practical oscillator, $|\beta A|$ is always adjusted to somewhat greater (about 0.5%) than unity. This ensures $|\beta A|$ not to fall below unity even after incidental variation in the characteristics of transistors and circuit parameters. As the condition for oscillation does not impose any restriction on the waveform, it does not require to be sinusoidal. When the signal generated is a sine wave, the device is called an *oscillator*. Oscillators can be classified as follows:

1. Phase-shift oscillator.
2. Resonant-circuit oscillator.
3. Wien-bridge oscillator.
4. Crystal oscillator.

The various signal-generating devices are given different names depending on their characteristics and uses. An oscillator with an additional capacity of amplitude modulation of the output signal and a wide range of tuning is called a *signal generator*. A device delivering a choice of different waveforms with adjustable frequencies over a wide range is called a *function generator*. The oscillator is the basic element for all of them.

6.1.1 Phase-Shift Oscillator

For the feedback system shown in Fig. 6.2, if the amplifier gain A is a constant and does not depend on frequency, one way of getting a unity feedback gain is to use a frequency-dependent feedback network having transfer function β, where $|\beta| = 1/A$ and the phase angle is 180°. If the loading effect of the feedback network on the amplifier is negligible, the amplifier introduces a phase shift of 180° and the feedback network introduces an additional phase shift. If the phase shift of the feedback network is 180° for just a single frequency, then at this frequency the total loop phase shift is exactly zero, that is, $\angle A\beta = 0$. If $|A\beta|$ is set equal to unity, the feedback system, with $X_s = 0$, oscillates at this particular frequency.

A simple lead or lag network, consisting of one R and one C, produces a phase shift of less than 90° for finite frequencies. Therefore, at least three simple RC stages are required for a phase shift of 180°. Consider an RC phase-shift network shown in Fig. 6.3.

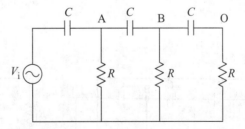

FIGURE 6.3 An RC phase-shift network.

Applying KCL to nodes A, B and O,

$$\left(2j\omega C + \frac{1}{R}\right)V_A - j\omega CV_B = j\omega CV_i \tag{6.3}$$

$$-j\omega CV_A + \left(2j\omega C + \frac{1}{R}\right)V_B - j\omega CV_o = 0 \tag{6.4}$$

$$-j\omega C V_B + \left(j\omega C + \frac{1}{R}\right) V_O = 0 \tag{6.5}$$

Manipulating the above equations, the transfer function

$$\frac{V_o}{V_i} = \beta(j\omega) = \frac{\omega^3 C^3 R^3}{\omega C R(\omega^2 C^2 R^2 - 5) + j(1 - 6\omega^2 C^2 R^2)} \tag{6.6}$$

The angle of shift of the transfer function $\beta(j\omega)$ is $180°$ when the imaginary part of the denominator is zero, that is,

$$1 - 6\omega^2 C^2 R^2 = 0$$

and hence the corresponding frequency,

$$\omega_o = \frac{1}{CR\sqrt{6}} \tag{6.7}$$

Substituting this frequency, that is, $\omega = \omega_o$, in Eq. (6.6), we get

$$\beta(j\omega) = -\frac{1}{29} \tag{6.8}$$

In other words, for the frequency ω_o given by Eq. (6.7), the gain of the transfer function,

$$|\beta| = \frac{1}{29} \tag{6.9}$$

and the phase shift,

$$\theta = 180° \tag{6.10}$$

Now, this circuit can be used as a feedback network to produce oscillation as shown in Figs. 6.4 and 6.5. The amplifier may be a field-effect transistor (FET) amplifier of Fig. 6.4 or a transistor amplifier as shown in Fig. 6.5.

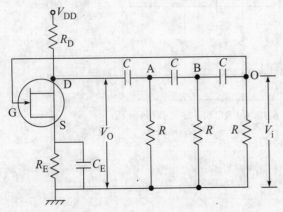

FIGURE 6.4 An FET amplifier with phase-shift network.

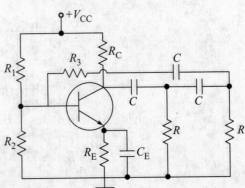

FIGURE 6.5 A transistor amplifier with phase-shift network.

For the FET amplifier of Fig. 6.4, assuming $R \gg R_D$ and neglecting the a.c. output resistance r_d of the FET, the midband amplifier gain $A = -g_m R_D$. Therefore, at the oscillation frequency $\omega_o = 1/R_C\sqrt{6}$ and for the loop gain to be unity,

$$\beta A = -\frac{1}{29} \times -g_m R_D = 1$$

Hence, $$g_m = \frac{29}{R_D} \tag{6.11}$$

Thus, to get oscillation, we must have $g_m \geq 29/R_D$.

In the amplifier of Fig. 6.5, the transistor does not have high input resistance, unlike the FET of Fig. 6.4. Since the feedback network should be loaded with a resistance equal to R, the resistance R_3 should be chosen such that the input resistance R_{in} be equal to R. The input resistance,

$$R_{in} = R_3 + \frac{R_1 R_2}{R_1 + R_2} \| h_{ie} = R$$

$$\therefore \quad R_3 = R - \frac{R_1 R_2}{R_1 + R_2} \| h_{ie}$$

If R_C is taken into consideration, then the oscillation frequency in Eq. (6.7) is modified as

$$\omega_o = \frac{1}{C\sqrt{6R^2 + 4R_c R}} \tag{6.12}$$

The phase-shift oscillator is suitable for the frequency range from a few hertz to several hundred kilohertz. For a large frequency variation, the three capacitors are usually varied simultaneously. This keeps the magnitude of β and βA constant.

6.1.2 Resonant-Circuit Oscillator

For determining the oscillation frequency, a resonant (LC) circuit is used in this type of oscillators. The resonant circuit constitutes the load impedance of the amplifier. The output developed across the LC circuit is fed back inductively to the input of the amplifier. Two circuits of this type of oscillators, one using an FET and another a transistor, are shown in Fig. 6.6(a) and (b), respectively. The drain of the FET in Fig. 6.6(a) has a tuned capacitor, whereas the collector of the transistor in Fig. 6.6(b) consists of a tuned load.

Under the assumption that the inductive coil is loss free, the resonant frequency,

$$\omega_r = \frac{1}{\sqrt{LC}} \tag{6.13}$$

and the impedance of the resonant circuit is large and purely resistive. In this case the output voltage is precisely 180° out of phase with the applied input voltage. The direction of the secondary winding is adjusted to give a phase shift of 180°. For this, the secondary winding is assumed to be unloaded. The total loop phase shift is, thus, exactly zero. This satisfies the phase-shift condition for oscillation. Since the transformer is considered to be unloaded, the ratio of amplitudes of secondary to primary voltages is M/L_1,

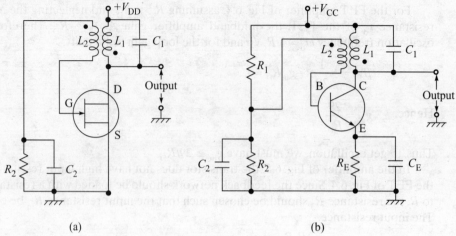

(a)   (b)

FIGURE 6.6 (a) A tuned drain FET oscillator and (b) a tuned collector transistor oscillator.

M being the mutual inductance. Hence, $\beta = M/L_1$, and as $-A\beta = 1$ for oscillation, the amplifier gain $A = L_1/M$. The oscillator of this type can oscillate at a very high frequency covering the frequency range of 100 kHz to 500 MHz.

Several oscillators of resonant type, represented in a general form, are depicted in Fig. 6.7(a). The active device may be a transistor, an FET, or an operational amplifier. Let us assume that the amplifier has a negative gain A_v, infinite input resistance, and output resistance R_o. A linear equivalent circuit of Fig. 6.7(a) is given in Fig. 6.7(b).

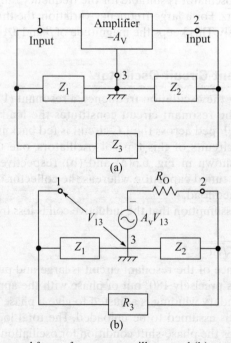

FIGURE 6.7 (a) A general form of resonant oscillator and (b) an equivalent circuit of (a).

From Fig. 6.7(b), the load impedance,

$$Z_L = \frac{Z_2(Z_1 + Z_3)}{Z_1 + Z_2 + Z_3}$$

The gain without feedback,

$$A = \frac{-A_v Z_L}{Z_L + R_o} \tag{6.14}$$

and the feedback gain,

$$\beta = \frac{-Z_1}{Z_1 + Z_3} \tag{6.15}$$

Hence, the loop gain,

$$-A\beta = \frac{-A_v Z_1 Z_2}{R_o(Z_1 + Z_2 + Z_3) + Z_2(Z_1 + Z_3)} \tag{6.16}$$

If the impedances Z_1, Z_2, and Z_3 are considered pure reactive (inductive or capacitive), then $Z_1 = jX_1$, $Z_2 = jX_2$, and $Z_3 = jX_3$. Hence, Eq. (6.16) can be written as

$$-A\beta = \frac{A_v X_1 X_2}{jR_o(X_1 + X_2 + X_3) - X_2(X_1 + X_3)} \tag{6.17}$$

For the loop phase shift to be zero,

$$X_1 + X_2 + X_3 = 0 \tag{6.18}$$

and

$$-A\beta = \frac{-A_v X_1}{X_1 + X_3} = \frac{A_v X_1}{X_2} \tag{6.19}$$

For $-A\beta = 1$, X_1 and X_2 must be of the same sign, and X_3 must be of the opposite sign. This means that both X_1 and X_2 must be either inductive or capacitive, and X_3 must be either capacitive or inductive.

Hartley Oscillator

In this oscillator, X_1 and X_2 are inductive and X_3 is capacitive. The oscillator using an FET amplifier is shown in Fig. 6.8(a). A transistor is used as the active device in Fig. 6.8(b). Hence, from Eq. (6.18),

$$\omega L_1 + \omega L_2 = \frac{1}{\omega C}$$

or

$$\omega = \omega_o = \frac{1}{\sqrt{LC}}$$

or

$$f_o = \frac{1}{2\pi\sqrt{LC}}, \text{ where } (L = L_1 + L_2) \tag{6.20}$$

Colpitts Oscillator

Unlike Hartley oscillator, in Colpitts oscillator X_1 and X_2 are capacitive and X_3 is inductive. The oscillator using an FET as the active device is shown in Fig. 6.9(a). A transistor

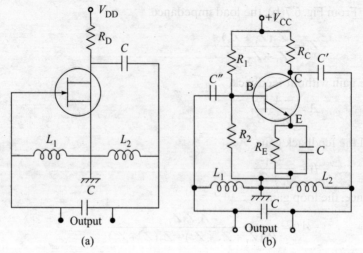

FIGURE 6.8 The Hartley oscillator using (a) an FET and (b) a transistor.

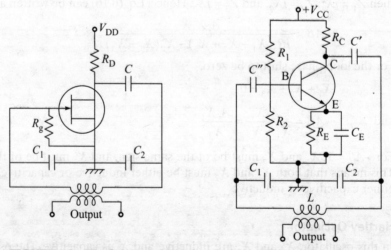

FIGURE 6.9 The Colpitts oscillator using (a) an FET and (b) a transistor.

is used as the active device in Fig. 6.9(b). Hence, from Eq. (6.18),

$$\omega L = \frac{1}{\omega C_1} + \frac{1}{\omega C_2}$$

or

$$\omega = \omega_o = \frac{1}{\sqrt{LC}}$$

or

$$f_o = \frac{1}{2\pi\sqrt{LC}} \text{ where } \left(C = \frac{C_1 C_2}{C_1 + C_2} \right) \quad (6.21)$$

Capacitors C' and C'' are blocking capacitors and have very low reactance at oscillation frequency and provide very high reactance to d.c. currents.

6.1.3 Wien-Bridge Oscillator

This oscillator consists of an ideal operational amplifier, resistances R_1 and R_2, and a feedback network as shown in Fig. 6.10(a). The feedback network consists of R and C. An equivalent circuit of Fig. 6.10(a) is drawn in Fig. 6.10(b). The gain of the amplifier without feedback,

$$A = \frac{V_o}{V_2} = \frac{R_1 + R_2}{R_1} = 1 + \frac{R_2}{R_1} \tag{6.22}$$

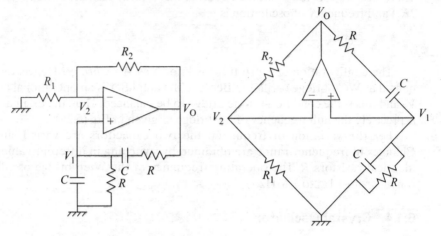

FIGURE 6.10 (a) The Wien-bridge oscillator and (b) an equivalent circuit of (a).

The impedance of parallel RC circuit,

$$Z_1 = \frac{R(1/j\omega C)}{R + (1/j\omega C)} = \frac{R}{1 + j\omega CR}$$

and the impedance of series RC circuit,

$$Z_2 = R + \frac{1}{j\omega C} = \frac{1 + j\omega CR}{j\omega C}$$

Hence, the voltage transfer function of the feedback network,

$$\beta(j\omega) = \frac{V_1}{V_o} = \frac{Z_1}{Z_1 + Z_2} = \frac{\dfrac{R}{1 + j\omega CR}}{\dfrac{R}{1 + j\omega CR} + \dfrac{1 + j\omega CR}{j\omega C}}$$

$$= \frac{j\omega CR}{1 - \omega^2 C^2 R^2 + j3\omega CR} \tag{6.23}$$

Since the gain A in Eq. (6.22) has zero phase shift, β must also have zero phase shift. This condition, which is essential for oscillation, is achieved at

$$\omega = \omega_o = \frac{1}{RC} \tag{6.24}$$

Thus, the feedback gain (in magnitude) is 1/3 and the phase angle is zero. For the loop gain to be unity

$$\beta A = \frac{A}{3} = \frac{1}{3}\left(1 + \frac{R_2}{R_1}\right) = 1$$

To satisfy the above condition,

$$R_2 = 2R_1 \tag{6.25}$$

Therefore, we conclude that the circuit of Fig. 6.10(a) oscillates when $R_2 = 2R_1$ (or $R_2 > 2R_1$) and frequency of oscillation is

$$\omega_o = \frac{1}{RC} \tag{6.26}$$

The configuration of Fig. 6.10(b) is known as *Wien bridge*. Hence, the oscillator is named as Wien-bridge oscillator. Because the amplifier is an operational amplifier, $V_1 = V_2$ and this is the condition for the bridge to be balanced. Thus, the oscillation frequency is precisely the null frequency of the bridge as given in Eq. (6.26).

For the variation of frequency, the two capacitors are varied simultaneously. Changes in frequency range are obtained by switching in different values for the two identical resistors R. The operating frequency of the Wien-bridge oscillator roughly ranges from 1 Hz to 1 MHz.

6.1.4 Crystal Oscillator

The frequency stability of the oscillators discussed earlier is usually poor because of the variations in temperature, humidity, and parameters of the transistor and the circuit components. Although RC oscillators can achieve frequency stabilities of about 0.1%, frequency stabilities of 0.01% are more typical of LC oscillators. In operations such as radio transmitters, it is essential that frequency of oscillation of the master oscillator be highly stable. For such a frequency, stable oscillator, a piezoelectric quartz crystal, may be used in place of the tuned circuit in the oscillator. Crystal oscillators can have frequency stabilities of 0.001% or even 0.0001%. The crystal oscillator is used as the master oscillator in radio transmitters, and as a local oscillator in radio receivers and telephones. Such oscillators are also used in digital wristwatches.

Quartz crystal has the property that if it is mechanically vibrated, it produces an a.c. voltage. Conversely, if an a.c. voltage is applied across the crystal, it vibrates at the frequency of applied voltage. This phenomenon is known as piezoelectric effect and the crystal is known as piezoelectric crystal. Thus, the quartz crystal forms an electromechanical system. The resonance frequency of this system depends on the dimensions of the crystal, orientation of the surfaces, and its mounting. The crystal is mounted between two plates. Plates have electrodes for applying the voltage.

The crystal mounted between two plates is shown in Fig. 6.11(a), whereas an equivalent electrical circuit is shown in Fig. 6.11(b). The electrical equivalent of the vibrational characteristics (mass, compliance, and friction) of the crystal is represented by inductance (L), capacitance (C), and resistance (R), respectively. These elements are connected in series. Further, since the quartz crystal is mounted between two plates, it forms an electrostatic capacitance (C') with crystal as the dielectric. The capacitance (C') is

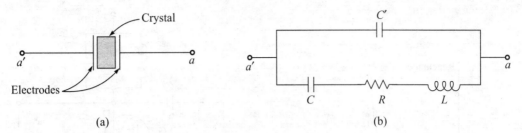

FIGURE 6.11 (a) A mounted quartz crystal and (b) its equivalent electrical circuit.

very much larger than C. Typical values for these elements for a 466-kHz crystal are $L = 3.1$ H, $C = 0.04$ pF, $R = 4.4$ kΩ, and $C' = 6$ pF. Corresponding Q is 2400.

For the circuit shown in Fig. 6.11(b), the resonance frequency of the series RLC circuit is given by

$$\omega_s = \frac{1}{\sqrt{LC}} \tag{6.27}$$

There is, however, an additional parallel resonance frequency due to parallel connection of C'. Since R is relatively small, it may be neglected for determining parallel resonance frequency. The admittance of the equivalent circuit,

$$Y = j\omega C' + \frac{1}{j\omega L + (1/j\omega C)}$$

After manipulating and arranging,

$$Y = \frac{j(\omega^2 LC' - C'/C - 1)}{\omega L - 1/\omega C} \tag{6.28}$$

At the parallel resonance frequency $Y = 0$, hence, for $\omega = \omega_p$, we must have

$$\omega_p^2 LC' - C'/C - 1 = 0$$

or

$$\omega_p = \sqrt{\frac{1}{L}\left(\frac{1}{C'} + \frac{1}{C}\right)} \tag{6.29}$$

As discussed earlier, $C' \gg C$; hence,

$$\frac{1}{C'} + \frac{1}{C} \approx \frac{1}{C}$$

The parallel resonance frequency, therefore, is given approximately by

$$\omega_p = \frac{1}{\sqrt{LC}} = \omega_s \tag{6.30}$$

This means that the parallel resonance frequency ω_p and the series resonance frequency ω_s are very near to each other. If the operating frequency ω lies between ω_p and ω_s ($\omega_s < \omega < \omega_p$), the reactance is inductive and outside this range it is capacitive. The variation of reactance is shown in Fig. 6.12.

For excellent frequency stability of oscillation, a piezoelectric quartz crystal may be used in place of the tuned circuit in the oscillator. In the circuit of Fig. 6.7, if a crystal is

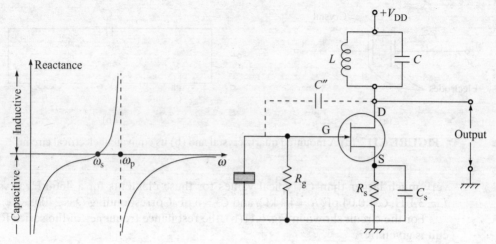

FIGURE 6.12 Variation of reactance with ω.

FIGURE 6.13 A crystal oscillator.

used for Z_1, a tuned LC combination for Z_2, and a capacitance C'' for Z_3, the resulting circuit is a crystal oscillator. This crystal oscillator is shown in Fig. 6.13. The active device is an FET. The capacitance C'' is equivalent to the sum of capacitances from drain to gate and the stray wiring capacitance.

To satisfy Eq. (6.20), the crystal reactance as well as that of the LC network in the crystal oscillator must be inductive. At the same time for the loop gain to be greater than unity, the crystal reactance can be too small. Hence, the oscillation frequency lies between ω_p and ω_s, but closer to ω_p. As $\omega_s \approx \omega_p$, the oscillator frequency is determined essentially by the crystal and not by other circuit parameters.

6.2 MULTIVIBRATORS

There are two types of multivibrators, namely, (1) *astable multivibrator* and (2) *monostable multivibrator*. The monostable multivibrator has one permanently stable state and one quasi-stable state. The astable multivibrator has both states quasi-stable. In monostable circuit a triggering signal is needed to transit from the stable state to the quasi-stable state. It remains in quasi-stable state for a long time in comparison with the transition time between states. However, it returns itself from the quasi-stable state to the stable state and no external signal is required for the reverse transition. For astable circuit, no external signal is required for transition from one state to another. It makes successive transitions from one quasi-stable state to another quasi-stable state.

Both astable and monostable circuits are extensively used in pulse circuits. The basic application of monostable circuit is where we need a fixed time interval and abrupt discontinuity in a voltage waveform. The astable circuit is an oscillator, as it is used as a square wave generator. Since it does not require a triggering signal, it is itself a basic source of fast waveforms.

6.2.1 Astable Multivibrator

An astable (free-running) multivibrator using npn transistor is shown in Fig. 6.14. A capacitance coupling is used between two stages. Neither transistor can remain permanently cutoff. The multivibrator circuit has two quasi-stable states and it makes periodic transitions between the two states. The output of the second-stage RC-coupled amplifier using transistor T_2 is fed back via C_1 to the input (base) of the first-stage transistor T_1. Similarly, the output of the first transistor T_1 is fed back via C_2 to the input (base) of the second-stage transistor T_2. In a symmetrical circuit (i.e., $R_1C_1 = R_2C_2$) the output waveform is a symmetrical square wave. When the time constant R_1C_1 is larger than the time constant R_2C_2, the output is a pulse train because the off time of T_1 is larger than the off time of T_2. The time period of the square wave is determined by the time constant of the symmetrical circuit. The time constant, when $R_1 = R_2 = R$ and $C_1 = C_2 = C$, is given by

$$T = 1.38 RC \tag{6.31}$$

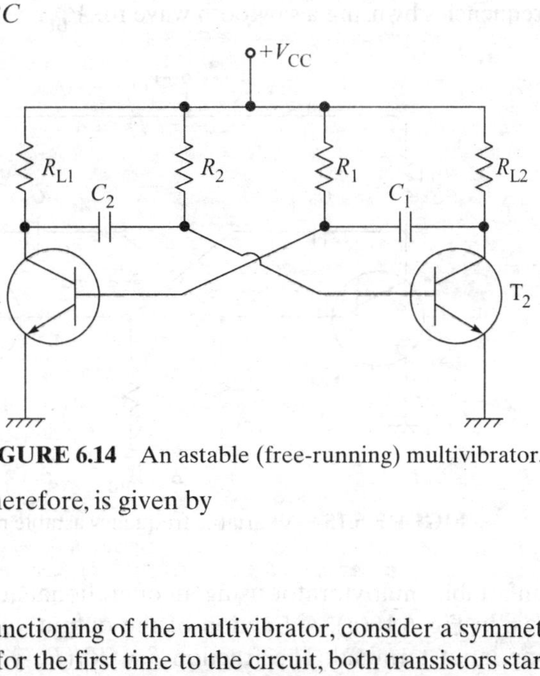

FIGURE 6.14 An astable (free-running) multivibrator.

and the frequency, therefore, is given by

$$f = \frac{0.7}{RC} \tag{6.32}$$

To explain the functioning of the multivibrator, consider a symmetrical circuit. When the power is applied for the first time to the circuit, both transistors start conducting. Considering small differences in the operating characteristics of the two transistors, one of the transistors (say T_1) conducts slightly more than T_2. Hence, the output voltage of T_1 falls more rapidly than that of T_2. This decrease in the output voltage of T_1 is applied to the R_2C_2 network. This negative change of voltage is applied to the input (base) of T_2 as the charge on a capacitor cannot change instantaneously. This decreases the output current of T_2, and hence increases the output voltage of T_2. This increase in the output voltage of T_2 is applied to the base of T_1 through R_1C_1 network. This increases the output current of T_1 even more heavily. The regenerative action continues until T_1 is saturated and T_2 is entirely cutoff. Since T_2 is cutoff, its output voltage is V_{CC} and C_1 charges to V_{CC} throw the low-resistance path between the base and emitter of T_1. With T_1 completely saturated its output voltage goes down to approximately zero. As the charge on C_2 cannot change instantaneously, the base voltage of T_2 is at $-V_{CC}$ bringing T_2 deep into cutoff.

Now, capacitor C_2 starts discharging through R_2 exponentially. When it gets discharged completely, it begins to charge up to V_{CC} with C_2 and R_2 junction at V_{CC} and output terminal of T_1 at zero. This suddenly places forward bias on T_2 forcing it to conduct. The output current of T_2 causes a fall in the output voltage. The negative change in voltage is fed back to the base of T_1, which in turn gets out of saturation. The regenerative action continues till T_1 is cutoff and T_2 is saturated. This completes a full cycle of operation. When T_2 is cutoff, the output voltage of the multivibrator is V_{CC}, and it is zero when T_2 is saturated.

For varying the frequency of the square wave, the variable capacitors are ganged together. The output frequency, then, can be changed by changing the value of C. Another method of varying the output frequency is as shown in Fig. 6.15. The base resistors in this circuit are returned to a separate power source, V_{BB}. The frequency can be changed by changing V_{BB}. In fact, the circuit of Fig. 6.15 can be swept through a range of different frequencies by using a sawtooth wave for V_{BB}.

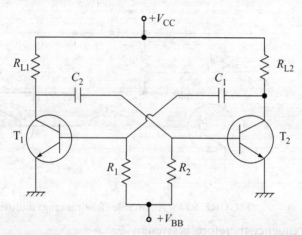

FIGURE 6.15 A variable frequency astable multivibrator.

Now, an astable multivibrator using an operational amplifier is discussed. The circuit is shown in Fig. 6.16(a). A fraction of the output voltage βv_o is feedback to the noninverting input terminal. The fraction $\beta = R_1/(R_1 + R_2)$. Hence, the differential input v_{in} is given by

$$v_{in} = v_c - \beta v_o \tag{6.33}$$

Assume that v_{in} is negative (i.e., $v_c < \beta v_o$) such that $v_o = +V_{CC}$. Now, the capacitor C charges towards V_{CC} through R with time constant RC. Hence, eventually the capacitor voltage v_c becomes larger than βv_o. It means that v_{in} becomes positive. When v_{in} is positive (i.e., $v_c > \beta v_o$), $v_o = -V_{CC}$, and hence the noninverting voltage becomes $-\beta v_o$. This means that the capacitor tends to charge towards $-V_{CC}$, again, with time constant RC. Therefore, eventually, v_c becomes more negative than the noninverting voltage. This input voltage v_{in} becomes negative again. The resulting output $v_o = +V_{CC}$ again, putting the circuit back into the originally assumed state. The process repeats and results in an output waveform as shown in Fig. 6.16(b).

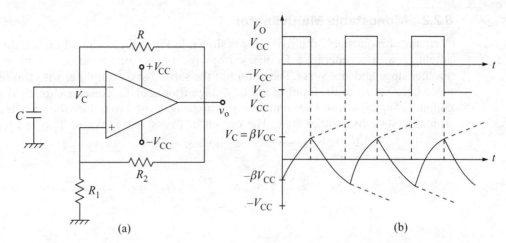

FIGURE 6.16 (a) An astable multivibrator and (b) associated waveforms.

When we want that the amplitude of the square wave should have a value less than V_{CC}, the output v_o is shunted to ground by two Zener diodes back-to-back and a resistance R_3 in series with the output terminal. In this case, the output square wave fluctuates between $+V_z$ to $-V_z$, where V_z is the break down voltage of the Zener diode.

This type of multivibrators, also called relaxation oscillators, are useful at lower frequencies, from 10 Hz to 10 kHz. At higher frequencies, the output is not ideally square wave due to the slow rate of the operational amplifier. As shown in Fig. 6.16(b), the output is ideally a square wave, whereas the capacitor voltage is a repetitive exponentially increasing and decreasing wave. The capacitor voltage may be approximately a triangular wave if a large capacitor is taken. As initial portion of the exponential curve approximates a straight line, the capacitor voltage may also be approximated as a triangular wave by taking small β. Thus, an astable multivibrator can produce a triangular as well as a square wave.

To derive an expression for the time period of the wave, let us consider the first half period from $t = 0$ to $t = T/2$. During this period, capacitor charges from $-\beta V_{CC}$ to $+\beta V_{CC}$ with a time constant of RC. The capacitor voltage is given by

$$v_c = V_{CC}\left(1 - (1+\beta)\, e^{-t/RC}\right)$$

Since at $t = T/2$, $v_c = \beta V_{CC}$,

$$\beta V_{CC} = V_{CC}\left(1 - (1+\beta)\, e^{-T/2RC}\right)$$

or

$$(1+\beta)e^{-T/2RC} = 1 - \beta$$

$\therefore$

$$T = 2RC\ \ln\frac{1+\beta}{1-\beta} = 2RC\ \ln\left(\frac{2R_1}{R_2} + 1\right) \qquad (6.34)$$

6.2.2 Monostable Multivibrator

A circuit of monostable multivibrator is shown in Fig. 6.17. Essentially, it is a two-stage amplifier with regenerative feedback from the output of one stage to the input of another stage and vice versa. The output of the second-stage amplifier with transistor T_2 is fed back via R_1 to the input of the first-stage transistor T_1. The capacitor C_1 is a small commutating capacitor. The commutating capacitor is used to reduce the transition time from one state to another state. The output of first-stage transistor T_1 is fed back via capacitor C_2 to the input of second-stage transistor T_2.

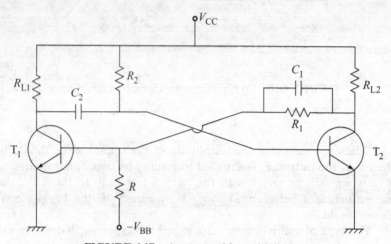

FIGURE 6.17 A monostable multivibrator.

Let us assume that the circuit parameters are so adjusted that it is in stable state. Under this condition transistor T_1 is cutoff and T_2 is saturated. Now, a negative trigger is applied to the input of T_2 or output of T_1. The single trigger applied to the base of T_2 is strong enough to bring T_2 completely below cutoff. The voltage at the output of T_2 rises to approximately equal to V_{CC}. Because of coupling between the output of T_2 and input of T_1, the first-stage transistor T_1 starts conducting. The transistor T_1 may be driven to saturation or it may operate within its active region. In either case, a current I_1 flows through resistance R_{L1}. The output voltage of the first stage drops abruptly by an amount $I_1 R_{L1}$. The input voltage of the second stage drops by the same amount because the capacitor voltage cannot change instantaneously. The multivibrator is now in the quasi-stable state.

The multivibrator remains in this quasi-stable state for only a finite time T because the base of T_2 is connected to V_{CC} via resistance R_2. Hence, the input voltage of T_2 rises and when it passes the cut-in voltage V_γ of T_2, a regenerative action takes place, turning T_1 off. Thus, the multivibrator returns to its initial stable state.

The delay time T may be varied either by changing the time constant $R_2 C_2$ or by adjusting the current I_1. The current I_1 is controlled by the base input current. This input current depends on V_{BB}. Hence, T may be varied by variation of V_{BB}.

A monostable multivibrator circuit using operational amplifier is obtained by modifying the circuit of the astable multivibrator of Fig. 6.16(a), as shown in Fig. 6.18(a). In the circuit of Fig. 6.18(a), a clamping diode, D_1, is added across the capacitor C and a

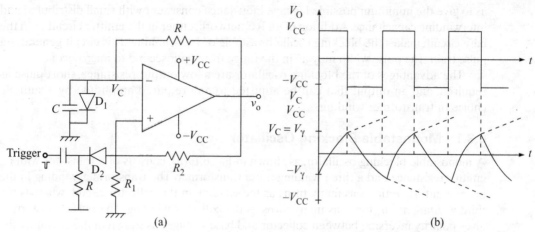

FIGURE 6.18 (a) A monostable multivibrator and (b) associated waveforms.

narrow negative triggering pulse is introduced through the diode, D_2, to the noninverting terminal. Now, assume that the circuit is in the stable state with the output voltage $v_o = +V_{CC}$. The capacitor cannot charge to βV_{CC} as in Fig. 6.16(a) because of the clamping diode. The capacitor is clamped at

$$V_c = V_\gamma \approx 0.6 \text{ V} \tag{6.35}$$

If the negative triggering pulse amplitude is greater than $\beta V_{CC} - V_\gamma$, then v_{in} becomes positive and the output switches to $-V_{CC}$. The capacitor, now, charges through R towards $-V_{CC}$ as D_1 is reverse biased. When v_c becomes more negative than $-\beta V_{CC}$, the differential input v_{in} is negative. Hence, the output switches to $+V_{CC}$. The capacitor, now, charges through R towards $+V_{CC}$ until v_c reaches V_γ and C is clamped again at V_γ. The output and capacitor voltage waveforms are shown in Fig. 6.18(b).

The pulse width T is given by

$$T = RC \ln \frac{1+(V_\gamma / V_{CC})}{1-\beta} \tag{6.36}$$

As $V_{CC} \gg V_\gamma$ and if $R_1 = R_2$ (i.e., $\beta = 0.5$), then $T = 0.69RC$.

6.3 BLOCKING OSCILLATORS

Blocking oscillator is a circuit in which the output of the active device is coupled to the input through a transformer, instead of an RC circuit. The coupling transformer is called a pulse transformer. The relative winding polarities are so chosen that the feedback is regenerative. The circuit can be made to generate a single pulse or a pulse train. Former circuit is a *monostable circuit* and later is an *astable circuit*. The blocking oscillator, therefore, consists of essentially an amplifier and a pulse transformer. The pulse transformer provides coupling between the input and the output of the amplifier. The pulse transformer is normally wound on a high permeability magnetic core. The winding is made in such a manner as to give the minimum possible leakage inductance consistent with small distributed and interwinding capacitance. Addition of an RC network, either in the emitter circuit or in the base circuit, makes the blocking oscillator astable or free-running. This circuit generates a pulse train. The pulse width may lie in the range from nanosecond to microsecond.

The advantages of the blocking oscillator are a low output resistance, short pulse as required, and an output that can be adjusted to any required amplitude by a suitable choice of transformer winding.

6.3.1 Monostable Blocking Oscillator

A monostable blocking oscillator, as shown in Fig. 6.19(a), consists of a transistor with an emitter resistance and a three winding pulse transformer. The transformer winding in the base circuit has n times as many turns as the winding in the collector circuit, whereas the third winding has n_1 times as many turns as the collector winding. The pulse transformer gives polarity inversion between collector and base voltages as shown in the circuit by the polarity dots. The third winding is connected to the load resistor R_L. The third winding direction is arbitrary and is chosen to obtain either a positive or a negative output pulse. The base supply V_{BB}, which is selected to be of the order of a few tenths of a volt, is added to avoid triggering by noise pulses and to prevent free-running. Since $V_{BB} \ll V_{CC}$ and does not basically affect the oscillator operation, it may be neglected in further discussion.

The equivalent circuit of Fig. 6.19(a) is shown in Fig. 6.19(b). The transformer is represented by an ideal transformer where we assumed that there is no leakage flux, no copper loss, and no magnetic loss. The magnetic inductance of the collector winding is L. The junction voltages V_{CE} and V_{BE} are neglected in the equivalent circuit for the sake of simplicity in calculation. Let us assume that a triggering pulse is applied momentarily to the collector to lower its voltage. Because of the polarity inversion, the base voltage rises by the transformer action. When the base-to-emitter voltage exceeds the cut-in voltage, the transistor comes out of cutoff and a collector current flows. It further decreases the collector voltage. In turn the base-to-emitter voltage increases. If the loop gain is greater than unity, the

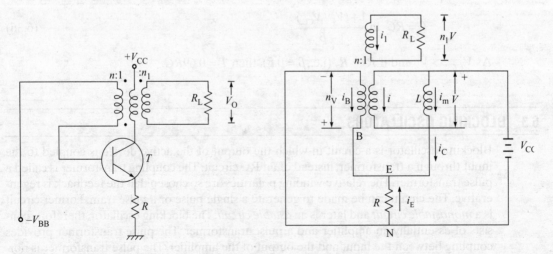

FIGURE 6.19 (a) A monostable blocking oscillator and (b) its equivalent circuit.

regenerative action continues till the transistor is saturated. During the saturation period, although the collector voltage is constant, the collector current continues to increase with a constant rate due to the inductance of the transformer. As the emitter current depends on the induced base voltage and the emitter resistance R, the emitter current is constant. Hence, the base current, which is equal to the emitter current minus the collector current, continues to decrease at a constant rate. When the base current is no longer sufficient, the transistor comes out of saturation and regeneration cuts off the transistor.

To determine the voltage and current waveforms, let us consider the equivalent circuit shown in Fig. 6.19(b). Applying KVL to loop BECNB gives

$$V + nV = V_{CC} \tag{6.37}$$

where V is the drop across the collector winding during the pulse. Hence,

$$V = \frac{V_{CC}}{n+1} \tag{6.38}$$

Applying KVL to loop BENB gives

$$nV = V_{EN} = (i_C + i_B)R$$

or

$$-i_E = i_C + i_B = \frac{nV}{R} = \frac{n}{n+1} \times \frac{V_{CC}}{R} \tag{6.39}$$

$$i_E = -\frac{n}{n+1} \times \frac{V_{CC}}{R} \tag{6.40}$$

As the sum of ampere-turns in the ideal transformer is zero, for the given polarity dots

$$(i_C - i_m) - ni_B + n_1 i_1 = 0 \tag{6.41}$$

But from the load circuit,

$$i_1 = -\frac{n_1 V}{R_L} = -\frac{n_1}{n+1} \times \frac{V_{CC}}{R_L} \tag{6.42}$$

and the magnetizing current is given by

$$L\frac{di_m}{dt} = V$$

or

$$i_m = \frac{Vt}{L} = \frac{t}{n+1} \times \frac{V_{CC}}{L} \tag{6.43}$$

Substituting for i_1 and i_m from Eqs. (4.42) and (4.43) in Eq. (4.41), we get

$$i_C - \frac{t}{n+1} \times \frac{V_{CC}}{L} - ni_B - \frac{n_1^2}{n+1} \times \frac{V_{CC}}{R_L} = 0 \tag{6.44}$$

Solving Eqs. (4.39) and (4.44)

$$i_C = \frac{V_{CC}}{(n+1)^2}\left(\frac{n^2}{R} + \frac{n_1^2}{R_L} + \frac{t}{L}\right) \tag{6.45}$$

$$i_B = \frac{V_{CC}}{(n+1)^2}\left(\frac{n}{R} - \frac{n_1^2}{R_L} - \frac{t}{L}\right) \tag{6.46}$$

From Eqs. (6.39), (6.45), and (6.46), it can be noted that the emitter current is constant, the collector current waveform is a trapezoidal with a positive slope, and the base current waveform is also trapezoidal, but with a negative slope during the pulse. These waveforms are shown in Fig. 6.20. The pulse width can be approximately obtained by taking $i_B = 0$, that is,

$$T_p = \frac{nL}{R} - \frac{n_1^2 L}{R_L} \tag{6.47}$$

At the termination of the pulse the magnetizing current i_m does not decay to zero when the transistor currents have dropped to zero, as the current through an inductor does not change instantaneously. The path of i_m is through the small capacitance of the transformer, which we have neglected for simplicity in analysis. Since the transformer capacitance is small, i_m decays rapidly and large voltage overshoots are induced at the collector, base, and load, as shown in Fig. 6.21. The overshoots must not be so large so as to break down the transistor. It is, therefore, important to provide adequate damping of the back swing in the voltages at $t = T_p$. It is absolutely essential for the operation of the blocking oscillator. Now, let us consider the base voltage waveform of Fig. 6.21(b). If the damping is not adequate, the back swing may oscillate in the positive direction as shown by the dotted line. In such a case, regeneration starts when the base voltage becomes slightly positive. The transistor re-enters the active region. The blocking oscillator is, then, free-running. The oscillator generates a continuous oscillation. The oscillation would be a very distorted sinusoid. In case the core losses of the transformer and the load resistor R_L are not able to provide

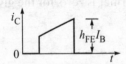

(a) Collector current waveform.

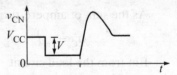

(a) Collector voltage waveform.

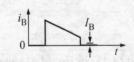

(b) Base current waveform.

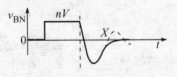

(b) Base voltage waveform.

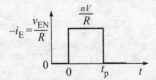

(c) Emitter current waveform.

FIGURE 6.20 Current waveforms.

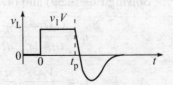

(c) Load voltage waveform.

FIGURE 6.21 Voltage waveforms.

adequate damping, an external resistor is connected to the transformer. This is essential for monostable operation of the blocking oscillator.

6.3.2 Astable Blocking Oscillator

One form of astable blocking oscillator may be obtained by adding an R_1C_1 network in the emitter circuit of the blocking oscillator shown in Fig. 6.19(a). This circuit is shown in Fig. 6.22(a). This circuit also differs from Fig. 6.19(a) in the reversal of the polarity of V_{BB}. The operation of the circuit of Fig. 6.22(a) is explained as follows. Let us assume that the capacitance voltage across C_1, as shown in Fig. 6.22(b), is initially v_1. Also assume that $v_1 > V_{BB} - V_\gamma$, where V_γ is the cut-in base-emitter voltage. Hence, the transistor is cutoff. The capacitor C_1 discharges through resistor R_1, and v_1 decreases exponentially with time constant C_1R_1. When $v_1 = V_{BB} - V_\gamma$, the base current starts flowing and the transistor comes out of cutoff. The collector current increases and the regenerative action begins. The regenerative action for the loop gain greater than unity quickly brings the transistor to saturation. The collector voltage waveform as shown in Fig. 6.22(b) and the base voltage waveform during T_p are similar to those in Fig. 6.21(b). During the period T_p, the capacitor is recharged to a voltage V, which is greater than v_1. The transistor comes out of saturation and the regeneration action brings it to cutoff mode. The transistor remains off for a period T_f. During the period T_f, the capacitor discharges to the voltage $V_{BB} - V_\gamma$ and the transistor comes in active region. At this point the cycle repeats and the time constant $T = T_p + T_f$. The waveforms of the collector voltage and the capacitor voltage are shown in Fig. 6.22 (b). The overshoots are for the reasons discussed while discussing the monostable blocking oscillator (see Section 6.2).

The period T_f is given by

$$T_f = C_1 R_1 \ln\left(\frac{V_1}{V_{BB} - V_\gamma}\right) \tag{6.48}$$

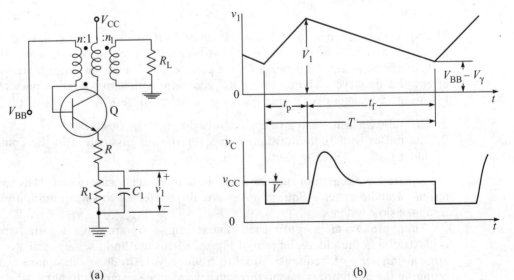

(a)   (b)

FIGURE 6.22 (a) An astable blocking oscillator and (b) associated waveforms.

The period T_p is calculated with the monostable blocking oscillator equivalent circuit of Fig. 6.19(b) after adding V_{BB}, C_1, and R_1. The time constant $C_1 R_1$ is often much greater than T_p. The approximate value of T_p may be given by

$$T_p \approx \frac{nL}{R} - \frac{n_1^2 L}{R_L} \tag{6.49}$$

6.4 SQUARE WAVE GENERATORS

Special shape waves are extensively used in the field of measurement as well as in other electronic works. Special waveforms commonly used are square waves, sawtooth waves, and pulse waves. Square wave generators are discussed in this section. Pulse wave generators are discussed in the next section, whereas the sawtooth generators are discussed in Chapter 8.

The main difference between a pulse generator and a square wave generator is in their duty cycle. The duty cycle is defined as follows:

$$\text{Duty cycle} = \frac{\text{Average value of the pulse over one cycle}}{\text{Peak value of the pulse}} \tag{6.50}$$

The average value of a pulse is given by

$$V_{av} = \frac{V_{peak} T_p}{T} \tag{6.51}$$

where V_{peak} is the peak value, T_p is the pulse width, and T is the time period. Hence,

$$\text{Duty cycle} = \frac{V_{av}}{V_{peak}} = \frac{\frac{V_{peak} T_p}{T}}{V_{peak}} = \frac{T_p}{T} \tag{6.52}$$

The square wave generators with equal on and off times, therefore, have duty cycle equal to 0.5 or 50% of the time period, irrespective of the wave frequency. The duty cycle of a pulse generator depends on the pulse width. Thus, a very short duration pulse gives a low-duty cycle. Therefore, a pulse generator with short-duration pulses has the following advantages over a square wave generator:

1. The pulse generator can give more power during its on period.
2. The pulses with short duration reduce the power dissipation in the component under test.

Square wave generators are used in low-frequency characteristic analysis, such as in testing of audio systems. Square waves are also preferred to short-duration pulses in testing of slow systems.

There are two main approaches of generating a square wave. The first approach is illustrated in the block diagram of Fig. 6.23. This method uses an oscillator generating a sine wave of frequency equal to frequency desired for the square wave. The output of the oscillator is passed through a wave-shaping circuit to obtain the desired

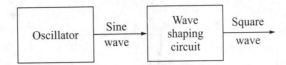

FIGURE 6.23 A block diagram for generating square waves.

waveform with desired frequency. In the second approach, square waves are directly generated without using an oscillator. Multivibrators and blocking oscillators are used in this method.

For generating square waves from sinusoidal waves, double-clipping circuits are used. Double-ended clipping circuits have already been discussed in Chapter 3. Clipping circuit using transistor and Schmitt trigger are also used for generating square waves from sinusoidal waves. Generating square waves from sinusoidal waves using the transistor clipping circuit and the Schmitt trigger are discussed in this section.

6.4.1 Transistor Clipping Circuit

When a transistor amplifier is overdriven to saturation, it can clip a sine wave. The transistor amplifier, shown in Fig. 6.24, certainly saturates and cuts off when a large amplitude sine wave is applied to it. During the positive half cycle of the input, the transistor saturates almost immediately and during the negative half cycle it cuts off. Thus, a good square wave is obtained at the output terminals.

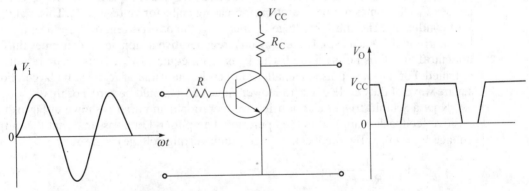

FIGURE 6.24 A transistor amplifier as a clipper.

6.4.2 Schmitt Trigger

A Schmitt trigger is an asymmetrical bistable multivibrator. It can be used to obtain square wave from a sinusoidal wave. The Schmitt trigger circuit is shown in Fig. 6.25.

At time $t = 0$, the input $V_i = 0$ and so there is no base current in T_1. The collector current is also zero and T_1 is cutoff. Transistor T_2 base is connected to a voltage divider formed by R_3. As T_1 is cutoff, the base voltage of T_2 is positive enough to force a base current which saturates T_2. With T_2 saturated the emitters of both T_1 and T_2 are at a voltage $V_E = V_{E1}$ and the output $V_o = V_{E1}$. Till $v_i < V_{E1}$, T_1 remains cutoff and T_2 saturated. The output remains at V_{E1}. When V_i crosses V_{E1}, the base current in T_1 starts to flow. Its

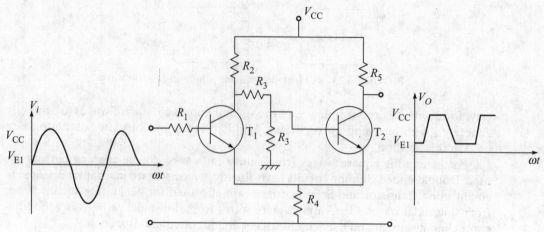

FIGURE 6.25 A Schmitt trigger circuit.

collector current increases, decreasing its collector voltage. This drop of collector voltage of T_1 results in decrease of base voltage of T_2, bringing it out of saturation and reducing V_E. The decrease in V_E further increases the collector current of T_1, dropping its collector voltage further. This regenerative action continues till T_1 is not saturated and T_2 is not cutoff. Now, $V_o = V_{CC}$ and $V_E = V_{E2}$, which is less than V_{E1}. Since the regenerative action is very quick, V_o reaches from V_{E1} to V_{CC} suddenly. Further increase in V_i only increases the base current of T_1. As T_2 is cutoff, V_o remains at V_{CC} until v_i is less than V_{E2}. Once v_i crosses V_{E2}, T_1 comes out of saturation, increasing collector voltage of T_1. This makes T_2 start conducting. The emitter voltage V_E increases, the base current of T_1 reduces and the base current of T_2 increases. There is again a regenerative action and it continues till T_1 is not cutoff and T_2 is not saturated. The V_o is again equal to V_{E1}. Thus a square pulse is obtained. For a square pulse of smaller duty cycle, the value of R_4 must be larger. For a square wave of duty cycle of 0.5, the lower value of V_o should be zero. For this $R_4 = 0$.

Now, a Schmitt trigger circuit using an operational amplifier is shown in Fig. 6.26. A positive feedback is applied to the operational amplifier. Let us assume that the output voltage $V_o = +V_{CC}$. The feedback voltage (noninverting voltage) is given by

$$V_n = \frac{R_1}{R_1 + R_2} V_o = \beta V_{CC} \tag{6.53}$$

where $\beta = R_1/(R_1 + R_2)$. If the input voltage $V_{in} = V_i - \beta V_{CC}$ is negative, then $V_o = +V_{CC}$. In contrast, if the input voltage $V_{in} = V_i - \beta V_{CC}$ is positive, then $V_o = -V_{CC}$. The transfer characteristic of the Schmitt trigger for an increasing inverting voltage V_i is shown in Fig. 6.27(a).

Next, let us assume that the output voltage V_o is $-V_{CC}$. In this case noninverting input voltage is given by

$$V_n = \frac{R_1}{R_1 + R_2} V_o = -\beta V_{CC} \tag{6.54}$$

If the input voltage $V_{in} = V_i + \beta V_{CC}$ is positive, then $V_o = -V_{CC}$. Conversely, if the input voltage $V_{in} = V_i + \beta V_{CC}$ is negative, then $V_o = +V_{CC}$. The transfer characteristic of the

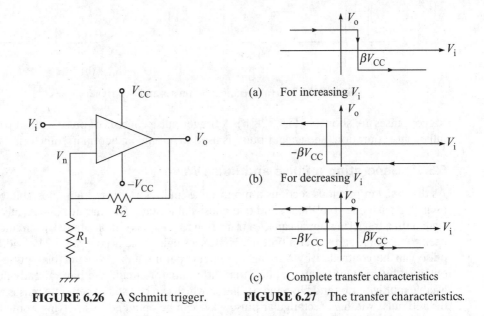

FIGURE 6.26 A Schmitt trigger.   **FIGURE 6.27** The transfer characteristics.

Schmitt trigger for a decreasing inverting voltage V_i is shown in Fig. 6.27(b). Combining, the two transfer characteristics of Fig. 6.27(a) and (b), a complete transfer characteristic, as shown in Fig. 6.27(c), is obtained. From the complete transfer characteristic it is clear that the output voltage of $+V_{CC}$ does not change to $-V_{CC}$ until the voltage applied to the inverting terminal V_i increases beyond βV_{CC}. Also, the output voltage of $-V_{CC}$ does not change to $+V_{CC}$ until voltage V_i decreases beyond $-\beta V_{CC}$. This phenomenon is known as *hysteresis*. The width of the hysteresis is the difference between the two threshold (or triggering) voltages βV_{CC} and $-\beta V_{CC}$. In this case, the width is $2\beta V_{CC}$. Therefore, the Schmitt trigger is a bistable device having two stable states outputs of $+V_{CC}$ and $-V_{CC}$. Thus, we see that a Schmitt trigger can convert a sinusoidal signal into a square signal. Also, the Schmitt trigger output is least affected by the noise in the sinusoidal input.

For improving the quality of the square wave, the sinusoidal voltage should be passed through an amplifier and then to the shaping circuit. The output of the shaping circuit is again passed through an amplifier and then to a next shaping circuit. This process may be repeated for still improving the quality of the square wave.

As discussed in Sections 6.2 and 6.3, a square wave can be generated directly by using an astable (free-running) multivibrator or a blocking oscillator.

6.5 PULSE AND PULSE WAVE GENERATORS

There are two types of pulses, namely, (1) rectangular pulses and (2) trigger pulses. The rectangular pulses are shown in Fig. 6.28(a). The rectangular pulses are unsymmetrical square waves in which the on period is of very short duration relative to the off duration. An ideal rectangular pulse consists of accurately horizontal and vertical segments. The

274 Chapter 6 / Wave Generators

(a)   (b)

FIGURE 6.28 (a) Rectangular pulses and (b) trigger pulses.

trigger pulses are shown in Fig. 6.28(b). A trigger pulse consists of pulse with step rise and fall of short duration. The trigger pulse is used to initiate an action in some other circuits.

6.5.1 Generating a Pulse and Pulse Wave

As discussed in previous sections a monostable multivibrator and a monostable blocking oscillator have one stable state and one quasi-stable state. The circuit remains in its stable state until a triggering signal causes a transition to the quasi-stable state. The circuit returns itself to its stable state after a time T_p. Hence, a single pulse is generated. The width of the pulse can be controlled by varying the circuit parameters. A rectangular pulse can be obtained by applying a square pulse to a differentiator circuit. We have already discussed astable multivibrator and blocking oscillator, which can be used to generate pulses with any desired pulse width. A rectangular pulse wave can be obtained by applying a square pulse wave to a differentiator circuit. The operation of a differentiator circuit is discussed next.

RC Differentiator Circuit

A trigger-type pulse can be obtained from a square pulse by means of a differentiating circuit. A typical differentiator consists of a small capacitor in series with a small resistor, as shown in Fig. 6.29. The operation of an RC circuit, as in Fig. 6.29, is discussed as follows.

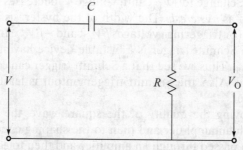

FIGURE 6.29 A differentiator (RC) circuit.

Step Input Voltage

Let a step input voltage be applied to the differentiator (RC) circuit of Fig. 6.29. A step voltage maintains the value zero for all $t < 0$ and a value equal to V for all $t > 0$. The voltage distribution of the circuit is given by

$$\frac{1}{CR}\int V_o \, dt + V_o = V$$

or

$$\frac{1}{CR}V_o + \frac{dV_o}{dt} = 0$$

After solving the differential equation,

$$V_o = Ve^{-t/RC} \tag{6.55}$$

The output voltage decays exponentially and for small time constant RC, it decays very rapidly as shown in Fig. 6.30.

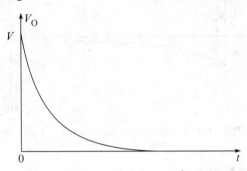

FIGURE 6.30 An output for step input.

Pulse Input Voltage

An ideal pulse is shown in Fig. 6.31(a) and can be considered as the sum of two step input voltages shown in Fig. 6.31(b) and (c), respectively. One step voltage is with value zero for $t < 0$ and V for $t > 0$, whereas the second step voltage value is zero for $t < T_p$ and $-V$ for $t > T_p$. Hence, the response of the differentiator at $t = T_p^-$ is given by

$$V_o = V_p = Ve^{-T_p/RC} \tag{6.56}$$

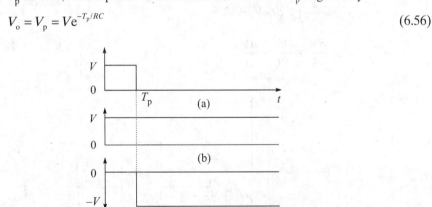

FIGURE 6.31 A pulse input.

At $t = T_p$, the input abruptly falls by the amount V. Since, the capacitor voltage cannot change instantaneously, the output voltage must also drop by the voltage V. Hence, at $t = T_p^+$, the output $V_o = V_p - V$. Since $V_p < V$, the output at $t = T_p^+$ becomes negative and then decays exponentially to zero. The output voltage for pulse with width $T_p \ll RC$ is as shown in Fig. 6.32(a). When $T_p \gg RC$, $V_p = 0$ and the output is as shown in Fig. 6.32(b). Thus, the differentiator can be used to obtain a rectangular pulse from a square pulse.

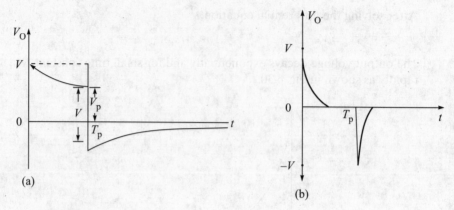

FIGURE 6.32 Pulse response (a) for $T_p \ll RC$ and (b) for $T_p \gg RC$.

Square Wave Input Voltage

A square wave, in general, remains at a constant value V' for an interval T_1 and for the next interval T_2 it remains at a constant value V''', as shown in Fig. 6.33. This waveform repeat with a period $T = T_1 + T_2$. We are interested in steady-state response of the differentiator for the square wave input. Before proceeding for obtaining the response, let us first find the average value of the steady-state response for a periodic input waveform. The differentiator circuit of Fig. 6.29 is governed by the equation

$$\frac{1}{RC}\int V_o dt + V_o = V_i$$

or

$$\frac{dV_i}{dt} = \frac{V_o}{RC} + \frac{dV_o}{dt}$$

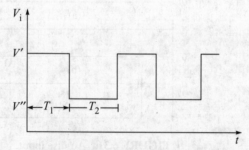

FIGURE 6.33 A square wave input.

Multiplying both sides by dt and then integrating,

$$\int_{t=0}^{t=T} dV_i = \frac{1}{RC}\int_{t=0}^{t=T} V_o dt + \int_{t=0}^{t=T} dV_o$$

or

$$V_i(T) - V_i(0) = \frac{1}{RC}\int_{t=0}^{t=T} V_o dt + V_o(T) - V_o(0)$$

Under steady state conditions, the output and input waveforms are periodic; hence,

$$V_i(T) = V_i(0) \text{ and } V_o(T) = V_o(0)$$

$$\therefore \int_0^T V_o dt = 0 \quad (6.57)$$

$$\therefore \int_0^T V_o dt = \text{Area under the output waveform over one cycle} = 0 \quad (6.58)$$

Thus, the positive area is equal to the negative area. Hence, it shows that the d.c. component in the output is zero.

The output of the RC circuit, like the pulse response, for a square wave input, in general, is shown in Fig. 6.34(a). In the special case, when $RC \gg T_1$ and T_2, the output is approximately a square wave as shown in Fig. 4.34(b), and when $RC \ll T_1$ and T_2, the output is as shown in Fig. 6.34(c).

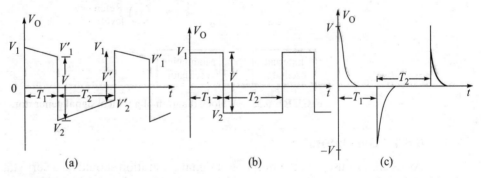

FIGURE 6.34 Output of RC circuit (a) in general, (b) when $RC \gg T_1$ and T_2, (c) when $RC \ll T_1$ and T_2.

From Fig. 6.34(a),

$$V_1' = V_1 e^{-T_1/RC} \qquad V_1' - V_2 = V \quad (6.59)$$

$$V_2' = V_2 e^{-T_1/RC} \qquad V_1 - V_2' = V' \quad (6.60)$$

The square wave of Fig. 6.34(a) is a symmetrical when $T_1 = T_2 = T/2$. Because of symmetry, $V_1 = -V_2$ and $V_1' = -V_2'$. Hence, for a symmetrical waveform, equations in Eq. (6.59) are identical with those in Eq. (6.60). Thus, we see that a differentiator can be used for obtaining a rectangular pulse train from a square wave.

6.6 SIGNAL GENERATORS

A signal generator is a device that generates radio frequency signals of accurately known characteristics. First, the frequency of the signal should be well known and stable.

Second, the amplitude should be controllable from very small to relatively large values. Third, the signal should be free of distortion. The signal generator has the capacity of amplitude modulation and a wide range of tuning. It covers the frequency range from a few hertz to many gigahertz. A typical signal generator, as shown in the block diagram of Fig. 6.35, consists of the following components:

1. An oscillator.
2. A modulating system.
3. An attenuator.
4. A reference voltage or power source.

The components of the signal generator are briefly described here.

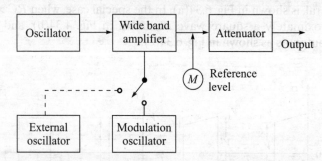

FIGURE 6.35 A block diagram of a typical signal generator.

6.6.1 Oscillators

An oscillators used for the purpose of signal generation should be a very stable LC oscillator. It must be tunable over the necessary frequency band. It must also generate a good sinusoidal waveform having no appreciable hum or noise modulation. The oscillator is designed to give an output that is as nearly constant as possible over a frequency range and from one band to other. In some cases automatic amplitude control is used. It should also have good inherent frequency stability.

Most of the general-purpose signal generators deliver amplitude-modulated signals in very high frequency and very low frequency ranges. In these signal generators, power amplifiers are used in conjunction with the oscillators. In this case, the amplitude modulation is applied to the amplifier instead of the oscillator. Most signal generators have modulating systems that provide either amplitude or frequency modulation. The application of the signal generator determines the type of modulation to be used.

6.6.2 Modulating System

Three types of modulating signals are used in amplitude modulation. They are sine wave, square wave, and pulse. The sine wave modulation is used for simulating a signal modulated by a voice or video signals. In microwave signal generators, square wave modulation is commonly used. For simulating signals, such as used in radar pulse communication systems, pulse modulation is common.

Two types of frequency modulations are widely used in signal generators. In the first type, a sine wave is used as modulating signal. It is used to simulate a signal that is modulated by a voice or a video wave. In the second type of frequency modulation, the instantaneous frequency of the oscillator is continuously swept over a wide band in order to trace out response curve. The signal generator of this type is known as sweep-frequency generator and is discussed in Section-6.9.

The modulation can be obtained either from an internally generated modulating signal or by a modulating signal applied from an external source (shown by dashed line in Fig. 6.35). The internal modulating system ordinarily provides modulation corresponding to standard test signals. More flexible and elaborate forms of modulation can be obtained by using external modulating signal. The detailed discussions on signal modulation and demodulation are given in Chapter 13.

6.6.3 Attenuator

An attenuator, unlike an amplifier, reduces signal level by a desired amount. The attenuation may be expressed as the ratio of the input power to the output power. The attenuation in decibel is expressed as the log of the ratio of the input power (P_i) to the output power (P_o) by the relationship

$$\bar{A} \text{ in decibel} = 10 \log\left(\frac{P_i}{P_o}\right) \quad (6.61)$$

When the signal voltage is concerned, the attenuation in decibel is expressed as the log of the ratio of the input voltage (V_i) to the output voltage (V_o) by the relationship

$$\bar{A} \text{ in decibel} = 20 \log\left(\frac{V_i}{V_o}\right) \quad (6.62)$$

If an input signal is passed through two attenuators in cascade, the output of the first attenuator is reduced by the ratio V_i / V_o', whereas the signal is further reduced by the ratio V_o' / V_o by the second attenuator. The total reduction in the output expressed in decibel is given by

$$\bar{A} \text{ in decibel} = 20 \log\left(\frac{V_i}{V_o}\right) = 20 \log\left(\frac{V_i}{V_o'}\right)\left(\frac{V_o'}{V_o}\right)$$
$$= 20 \log\left(\frac{V_i}{V_o'}\right) + 20 \log\left(\frac{V_o'}{V_o}\right) = \bar{A}_1 \text{ in decibel} + \bar{A}_2 \text{ in decibel} \quad (6.63)$$

Hence, the total attenuation in decibel of two attenuators in cascade is the sum of the decibel attenuation of each attenuator. The result can be extended for more than two attenuators in cascade. Hence, in general, the total attenuation in decibel for n attenuators in cascade is given by

$$\bar{A} \text{ in decibel} = \bar{A}_1 \text{ in decibel} + \bar{A}_2 \text{ in decibel} + \ldots + \bar{A}_n \text{ in decibel} \quad (6.64)$$

where $\bar{A}_i$ is attenuation in decibel for the ith attenuator $(i = 1, 2, \ldots, n)$.

The resistive π attenuator can be designed with standard components up to about 20 dB and for frequency up to 100 MHz. The capacitive reactances alter the attenuation for higher frequencies. This can be eliminated to some extent by cascading lower values attenuators for obtaining higher value of attenuations. Code-operated switches can be used with cascaded attenuators for providing attenuations in steps by manipulating the switches in a simple binary sequence. For example, four-switch cascaded attenuators of 1, 2, 4, 8 dB could provide attenuations of 0 to 15 in 1-dB steps. This technique can be used to have up to 100-dB attenuations in 1-dB steps by using seven-switch attenuators. The detailed discussion on attenuator circuits is given in Chapter 7.

6.6.4 Reference Source

In low-cost signal generators, metering circuit is employed to the input of the attenuator and the level is set by a manual adjustment. In modern signal generators, automatic level control is employed to set the level at the input of the attenuator. For this a reference source is required. To sense the voltage level at the input of the attenuator, an accurate and broadband voltage measuring device is used. For this either a transistor voltmeter or a thermocoupled meter can be used.

Many signal generators employ a precise frequency readout. Old signal generators employed precision dials and mechanical dial drives with hand-calibrated dial plates. In more expensive signal generators crystal oscillators were used to periodically check the dial calibration. Presently, the frequency counter is used, as it is simple and very accurate.

6.6.5 Audio Signal Generators

Audio frequency signal generators are employed in determining the variation of amplification with the frequency of an audio amplifier. The oscillator used in an audio signal generator is an audio frequency oscillator, typically phase-shift type. The output of the oscillator is applied to an adjustable resistance attenuator. A transistor voltmeter is commonly used as a reference level indicator.

6.7 FUNCTION GENERATORS

A function generator is a versatile instrument, which delivers signals of different waveforms with adjustable frequency over a wide range. The most common waveforms generated are the sine, square, triangular, and sawtooth waves. The frequency range is from a fraction of hertz to several kilohertz. The special features of the function generator may be listed as follows:

1. Various outputs may be available at the same time.
2. It has the capability to phase lock to an external signal source.
3. It phase locks to a frequency standard and then gives a waveform, the frequency of which is same as standard frequency and has the same accuracy and stability.

Since the low frequency of an RC oscillator is limited, a different approach is followed. The block diagram of the function generator is shown in Fig. 6.36. This function generator delivers sine, triangular, and square waves. The frequency range is from 0.01 Hz to 100 kHz. A frequency dial on the front panel governs the frequency control network. Alternatively, the frequency control network is governed by an externally applied control

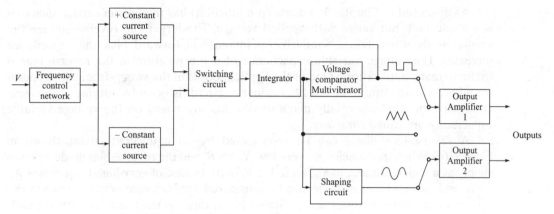

FIGURE 6.36 A schematic diagram of a function generator.

voltage. The frequency-control voltage regulates the two constant current sources. Positive current source supplies a constant current to an integrator, which in turn gives an output voltage with time. The slope of the integrator output increases or decreases depending on the increase or decrease in the current supplied by the positive current source. The integrator output is connected to the inverting terminal of the Schmitt trigger circuit of Fig. 6.26. When this voltage reaches a predetermined voltage, the Schmitt trigger changes its state. The predetermined voltage depends on the value of β. This change in state operates the switching circuit, which turns off the positive current source and turns on the negative current source, which was previously off.

As the negative current source supplies a reverse current to the integrator, the integrator output decreases linearly with time. The Schmitt trigger, which operates as a comparator, again, changes its state at a predetermined voltage level on the negative slope of the output voltage. This changes the switching position, cutting off the negative current source and switching on the positive current source again. This cycle is repeated.

The output of the integrator is a triangular wave, whereas the output of the comparator is a square wave. The frequencies of the waves are same and determined by the magnitude of the constant current sources. The shaping circuit converts the triangular wave into a sine wave with less than 1% distortion. The two output amplifiers can be connected to any two of the outputs simultaneously.

6.8 SWEEP-FREQUENCY GENERATOR

Sine wave signal generators discussed earlier have been designed to generate a voltage sine wave whose frequency is known and stable. In applications, for example, measuring frequency response of amplifiers, filters, and other networks, a sweeping source of frequency is required. A sweep-frequency generator differs from a single-frequency generator (oscillator) in the sense that the former is capable of being electronically tuned. The solid-state variable capacitance diode is used for electronically tuning the sweep-frequency generator. The value of capacitance of the diode depends on the reverse bias-voltage applied to the diode. Hence, the frequency is ultimately controlled by the voltage. Hence, the sweep-frequency generator is also known as *voltage-controlled oscillator*.

As discussed in Chapter 1, a diode (p-n junction) has the barrier capacitance that is not constant, but varies with applied voltage. The larger the reverse voltage the smaller is the capacitance. Similarly, for increase in forward bias, the capacitance increases. The voltage variable capacitance of a p-n junction in the reverse bias is useful in many applications. One of the applications is in the sweep-frequency generator where voltage tuning of the LC resonance circuit is done by varying the reverse voltage of the diode. Specially made diodes that are based on the voltage variable capacitance are called *varactors*.

A reverse-bias diode can be represented by an equivalent circuit shown in Fig. 6.37(a). When frequencies are very low, $X_C \gg R_r$ and the reverse-bias diode behaves simply as a high resistance as shown in Fig. 6.37(b). In case of very high frequencies, $X_C \ll R_r$ and the reverse-bias diode may be considered a capacitance only, as shown in Fig. 6.37(c). As discussed above, the capacitance of the diode is not fixed. Its value depends on the magnitude of the reverse voltage applied across the diode, as shown in Fig. 6.38. For very low reverse current, silicon is used in varactors. The capacitance in reverse direction is approximately given by

$$C_d = \frac{K}{\sqrt{0.7 + V}} \tag{6.65}$$

where K is the constant and V is the reverse voltage.

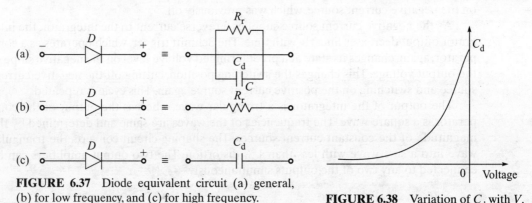

FIGURE 6.37 Diode equivalent circuit (a) general, (b) for low frequency, and (c) for high frequency.

FIGURE 6.38 Variation of C_d with V.

A sweep-frequency generator is made by using a variable capacitance diode in the filter circuit of an LC resonant oscillator, as shown in Fig. 6.39. A sweep-voltage generator, discussed in Chapter 8, is used to control the capacitance of the varactor.

Because of the nonlinear relation between the diode capacitance and the sweep voltage, the sweep voltage and frequency relationship is nonlinear. The amount of non-linearity depends on the type of oscillator used and to a great extent on the frequency range covered by the generator. For narrower frequency range, the relationship between the voltage and the frequency is more linear. Generally, there is a limit of two to one of the maximum to minimum frequency of any sweep-frequency generator.

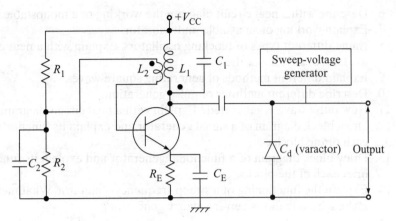

FIGURE 6.39 A sweep-frequency generator.

A broadband sweep-frequency generator is made by mixing a fixed-frequency oscillator with a sweep-frequency generator at a lower frequency well above the required frequency band. For an example a 0- to 400-MHz signal is generated by mixing 500- to 900-MHz sweep-frequency generator with a fixed 500-MHz oscillator. The output frequency range covers from 0 to 400 MHz, whereas the maximum-to-minimum frequency ratio of the sweep-frequency generator is less than two. The block diagram is shown in Fig. 6.40.

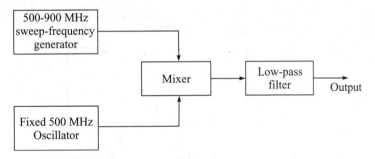

FIGURE 6.40 A wideband sweep-frequency generator.

6.9 REVIEW QUESTIONS

1. Name various types of oscillators being used and explain with a neat circuit diagram the working of any one of them.
2. Draw the circuit of phase-shift oscillator. What determines the frequency of oscillation?
3. What are differences between the Hartley and Colpitts oscillators? Explain their working principles.
4. Explain the oscillation conditions in a Wien-bridge oscillator.
5. Describe in brief a crystal oscillator and mention its merits and demerits.

6. Describe with a neat circuit diagram the working of a monostable multivibrator.
7. Explain working of an astable multivibrator.
8. Name different types of blocking oscillators. Explain with a neat circuit diagram the working of any one of them.
9. Explain different methods of generating square waves.
10. Describe different methods of pulse generation.
11. How pulse wave is generated? Explain with a neat circuit diagram.
12. Draw block diagram of a signal generator and explain its function describing in brief each of the blocks.
13. Draw block diagram of a functional generator and explain its function describing in brief each of the blocks.
14. Explain the functioning of a sweep-frequency generator. What method is adopted to make a broadband sweep-frequency generator?

6.10 SOLVED PROBLEMS

1. A phase-shift oscillator is shown in Fig. 6.41. (a) For given $C = 0.001$ µF and oscillation frequency $= 10$ kHz, determine R and the minimum value of R_F. (b) For $R = 1.5$ kΩ and oscillation frequency $= 1.2$ kHz, determine C and the minimum value of R_F.

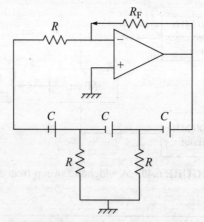

FIGURE 6.41 A phase-shift oscillator.

Solution

(a) The oscillation frequency of phase-shift oscillator,

$$f = \frac{1}{2\pi RC\sqrt{6}}$$

or $R = \dfrac{1}{2\pi fC\sqrt{6}} = \dfrac{10^6}{2\pi \times 10\sqrt{6}}$

   $= 6.5 \text{ k}\Omega$

Since $\beta A = \dfrac{1}{29} \times \dfrac{R_F}{R} = 1$

or $R_F = 29R = 29 \times 6.5 = \mathbf{188.5 \ k\Omega}$

(b) $C = \dfrac{1}{2\pi f R\sqrt{6}} = \dfrac{10^{-6}}{2\pi \times 1.2 \times 1.5 \sqrt{6}}$

$= \mathbf{0.0361 \ \mu F}$

$R_F = 29R = 29 \times 1.5 = \mathbf{43.5 \ k\Omega}$

2. (a) For Fig. 6.4(a), $g_m = 5 \ m\Omega^{-1}$ and $R = 100 \ k\Omega$, determine the value of C and minimum value of R_D, if the oscillation frequency is 1 kHz. (b) Determine the minimum value of g_m and the frequency of oscillation, if $R = 100 \ k\Omega$, $C = 0.001 \ \mu F$, and $R_D = 5 \ k\Omega$.

Solution

(a) The midband gain of the FET amplifier $A = g_m R_D$. Hence, for oscillation,

$$\beta A = (-1/29) \times (-g_m R_D) = 1$$

Hence, $R_D = \dfrac{29}{g_m} = \dfrac{29}{5} = \mathbf{5.8 \ k\Omega}$

$C = \dfrac{1}{2\pi f R \sqrt{6}} = \dfrac{10^{-6}}{2\pi \times 1.0 \times 100 \sqrt{6}} = \mathbf{0.0065 \ \mu F}$

(b) $g_m = \dfrac{29}{R_D} = \dfrac{29}{5 \times 10^3} = \mathbf{5.8 \ m\Omega^{-1}}$

$f = \dfrac{1}{2\pi RC \sqrt{6}} = \dfrac{10^6}{2\pi \times 100 \times 1.0 \sqrt{6}} = \mathbf{0.649 \ k\Omega}$

3. For the phase-shift oscillator of Fig. 6.5, $R_1 = 27 \ k\Omega$, $R_2 = 3 \ k\Omega$, $R = 1 \ k\Omega$, $h_{ie} = 0.5 \ k\Omega$, and $f = 1$ kHz. (a) Determine the value of C and R_s, if $R_C = 1.2 \ k\Omega$. (b) Determine the value of R_C, if $C = 0.047 \ \mu F$.

Solution

The frequency is given by

(a) $f = \dfrac{1}{2\pi RC \sqrt{6 + 4R_C / R}}$

$C = \dfrac{1}{2\pi f R \sqrt{6 + 4R_C / R}} = \dfrac{10^{-6}}{2\pi \times 1.0 \times 1.0 \sqrt{6 + 4 \times 1.2 / 1}}$

$= \mathbf{0.048 \ \mu F}$

$R_s = R - \dfrac{R_p h_{ie}}{R_p + h_{ie}}$

where $R_p = \dfrac{R_1 R_2}{R_1 + R_2} = \dfrac{27 \times 3}{27 + 3} = 2.7 \text{ k}\Omega$

Hence, $R_s = 1 - \dfrac{2.7 \times 0.5}{2.7 + 0.5} = 578 \text{ }\Omega$

(b) $\sqrt{6 + 4R_C/R} = \dfrac{1}{2\pi f RC} = \dfrac{1}{2\pi \times 1 \times 1 \times 0.047} = 3.39$

or $R_C = \dfrac{(3.39)^2 - 6}{4} = 1.367 \text{ k}\Omega$

4. For the Wien-bridge oscillator shown in Fig. 6.10(a), the parallel RC and series RC combinations are interchanged. (a) Show that feedback transfer function is given by

$$\beta = \beta(j\omega) = \dfrac{1 - \omega^2 R^2 C^2 + 2j\omega RC}{1 - \omega^2 R^2 C^2 + 3j\omega RC}$$

(b) Find the relation between R_1 and R_2 for which the oscillation frequency is $f_o = 1/RC$.

Solution

(a) The impedance of series RC circuit,

$$Z_1 = R + \dfrac{1}{j\omega C} = \dfrac{1 + j\omega CR}{j\omega C}$$

and the impedance of parallel RC circuit,

$$Z_2 = \dfrac{R(1/j\omega C)}{R + (1/j\omega C)} = \dfrac{R}{1 + j\omega CR}$$

Hence, the voltage transfer function of the feedback network,

$$\beta(j\omega) = \dfrac{V_1}{V_o} = \dfrac{Z_1}{Z_1 + Z_2} = \dfrac{\dfrac{1 + j\omega CR}{j\omega C}}{\dfrac{1 + j\omega CR}{j\omega C} + \dfrac{R}{1 + j\omega CR}}$$

$\therefore$ $\beta = \beta(j\omega) = \dfrac{1 - \omega^2 R^2 C^2 + 2j\omega RC}{1 - \omega^2 R^2 C^2 + 3j\omega RC}$

(b) Substituting for $\omega = 1/RC$ in the above equation, $\beta = 2/3$. For satisfying the oscillation condition,

$$\beta A = \dfrac{2}{3}\left(\dfrac{R_1 + R_2}{R_1}\right) = 1$$

$\therefore$ $R_1 = 2R_2$

5. (a) For the Wien bridge shown in Fig. 6.6, if $R = R_1 = 10$ kΩ, $f_o = 10$ kHz, determine the value of C and the minimum value of R_2. (b) Determine the value of R and the maximum value of R_1, if $R_2 = 10$ kΩ, $f_o = 1$ kHz, and $C = 0.02$ µF.

Solution

For the oscillation frequency,

$$C = \frac{1}{2\pi fR} = \frac{10^{-6}}{2\pi \times 10 \times 10} = 0.00159 \text{ µF}$$

The minimum value of R_2 is given by

$$R_2 = 2R_1 = 2 \times 10 = \mathbf{20 \text{ k}\Omega}$$

(b) For the oscillation frequency

$$R = \frac{1}{2\pi fC} = \frac{10^3}{2\pi \times 1.0 \times 0.02} = \mathbf{7.96 \text{ k}\Omega}$$

The maximum value of R_1 is given by

$$R_1 = \frac{R_2}{2} = \frac{10}{2} = \mathbf{5 \text{ k}\Omega}$$

6. Consider the FET Hartley oscillator shown in Fig. 6.7(a). For the circuit $L_1 = 85$ µH and $L_2 = 105$ µH. (a) Determine the oscillation frequency and the minimum value of g_m, if $C = 200$ pF and $R_D = 2$ kΩ. (b) For given $f_o = 500$ kHz and $R_D = 1$ kΩ, determine the value of capacitance and the minimum value of g_m.

Solution

(a) The oscillation frequency is given by

$$f_o = \frac{1}{2\pi\sqrt{LC}}$$

$$= \frac{1}{2\pi\sqrt{190 \times 10^{-6} \times 200 \times 10^{-12}}}$$

$$= \mathbf{816 \text{ kHz}} \qquad (L = L_1 + L_2)$$

From Eq. (6.19), for $A_v = g_m R_D$, $X_1 = L_1$ and $X_2 = L_2$,

$$\frac{g_m R_D L_1}{L_2} = 1$$

$$g_m = \frac{L_2}{L_1 R_D} = \frac{105}{85 \times 2 \times 10^3} = \mathbf{0.62 \text{ m}\Omega^{-1}}$$

(b) The value of C is given by

$$C = \frac{1}{4\pi^2 f_o^2 L} = \frac{1}{4\pi^2 (500 \times 10^3)^2 \times 190 \times 10^{-6}}$$

$$= \frac{10^{-6}}{190\pi^2} = 533.3 \text{ pF}$$

The minimum value of

$$g_m = \frac{L_2}{L_1 R_D} = \frac{105}{85 \times 1} = 1.235 \text{ m}\Omega^{-1}$$

7. For the circuit of FET Colpitts oscillator shown in Fig. 6.7(a), $C_1 = 1250$ pF and $C_2 = 2450$ pF, and $R_G = 10$ MΩ. (a) Determine the oscillation frequency and the minimum value of g_m, if $L = 50$ µH and $R_D = 1$ kΩ. (b) For given $f_o = 500$ kHz and $g_m = 2$ mΩ^{-1}, determine the value of L and minimum value of R_D.

Solution

(a) The impedances

$$Z_1 = \frac{R_G}{1 + j\omega C_1 R_G} \approx \frac{1}{j\omega C_1} \qquad (\text{As } \omega C_1 R_G \gg 1)$$

$$Z_2 = \frac{1}{j\omega C_2} \text{ and } Z_3 = \omega L$$

The oscillation frequency is given by

$$f_o = \frac{1}{2\pi\sqrt{LC}}$$

where $C = \frac{C_1 C_2}{C_1 + C_2} = \frac{1250 \times 2450}{1250 + 2450} = 828 \text{ pF}$

$$f_o = \frac{1}{2\pi\sqrt{50 \times 10^{-6} \times 828 \times 10^{-12}}}$$

$$= 782 \text{ kHz}$$

From Eq. (6.19), for $A_v = g_m R_D$, $X_1 = \frac{1}{\omega C_1}$ and $X_2 = \frac{1}{\omega C_2}$,

$$\frac{g_m R_D C_2}{C_1} = 1$$

$$g_m = \frac{C_1}{C_2 R_D} = \frac{1250}{2450 \times 1} = 0.51 \text{ m}\Omega^{-1}$$

(b) The value of L is given by

$$L = \frac{1}{4\pi^2 f_o^2 C} = \frac{1}{4\pi^2 (500 \times 10^3)^2 \times 828 \times 10^{-12}}$$

$$= 122.4 \text{ µF}$$

The minimum value of

$$R_D = \frac{C_1}{C_2 g_m} = \frac{1250}{2450 \times 2} = 0.255 \text{ k}\Omega$$

8. The electric equivalent circuit of a quartz crystal, as shown in Fig. 6.11(b), has the values $R = 2$ kΩ, $L = 5.5$ H, $C = 0.08$ pF, and $C' = 12$ pF. Determine (a) the series resonance frequency ω_s, (b) the parallel resonance frequency ω_p, and (c) the Q factor at the series resonance frequency.

Solution

(a) The series resonance frequency,

$$\omega_s = \frac{1}{\sqrt{LC}} = \frac{10^6}{\sqrt{5.5 \times 0.08}} = 1.5 \text{ MHz}$$

(b) The parallel resonance frequency,

$$\omega_p = \frac{1}{\sqrt{LC_{eq}}}$$

where $C_{eq} = \frac{CC'}{C+C'} = \frac{0.08 \times 12}{0.08 + 12} = 0.0795$ pF

$$\omega_p = \frac{10^6}{\sqrt{5.5 \times 0.0795}} = 1.5123 \text{ MHz}$$

(c) The Q factor at the series resonance frequency,

$$Q = \frac{\omega_s L}{R} = \frac{1.5 \times 10^6 \times 5.5}{2} = 4125000$$

9. A multivibrator is shown in Fig. 6.16(a). (a) Determine oscillation frequency f_o, if $R = 91$ kΩ, $R_1 = R_2 = 10$ kΩ, and $C = 0.05$ µF. (b) For given values of $f_o = 150$ Hz, $C = 0.1$ µF and $R_1 = R_2 = 10$ kΩ, determine the value of R. (c) When $R = 15$, $R_1 = R_2 = 20$, and $f_o = 100$ Hz, find the value of C.

Solution

The time period is given by

$$T = 2RC \ln\left(\frac{2R_1}{R_2} + 1\right) \text{ and } f_o = \frac{1}{T}$$

(a) $T = 2 \times 91 \times 10^3 \times 0.05 \times 10^{-6} \ln\left(\frac{2 \times 10}{10} + 1\right) = 10$ ms

$$f_o = \frac{1}{10 \times 10^{-3}} = 100 \text{ Hz}$$

(b) $T = \dfrac{1}{f_o} = \dfrac{1}{150} = 6.667$ ms

$6.667 \times 10^{-3} = 2R \times 0.1 \times 10^{-6} \ln\left(\dfrac{2 \times 10}{10} + 1\right) = 0.22R \times 10^{-6}$

$R = \dfrac{6.667 \times 10^3}{0.22} = \mathbf{30.3\ k\Omega}$

(c) $T = \dfrac{1}{f_o} = \dfrac{1}{100} = 10.0$ ms

$10.0 \times 10^{-3} = 2C \times 15 \times 10^3 \ln\left(\dfrac{2 \times 20}{20} + 1\right) = 33.0C \times 10^3$

$C = \dfrac{10.0 \times 10^{-3}}{33.0 \times 10^3} = \mathbf{0.303\ \mu F}$

10. The blocking oscillator shown in Fig. 6.19(a) uses a silicon transistor with $V_{CE} = 0.3$ V and $V_{BE} = 0.7$ V, and $h_{fe} = 50$. The transformer turns are $n = 2$ and $n_1 = 1$, and $L = 3$ mH. Calculate (a) the pulse amplitude at the collector, (b) the collector current i_C, (c) the base current i_B, and (d) the pulse width. Take $R = 1.5$ kΩ, $R_L = 2.5$ kΩ, and $V_{CC} = 10$ V.

Solution

(a) From the collector circuit the pulse amplitude at the collector,

$$I_C = h_{fe} I_B = \dfrac{V_{CC} - V_{CE}}{(h_{fe} + 1)R} \times h_{fe}$$

$$= \dfrac{10 - 0.3}{51 \times 1.5} \times 50 = \mathbf{6.34\ mA}$$

The equivalent circuit is shown in Fig. 6.42. Applying KVL to loop BECNB

$$nV + V = V_{CC} + V_{BE} - V_{CE}$$

or $V = \dfrac{10 + 0.7 - 0.3}{2 + 1} = 3.467$ V

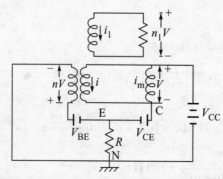

FIGURE 6.42 An equivalent circuit for Solved Problem 10.

Applying KVL to loop BENB

$$nV - V_{BE} = (i_C + i_B)R$$

or $\quad -i_E = i_C + i_B = \dfrac{nV - V_{BE}}{R} = \dfrac{2 \times 3.467 - 0.7}{1.5} = 4.156 \text{ mA}$ \hfill (1)

From load circuit,

$$i_1 = -\dfrac{n_1 V}{R_L} = -\dfrac{1 \times 3.467}{2.5} = -1.387 \text{ mA}$$

For magnetizing circuit,

$$i_m = \dfrac{Vt}{L} = \dfrac{3.467t}{3 \times 10^{-3}} = 1155.7 \text{ mA}$$

Total ampere turn is zero and given by

$$(i_C - i_m) - n i_B + n_1 i_1 = 0$$

or $\quad i_C - 2i_B = 1.387 + 1155.7t$ \hfill (2)

(b) Solving for i_C from (1) and (2),

$$i_C = \mathbf{3.233 + 385t}$$

(c) Substituting for i_C in (1),

$$i_B = \mathbf{0.923 - 385t}$$

(d) At T_p, $i_B \approx 0$, hence the pulse width,

$$T_p = \dfrac{0.923}{385} = \mathbf{2.397 \text{ ms}}$$

11. For the Schmitt trigger shown in Fig. 6.26, suppose that $R_1 = 5$ kΩ, $R_2 = 20$ kΩ, and $V_{CC} = 12$ V. Find the hysteresis width.

Solution

The feedback gain,

$$\beta = \dfrac{R_1}{R_1 + R_2} = \dfrac{5}{5 + 20} = 0.2$$

Hence, the hysteresis width, from Fig. 6.27,

$$H_w = 2\beta V_{CC} = 2 \times 0.2 \times 12 = \mathbf{4.8 \text{ V}}$$

6.11 EXERCISES

1. A phase-shift oscillator is shown in Fig. 6.41. (a) For given $C = 0.01$ µF and $R = 10$ kΩ, determine the oscillation frequency and the minimum value of R_F. (b) For $R_F = 200$ kΩ and $C = 0.05$ µF, determine R and the oscillation frequency.

2. (a) For Fig. 6.4, g_m = 10 kΩ⁻¹ and C = 0.001 µF, determine the value of R and the minimum value of R_D, if the oscillation frequency is 1 kHz. (b) Determine the minimum value of g_m and C, if R = 100 kΩ, frequency of oscillation = 1 kHz, and R_D = 5 kΩ.

3. For the phase-shift oscillator of Fig. 6.5, R_1 = 2.5 kΩ, R_2 = 2.5 kΩ, R_C = 1.5 kΩ, h_{ie} = 1.1 kΩ, and f = 1 kHz. (a) Determine the value of R and R_s, if C = 0.05 µF. (b) Determine the value of C and R_s, if R = 1.3 kΩ.

4. (a) For the Wien bridge shown in Fig. 6.6, if R = R_1 = 5 kΩ, C = 0.02 µF, determine the value of C and minimum value of R_2. (b) Determine the value of C and the maximum value of R_1, if R_2 = 10 kΩ, f_o = 1 kHz, and R = 10 kΩ.

5. An oscillator circuit is shown in Fig. 6.43. (a) Show that the voltage transfer function for the RC is as given below:

$$\beta = \frac{1}{3 + j(\omega CR - 1/\omega CR)}$$

(b) Find the oscillation frequency f_o. Also, derive relationship between R_1 and R_2 for oscillation frequency.

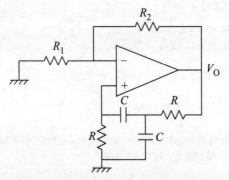

FIGURE 6.43 Circuit for question 5.

6. For the FET Hartley oscillator shown in Fig. 6.7(a) L_1 = 120 µH and L_2 = 150 µH. (a) Determine the oscillation frequency and the minimum value of g_m, if C = 500 pF and R_D = 1 kΩ. (b) For given f_o = 750 kHz and g_m = 5 mΩ⁻¹, determine the value of C and the minimum value of R_D.

7. For the circuit of FET Colpitts oscillator shown in Fig. 6.8(a), C_1 = 1200 pF, C_2 = 1500 pF, and R_G = 10 MΩ. (a) Determine the oscillation frequency and the minimum value of g_m, if L = 75 µH and R_D = 3 kΩ. (b) For given f_o = 450 kHz and g_m = 5 mΩ⁻¹, determine the value of L and the minimum value of R_D.

8. The electric equivalent circuit of a quartz crystal, as shown in Fig. 6.11(b), has the values R = 1.5 kΩ, L = 4.5 H, C = 0.04 pF, and C' = 10 pF. Determine (a) the series resonance frequency ω_s, (b) the parallel resonance frequency ω_p, and the Q factor at the series resonance frequency.

9. A multivibrator is shown in Fig. 6.16(a). (a) Determine oscillation frequency f_o, if $R = 55$ kΩ, $R_1 = R_2$ 15 kΩ, and $C = 0.04$ µF. (b) For given values of $f_o = 100$ Hz, $C = 0.05$ µF $R = 20$ kΩ, and $R_2 = 10$ kΩ, determine the value of R_1. (c) When $R = 20$ kΩ, $R_1 = 15$ kΩ, $C = 0.05$ µF, and $f_o = 100$ Hz, find the value of R_2.

10. Consider the blocking oscillator shown in Fig. 6.19(a). The transformer turns are $n = 3$ and $n_1 = 2$, and $L = 3$ mH. Calculate (a) the pulse amplitude at the collector, (b) the collector current i_C, (c) the base current i_B, and (d) the pulse width. Take $R = 2$ kΩ, $R_L = 3.5$ kΩ, and $V_{CC} = 15$ V.

11. For the Schmitt trigger shown in Fig. 6.26, suppose that $R_1 = 10$ kΩ, $R_2 = 25$ kΩ, and $V_{CC} = 15$ V. Find the hysteresis width.

7

ATTENUATORS AND FILTERS

Outline

7.1 Attenuators   296
7.2 Symmetrical Attenuators   300
7.3 Padding Sources and Loads   308
7.4 Passive Filters   310
7.5 RC Filters   312
7.6 LC Filters   316
7.7 Active Filters   319
7.8 Active Resonant Bandpass Filter   325
7.9 Review Questions   330
7.10 Solved Problems   330
7.11 Exercises   341

7.1 ATTENUATORS

Unlike an amplifier an attenuator is used to reduce the signal level by a desired amount. The ordinary voltage divider, shown in Fig. 7.1, is one of the simplest types of attenuators. The voltage gain of the network is the ratio of the output voltage to the input voltage. From Fig. 7.1,

$$V_o = IR_2$$

But

$$I = \frac{V_i}{R_1 + R_2}$$

$$\therefore \quad V_o = \frac{R_2}{R_1 + R_2} V_i \tag{7.1}$$

The gain of the voltage divider,

$$A = \frac{V_o}{V_i} = \frac{R_2}{R_1 + R_2} \tag{7.2}$$

From Eq. (7.2) it is clear that the gain of the attenuator is always less than unity. The gain expressed in decibel (dB) is

$$A_{dB} = 20 \log_{10} A \tag{7.3}$$

Since A is less than unity, A_{dB} is always negative. For example, if $R_1 = 9$ kΩ and $R_2 = 1$ kΩ, the gain $A = 0.1$, and $A_{dB} = -20$ dB.

The term *attenuation*, which is more commonly used for attenuators, is the reciprocal of the gain. The attenuation is, therefore, defined as

$$\overline{A} = \frac{V_i}{V_o} = \frac{1}{A} = \frac{R_1 + R_2}{R_2} \tag{7.4}$$

and

$$\overline{A}_{dB} = -A_{dB} \tag{7.5}$$

Hence, $\overline{A}$ is always greater than unity and $\overline{A}_{dB}$ is always positive. For the above example, $\overline{A} = 10$ and $\overline{A}_{dB} = 20$ dB. Thus, the attenuation $\overline{A}_{dB}$ is negative of gain A_{dB}.

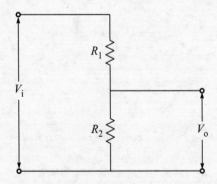

FIGURE 7.1 A resistance attenuator.

Now, let us consider the operational amplifier (OP AMP) of Fig. 5.8(a) and Eq. (5.19). If $Z_f = R_f$ and $Z_s = R_s$, the output of the amplifier is given by

$$V_o = \frac{R_f}{R_s} V_s \tag{7.6}$$

If $R_f < R_s$, the OP AMP works as an attenuator.

As far as d.c. or low-frequency a.c. input signals are concerned, the voltage divider of Fig. 7.1 works satisfactorily. At higher frequencies, however, the reactance due to stray capacitance C_2 that shunts R_2 becomes appreciable and affects the output. The voltage divider with the stray capacitance is shown in Fig. 7.2(a). The capacitance C_2 may be, for example, the input capacitance of the next stage of amplifier. The Thevenin's equivalent of Fig. 7.2(a) is shown in Fig. 7.2(b). The resistance R is equal to the parallel combination of R_1 and R_2 and is given by

$$R = \frac{R_1 R_2}{R_1 + R_2} \tag{7.7}$$

and the Thevenin's voltage equivalent is given by

$$V_i' = \frac{R_2}{R_1 + R_2} V_i \tag{7.8}$$

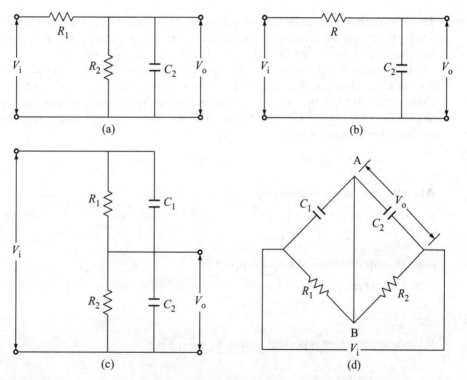

FIGURE 7.2 An attenuator. (a) An actual circuit considering stray capacitance, (b) equivalent circuit of (a), (c) compensated attenuator, and (d) circuit of (c) redrawn as a bridge.

Both resistances R_1 and R_2 are required to be large so that the input resistance R [Fig. 7.2(b)] of the attenuator may be large enough to prevent loading down of the input signal. Because of the capacitance C_2, the output of the attenuator does not follow the input instantaneously. The output, for a step input, is given by

$$V_o = V_i \left(1 - e^{-t/RC_2}\right) \tag{7.9}$$

The rise time is the interval of time required for the step response of a system to go from 10% to 90% of its final value. The time required for V_o to reach 10% of it final value is $0.1\,RC_2$, and the time to reach 90% of its final value is $2.3RC_2$. Hence, the rise time is $2.2RC_2$. For example, if R_1 and R_2 are taken equal to 1 MΩ and C_2 = 15 pF, then the rise time for V_o in Fig. 7.2(b) for a step input is 2.2 × 0.5 × 15 = 16.5 μs. This rise time is large and is unacceptable. At higher frequencies the output of the attenuator reduces appreciably. This effect, thus, limits the frequency range of the attenuator. To make the attenuator frequency independent, it may be compensated by shunting resistance R_1 by a capacitance C_1 as shown in Fig. 7.2(c). The circuit of Fig. 7.2(c) is redrawn as a bridge in Fig. 7.2(d), the resistances and capacitances forming its four arms. For $R_1C_1 = R_2C_2$, the bridge is balanced and no current flows through the branch connecting the points A and B. Hence, under the condition $R_1C_1 = R_2C_2$, the output

$$V_0 = \frac{R_2}{R_1 + R_2} V_i = \frac{C_1}{C_1 + C_2} V_i = \overline{A} V_i \tag{7.10}$$

Thus, the output is independent of frequency. In practice, C_1 is made adjustable and the final adjustment for compensation is made experimentally by the method of square wave testing as the condition $R_1C_1 = R_2C_2$ must be satisfied precisely.

Let us consider the step response of the attenuator if the compensation is incorrect. As the input changes abruptly by V_i at $t = 0$, the voltage across C_1 and C_2 must also change abruptly. But this is possible if an infinite current flows through the capacitances at $t = 0$ for an infinitesimal time. Hence, a finite charge is developed across each capacitor. The charge is given by

$$q = \int_{0-}^{0+} i\, dt$$

At $t = 0+$, the voltage is given by

$$V_i = \frac{q}{C_1} + \frac{q}{C_2} = \frac{C_1 + C_2}{C_1 C_2} q \tag{7.11}$$

and the output voltage at $t = 0+$ is given by

$$V_o(0+) = \frac{q}{C_2} = \frac{C_1}{C_1 + C_2} V_i \tag{7.12}$$

When compensation is perfect, that is, $C_1 = \dfrac{C_2 R_2}{R_1}$,

$$V_o(0+) = \overline{A} V \tag{7.13}$$

The output, therefore, is a step output.

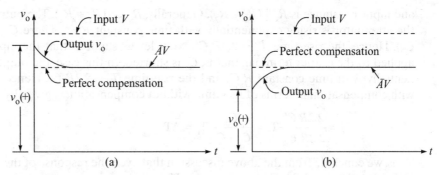

FIGURE 7.3 Step response (a) when $C_1 > C_2 R_2 / R_1$ and (b) when $C_1 < C_2 R_2 / R_1$.

When the attenuator is overcompensated, that is, $C_1 > \dfrac{C_2 R_2}{R_1}$,

$$V_o(0+) > \overline{A} V \qquad (7.14)$$

The output, therefore, is initially spiked step voltage as shown in Fig. 7.3(a).

When the attenuator is undercompensated, that is, $C_1 < \dfrac{C_2 R_2}{R_1}$,

$$V_o(0+) < \overline{A} V \qquad (7.15)$$

The output, therefore, is initially rounded step voltage as shown in Fig. 7.3(b).

In the above analysis it has been taken that an infinite current flows through the capacitors for an infinitesimal time, which never happens in practice. The reason for reaching to the conclusion of infinite current for infinitesimal time is the assumption that the input source has zero impedance. Now, we take the input source resistance R_s and study the effect on the step response of the compensated attenuator. The circuit of compensated attenuator with source resistance R_s is shown in Fig. 7.4(a). The branch joining the points A and B in Fig. 7.4(a) can be removed, as no current flows through the branch for the compensated attenuator. The Thevenin's equivalent of Fig. 7.4(a) is shown in Fig. 7.4(b). The equivalent input voltage,

$$V_i' = \dfrac{R_1 + R_2}{R_s + R_1 + R_2} V_i$$

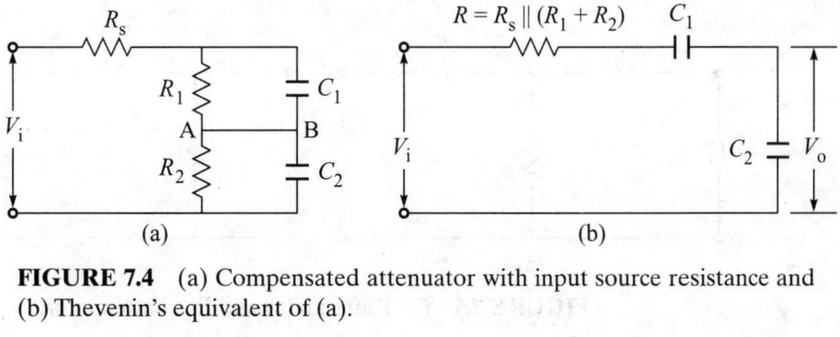

FIGURE 7.4 (a) Compensated attenuator with input source resistance and (b) Thevenin's equivalent of (a).

and input resistance is $R_s \parallel (R_1 + R_2)$. Generally, $R_s \ll (R_1 + R_2)$. The output voltage of the attenuator is rising exponentially with time constant $R_s C'$, where $C' = C_1 C_2/(C_1 + C_2)$. Hence, the rise time $T'_s = 2.2 R_s C'$. Now, let us suppose that the input is directly applied to the output terminal, that is, C_1 is shorted. In this case, the output rises exponentially with time constant $R_s C_2$ and the rise time $T_s = 2.2 R_s C_2$. Hence, the rise time with compensation in terms of rise time without compensation is given by

$$T'_r = \frac{2.2 R_s C'}{2.2 R_s C_2} T_r = \frac{C_1}{C_1 + C_2} T_r = \overline{A} T_r \qquad (7.16)$$

Thus, we conclude from the above discussion that even the response of the compensated attenuator is not an ideal step response. However, an improvement in rise time does result if the attenuator is compensated. For example, if the output of the attenuator is one-tenth the input, the rise time of the output using the compensated attenuator is one-tenth of what it would be without compensation.

7.2 SYMMETRICAL ATTENUATORS

In addition to introducing attenuation of the input signal, an attenuator is used for impedance matching between the input and output terminals. An attenuator is inserted between a source with source resistance R_s and a load R_L, as shown in Fig. 7.5. For the impedance matching, the input resistance of the attenuator must be equal to the source resistance R_s and its output resistance must be equal to the load resistance.

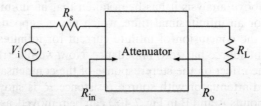

FIGURE 7.5 An attenuator for impedance matching.

There are two forms of symmetrical attenuators, namely, T attenuator and π attenuators shown in Fig. 7.6. These networks are symmetrical about a vertical centre line. Since the input and the output have a common terminal, the networks of Fig. 7.6 are referred to as unbalanced symmetrical attenuators.

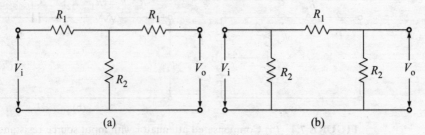

FIGURE 7.6 (a) T attenuator and (b) π attenuator.

The input resistance R_{in} of the attenuator of Fig. 7.5, looking into the attenuator, depends on the value of the load resistance R_L. For a particular value of R_L, the input resistance R_{in} is equal to R_L. This value of R_L is called characteristic or image resistance of the attenuator. Every attenuator has a characteristic resistance R_{ch}, and so an attenuator should normally be loaded in its characteristic resistance. If an attenuator with $R_{in} = R_{ch}$ is inserted between a source with $R_s = R_{ch}$ and a load $R_L = R_{ch}$, there is an impedance matching between the source and load of the attenuator. If R_{is} is the input resistance of the attenuator with output terminals shorted and R_{io} is its input resistance with output terminals open, the characteristic resistance is given by

$$R_{ch} = \sqrt{R_{is} R_{io}} \tag{7.17}$$

7.2.1 Symmetrical T Attenuators

A symmetrical T attenuator has been shown again in Fig. 7.6(a). Let us define

$$m = \frac{R_2}{R_1} \tag{7.18}$$

or $\quad R_2 = mR_1 \tag{7.19}$

The T attenuator is redrawn in Fig. 7.7(a). The input resistance when the output is shorted is given by

$$R_{is} = R_1 + \frac{mR_1}{(1+m)} \tag{7.20}$$

and the input resistance when output is open is given by

$$R_{io} = (1+m)R_1 \tag{7.21}$$

Hence, the characteristic resistance of the T attenuator,

$$R_{ch} = \sqrt{\left(R_1 + \frac{mR_1}{(1+m)}\right)(1+m)R_1}$$

or $\quad R_{ch} = R_1\sqrt{(1+2m)} \tag{7.22}$

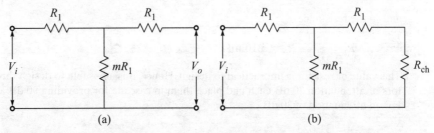

FIGURE 7.7 (a) Symmetrical T attenuator and (b) the attenuator loaded with R_{ch}.

The attenuator loaded with its characteristic resistance R_{ch} is shown in Fig. 7.7(b). The attenuation under the condition of Fig. 7.7(b) is another parameter required to be known. The attenuation of the attenuator loaded with the characteristic resistance,

$$\overline{A} = \frac{V_i}{V_o} = \frac{1+m+\sqrt{1+2m}}{m} \tag{7.23}$$

Therefore, for the given R_1 and R_2 of the T attenuator, its characteristic resistance and attenuation when loaded with its characteristic resistance can be calculated by using Eqs. (7.22) and (7.23), respectively.

The problem of designing a T attenuator is to find the values of R_1 and R_2 for given characteristic resistance R_{ch} and attenuation $\overline{A}$. By solving Eqs. (7.22) and (7.23) simultaneously, R_1 and R_2 are obtained as follows:

$$R_1 = \frac{\overline{A}-1}{\overline{A}+1} R_{ch} \tag{7.24}$$

$$R_2 = \frac{2\overline{A}}{\overline{A}^2 - 1} R_{ch} \tag{7.25}$$

From Eqs. (7.24) and (7.25), it is clear that, for a given attenuation $\overline{A}$, R_1 and R_2 are directly proportional to R_{ch}. If R_1 and R_2 are known for $R_{ch} = R_{ch1}$, then R_1 and R_2 for $R_{ch} = R_{ch2}$ can be obtained by multiplying each R_1 and R_2 by R_{ch2}/R_{ch1}.

7.2.2 Cascading T Attenuators

Equation (7.25) reveals that for higher values of attenuation the value of R_2 becomes very small. Construction of very small resistances is difficult as well as uneconomical. Hence, it is convenient to use a number of attenuators of small attenuations in cascade in place of a single attenuator of given high-value of attenuation.

Example 7.1

Design a T attenuator for attenuation of 90 dB and characteristic resistance of 50 Ω.

Solution

For an attenuation of 90 dB,

$$\overline{A} = 31622.78$$

and so the resistance

$$R_2 = \frac{2\overline{A}}{\overline{A}^2 - 1} R_{ch} = 3.16 \text{ m}\Omega$$

This value of resistor is impractical to design. Hence, it is advisable to design three attenuators of attenuation 30 dB each and place them in cascade for providing 90 dB attenuation. For an attenuation of 30 dB,

$$\overline{A} = 31.623$$

and so the resistances

$$R_1 = \frac{\overline{A}-1}{\overline{A}+1} R_{ch} = 46.9 \; \Omega$$

$$R_2 = \frac{2\overline{A}}{\overline{A}^2 - 1} R_{ch} = 3.17 \; \Omega$$

Three 30-dB and 50-Ω attenuators in cascade are shown in Fig. 7.8. We see that the resistance level is preserved from the load to the source. Starting from the load end, we find that the input resistance of the third attenuator is 50 Ω. Hence, all attenuators are loaded correctly in characteristic resistance.

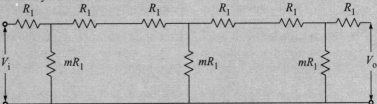

FIGURE 7.8 Three T attenuators in cascade.

7.2.3 Symmetrical π Attenuators

A symmetrical π attenuator has been shown in Fig. 7.6(b). The π attenuator is, sometimes, preferred to the T attenuator. Like in T attenuator let us define

$$m = \frac{R_2}{R_1} \tag{7.26}$$

or $\quad R_2 = mR_1 \tag{7.27}$

The π attenuator is redrawn in Fig. 7.9(a). The input resistance when the output is shorted is given by

$$R_{is} = \frac{mR_1}{(1+m)} \tag{7.28}$$

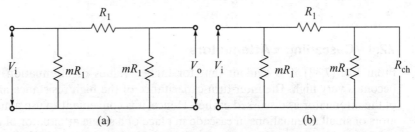

FIGURE 7.9 (a) Symmetrical π attenuator and (b) the attenuator loaded with R_{ch}.

and the input resistance when the output is open is given by

$$R_{io} = \frac{m(1+m)R_1}{(1+2m)} \tag{7.29}$$

Hence, the characteristic resistance of the π attenuator,

$$R_{ch} = \sqrt{\left(\frac{mR_1}{1+m}\right)\left(\frac{m(1+m)R_1}{1+2m}\right)}$$

or

$$R_{ch} = \frac{m}{\sqrt{1+2m}} R_1 \tag{7.30}$$

The attenuator loaded with its characteristic resistance R_{ch} is shown in Fig. 7.9(b). The attenuation under the condition of Fig. 7.9(b) is another parameter required to be known. The attenuation of the attenuator loaded with its characteristic resistance,

$$\overline{A} = \frac{V_i}{V_o} = \frac{1+m+\sqrt{1+2m}}{m} \tag{7.31}$$

Therefore, for a given R_1 and R_2 of the π attenuator, its characteristic resistance and attenuation when loaded with its characteristic resistance can be calculated by using Eqs. (7.30) and (7.31), respectively.

The problem of designing a π attenuator is to find the values of R_1 and R_2. For given characteristic resistance R_{ch} and attenuation $\overline{A}$, the π attenuator can be designed by using Eqs. (7.30) and (7.31), simultaneously. R_1 and R_2 are given by

$$R_1 = \frac{\overline{A}^2 - 1}{2\overline{A}} R_{ch} \tag{7.32}$$

$$R_2 = \frac{\overline{A}+1}{\overline{A}-1} R_{ch} \tag{7.33}$$

From Eqs. (7.32) and (7.33), it is clear that, for a given attenuation $\overline{A}$, R_1 and R_2 are directly proportional to R_{ch}. If R_1 and R_2 are known for $R_{ch} = R_{ch1}$, then R_1 and R_2 for $R_{ch} = R_{ch2}$ can be obtained by multiplying each R_1 and R_2 by R_{ch2}/R_{ch1}.

7.2.4 Cascading π Attenuators

From Eq. (7.32) it is evident that for higher values of attenuation the value of R_1 becomes very high. The inter-turn capacitance of the high resistance affects the output of the attenuator as discussed earlier. Hence, it is convenient to use a number of attenuators of small attenuations in cascade in place of a single attenuator of given high-value of attenuation.

Example 7.2

Design a π attenuator for attenuation of 90 dB and characteristic resistance of 50 Ω.

Solution

For an attenuation of 90 dB,

$$\overline{A} = 31622.78$$

and so the resistance

$$R_1 = \frac{\overline{A}^2 - 1}{2\overline{A}} R_{ch} = \frac{(31622.78)^2 - 1}{2 \times 31622.78} \times 50 = 790.6 \text{ k}\Omega$$

This value of resistor is very high. Hence, it is advisable to design three attenuators of attenuation 30 dB each and place them in cascade for providing 90 dB attenuation. For an attenuation of 30 dB,

$$\overline{A} = 31.623$$

and so the resistances

$$R_1 = \frac{\overline{A}^2 - 1}{2\overline{A}} R_{ch} = \frac{(31.623)^2 - 1}{2 \times 31.623} \times 50 = 790.6 \text{ }\Omega$$

$$R_2 = \frac{\overline{A} + 1}{\overline{A} - 1} R_{ch} = \frac{31.623 + 1}{31.623 - 1} \times 50 = 53.26 \text{ }\Omega$$

Three 30-dB and 50-Ω attenuators in cascade are shown in Fig. 7.10. We see that the resistance level is preserved from the load to the source. Starting from the load end, we find that the input resistance of the third attenuator is 50 Ω. Hence, all attenuators are loaded correctly in characteristic resistance.

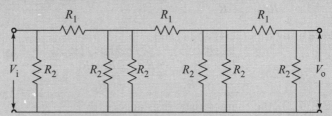

FIGURE 7.10 Three π attenuators in cascade.

7.2.5 Bridged T Attenuators

The bridged T attenuator is an important type of resistance attenuator and is shown in Fig. 7.11(a). As the input and output terminals are bridged by the resistance R_3, this attenuator is known as bridged attenuator. Let us choose the resistances such that

$$R_1 = \sqrt{R_2 R_3} \tag{7.34}$$

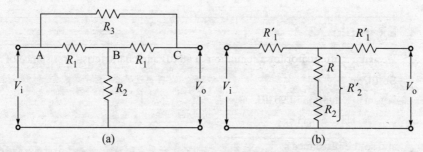

FIGURE 7.11 (a) A bridged T attenuator and (b) equivalent of (a) after delta-star conversion.

After delta-star conversion, Fig. 7.11(a) changes into a T attenuator as shown in Fig. 7.11(b). The T attenuator resistances are given as

$$R'_1 = \frac{R_1 R_3}{2R_1 + R_3} \tag{7.35}$$

$$R'_2 = R_2 + R = R_2 + \frac{R_1^2}{2R_1 + R_3}$$

$$= \frac{2R_1(R_1 + R_2)}{2R_1 + R_3} \tag{7.36}$$

As Fig. 7.11(b) is a T attenuator, Eqs. (7.22) and (7.23) for characteristic resistance and attenuation, respectively, when the attenuator is loaded with its characteristic resistance, can be applied. The parameter m of the T attenuator [Fig. 7.11(b)] is defined as

$$m = \frac{R'_2}{R'_1} = \frac{2(R_1 + R_2)}{R_3} \tag{7.37}$$

and hence using Eqs. (7.34) and (7.37)

$$\sqrt{1+2m} = \sqrt{\frac{R_3^2 + 4R_1 R_3 + 4R_1^2}{R_3^2}} = \frac{2R_1 + R_3}{R_3} \tag{7.38}$$

From Eq. (7.22), the characteristic resistance,

$$R_{ch} = R'_1 \sqrt{1+2m} = \frac{R_1 R_3}{2R_1 + R_3} \times \frac{2R_1 + R_3}{R_3} = R_1 \tag{7.39}$$

From Eq. (7.23), the attenuation,

$$\overline{A} = \frac{V_i}{V_o} = \frac{1 + m + \sqrt{1+2m}}{m}$$

Substituting from Eqs. (7.37) and (7.38) and manipulating,

$$\overline{A} = \frac{R_1 + R_2}{R_2} \tag{7.40}$$

To design a bridged T attenuator, for given R_{ch} and $\overline{A}$, the parameters $R_1, R_2,$ and R_3 can be obtained by solving simultaneously Eqs. (7.34), (7.39), and (7.40). The parameters are given by

$$R_1 = R_{ch} \tag{7.41}$$

$$R_2 = \frac{R_{ch}}{\overline{A} - 1} \tag{7.42}$$

and $\qquad R_3 = (\overline{A} - 1)R_{ch} \tag{7.43}$

From Eqs. (7.41) to (7.43), it is clear that R_1 is independent of $\overline{A}$, and only R_2 and R_3 are function of $\overline{A}$. This is an important property and is used for designing a variable attenuator.

7.2.6 Variable Attenuators

Commercial instruments are often required to have variable attenuators. A variable attenuator provides freedom of selection of different attenuations while keeping the characteristic resistance unchanged. This can be achieved by using ganged rheostats for the resistors in the attenuator. In attenuators of T and π types, all three resistances are required to vary according to the design equations. In bridged T attenuator, R_1 is not affected by the variation of $\overline{A}$, and so it may be fixed at the value of R_{ch}. In bridged T attenuator, it is, therefore, required only to vary resistors R_2 and R_3 according to Eqs. (7.38) and (7.39). A variable bridged T attenuator is shown in Fig. 7.12. Resistors R_2 and R_3 are ganged together. As R_2 is inversely proportional to $\overline{A}$ and R_3 is directly proportional to $\overline{A}$, linear rheostats do not track properly. However, logarithmic rheostats can be ganged together in accordance with the design equations. Thus, we can design a continuously variable attenuator with constant characteristic resistance.

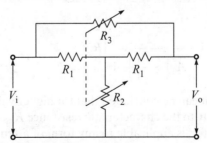

FIGURE 7.12 A variable attenuator.

Because the tracking of R_2 and R_3 is not perfect, some deviation in the value of characteristic resistance is possible. However, the reasonable accurate value of R_{ch} over the range of adjustment is provided by the variable bridged T attenuator.

For more precise values of R_{ch} and $\overline{A}$, the usual procedure is to build a step attenuator. A step attenuator means the variation in step; for example, a step attenuator can be designed to cover from 0 to 10 dB in steps of 3 dB.

A popular way of building a step attenuator is to build a number of simple T attenuators with different values of attenuation. By means of a proper switching arrangement, various combinations of attenuators can be cascaded to produce the desired amount of attenuation.

7.3 PADDING SOURCES AND LOADS

As we have seen, it is desirous that the source and load resistances be equal to the characteristic resistance of the attenuator. Under this condition, impedance matchings occur at both source and load terminals. Impedance matching also provides precise attenuation $\overline{A}$ to the input signal. However, it may require to use an attenuator in a unmatched situation.

There are cases where a signal source resistance (or Thevenin's equivalent) is unknown or it varies because of temperature, ageing, or any other reasons. If we have to use this source in a system with its input resistance R_i, we must stabilize the source resistance to R_i. A way of doing so is by cascading an attenuator with characteristic resistance $R_{ch} = R_i$. If the attenuator has enough attenuation, the Thevenin's resistance, looking back into the attenuator, is very close to R_{ch}. The price paid for this is the loss of some amount of signal level. But this is generally done in good commercial sources where it is far better to have a fixed source resistance with less signal than to have more signal with an unreliable source resistance. This use of attenuator is commonly known as padding a source. That is why attenuators are often called *pads*. Similarly, an attenuator can be used with an uncertain load for matching with a source of known resistance.

A source with resistance R_s padded with a T attenuator is shown in Fig. 7.13. The relation between the output resistance R_o and the characteristic resistance R_{ch} of the attenuator is derived here. The output resistance of the circuit of Fig. 7.13 is given by

$$R_o = R_1 + \frac{R_2(R_1 + R_s)}{R_2 + R_1 + R_s} \tag{7.44}$$

Now, let us define $R_s = kR_{ch}$ (7.45)

Substituting for R_1, R_2, and R_s from Eqs. (7.24), (7.25), and (7.45), respectively, in Eq. (7.44) and manipulating, the output resistance is given by

$$R_o = \left(1 + \frac{2}{\overline{A}^2\left(\frac{k+1}{k-1}\right) - 1}\right) R_{ch} \tag{7.46}$$

From Eq. (7.46), it can be concluded that for higher values of $\overline{A}$, the output resistance R_o approaches closely to the characteristic resistance R_{ch}. Equation (7.46), although derived for the T attenuator, is applicable to any form of attenuators.

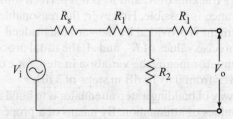

FIGURE 7.13 A source padded with a T attenuator.

Example 7.3

For the circuit of Fig. 7.13, $k = 3$ and $R_{ch} = 50\ \Omega$. Find the output resistance (a) for $\overline{A} = 2$ and (b) for $\overline{A} = 10$.

Solution

(a) For $\overline{A} = 2$, from Eq. (7.46),

$$R_o = \left(1 + \frac{2}{4 \times 2 - 1}\right) \times 50 = 64.3\ \Omega$$

(b) For $\overline{A} = 10$, from Eq. (7.46),

$$R_o = \left(1 + \frac{2}{100 \times 2 - 1}\right) \times 50 = 50.5\ \Omega$$

From the above example, the Thevenin's equivalent source resistance, even for $R_s = 3R_{ch}$ and for $\overline{A} = 10$ is 50.5, which is very close to the characteristic resistance R_{ch}. But the price paid for this is the loss of signal level by a factor of 10.

Now, consider the case where R_s is not known. For this, two limiting values of R_o can be obtained. The source resistance R_s can have any value between $R_s = 0$ and $R_s = \infty$.

For $R_s = 0$, $k = 0$ and Eq. (7.46) reduces to

$$R_o = \left(1 - \frac{2}{\overline{A}^2 + 1}\right) R_{ch} \tag{7.47}$$

For $R_s = \infty$, $k = \infty$ and Eq. (7.46) reduces to

$$R_o = \left(1 + \frac{2}{\overline{A}^2 - 1}\right) R_{ch} \tag{7.48}$$

Example 7.4

Find the Thevenin's equivalent source resistance for two limiting values of R_s if $R_{ch} = 50\ \Omega$, for (a) $\overline{A} = 2$ and (b) $\overline{A} = 10$.

Solution

(a) $\overline{A} = 2$

(i) For $R_s = 0$,

$$R_o = \left(1 - \frac{2}{2^2 + 1}\right) \times 50 = 30\ \Omega$$

(ii) For $R_s = \infty$,

$$R_o = \left(1 + \frac{2}{2^2 - 1}\right) \times 50 = 83.33\ \Omega$$

(b) $\overline{A} = 10$.
  (i) For $R_s = 0$,
  $$R_o = \left(1 - \frac{2}{10^2 + 1}\right) \times 50 = \mathbf{49\ \Omega}$$
  (ii) For $R_s = \infty$,
  $$R_o = \left(1 + \frac{2}{10^2 - 1}\right) \times 50 = \mathbf{51\ \Omega}$$

Finally, from the above example, we conclude that if a pad of $\overline{A} = 10$ (or 20 dB) is used, the padded source (or load) resistance is very close to the value of R_{ch}.

7.4 PASSIVE FILTERS

Filters are used to separate some sinusoidal frequencies from others. A filter is normally placed between a source and a load. If the input source produces a signal containing a number of sinusoidally varying signals of different frequencies, some frequencies are passed through the filter to the load with very little attenuation, whereas other frequencies are heavily attenuated.

Filters are divided into the following four groups:

1. Low-pass filter.
2. High-pass filter.
3. Bandpass filter.
4. Bandstop filter.

Ideal responses of four types of filters are shown in Fig. 7.14.

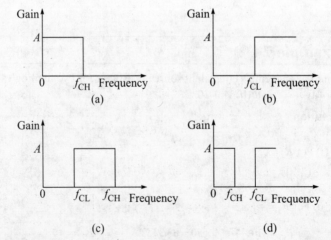

FIGURE 7.14 Ideal responses of (a) low-pass filter, (b) high-pass filter, (c) bandpass filter, and (d) bandstop filter.

7.4.1 Low-Pass Filter

In an ideal low-pass filter, the gain (V_o/V_i) is a constant A from zero frequency up to a cutoff frequency f_c and zero above the cutoff frequency [Fig. 7.14(a)]. However, the response of a real low-pass filter is not as shown in Fig. 7.14(a). In a real low-pass filter, the change in gain from A to zero is not instantaneous. It is as shown in Fig. 7.15(a). The cutoff frequency of a practical low-pass filter is defined as that frequency at which the voltage gain drops from A to $0.707A$ (or by 3 dB).

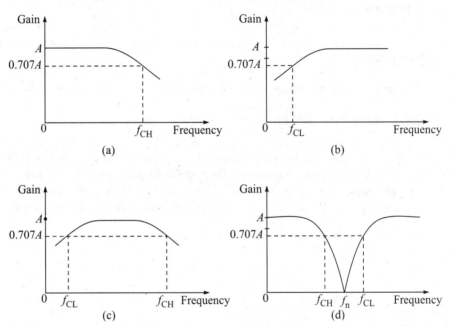

FIGURE 7.15 Real responses of (a) low-pass filter, (b) high-pass filter, (c) bandpass filter, and (d) bandstop filter.

7.4.2 High-Pass Filter

An ideal high-pass filter has zero gain from zero frequency up to a cutoff frequency f_c. Above the cutoff frequency, the gain is a constant A. The response of the ideal high-pass filter is shown in Fig. 7.14(b). Like low-pass filter the response of a practical high-pass filter differs from that of an ideal high-pass filter of Fig. 7.14(b). Unlike the ideal high-pass filter, the change in gain from zero to a constant A is not instantaneous. The response of a real filter is shown in Fig. 7.15(b). The cutoff frequency of a practical high-pass filter is also defined as the frequency at which the voltage gain rises from zero to $0.707A$ (or by 3 dB).

7.4.3 Bandpass Filter

A bandpass filter has two cutoff frequencies. As shown in Fig. 7.14(c) the gain of an ideal bandpass filter is a constant A between the cutoff frequencies and zero outside of these frequencies. Like low-pass and high-pass filters, the changes of gain at cutoff frequencies are not instantaneous in a practical filter. The response of the real bandpass filter is shown in Fig 7.15(c). The definition of cutoff frequencies is same as that in the low- and high-pass filters.

7.4.4 Bandstop Filter

A bandstop filter, like the bandpass filter, has two cutoff frequencies. As shown in Fig. 7.14(d) the gain of an ideal bandstop filter is a constant A outside the cutoff frequencies and zero between the cutoff frequencies. The changes of gain at cutoff frequencies, again, are not instantaneous in a practical filter. The response of the real bandstop filter is shown in Fig. 7.15(d).

7.5 RC FILTERS

The simplest and most economical filters are RC filters using resistance and capacitance.

7.5.1 Low-Pass RC Filter

A low-pass RC filter is shown in Fig. 7.16. At very low frequencies, the capacitive reactance is very high and the output voltage V_o is approximately equal to V_i. The gain, thus, is unity. At very high frequencies, the capacitive reactance is very low, and hence the output voltage is very small. Thus, this filter attenuates heavily at high frequencies. The passband gain of the filter is unity. The cutoff frequency for the filter is given by

$$f_{cH} = \frac{1}{2\pi RC} \tag{7.49}$$

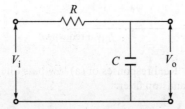

FIGURE 7.16 RC low-pass filter.

Due to load resistance R_L across the output terminals and source resistance R_s, the passband gain and cutoff frequency are given as follows

$$A = \frac{R_L}{R_s + R + R_L} \tag{7.50}$$

$$f_{cH} = \frac{1}{2\pi R_{eq} C} \tag{7.51}$$

where $$R_{eq} = \frac{(R_s + R)R_L}{R_s + R + R_L} \tag{7.52}$$

The ideal response of the RC low-pass filter is as shown in Fig. 7.14(a) and real response as in Fig. 7.15(a).

7.5.2 High-Pass RC Filter

A simple high-pass RC filter is shown in Fig. 7.17. The response is as shown in Fig. 7.15(b). The cutoff frequency is given by

$$f_{cL} = \frac{1}{2\pi RC} \tag{7.53}$$

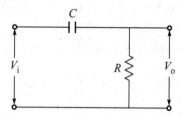

FIGURE 7.17 RC high-pass filter.

If load and source resistances are R_L and R_s, respectively, the passband gain and cutoff frequency are given by Eqs. (7.54) and (7.55), respectively.

$$\text{Gain } A = \frac{RR_L}{R_s(R+R_L)+RR_L} \tag{7.54}$$

$$f_{cL} = \frac{1}{2\pi R_{eq} C} \tag{7.55}$$

where

$$R_{eq} = R_s + \frac{RR_L}{R+R_L} \tag{7.56}$$

7.5.3 Bandpass RC Filter

A simple bandpass filter is constructed by cascading low-pass filter and high-pass filter. An RC bandpass filter is shown in Fig. 7.18. If parameters are appropriately selected, the filter has a lower cutoff frequency f_{cL} and an upper cutoff frequency f_{cH}. These frequencies are given by

$$f_{cL} = \frac{1}{2\pi R_2 C_2} \tag{7.57}$$

$$f_{cH} = \frac{1}{2\pi R_1 C_1} \tag{7.58}$$

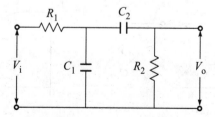

FIGURE 7.18 RC bandpass filter.

In the passband the capacitor C_1 may be considered open and C_2 shorted. Hence, in bandpass the filter behaves as a simple voltage divider. The gain under bandpass is given by

$$A = \frac{R_2}{R_1 + R_2} \tag{7.59}$$

If load and source resistances are R_L and R_s, respectively, the passband, gain and cutoff frequencies are given by Eqs. (7.60), (7.61), and (7.63), respectively.

$$A = \frac{R_2 R_L}{(R_s + R_1)(R_2 + R_L) + R_2 R_L} \tag{7.60}$$

$$f_{cL} = \frac{1}{2\pi R_{eq} C_2} \tag{7.61}$$

where

$$R_{eq} = \frac{R_2 R_L}{R_2 + R_L} \tag{7.62}$$

$$f_{cH} = \frac{1}{2\pi R_{eq} C_1} \tag{7.63}$$

where

$$R_{eq} = R_s + R_1 \tag{7.64}$$

If the cutoff frequencies are closed together, then too much interaction takes place between the two sections of the filter. The circuit, therefore, is not suitable for narrow band applications. Because of this reason, the circuit is designed for the situation where the higher cutoff frequency is much larger than the lower cutoff frequency. It is recommended, therefore, that the cutoff frequencies be at least one decade apart, that is, $f_{cH}/f_{cL} > 10$. To satisfy the above condition, $R_2 C_2 > R_1 C_1$, and if $C_1 = C_2$, $R_2 > R_1$.

7.5.4 Bandstop RC Filter

A bandstop RC filter uses either a Wien bridge or a twin T circuit. The Wien-bridge circuit is used in a notch filter. A notch filter completely attenuates out one particular frequency. The twin T circuit has two T circuits in parallel.

A special form of the Wien bridge is shown in Fig. 7.19(a). At very low and very high frequencies, the bridge is unbalanced. At very low frequency, the capacitors may be

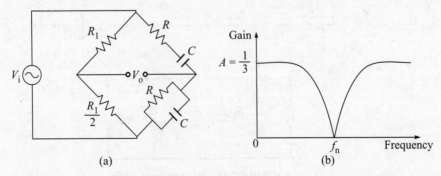

FIGURE 7.19 (a) The Wien-bridge bandstop filter and (b) the response of the filter.

treated as open circuited and there is no current through the resistor R. The output voltage is equal to the voltage drop across $R_1/2$. Hence, the output voltage

$$V_o = \frac{V_i}{3} \tag{7.65}$$

At very high frequency the capacitors may be treated as short circuited. The output voltage is again equal to the voltage drop across the resistor $R_1/2$. The output voltage is again given by Eq. (7.65). At a frequency f_n, between very high and very low frequencies, the bridge may be balanced. The frequency at which the bridge is balanced is called notch frequency f_n. Hence, the output at the notch frequency is zero. When the bridge is balanced,

$$\frac{R_1}{R + \dfrac{1}{j\omega_n C}} = \frac{\dfrac{R_1}{2}}{\dfrac{R}{1 + j\omega_n CR}}$$

or

$$R + \frac{1}{j\omega_n C} = \frac{2R}{1 + j\omega_n CR}$$

or

$$\omega_n^2 = \frac{1}{R^2 C^2}$$

$$\therefore \quad f_n = \frac{1}{2\pi RC} \tag{7.66}$$

Thus, the notch frequency can be calculated from Eq. (7.66). The response of this filter is shown in Fig. 7.19(b).

A bandstop RC filter using twin T circuit is shown in Fig. 7.20. The circuit behaves very much same as the Wien bridge. At very low frequencies, the capacitors may be treated open circuited, and hence the gain is unity. At very high frequencies, the capacitors may be treated short circuited, and hence the gain is unity. Thus, the passband gain is approximately equal to unity. The response is as shown in Fig. 7.19(b), but with $A = 1$.

The nodal equations for the nodes A, O, and B are

At node A: $\left(\dfrac{2}{R} + j2\omega C\right) V_A - \dfrac{1}{R} V_O = \dfrac{1}{R} V_i$ \hfill (7.67)

At node O: $-\dfrac{1}{R} V_A + \left(\dfrac{1}{R} + j\omega C\right) V_O - j\omega C V_B = 0$ \hfill (7.68)

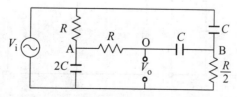

FIGURE 7.20 A bandstop RC filter using twin T circuit.

At node B: $-j\omega C V_O + \left(\dfrac{2}{R} + j2\omega C\right) V_B = j\omega C V_i$ \hfill (7.69)

Solving Eqs. (7.67) to (7.69) for V_O

$$\left(\dfrac{1}{R^2} + j\dfrac{4\omega C}{R} - \omega^2 C^2\right) V_O = \left(\dfrac{1}{R^2} - \omega^2 C^2\right) V_i \quad (7.70)$$

At notch frequency, $V_O = 0$; hence, the notch frequency is given by

$$\left(\dfrac{1}{R^2} - \omega^2 C^2\right) = 0$$

or $\quad \omega = \dfrac{1}{RC}$

$\therefore \quad f_n = \dfrac{1}{2\pi RC}$ \hfill (7.71)

The twin T circuit RC filter has a distinct advantage over the Wien-bridge filter in that it has a common input and output terminal. However, in tuning the twin T circuit, RC filter needs to gang together three resistors and three capacitors, whereas in Wien-bridge filter only two capacitors need to be ganged. The Wien-bridge filter is, therefore, preferred where a tuning notch filter is needed despite of the lack of a common input and output terminal.

7.6 LC FILTERS

LC filters are closer to the ideal filters and their responses are also closer to that of the ideal filters. However, inductors are generally expensive and usually not available in precise values. In addition, inductors are usually available in odd shapes and are not used in printed circuit boards.

7.6.1 Low-Pass LC Filter

Low-pass LC filters are shown in Fig. 7.21. The low-pass LC filter of Fig. 7.21(a) is obtained by replacing the resistor of Fig. 7.16 by an inductor of inductance L. Figures 7.21(b) and (c) are a T network and a π network, respectively. For very low frequencies, the inductors

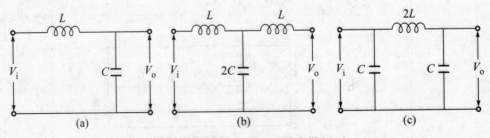

FIGURE 7.21 Low-pass LC filters.

may be treated shorted whereas the capacitors open. Hence, the signals of low frequencies pass through the filter without appreciable attenuation. At very high frequencies, the capacitors may be treated shorted and the inductors open. Thus, the situations reverse and the high-frequency signals are heavily attenuated. The passband gain is unity.

The characteristic resistance for the filters of Fig. 7.21(a) is given by

$$R_{ch} = \sqrt{R_{is}R_{io}} = \sqrt{j\omega L\left(j\omega L + \frac{1}{j\omega C}\right)}$$

For passband, $(j\omega L)^2$ is very small and may be neglected. Hence,

$$R_{ch} = \sqrt{\frac{L}{C}} \tag{7.72}$$

Similarly, it can be shown that the characteristic resistances for the filters of Fig. 7.21(b) and (c) are same as given by Eq. (7.72). For the filter to function properly, the source and load resistances must be equal to the characteristic resistance R_{ch}. The cutoff frequency f_{cH} of the low-pass LC filter can be given by

$$f_{cH} = \frac{1}{2\pi\sqrt{LC}} \tag{7.73}$$

For a given values of R_{ch} and f_{cH}, the filter parameters are given by

$$L = \frac{R_{ch}}{2\pi f_{cH}} \tag{7.74}$$

$$C = \frac{1}{2\pi f_{cH} R_{ch}} \tag{7.75}$$

7.6.2 High-Pass LC Filter

High-pass LC filters are shown in Fig. 7.22. The high-pass LC filter of Fig. 7.22(a) is obtained by replacing the resistor of Fig. 7.17 by an inductor of inductance L. Figures 7.22(b) and (c) are a T network and a π network, respectively, obtained by interchanging capacitors and inductors in Fig. 7.21(b) and (c), respectively. For very low frequencies, the inductors may be treated shorted whereas the capacitors open. Hence, the signals of low frequencies are heavily attenuated. At very high frequencies, the capacitors may be treated shorted and the inductors open. Thus, the situations reverse and the high-frequency signals pass through the filter without appreciable attenuation.

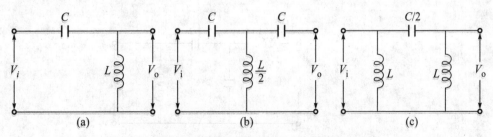

FIGURE 7.22 High-pass LC filters.

The characteristic resistance for the filters of Fig. 7.22(a) is given by

$$R_{ch} = \sqrt{R_{is} R_{io}} = \sqrt{\frac{1}{j\omega C}\left(j\omega L + \frac{1}{j\omega C}\right)}$$

For passband, $(1/j\omega C)^2$ is very small and may be neglected. Hence,

$$R_{ch} = \sqrt{\frac{L}{C}} \tag{7.76}$$

Similarly, it can be shown that the characteristic resistance for the filters of Fig. 7.22(b) and (c) is same as shown in Eq. (7.76). For the filter to function properly, the source and load resistances must be equal to the characteristic resistance R_{ch}. The cutoff frequency f_{cL} of the high-pass LC filter can be given by

$$f_{cL} = \frac{1}{2\pi\sqrt{LC}} \tag{7.77}$$

For given values of R_{ch} and f_{cH}, the filter parameters are given by

$$L = \frac{R_{ch}}{2\pi f_{cL}} \tag{7.78}$$

$$C = \frac{1}{2\pi f_{cL} R_{ch}} \tag{7.79}$$

7.6.3 Bandpass LC Filter

A simple bandpass filter is constructed by cascading the low-pass LC filter of Fig. 7.21(a) and the high-pass LC filter of Fig. 7.22(a). The circuit can also be obtained, simply, by replacing resistors in Fig. 7.18 by inductors. The bandpass LC filter is shown in Fig. 7.23. The passband gain is approximately unity.

$$f_{cL} = \frac{1}{2\pi\sqrt{L_2 C_2}} \tag{7.80}$$

$$f_{cH} = \frac{1}{2\pi\sqrt{L_1 C_1}} \tag{7.81}$$

The parameters are designed by using the formulas of low and high-pass filters.

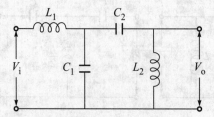

FIGURE 7.23 A bandpass LC filter.

7.6.4 Bandstop LC Filter

The circuit is obtained by replacing the resistors of Fig. 7.20 by inductors and is drawn as in Fig. 7.24.

The notch frequency is given by

$$f_n = \frac{1}{2\pi\sqrt{LC}} \tag{7.82}$$

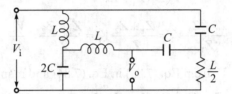

FIGURE 7.24 A bandstop LC filter.

7.7 ACTIVE FILTERS

The passive filters discussed in the preceding sections have some serious drawbacks. One drawback of the passive filters is their inability to have a gain greater than unity. This is because a passive filter cannot add energy to a signal. Some filter configurations need inductors as their elements. Inductors are generally expensive and are usually available in odd shapes. This is another drawback of the passive filters. An active filter uses the operational amplifier (OP AMP) as the active element and only resistors and capacitors for the passive elements. Let us consider the circuit of Fig. 7.25(a) using OP AMP as the active element. The circuit is a generalized active filter configuration of second order. The closed loop gain from inverting input to the output is given below,

$$A_{vf} = \frac{V_o}{V_1} = \frac{R_1 + R_2}{R_1} \tag{7.83}$$

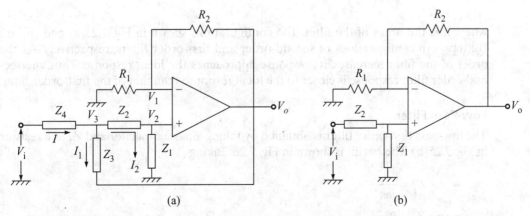

FIGURE 7.25 A generalized active filter configuration of (a) second order and (b) first order.

From Fig. 7.25,
$$I = I_1 + I_2$$

or
$$\frac{V_i - V_3}{Z_4} = \frac{V_3 - V_o}{Z_3} + \frac{V_3}{Z_1 + Z_2}$$

or
$$\frac{V_i}{Z_4} = \frac{(Z_1 + Z_2 + Z_3)Z_4 + Z_3(Z_1 + Z_2)}{Z_3 Z_4 (Z_1 + Z_2)} V_3 - \frac{V_o}{Z_3} \tag{7.84}$$

But
$$V_3 = \frac{Z_1 + Z_2}{Z_1} V_2 = \frac{Z_1 + Z_2}{Z_1} V_1 = \frac{Z_1 + Z_2}{Z_1} \times \frac{V_o}{A_{vf}} \tag{7.85}$$

Substituting for V_3 from Eq. (7.85) in Eq. (7.84) and manipulating, we get

$$V_i = \frac{(Z_1 + Z_2 + Z_3)Z_4 + Z_3(Z_1 + Z_2) - Z_1 Z_4 A_{vf}}{Z_1 Z_3 A_{vf}} V_o$$

$$\frac{V_o}{V_i} = \frac{Z_1 Z_3 A_{vf}}{Z_3(Z_1 + Z_2 + Z_4) + Z_2 Z_4 + Z_1 Z_4 (1 - A_{vf})} \tag{7.86}$$

When we take $Z_4 = 0$ and $Z_3 = \infty$, Fig. 7.25(a) reduces as Fig. 7.25(b) and Eq. (7.86) reduces to Eq. (7.87).

$$\frac{V_o}{V_i} = \frac{Z_1 A_{vf}}{Z_1 + Z_2} \tag{7.87}$$

7.7.1 Butterworth Filter

A general Butterworth filter gain A is of the form given by

$$A = \frac{A_{vf}}{\sqrt{1 + \left(\frac{\omega}{\omega_o}\right)^{\pm 2n}}} \tag{7.88}$$

where n is the order of the filter. The configurations shown in Fig. 7.25(a) and (b) are Butterworth configurations of second-order and first-order filters, respectively. As the order of the filter increases, its response approaches the ideal response. Thus, the second-order filter response is closer to the ideal response than that of the first-order filter.

Low-Pass Filter

The low-pass first-order filter is obtained by taking Z_1 as a capacitor and Z_2 as a resistor in Fig. 7.25(b). The circuit is shown in Fig. 7.26. Taking

$$Z_1 = \frac{1}{j\omega C}$$

$$Z_2 = R$$

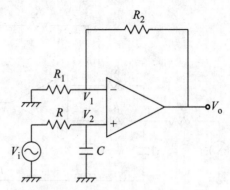

FIGURE 7.26 Fist-order low-pass filter.

The gain of the filter from Eq. (7.87),

$$A = \frac{V_o}{V_i} = \frac{\frac{1}{j\omega C} A_{vf}}{\frac{1}{j\omega C} + R}$$

$$= \frac{A_{vf}}{1 + j\omega RC} \tag{7.89}$$

where $\quad A_{vf} = \dfrac{R_1 + R_2}{R_1}$

At the cutoff frequency, $|A| = 0.707 A_{vf} = \dfrac{A_{vf}}{\sqrt{2}}$.

Hence, $\quad \omega_{cH} = \dfrac{1}{RC}$

and $\quad f_{cH} = \dfrac{1}{2\pi RC} \tag{7.90}$

For second-order filter, Z_1 and Z_3 [in Fig. 7.25 and Eq. (7.86)] are taken as capacitors and Z_2 and Z_4 as resistors. Thus, the second-order low-pass Butterworth filter is as shown in Fig. 7.27. Hence,

$$Z_1 = Z_3 = \frac{1}{j\omega C}$$

and $\quad Z_2 = Z_4 = R$

The gain of the second-order filter is obtained by substituting the above values of Z_1, Z_2, Z_3, and Z_4 in Eq. (7.86) and the gain is given by

$$A = \frac{V_o}{V_i} = \frac{A_{vf}}{1 - \omega^2 R^2 C^2 + j\omega RC(3 - A_{vf})} \tag{7.91}$$

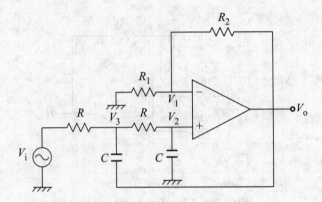

FIGURE 7.27 Second-order low-pass filter.

Eq. (7.90) represents a second-order filter for which damping ratio is $(3 - A_{vf})/2$. For the damping ratio of 0.707, the amplifier gain $A_{vf} = 1.586$. Hence, Eq. (7.91) reduces as

$$A = \frac{V_o}{V_i} = \frac{A_{Vf}}{1 - \omega^2 R^2 C^2 + j\sqrt{2}\omega RC} \qquad (7.92)$$

$$\therefore \quad |A| = \frac{A_{Vf}}{\sqrt{(1 - \omega^2 R^2 C^2)^2 + (\sqrt{2}\omega RC)^2}}$$

$$= \frac{A_{Vf}}{\sqrt{1 + (\omega^2 R^2 C^2)^2}}$$

This is the Butterworth filter gain for $n = 2$.
 For 3-dB gain,
$$1 + (\omega RC)^4 = 2$$

or $\quad \omega = \dfrac{1}{RC}$

$$\therefore \quad f_{cH} = \frac{1}{2\pi RC} \qquad (7.93)$$

High-Pass Filter

The high-pass first-order filter is obtained by taking Z_1 as a resistor and Z_2 as a capacitor in Fig. 7.25(b). The circuit is shown in Fig. 7.28. Taking

$$Z_1 = R$$

$$Z_2 = \frac{1}{j\omega C}$$

Sec. 7.7 / Active Filters

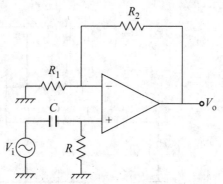

FIGURE 7.28 First-order high-pass filter.

The gain of the filter from Eq. (7.87),

$$A = \frac{V_o}{V_i} = \frac{RA_{vf}}{R + \dfrac{1}{j\omega C}} = \frac{A_{vf}}{1 - j(1/\omega RC)} \quad (7.94)$$

where $\quad A_{vf} = \dfrac{R_1 + R_2}{R_1}$

At the cutoff frequency, $|A| = 0.707 = \dfrac{A_{vf}}{\sqrt{2}}$

Hence, $\quad \omega_{cL} = \dfrac{1}{RC}$

and $\quad f_{cL} = \dfrac{1}{2\pi RC} \quad (7.95)$

For second-order high-pass filter, Z_1 and Z_3 [in Fig. 7.25(a) and Eq. (7.86)] are taken as resistors and Z_2 and Z_4 as capacitors. Thus, the second-order high-pass Butterworth filter is as shown in Fig. 7.29. Hence,

$$Z_1 = Z_3 = R$$

and $\quad Z_2 = Z_4 = \dfrac{1}{j\omega C}$

The gain of the second-order filter is obtained by substituting the above values of $Z_1, Z_2, Z_3,$ and Z_4 in Eq. (7.86). Thus, the gain is given by

$$A = \frac{V_o}{V_i} = \frac{A_{vf}}{1 - (1/\omega^2 R^2 C^2) - j(1/\omega RC)(3 - A_{vf})} \quad (7.96)$$

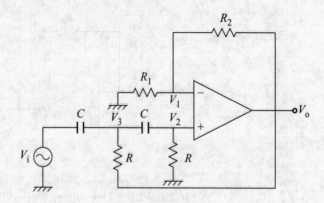

FIGURE 7.29 Second-order high-pass filter.

As in low-pass pass filter, for the damping ratio of 0.707 (i.e., $A_{vf} = 1.586$), Eq. (7.96) reduces as

$$A = \frac{V_o}{V_i} = \frac{A_{Vf}}{1 - (1/\omega^2 R^2 C^2) - j\sqrt{2}(1/\omega RC)} \quad (7.97)$$

$$\therefore \quad |A| = \frac{A_{Vf}}{\sqrt{\left[1 - (1/\omega^2 R^2 C^2)\right]^2 + (\sqrt{2}/\omega RC)^2}}$$

$$= \frac{A_{Vf}}{\sqrt{1 + (1/\omega^2 R^2 C^2)^2}}$$

For 3-dB gain,

$$1 + (1/\omega RC)^4 = 2$$

or

$$\omega = \frac{1}{RC}$$

$$\therefore \quad f_{cL} = \frac{1}{2\pi RC} \quad (7.98)$$

Bandpass Filter

A bandpass active filter, like a bandpass passive filter, is obtained by cascading a low-pass active filter with cutoff frequency f_{cH} and a high-pass active filter with cutoff frequency f_{cL}, provided $f_{cH} > f_{cL}$ as shown in Fig. 7.14(c).

Bandstop Filter

A bandstop active filter is obtained by paralleling a high-pass filter with cutoff frequency f_{cL} with a low-pass filter with cutoff frequency f_{cH}, as shown in Fig. 7.30. Here, it is required that $f_{cH} < f_{cL}$, as in Fig. 7.14(d).

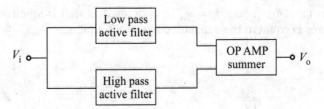

FIGURE 7.30 An active bandstop filter.

7.8 ACTIVE RESONANT BANDPASS FILTER

A single LC resonant circuit can be used to obtain an approximate narrow band characteristic. A single Butterworth filter cannot get the narrow band characteristic. The LC resonant filter has a response that peaks at a centre frequency (called resonant frequency) f_o and drops off with frequency on both sides of f_o, symmetrically. A basic LC resonant filter, as shown in Fig. 7.31, is a second-order circuit.

From Fig. 7.31,

$$V'_i = \frac{V_i R}{R + j\left(\omega L - \dfrac{1}{\omega C}\right)}$$

and

$$A_{vf} = \frac{R_1 + R_2}{R_1}$$

Hence,

$$\frac{V_o}{V'_i} = A_{vf}$$

and

$$A = \frac{V_o}{V'_i} \times \frac{V'_i}{V_i} = \frac{V_o}{V_i} = \frac{A_{vf} R}{R + j\left(\omega L - \dfrac{1}{\omega C}\right)} \tag{7.99}$$

Also,

$$A = \frac{V_o}{V_i} = \frac{A_{vf} R}{R + \left(j\omega L + \dfrac{1}{j\omega C}\right)} = \frac{j\omega(R/L) A_{vf}}{(j\omega)^2 + j\omega(R/L) + 1/LC} \tag{7.100}$$

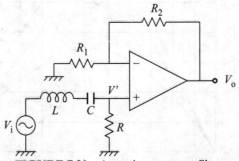

FIGURE 7.31 An active resonant filter.

The resonant frequency $f_o = \omega_o/2\pi$ is defined as that frequency at which the inductive reactance is equal to the capacitive reactance, that is,

$$\omega_o L = \frac{1}{\omega_o C}$$

or $\qquad \omega_o = \dfrac{1}{\sqrt{LC}} \qquad$ (7.101)

The quality factor or Q factor is given by

$$Q = \frac{\omega_o L}{R} = \frac{1}{\omega_o CR} = \frac{1}{R}\sqrt{\frac{L}{C}} \qquad (7.102)$$

Eq. (7.100) can, also, be written as

$$A = \frac{V_o}{V_i} = \frac{j\omega\left(\dfrac{\omega_o}{Q}\right)A_{vf}}{(j\omega)^2 + j\omega\left(\dfrac{\omega_o}{Q}\right) + \omega_o^2} \qquad (7.103)$$

$$\therefore \qquad |A| = \frac{A_{vf}}{\sqrt{1 + Q^2\left(\dfrac{\omega}{\omega_o} - \dfrac{\omega_o}{\omega}\right)^2}} \qquad (7.104)$$

The gain magnitude given in Eq.(7.104) in decibel is plotted against the normalized frequency on log scale in Fig. 7.32 for different values of Q.

From Fig. 7.32, it is seen that for every frequency $\omega' < \omega_o$, there exists a frequency $\omega'' > \omega_o$ for which $|A|$ has the same value. Hence, it can be shown that $\omega_o^2 = \omega' \, \omega''$.

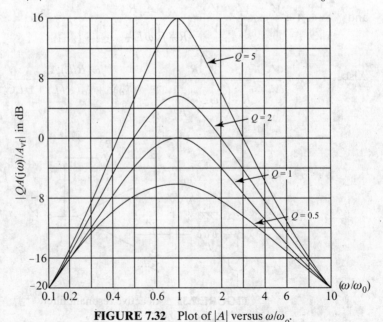

FIGURE 7.32 Plot of $|A|$ versus ω/ω_o.

7.8.1 Bandwidth of Passband

If ω_1 and ω_2 are two frequencies at equal distance on both sides from ω_0 for which the gain drops by 3 dB from its value A_{vf} at ω_0. Then the bandwidth of the filter is given by

$$B = \frac{\omega_2 - \omega_1}{2\pi} = \frac{1}{2\pi}\left(\omega_2 - \frac{\omega_0^2}{\omega_2}\right) \tag{7.105}$$

Since ω_2 is a cutoff frequency for a particular value of Q, the gain at ω_2 is given by

$$|A| = \frac{A_{vf}}{\sqrt{1 + Q^2\left(\frac{\omega_2}{\omega_0} - \frac{\omega_0}{\omega_2}\right)^2}} = \frac{A_{vf}}{\sqrt{2}}$$

which gives

$$Q\left(\frac{\omega_2}{\omega_0} - \frac{\omega_0}{\omega_2}\right) = 1 = \frac{Q}{\omega_0}\left(\omega_2 - \frac{\omega_0^2}{\omega_2}\right) \tag{7.106}$$

Solving Eqs. (7.105) and (7.106), the bandwidth,

$$B = \frac{\omega_0}{2\pi Q} = \frac{f_0}{Q} \tag{7.107}$$

The bandwidth, alternatively, can be given in terms of filter parameter by substituting for Q from Eq. (7.101),

$$B = \frac{\omega_0}{2\pi} \times \frac{1}{\frac{\omega_0 L}{R}} = \frac{R}{2\pi L} \tag{7.108}$$

7.8.2 Active RC Bandpass Filter

An active bandpass filter that can replace the active resonant filter of Fig. 7.31 is suggested here. The gain, given by Eq. (7.100) for the RLC filter of Fig. 7.31, can be obtained with the multiple feedback circuit shown in Fig. 7.33(a). This circuit uses two capacitors, three resistors, and one OP AMP. Equivalent circuit of Fig. 7.33(a) is shown in Fig. 7.33(b). The merit of this circuit is that it avoids the use of an inductor. The problems of using an inductor has already been discussed earlier.

In the circuit of Fig. 7.33(b), the equivalent voltage source and its series resistance are given by

$$V_i' = \frac{V_i R_2}{R_1 + R_2} \tag{7.109}$$

and

$$R' = \frac{R_1 R_2}{R_1 + R_2} \tag{7.110}$$

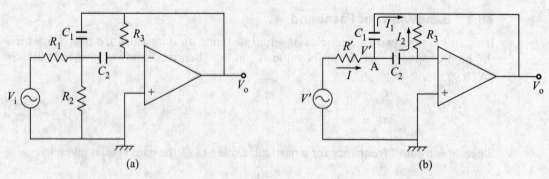

FIGURE 7.33 (a) An active RC resonant filter and (b) its equivalent circuit.

At node A,
$$I = I_1 + I_2$$

or
$$\frac{V_i' - V'}{R'} = \frac{V' - V_o}{\frac{1}{j\omega C_1}} + \frac{-V_o}{R_3}$$

or
$$\frac{V'}{R'} = \left(\frac{1}{R'} + j\omega C_1\right)V' - \left(\frac{1}{R_3} + j\omega C_1\right)V_o$$

$$\therefore \quad V' = \frac{I_2}{j\omega C_2} = -\frac{V_o}{j\omega C_2 R_3}$$

$$\therefore \quad \frac{V_i}{R_1} = \left(\frac{1}{R'} + j\omega C_1\right) \times \left(\frac{-V_o}{j\omega C_2 R_3}\right) - \left(\frac{1}{R_3} + j\omega C_1\right)V_o$$

Arranging the above equation, we get

$$A = \frac{V_o}{V_i} = \frac{-j\omega/C_1 R_1}{(j\omega)^2 + j\omega\left(\dfrac{C_1 + C_2}{C_1 C_2 R_3}\right) + \dfrac{1}{C_1 C_2 R_3 R'}} \tag{7.111}$$

Equating coefficients of Eqs. (7.100), (7.103), and (7.110), we obtain the following relations:

$$-R_1 C_1 = \frac{L}{RA_{vf}} = \frac{Q}{\omega_o A_{vf}} \tag{7.112}$$

$$\frac{C_1 C_2}{C_1 + C_2} R_3 = \frac{L}{R} = \frac{Q}{\omega_o} \tag{7.113}$$

$$C_1 C_2 R_3 R' = LC = \frac{1}{\omega_o^2} \tag{7.114}$$

Hence, we conclude that for designing the active RC bandpass filter equivalent to the resonant (RLC) bandpass filter, it is only required to have real positive values of R_1, R', R_3, C_1, and C_2, which satisfy Eqs. (7.112) to (7.114). Since there are only three equations for solving five parameters, two of them may be chosen arbitrarily.

Example 7.5

For the data $-A_{vf} = 50$, 3-dB bandwidth $B = 15$ Hz, and $f_o = 150$ Hz, design (a) an active resonant (RLC) filter and (b) a second-order RC bandpass filter.

Solution

From Eq. (7.107), $Q = 150/15 = 10$ and $\omega_o = 2\pi f_o = 2\pi \times 150 = 942.5$ rad/s.

(a) As far as RLC bandpass filter is concerned, there are two independent equations to solve for three parameters. Hence, let us assume $C = 0.1$ µF. From Eq. (7.114),

$$L = \frac{1}{\omega_o^2 C} = \frac{1}{(942.5)^2 \times 0.1 \times 10^{-6}} = 11.26 \text{ H}$$

The value of inductance is unreasonably large.

From Eq. (7.112) or (7.113),

$$R = \frac{\omega_o L}{Q} = \frac{942.5 \times 11.26}{10} = 1061 \text{ } \Omega$$

(b) Let us choose,

$$C_1 = C_2 = 0.1 \text{ µF}$$

From Eq.(7.112),

$$R_1 = \frac{Q}{-A_{vf} \omega_o C_1} = \frac{10}{50 \times 942.5 \times 0.1 \times 10^{-6}} = 2122 \text{ } \Omega$$

From Eq.(7.113),

$$R_3 = \frac{Q}{\omega_o \left(\frac{C_1 C_2}{C_1 + C_2} \right)} = \frac{10}{942.5 \left(\frac{0.1 \times 0.1}{0.2} \right) \times 10^{-6}} = 212 \text{ k}\Omega$$

From Eq.(7.114),

$$R' = \frac{1}{\omega_o^2 \times C_1 C_2 R_3} = \frac{1}{(942.5)^2 \times (0.1 \times 10^{-6})^2 \times 212 \times 10^3} = 531 \text{ } \Omega$$

7.9 REVIEW QUESTIONS

1. Name different types of attenuators. Define the following terms used with an attenuator:
   a. Attenuation.
   b. Characteristic resistance.
2. How is a potential divider-type attenuator compensated against the stray capacitance?
3. Derive expressions for characteristic resistance and attenuation of a T attenuator.
4. Derive expressions for characteristic resistance and attenuation of a π attenuator.
5. Derive expressions for characteristic resistance and attenuation of a bridged T attenuator.
6. Why is it essential to use a number of attenuators in cascade in place of a single high attenuation attenuator?
7. Derive expressions for cutoff frequencies of an RC (i) low-pass filter and (ii) high-pass filter.
8. Draw different circuits for low-pass filters. Derive expressions for characteristic resistance and cutoff frequency for each of them.
9. Draw different circuits for high-pass filters. Derive expressions for characteristic resistance and cutoff frequency for each of them.
10. Describe with neat circuit diagram the working of a Wien-bridge bandstop filter. Derive expression for the notch frequency.
11. Why is an inductor not a good choice as a parameter in a filter?
12. What are the advantages of the active filter in comparison with the passive filter?
13. Describe the function of the second-order Butterworth low-pass and high-pass filters.
14. Describe a resonant bandpass filter. Suggest an RC bandpass filter.

7.10 SOLVED PROBLEMS

1. A voltage divider shown in Fig. 7.1 is used as an attenuator. The resistances $R_1 = 75$ kΩ and $R_2 = 15$ kΩ. Find the voltage gain and the attenuation in decibel. If the transistor and stray capacitance C_2 across R_2 is of 50 µF, find the value of compensating capacitance C_1 across R_1.

   Solution

   The gain of the attenuator,

   $$A = \frac{R_2}{R_1 + R_2} = \frac{15}{75+15} = \frac{1}{6} = 0.167$$

   and the gain in decibel,

   $$A_{dB} = 20 \log_{10} A = 20 \log_{10} \left(\frac{1}{6}\right)$$

   $$= -20 \log_{10} 6 = -15.563 \text{ dB}$$

The attenuation,

$$\overline{A} = \frac{1}{A} = 6$$

and the attenuation in decibel,

$$A_{dB} = 20 \log_{10} \overline{A} = 20 \log_{10} 6 = \textbf{15.563 dB}$$

The capacitance of compensating capacitor,

$$C_1 = \frac{R_2}{R_1} C_2 = \frac{15 \times 50}{75} = \textbf{10 μF}$$

2. Find the characteristic resistance R_{ch} of the T attenuator shown in Fig. 7.6(a), if R_1 = 21 Ω and R_2 = 31 Ω. Prove that the input resistance of the attenuator is equal to R_{ch} when it is loaded by its characteristic resistance.

Solution

The characteristic resistance,

$$R_{ch} = \sqrt{R_{is} R_{io}}$$

The input resistance when the output is short circuited,

$$R_{is} = 21 + 21 \| 31$$

$$= 21 + \frac{21 \times 31}{21 + 31} = 33.5 \, \Omega$$

The input resistance when the output is open circuited,

$$R_{io} = 21 + 31 = 52 \, \Omega$$

$$\therefore \quad R_{ch} = \sqrt{52 \times 33.5} = \textbf{41.7 Ω}$$

When the attenuator is loaded with R_{ch}, the input resistance,

$$R_{in} = 21 + 31 \| (21 + 41.7)$$

$$= 21 + \frac{31 \times 62.7}{31 + 62.7} = \textbf{41.7 Ω (QED)}$$

3. Find characteristic resistance and attenuation of the T attenuator of Fig. 7.6(a), when R_1 = 25 Ω and R_2 = 35 Ω. Determine the values of the characteristic resistance and the attenuation if the resistances of the attenuator are reduced by a factor of 10.

Solution

The ratio,

$$m = \frac{R_2}{R_1} = \frac{35}{25} = 1.4$$

From Eq. (7.22), the characteristic resistance,

$$R_{ch} = R_1 \sqrt{1+2m}$$

$$= 25\sqrt{1+2\times 1.4} = \mathbf{48.734\ \Omega}$$

From Eq. (7.23), the attenuation,

$$\overline{A} = \frac{1+m+\sqrt{1+2m}}{m}$$

$$= \frac{1+1.4+\sqrt{1+2\times 1.4}}{1.4} = 3.1$$

$$\overline{A}_{dB} = 20 \log_{10} 3.1 = \mathbf{9.83\ dB}$$

When the resistances are reduced by a factor of 10, $R_1 = 2.5\ \Omega$ and $R_2 = 3.5\ \Omega$. Hence, the ratio m remains the same, that is, $m = 1.4$. From Eq. (7.23) the attenuation depends only on the ratio of the two resistances and not on their values; hence, the attenuation remains the same. From Eq. (7.22), it is also clear that the characteristic resistance is reduced by the same factor of 10. Therefore,

$$R_{ch} = \mathbf{48.734\ \Omega}$$

$$\overline{A} = 3.1 \text{ or } \overline{A}_{dB} = 9.83 \text{ dB}$$

4. Design a T attenuator with characteristic resistance of 45 Ω and attenuation of 20 dB.

Solution

$$\overline{A} = 20 \text{ dB} \quad \text{or} \quad \overline{A} = 10$$

From Eq. (7.24),

$$R_1 = \frac{\overline{A}-1}{\overline{A}+1} \times R_{ch} = \frac{10-1}{10+1} \times 45 = \mathbf{36.8\ \Omega}$$

From Eq.(7.25),

$$R_2 = \frac{2\overline{A}}{\overline{A}^2 - 1} \times R_{ch} = \frac{2\times 10}{10^2 - 1} \times 45 = \mathbf{9.1\ \Omega}$$

5. Design an 80-dB, 350-Ω attenuator using four sections of T attenuators in cascade.

Solution

Let us choose all four sections identical. Hence, each section attenuation

$$\overline{A} = 20 \text{ dB} \quad \text{or} \quad \overline{A} = 10$$

From Eq. (7.24)

$$R_1 = \frac{\overline{A}-1}{\overline{A}+1} \times R_{ch} = \frac{10-1}{10+1} \times 350 = \mathbf{286.36\ \Omega}$$

From Eq. (7.25)

$$R_2 = \frac{2\overline{A}}{\overline{A}^2-1} \times R_{ch} = \frac{2 \times 10}{10^2-1} \times 350 = \mathbf{70.7\ \Omega}$$

The cascaded attenuators are shown in Fig. 7.34.

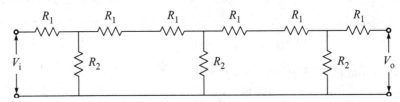

FIGURE 7.34 Attenuator for Solved Problem 5.

6. (a) For the π attenuator shown in Fig. 7.6(b), $R_1 = 1450\ \Omega$ and $R_2 = 350\ \Omega$. Find the characteristic resistance and the attenuation of the attenuator. (b) Determine the resistances of the attenuator if the desired characteristic resistance and attenuation are 495 Ω and 20 dB, respectively.

Solution

(a) The ratio,

$$m = \frac{R_2}{R_1} = \frac{350}{1450} = 0.24$$

From Eq. (7.30), the characteristic resistance,

$$R_{ch} = \frac{m}{\sqrt{1+2m}} R_1$$

$$= \frac{0.24}{\sqrt{1+2\times 0.24}} 1450 = \mathbf{286\ \Omega}$$

From Eq. (7.31), the attenuation,

$$\overline{A} = \frac{1+m+\sqrt{1+2m}}{m}$$

$$= \frac{1+0.24+\sqrt{1+2\times 0.24}}{0.24} = \mathbf{10.235}$$

(b) $\overline{A} = 20$ dB or $\overline{A} = 10$

From Eq.(7.32),
$$R_1 = \frac{\overline{A}^2 - 1}{2\overline{A}} \times R_{ch} = \frac{10^2 - 1}{2 \times 10} \times 495 = \mathbf{2450.25\ \Omega}$$

From Eq. (7.33),
$$R_2 = \frac{\overline{A} + 1}{\overline{A} - 1} \times R_{ch} = \frac{10 + 1}{10 - 1} \times 495 = \mathbf{605\ \Omega}$$

7. Design a 40-dB, 450-Ω bridged T attenuator.

Solution

$\overline{A} = 40$ dB or $\overline{A} = 100$

From Eq.(7.37),
$R_1 = R_{ch} = \mathbf{450\ \Omega}$

From Eq.(7.38),
$$R_2 = \frac{R_{ch}}{\overline{A} - 1} = \frac{450}{100 - 1} = \mathbf{4.54\ \Omega}$$

From Eq.(7.39),
$$R_3 = (\overline{A} - 1)R_{ch} = (100 - 1) \times 450 = \mathbf{44.55\ k\Omega}$$

8. For the RC low-pass filter of Fig. 7.16, $R = 1$ kΩ and $C = 0.1$ µF. Find the passband gain and cutoff frequency, when (a) $R_s = 0$ and no load, (b) $R_s = 0$ and $R_L = 2$ kΩ, and (c) $R_s = 3$ kΩ and $R_L = 5$ kΩ.

Solution

(a) The passband gain is always unity in this condition, that is,
$A = 1$

The cutoff frequency is given by
$$f_{cH} = \frac{1}{2\pi RC} = \frac{1}{2\pi \times 1 \times 10^3 \times 0.1 \times 10^{-6}} = \mathbf{1.59\ kHz}$$

(b) The passband gain,
$$A = \frac{R_L}{R + R_L} = \frac{2}{1 + 2} = \mathbf{0.667}$$

The cutoff frequency is given by
$$f_{cH} = \frac{1}{2\pi(R \parallel R_L)C}$$

where $R \| R_L = \dfrac{RR_L}{R+R_L} = \dfrac{1 \times 2}{1+2} = 0.667 \text{ k}\Omega$

Hence, $f_{cH} = \dfrac{1}{2\pi \times 0.667 \times 10^3 \times 0.1 \times 10^{-6}} = \mathbf{2.39 \text{ kHz}}$

(c) The passband gain,

$$A = \dfrac{R_L}{R_s + R + R_L} = \dfrac{5}{3+1+5} = \mathbf{0.5556}$$

The cutoff frequency is given by

$$f_{cH} = \dfrac{1}{2\pi[(R_s+R) \| R_L]C}$$

where $(R_s+R) \| R_L = \dfrac{(R_s+R)R_L}{R_s+R+R_L} = \dfrac{(3+1) \times 2}{3+1+2} = 1.333 \text{ k}\Omega$

Hence, $f_{cH} = \dfrac{1}{2\pi \times 1.333 \times 10^3 \times 0.1 \times 10^{-6}} = \mathbf{1.194 \text{ kHz}}$

10. Find the passband gain and the cutoff frequency of a high-pass RC filter with $R = 125$ kΩ, $C = 550$ pF, $R_s = 15$ kΩ, and $R_L = 145$ kΩ.

Solution

The high passband gain,

$$A = \dfrac{R \| R_L}{R_s + R \| R_L}$$

where $R \| R_L = \dfrac{RR_L}{R+R_L} = \dfrac{125 \times 145}{125+145} = 67.13 \text{ k}\Omega$

Hence, $A = \dfrac{67.13}{15+67.13} = \mathbf{0.82}$

The cutoff frequency is given by

$$f_{cH} = \dfrac{1}{2\pi[R_s + R \| R_L]C}$$

$$= \dfrac{1}{2\pi \times (15+67.13) \times 10^3 \times 550 \times 10^{-12}} = \mathbf{3.523 \text{ kHz}}$$

11. In the bandpass RC filter shown in Fig. 7.18, $R_1 = 12$ kΩ, $R_2 = 1.2$ MΩ, $C_1 = 150$ pF, and $C_2 = 0.02$ μF. Calculate the passband gain, and the lower and higher cutoff frequencies.

Solution

The bandpass gain,

$$A = \frac{R_2}{R_1 + R_2} = \frac{1.2}{0.012 + 1.2} = 0.99$$

The lower cutoff frequency is given by

$$f_{cL} = \frac{1}{2\pi R_2 C_2} = \frac{1}{2\pi \times 1.2 \times 10^6 \times 0.02 \times 10^{-6}} = 6.63 \text{ Hz}$$

The higher cutoff frequency is given by

$$f_{cH} = \frac{1}{2\pi R_1 C_1} = \frac{1}{2\pi \times 12 \times 10^3 \times 150 \times 10^{-12}} = 88.42 \text{ kHz}$$

12. The circuit shown in Fig. 7.35 uses an ideal OP AMP. (a) Find the passband gain, the frequency f_o, and the damping ratio. (b) Using this circuit design, find a second-order Butterworth low-pass filter with cutoff frequency $f_{ch} = 1$ kH and low-frequency gain equal to -1.

Solution

At node A,

$$I = I_1 + I_2 + I_3$$

or

$$\frac{V_i - V'}{R_3} = \frac{V' - V_o}{R_2} - j\omega C_1 V_o + j\omega C_2 V'$$

or

$$\frac{V_i}{R_3} = \frac{R_2 + R_3 + j\omega C_2 R_2 R_3}{R_2 R_3} V' - \frac{1 + j\omega C_1 R_2}{R_2} V_o$$

But $\quad V' = I_2 R_1 = -j\omega C_1 R_1 V_o$.

Hence,

$$\frac{V_i}{R_3} = -\left(\frac{R_2 + R_3 + j\omega C_2 R_2 R_3}{R_2 R_3} j\omega C_1 R_1 + \frac{1 + j\omega C_1 R_2}{R_2} \right) V_o$$

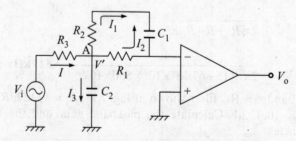

FIGURE 7.35 Circuit for Solved Problem 12.

$$V_i = -\left(\frac{(j\omega)^2 C_1 C_2 R_1 R_2 R_3 + j\omega C_1(R_1R_2 + R_2R_3 + R_3R_1) + R_3}{R_2}\right) V_o$$

$$\therefore \quad A = \frac{V_o}{V_i} = \frac{-R_2/R_3}{(j\omega)^2 C_1 C_2 R_1 R_2 + j\omega \dfrac{C_1}{R_3}(R_1R_2 + R_2R_3 + R_3R_1) + 1} \quad (7.115)$$

(a) Passband gain,

$$A = -\frac{R_2}{R_3} \quad (7.116)$$

The frequency,

$$f_o = \frac{\omega_o}{2\pi} = \frac{1}{2\pi\sqrt{C_1 C_2 R_1 R_2}} \quad (7.117)$$

From Eq. (7.115), the damping ratio δ is given by

$$\frac{2\delta}{\omega_o} = \frac{C_1(R_1R_2 + R_2R_3 + R_3R_1)}{R_3}$$

$$\therefore \quad \delta = \frac{C_1(R_1R_2 + R_2R_3 + R_3R_1)}{2R_3\sqrt{C_1 C_2 R_1 R_2}} \quad (7.118)$$

(b) For Butterworth filter, $\delta = \dfrac{1}{\sqrt{2}}$. Also, let

$$2R_1 = R_2 = R_3 = R$$

and $\quad xC_1 = C_2 = C$

Substituting for damping ratio, resistors, and capacitors in Eq. (7.118),

$$\frac{1}{\sqrt{2}} = \frac{\sqrt{2}}{\sqrt{x}}$$

Hence, $x = 4$

Substituting capacitors and resistors values in (7.115),

$$A = \frac{V_o}{V_i} = \frac{-1}{\left[1 - \left(\dfrac{\omega CR}{2\sqrt{2}}\right)^2\right] + j\sqrt{2}\left(\dfrac{\omega CR}{2\sqrt{2}}\right)}$$

and $\quad |A| = \dfrac{1}{\sqrt{1 + \left(\dfrac{\omega^2 C^2 R^2}{8}\right)^2}}$

For $|A| = 3$ dB,

$$\omega_{cH} = \frac{2\sqrt{2}}{CR} = \frac{1}{\sqrt{C_1 C_2 R_1 R_2}}$$

and the cutoff frequency

$$f_{cH} = \frac{1}{2\pi \sqrt{C_1 C_2 R_1 R_2}} \qquad (7.119)$$

Let us choose $C_1 = 0.05$ μF, then $C_2 = 4 \times 0.05 = 0.2$ μF. For $f_{cH} = 1$ kHz,

$$1 \times 10^3 = \frac{10^6}{2\pi \sqrt{0.05 \times 0.2 \times (R^2/2)}}$$

$$R = \frac{10^3}{2\pi \sqrt{0.05 \times 0.1}} = 2250 \text{ } \Omega$$

Hence, $R_1 = \mathbf{1125 \text{ } \Omega}$

and $\quad R_2 = \mathbf{2250 \text{ } \Omega}$

13. For the low-pass filter shown in Fig. 7.35, determine the damping ratio, cutoff frequency, and the passband gain of the filter if $R_1 = R_2 = R_3 = 2$ kΩ and $C_1 = C_2 = 0.2$ μF.

Solution

From Eq. (7.119), the cutoff frequency,

$$f_{cH} = \frac{1}{2\pi \sqrt{C_1 C_2 R_1 R_2}}$$

$$= \frac{1}{2\pi \sqrt{(0.2 \times 10^{-6})^2 (2 \times 10^3)^2}} = \mathbf{398 \text{ Hz}}$$

The damping ratio is given by Eq. (7.118) as

$$\delta = \frac{C_1 (R_1 R_2 + R_2 R_3 + R_3 R_1)}{2 R_3 \sqrt{C_1 C_2 R_1 R_2}}$$

$$= \frac{0.2 \times 10^{-6} \times 3 \times 2 \times 2 \times 10^6}{2 \times 2 \times 10^3 \times 0.2 \times 10^{-6} \times 2 \times 10^3} = \mathbf{1.5}$$

The gain is given by

$$A = \frac{-R_2}{R_3} = \mathbf{-1}$$

14. For the low-pass filter shown in Fig. 7.36, determine the passband gain and cutoff frequency.

Solution

The impedances are

$$Z_1 = R_1$$

and

$$Z_2 = \frac{1}{\frac{1}{R_2} + j\omega C} = \frac{R_2}{1 + j\omega C R_2}$$

The voltage gain of the network,

$$A = \frac{-Z_2}{Z_1} = \frac{-R_2/R_1}{1 + j\omega C R_2}$$

The passband gain,

$$A_o = -\frac{R_2}{R_1} \tag{7.120}$$

The gain magnitude,

$$|A| = \frac{R_2/R_1}{\sqrt{1 + (\omega C R_2)^2}}$$

At the cutoff frequency, the gain magnitude,

$$|A|_{cH} = \frac{R_2/R_1}{\sqrt{2}}$$

Hence, the cutoff frequency,

$$f_{cH} = \frac{\omega_{cH}}{2\pi} = \frac{1}{2\pi C R_2} \tag{7.121}$$

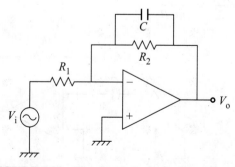

FIGURE 7.36 The circuit for Solved Problem 14.

15. For the high-pass filter shown in Fig. 7.37, determine the passband gain and the cutoff frequency.

Solution

The impedances are

$$Z_1 = R_1 + \frac{1}{j\omega C} = \frac{1 + j\omega C R_1}{j\omega C}$$

and $\quad Z_2 = R_2$

The voltage gain of the network,

$$A = \frac{-Z_2}{Z_1} = \frac{-j\omega C R_2}{1 + j\omega C R_1}$$

$$= \frac{-R_2/R_1}{1 + \dfrac{1}{j\omega C R_1}}$$

The passband gain,

$$A_o = -\frac{R_2}{R_1} \tag{7.122}$$

The gain magnitude,

$$|A| = \frac{R_2/R_1}{\sqrt{1 + (1/\omega C R_1)^2}}$$

At the cutoff frequency, the gain magnitude,

$$|A|_{cH} = \frac{R_2/R_1}{\sqrt{2}}$$

Hence, the cutoff frequency,

$$f_{cH} = \frac{\omega_{cH}}{2\pi} = \frac{1}{2\pi C R_1} \tag{7.123}$$

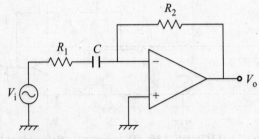

FIGURE 7.37 The circuit for Solved Problem 15.

7.11 EXERCISES

1. A voltage divider type attenuator has resistances $R_1 = 55$ kΩ and $R_2 = 6.5$ kΩ. Find the voltage gain and attenuation in decibel. If the stray and transistor capacitances together across the 6.5-kΩ resistor is 100 µF, find the value of trimer to be connected across R_1 for compensating the effect of above capacitances.

2. Find the characteristic resistance and attenuation of a symmetrical T attenuator with resistances $R_1 = 27$ Ω and $R_2 = 36$ Ω. Show that the input resistance of the attenuator is equal to the characteristic resistance when it is loaded with its characteristic resistance.

3. Design a symmetrical T attenuator if the characteristic resistance and attenuation needed are 500 Ω and 20 dB, respectively.

4. Design a 70-dB, 550-Ω T attenuator. Choose a suitable number of T sections so that they can be cascaded to get the above attenuator. Design each section. Show the cascaded attenuator.

5. Find the characteristic resistance and attenuation of a π attenuator with resistances $R_1 = 2$ kΩ and $R_2 = 0.4$ kΩ. Find the values of R_1 and R_2 if we need a 20-dB, 50-Ω π attenuator.

6. Design a 20-dB, 250-Ω bridged T attenuator.

7. Calculate the passband gain and cutoff frequency of a low-pass RC filter when (a) $R = 10$ kΩ, $C = 450$ pF, $R_s = 0$, and $R_L = \infty$; (b) $R = 10$ kΩ, $C = 450$ pF, $R_s = 0$, and $R_L = 50$ kΩ; and (c) $R = 10$ kΩ, $C = 450$ pF, $R_s = 5$ kΩ, and $R_L = 50$ kΩ.

8. Find the passband gain and cutoff frequency of a high-pass RC filter with $R = 65$ kΩ, $C = 350$ pF, $R_s = 15$ kΩ, and $R_L = 125$ kΩ.

9. The elements of the bandpass filter shown in Fig. 7.18 are $R_1 = 10$ kΩ, $C_1 = 250$ pF, $R_2 = 1.5$ MΩ, and $C_2 = 0.1$ µF. Calculate the passband gain, and the lower and upper cutoff frequencies.

10. In the active filter shown in Fig. 7.35, calculate the values of C_1 and C_2, if $R_1 = 1$ kΩ and $R_2 = R_3 = 2$ kΩ, and the required values of damping ratio and cutoff frequency are 0.707 and 1 kHz, respectively.

11. For the active low-pass filter shown in Fig. 7.36, $R_1 = 1.5$ kΩ, $R_2 = 2.5$ kΩ, and $C = 0.1$ µF. Calculate the passband gain and the cutoff frequency.

12. For the active high-pass filter shown in Fig. 7.37, $R_1 = 15$ kΩ, $R_2 = 25$ kΩ, and $C = 0.01$ µF. Calculate the passband gain and the cutoff frequency.

13. Design an active resonant (RLC) filter for $A_{vf} = -50$, 3-dB bandwidth $B = 15$ Hz, and $f_o = 150$ Hz.

14. Design a second-order RC bandpass filter for $A_{vf} = -50$, 3-dB bandwidth $B = 15$ kHz, and $f_o = 150$ kHz.

15. Find the expressions for central (resonant) frequency ω_o, the bandwidth BW, the quality factor Q, and the gain at $\omega = \omega_o$, when $C_1 = C_2 = C$.

8

CATHODE-RAY OSCILLOSCOPE

Outline

8.1 Introduction 344

8.2 Cathode-Ray Tube 345

8.3 Focusing Devices 347

8.4 Deflecting Forces 349

8.5 Vertical Deflection System 352

8.6 Horizontal Deflection System 355

8.7 Synchronization of Sweep 358

8.8 Methods of Improving Sweep Linearity 360

8.9 CRO Block Diagram 362

8.10 Applications of Cathode-Ray Oscilloscopes 364

8.11 Special Purpose CROs 371

8.12 Review Questions 377

8.13 Solved Problems 378

8.14 Exercises 382

8.1 INTRODUCTION

Cathode-ray oscilloscopes are very important instruments used in electrical and electronic measurements. They are commonly known as CROs. The CROs use cathode rays to produce a visual image of an electrical signal on their screens. The cathode rays have the following properties:

1. They are deflected by electrical and magnetic fields.
2. They produce fluorescence when strike on a fluorescent screen.

Following are some applications of the CRO:

(a) Observation of waveform of an electrical signal.
(b) Observation of phenomena and other time-varying electrical signals from a very low frequency range to a very high frequency range.
(c) Measurement of voltage and current.
(d) Measurement of frequency and phase angle of a.c. voltages and currents.

A general-purpose CRO consists of the following constituents:

1. Cathode-ray tube (CRT).
2. Deflection systems:
   (a) Vertical deflection system.
   (b) Horizontal deflection system.
3. Time-base generator.
4. Power supply.
5. Delay line.

The CRT is the heart of the CRO and the remaining part of the CRO consists of circuitry to operate the CRT. The CRT produces a sharply focused and accelerated beam of electrons that moves with a very high velocity from its source and strikes the screen. The screen is coated with fluorescent material. The beam of electrons is produced by the cathode, and focused and accelerated by the focusing and accelerating electrodes, respectively. The fall of an electron beam at very high velocity causes the screen to glow at a small spot. Before striking the screen, the electron beam passes through a set of vertical deflection plates and a set of horizontal deflection plates. When voltage is applied to the deflection plates, the electron beam moves either in vertical or in horizontal direction depending on whether the voltage is applied to the vertical deflection plates or the horizontal deflection plates. The vertical and horizontal movements of the electron beam are independent of each other. Hence, the spot on the screen can be positioned anywhere on the screen. It is only required to simultaneously apply appropriate vertical and horizontal voltages to the vertical and horizontal deflection plates, respectively.

When a periodic signal is applied to the vertical deflection plates and a sawtooth signal of same period is applied to the horizontal deflection plates, the spot created by the electron beam traces an image of the signal applied to the vertical deflection plates on the screen.

8.2 CATHODE-RAY TUBE

There are various types of CRTs depending on the purpose for which they are designed. However, it essentially, as shown in Fig. 8.1, consists of the following constituents:

1. *Electron gun.* It produces a focused and accelerated electron beam. The electrons are produced by the cathode, and they are focused and accelerated in the form of a beam by the focusing and accelerating electrodes.
2. *Deflecting systems.* There are vertical and horizontal deflecting systems. A deflecting system may be either electrostatic or electromagnetic. They are used to deflect the electron beam in either vertical and horizontal direction.
3. *Fluorescent screen.* The screen is coated with a fluorescent material. The electron beam strikes the screen causing a well-defined visual spot.

8.2.1 Electron Gun

It consists of a source of electrons, and accelerating and focusing arrangements. The source of electrons consists of, in general, an indirectly heated cathode K, as shown in Fig. 8.1, to give out streams of electrons. The cathode is coated with barium and strontium oxides. The heater H heats up the cathode for emission of electrons. The heat shield HS is placed outside and coaxial with the cathode to avoid wastage of heat. The control electrode (or grid) CG is in the form of a metal cylinder having a metal diaphragm. The diaphragm has a small aperture in front of the cathode. As the control electrode is negatively biased, it repels electrons from all sides and concentrates them in a fine beam. The first anode A_1 consists of a cylinder containing a number of diaphragms having narrow apertures. This positively biased anode is placed in front of the control electrode and is used to accelerate the electrons. Hence, it is also called accelerating anode. Another accelerating anode is A_3, which is maintained at high positive potential with respect to

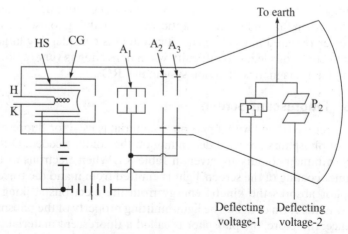

FIGURE 8.1 A cathode-ray tube.

the cathode. The electrode A_2 is kept at negative potential with respect to anode A_1 and A_2. By varying this negative potential the convergence of the beam can be varied. The electrode A_2 is known as focusing electrode. The combination of A_1, A_2, and A_3 is regarded as electron lens.

The focusing arrangements may be either electrostatic or electromagnetic. In the electrostatic focusing, an electrostatic field is set up between two conductors in such a way that electrons passing through the electrostatic field are deflected towards the axis. The electromagnetic focusing utilizes the fact that an electron entering a constant magnetic field perpendicular to its path is deflected and moved in a circular path. The magnetic field setup is parallel to the axis of the tube. The electrons travelling parallel to the tube axis are not affected by the magnetic field, whereas those having a component of velocity away from the axis move in a spiral path, which finally returns them to the axis. The magnetic focusing with the magnetic deflection is, generally, used in televisions and finds little application in CROs. The electrostatic and electromagnetic focusing are discussed in detail in Section-8.3.

8.2.2 Deflecting Systems

Like the focusing arrangement, the deflecting system may be electrostatic or electromagnetic. In Fig. 8.1, P_1 and P_2 are two pairs of parallel plates known as horizontal or X plates and vertical or Y plates, respectively. If the horizontal (X) pair of deflecting plates P_1 is charged, the electric field produced between the plates deflects the electron beam to the right or left depending on the direction of the electric field. Thus, a horizontal line is traced when the charging voltage is alternating. Similarly, the vertical (Y) pair of deflecting plates P_2, when charged with alternating voltage, causes the electron beam to trace a vertical line.

In the electromagnetic system, the deflection is obtained by producing a magnetic field perpendicular to the axis of the tube by means of coils outside the tube. The electron beam entering the field moves in a circular path and comes out from the field with its path deflected in a plane perpendicular to the field. Two fields, perpendicular to each other, are used to produce deflection in the horizontal and vertical directions. The advantage of the electromagnetic system is that a greater deflection of the beam can be produced than that in the electrostatic system, enabling to use a shorter tube. However, the power required in electromagnetic system is considerably large. The inductance of the deflecting coils leads to difficulty when deflection is very rapid. These facts limit the use of electromagnetic deflection system in CROs.

8.2.3 Fluorescent Screen

A glass screen coated with a phosphor material is used for fluorescent screen. Different types of phosphors are used depending on the colour needed. Different types of phosphors with their colours are given in Table-8.1. When electrons with high velocity strike the inner coating of the screen, light is emitted from it and the trace is made visible. The phosphor absorbs the kinetic energy from the electrons striking it and gives up the energy in the form of light. The light-emitting property of the phosphor is known as fluorescence; therefore, the phosphor is called a fluorescent material. The second property of the phosphor is to continue emission of light for some time even after the source of

TABLE-8.1 Fluorescent materials, their colours, and persistence.

Types of phosphors	Trace colours	Persistence	Applications
P_1	Green	Medium	General-purpose CRO
P_2	Yellow-green	Medium	CRO for observation of low- and medium-speed signals
P_4	White	Medium-short	Television
P_5	Blue	Very short	Photography
P_7	White-green	Long	Radar
P_{11}	Blue	Medium-short	Photography
P_{12}	Orange	Long	Radar
P_{31}	Green	Medium-short	General-purpose CRO

excitation is cutoff. This property is known as phosphorescence. The length of time for which phosphorescence occurs is a measure of persistence of the fluorescent materials. If persistence lasts for a few microseconds, it is called short persistence. It is called medium persistence, if it persists for a few milliseconds, and long persistence, if it persists for a few seconds to several minutes.

8.3 FOCUSING DEVICES

Because of the mutual repulsion between the electrons, the beam coming out of the accelerating electrodes has a tendency to spread from the axis of the CRT. To prevent this spread and to bring the beam to a sharp focus at the screen, some focusing devices are required. The focusing device may be either electrostatic or electromagnetic.

8.3.1 Electrostatic Focusing

Three anodes A_1, A_2, and A_3 of the electron gun are the preaccelerating anode, the focusing anode, and the accelerating anode, respectively. These three anodes together form an electron lens system for focusing the electron beam into a spot on the fluorescent screen. As discussed earlier A_1 and A_3 are connected to a positive potential of the order of 1.5 kV or higher voltage supplied from the CRT high-voltage supply. The anode A_2 is connected to a lower potential of, say, 500 V.

We know that an electron in an electric field experiences a force. The operation of electrostatic focusing device utilizes this effect of an electrostatic field on an electron. The force experienced by the electron is opposite in direction and equal in magnitude to 1.602×10^{-19} times the electrostatic field. This condition is valid only when the electrostatic field is of uniform intensity; but this is not true in practice. An electrostatic field between two rectangular plates when a voltage V is applied across the plate is shown in Fig. 8.2. The internal repulsive force between the electric field lines causes spreading of the gap between two lines, resulting in a curvature of the field at the edge of the plates, as shown in Fig. 8.2. For the large size of plates, the electric field may be considered of uniform intensity. Thus, the force on the electron is in the direction

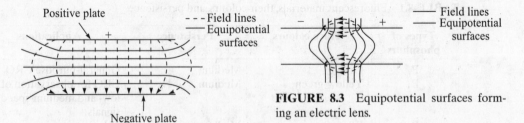

FIGURE 8.2 Equipotential surfaces in an electric field.

FIGURE 8.3 Equipotential surfaces forming an electric lens.

perpendicular to the equipotential surfaces. The equipotential surfaces are shown by solid lines in Fig. 8.2.

Let us consider two cylinders placed end to end, as shown in Fig. 8.3. If a potential difference is applied between the cylinders, the electric field developed between them is not uniform because of the reasons mentioned above. The equipotential surfaces are shown by solid lines and the field is shown by dashed line in Fig. 8.3. The equipotential surfaces as shown in Fig. 8.3 form an electron lens and follow the same law for bending of electron beam as for the bending of light beam at an optical lens. Thus, two electron lenses are formed in the electrostatic focusing in a CRT, one between the electrodes A_1 and A_2 and another between the electrodes A_2 and A_3.

Now, the focusing of the electron beam can be explained as follows. The electrons emitted by the cathode are slightly divergent. The electrons that enter the electric field between the electrodes A_1 and A_2 at an angle other than normal to the equipotential surfaces are refracted towards the normal. The electron beam, thus, tends to become parallel to the axis of the CRT. This approximately parallel beam of electrons enters the electric field between the electrodes A_2 and A_3 (the second electron lens) and is refracted once again, and becomes slightly convergent and focused on the screen at the centre of the CRT axis. The focal length of this double concave lens is controlled by varying the voltage on the focusing electrode A_2. The focal point of the beam, hence, can be moved along the axis of the CRT.

8.3.2 Electromagnetic Focusing

The axis of the electromagnetic coil coincides with the axis of the CRT. The coil may extend up to the complete length of the tube or it may be concentrated in a smaller region. The first type of the coil is used in the *image orthicon* camera tube in television, whereas the second type is used in the picture tube of television receiver, in radar indicator, and in general-purpose CRO. Figure 8.4 illustrates the principle of the magnetic focusing.

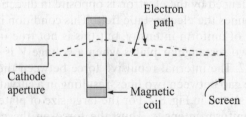

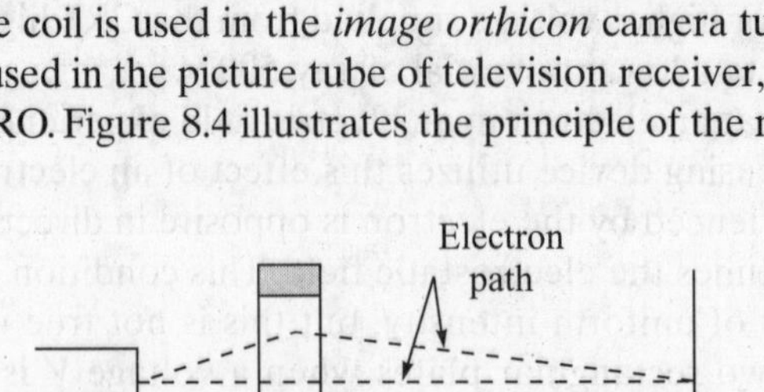

FIGURE. 8.4 Magnetic focusing.

The magnetic coil surrounds the CRT tube in such a way that the lines of magnetic field are uniformly distributed and are parallel to the axis of the tube. The electrons moving parallel to the axis of the tube are not affected by the magnetic field, whereas the electrons moving at an angle to the axis experience a force. The direction of the force experienced by the electrons is perpendicular to both the direction of electron motion and to the magnetic field. Thus, two forces are acting on the electrons. The first is the attractive force of the anode, which causes the electrons to move forward and the second is due to the magnetic field, causing the side motion. Hence, the electrons move in a spiral path, which finally returns them to the axis of the tube.

8.4 DEFLECTING FORCES

In Section-8.2, it has been mentioned that there are two methods of producing deflection in the electron beam. Like focusing method, the two deflecting methods are (1) electrostatic deflection and (2) electromagnetic deflection.

8.4.1 Electrostatic Deflection

The theory of electrostatic deflection is discussed here. Let

m = mass of an electron in kilograms.

q = charge of an electron in coulombs.

V = accelerating voltage in volts, that is, potential difference between the cathode and the last anode.

E = potential between the deflecting plates in volts.

l = axial length of the deflecting plates in metres.

L = distance from the centre of the deflecting plates to the screen in metres.

S = spacing between the two plates.

The deflection of the electron is shown in Fig. 8.5. If v is the velocity of the electron beam in the axial direction, then

$$\frac{1}{2}mv^2 = qV$$

or

$$v^2 = \frac{2qV}{m} \tag{8.1}$$

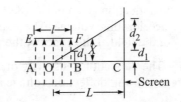

FIGURE 8.5 Electrostatic deflection.

The force acting on an electron in the direction perpendicular to the axis,

$$F = \frac{Eq}{s}$$

and the acceleration in this direction,

$$a = \frac{F}{m} = \frac{Eq}{ms} \tag{8.2}$$

The distance travelled by the electron in the direction perpendicular to the axis at the time it leaves the deflecting plate,

$$d_1 = \frac{1}{2} at^2 \tag{8.3}$$

where t is the time for which force F is acting on the electron and is given by

$$t = \frac{l}{v} \tag{8.4}$$

Hence, substituting for a and t from Eqs. (8.2) and (8.4), respectively, in Eq. (8.3),

$$d_1 = \frac{Eql^2}{2msv^2} \tag{8.5}$$

The velocity v' is in the direction perpendicular to the axis and is acquired in time t,

$$v' = \frac{d_1}{t/2} = \frac{Eql}{msv} \tag{8.6}$$

Hence, the distance d_2, in Fig. 8.5, is given by

$$d_2 = v't' = \frac{Eql}{msv} \times \frac{L - l/2}{v} \tag{8.7}$$

where t' is the time taken by the electron in moving the distance between B and C. The total distance travelled by the electron, therefore, is given by

$$d = d_1 + d_2 = \frac{EqlL}{msv^2} \tag{8.8}$$

Substituting for v^2 from Eq. (8.1) in Eq. (8.8),

$$d = \frac{ElL}{2sV} \tag{8.9}$$

The expression for the deflection given in Eq. (8.9) shows that the deflection is inversely proportional to the accelerating voltage. The expression is approximate, as we have ignored the fringing of the field at the edges of the plates and nonuniformity of the field.

Deflection sensitivity. The deflection sensitivity of the CRT is defined as the deflection, in metre, on the screen per unit deflection voltage. Thus the deflection sensitivity,

$$S = \frac{d}{E} = \frac{lL}{2sV} \text{ m/V} \tag{8.10}$$

Deflection factor. The deflection factor is defined as the reciprocal of the deflection sensitivity. Thus the deflection factor,

$$G = \frac{1}{S} = \frac{2sV}{lL} \text{ V/m} \tag{8.11}$$

From Eqs. (8.10) and (8.11), it is evident that the sensitivity of the CRT is independent of the deflection voltage. It varies inversely with the accelerating voltage. Thus, a high-deflection voltage is required for a given deflection, if the accelerating voltage is high. This is a requirement for a highly accelerated electron beam. A highly accelerated beam has more kinetic energy and so produces a brighter image on the screen. A typical value of sensitivity range is from 0.1 mm/V to 1.0 mm/V.

8.4.2 Electromagnetic Deflection

Let us assume that an electron beam travelling along the axis of the tube having a velocity v enters a magnetic field of flux density B. The field is assumed to be uniform and covers an axial distance l, as shown in Fig. 8.6. The force acting on the electron,

$$F = Bqv \tag{8.12}$$

where q is the charge of an electron in coulombs. The force acts in a direction perpendicular to the path of the electron flow. The path is an arc of a circle of radius,

$$r = \frac{mv}{qB} \tag{8.13}$$

where m is the mass of an electron in kilograms. The electron emerges from the field and travels in a direction inclined at an angle θ to the axis and strikes the screen at S. If the total angular deflection is small, then

$$\text{arc PQ} = l$$

and

$$\theta = \frac{l}{r} = \frac{lqB}{mv} \tag{8.14}$$

The displacement of the electron beam,

$$d = L \tan \theta \approx L \theta$$

or

$$d = \frac{LlqB}{mv}$$

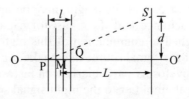

FIGURE 8.6 Electromagnetic deflection.

After substituting the value of v from Eq. (8.1),

$$d = LlB\sqrt{\frac{q}{2mV}} \qquad (8.15)$$

The expression is again approximate, as the variation of field at the edges is neglected.
Magnetic sensitivity. The magnetic sensitivity is defined as the deflection, in metres, on the screen caused by a magnetic flux density of 1 Wb/m², that is, the magnetic sensitivity,

$$S_m = \frac{d}{B} = Ll\sqrt{\frac{q}{2mV}} \qquad (8.16)$$

Alternatively, the sensitivity S_m is defined as the deflection, in millimetres, on the screen, when one milliampere current flows through the magnetic deflection coil. This is expressed in mm/mA.

8.5 VERTICAL DEFLECTION SYSTEM

The vertical deflection system, generally, performs the following function:

1. Faithfully reproduces the input signal waveform.
2. Provides a buffer between the input source and the CRT deflection plates.

The various elements of the system are as shown in Fig. 8.7. The different elements given in the block diagram are discussed in this section.

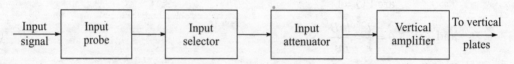

FIGURE 8.7 A block diagram of various elements.

8.5.1 Input Probe

The input terminals of a CRO are connected through a shield cable to the signal source to avoid the effect of stray fields such as due to nearby power line. The capacitance of the cable, which increases with its length, may be appreciably high if a long cable is used. The cable capacitance and the transistor capacitance of the vertical amplifier together may be high. The combination of this high input capacitance with the high output impedance of the signal source may make it impossible to get faithful observations of the signal waveforms, particularly with fast waveforms. An input probe is used to connect the vertical amplifier to the input signal source without loading the source. A general-purpose probe is shown in Fig. 8.8. It consists of a series resistance R_1

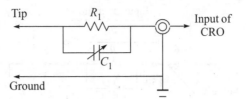

FIGURE 8.8 A CRO probe.

shunted with a variable capacitor C_1. The resistor provides attenuation which helps in getting faithful observation of the signal waveform. The capacitor provides compensation at high frequency. The attenuator compensation is discussed in detail while discussing the input attenuator element of the vertical deflection system.

Both R_1 and C_1 contained in the probe body. At one end is the probe tip for connection to the input and the other end is connected through a length of shielded cable to the CRO. The ground connector is connected to the ground of the input source.

8.5.2 Input Selector

It is essentially a three-position switch. The three positions are a.c., ground, and d.c. In the a.c. position of the input selector, the signal is connected through a capacitor to an attenuator. The capacitor blocks the d.c. component of the input waveform if it is d.c. biased. At the d.c. position the signal voltage is directly connected to the attenuator. Thus, both a.c. and d.c. components are applied to the attenuator. This mode is useful when total instantaneous values of the input signal are required to be absorbed or measured. The ground position is the intermediate position between a.c. and d.c., and is useful from safety point of view. When brought in ground position, any charge in the input attenuator is removed by momentarily grounding the attenuator input.

8.5.3 Input Attenuator

It is fundamentally an RC voltage divider and is controlled by the range selection switch on the panel. A typical setting of the switch may be 0.1, 0.2, 0.5, 1, 2, 5, 10, 20, and 50 V/div. So, the maximum attenuation of the signal is at 50 V/div setting. For linear operation of the CRO over the specified frequency range (say, d.c. to 25 MHz), the attenuation of the signal must be independent of frequency. This requires a compensated attenuator. The attenuators have already been discussed in Section-7.1. An attenuator and the input stage of the vertical amplifier are shown in Fig. 8.9. The input impedance of the amplifier is represented by a parallel combination of resistor R_i and a capacitor C_i. The resistor R_a is the attenuation resistor and C_a is the compensating capacitor.

The volt per division selector switches S_1 and S_2 are in synchronism. When the switches are at the uppermost position, the signal V_i is directly connected to the input of the vertical amplifier, and so there is no attenuation in the signal. From the different switch positions in the Fig. 8.9, it is the position corresponding to 0.1 V/div. The sensitivity of the deflection system in this case is maximum. With any of the positions 2 to 9,

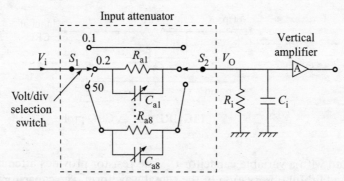

FIGURE 8.9 Attenuator with input stage of vertical amplifier.

which correspond to 0.2, 0.5, 1, 2, 5, 10, 20, and 50 V/div, respectively, the corresponding attenuator $R_a C_a$ is connected into the circuit. The voltage division takes place. The voltage ratio V_o/V_i is equal to the ratio of the amplifier input impedance and the total circuit impedance. To make the attenuator independent of frequency of the signal, this ratio of impedances is kept constant by making time constants $R_{aj} C_{aj}$ ($j = 1, 2, ..., 8$) equal to the time constant of $R_i C_i$ (Section-7.1).

8.5.4 Vertical Amplifier

The input attenuator functions as an input selector and gives a constant output voltage. This constant voltage (for any position of switches S_1 and S_2) is fed to the vertical amplifier unit. The vertical amplifier has a constant gain. It is easier to design an amplifier with a constant gain to meet and maintain the requirements of stability and bandwidth. In general, the vertical amplifier consists of two major circuit blocks, namely, preamplifier and main amplifier, as shown in Fig. 8.10. The input stage, the first element of the preamplifier, often consists of a field-effect transistor (FET) source follower. The high input impedance of the source follower virtually isolates the amplifier from the attenuator. The second element of the preamplifier is the phase inverter, which provides two antiphase output signals. These two antiphase signals are required to operate the push-pull output amplifier. Sometimes, BJT (bipolar junction transistor) emitter follower forms a buffer element between the input stage and the phase inverter. It acts as an impedance transformer to match the medium impedance of the FET output to the low impedance input of the phase inverter.

The main vertical amplifier consists of a push-pull output amplifier. The output amplifier delivers equal signal voltages of opposite polarity to the vertical deflection plates. Push-pull amplifiers are used because they improve the deflection linearity of the CRT.

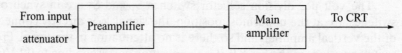

FIGURE 8.10 Two major circuit blocks of the vertical amplifier.

8.6 HORIZONTAL DEFLECTION SYSTEM

The CRO is most widely used for the observation of a periodic voltage waveform. For example, consider the waveform of an alternating voltage of 100 Hz. When this voltage is applied to the vertical plates of a CRO, the luminous spot on the screen moves vertically in a straight line with simple harmonic motion with a frequency of 100 Hz. Now, if a voltage, increasing at a constant rate from zero, is applied to the horizontal plates, the luminous spot, due to the voltage applied to the vertical plates, moves at a constant speed across the screen. Thus, the displacement produced by the vertical plates is spread out in the horizontal direction. If the voltage applied to the horizontal plates sweeps the spot in 0.01 s, a single cycle of the test voltage applied to the vertical plates appears on the screen. After one sweep the voltage applied to the horizontal plates falls to zero instantaneously. By repeating the sweep the same trace of the luminous spot can be produced again and again giving the impression of a stationary wave. If the sweep time is 0.02 s, two cycles of the test voltage wave appear on the screen. This type of voltage applied to the horizontal plates is called the *sweep-* or *time-base voltage*. The time-base voltage generator is inbuilt in the CRO.

In some applications such as in the measurements of frequency and phase of an a.c. signal, two signals of same nature are compared, and so an external voltage signal is applied to the horizontal system as well, instead of a time-base voltage. Hence, this requires the horizontal deflection system to be of the same class as the vertical deflection system. The block diagram description of the horizontal deflection system is shown in Fig. 8.11.

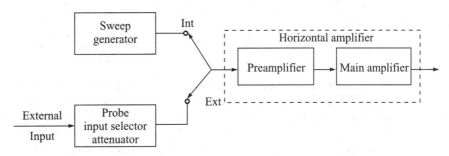

FIGURE 8.11 A block diagram of the horizontal deflection system.

The horizontal deflection system is connected by a two-way switch which connects the horizontal amplifier either to 'int' position or to 'ext' position. At 'int' position the horizontal amplifier is connected to the internal time-base generator, whereas at 'ext' position the horizontal amplifier is connected to the input attenuator and then to input selector switch. The selector switch is similar to the selector switch of the vertical deflection system and is used to select the proper input signal. Again, the input signal is applied through a probe. The probe, input selector, and input attenuator are similar to those used in the vertical deflection system.

8.6.1 Time-Base Generator

All the circuits used for the generation of time base fundamentally depend on the charging of a capacitor. The voltage across a capacitor is given by

$$V_C = \frac{1}{C}\int I\,dt$$

Hence,
$$\frac{dV_C}{dt} = \frac{I}{C} \tag{8.17}$$

where I and C are current and capacitance, respectively, and t represents time. Thus, for the constant capacitor current, the voltage across the capacitor increases at a constant rate. The basic sweep circuit is shown in Fig. 8.12(a). Initially, the switch S is closed and the output voltage V_o is zero. When the switch is opened, the capacitor starts charging and gives an exponentially rising voltage from zero, as shown in Fig. 8.12(b). In practice, the function of the switch S is performed by an electronic device, such as a unijunction transistor (UJT), a thyristor, a gas thyraton. The sweep voltage in Fig. 8.12(b) is not linear because the charging current is not linear.

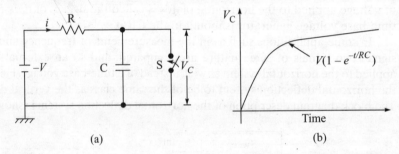

FIGURE 8.12 (a) A basic sweep circuit and (b) exponential output of the circuit.

An electronic switching circuit is shown in Fig. 8.13(a). A UJT is used as the switch in this circuit. When the supply V_{yy} is switched on, the capacitor C starts charging through the resistor R. The emitter voltage of the UJT rises towards V_{YY}. When this voltage reaches to the peak voltage V_p of the device, it turns on and the capacitor starts discharging through it to the valley voltage V_v of the UJT. The valley voltage is

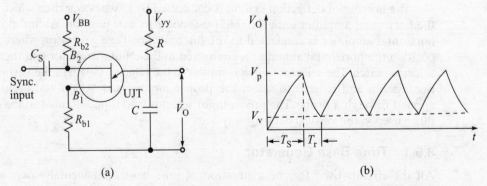

FIGURE 8.13 (a) A switching circuit with UJT and (b) sawtooth waveform.

the minimum voltage required to maintain the conduction of the device. When the capacitor voltage reaches the valley voltage, the device turns off and the capacitor starts charging again, repeating the cycle. The sweep (sawtooth) waveform is shown in Fig. 8.13(b).

The waveform generated above is repetitive. The next sweep is initiated immediately at the termination of the previous sweep. It does not wait for an external signal for initializing the switch. This mode of operation is known as *free-running, astable,* or *recurrent mode.* In practice, the frequency is varied through R and C. Continuous variation of frequency is obtained by varying R, and the ranges are changed by changing the value of C.

For a stationary pattern of n cycle of a periodic signal of frequency f_v applied to the vertical plates,

$$f_v = nf_s \tag{8.18}$$

where f_s is the sweep frequency. A portion of the last cycle of the vertical signal occurs during the return time T_r shown in Fig. 8.13(b). If $T_r \ll T_s$, the return trace on the screen is very faint or even invisible. This disappearance of the return trace is desirable for the clarity of the pattern. To reduce the return time, the capacitor is kept small. However, a small capacitor increases the nonlinearity of the sweep. A circuit employing an additional npn transistor, as shown in Fig. 8.14, may be used to reduce T_r. When T_1 starts conducting, the emitter-base junction of T_2 becomes forward biased and it conducts. Hence, the discharge of the capacitor C is hastened.

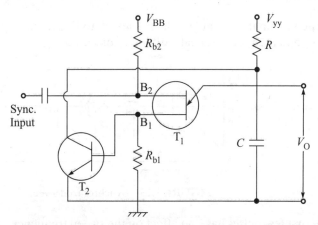

FIGURE 8.14 A circuit to reduce the return time T_r.

Where T_r is not very small in comparison with T_s, the return trace may be made invisible by turning off or blanking the beam during retrace. For this, the voltage developed at B_2 of the UJT, which is a negative voltage spike, is applied to the grid of the CRT. When f_v is only slightly different from nf_s, the waveform drifts slowly along the screen. To overcome this, the vertical signal is applied to B_2 of the UJT. This process is termed as synchronization. The synchronization method is discussed in Section-8.7.

8.6.2 Horizontal Amplifier

Contrary to the vertical amplifier, the horizontal amplifier has to generally process a sweep signal with fairly high amplitude and relatively slow rise time. Hence, the horizontal amplifier performance requirements are of lower order. However, because of the smaller horizontal deflection sensitivity of the CRT than the vertical deflection sensitivity, the gain of the horizontal amplifier is relatively higher.

As discussed earlier, in some applications, the CRO works in X-Y mode and not in Y-t mode. In X-Y mode, The Y-input signal is applied to the vertical plates as usual, whereas the horizontal time base is replaced by an external signal that is applied, as shown in Fig. 8.11, by changing the sweep selector switch to position 'ext' from the position 'int'.

8.7 SYNCHRONIZATION OF SWEEP

Synchronization of a sweep signal to an external signal is possible if the external signal is applied to the sync. input in Fig. 8.13(a) and Fig. 8.14 in such a way as to lower the peak voltage V_p of the UJT. A negative voltage applied at the base 2 lowers V_p. Thus, the sweep ends prematurely. The situation is illustrated in Fig. 8.15. A train of negative sync. pulses is superimposed on the peak voltage V_p. The retrace time T_r is assumed to be negligibly small. For synchronization, the following two conditions must be satisfied:

1. The time interval between the pulses T must be less than the natural period of the sweep.
2. The pulse amplitude is at least large enough to bridge the gap between the peak voltage V_p and the required peak voltage V_p' of the sweep.

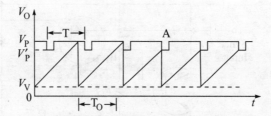

FIGURE 8.15 Synchronized sweep.

The first few pulses have no effect on the sweep frequency, and the sweep generator continues to run unsynchronized at the natural sweep frequency. The capacitor is stopped charging prematurely by a sync. pulse, which occurs at the moment where the sweep voltage equals the momentarily peak voltage V_p' of the UJT, the point A in Fig. 8.15. At this instant the capacitor discharges rapidly and the ramp is terminated. The sweep comes in synchronism.

A sinusoidal sync. signal of sufficient amplitude can also be applied instead of a train of negative sync. pulses for the sweep synchronization. The sine wave sync. signal can either shorten the natural period like negative sync. pulses or lengthen the natural period of the sawtooth.

8.7.1 Triggered Sweep

A waveform to be traced on the CRO may not be periodic. It occurs at irregular intervals. In such a case it is desirable that instead of free-running sweep circuit, the circuit should remain quiescent and wait to be initiated by the waveform under observation. Even if the waveform is periodic, the interesting part of the waveform may be short in time duration in comparison with the period of the waveform, such as a narrow pulse. To study the pulse form in detail, the display of the pulse should be stretched out in time, suppressing the display of the remaining part of the waveform. For this, it is required that the sweep be set for a very short duration comparable with the time duration between the pulses. The sweep circuit remains quiescent until it is initiated by the pulse. Hence, the laboratory-type CROs are usually provided with a time base with a triggering circuit. The triggering circuit provides triggering pulses to initiate the sweep. This type of sweep is known as *triggered sweep*.

For triggered sweep, the circuit of Fig. 8.13(a) is modified as in Fig. 8.16. With the mechanical switch S open, position B, the clamping voltage V_R of the clamping diode D is determined by the resistors R_1 and R_2 and is given by

$$V_R = \frac{R_2}{R_1 + R_2} V_{YY} \tag{8.19}$$

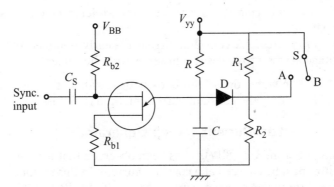

FIGURE 8.16 Modified circuit for a triggered sweep.

The voltage V_R is made less than the peak voltage V_p. When the capacitor voltage rises to V_R, the diode D conducts and prevents further rise of capacitor voltage. Thus, the capacitor voltage is clamped to V_R, and the peak voltage V_p of the UJT never reaches. The capacitor voltage remains constant at V_R. Now, a negative signal (pulse) is applied at the sync. input terminal of the UJT circuit. If the sync. input lowers the peak voltage V_p below V_R even momentarily, the capacitor discharges quickly in time T_r through the UJT. After reaching to V_v the capacitor charges again towards V_{YY}, reaching up to V_R. The output waveform of the triggered sweep generator is shown in Fig. 8.17.

The trigger pulse first initiates the retrace before the generation of the sweep starts. Thus, the initial part of the waveform to be traced is lost in the interval T_r. To avoid this, a sufficient signal delay is provided by the vertical delay line, discussed later in Section-8.9. When the switch S is closed, the normal sweep is generated.

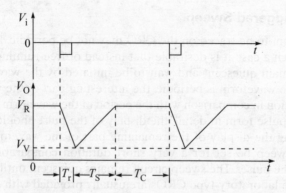

FIGURE 8.17 Output of a triggered sweep generator.

8.8 METHODS OF IMPROVING SWEEP LINEARITY

The methods of improving the linearity of sweep are as follows:

1. *Constant-current charging.* The capacitor is charged from a constant-current source.
2. *Miller circuit.* It uses an operational integrator, which changes the step signal into a ramp waveform.
3. *Phantastron circuit.* It is a version of the Miller circuit and uses a pulse instead of a step input.
4. *Bootstrap circuit.* A constant current is nearly obtained by maintaining a constant voltage across a fixed resistor in series with the capacitor.

It is beyond the scope of this book to deal with all the above methods of sweep generation. Here, only the constant-current method is discussed.

8.8.1 Constant-Current Sweep Generator

In Section-8.6 and Eq. (8.17), it has been observed that for obtaining a linear sweep, the capacitor must be charged at a constant current. In this method, therefore, the capacitor is charged through a constant-current source. The collector current of a transistor in the common-base configuration is nearly constant, except for very small values of V_{CB}, when the emitter current is held constant. This characteristic of the transistor is employed, in this method, to generate a quite linear ramp by causing a constant-current charging of the capacitor, as shown in Fig. 8.18(a).

A modified version of the circuit of Fig. 8.18(a) is shown in Fig. 8.18(b). This circuit employs a single supply V_{YY} and provides considerable temperature compensation. The emitter voltage V_{EE} is given by

$$V_{EE} = V_Z + V_D \tag{8.20}$$

where V_Z and V_D are the voltages across the Zener diode D_Z and the diode D, respectively. The collector voltage V_{CC} is equal to the drop across the resistor R, that is,

$$V_{CC} = V_{YY} - V_{EE} \tag{8.21}$$

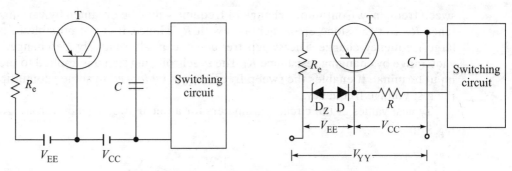

FIGURE 8.18 Constant-current charging schemes.

The diode D is made of the same material as the transistor. Hence, the diode serves to compensate for the temperature-dependent emitter-to-base voltage, V_{EB}. Voltages V_D and V_{EB} are always equal and opposite. The voltage across R_e remains equal to V_Z. The emitter current is given by

$$I_e = \frac{V_Z}{R_e} \tag{8.22}$$

The emitter current is constant provided that a temperature-compensated Zener diode is used.

A circuit generating quite linear sweep employing the principle of constant-current charging is illustrated in Fig. 8.19. The constant charging current is obtained by using a transistor T_2 in common-base configuration. Resistor R_o and capacitor C_o are used for the purpose of isolating d.c. component of voltage to avoid a steady deflection of the beam. Capacitors C_1, C_2, and C_3 are for different ranges of the

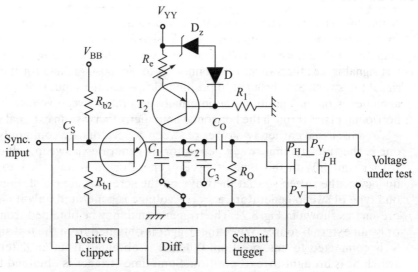

FIGURE 8.19 A constant-current charging sweep generator.

sweep frequency. Continuous change of frequency may be obtained by varying resistance R_e. But the slope error increases with R_e. Hence, it is not possible to have a large frequency change. The sweep frequency can, alternatively, be changed for a wide range by changing resistance R_1. The synchronizing signal is applied to the sync. input terminal. It enables the sweep frequency to be *locked* to some submultiples of the test voltage frequency.

Typical values of the circuit parameters for a linear sweep generator are given in Fig. 8.20.

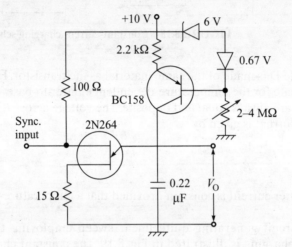

FIGURE 8.20 A typical linear sweep generator.

8.9 CRO BLOCK DIAGRAM

A simplified block diagram of a CRO is shown in Fig. 8.21. There are vertical and horizontal attenuators and amplifiers so that signals can be adjusted to give the desired amplitude to the trace. The voltage under test is applied to the vertical plates. When the test signal is a.c., the switch S_1 is connected to 'ac' position, and for the d.c.-biased test signal the switch S_1 is brought to position 'dc'. When the input switch of the horizontal amplifier is on 'int' position, the internally generated sweep voltage is applied to the horizontal plates through the horizontal amplifier, otherwise an external signal is applied.

As discussed earlier, two types of sweep generators are commonly used. The first one is the free-running sweep generator. It generates a sweep voltage without any trigger signal. With this type of sweep generators, it is necessary to adjust the frequency of the sweep so that the wave on the screen is almost stationary. In the second type of sweep generators, a sweep voltage is generated only if a trigger signal is present, as shown in Fig. 8.21. The trigger signal may be obtained from the test signal or by an external source. When the trigger is obtained from the test signal, the switch S_2 is connected to 'int' position. When it is obtained from an external source, the switch S_2 is brought to 'ext' position. Sometimes trigger is obtained from the power supply to the CRO. Under this condition, switch S_2 is brought to the 'line' position.

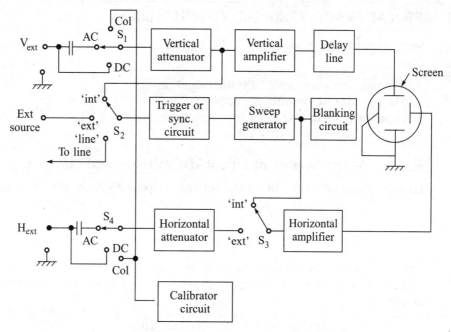

FIGURE 8.21 A simplified block diagram of a CRO.

8.9.1 Blanking Circuit

The blanking circuit shuts off the electron stream while the trace returns to its original position, so that the return path does not appear on the screen. For this, the voltage developed at B_2 of the UJT of the time base generator, which is a negative voltage spike, is applied to the grid of the CRT.

8.9.2 Delay Line

All the circuits in the CRO cause a certain amount of delay in transmitting the signal voltage to the deflection plates. The horizontal time-base signal is initiated or triggered by a portion of the test signal applied to the vertical plates. Generating and shaping of a trigger pulse in the horizontal circuit takes time (about 80 ns). For the operator to observe the leading edge of the signal waveform to be traced, the vertical plate signal must be delayed by at least the same amount of time. The vertical delay line does this function. The delay line may be of the lumped-parameter type or distributed-parameter type. The lumped-parameter delay line is made of a number of cascaded symmetrical LC network. The distributed-parameter delay line consists of a specially made coaxial cable with a high value of inductance per unit length.

8.9.3 Power Supply

The power supply consists of high-voltage and low-voltage sections. The high-voltage section is used to operate the CRT, and the low-voltage section supplies the electronic circuitry of the oscilloscope.

8.10 APPLICATIONS OF CATHODE-RAY OSCILLOSCOPES

Some important applications of a CRO going to be discussed in this section are given below:

1. Measurements of sinusoidal voltages and currents.
2. Measurements of frequency and phase.
3. Determination of B-H loops for magnetic materials.
4. Measurement of dielectric loss.

8.10.1 Measurements of Sinusoidal Voltages and Currents

For the measurement of a sinusoidal voltage, the peak-to-peak value is obtained from its trace on the screen of the CRO. The r.m.s. value of the voltage is given by

$$V_{r.m.s.} = \frac{\text{Peak-to-peak value of voltage}}{2\sqrt{2}} \tag{8.23}$$

The value of current can be determined by passing the current through a known resistance R and measuring the voltage drop across the resistance. The r.m.s. value of current is given by

$$I_{r.m.s.} = \frac{\text{Peak-to-peak value of voltage drop}}{2\sqrt{2}\ R} \tag{8.24}$$

8.10.2 Measurement of Frequency and Phase

The frequency to be measured is compared with a known frequency. For this a sinusoidal voltage of known frequency is applied to the horizontal plates and the sinusoidal voltage of unknown frequency is applied to the vertical plates. If the ratio of the two frequencies is either an integer or a ratio of two integers such as 3/2, 5/4, etc., a clearly defined and distinctive pattern appears on the screen. These patterns are known a *Lissajous figures*. The Lissajous figures are discussed first.

Lissajous Figures

Consider a special case where two voltages applied to the vertical and horizontal plates have equal frequency and same phase. Let the voltage applied to the vertical plates, also called the Y plates, be

$$e_y(t) = E_y \sin(2\pi f)t \tag{8.25}$$

and the voltage to the horizontal plates, also known as the X plates, be

$$e_x(t) = E_x \sin(2\pi f)t \tag{8.26}$$

where E_y and E_x are the amplitudes of the two voltages. The deflections of the luminous spot in the vertical and horizontal directions are given, respectively, by

$$D_y = Ke_y(t) = KE_y \sin(2\pi f)t \tag{8.27}$$

$$D_x = Ke_x(t) = KE_x \sin(2\pi f)t \tag{8.28}$$

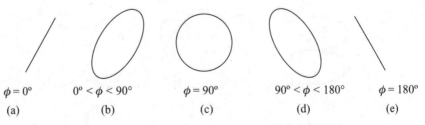

FIGURE 8.22 Patterns for inputs of same frequency, but different phase.

where K is a constant. Hence, the ratio of the two deflections,

$$\frac{D_y}{D_x} = \frac{E_y}{E_x} = m \qquad (m \text{ is a constant})$$

or $\quad D_y = mD_x \qquad (8.29)$

Equation (8.29) is an equation for a straight line passing through the origin and making an angle $\theta = (\tan^{-1} m)$ with the horizontal axis (or x-axis). The pattern of the trace is shown in Fig. 8.22(a).

When the phase difference between the two voltages is 180°, the constant m is negative and the straight line tilts, as shown in Fig. 8.22(e). It makes an angle θ with the negative x-axis. When $E_y = E_x$, $\theta = 45°$.

When $E_y = E_x = E$ and the phase difference (ϕ) between the two voltages is 90°,

$$e_y(t) = E_y \sin (2\pi f)t$$

$$e_x(t) = E_x \cos (2\pi f)t$$

and the respective deflections are

$$D_y = KE \sin (2\pi f)t$$

$$D_x = KE \cos (2\pi f)t$$

Hence, the resultant deflection

$$D = \sqrt{D_y^2 + D_x^2} = KE\sqrt{\sin^2 (2\pi f)t + \cos^2 (2\pi f)t}$$

$$= KE \text{ (a constant)} \qquad (8.30)$$

and the pattern is a circle as shown in Fig. 8.22(c). When the phase difference (ϕ) is other than 90°, the pattern is an ellipse, as shown in Fig. 8.22(b) and (d). If the voltages are not equal, but $\phi = 90°$, then also the pattern is an ellipse. If $E_y > E_x$, the major axis is in vertical direction, and if $E_y < E_x$, the major axis is in the horizontal direction.

Until now, we considered the cases of equal frequencies. When the frequencies of two signals are not equal, the Lissajous figures obtained are not as simple as shown in Fig. 8.22. The Lissajous figures are as shown in Fig. 8.23.

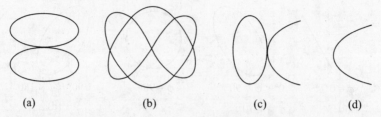

(a) (b) (c) (d)

FIGURE 8.23 The Lissajous figures when frequencies are not equal.

Lissajous figures are of two types: closed Lissajous figures and open Lissajous figures. The closed Lissajous figures have no break or discontinuity. This means that they have no loose or free end. Examples of closed Lissajous figures are those shown in Fig. 8.22(b), (c), and (d), and Fig. 8.23(a) and (b). The open Lissajous figures are simply steady traces and have free ends. Examples of open Lissajous figures are those shown in Fig. 8.22(a) and (e), and Fig. 8.23(c) and (d).

For closed Lissajous figures the ratio of vertical frequency to horizontal frequency is equal to the ratio of the number of positive y peaks to the number of positive x peaks. This means that

$$\frac{f_y}{f_x} = \frac{\text{Number of positive y peaks}}{\text{Number of positive x peaks}} \tag{8.31}$$

To illustrate the above statement, consider the Lissajous figures shown in Fig. 8.24. In Fig. 8.24(a), the number of positive y peaks is 2 and the number of positive x peaks is 1. Hence,

$$\frac{f_y}{f_x} = \frac{2}{1} \tag{8.32}$$

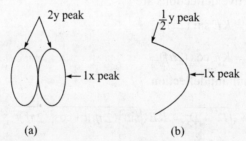

(a) (b)

FIGURE 8.24 (a) Close and (b) open Lissajous figures.

In case of open Lissajous figure (pattern), free ends are counted as ½ instead of 1. For example, the pattern shown in Fig. 8.24(b), there is no closed y peak, but a free end. Hence, the positive y peak taken is ½. The positive x peak in this pattern is 1. Hence,

$$\frac{f_y}{f_x} = \frac{1/2}{1} = \frac{1}{2} \tag{8.33}$$

Measurement of Frequency

The voltage whose frequency is to be measured is applied to the Y plates, whereas a voltage of same amplitude, but with varying known frequency, is applied to the X plates. The known frequency is varied until a pattern, shown in Fig. 8.22, is obtained. In this case the unknown frequency, that is, the frequency of the voltage applied to the Y plates, is equal to the known frequency. When it is not possible to adjust the known frequency equal to the unknown frequency, the known frequency is adjusted to a multiple or submultiple of the unknown frequency. The ratio of the two frequencies can be obtained from the pattern observed and by using the method explained above. As f_x is known, the unknown frequency can be known. For example, for the pattern of Fig. 8.24(a), the unknown frequency is twice of the known frequency, whereas for the pattern of Fig. 8.24(b), the unknown frequency is half of the known frequency.

Measurement of Phase Angle

To determine the phase between the two sinusoidal voltages of equal amplitude and frequency, as shown in Fig. 8.25, the two voltages are applied to the vertical and horizontal plates of the CRO. As discussed earlier, the pattern obtained, in this case, is an ellipse, as shown in Fig. 8.26. Let the two voltages be expressed as

$$e_x = E \sin (2\pi f)t$$
$$e_y = E \sin (2\pi f t + \phi)$$

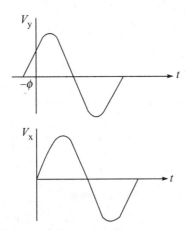
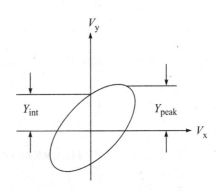

FIGURE 8.25 Two voltage waves.   **FIGURE 8.26** Pattern as an ellipse.

The deflections in X and Y directions are

$$D_x = KE \sin (2\pi f)t$$
$$D_y = KE \sin (2\pi f t + \phi)$$

For $t = 0$,
$$D_x = 0$$
and
$$D_y = KE \sin \phi$$

Thus, the D_y at $t = 0$ is the deflection in the y-axis, while the deflection in x-axis is zero, that is, the intercept on the y-axis is as shown by Y_{int} in Fig. 8.26. Hence,

$$Y_{int} = KE \sin \phi \tag{8.34}$$

The peak deflection in Y direction, as shown in Fig. 8.26, is

$$Y_{peak} = KE \tag{8.35}$$

From Eqs. (8.34) and (8.35),

$$\sin \phi = \frac{Y_{int}}{Y_{peak}} \tag{8.36}$$

and

$$\phi = \sin^{-1} \left(\frac{Y_{int}}{Y_{peak}} \right) \tag{8.37}$$

An alternate method of determine the phase difference of two sinusoidal voltages is the use of a dual-beam CRO. The dual-beam CRO has two independent but identical vertical plates. The dual-beam CRO is discussed in the next section. The two voltage waves whose phase difference is to be determined are applied to the two vertical plates. The origins of the two traces are made to coincide. The phase angle between the two waves can be read directly from the traces, as shown in Fig. 8.27.

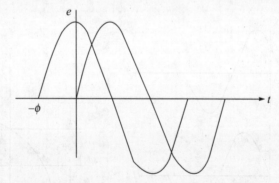

FIGURE 8.27 Two traces on a dual-beam CRO.

8.10.3 Determination of B-H Loop for Magnetic Materials

The connections of the experimental setup are as shown in Fig. 8.28. A laminated ring specimen is preferred. The voltage drop across the resistor R_1 is proportional to the magnetizing force, which is proportional to the magnetizing current. This voltage is applied to the horizontal plates. In the secondary side, the induced voltage e_2 in magnitude is given by

$$|e_2| = N_2 \frac{d\phi}{dt}$$

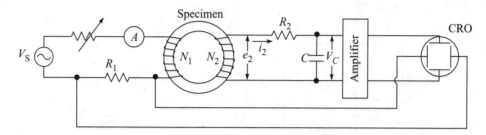

FIGURE 8.28 Determination of B-H loop of magnetic materials.

where ϕ is the instantaneous value of flux. The instantaneous value of secondary current,

$$i_2 = \frac{e_2}{R_2} = \frac{N_2}{R_2}\frac{d\phi}{dt} \qquad \text{(Assuming } R_2 \gg 1/\omega C\text{)}$$

and

$$v_c = \frac{q}{C} = \frac{1}{C}\int i_2\, dt$$

$$= \frac{1}{C}\int \frac{N_2}{R_2}\frac{d\phi}{dt}\, dt = \frac{N_2}{R_2 C}\int d\phi$$

$$= \frac{N_2}{R_2 C}\phi \qquad (8.38)$$

Thus, the voltage v_c is proportional to the flux, and the flux is proportional to the flux density. Hence, the voltage v_c is proportional to the flux density. The voltage v_c is first amplified and then applied to the vertical plates of the CRO. Thus, the B-H loop is observed on the screen of the CRO.

8.10.4 Measurement of Dielectric Loss

Figure 8.29 illustrates the circuit for measuring dielectric loss of a dielectric material. The capacitor C_s in the circuit is formed by using dielectric material under test. The capacitor C_o is a loss-free capacitor with a known capacitance. The capacitors C_s and C_o in series are connected to the supply with known frequency. The voltage across the capacitor C_s is applied to the X plates, whereas that across the capacitor C_o is applied to the Y plates. The voltage across C_o is proportional to the integral of current through the capacitors. If the supply voltage is sinusoidal, the Lissajous figure on the screen is an ellipse.

Let K_x and K_y are deflection sensitivities in centimetre per volt of X and Y plates, respectively. V_s and V_o are r.m.s. voltages across C_s and C_o, respectively. The r.m.s. value of current through the capacitors is I. The current taken by the deflection plates may be neglected. The phasor diagram showing current and voltages is shown in Fig. 8.30.

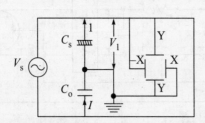

FIGURE. 8.29 A circuit for measurement of dielectric loss.

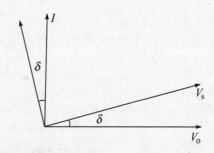

FIGURE 8.30 A vector diagram showing current and voltages.

The instantaneous values of voltages,

$$v_s = \sqrt{2}V_s \sin \omega t$$

$$v_o = \sqrt{2}V_o \sin(\omega t - \delta)$$

The deflection in the X direction,

$$D_x = K_x v_s = \sqrt{2}K_x V_s \sin \omega t$$

and that in the Y direction,

$$D_y = K_y v_o = \sqrt{2}K_y V_o \sin(\omega t - \delta)$$

The area of the ellipse shown in Fig. 8.31,

$$A = \int D_y dD_x$$

$$\therefore \quad dD_x = \sqrt{2}K_x V_s \omega \cos \omega t \, dt$$

$$\therefore \quad A = \int_0^T \sqrt{2}K_y V_o \sin(\omega t - \delta) \sqrt{2}K_x V_s \omega \cos \omega t \, dt \tag{8.39}$$

where T is the time period.

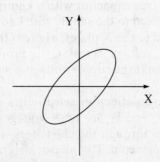

FIGURE 8.31 Pattern on the screen.

After integration of Eq. (8.39),

$$A = 2\pi K_x K_y V_s V_o \sin \delta \tag{8.40}$$

But $\quad V_o = \dfrac{I}{\omega C_o}$

Substituting for V_o in Eq. (8.40),

$$A = \dfrac{2\pi K_x K_y}{\omega C_o} V_s I \sin \delta \tag{8.41}$$

From Fig. 8.30, the dielectric loss per cycle,

$$W = V_s I \cos(90° - \delta) = V_s I \sin \delta \tag{8.42}$$

From Eqs. (8.41) and (8.42),

$$A = \dfrac{2\pi K_x K_y}{\omega C_o} W = KW \tag{8.43}$$

where $\quad K = \dfrac{2\pi K_x K_y}{\omega C_o} \tag{8.44}$

Thus, the area of the trace on the screen (i.e., ellipse) is proportional to the dielectric loss. The dielectric loss per cycle,

$$W = \dfrac{\text{Area of the ellipse}}{K} \tag{8.45}$$

8.11 SPECIAL PURPOSE CROs

In the preceding sections we have discussed in detail about the working principle, construction, and applications of a general purpose CRO. Some of the special purpose CROs discussed in this section are as follows:

1. Dual-trace CRO.
2. Dual-beam CRO.
3. Storage CRO.
4. Digital storage CRO
5. Sampling CRO.

8.11.1 Dual-Trace CRO

A dual-trace CRO has two identical vertical input circuits marked as channel A and channel B. A block diagram of the dual-trace CRO is shown in Fig. 8.32. Both the channels have their own preamplifiers and delay lines. The outputs of the preamplifiers, through the delay lines, are fed to an electronic switch. One function of the electronic switch is to alternatively connect the input terminals of the main vertical amplifier to the output of channel A and channel B.

372 Chapter 8 / Cathode-Ray Oscilloscope

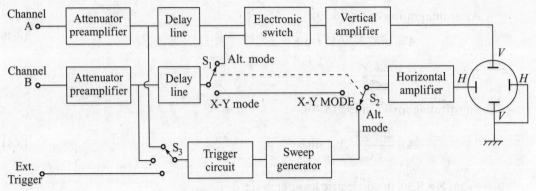

FIGURE 8.32 A dual-trace CRO.

Another function of the electronic switch is to select different modes of display. Electronic switch circuit operates the mode selection switches S_1 and S_2 in synchronism. When the mode selector is in 'alt' (alternate) mode, the electronic switch works in a normal way and alternatively connects channel A and channel B to vertical main amplifier. The switching rate of the electronic switch is synchronized to the sweep rate so that switching takes place at the start of each new sweep. As each channel has its own calibrated input attenuator and vertical position control, the amplitudes of the input signals can be adjusted independently and the traces can be positioned separately on the screen. The sweep trigger signal, as shown in Fig. 8.32, is available from channel A and channel B and it is picked off before the delay line. This maintains the correct phase relationship between the two input signals.

In the X-Y mode of operation, the horizontal amplifier is connected to channel B instead of the sweep generator. As both preamplifier blocks are identical, accurate X-Y measurements can be made.

Other modes of operations (not shown in the figure) are the chopper mode and the added mode. In the chopper mode, the electronic switch is free-running at the rate of 100 to 500 kHz and is entirely independent of the frequency of the sweep. When the chopping rate is much faster than the sweep rate, the electronic switch successively connects small segments of channel A and channel B inputs to the main vertical amplifier. The individual small segments of both the inputs fed to the main vertical amplifier reconstitutes the original signals of channel A and channel B on the screen, without any visual interruptions. When the sweep rate and chopping rate are approximately equal, the continuity of the images displayed is lost. The alternate mode of operation is preferred in this case.

In added mode of operation, the sum of the two signals is displayed as a single image. With polarity inversion switches in both the channels (not shown in the figure), it is possible to display A + B, A – B, B – A, and –A – B. In addition, a dual-trace CRO can be used to display either channel A or channel B like a conventional CRO.

8.11.2 Dual-Beam CRO

Unlike the dual-trace CRO a dual-beam CRO has two independent but identical vertical systems including the two sets of vertical plates. Hence, the two input signals

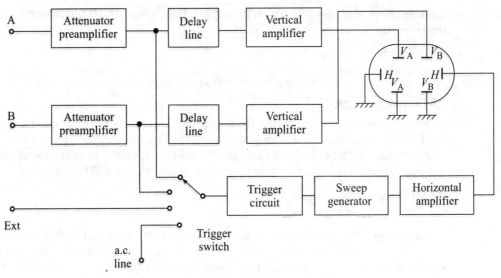

FIGURE 8.33 A dual-beam CRO.

are fed to the dual-beam CRO through channels A and B, and two separate images are displayed on the screen. The block diagram of the dual-beam CRO is shown in Fig. 8.33.

A special CRT is used for the dual-beam CRO. The CRT produces two completely separate electron beams. The two beams are independently deflected in the vertical direction. There are two methods of getting separate beams. In some dual-beam CROs, the output of a single electron gun is mechanically split into two separate beams. This is called split beam technique. Another type of dual-beam CRO uses a CRT that has two separate electron guns.

Like a conventional CRO, the dual-beam CRO also consists of only one horizontal system. The sweep generator drives one set of horizontal plates. Thus, two beams are swept across the screen at the same rate. The trigger circuit can be actuated by channel A, channel B, an external signal, or the supply line.

The dual-beam CROs are not as versatile as the dual-trace CROs because the later can be operated in a number of modes. But the importance of the dual-beam CRO is where it is to display two related phenomena, such as the input and output of a system for comparison.

8.11.3 Storage CRO

The persistence of the phosphor in the conventional CROs ranges from a few milliseconds to several seconds. The image on the screen disappears after the removal of the input signal. Hence, it not possible to make a real-time observation of the one-time events which occur only once. The images in this case disappear from the screen after a relatively short period. A storage CRO has a property to retain the display for much longer period. The image persists for several hours after the image is first written on the phosphor. The retention feature of the storage CRO can be utilized in displaying the waveform of a very low frequency signal. Following two techniques are utilized in the construction of the storage CROs:

1. Analogue technique.
2. Digital technique.

Analogue Storage CRO

In analogue technique a specially designed CRT is used. In an analogue storage CRO, the CRT consists of two electron guns. One is the regular electron gun similar to that in a conventional CRO called *write gun* and another is a special gun called *flood gun*. There are two storage techniques used in the design of the storage CRT:

1. Mesh storage technique.
2. Phosphor storage technique.

Mesh Storage CRT In a mesh storage CRT, a dielectric material is deposited on a storage mesh as the storage target. The storage mesh is a cross wire mesh. The storage mesh is placed between the deflection plates and the phosphor screen, as shown in Fig. 8.34. The mesh storage CRT based on the dielectric mesh target was first developed by Dr. Andrew Haeff in 1947.

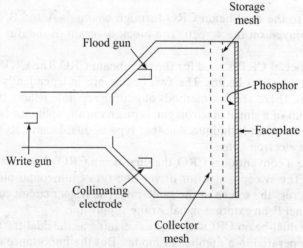

FIGURE 8.34 A mesh storage CRT.

A mesh storage CRT contains a dielectric deposited storage mesh, a collector mesh, a flood gun, and a collimator. These are in addition to all the elements of a standard CRT. The storage target is a thin deposition of a dielectric material on the storage mesh. The dielectric material may be magnesium fluoride. The dielectric material used has a characteristic of emitting secondary emission when bombarded by electrons of sufficient energy. The flood gun is grounded, whereas the collector mesh is biased at a positive potential higher than the storage mesh.

When the write gun electrons hit the storage mesh, secondary emission results. Thus, the areas of the storage surface hit by the beam electrons release electrons due to secondary emission and become positively charged. The secondary electrons are collected either by the collector mesh or by the display phosphor target. Thus, the write gun beam deflection pattern is traced on the storage surface as a positive charge pattern. The high insulation of the dielectric material prevents the loss of charge for a considerable period. Thus, the pattern is effectively stored.

To view the pattern stored on the storage mesh, the flood gun is turned on after the write gun is turned off. The flood gun emits a great flood of electrons. These electrons are attracted towards the collector mesh, as it is biased more positive than the deflection region. The collimator is a conductive coating on the CRT envelope with an applied potential. It helps to align the flood beam so that the electrons approach the storage target perpendicularly. Some electrons penetrate beyond the collector mesh. These electrons encounter the positive charged regions, that is, stored trace regions. Hence, these electrons are allowed to pass through the display phosphor, as shown in Fig. 8.35. The electrons encountering the negative charged regions, that is, without trace regions, are repelled back to the collector mesh, as shown in Fig. 8.35. Thus, the trace of the write beam on the storage surface reappears on the CRT display phosphor.

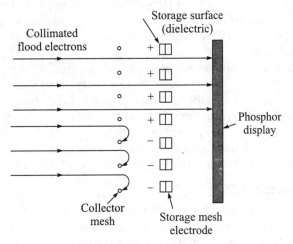

FIGURE 8.35 Flow of flood electrons.

A typical mesh storage CRT stores a trace for an hour or more and the trace may be displayed at bright intensity for at least a minute. The intensity fades out because ions generated by flood gun charge in other regions of the storage surface and hence the entire display surface consequently appears to be written. This is called fading positive. To erase the storage surface of the traces, a momentary-contact ERASE button is provided. It biases the storage mesh to the potential of collector mesh.

Phosphor Storage CRT Phosphor storage CRT uses a thin layer of scattered phosphor particles to serve as both the storage element and the display element. The storage element has a bistable property, which means that a trace is either stored or not stored. The material used is a P_1 phosphor doped for good secondary emission characteristic. The deposition of scattered phosphor particles is shallow enough so that the surface is not electrically continuous as shown in Fig. 8.36. The layer may be more than one particle thick, but it should be less than the thickness beyond which no storage is possible. The

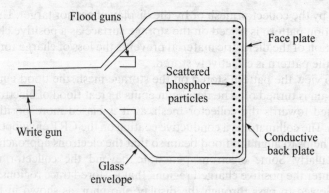

FIGURE 8.36 A phosphor storage CRT.

controlling electrode is the conductive backplate which is a transparent film deposited on the inside surface of the faceplate, before the phosphor is deposited. An operating voltage ranging from 100 to 200 V on the electrode gives a stable storage characteristic. A voltage below 100 V uniformly erases the target and that above 200 V uniformly writes the target.

The flood gun uniformly bombards the entire CRT screen with low-energy electrons. The phosphor particles struck by the low-energy electrons take on fairly low-level charge. The particles not hit by the electrons remain in a 'no-charge' condition. When a trace is to be recorded, the writing gun is turned on and high-energy electrons strike the screen forming an image. The phosphor particles that are struck by the high-energy electrons possess a considerable amount of charge and attract additional flood gun electrons, which retain the image. For erasing the screen, the controlling electrode is grounded.

Digital Storage CRO

A digital storage oscilloscope uses a conventional CRO. However, the analogue input signal is digitized using an A/D converter, and the digital signal is stored in a memory. When it is desired the digital signal is converted into an analogue signal using a D/A converter, and the reconstructed signal is applied to the vertical channel of the CRO. The block diagram of the digital storage CRO is shown in Fig. 8.37. The A/D and D/A converters and memory are discussed in Chapter 11.

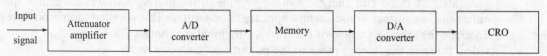

FIGURE 8.37 A block diagram of the digital storage oscilloscope.

8.11.4 Sampling CRO

The bandwidth of the vertical amplifier limits the frequency range of signals that can be displayed on a conventional CRO. Sampling CRO is used to improve high-frequency performance. The sampling technique is used in the sampling CRO to slow down the

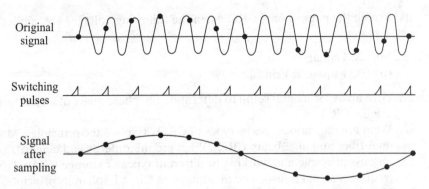

FIGURE 8.38 Original signal, switching pulses, and signal after sampling.

high frequency of the input signal many thousands of time. Sampling is a time-stretching process and consists of a sampler and a tuned amplifier. A simple sampler consists of an electronic switch and a storage capacitor. When the switch is closed momentarily, the capacitor is charged to the instantaneous value of the input voltage and holds this value till the switch does not close again. Samples of the input waveform are taken from successive cycles with one sample in one cycle. Each sample is slightly delayed with respect to the preceding sample, as shown in Fig. 8.38. The constructed waveform that is displayed on the CRO screen is a composite waveform made up of the samples taken from successive cycles of the input signal.

8.12 REVIEW QUESTIONS

1. Explain clearly about the main components of a CRT. Discuss briefly how the intensity, focusing, Y position, and volts per division adjustments are incorporated.
2. Discuss both electrostatic and electromagnetic focusing devices. What are their advantages and disadvantages?
3. Derive the expression of deflection in the electron beam in an electrostatic deflection method.
4. Derive the expression of deflection in the electron beam in an electromagnetic deflection method.
5. With a block diagram show the different components of the vertical deflection system of a CRO. Discuss each of the components.
6. Describe the constructional details of a sweep generator used in a CRO. How is the sweep linearity of the sweep improved?
7. How is sweep synchronized? Discuss the method of synchronization.
8. Discuss how a CRO can be used to measure the frequency of a sinusoidal signal. What is Lissjous figure?
9. Explain in short how the phase angle between two sinusoidal voltages can be determined with the help of a CRO.

10. Sketch the pattern on a CRO when the following voltages are applied on the X and Y plates:

    (a) X: $V\sin\omega t$; Y: $V\cos\omega t$
    (b) X: $V\sin\omega t$; Y: $V\sin 2\omega t$

11. Give an experimental setup to determine the phase angle of an unknown impedance using a CRO.
12. What are the various special types of CRO? Explain the principle and working of the dual-trace and dual-beam CROs. What are their merits and demerits?
13. Discuss principle and working of different types of storage CROs. How a very high frequency signal is displayed in a sampling CRO. Explain its principle.
14. How is B-H loop of a magnetic material displayed on the CRO. Explain the method?
15. How is dielectric loss of a dielectric material known using a CRO. Explain the method?

8.13 SOLVED PROBLEMS

1. An electrostatically deflected CRT has deflecting plates 2.5 cm long and 7.0 mm apart. The distance of the screen from the centre of the deflecting plates is 55 cm. Find (i) the velocity of the beam and (ii) the deflecting sensitivity of the tube. The accelerating voltage is 2.0 kV, charge on an electron is 1.6×10^{-19} C, and mass of an electron is 9.1×10^{-31} kg.

   Solution

   From Eq. (8.1) the velocity of the beam,

   $$v = \sqrt{\frac{2qV}{m}} = \sqrt{\frac{2 \times 1.6 \times 10^{-19} \times 2.0 \times 10^3}{9.1 \times 10^{-31}}}$$

   $$= 2.65 \times 10^7 \text{ m/s}$$

   From Eq. (8.10) the deflection sensitivity,

   $$S = \frac{lL}{2sV} = \frac{2.5 \times 10^{-2} \times 55 \times 10^{-2}}{2 \times 7.0 \times 10^{-3} \times 2.0 \times 10^3}$$

   $$= \textbf{0.357 mm/V}$$

2. Following are the given data for an electrostatic deflected CRT:

   Length of the deflecting plates = 1.5 cm

   Spacing of the deflecting plates = 0.55 cm

   Distance of the screen from the centre of the plates = 25 cm

   Calculate the approximate voltage between the plates to give a deflection of 7.5 mm on the screen. The accelerating voltage is 300 V.

Solution

From Eq. (8.9), the deflection,

$$d = \frac{ElL}{2sV}$$

Hence, the potential difference between the deflecting plates

$$E = \frac{2sVd}{lL} = \frac{2 \times 0.55 \times 10^{-2} \times 300 \times 7.5 \times 10^{-3}}{1.5 \times 10^{-2} \times 25 \times 10^{-2}}$$

$$= \mathbf{6.6\ V}$$

3. Following are the given data for an electrostatic deflected CRT:

   Length of the deflecting plates = 2.5 cm
   Spacing of the deflecting plates = 0.5 cm
   Distance of the screen from the centre of the plates = 25 cm

   The electron beam is accelerated by a potential difference of 2.5 kV and is centrally projected between the plates. Calculate the deflection on the screen and the deflecting voltage. Also, calculate the deflection sensitivity and factor of the CRT.

Solution

Deflection of the beam on the screen is shown in Fig. 8.39. The beam projected backwards cuts the horizontal axis at O at the centre of the deflecting plates. Hence, from the similar triangles OCD and OBA,

$$\frac{CD}{BA} = \frac{OC}{OB}$$

Hence, the deflection,

$$d = CD = \frac{OC \times BA}{OB} = \frac{L \times s/2}{l/2} = \frac{Ls}{l}$$

$$= \frac{25 \times 10^{-2} \times 0.5 \times 10^{-2}}{2.5 \times 10^{-2}} = \mathbf{5.0\ cm}$$

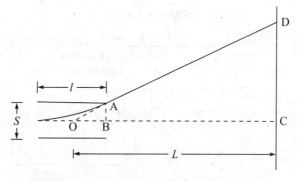

FIGURE 8.39 Schematic for Solved Problem 3.

From Eq. (8.9), the deflection voltage,

$$V_d = \frac{2dVs}{lL} = \frac{2 \times 5.0 \times 10^{-2} \times 2.5 \times 10^3 \times 0.5 \times 10^{-2}}{2.5 \times 10^{-2} \times 25 \times 10^{-2}}$$

$$= 200 \text{ V}$$

The deflection sensitivity,

$$S = \frac{d}{V_d} = \frac{5.0 \times 10^{-2}}{200} = \mathbf{0.25 \text{ mm}}$$

The deflection factor,

$$D = \frac{1}{s} = \frac{1}{0.25} = \mathbf{4 \text{ V/mm}}$$

4. The plate lengths and spacing between the vertical and horizontal plates are 2.5 cm and 1.5 cm, respectively. Both pairs of the plates are at the same potential of 110 V (r.m.s.). Find the length of the line produced on the screen at a distance of 30 cm from the centre of near plates. Accelerating voltage may be taken as 1.5 kV. The distance between the two pairs of plates is 3.5 cm.

Solution

Fig. 8.40 may be referred.

From Eq. (8.9), the deflection

$$d = \frac{ElL}{2sV}$$

Hence, the vertical deflection,

$$d_V = \frac{ElL_V}{2sV} = \frac{2\sqrt{2} \times 110 \times 2.5 \times 10^{-2} \times 33.5 \times 10^{-2}}{2 \times 1.5 \times 10 \times 10^{-2} \times 1.5 \times 10^3}$$

$$= \mathbf{5.79 \text{ cm}}$$

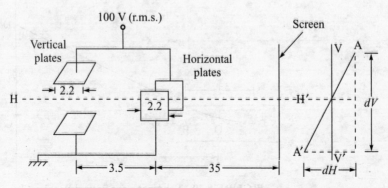

FIGURE 8.40   Schematic for Solved Problem 4.

and the horizontal deflection,

$$d_H = \frac{EIL_H}{2sV} = \frac{2\sqrt{2} \times 110 \times 2.5 \times 10^{-2} \times 30.0 \times 10^{-2}}{2 \times 1.5 \times 10 \times 10^{-2} \times 1.5 \times 10^3}$$

$$= 5.18 \text{ cm}$$

From Fig. 8.40, the length of the image (line AA'),

$$m = \sqrt{d_V^2 + d_H^2} = \sqrt{(5.79)^2 + (5.18)^2}$$

$$= 7.7 \text{ cm}$$

5. A magnetically deflected CRT has an accelerating voltage of 11 kV. The screen is situated at a distance of 35 cm from the centre of the deflecting coil. The effective length over which the coil assumed to give uniform field is 5.5 cm. Calculate the flux density required to give a deflection of 10 cm. Let us take $q = 1.6 \times 10^{-19}$ C and $m = 9.1 \times 10^{-31}$ kg, where q is the charge and m is the mass of an electron.

Solution

From Eq. (8.15), the deflection,

$$d = LlB\sqrt{\frac{q}{2mV}}$$

or $$B = \frac{d}{Ll}\sqrt{\frac{2mV}{q}} = \frac{10 \times 10^{-2}}{35 \times 10^{-2} \times 5.5 \times 10^{-2}}\sqrt{\frac{2 \times 9.1 \times 10^{-31} \times 11 \times 10^3}{1.6 \times 10^{-19}}}$$

$$= 1.84 \text{ mWb/m}^2$$

6. When comparing two frequencies the Lissajous figures obtained are shown in Fig. 8.41. Determine the ratio of the two frequencies.

Solution

The ratio of frequencies is given by

$$\frac{f_V}{f_H} = \frac{\text{Number of positive V peaks}}{\text{Number of positive H peaks}}$$

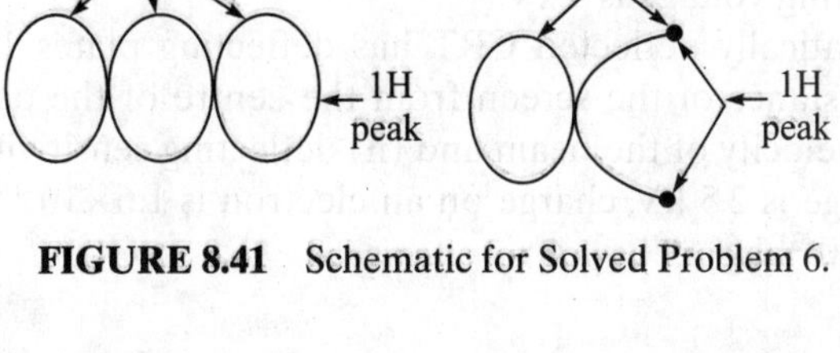

FIGURE 8.41 Schematic for Solved Problem 6.

(a) Number of positive H peaks = 1
Number of positive V peaks = 3

$$\frac{f_V}{f_H} = \frac{3}{1}$$

(b) Number of positive H peaks = 1
Number of positive V peaks = $1\frac{1}{2}$

$$\frac{f_V}{f_H} = \frac{1\frac{1}{2}}{1} = \frac{3}{2}$$

7. Two sinusoidal voltage waves of same amplitude and frequency but of different phases are applied to X plates and Y plates. The pattern obtained on the screen is an ellipse. The peak Y deflection and its intercept on the y-axis are 5 and 4 divisions, respectively. Find the relative phase angle when the major axis of the ellipse makes (i) positive slope and (ii) negative slope with the x-axis.

Solution

From Eq. (8.37), the relative phase,

$$\phi = \sin^{-1}\left(\frac{Y_{intercept}}{Y_{peak}}\right)$$

(i) $\phi = \sin^{-1}\left(\frac{4}{5}\right) = \mathbf{53°}$

(ii) $\phi = 180° - 53° = \mathbf{127°}$

8.14 EXERCISES

1. Following are the given data for an electrostatic deflected CRT:
   Length of the deflecting plates = 2.0 cm
   Spacing of the deflecting plates = 0.5 cm
   Distance of the screen from the centre of the plates = 30 cm
   Calculate the deflection when voltage applied between the vertical plates is 100 V. The accelerating voltage is 2 kV.

2. An electrostatically deflected CRT has deflecting plates 1.5 cm long and 5.0 mm apart. The distance of the screen from the centre of the deflecting plates is 50 cm. Find (i) the velocity of the beam and (ii) deflecting sensitivity of the tube. The accelerating voltage is 2.5 kV, charge on an electron is 1.6×10^{-19} C, and mass of an electron is 9.1×10^{-31} kg.

3. In an electrostatically deflected CRT, the length of the deflecting plates is 2.0 cm and the spacing of the deflecting plates is 5.0 mm. The distance of the screen from the centre of the plates is 35 cm. The electron beam is accelerated by a potential difference of 2 kV and is centrally projected between the plates. Calculate the deflection on the screen and the deflecting voltage. Also, calculate the deflection sensitivity and factor of the CRT.
4. A magnetically deflected CRT has an accelerating voltage of 5 kV. The screen is situated at a distance of 27 cm beyond the deflecting coil. The effective length over which the coil assumed to give uniform field is 6.0 cm. Calculate the flux density required to give a deflection of 7.5 cm. Let us take $q = 1.6 \times 10^{-19}$ C and $m = 9.1 \times 10^{-31}$ kg, where q is the charge and m is the mass of an electron.
5. Two sinusoidal voltage waves of equal amplitude and same frequency but of different phases are applied to X plates and Y plates. The pattern obtained on the screen is an ellipse. The peak Y deflection and its intercept on the y-axis are 4 and 3 divisions. Find the relative phase angle when the major axis of the ellipse makes (i) positive slope and (ii) negative slope with the x-axis.

9

ELECTRICAL POWER SUPPLIES

Outline

9.1 Introduction 386
9.2 Controlled Rectifiers 387
9.3 Inverters 417
9.4 A.C. Voltage Controllers 433
9.5 D.C. Voltage Regulators 444
9.6 Choppers 450
9.7 Switched Mode Power Supply 457
9.8 Uninterrupted Power Supply 462
9.9 Review Questions 464
9.10 Solved Problems 466
9.11 Exercises 487

9.1 INTRODUCTION

In modern age, electrical power supply plays an important role in our life. In addition to domestic establishment, all modern equipments, industries, research laboratories, computer systems, and commercial establishment need electrical power supply. However, the power requirements of different users vary. They may differ in the type of supply (such as d.c. or a.c.), the magnitude of the supply voltage, the amount of power, the quality of power supply, etc. The power supplied by the electricity board is a.c. power. The general power supply may not be always available. Also, the quality of the power supplied may not be good enough for certain requirements. Especially, in research laboratories some equipments or experimental setups may require d.c. supply. Therefore, an additional power supply is needed.

We generally need an a.c. power supply as a standby source, when mains power supply is not available. This a.c. power supply is known as uninterrupted power supply (UPS).

Power supply specifications differ according to their applications. However, the specifications that are generally common for all applications are listed below:

1. Isolation between the source and the load.
2. Compactness in size.
3. High efficiency.
4. Low harmonic distortion.
5. Good voltage regulation.

As mentioned above, some applications require d.c. supply, which needs to be obtained from the a.c. mains supply. The conversion from a.c. to d.c. is known as rectification. Different rectifier circuits using diodes have been discussed in Chapter 3. Rectifiers using diodes are known as uncontrolled rectifiers because the output d.c. voltage is fixed for a given a.c. supply. For low-power devices in which a fixed voltage is required, the diode rectifiers are used. Such rectifiers have the advantages of low cost and simple circuitry.

In high-power devices, rectifiers with thyristors are used because thyristors are available in high-voltage, high-current ratings. These rectifiers are known as controlled rectifiers. These are called controlled rectifiers because their output voltages can be controlled by changing the firing angle of the thyristors. Depending on whether the a.c. input is single phase or three phase, a controlled rectifier is called single-phase or three-phase rectifier. High-power rectifiers are mostly three-phase rectifiers. A controlled rectifier is also known as a.c. to d.c. converter.

There are cases where we require a.c. from d.c. The device used for this purpose is known as an inverter. An inverter converts d.c. supply into a.c. supply. The output voltage and frequency can be varied as per the requirement of the load. The d.c. input supply may be a battery, solar cells, fuel cells, etc. In small- and medium-size inverters are used in domestic and commercial installations as a source of standby electric supply. In industrial installations, inverters are used for variable-speed a.c. drives, induction heating, etc. In high-voltage direct current (HVDC) power supply, a.c. is converted into d.c. by controlled rectifiers at the a.c. (generating) end of the power line. The d.c. voltage is transmitted over the HVDC line. The d.c. voltage at the receiving end is converted into the a.c. voltage by an inverter and a.c. is supplied to the consumers.

In industrial heating, for illumination level controller, speed control of induction motors, etc., variable voltage supply at constant frequency is required. The device used for this purpose is known as an a.c. voltage regulator. The a.c. voltage level can be varied by conventional methods such as by an autotransformer, tap changing transformer, and saturable reactor. However, the modern method uses a.c. regulators by using thyristors and triacs. These regulators are more popular because of high efficiency, fast control, and compact size. However, these regulators introduce high level of harmonics. Active filters may be used to minimize the harmonic level.

9.2 CONTROLLED RECTIFIERS

As mentioned above, controlled rectifiers (or converters) employ thyristors as active devices unlike uncontrolled rectifiers, which employ diodes. If the a.c. input is single phase, the converter is called a single-phase converter, whereas it is called a three-phase converter when the input is a three-phase a.c. supply. The single-phase converter like uncontrolled rectifier may be half-wave converter or full-wave converter. Half-wave converters use only one thyristor as an active device. They introduce a d.c. component in the supply transformer, which results in saturation of the transformer. They also introduce harmonics in the supply line. These disadvantages are minimized in single-phase and three-phase full-wave converters. In practice, there are two configurations for full-wave converters:

1. Midpoint converter.
2. Bridge converter.

A midpoint converter uses an input transformer with central tapped secondaries. Thus, there are two secondary windings for each input phase. The single-phase midpoint full-wave converter uses two thyristors, whereas the three-phase full-wave converter uses six thyristors (two thyristor per phase). The bridge converter uses thyristors in the form of a bridge circuit. The single-phase full-wave bridge converter uses four thyristors, whereas the three-phase full-wave bridge converter uses six thyristors.

The single-phase or three-phase full-wave bridge converters may be of two types:

1. Full-controlled converter or full converter.
2. Half-controlled converter or semiconverter.

A full converter uses thyristors only as active devices and there is wider control over the level of output voltage. A semiconverter uses a mixture of diodes and thyristors and there is a limited control over the level of output voltage.

9.2.1 Single-Phase Half-Wave Converters

Like half-wave uncontrolled rectifier, the half-wave converter uses only one thyristor. The converter may be connected to a resistive load, an inductive load consisting of resistance and inductance, and a load consisting of a resistance, an inductance, and a back e.m.f. The third type of load may be written as RLE load and an example of this type of load is a d.c. motor. The converter is used as a variable voltage d.c. source in the armature-controlled d.c. shunt motor. The thyristor conducts during the positive half of the cycle when it is triggered at an angle α by applying a positive pulse to the gate. The thyristor blocks during the negative half of the cycle. During nonconducting period, the voltage drop across the thyristor is same as the input voltage, whereas during the conduction period, the voltage drop across the thyristor is very small, only about 1 V.

When the load is inductive, that is, a combination of resistance and inductance, the load current lags the load voltage. When the thyristor starts conducting, the load voltage instantaneously becomes equal to the input voltage. However, the load current does not change instantaneously. It increases slowly, resulting in a wave shape different from that obtained in the case of pure resistive load. When the input voltage becomes zero at $\omega t = \pi$, the load voltage also becomes zero. But the load current does not become zero and tends to continue beyond $\omega t = \pi$. This results in a distorted current wave shape and a poor power factor. To improve the wave shape and power factor, a *commutating diode* (also known as *freewheeling* or *by-pass diode*) is connected across the load. Following four types of loads are discussed here:

1. Resistive load.
2. Inductive load.
3. Inductive load with commutating diode.
4. Inductive load and back e.m.f.

Resistive (R) Load The circuit for the single-phase half-wave converter connected to a resistive load is shown in Fig. 9.1. The input voltage, the triggering pulse, the output voltage, the output current, and the voltage across the thyristor are shown in Fig. 9.2. The average (d.c.) output voltage,

$$V_{d.c.} = \frac{1}{2\pi} \int_\alpha^\pi V_m \sin \omega t \, d\omega t$$

$$= \frac{1}{2\pi} \left| -V_m \cos \omega t \right|_\alpha^\pi$$

$$= \frac{V_m}{2\pi}(1 + \cos \alpha) \tag{9.1}$$

The average load current,

$$I_{d.c.} = \frac{V_{d.c.}}{R} = \frac{V_m}{2\pi R}(1 + \cos \alpha) \tag{9.2}$$

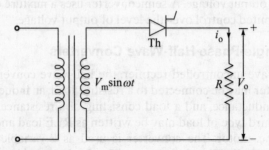

FIGURE 9.1 A half-wave converter with R load.

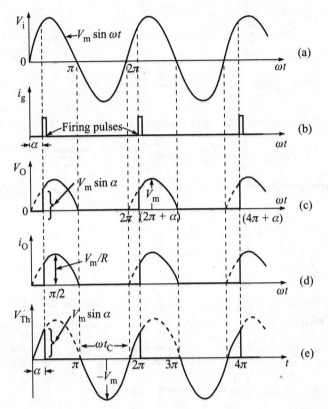

FIGURE 9.2 Waveforms of (a) input voltage, (b) firing pulse, (c) output voltage, (d) output current, and (e) voltage drop across thyristor.

The r.m.s. output voltage,

$$V_{r.m.s.}^2 = \frac{1}{2\pi}\int_\alpha^\pi V_m^2 \sin^2 \omega t \, d\omega t = \frac{V_m^2}{4\pi}\int_\alpha^\pi (1-\cos 2\omega t) \, d\omega t$$

$$= \frac{V_m^2}{4\pi}\left((\pi-\alpha) + \frac{\sin 2\alpha}{2}\right)$$

Hence, $\quad V_{r.m.s.} = \frac{V_m}{2}\sqrt{\left(\frac{\pi-\alpha}{\pi} + \frac{\sin 2\alpha}{2\pi}\right)}$ \hfill (9.3)

The r.m.s. load current,

$$I_{r.m.s.} = \frac{V_{r.m.s.}}{R} = \frac{V_m}{2R}\sqrt{\left(\frac{\pi-\alpha}{\pi} + \frac{\sin 2\alpha}{2\pi}\right)} \qquad (9.4)$$

From Eqs. (9.1) and (9.3), it is evident that the average and r.m.s. output voltages depend on the firing angle α. Hence, for $\alpha = \pi$, both the average and r.m.s. output voltages are equal to zero, whereas for $\alpha = 0$, both the average and r.m.s. output voltages are same as in the uncontrolled rectifiers, and are given by

$$V_{d.c.} = \frac{V_m}{\pi} \qquad (9.5)$$

and
$$V_{r.m.s.} = \frac{V_m}{2} \tag{9.6}$$

The total output power,

$$P_o = \frac{V_{r.m.s.}^2}{R} = \frac{V_m^2}{4R}\left(\frac{(\pi-\alpha)}{\pi} + \frac{\sin 2\alpha}{2\pi}\right) \tag{9.7}$$

As the input current is same as the r.m.s. load current, the input VA,

$$VA_{input} = \frac{V_m}{\sqrt{2}} \times \frac{V_{r.m.s.}}{R} = \frac{V_m^2}{2\sqrt{2}R}\sqrt{\left(\frac{\pi-\alpha}{\pi} + \frac{\sin 2\alpha}{2\pi}\right)} \tag{9.8}$$

The input power factor,

$$pf = \frac{P_o}{VA_{input}} = \frac{1}{\sqrt{2}}\sqrt{\left(\frac{\pi-\alpha}{\pi} + \frac{\sin 2\alpha}{2\pi}\right)} \tag{9.9}$$

The peak inverse voltage across the thyristor,

$$V_{peak} = V_m \tag{9.10}$$

Example 9.1

A single-phase half-wave converter is shown in Fig. 9.3. The firing angle $\alpha = \pi/4$. (a) Find $V_{d.c.}, I_{d.c.}, V_{r.m.s.},$ and $I_{r.m.s.}$. (b) Determine $P_{d.c.}$ and $P_{a.c.}$. (c) Also determine rectification efficiency, form factor, ripple factor, peak inverse voltage across the thyristor, VA rating of the transformer, and transformer utilization factor.

Solution

(a) The peak input voltage to the thyristor circuit,

$$V_m = \frac{\sqrt{2} \times 230}{2} = 162.6 \text{ V}$$

From Eq. (9.1),

$$V_{d.c.} = \frac{162.6}{2\pi}(1+\cos \pi/4)$$

$$= 44.2 \text{ V}$$

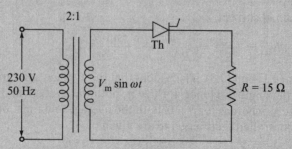

FIGURE 9.3 Circuit for Example 9.1

From Eq. (9.2),
$$I_{d.c.} = \frac{V_{d.c.}}{R} = \frac{44.2}{15} = \mathbf{2.95\ A}$$

From Eq. (9.3),
$$V_{r.m.s.} = \frac{V_m}{2}\sqrt{\left(\frac{\pi-\alpha}{\pi}+\frac{\sin 2\alpha}{2\pi}\right)} = \frac{162.6}{2}\sqrt{\left(\frac{3}{4}+\frac{1}{2\pi}\right)}$$
$$= \mathbf{77.52\ V}$$

From Eq. (9.4),
$$I_{r.m.s.} = \frac{V_{r.m.s.}}{R} = \frac{77.52}{15} = \mathbf{5.17\ A}$$

(b) $P_{d.c.} = V_{d.c.}I_{d.c.} = 44.2 \times 2.95 = \mathbf{130.39\ W}$

and $\quad P_{a.c.} = V_{r.m.s.} \times I_{r.m.s.} = 77.52 \times 5.17 = \mathbf{400.78\ W}$

(c) The rectification efficiency,
$$\eta_r = \frac{P_{d.c.}}{P_{a.c.}} = \frac{130.39}{400.78} = 0.325$$
$$= \mathbf{32.5\%}$$

Form factor,
$$F = \frac{V_{r.m.s.}}{V_{d.c.}} = \frac{77.52}{44.2} = \mathbf{1.754}$$

Ripple factor,
$$\gamma = \sqrt{F^2-1} = \sqrt{(1.754)^2-1} = \mathbf{1.44}$$

Peak inverse voltage across the thyristor,
$$V_{peak} = V_m = \mathbf{162.6\ V}$$

The transformer secondary current is same as the r.m.s. value of the load current. Hence, the VA rating of the transformer is given by
$$VA = \frac{230}{2} \times 5.17 = \mathbf{594.55\ VA}$$

The transformer utilization factor,
$$UF = \frac{P_{d.c.}}{VA} = \frac{130.39}{594.55} = \mathbf{0.22}$$

Inductive (RL) Load In case of resistive load, the load current and voltage are in phase. At the firing angle α, the load voltage instantaneously becomes equal to the input voltage at $\omega t = \alpha$. The load current also changes instantaneously. At $\omega t = \pi$, both the load voltage and the current become zero, simultaneously. The circuit of a single-phase

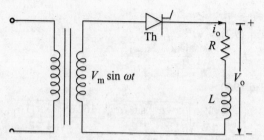

FIGURE 9.4 Half-wave converter with RL load.

half-wave converter with inductive load is shown in Fig. 9.4. The input voltage is a sinusoidal signal and is shown in Fig. 9.5(a). At the firing angle α, that is, $\omega t = \alpha$, the thyristor starts conducting and the load voltage instantaneously becomes equal to the input voltage. However, the load current does not change instantaneously; it increases slowly. After some time the output current reaches to a maximum value, and then decreases. When the input voltage becomes zero at $\omega t = \pi$, the load voltage also becomes zero. But the load current does not become zero and tends to continue beyond $\omega t = \pi$. After $\omega t = \pi$, the thyristor becomes reverse biased, but it does not turn off, as the load current is not below the holding current. The load current reduces to zero after some time, say at an angle β, which is greater than π. The thyristor is turned off at $\omega t = \beta$, as its anode is

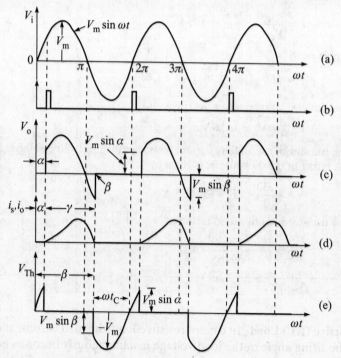

FIGURE 9.5 Waveforms of (a) input voltage, (b) firing pulse, (c) output voltage, (d) output current, and (e) voltage drop across thyristor.

already reverse biased and the output current is equal to zero. The angle β is called the *extinction angle* and $\gamma = \beta - \alpha$ is called the *conduction angle*. The cycle is again repeated at $\omega t = 2\pi + \alpha$. The input voltage, the triggering pulse, the output voltage, the output current, and the voltage across the thyristor are shown in Fig. 9.5. During the period when the thyristor is conducting, the KVL equation of the circuit is given by

$$Ri_o + L\frac{di_o}{dt} = V_m \sin \omega t$$

or
$$\frac{di_o}{dt} + \frac{R}{L}i_o = \frac{V_m}{L}\sin \omega t \tag{9.11}$$

The transient response,

$$i_o^t = Ke^{-(R/L)t} \tag{9.12}$$

where K is a constant. The steady state response,

$$i_o^{ss} = A\cos\omega t + B\sin\omega t \tag{9.13}$$

where A and B are constants. Differentiating Eq. (9.13),

$$\frac{di_o^{ss}}{dt} = -A\omega\sin\omega t + B\omega\cos\omega t \tag{9.14}$$

Substituting from Eqs. (9.13) and (9.14) in Eq. (9.11), and arranging,

$$\left(\frac{AR}{L} + B\omega\right)\cos\omega t + \left(\frac{BR}{L} - A\omega\right)\sin\omega t = \frac{V_m}{L}\sin\omega t$$

Hence,
$$\left(\frac{AR}{L} + B\omega\right) = 0$$

and
$$\left(\frac{BR}{L} - A\omega\right) = \frac{V_m}{L}$$

Solving for A and B,

$$A = \frac{-\omega L V_m}{R^2 + \omega^2 L^2}$$

$$B = \frac{RV_m}{R^2 + \omega^2 L^2}$$

Substituting these values of A and B in Eq. (9.13),

$$i_o^{ss} = \frac{-\omega L V_m}{R^2 + \omega^2 L^2}\cos\omega t + \frac{RV_m}{R^2 + \omega^2 L^2}\sin\omega t$$

$$= \frac{V_m}{\sqrt{R^2 + \omega^2 L^2}}\sin(\omega t - \theta) \tag{9.15}$$

where $\theta = \tan^{-1}(\omega L/R)$.

Hence, the total response,

$$i_o = i_o^t + i_o^{ss} = Ke^{-(R/L)t} + \frac{V_m}{\sqrt{R^2 + \omega^2 L^2}} \sin(\omega t - \theta)$$

At $\omega t = \alpha$, $i_o = 0$, then

$$0 = Ke^{-(R/L)(\alpha/\omega)} + \frac{V_m}{\sqrt{R^2 + \omega^2 L^2}} \sin(\alpha - \theta)$$

Hence, $\quad K = \left(-\frac{V_m}{\sqrt{R^2 + \omega^2 L^2}} \sin(\alpha - \theta)\right) e^{(R/L)(\alpha/\omega)}$ (9.16)

and the total dynamic response,

$$i_o = \left(-\frac{V_m}{\sqrt{R^2 + \omega^2 L^2}} \sin(\alpha - \theta)\right) e^{-(R/L)(t - \alpha/\omega)}$$

$$+ \frac{V_m}{\sqrt{R^2 + \omega^2 L^2}} \sin(\omega t - \theta) \quad (9.17)$$

When $\omega t = \beta$, the load current $i_o = 0$; hence,

$$\sin(\beta - \theta) = \sin(\alpha - \theta) e^{-(R/\omega L)(\beta - \alpha)} \quad (9.18)$$

The extinction angle β can be obtained by solving Eq. (9.18). If the extinction angle is known, the average output voltage is given by

$$V_{d.c.} = \frac{1}{2\pi} \int_\alpha^\beta V_m \sin \omega t \, d\omega t = \frac{1}{2\pi} \left| -V_m \cos \omega t \right|_\alpha^\beta$$

$$= \frac{V_m}{2\pi} (\cos\alpha - \cos\beta) \quad (9.19)$$

The average load voltage across the inductor is zero; hence, the average current,

$$I_{d.c.} = \frac{V_{d.c.}}{R} = \frac{V_m}{2\pi R} (\cos\alpha - \cos\beta) \quad (9.20)$$

The r.m.s. output voltage,

$$V_{r.m.s.}^2 = \frac{1}{2\pi} \int_\alpha^\beta V_m^2 \sin^2 \omega t \, d\omega t = \frac{V_m^2}{4\pi} \int_\alpha^\beta (1 - \cos 2\omega t) \, d\omega t$$

$$= \frac{V_m^2}{4\pi} \left((\beta - \alpha) + \frac{\sin 2\alpha - \sin 2\beta}{2}\right)$$

Hence, $\quad V_{r.m.s.} = \frac{V_m}{2\sqrt{\pi}} \sqrt{\left((\beta - \alpha) + \frac{\sin 2\alpha - \sin 2\beta}{2}\right)}$ (9.21)

The load current flows during the period $\alpha < \omega t < \beta$. The same current flows through the source. The r.m.s. value of load current can be obtained from Eq. (9.17). The power consumed in the load is $I_{r.m.s.}^2 R$. The power consumed in the load is supplied by the source during $\alpha < \omega t < \pi$, as both the output voltage and the current are positive. During the period $\pi < \omega t < \beta$, the output voltage is negative and the output current is positive. The energy stored in the inductor is, therefore, returned to the source during this period.

Inductive Load with Commutating Diode As mentioned above, to improve the wave shape and power factor, a commutating diode is connected across the load. The circuit of a single-phase half-wave converter with inductive (RL) load and commutating diode is shown in Fig. 9.6. At $\omega t = \alpha$, the thyristor is fired and starts conducting. The output voltage instantaneously rises to the input voltage. Due to inductance, the load current does not rise instantaneously, but it rises gradually to a maximum value. The diode is reverse biased and the load current flows through the thyristor. At $\omega t = \pi$, the input voltage is zero and so is the output voltage. But the output current continues beyond $\omega t = \pi$. The output voltage reverses from positive to negative value. The diode CD becomes forward biased and starts conducting. The load current is immediately transferred to the diode. The thyristor current becomes zero. As the output current (also the thyristor current) is zero and the anode voltage becomes reverse biased, the thyristor is turned off at $\omega t = \pi$. During the interval from $\omega t = \pi$ to $\omega t = 2\pi + \alpha$, the load current continues to flow through the commuting diode. At $\omega t = 2\pi + \alpha$, the thyristor is turned on again by the firing pulse, as the thyristor is forward biased. The above cycle is repeated after $\omega t = 2\pi + \alpha$. The use of commutating diode improves the wave form of the load current and power factor. It also improves the circuit performance. Different waves are shown in Fig. 9.7.

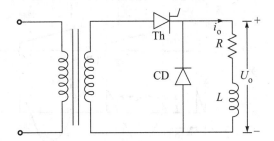

FIGURE 9.6 A circuit with RL load and CD diode.

The expression of the load current can be derived as it has been done for the inductive load without commutating diode. The only difference is that the current in this case is not zero at $\omega t = \alpha$. For the period when CD is not conducting, the total response of the circuit of Fig. 9.6, as in case of Fig. 9.4, is given by

$$i_o = Ke^{-(R/L)t} + \frac{V_m}{\sqrt{R^2 + \omega^2 L^2}} \sin(\omega t - \theta)$$

Let at $\omega t = \alpha, i_o = I_o$, then

$$I_o = Ke^{-(R/L)(\alpha/\omega)} + \frac{V_m}{\sqrt{R^2 + \omega^2 L^2}} \sin(\alpha - \theta)$$

Hence, $$K = \left(I_o - \frac{V_m}{\sqrt{R^2 + \omega^2 L^2}} \sin(\alpha - \theta)\right) e^{(R/L)(\alpha/\omega)} \qquad (9.22)$$

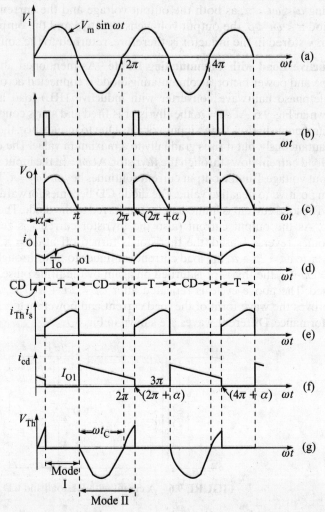

FIGURE 9.7 Waveforms of (a) input voltage, (b) firing pulse, (c) output voltage, (d) output current, (e) thyristor current, (f) diode current, and (g) voltage drop across thyristor.

and the total dynamic response,

$$i_o = \left(I_o - \frac{V_m}{\sqrt{R^2 + \omega^2 L^2}} \sin(\alpha - \theta)\right) e^{-(R/L)(t - \alpha/\omega)}$$

$$+ \frac{V_m}{\sqrt{R^2 + \omega_2 L^2}} \sin(\omega t - \theta) \tag{9.23}$$

where $\theta = \tan^{-1}(\omega L/R)$.

During the interval from $\omega t = \pi$ to $\omega t = 2\pi + \alpha$, the thyristor is nonconducting and the commutating diode is conducting. During this interval, the KVL equation

$$Ri_o + L\frac{di_o}{dt} = 0 \tag{9.24}$$

The solution of Eq. (9.24) is of the form given as follows:
$$i_o = Ke^{-(R/L)t} \qquad (9.25)$$
where K is a constant. Let at $\omega t = \pi, i_o = I'_o$. Then

$$I'_o = Ke^{-(R/L)(\pi/\omega)}$$

or $\qquad K = I'_o e^{(R/L)(\pi/\omega)}$

Hence, $\qquad i_o = I'_o\, e^{-(R/L)(t-\pi/\omega)} \qquad (9.26)$

It is evident from the above discussions that the source supplies power to the load resistance during $\alpha < \omega t < \pi$. During the interval from $\omega t = \pi$ to $\omega t = 2\pi + \alpha$, the energy stored in the inductor is supplied to the load resistance. The average values of voltage and current are same as given in Eqs. (9.1) and (9.2).

When the inductive load is without the commutating diode, as in Fig. 9.4, the power consumed in the load is supplied by the source during $\alpha < \omega t < \pi$, as both the output voltage and the output current are positive. During the period $\pi < \omega t < \beta$, the output voltage is negative and the output current is positive. The energy stored in the inductor is, therefore, returned to the source during this period. When commutating diode CD is used in conjunction with the inductive load, as shown in Fig. 9.6, the energy stored in the inductor is absorbed by the load resistance in the period $\pi < \omega t < 2\pi + \alpha$. It can, therefore, be concluded that the power delivered to the load resistance is more when CD is used (Fig. 9.6). As volt ampere is almost same in both circuits of Figs. 9.4 and 9.6, the input power factor is improved when CD is used with the inductive load.

It can also be seen from Figs. 9.5 and 9.7 that the load current waveform is improved when CD is used. Thus, the advantages of using CD are as listed below:

1. Input power factor is improved.
2. Load current waveform is improved.
3. Rectification efficiency is improved.

The commutating diode used in the circuit of Fig. 9.6 prevents the output voltage from becoming negative. Whenever the output voltage tends to become negative, the commutating diode starts conducting, making the thyristor current and the output voltage zero. This results in turning off the thyristor.

The supply current in single-phase half-wave converters taken from the supply is always unidirectional. Thus, these converters introduce a d.c. component into the supply line. The d.c. component may lead to saturation of the supply transformer. Harmonics are also introduced in the supply. These are undesirable.

Inductive Load and Back E.M.F The circuit of a single-phase half-wave converter with RLE load (a load consisting of a resistance, an inductance, and a back e.m.f.) is shown in Fig. 9.8. The back e.m.f. in the load may be due to a battery or a d.c. motor. The firing pulse must be applied to the gate of the thyristor when its anode is forward biased. Hence, the minimum value of firing angle is obtained from the relation given in Eq. (9.27).

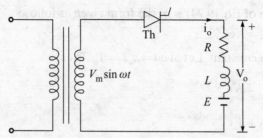

FIGURE 9.8 Half-wave converter with RLE load.

$$V_m \sin \omega t = E \tag{9.27}$$

Hence, the minimum value of firing angle,

$$\alpha_{min} = \sin^{-1}\left(\frac{E}{V_m}\right) \tag{9.28}$$

and the maximum value of firing angle,

$$\alpha_{max} = \pi - \alpha_{min} \tag{9.29}$$

These angles are shown in Fig. 9.9(a). During the interval, the load current is zero, that is, the thyristor is nonconducting, the output voltage is equal to E. When the load current is not zero, that is, the thyristor is conducting, the output voltage follows the input voltage

For the circuit of Fig. 9.8, the KVL equation is given by

$$Ri_o + L\frac{di_o}{dt} = V_m \sin \omega t - E \tag{9.30}$$

The difference between Eqs. (9.11) and (9.30) is that in Eq. (9.11) there is only one input $V_m \sin \omega t$, whereas in Eq. (9.30), there is an additional input $-E$. Hence, the transient response is same as given in Eq. (9.12). As there are two inputs, there are two steady-state values. As the first input is same as in Eq. (9.11), the first steady-state solution is same and is given in Eq. (9.15). The second steady-state solution is $-(E/R)$. Hence, the total response is given by

$$i_o = i_o^t + i_{o1}^{ss} + i_{o2}^{ss} = Ke^{-(R/L)t} + \frac{V_m}{\sqrt{R^2 + \omega^2 L^2}} \sin(\omega t - \theta) - \frac{E}{R}$$

At $\omega t = \alpha$, that is, $t = \alpha/\omega$, $i_o = 0$. Hence,

$$K = \left(\frac{E}{R} - \frac{V_m}{\sqrt{R^2 + \omega^2 L^2}} \sin(\alpha - \theta)\right) e^{(R/L)(\alpha/\omega)}$$

and the current is given by

$$i_o = \frac{V_m}{\sqrt{R^2 + \omega_2 L^2}} \sin(\omega t - \theta) - \left(\frac{V_m \sin(\alpha - \theta)}{\sqrt{R^2 + \omega^2 L^2}}\right) e^{-(R/L)(t - \alpha/\omega)}$$

$$- \frac{E}{R}\left(1 - e^{-(R/L)(t - \alpha/\omega)}\right) \tag{9.31}$$

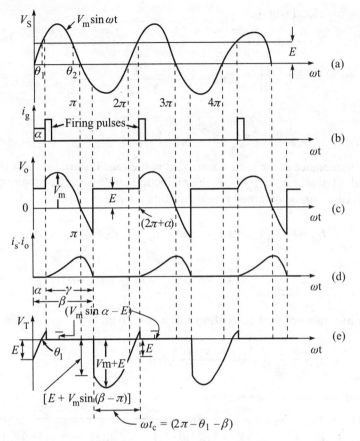

FIGURE 9.9 Waveforms of (a) input voltage, (b) firing pulse, (c) output voltage, (d) output and supply current, and (e) voltage drop across thyristor.

where $\theta = \tan^{-1}(\omega L/R)$. Equation (9.31) is applicable for the interval $\alpha < \omega t < \beta$. The extinction angle β depends on the load e.m.f. E, firing angle α, and the load impedance angle θ. The average voltage across the inductor is zero; hence, the average current is given by

$$I_{d.c.} = \frac{1}{2\pi R}\int_{\alpha}^{\beta}(V_m \sin\omega t - E)\,d\omega t = \frac{1}{2\pi R}\left|-V_m \cos\omega t - E\omega t\right|_{\alpha}^{\beta}$$

$$= \frac{1}{2\pi R}(V_m(\cos\alpha - \cos\beta) - E(\beta - \alpha)) \qquad (9.32)$$

Since the conduction angle $\gamma = \beta - \alpha$,

$$I_{d.c.} = \frac{1}{2\pi R}(V_m[\cos\alpha - \cos(\gamma + \alpha)] - E\gamma) \qquad (9.33)$$

$$= \frac{1}{\pi R}\left[V_m \sin\left(\alpha + \frac{\gamma}{2}\right)\sin\frac{\gamma}{2} - \frac{E\gamma}{2}\right] \qquad (9.34)$$

The average load voltage,

$$V_{d.c.} = E + I_{d.c.}R$$

$$= E + \frac{1}{\pi}\left[V_m \sin\left(\alpha + \frac{\gamma}{2}\right)\sin\frac{\gamma}{2} - \frac{E\gamma}{2}\right]$$

$$= E\left(1 - \frac{\gamma}{2\pi}\right) + \frac{V_m}{\pi}\sin\left(\alpha + \frac{\gamma}{2}\right)\sin\frac{\gamma}{2} \quad (9.35)$$

If the inductance in the load is zero, that is, the load is pure resistive, then the thyristor is turned off at $\omega t = \pi - \alpha_{min}$. Now, the extinction angle $\beta < \pi$. The average value of current, in this case, can be obtained by substituting $\beta = \pi - \alpha_{min}$ in Eq. (9.32). Hence,

$$I_{d.c.} = \frac{1}{2\pi R}(V_m(\cos\alpha - \cos(\pi - \alpha_{min})) - E(\pi - \alpha_{min} - \alpha))$$

$$= \frac{1}{2\pi R}(V_m(\cos\alpha + \cos\alpha_{min}) - E(\pi - (\alpha_{min} + \alpha))) \quad (9.36)$$

The r.m.s. value of load current, with $L = 0$, is given by

$$I_{r.m.s.}^2 = \frac{1}{2\pi}\int_\alpha^\beta \left(\frac{V_m \sin\omega t - E}{R}\right)^2 d\omega t$$

After integrating and simplifying,

$$I_{r.m.s.} = \frac{1}{2R\sqrt{\pi}}\sqrt{(V_m^2 + 2E^2)(\beta - \alpha) - \frac{V_m^2}{2}(\sin 2\beta - \sin 2\alpha) - 4V_m(\sin\beta - \sin\alpha)} \quad (9.37)$$

The power delivered to the load,

$$P = I_{r.m.s.}^2 R + EI_{d.c.} \quad (9.38)$$

and the supply power factor,

$$\text{pf} = \frac{P}{VA_{input}} = \frac{I_{r.m.s.}^2 R + EI_{d.c.}}{V_i I_{r.m.s.}} \quad (9.39)$$

The waveforms of input voltage, firing pulse, output voltage, load and supply current, and voltage drop across the thyristor are shown in Fig. 9.9.

9.2.2 Single-Phase Full-Wave Converters

The supply current in single-phase half-wave converters is always unidirectional. Thus, these converters introduce a d.c. component into the supply line. The d.c. component may lead to the saturation of the supply transformer. Harmonics are also introduced in the supply. These are undesirable. These shortcomings can be removed to some extent by the use of single-phase full-wave converters.

Resistive (R) Load A single-phase full-wave converter using central tapped transformer is shown in Fig. 9.10. This circuit is known as *M-2 connection* because it uses the midpoint of secondary of the supply transformer, and the number of load current pulses per cycle is two. The load connected to the converter is purely resistive. Thyristor Th_1 conducts during $\alpha < \omega t < \pi$ in the positive half cycle of the input signal. Thyristor Th_2 conducts in the interval from $\omega t = \pi + \alpha$ to $\omega t = 2\pi$. The output voltage is controlled by controlling the firing angle α. Since the load is a pure resistance, the output current wave shape is in phase with and similar to that of the output voltage. The output voltage and current, and the voltage drops across the thyristors are shown in Fig. 9.11. The output voltage and current are always positive.

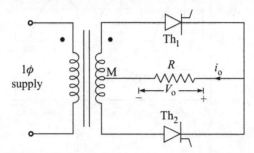

FIGURE 9.10 Full-wave converter with R load.

Let us assume that the turns ratio from primary to each secondary is unity. The average output voltage is given by

$$V_{d.c.} = \frac{1}{\pi} \int_\alpha^\pi V_m \sin \omega t \, d\omega t$$

$$= \frac{1}{\pi} \left| -V_m \cos \omega t \right|_\alpha^\pi$$

$$= \frac{V_m}{\pi}(1+\cos\alpha) \tag{9.40}$$

The average load current,

$$I_{d.c.} = \frac{V_{d.c.}}{R} = \frac{V_m}{\pi R}(1+\cos\alpha) \tag{9.41}$$

The r.m.s. output voltage,

$$V_{r.m.s.}^2 = \frac{1}{\pi} \int_\alpha^\pi V_m^2 \sin^2 \omega t \, d\omega t = \frac{V_m^2}{2\pi} \int_\alpha^\pi (1-\cos 2\omega t) \, d\omega t$$

$$\therefore \quad V_{r.m.s.} = V_m \sqrt{\left(\frac{\pi-\alpha}{2\pi} + \frac{\sin 2\alpha}{4\pi}\right)} \tag{9.42}$$

The r.m.s. load current,

$$I_{r.m.s.} = \frac{V_{r.m.s.}}{R} = \frac{V_m}{R} \sqrt{\left(\frac{\pi-\alpha}{2\pi} + \frac{\sin 2\alpha}{4\pi}\right)} \tag{9.43}$$

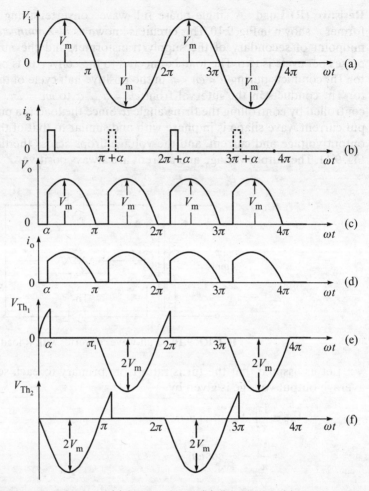

FIGURE 9.11 Waveforms of (a) input voltage, (b) firing pulse, (c) output voltage, (d) output current, and (e) & (f) voltage drops across thyristor.

Inductive (RL) Load The circuit of a single-phase full-wave converter using central tapped transformer and supplying an RL load is shown in Fig. 9.12. At $\omega t = \alpha$, the thyristor Th_1 starts conducting, and the output voltage immediately follows the input voltage. The output current lags the output voltage and rises slowly to a maximum value depending on the time constant L/R. At $\omega t = \pi$, Th_1 is reversed biased, but as the current is not

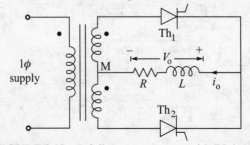

FIGURE 9.12 A full-wave converter with RL load.

below the holding current it does not turned off. The current continues to flow beyond $\omega t = \pi$. At $\omega t = \pi + \alpha$, the thyristor Th_2 starts conducting through the load and the current is transferred from Th_1 to Th_2. The thyristor Th_1 is turned off. The load current starts rising again. Th_2 conducts until $\omega t = 2\pi + \alpha$. The cycle repeats at $\omega t = 2\pi + \alpha$. The load current never reaches zero. If the load is highly inductive, the load current is almost constant. The waveforms of input voltage, firing pulses, load voltage and current, thyristor currents, and voltage drops across them are shown in Fig. 9.13.

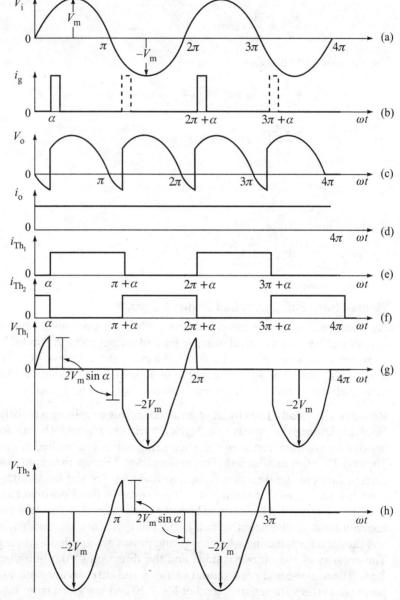

FIGURE 9.13 Waveforms of (a) input voltage, (b) firing pulses, (c) load voltage, (d) load current, (e) & (f) thyristor currents, and (g) & (h) voltage drops across thyristors.

The average value of output voltage,

$$V_{d.c.} = \frac{1}{\pi}\int_{\alpha}^{\pi+\alpha} V_m \sin\omega t\, d\omega t = \frac{1}{\pi}\left| -V_m \cos\omega t\right|_{\alpha}^{\pi+\alpha}$$

$$= \frac{2V_m}{\pi}\cos\alpha \tag{9.44}$$

The average load current,

$$I_{d.c.} = \frac{V_{d.c.}}{R} = \frac{2V_m}{\pi R}\cos\alpha \tag{9.45}$$

The r.m.s. output voltage,

$$V_{r.m.s.}^2 = \frac{1}{\pi}\int_{\alpha}^{\pi+\alpha} V_m^2 \sin^2\omega t\, d\omega t$$

$$= \frac{V_m^2}{2\pi}\int_{\alpha}^{\pi+\alpha}(1-\cos 2\omega t)\, d\omega t = \frac{V_m^2}{2}$$

$$\therefore \quad V_{r.m.s.} = \frac{V_m}{\sqrt{2}} \tag{9.46}$$

The r.m.s. load current,

$$I_{r.m.s.} = \frac{V_{r.m.s.}}{R} = \frac{V_m}{\sqrt{2}R} \tag{9.47}$$

Single-Phase Full-Controlled Bridge Converter

As mentioned earlier, a full-controlled bridge converter uses only thyristors as active devices and has wider control over the level of output voltage. This bridge converter is also known as full converter. The circuit of a full converter is known as *B-2 connection* because it uses thyristors in the form of a bridge, and the number of load current pulses per cycle is two. Single-phase full converters with resistive load and RLE load are discussed here.

Resistive (R) Load The circuit of a full converter supplying a resistive load is shown in Fig. 9.14. During the positive half cycle, thyristors Th_1 and Th_2 are forward biased, and hence they are triggered at $\omega t = \alpha$ when firing pulses are applied at their gates. Thyristors Th_1 and Th_2 start conducting. The load voltage follows the input voltage and the load current flows in the direction from a to b. At $\omega t = \pi$, the load voltage, and current are zero. Because at $\omega t = \pi$ the input voltage reverses from positive to negative value, the thyristors Th_1 and Th_2 are reverse biased and so these thyristors are turned off by natural commutation. In the negative half cycle, the thyristors Th_3 and Th_4 are forward biased and they start conducting when triggering pulses are applied to their gates at $\omega t = \pi + \alpha$. The polarity of the output voltage and the direction of the output current remain the same. The average and r.m.s. output voltages and currents are same as in the case of mid-point converter with resistive load of Fig. 9.10 and are given from Eq. (9.40) to (9.43).

The waveforms of different voltages and currents are shown in Fig. 9.15.

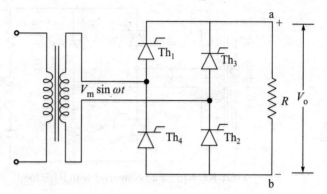

FIGURE 9.14 Full-bridge converter with R load.

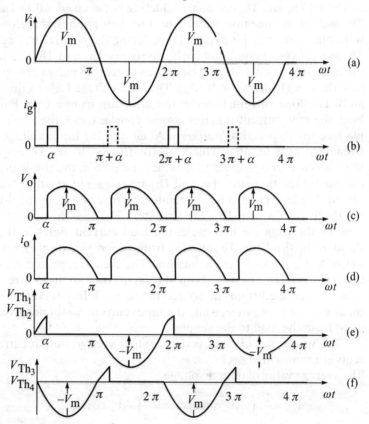

FIGURE 9.15 Waveforms of (a) input voltage, (b) firing pulses, (c) load voltage, (d) load current, and (e) & (f) voltage drops across thyristors.

RLE Load The circuit of a full-controlled bridge converter supplying an RLE load is shown in Fig. 9.16. The load current may be assumed to be continuously flowing over the working range. This means that the load is always connected to the a.c. voltage source through the thyristors. Since the thyristors Th_3 and Th_4 are conducting when

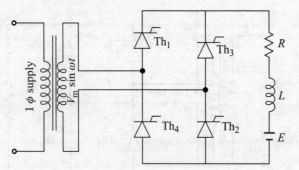

FIGURE 9.16 Full converter with RLE load.

thyristors Th_1 and Th_2 are nonconducting between $\omega t = 0$ and $\omega t = \alpha$, the cathodes of Th_1 and Th_2 are negative with respect to their anodes; therefore, Th_1 and Th_2 are forward biased during this period. Thus, during the positive half cycle, thyristors Th_1 and Th_2 are forward biased. Similarly, thyristors Th_3 and Th_4 are forward biased during the complete negative half cycle. When triggering pulses are applied at $\omega t = \alpha$, thyristors Th_1 and Th_2 start conducting. The load voltage follows the input voltage immediately. The load current flows in the direction from a to b. For the highly inductive load, the load current remains almost constant and, therefore, may be assumed ripple-free for all practical purposes. At $\omega t = \pi$, the input voltage reverses its direction from positive to negative. Hence, the thyristors Th_1 and Th_2 are reverse biased. But as the load current continues to flow beyond $\omega t = \pi$, the thyristors Th_1 and Th_2 are conducting till the thyristors Th_3 and Th_4 are triggered at $\omega t = \pi + \alpha$, as they are forward biased. The load current is immediately transferred from the thyristors Th_1 and Th_2 to the thyristors Th_3 and Th_4. The thyristors Th_1 and Th_2 turn off by natural commutation. In the negative half cycle, the load current flows in the same direction, but through the thyristors Th_3 and Th_4 from $\omega t = \pi + \alpha$ to $\omega t = 2\pi + \alpha$. The polarity of the output voltage also remains the same. The cycle is repeated at $\omega t = 2\pi + \alpha$. During the interval $\omega t = \alpha$ to $\omega t = \pi$, both the input voltage and current are positive; hence, power is supplied from the source to the load. In interval $\omega t = \pi$ to $\omega t = \pi + \alpha$, the input voltage is negative, while the input current is still positive. Hence, power is supplied from the load to the supply.

The waveforms of input voltage, output voltage, output current, and thyristors currents are shown in Fig. 9.17.
The average value of output voltage,

$$V_{d.c.} = \frac{1}{\pi} \int_{\alpha}^{\pi+\alpha} V_m \sin\omega t \, d\omega t = \frac{1}{\pi} \left| -V_m \cos\omega t \right|_{\alpha}^{\pi+\alpha}$$

$$= \frac{2V_m}{\pi} \cos\alpha \qquad (9.48)$$

The average load current,

$$I_{d.c.} = \frac{V_{d.c.} - E}{R} = \frac{2V_m \cos\alpha - \pi E}{\pi R} \qquad (9.49)$$

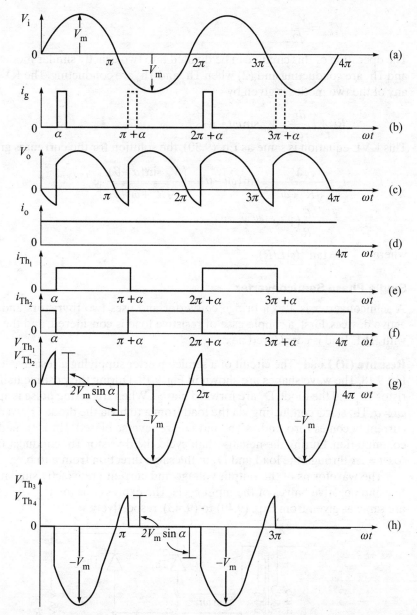

FIGURE 9.17 Waveforms of (a) input voltage, (b) firing pulses, (c) load voltage, (d) load current, (e) & (f) thyristor currents, and (g) & (h) voltage drops across thyristors.

The r.m.s. output voltage,

$$V_{r.m.s.}^2 = \frac{1}{\pi}\int_{\alpha}^{\pi+\alpha} V_m^2 \sin^2 \omega t \, d\omega t$$

$$= \frac{V_m^2}{2\pi}\int_{\alpha}^{\pi+\alpha}(1-\cos 2\omega t)\,d\omega t = \frac{V_m^2}{2}$$

$$\therefore \qquad V_{\text{r.m.s.}} = \frac{V_m}{\sqrt{2}} \qquad (9.50)$$

The operation of this circuit can be divided into two exactly similar modes: (1) when Th_1 and Th_2 are conducting and (2) when Th_3 and Th_4 are conducting. The KVL equation for any of the two modes is given by

$$Ri_o + L\frac{di_o}{dt} = V_m \sin\omega t - E \qquad (9.51)$$

This KVL equation is same as Eq. (9.30), the solution for the current is given by

$$i_o = \frac{V_m}{\sqrt{R^2 + \omega_2 L^2}} \sin(\omega t - \theta) - \left(\frac{V_m \sin(\alpha - \theta)}{\sqrt{R^2 + \omega^2 L^2}}\right) e^{-(R/L)(t-\alpha/\omega)}$$

$$- \frac{E}{R}\left(1 - e^{-(R/L)(t-\alpha/\omega)}\right) \qquad (9.52)$$

where $\theta = \tan^{-1}(\omega L/R)$.

Single-Phase Semiconverter

A semiconverter is also a bridge converter and uses two thyristors and two diodes as active devices. First, a simple case of resistive load is considered, and the semiconverter with RLE load is considered next.

Resistive (R) Load The circuit of a semiconverter supplying a resistive load is shown in Fig. 9.18. The wave shapes are shown in Fig. 9.19. During the positive half cycle, the thyristor Th_1 and the diode D_1 are forward biased. When triggering pulse is applied to Th_1 at $\omega t = \alpha$, Th_1 starts conducting via the load from a to b via the diode D_1. At $\omega t = \pi$, the load current becomes zero, and as Th_1 and D_1 are reverse biased, Th_1 is turned off by natural commutation. During the negative half cycle, the thyristor Th_2 conducts, from $\omega t = \pi + \alpha$ to $\omega t = 2\pi$ through the load and D_2 in the same direction from a to b.

The waveforms of the output voltage and current are exactly similar for both positive and negative halves of the input cycle. The expressions for $V_{\text{d.c.}}$, $I_{\text{d.c.}}$, $V_{\text{r.m.s.}}$, and $I_{\text{r.m.s.}}$ are same as given from Eqs. (9.40) to (9.43), respectively.

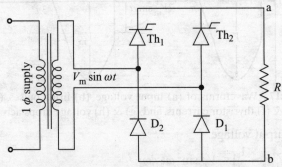

FIGURE 9.18 Semiconverter with R load.

RLE Load The circuit of the semiconverter with RLE load is shown in Fig. 9.20. This circuit differs from the circuit of Fig. 9.18 in the sense that a commutating diode CD is used

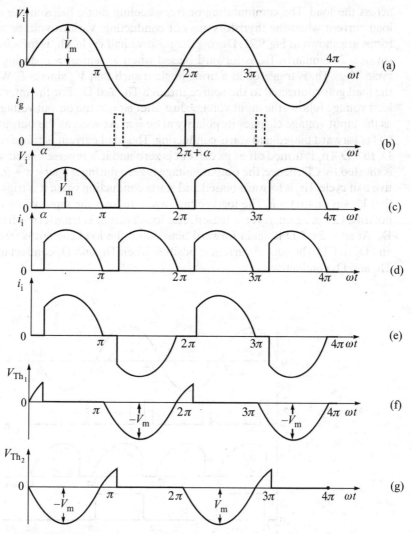

FIGURE 9.19 Waveforms of (a) input voltage, (b) triggering pulses, (c) output voltage, (d) output current, (e) input current, and (f) & (g) thyristor voltages.

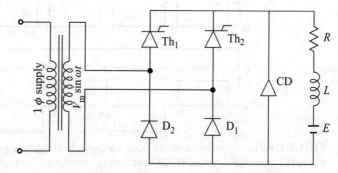

FIGURE 9.20 Semiconverter with RLE load.

across the load. The commutating or freewheeling diode helps in the conduction of the load current when the thyristors are not conducting. Various voltage and current waveforms are shown in Fig. 9.21. During the positive half cycle, Th_1 is forward biased when $V_m \sin \omega t > E$. Similarly, Th_2 is forward biased when $V_m \sin \omega t > E$ during the negative half cycle. Thus, Th_1 is triggered at a firing angle α such that $V_m \sin \alpha > E$. When Th_1 turns on, the load gets connected to the source through Th_1 and D_1. For the interval $\alpha < \omega t < \pi$, the load voltage follows the input voltage. Just after $\omega t = \pi$, the output voltage tends to reverse as the input voltage changes its polarity at $\omega t = \pi$. As soon as the output voltage reverses, CD is forward biased and starts conducting. The load current is transferred from Th_1 and D_1 to CD. Th_1 is turned off as its current is zero and it is reverse biased at $\omega t = \pi$. The load is shorted by CD; hence, the output voltage is zero during $\pi < \omega t < \pi + \alpha$. During the negative half cycle, Th_2 is forward biased and starts conducting when it is triggered at $\pi + \alpha$ such that $V_m \sin(\pi + \alpha) > E$. The load voltage again follows the input voltage, but with reverse polarity. CD is, again, reverse biased. The load current is transferred from CD to Th_2 and D_2. At $\omega t = 2\pi$, CD is again forward biased and the load current is transferred from Th_2 and D_2 to CD. The source current is positive when Th_1 and D_1 conduct and negative when Th_2 and D_2 conduct.

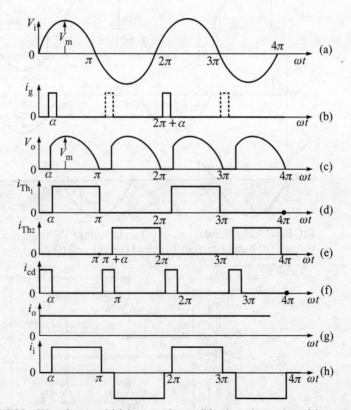

FIGURE 9.21 Waveforms of (a) input voltage, (b) triggering pulses, (c) output voltage, (d) & (e) thyristor currents, (f) CD current, (g) output current, and (h) input current.

During the conduction through the thyristor and the diode, the source supplies the energy to load. The energy is partially dissipated as heat in the resistance, partially stored in the inductor, and partially stored as electrical energy in the load e.m.f. (such as a battery). During the freewheeling period the energy stored in the inductor is partially dissipated as heat in the resistor and partially stored in the load e.m.f. No energy is fed back to the source during freewheeling period.

The average and r.m.s. values of load voltage are same as given in Eqs. (9.40) and (9.42), respectively. The operation of semiconverter can be divided into two modes. The first mode is when CD conducts ($0 < \omega t < \alpha$) and the second mode is when thyristor and diode conduct ($\alpha < \omega t < \pi$). During ($0 < \omega t < \alpha$) the KVL equation is given by

$$Ri_{o1} + L\frac{di_{o1}}{dt} = -E \tag{9.53}$$

With the initial condition, at $\omega t = 0$, $i_{o1} = I_o$, the solution of Eq. (9.53) is

$$i_{o1} = I_o e^{-(R/L)t} - \frac{E}{R}\left(1 - e^{-(R/L)t}\right) \tag{9.54}$$

At $\omega t = \alpha$, let $i_{o1} = I'_o$, then

$$I'_o = I_o e^{-(R/L)(\alpha/\omega)} - \frac{E}{R}\left(1 - e^{-(R/L)(\alpha/\omega)}\right) \tag{9.55}$$

During $\alpha < \omega t < \pi$, the KVL equation is given by

$$Ri_{o2} + L\frac{di_{o2}}{dt} = V_m \sin\omega t - E \tag{9.56}$$

As done earlier, the solution of above differential equation, with initial condition $i_{o2} = I'_o$, at $\omega t = \alpha$ [Eq. (9.55)], is given by

$$i_{o2} = \frac{V_m}{\sqrt{R^2 + \omega_2 L^2}} \sin(\omega t - \theta) + \left(I'_o - \frac{V_m \sin(\alpha - \theta)}{\sqrt{R^2 + \omega^2 L^2}}\right)e^{-(R/L)(t-\alpha/\omega)}$$

$$- \frac{E}{R}\left(1 - e^{-(R/L)(t-\alpha/\omega)}\right), \qquad \text{for } i_{o2} \geq 0 \tag{9.57}$$

where $\theta = \tan^{-1}(\omega L/R)$.

9.2.3 Three-Phase Full-Controlled Bridge Converters

The circuit of a three-phase full-controlled bridge converter is shown in Fig. 9.22. This converter uses six thyristors. The load connected is an RLE load. As per Fig. 9.22, the odd numbered thyristors Th_1, Th_3, and Th_5 form the positive group because they are turned on when supply voltages are positive. The even numbered thyristors Th_2, Th_4, and Th_6 form the negative group because they are turned on when supply voltages are negative. The firing frequency is six times the supply frequency. This means that in one cycle of supply, six thyristors are fired one by one in sequence at an interval of $\pi/3$. There are six pulses of output in one cycle of input wave. Therefore, this converter is also known as six pulse converter. The three-phase input voltage waves are shown in Fig. 9.23. The firing angle α of Th_1 is measured from $\pi/6$ with reference to the positive half cycle of voltage of phase A. Each thyristor conducts for $2\pi/3$. The firing sequence of thyristors is Th_1, Th_2, Th_3, Th_4, Th_5, and Th_6. Two thyristors conduct at one time.

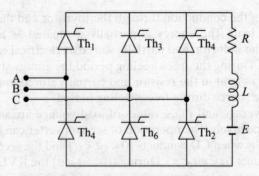

FIGURE 9.22 Three-phase full converter.

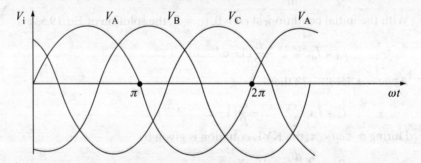

FIGURE 9.23 Three-phase voltage waves.

Commutation occurs every $\pi/3$ radian, that is, six times in one cycle. Commutation is natural. For phase sequence ABC and thyristors numbered as in Fig. 9.22, for $\alpha = 0$, the pairs of thyristors that conduct at a time (as shown in Fig. 9.24) are in the sequence $Th_1 Th_2$, $Th_2 Th_3$, $Th_3 Th_4$, $Th_4 Th_5$, $Th_5 Th_6$, and $Th_6 Th_1$.

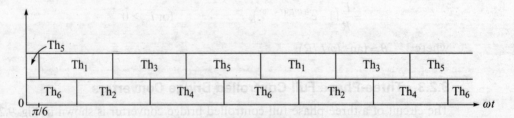

FIGURE 9.24 Conduction period for $\alpha = 0$.

If the firing angle $\alpha \neq 0$, the triggerings of thyristors are delayed by an angle α from the instants shown in Figs 9.23 and 9.24. The conduction period of each thyristor remains the same, that is, $2\pi/3$ radian. The output voltage is shown in Fig. 9.25.

The three-phase voltages can be expressed as follows:

$$v_a = V_m \sin \omega t$$
$$v_b = V_m \sin (\omega t - 2\pi/3)$$
and
$$v_c = V_m \sin (\omega t - 4\pi/3)$$

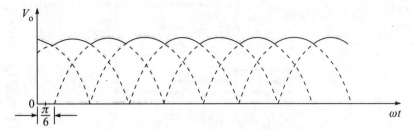

FIGURE 9.25 Output voltage variation.

The line voltages, hence, can be written as

$$v_{ab} = v_a - v_b = \sqrt{3}V_m \sin(\omega t + \pi/6)$$

$$v_{bc} = v_b - v_c = \sqrt{3}V_m \sin(\omega t - \pi/2)$$

$$v_{ca} = v_c - v_a = \sqrt{3}V_m \sin(\omega t + 5\pi/6)$$

The average output voltage,

$$V_{d.c.} = \frac{3}{\pi} \int_j^k \sqrt{3}V_m \sin(\omega t + \pi/6)\, d\omega t = \frac{3\sqrt{3}}{\pi} V_m \cos\alpha \qquad (9.58)$$

The r.m.s. output voltage,

$$V_{r.m.s.} = \left[\frac{3}{\pi} \int_j^k 3V_m^2 \sin^2(\omega t + \pi/6)\, d\omega t\right]^{\frac{1}{2}}$$

$$= \sqrt{3}V_m \sqrt{\frac{1}{2} + \frac{3\sqrt{3}}{4\pi}\cos 2\alpha} \qquad (9.59)$$

where $k = (\pi/2) + \alpha$ and $j = (\pi/6) + \alpha$.

When $\alpha \leq \pi/3$, the instantaneous output voltage has positive part only. For $\alpha > \pi/3$, the instantaneous output voltage has negative part also. The thyristor currents, however, are always positive. Therefore, the output current is positive only.

Three-Phase Semiconverter

The circuit of a three-phase semiconverter is shown in Fig. 9.26. It differs from the three-phase full converter in the sense that the negative-group thyristor in the full converter is replaced by three diodes. A freewheeling diode is also used. The three thyristors that belong to the positive group are numbered as Th_1, Th_2, and Th_3. The diodes that are in the negative group are numbered as D_1, D_2, and D_3. The thyristors are fired at intervals of $2\pi/3$ radian. Hence, the firing frequency is three times the supply frequency. At one time one thyristor and one diode conduct. The wave shape of the output voltage depends on the firing angle. The input voltage, conduction period, and output voltage for $\alpha = 0$ and $\alpha = \pi/3$ are shown in Fig. 9.27 and Fig. 9.28, respectively.

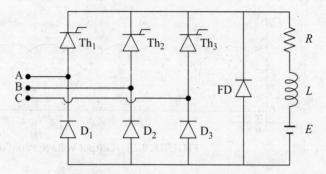

FIGURE 9.26 Three-phase semiconverter.

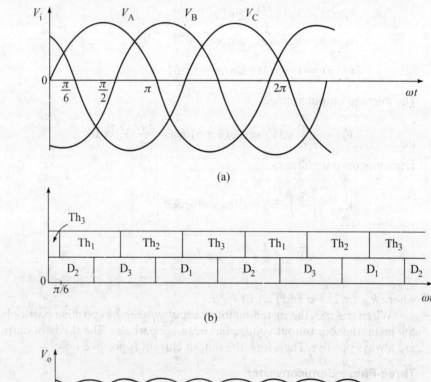

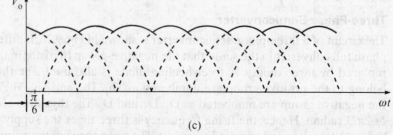

FIGURE 9.27 Waveforms of (a) phase input voltages, (b) conduction period and (c) output voltage (for $\alpha = 0$).

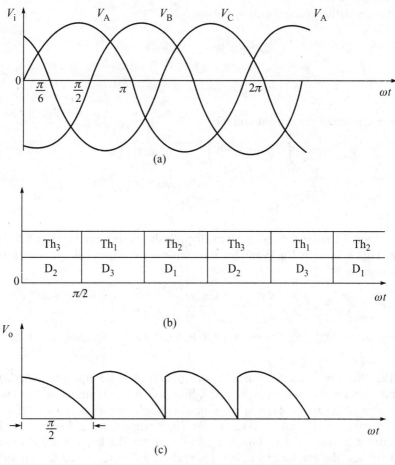

FIGURE 9.28 Waveforms of (a) phase input voltages, (b) conduction period, and (c) output voltage (for $\alpha = \pi/3$).

The three-phase voltages can be expressed as follows:

$$v_a = V_m \sin \omega t$$
$$v_b = V_m \sin(\omega t - 2\pi/3)$$

and
$$v_c = V_m \sin(\omega t - 4\pi/3)$$

The line voltages, hence, can be written as

$$v_{ab} = v_a - v_b = \sqrt{3}V_m \sin(\omega t + \pi/6)$$
$$v_{bc} = v_b - v_c = \sqrt{3}V_m \sin(\omega t - \pi/2)$$
$$v_{ca} = v_c - v_a = \sqrt{3}V_m \sin(\omega t + 5\pi/6)$$

When $\alpha \leq \pi/3$, the average output voltage

$$V_{d.c.} = \frac{3}{2\pi}\left(\int_{(\pi/6)+\alpha}^{\pi/2} v_{ab}\,d\omega t + \int_{\pi/2}^{(5\pi/6)+\alpha} v_{a.c.}\,d\omega t\right)$$

$$= \frac{3\sqrt{3}V_m}{2\pi}(1+\cos\alpha) \tag{9.60}$$

The r.m.s. value of the output voltage,

$$V_{r.m.s.} = \left[\frac{3}{2\pi}\left(\int_{(\pi/6)+\alpha}^{\pi/2} v_{ab}^2\,d\omega t + \int_{\pi/2}^{(5\pi/6)+\alpha} v_{a.c.}^2\,d\omega t\right)\right]^{0.5}$$

$$= \sqrt{3}V_m\left[\frac{3}{4\pi}\left(\frac{2\pi}{3}+\sqrt{3}\cos^2\alpha\right)\right]^{0.5} \tag{9.61}$$

When $\alpha > \pi/3$, the output voltage waveform is discontinuous. However, the average output voltage is same as Eq. (9.60) for the continuous output voltage waveform. The r.m.s. value is given by

$$V_{r.m.s.} = \sqrt{3}V_m\left[\frac{3}{4\pi}\left(\pi - \alpha + \frac{\sin 2\alpha}{2}\right)\right]^{0.5} \tag{9.62}$$

The operation of the semiconverter can be explained as follows:

(i) For $\alpha = 0$:

The thyristors Th_1, Th_2, and Th_3 are triggered at ωt equal to $\pi/6$, $5\pi/6$, and $3\pi/2$, respectively. For $0 < \omega t < \pi/6$, v_c is the highest positive instantaneous input voltage and v_b is the highest negative instantaneous input voltage. Hence, as per Fig. 9.26, the conducting elements are Th_3 and D_2. At $\omega t = \pi/6$, Th_1 is triggered, and because the highest positive instantaneous input voltage is v_a, and v_b is still the highest negative instantaneous input voltage, the conducting pair is Th_1 and D_2. Just after $\omega t = \pi/2$, v_a is still the highest positive instantaneous input voltage, but v_c tends to be the highest negative instantaneous input voltage. Hence, D_3 becomes most forward biased and conducting pair is Th_1 and D_3. At $\omega t = 5\pi/6$, v_b tends to be the highest positive instantaneous input voltage and v_c still continues to be the highest negative instantaneous input voltage. Hence, when Th_2 is triggered at $\omega t = 5\pi/6$, the conducting pair is Th_2 and D_3. At $\omega t = 7\pi/6$, v_b is still the highest positive instantaneous input voltage, but v_a tends to be the highest negative instantaneous input voltage. Hence, the new conducting pair is Th_2 and D_1. At $\omega t = 3\pi/2$, v_c tends to be the highest positive instantaneous input voltage, and v_a still continues to be the highest negative instantaneous input voltage. Hence, when Th_3 is triggered at $\omega t = 3\pi/2$, the new conducting pair is Th_3 and D_1. At $\omega t = 11\pi/6$, v_c is still the highest positive instantaneous input voltage, but v_b tends to be the highest negative instantaneous input voltage. Hence, the new conducting pair is Th_3 and D_2. The period of conduction and the output voltage with phase input voltages are shown in Fig. 9.27. For $\alpha = 0$, the freewheeling diode does not play any role.

(ii) For $\alpha = \pi/3$:

The thyristors Th_1, Th_2, and Th_3 are triggered at ωt equal to $\pi/2$, $7\pi/6$, and $11\pi/6$, respectively. For $0 < \omega t < \pi/6$, v_c is the highest positive instantaneous input voltage and v_b

is the highest negative instantaneous input voltage. As Th_3 is in the conducting mode, the conducting elements are Th_3 and D_2. The output voltage is v_{cb}. For $\pi/6 < \omega t < \pi/2$, v_a is the highest positive instantaneous input voltage and v_b is still the highest negative instantaneous input voltage. As Th_1 is not triggered as yet, the conducting elements are still Th_3 and D_2. The output voltage is still v_{cb}. At $\omega t = \pi/2$, Th_1 is triggered, and as the highest positive instantaneous input voltage is v_a and v_c tends to be the highest negative instantaneous input voltage, the conducting pair is Th_1 and D_3. The same pair of Th_1 and D_3 continues to conduct until $\omega t = 7\pi/6$. The output voltage from $\omega t = \pi/2$ to $\omega t = 7\pi/6$ is $v_{a.c.}$. At $\omega t = 7\pi/6$, Th_2 is triggered and as the highest positive instantaneous input voltage is v_b and v_a tends to be highest negative instantaneous input voltage, the conducting pair is Th_2 and D_1. The same pair of Th_2 and D_1 continues to conduct until $\omega t = 11\pi/6$. The output voltage from $\omega t = 7\pi/6$ to $\omega t = 11\pi/6$ is v_{ba}. Freewheeling diode does not come into action again. The conduction period and the output voltage with phase input voltages are shown in Fig. 9.28.

From Fig. 9.27, it is clear that voltage pulses v_{ab}, v_{bc}, and v_{ca} in the output voltage, for $\alpha = 0$, are eliminated and do not appear in the output voltage of Fig. 9.28 for $\alpha = \pi/3$. The ripple in this output voltage, therefore, is increased. It can be observed that, for $\alpha > \pi/3$, freewheeling diode comes into action when the output voltage is negative. The output voltage wave is more and more discontinuous and distorted as firing angle is increased.

Three-phase converters have some merits over single-phase converters. A comparison between them are given in Table-9.1.

Table-9.1 Comparison of three-phase and single-phase converters.

Three-phase converters	Single-phase converters
Line current has less distortion; hence, input power factor is higher.	Line current has significant distortion; hence, line power factor is poorer.
Ripple in d.c. output current is less; hence, needs smaller filter.	Ripple in d.c. output current is high; hence, needs larger filter.
The voltage variation from no load to full load is small, and voltage regulation is less.	Voltage drop in the circuit is high resulting in poor voltage regulation.
Firing circuit is more complicated.	Firing circuit is less complicated.
Same size converter can handle high power.	Same size converter can handle small power.

9.3 INVERTERS

An inverter is a device used for converting d.c. supply into a.c. supply. The output voltage and its frequency can be varied as per the requirement of the load. Some applications of inverters are as variable frequency supply for speed control of a.c. drives, in induction heating, as standby power supply for aircraft, as UPS for computer, as receiving end of HVDC transmission lines for converting d.c supply into a.c. supply, etc. The d.c. input supply to the inverter may come from an existing power supply network, an alternator through a rectifier, a battery, solar cells, fuel cells, or magnetohydrodynamic generator.

Inverters can be broadly classified on the basis of the type of input source. Thus, there are two types of inverters, namely, voltage-source inverters and current-source inverters. The voltage-source inverters convert energy from a d.c. voltage source, such as a battery or a fixed d.c. voltage. The current-source inverters use a d.c. current source as the input source of energy. If the inverter output is single phase, it is called a single-phase inverter, and when the output is three phase, the inverter is called a three-phase inverter. On the basis of the connections of the semiconductor devices and commutating elements, inverters are classified as series inverters, parallel inverters, and bridge inverters.

9.3.1 Single-Phase Series Inverters

In a series inverter, commutating elements L and C are connected in series with the load resistance R. If the load has some amount of inductance or capacitance, these can be combined with the commutating inductance L and capacitance C. A series inverter with thyristors and commutating L and C is shown in Fig. 9.29. The commutating L and C are chosen such that the RLC series circuit forms an underdamped circuit. The condition for the circuit to be underdamped is

$$R^2 < \frac{4L}{C} \tag{9.63}$$

Two thyristor are turned on appropriately so that the output voltage of desired frequency can be obtained.

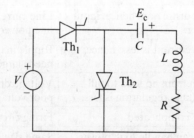

FIGURE 9.29 A basic series inverter.

The operation of the series inverter of Fig. 9.29 is described as follows:

Mode-1: *Thyristor Th_1 on and Th_2 off.*

The equivalent circuit is shown in Fig. 9.30(a). Since the thyristor Th_1 is forward biased by the d.c. input voltage, it starts conducting when an external pulse is applied at $t = 0$. A current flows in the load resistance R through Th_1, L, and C. For the series RLC circuit of Fig. 9.30(a), the current i is given by

$$Ri + L\frac{di}{dt} + \frac{1}{C}\int i\,dt = V \qquad \text{[with } v_C(0) = -E_C\text{]} \tag{9.64}$$

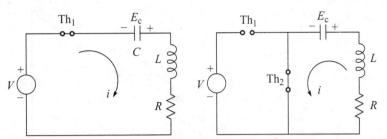

FIGURE 9.30 An equivalent circuit of Fig. 9.29 with (a) Th_1 on, Th_2 off and (b) Th_1 off, Th_2 on.

and the solution of Eq. (9.65) is

$$i(t) = \frac{V + E_C}{\omega_r L} e^{-\xi t} \sin \omega_r t \qquad (9.65)$$

where damping factor (ξ) and damped frequency (ω_r) of the circuit are given by

$$\xi = \frac{R}{2L} \qquad (9.66)$$

$$\omega_r = \sqrt{\frac{1}{LC} - \left(\frac{R}{2L}\right)^2} \qquad (9.67)$$

The resonance frequency

$$\omega_o = \sqrt{\omega_r^2 + \xi^2} = \frac{1}{\sqrt{LC}} \qquad (9.68)$$

The load current is not constant but varies as given by Eq. (9.65) and as shown in Fig. 9.32(b). The load current rises to a maximum value and decays to zero at point 'a' in Fig. 9.32(b). At point 'a', as the load current tends to reverse, the thyristor Th_1 is turned off. The voltage across the capacitor is given by

$$v_C = V - (V + E_C) \frac{\omega_o}{\omega_r} e^{-\xi t} \cos(\omega_r t - \theta) \qquad (9.69)$$

where, from Fig. 9.31, $\theta = \cos^{-1}(\omega_r/\omega_o)$. The current $i(t)$ is zero, when $\omega_r t = \pi$, that is, $t = (\pi/\omega_r)$. Hence, the capacitor voltage at point 'a', that is, $t = (\pi/\omega_r)$, from Eq. (9.70), is given by

$$V_{Ca} = V + (V + E_C) e^{-\pi \xi/\omega_r} \qquad (9.70)$$

Mode-2: *Thyristor Th_1 and Th_2 both are off.*

The time interval between the instant Th_1 is turned off and the instant Th_2 is turned on, 'ab' in Fig. 9.32, must be greater than the time T_{off} of the thyristor. This is necessary to ensure that the stored charge in Th_1 is reduced to zero. Th_1 is in completely off state at the point 'b'. The voltage across the inductor during this mode is zero. The left side of the capacitor is positive.

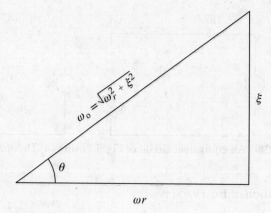

FIGURE 9.31 Relationship between ω_o, ω_r, and ξ.

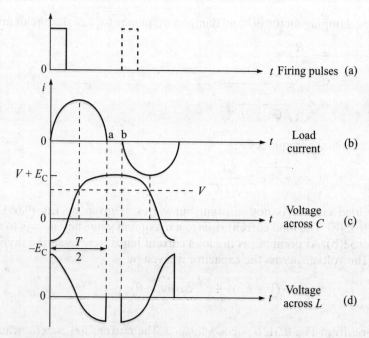

FIGURE 9.32 (a) Triggering pulses, and (b) current, and (c) voltage waveforms for series inverter.

Mode-3: *Thyristor Th_1 off and Th_2 on.*

The equivalent circuit is shown in Fig. 9.30(b). The time origin is taken when Th_2 is turned on. The direction of current is reversed. The discharging current, that is, the load current during this mode is given by

$$Ri + L\frac{di}{dt} + \frac{1}{C}\int i\,dt = V_{Ca} \tag{9.71}$$

and the solution of Eq. (9.71) is identical to that given in Eq. (9.65) by substituting V_{Ca} equal to $V + E_C$. Hence,

$$i(t) = \frac{V_{Ca}}{\omega_r L} e^{-\xi t} \sin \omega_r t \tag{9.72}$$

For the peak value of positive current, given by Eq. (9.65), to be equal to the negative peak value of current, V_{Ca} must equal to $V + E_C$. Hence, the C and L should be taken accordingly. The capacitor voltage, however, should not be more than the blocking voltage of the thyristor. The initial capacitor voltage E_C should be sufficient to turn off the thyristor in a time less than T_{off}.

The waveforms of load current, voltage across capacitor, and voltage across inductor are shown in Fig. 9.32. The source current flows only during the positive half cycle of the load current.

The basic series inverter discussed is very simple. It, however, suffers from the following drawbacks:

1. For a load power, the load current is taken from the supply only during the positive half cycle. Hence, the source current rating of the d.c. source is high. The source current has high harmonic content.
2. The inverter maximum operating frequency is limited because the operating frequency has to be less than the damped frequency of the circuit.
3. When the operating frequency is much lower than the damped frequency, the load voltage waveform is considerably distorted because T_{off} is large compared to the thyristor on time.
4. The amplitude and duration of the load current in each half cycle depends on the load circuit parameters. Hence, the output regulation of the inverter is poor.
5. Since the load current flows continuously through the commutating components, the ratings of the commutating components is high.

The drawbacks 3, 4, 5 are due to the nature of the circuit. However, drawbacks 1 and 2 can be removed by modifying the circuit, as discussed below.

Modified Series Inverters

A modified series inverter is shown in Fig. 9.33. It has two mutually coupled inductors L_1 and L_2, exactly identical and having equal inductance L. When the thyristor Th_1 is turned on at $t = 0$, the current i_1, in the loop consisting of V, Th_1, L_1, C, and R, begins to increase in clockwise direction. The rate of change of current is positive, and hence a voltage is induced in L_1 with the upper terminal positive. Since both inductors are mutually coupled, same voltage, as in L_1, is induced in L_2 with the upper terminal positive. This induced voltage adds to the initial capacitor voltage E_C in reverse biasing Th_2. After reaching the peak value, the current i_1 decreases. Hence, the rate of change of current is negative and the polarity of the induced voltages in L_1 is reversed. When the current i_1 is nearing zero, the voltage across C is somewhat less than $V + E_C$, with the right plate positive. Th_2 is forward biased and it can be turned on. Th_1 is reverse biased now and is turned off. The current i_2, in the loop consisting Th_2, L_2, R, and C, begins to increase in clockwise direction. Near the end of the negative half cycle, a similar action makes Th_2 reverse biased and turns it off. The cycle repeats. Since the excitation voltage and loop parameters in both halves of the cycle are same, $i_1 = i_2$. Thus, the time T_{off} can be avoided giving a higher output frequency. The modified series inverter removes drawback 2.

The danger of short circuit of d.c. source in series inverter of Fig. 9.29 is avoided in the modified series inverter because of the presence of inductors L_1 and L_2. However, the load draws power from the d.c. source in the positive half cycle only in the modified series inverter of Fig. 9.33. This drawback can be removed by further modifying the circuit of Fig. 9.33 as shown in Fig. 9.34.

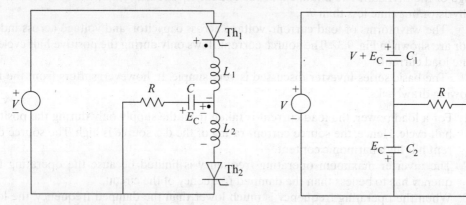

FIGURE 9.33 Modified series inverter.   **FIGURE 9.34** Half-bridge series inverter.

The circuit of Fig. 9.34 uses two capacitors in place of one in Fig. 9.33. The capacitances of the two capacitors are equal, that is, $C_1 = C_2 = C$. Also, $L_1 = L_2 = L$. In the circuit of Fig. 9.34, the power is drawn from the d.c. source during both halves of cycle. The operation of the circuit is explained as follows:

The initial voltage across the capacitor C_2 is assumed to be E_C with lower plate positive. Since two capacitors in series are connected across the d.c. source,

$$V_{C1} + V_{C2} = V$$

Hence, $V_{C1} = V + E_C$ $(\because V_{C2} = -E_C)$

Thus, the capacitor C_1 is charged to voltage $V + E_C$ with the upper plate positive. When Th_1 is turned on at $t = 0$, the load current i_o has two components i_1 and i_2. The current i_1 flows from the d.c. source, through Th_1, L_1, R, and C_2. The current i_2 is the discharge current of C_1 and flows through Th_1, L_1, and R. The driving voltage in both loops is $V + E_C$. The circuit elements in both loops are also same; hence, $i_1 = i_2$. At the end of the positive half cycle, the load current is zero, and as the rate of change of current is negative, the voltage developed across L_1 is reversed and Th_1 is reverse biased. Hence, Th_1 is turned off. The voltage induced across L_2 is also reversed, making Th_2 forward biased and which can be tuned on. The voltages across the capacitors are exchanged. Hence, $V_{C1} = E_C$ and $V_{C2} = V + E_C$. An exactly similar operation occurs in the negative half cycle. In each half of the cycle, one half of the load current is supplied by the d.c. source and another half by one of the capacitors. The maximum forward off state voltage across each thyristor is $V + E_C$ and the peak reverse voltage is E_C. The waveforms are shown in Fig. 9.35.

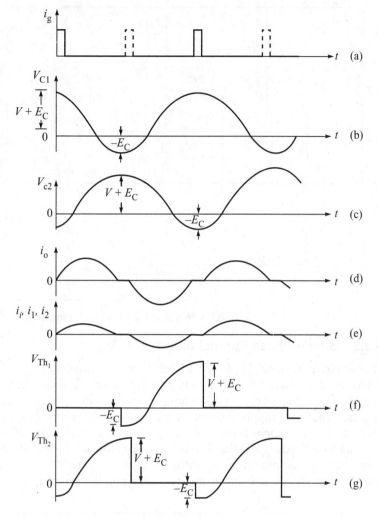

FIGURE 9.35 (a) Triggering pulses, (b) & (c) capacitor voltages, (d) load current, (e) input current, and (f) & (g) voltages across thyristors.

9.3.2 Three-Phase Series Inverter

Three-phase series inverter is shown in Fig. 9.36. Basically, it is a combination of the three single-phase inverters of Fig. 9.34. So, there are six thyristors. All thyristors are turned on in a proper sequence and at proper instants of time to give a three-phase output. The load is three resistors of resistance R connected in star. The capacitances of capacitors C_1 and C_2 are taken large enough to establish a constant neutral voltage. Each phase operates independently. The capacitor C in series with R forms an underdamped circuit with centre tapped inductor to provide commutation as in the single-phase inverter. The sequence of triggering the thyristors is 1, 6, 2, 4, 3, and 5. Thus, the firing frequency is six times the output frequency. Therefore, the interval between two successive firing is $T/6$, where T is the time period of the output wave.

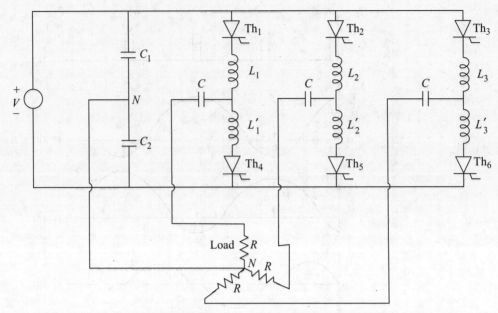

FIGURE 9.36 A three-phase series inverter.

9.3.3 Single-Phase Parallel Inverter

A basic circuit for a single-phase parallel inverter is shown in Fig. 9.37. In addition to two thyristors, the circuit consists of an inductor, an output transformer, and a commutating capacitor. The transformer primary has two windings of equal number of turns connected in series. The input supply is connected across the series combination of the inductor L, thyristor Th_1, and upper primary winding. The transformer turn ratio from each primary winding to secondary winding is unity. Due to inductor L, the source current is almost constant. The commutating capacitor C comes in parallel with the load through the transformer. That is why this inverter is called parallel inverter.

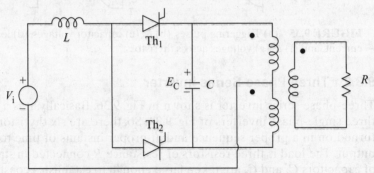

FIGURE 9.37 A single-phase parallel inverter.

The operation of the parallel inverter is explained as follows:

Mode-1: *Th_1 is on and Th_2 is off.*

With the start of conduction of Th_1, the current starts developing. The current produces a magnetic flux that links both the primary windings and the secondary winding.

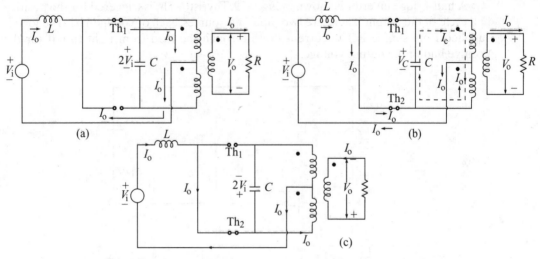

FIGURE 9.38 (a) Mode-1, (b) mode-2, and (c) under steady state.

As a result a voltage V_i is induced in the upper and lower primary windings and the secondary winding. The total voltage across the primary windings is $2V_i$. This primary voltage charges the capacitor C to a voltage $2V_i$ with the upper plate positive, as shown in Fig. 9.38(a). The thyristor Th_2, now, is forward biased through Th_1 by the capacitor voltage $2V_i$. When steady-state is reached, a current I_o flows through source, L, Th_1, and the upper half of primary winding. The current flow is shown in Fig. 9.38(a). Under steady state, $v_o = V_i$, $v_c = v_{Th2} = 2V_i$, $i_o = I_o$, $v_{Th1} = 0$, $i_c = 0$, $i_{Th1} = I_o$, and $i_{Th2} = 0$. During the transient state, currents and voltages vary exponentially.

Mode-2: *Th_1 is off and Th_2 is on.*

As Th_2 is forward biased, when a triggering pulse is applied to its gate, it turns on. Just after Th_2 is on, Th_1 is reverse biased by the capacitor voltage $2V_i$. It is, therefore, turned off. The steady state current I_o, now, flows through Th_2, the lower half of primary winding, the source, and L, as shown in Fig. 9.38(b). At the same time, capacitor discharges through the primary windings. The capacitor current i_c is also shown by a dotted line in Fig. 9.38(b). The capacitor current i_c decays exponentially from $-I_o$ to 0. The charging current of capacitor is taken as positive. Hence, the discharge current is negative. The capacitor current flows till the polarity of the capacitor voltage is not reversed, that is, from $+2V_i$ to $-2V_i$. The load voltage changes from V_i to $-V_i$. Under steady state, the current distribution is as shown in Fig. 9.38(c).

The conditions of Fig. 9.38(a) and Fig. 9.38(c) are identical, and applying a triggering pulse to Th_1, the same changes, as in mode-2, occur but in the reverse direction, and the cycle is repeated.

9.3.4 Single-Phase Half-Bridge Inverter

Bridge inverters are very popular and are commonly used in d.c. to a.c. conversion. A bridge circuit for a single-phase inverter can easily be extended for three-phase inverters. Moreover, an output transformer is not essential in a bridge inverter. Single-phase bridge inverters are of two types, namely, (i) half-bridge inverters and full-bridge inverters.

A half-bridge inverter is shown in Fig. 9.39. Thyristor Th_1 is triggered for the positive half cycle of the output. There are two input d.c. sources, each of voltage $V_i/2$. The load is supplied from source 1. The current flows through the load from right to left for the positive half of the output voltage.

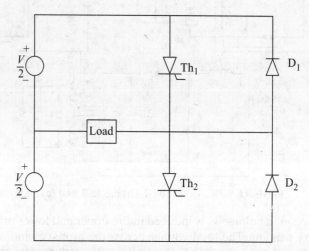

FIGURE 9.39 A single-phase half-bridge inverter.

Th_1 conducts for the period $T/2$ (i.e., $0 < t < T/2$), where T is the time period of the output voltage. During this period the output voltage is $+V_i/2$. At $T/2$, thyristor Th_1 is commutated and Th_2 is turned on. For the duration $T/2 < t < T$, Th_2 conducts and the source 2 supplies the load. The direction of current reverses and, so, flows from left to

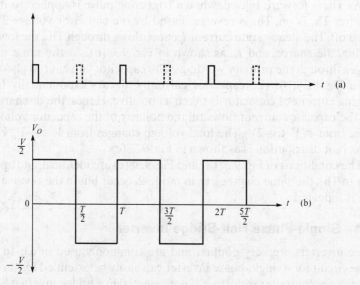

FIGURE 9.40 (a) Triggering pulses and (b) wave shape of output voltage.

right. The output voltage is $-V_i/2$. Thus, the negative half of the output is obtained. The gate pulses for Th_1 (by full line) and Th_2 (by dotted line) and the voltage output are shown in Fig. 9.40. The output current waveform is similar to the voltage waveform. Diodes D_1 and D_2 provide freewheeling operation. For resistive load, these diodes do not come into operation. These diodes are needed for reactive loads. The frequency of the output voltage depends on the gating frequency of each thyristor. Thus, the frequency

$$f = \frac{1}{T} \text{ Hz} \quad \text{and} \quad \omega = \frac{2\pi}{T} \text{ rad/s} \tag{9.73}$$

9.3.5 Single-Phase Full-Bridge Inverter

A full-bridge inverter is shown in Fig. 9.41. The full-bridge inverter uses only single d.c. source of voltage V_i. There are four thyristors and four diodes. In Fig. 9.41, thyristors Th_1 and Th_2 are triggered simultaneously, whereas Th_3 and Th_4 are fired together. During $0 < t < T/2$, Th_1 and Th_2 conduct and current flows from left to right. At $t = T/2$, both Th_1 and Th_2 are turned off together and Th_3 and Th_4 are turned on together. Both Th_3 and Th_4 conduct for the duration $T/2 < t < T$. During this period the load current flows from right to left and the output voltage reverses. Thus, the direction of load current and voltage reverses. Like half-bridge inverter, diodes are freewheeling diodes and come in operation when the load is reactive. The triggering pulses and waveforms of output voltage and current for the resistive load are shown in Fig. 9.42.

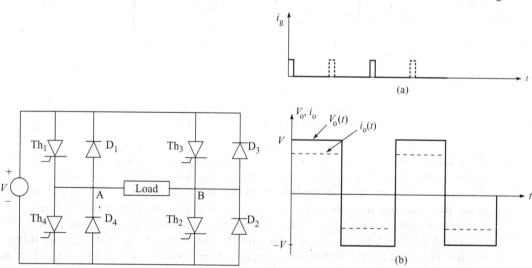

FIGURE 9.41 A single-phase full-bridge inverter.   **FIGURE 9.42** (a) Triggering pulses, (b) $v_o(t)$ and $i_o(t)$.

The half-bridge inverter requires a three-wire source. This drawback of the half-bridge inverter is eliminated in the full-bridge inverter. However, a full-bridge inverter uses four thyristors and four diodes. Hence, it is costlier than the half-bridge inverter. This drawback is compensated as the output voltage is double and the power output is four times compared to corresponding values in the half-bridge inverter.

Output Waves for Reactive Loads

As said earlier, when the load is resistive, the load current is also square wave and in phase with the voltage. However, for reactive load the load current does not change instantaneously and it is not a square wave. The load current is also not in phase with the voltage. The diodes connected in antiparallel with thyristors allow the current to flow when the thyristors are turned off. During freewheeling, the energy is fed back to the d.c. source, the diodes are also called feedback diodes.

Figure 9.43 shows an inverter load consisting of resistance (R), inductance (L), and capacitance (C). The voltage V is the output voltage of the inverter. In case of half-bridge inverter $V = (V_i/2)$, and in case of full-bridge inverter $V = V_i$. The KVL equations are

$$V = Ri_o + L\frac{di_o}{dt} + \frac{1}{C}\int i_o dt \qquad \text{when } 0 < t < \frac{T}{2}$$

$$-V = Ri_o + L\frac{di_o}{dt} + \frac{1}{C}\int i_o dt \qquad \text{when } \frac{T}{2} < t < T$$

The solution of the above equations depends on the relative values of R, L, and C. The variation of load currents for the following three conditions of the load are considered.

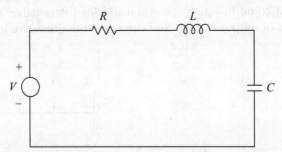

FIGURE 9.43 An inverter with RLC load.

When Load Consists of Only R and L Just before the instant $t = 0$, thyristors Th$_3$ and Th$_4$ (in Fig. 9.39, Th$_2$) are conducting, and the load current flows from B to A (taken negative). The current is shown as $-i_o$ in Fig 9.44(b). When Th$_3$ and Th$_4$ are turned off at $t = 0$, the inductance does not allow the current to become zero and change direction. Hence, freewheeling diodes D$_1$ and D$_2$ (D$_1$ in half-bridge inverter) start conducting and the current is fed back to the d.c. source. At $t = 0$, Th$_3$ and Th$_4$ (Th$_2$ in half-bridge inverter) are turned off, and Th$_1$ and Th$_2$ (Th$_1$ in half-bridge inverter) are turned on, but do not conduct as they are reverse biased by the voltage drops across the diodes. The current decreases and becomes zero at $t = t_1$ in Fig. 9.44(b). Th$_1$ and Th$_2$ (Th$_1$ in half-bridge inverter) start conducting as they are, now, forward biased. The load current flows from A to B (in positive direction) and increases from zero value linearly, as shown in Fig 9.44(b). At $t = T/2$, Th$_1$ and Th$_2$ are turned off and Th$_3$ and Th$_4$ are turned on, but the current does not change to zero. The freewheeling diodes D$_3$ and D$_4$ (D$_2$ in half-bridge inverter) start conducting and current is fed back to the source. As Th$_3$ and Th$_4$ are reverse biased by the voltage drops across the diodes, they do not conduct. At $t = t_2$, the current becomes zero and Th$_3$ and Th$_4$

Sec. 9.3 / Inverters   429

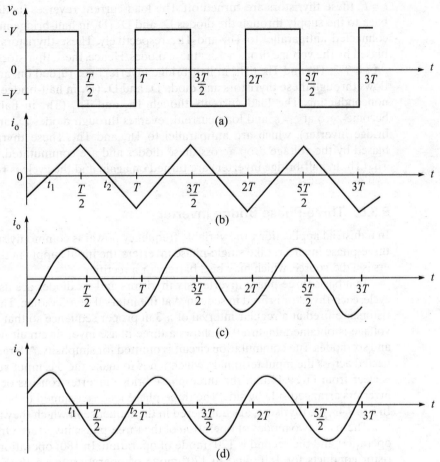

FIGURE 9.44 Wave shapes of (a) load voltage, and current when load is (b) RL, (c) RLC overdamped, and (d) RLC underdamped.

(Th_2 in half-bridge inverter) start conducting as they are forward biased now. Th_3 and Th_4 conduct until $t = T$ when they are turned off and Th_1 and Th_2 are turned on. The cycle is repeated. The wave shape of the load current is shown in Fig. 9.44(b).

When Load is RLC Overdamped The sequence of conduction of thyristors and diodes is the same as that in the RL load. However, the wave shape is not same as in the case of the RL load due the nature of the load. The wave shape of the load current in this case is shown in Fig. 9.44(c).

When Load is RLC Underdamped Because of the nature of the load, one group of thyristors (one thyristor in half-bridge inverter) is turned off before the other group of thyristors (other thyristor in half-bridge inverter) is turned on. Hence, just after $t = 0$, Th_1 and Th_2 (Th_1 in half-bridge inverter) are conducting through the load. The load current is positive and $i_o = I_o$, as shown in Fig. 9.44(d). As the load current reduces to zero at

$t = t_1$, these thyristors are turned off. The load current reverses its direction and flows back to the supply through the diodes D_1 and D_2 (D_1 in half-bridge inverter), which are connected antiparallel to Th_1 and Th_2, respectively. These thyristors are, now, reverse biased by the voltage drops across these diodes. Hence, these thyristors are commutated. At $t = T/2$, Th_3 and Th_4 (Th_2 in half-bridge inverter) are turned on, and so load current flows through these thyristors and diode D_1 and D_2 (D_1 in half-bridge inverter) become nonconducting. The load currents though Th_3 and Th_4 (Th_2 in half-bridge inverter) become zero at $t = t_2$ and load current reverses through diodes D_3 and D_4 (D_2 in half-bridge inverter), which are antiparallel to Th_3 and Th_4. These thyristors are reverse biased by the voltage drop across these diodes and are commutated. At $t = T$, Th_1 and Th_2 (Th_1 in half-bridge inverter) are turned on again and the cycle is repeated.

9.3.6 Three-Phase Bridge Inverter

In industrial applications, the variable frequency power is commonly supplied through a three-phase inverter. Like single-phase inverters, the input supply in three-phase inverters is a d.c. source, which may be a battery or a rectifier.

In a basic three-phase inverter, six thyristors and six diodes are used. Hence, in one cycle, each thyristor is fired in sequence at the interval of $\pi/3$ radian. This means that thyristors are fired at a regular interval of $\pi/3$ in proper sequence so that a three-phase a.c. voltage is obtained. Figure 9.45 shows a three-phase inverter circuit using six thyristors and six diodes. The commutation circuit is omitted for simplicity. A large capacitor is connected across the input terminals, which tends to make the d.c. input voltage constant. It is clear from Fig. 9.44 that the three-phase bridge inverter consists of three half-bridge inverters arranged side by side. The three-phase load is assumed to be connected in star. In Fig. 9.45, the thyristors are numbered in the sequence in which they are turned on.

There are two modes of operation of the three-phase inverters. One is 180° mode of operation and the second is 120° mode of operation. In 180° operation mode, each thyristor conducts for 180°, and in 120° mode of operation, each thyristor conducts for

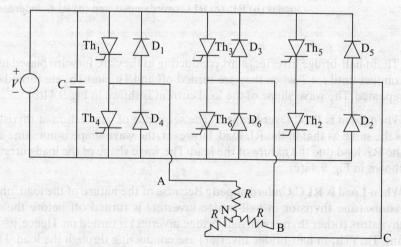

FIGURE 9.45 Three-phase bridge inverter using thyristors.

120°. In 180° mode of operation, three thyristors conduct at one time, and in 120° mode, only two thyristors conduct at one time. The conducting patterns of the two modes are shown in Fig. 9.46.

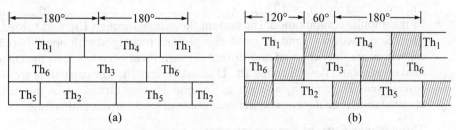

FIGURE 9.46 Conducting patterns of (a) 180° mode and (b) 120° mode.

9.3.7 Current Source Inverters

The input source in a current source inverter is a d.c. current source. In a current source, current drawn from the source is constant. Hence, the load current remains constant with the change of load on the inverter. The load voltage changes with the variation of load. A voltage source in series with a large inductor behaves as a constant current source. The advantages of the current source inverters are listed below:

1. Because of constant input current, the problems of misfiring of devices and short circuiting of source are not there in a current source inverter.
2. Peak current of device is limited.
3. Commutation circuit is simple.
4. Freewheeling diodes are not needed.

Single-Phase Current Source Inverter

The circuit of a single-phase current source inverter is shown in Fig. 9.47. The inverter is fed from a variable voltage source in series with a large inductor. There are four thyristors, four diodes, and two commutating capacitors. Diodes isolate the capacitors from the load. Hence, the capacitors are prevented from discharging through the load.

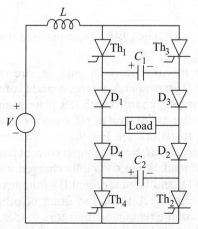

FIGURE 9.47 Single-phase current source inverter.

Let us assume that thyristors Th_1 and Th_2 are conducting. Capacitors C_1 and C_2 are charged fully with the polarity, as shown in Fig. 9.47, and the current is not flowing through C_1 and C_2. The current, hence, is flowing through Th_1, D_1, load, D_2, and Th_2. When Th_3 and Th_4 are turned on, Th_1 and Th_2 are reverse biased by the capacitor voltages and turned off by the impulse commutation. The current flows, now, through Th_3, C_1, D_1, load, D_2, C_2, and Th_4. The capacitors C_1 and C_2 are discharged and recharged with opposite polarity. When capacitors are fully charged and current through them becomes zero, the current through load reverses and flows through Th_3, D_3, load, D_4 and Th_4. The capacitors are, now, ready to reverse bias Th_3 and Th_4 as soon as Th_1 and Th_2 are turned on in the next cycle. The commutation time depends on the magnitude of load current and voltage.

Three-Phase Current Source Inverter

The circuit of a three-phase current source inverter is shown in Fig. 9.48. It is an extension of the single-phase current source inverter of Fig. 9.47. It employs six thyristors, six diodes, and six capacitors. Terminals A, B, and C are the output terminals. Since there are six thyristors, they are triggered in a sequence at an interval of $\pi/3$ radian. For 120° mode of operation, two thyristors conduct at one time. The operation for this mode is explained as follows.

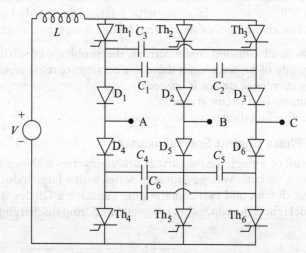

FIGURE 9.48 Three-phase current source inverter.

Let us assume that thyristors Th_1 and Th_6 are conducting. The load current flows through Th_1, diode D_1, terminal A, phases A and C of the load, terminal C, diode D_6, and Th_6. Capacitor C_1 is fully charged with left plate positive. At the end of the conduction period of Th_1 and after an interval of 60° from the triggering of Th_6, Th_2 is triggered. For firing sequence of thyristors, see Fig. 9.46(b). The capacitor voltage reverse biases the thyristor Th_1 and turns it off. Now the input current that is constant flows through Th_2, C_1, D_1, and phase A of load. When C_1 is fully charged with opposite polarity, the current is shifted from D_1 to D_2 and from terminal B of the inverter. Thus, the full-load current is transferred from terminal A to B. The firing of other thyristors are as shown in Fig. 9.46(b). The actions of the remaining thyristors, diodes, and capacitors are similar to those discussed above. The load is not shown in Fig. 9.48.

9.4 A.C. VOLTAGE CONTROLLERS

As discussed in Section-9.1, an a.c. voltage controller gives a variable voltage supply at a constant frequency. A.C. voltage controllers are classified as single-phase and three-phase voltage controllers. Furthermore, these types may be either half wave or full wave. Input to an a.c. voltage controller is from an a.c. voltage source. Therefore, it is line commutated and there is no need of force commutation. An a.c. voltage controller is also called an a.c. voltage regulator.

Two types of controls are used in a.c. voltage controllers, namely, integral cycle control and phase control. In the first type of control, the a.c. supply is connected to the load for certain number of cycles and is disconnected from the load for another certain number of cycles. Each on period and off period consists of an integral number of cycles. Thyristors are turned on by gate pulses at zero voltage crossing of input voltage. In phase control the thyristors are turned on at such a time that the load is connected to the a.c. supply for a part of each half cycle or a part of each cycle. The first type is called as half-wave a.c. voltage controller and the second as full-wave a.c. voltage controller.

The integral cycle-control type of a.c. voltage controllers has very limited applications. It has applications where the loads have mechanical inertia and high thermal constant, such as industrial heating and speed control of motors. Most a.c. voltage controllers being used are phase-control type. Hence, integral cycle-controlled a.c. voltage controllers for resistive loads are discussed in this section. The phase-control type of a.c. voltage controllers are discussed in succeeding sections.

9.4.1 Integral Cycle-Control A.C. Voltage Controllers

A single-phase circuit for an integral control a.c. voltage controller is shown in Fig. 9.49. The load considered is resistive. The gate pulses, input voltage, and output voltage are shown in Fig. 9.50. The pulses applied during the positive half cycle is shown by full lines, whereas pulses applied during the negative half cycle is shown by dotted lines. Let us

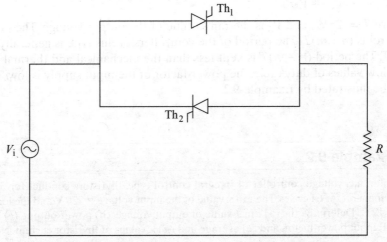

FIGURE 9.49 An integral cycle-control a.c. controller.

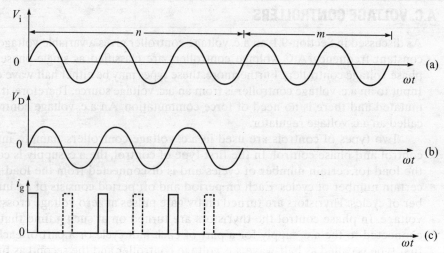

FIGURE 9.50 For integral control (a) input voltage, (b) output voltage, and (c) gate pulses.

assume that the controller is on for n cycles and off for m cycles. The sequence is repeated. Hence, the duty cycle of the controller,

$$\delta = \frac{n}{n+m} \tag{9.74}$$

Let the input voltage,

$$V_i(t) = \sqrt{2}V \sin \omega t$$

The r.m.s. value of output voltage,

$$v_o(t) = \sqrt{\delta \frac{2}{T} \int_0^T V^2 \sin^2 \omega t \, d\omega t}$$

$$= V\sqrt{\delta} \tag{9.75}$$

where $T = (2\pi/\omega)$ and V_o is the r.m.s. value of the output voltage. The period of on-off control is $(n+m)T$. The period of the controllers of this type is generally kept equal to $100T$. The period $(n+m)T$ is kept less than the mechanical and thermal time constant. For low values of duty cycle, the power factor of the input supply is low. This statement can be illustrated by Example-9.2.

Example 9.2

In an a.c. voltage controller of integral control type, thyristors conduct for 36 cycles and remain off for 64 cycles. The r.m.s. value of the input voltage is 220 V and the load resistance is 12 Ω. Determine the (a) r.m.s. value of output voltage, (b) power output, (c) power input, (d) input power factor, and (e) average and r.m.s. values of thyristor current. Neglect losses.

Solution

(a) Duty cycle,
$$\delta = \frac{n}{n+m} = \frac{36}{36+64} = 0.36$$

Hence, the r.m.s. value of output voltage,
$$V_o = 220 \times \sqrt{0.36} = \mathbf{132\ V}$$

(b) The power output,
$$P_o = \frac{V_o^2}{R} = \frac{132^2}{12} = \mathbf{1452\ V}$$

(c) Since losses are neglected, the input power
$$P_i = \mathbf{1452\ W}$$

(d) The r.m.s. value of load current,
$$I_o = \frac{132}{12} = \mathbf{11\ A}$$
$$I_i = \mathbf{11\ A}$$

and   input $VA = 220 \times 11 = \mathbf{2420\ VA}$
The input power factor,
$$\text{pf} = \frac{P}{VA} = \frac{1452}{2420} = \mathbf{0.6}$$

(e) Peak value of thyristor current,
$$I_{Th}^m = \frac{\sqrt{2} \times 220}{12} = \mathbf{25.93\ A}$$

The thyristor current flows for only one half of a cycle, that is, during the positive half, upper thyristor conducts, whereas during the negative half, current flows through the lower thyristor. Hence, the average value of current through each thyristor,

$$I_{av} = \frac{\delta}{T} \int_0^{T/2} 25.93 \sin \omega t\, dt$$

$$= \frac{25.93 \times 0.36}{\pi} = \mathbf{2.97\ A}$$

and the r.m.s. value of thyristor current,

$$I_{Th} = \sqrt{\frac{\delta}{T} \int_0^{T/2} 25.93^2 \sin^2 \omega t\, dt}$$

$$= \frac{25.93 \sqrt{0.36}}{2} = \mathbf{7.78\ A}$$

9.4.2 Single-Phase Phase-Control A.C. Voltage Controller

As mentioned earlier, there are two types of phase-control a.c. voltage controllers: half-wave controller and full-wave controller. In half-wave controller, a thyristor and a diode are connected in antiparallel, whereas in full-wave controller, two thyristors are connected in antiparallel.

Half-Wave Phase-Control A.C. Voltage Controller

The circuit of half-wave phase-control a.c. voltage controller is shown in Fig. 9.51. A thyristor Th and a diode D in antiparallel combination are connected in series with the a.c. source and the load. Here, the resistive load is considered. During the positive half cycle, thyristor Th is turned on at a phase angle α. It conducts up to $\omega t = \pi$. During the negative half cycle, the diode D is positive biased and conducts during whole of the negative half cycle. The triggering pulses, input and output voltages, and output current are shown in Fig. 9.52. As the load is resistive, the load current wave is exactly similar to the load voltage wave.

The average output or load voltage,

$$V_{av} = \frac{1}{T} \int_{\alpha/\omega}^{T} V_m \sin \omega t \, dt$$

$$= \frac{V_m}{2\pi}(\cos \alpha - 1) \qquad (9.76)$$

The r.m.s. value of output voltage,

$$V_o = \sqrt{\frac{1}{T} \int_{\alpha/\omega}^{T} V_m^2 \sin^2 \omega t \, dt}$$

$$= \frac{V_m}{\sqrt{2\pi}} \sqrt{\left(2\pi - \alpha + \frac{\sin 2\alpha}{2}\right)} \qquad (9.77)$$

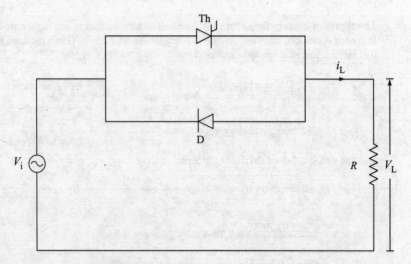

FIGURE 9.51 Half-wave a.c. voltage controller.

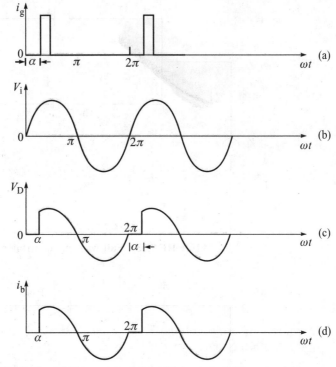

FIGURE 9.52 (a) Triggering pulses, (b) input voltage wave, (c) output voltage wave, and (d) output current wave.

Figure 9.52 shows that the positive and negative half cycles are not identical for both output voltage and current waveforms. Hence, a d.c. component is introduced in the supply and load current, which is not desirable. Because of this reason, half-wave controller is not very popular. In this type of controller only positive half cycle is controlled; hence, it is also called unidirectional controller.

Full-Wave Phase-Control A.C. Voltage Controller

In this type of controllers, both positive and negative half cycles are phase controlled. Due to this reason, this type of controller is also known as bidirectional controller or regulator. The full-wave controller, therefore, is discussed first with resistive load and then with RL load.

With R Load The circuit of the full-wave phase-control a.c. voltage controller is shown in Fig. 9.53. It uses two thyristors in antiparallel. Thyristor Th_1 conducts in the positive half cycle when triggered at $\omega t = (n\pi + \alpha)$, where $n = 0, 2, 4, \ldots$. Thyristor Th_2 conducts in the negative half cycle when triggered at $\omega t = (n\pi + \alpha)$, where $n = 1, 3, 5, \ldots$. The phase angle α is same for both halves. A single-phase full-wave controller using a single triac in place of two thyristor is shown in Fig. 9.54.

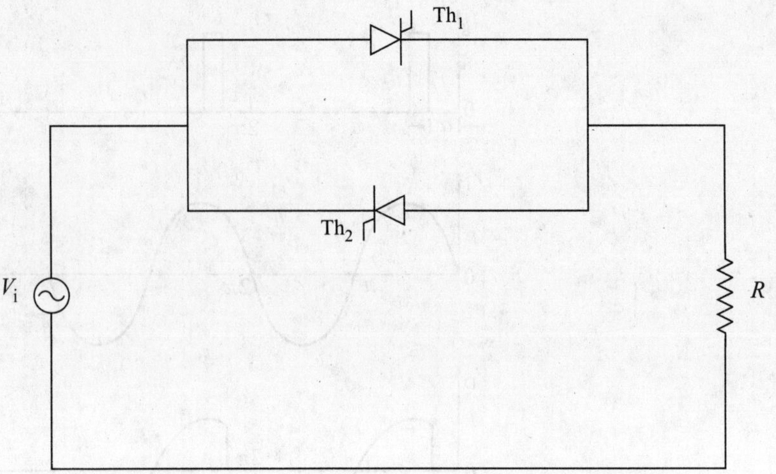

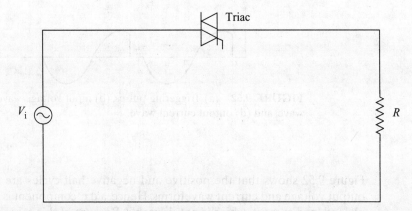

FIGURE 9.53 Full-wave a.c. controller using thyristors.

FIGURE 9.54 Full-wave a.c. controller using triac.

The average output voltage,

$$V_{av} = \frac{2}{T}\int_{\alpha/\omega}^{T/2} V_m \sin\omega t\, dt$$

$$= \frac{V_m}{\pi}(1+\cos\alpha) \tag{9.78}$$

The r.m.s. value of voltage,

$$V_o = \sqrt{\frac{2}{T}\int_{\alpha/\omega}^{T/2} V_m^2 \sin^2\omega t\, dt}$$

$$= \frac{V_m}{\sqrt{2\pi}}\sqrt{\left(\pi-\alpha+\frac{\sin 2\alpha}{2}\right)} \tag{9.79}$$

Triggering pulses, input voltage, output voltage, and current waveforms are shown in Fig. 9.55. Triggering pulses for Th_1 are shown by full lines, whereas that for Th_2 are shown by dotted lines.

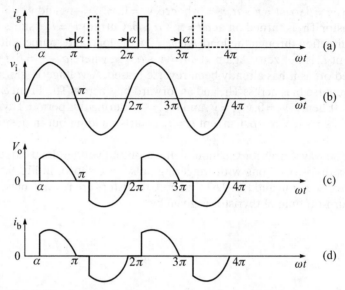

FIGURE 9.55 (a) Triggering pulses, (b) input voltage wave, (c) output voltage wave, and (d) output current wave.

With RL Load The circuit of a full-wave phase-control controller with RL load is shown in Fig. 9.56. When Th_1 is turned on at $\omega t = \alpha$ in the positive half cycle, the current does not change instantaneously as in the case of resistive load due to inductance. The current i_o starts building up through the load. At $\omega t = \pi$, the load and source voltages are

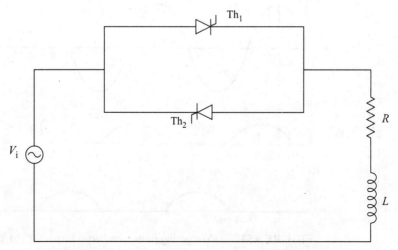

FIGURE 9.56 Full-wave a.c. voltage controller with RL load.

zero, but the current is not zero due to inductance. Let at $\omega t = \beta$, $(\beta > \pi)$, the current becomes zero. Angle β is called the extinction angle. When $\omega t > \pi$, Th_1 is reverse biased, but does not turn off, as i_o is not zero. At $\omega t = \beta$, Th_1 is turned off as i_o is zero and a voltage $V_m \sin \beta$ at once appear as reverse bias across Th_1 and as forward bias across Th_2. From $\omega t = \beta$ to $\omega t = \pi + \alpha$, $i_o = 0$; hence, $v_o = 0$, $v_{Th1} = -v_s$, and $v_{Th2} = v_s$ during this period. Thyristor Th_2 is turned on at $\omega t = \pi + \alpha > \beta$. Current $i_o = i_{Th2}$ starts building but in negative direction through the load. At $\omega t = 2\pi$, the load and source voltages are zero, but the current i_o is not zero. Again at $\omega t = \pi + \alpha + \gamma$, where $\alpha + \gamma = \beta$, $i_o = i_{Th2} = 0$ and Th_2 is turned off as it has already been reverse biased. A voltage $V_m \sin(\pi + \beta)$ at once appear as a reverse bias across Th_2 and as forward bias across Th_1. From $\omega t = \pi + \beta$ to $\omega t = 2\pi + \alpha$, $i_o = 0$; hence, $v_o = 0$, $v_{Th1} = v_s$, and $v_{Th2} = -v_s$ during this period. Thyristor Th_1 is turned on at $\omega t = 2\pi + \alpha > \pi + \beta$. Current $i_o = i_{Th1}$ starts building but in positive direction through the load.

The waveforms for the input voltage, output voltage, load current, and voltage drops across thyristors along with triggering pulses are shown in Fig. 9.57. As it is clear from above discussion and Fig. 9.57 that both thyristors are reverse biased for π radian; hence, the turn-off time of thyristor is given by

$$t_{off} = \frac{\pi}{\omega} \tag{9.80}$$

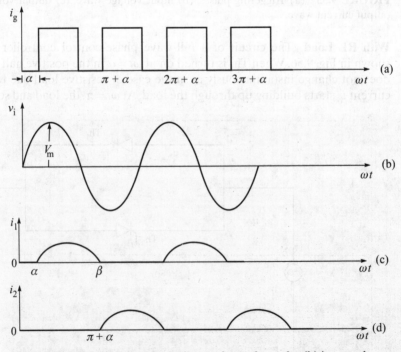

FIGURE 9.57 (a) Triggering pulses, and waveforms for (b) input voltage, (c) & (d) currents through thyristors.

The load current is dependent on the load impedance and the load angle ϕ. The impedance of the RL load,

$$Z = \sqrt{R^2 + \omega^2 L^2} \tag{9.81}$$

and the load angle $\phi = \tan^{-1}\left(\dfrac{\omega L}{R}\right)$ \hfill (9.82)

The KVL equation for the circuit of Fig. 9.56, when Th_1 is conducting,

$$Ri_o + L\frac{di_o}{dt} = V_m \sin \omega t$$

Solution of above equation is given by

$$i_o = \frac{V_m}{Z}\sin(\omega t - \phi) + Ke^{-Rt/L} \tag{9.83}$$

The constant K is found from the initial condition, that is, $i_o = 0$ at $\omega t = \alpha$. Hence,

$$0 = \frac{V_m}{Z}\sin(\alpha - \phi) + Ke^{-R\alpha/\omega L}$$

and
$$K = -\frac{V_m}{Z}\sin(\alpha - \phi)e^{R\alpha/\omega L}$$

Substituting this value of K in Eq. (9.83),

$$i_o = \frac{V_m}{Z}\left[\sin(\omega t - \phi) - \sin(\alpha - \phi)e^{\frac{R}{L}\left(\frac{\alpha}{\omega} - t\right)}\right] \tag{9.84}$$

The current i_o becomes zero again at $\omega t = \beta$. Substituting $i_o = 0$ and $\omega t = \beta$ in Eq. (9.84), we obtain the following relation between β and α for different values of ϕ:

$$\sin(\beta - \phi) = \sin(\alpha - \phi)e^{\frac{R}{L}\left(\frac{\alpha}{\omega} - \frac{\beta}{\omega}\right)} \tag{9.85}$$

The value of β can be obtained from Eq. (9.85) for a given value of ϕ and α. Also, the triggering pulse is required to be applied after the current through the conducting thyristor is zero, that is, $\alpha > (\beta - \pi)$. For the resistive load, from Eq. (9.85), $\beta = 0$. Hence, a short pulse at any value of α can trigger the thyristor. In case of RL load, α is dependent on β, but β changes with the load ϕ given by Eq. (9.82). Hence, the exact value of α is not known and a short pulse may not be sufficient to trigger the thyristor. To overcome this difficulty, a rectangular pulse of sufficient width must be applied for triggering the thyristor.

The conduction period of each thyristor is $(\beta - \alpha)$. Hence, the average value of each thyristor current,

$$I_{Th}^{av} = \frac{1}{T}\int_{\alpha/\omega}^{\beta/\omega} i_o \, dt$$

$$= \frac{V_m}{TZ}\int_{\alpha/\omega}^{\beta/\omega} [\sin(\omega t - \phi) - \sin(\alpha - \phi)e^{(R/L)(\alpha/\omega - t)}]dt \tag{9.86}$$

and the r.m.s. value of each thyristor current,

$$I_{Th} = \frac{1}{T}\int_{\alpha/\omega}^{\beta/\omega} i_o^2 dt$$

$$= \frac{V_m}{Z}\sqrt{\frac{1}{T}\int_{\alpha/\omega}^{\beta/\omega}[\sin(\omega t-\phi)-\sin(\alpha-\phi)e^{(R/L)(\alpha/\omega-t)}]^2 dt} \qquad (9.87)$$

Since the current through both the thyristors are equal in magnitude, the r.m.s. value of load current is given by

$$I_o = \sqrt{2I_{Th}^2} \qquad (9.88)$$

The r.m.s. value of output voltage is given by,

$$V_o = \sqrt{\frac{2}{T}\int_{\alpha/\omega}^{\beta/\omega} V_m^2 \sin^2\omega t\, dt}$$

$$= \frac{V_m}{\sqrt{2\pi}}\sqrt{\left(\beta-\alpha+\frac{\sin 2\alpha-\sin 2\beta}{2}\right)} \qquad (9.89)$$

From the above analysis, we get the following observations:
(a) If $\alpha = \phi$, from Eq. (9.85), we conclude that

$$\sin(\beta-\alpha) = 0$$

$$\therefore \quad \beta-\alpha = \text{conduction angle} = \pi \qquad (9.90)$$

(b) The conduction angle $\gamma = \beta - \alpha$ cannot exceed π, Hence,

$$\phi \leq \alpha \leq \pi \qquad (9.91)$$

(c) For $\alpha < \phi$, the thyristor does not conduct, as during that time it is reverse biased.

9.4.3 Three-Phase A.C. Voltage Controller

Like single half-wave controller, a three-phase half-wave controller can also be used. The half-wave controller uses three thyristors and three diodes. However, the output of such controller contains high harmonics. Half-wave circuit of a three-phase controller, therefore, is not used in real practice.

A three-phase full-wave controller circuit is shown in Fig. 9.58. This circuit uses six thyristors, two for each phase. Let the voltages be given by

$$V_R = V_m \sin\omega t$$

$$V_Y = V_m \sin(\omega t - 120°)$$

$$V_B = V_m \sin(\omega t + 120°)$$

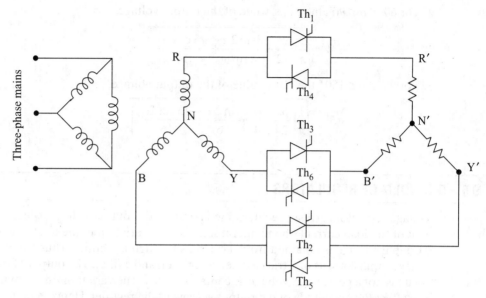

FIGURE 9.58 Three-phase full-wave a.c. voltage controller.

The line input voltages are

$$V_{YR} = V_R - V_Y = \sqrt{3}V_m \sin(\omega t + 30°)$$

$$V_{BY} = V_Y - V_B = \sqrt{3}V_m \sin(\omega t - 90°)$$

$$V_{RB} = V_B - V_R = \sqrt{3}V_m \sin(\omega t + 150°)$$

The following results are obtained when the three-phase controller circuit is analyzed:

1. For $0 \leq \alpha \leq 60°$, two thyristors conduct during some period and three thyristors during another interval. The conditions alternate between two and three conducting thyristors.
2. For $60° \leq \alpha \leq 90°$, two thyristors conduct at one time.
3. For $90° \leq \alpha \leq 150°$, two thyristors conduct at one time, but during some intervals no thyristor conducts.
4. For $150° \leq \alpha$, the output voltage is zero.

Hence, the controlling interval is given by

$$0 \leq \alpha \leq 150° \tag{9.92}$$

The r.m.s. value of the output voltage depends on the value of triggering angle α. Hence, the r.m.s. value of the output voltage for the above first three conditions can be obtained as follows:

1. For $0 \leq \alpha \leq 60°$, the r.m.s. value of the output voltage,

$$V_o = \sqrt{3}V_m \sqrt{\frac{1}{\pi}\left(\frac{\pi}{6} - \frac{\alpha}{4} + \frac{\sin 2\alpha}{8}\right)} \tag{9.93}$$

2. For $60° \leq \alpha \leq 90°$, the r.m.s. value of the output voltage,

$$V_o = \sqrt{3}V_m \sqrt{\frac{1}{\pi}\left(\frac{\pi}{12} + \frac{3\sin 2\alpha}{16} + \frac{\sqrt{3}\cos 2\alpha}{16}\right)} \qquad (9.94)$$

3. For $90° \leq \alpha \leq 150°$, the r.m.s. value of the output voltage,

$$V_o = \sqrt{3}V_m \sqrt{\frac{1}{\pi}\left(\frac{5\pi}{24} - \frac{\alpha}{4} + \frac{\sin 2\alpha}{16} + \frac{\sqrt{3}\cos 2\alpha}{16}\right)} \qquad (9.95)$$

9.5 D.C. VOLTAGE REGULATORS

Voltage regulators give a predetermined constant d.c. output voltage, which is independent of the load current, the temperature, and any variation in the a.c. line voltage. The d.c. input voltage is obtained from a.c. mains through a rectifier. Thus, the unregulated voltage input consists of a transformer, a rectifier, and a filter. The output of the rectifier circuit is not a perfect d.c. voltage. It contains ripples. The variations in a.c. mains voltage also affect the magnitude of d.c. output voltage of the rectifier. Hence, a regulator circuit follows the rectifier circuit to get a stable d.c. voltage.

There are three reasons for using a voltage regulator. The first reason is that an unregulated supply has poor regulation, and hence the output voltage is not constant as the load varies. The second reason is that the d.c. output of the unregulated supply varies with the a.c. input. The third reason is that the d.c. output voltage varies with the change in temperature.

Voltage regulators can be grouped on the principle used for keeping the output voltage constant with varying load. The first group uses a Zener diode and an operational amplifier and a transistor. The conduction of the transistor varies with the variation in the output voltage such that the output voltage remains constant. There are two types of this group of regulators, namely, *shunt regulator* and *series regulator*. The second group of regulators uses pulse width modulation (PWM) to achieve voltage regulation. The magnitude of the pulse turns on or turns off the control transistor or metal oxide semiconductor field-effect transistor (MOSFET). This group of regulators is called *switching regulators*. There are three types of this group of regulators, namely, *buck regulator, boost regulator*, and *buck-boost regulator*.

9.5.1 Shunt Voltage Regulator

The circuit of a shunt regulator is shown in Fig. 9.59. The d.c. input voltage is obtained from an a.c. mains through a bridge rectifier and a filter. In this regulator, an operational amplifier is used as a voltage comparator. The constant d.c. voltage V_z of the Zener diode is applied to the inverting terminal of the operational amplifier. Thus, the inverting input is kept at a constant voltage. The feedback voltage obtained from voltage divider R_1 and R_2 is fed to the noninverting terminal. Thus, the feedback voltage is compared with the Zener voltage. The output voltage of the operational amplifier is used to drive the shunt transistor T, as the output is the input to the base of the transistor. The feedback helps in maintaining the output voltage constant despite variations

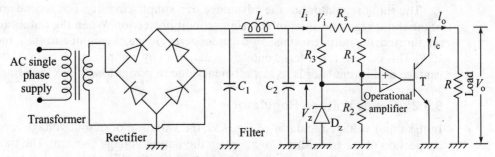

FIGURE 9.59 A shunt voltage regulator.

in the input voltage and load. The output voltage V_o can be changed by varying the feedback gain β.

The action of the shunt regulator can be explained as follows. When the output voltage increases, the feedback to the noninverting input of the operational amplifier increases, resulting an increase in its output voltage. The output voltage of the operational amplifier drives the base of the controlling transistor T, increasing the collector current. Finally, it increases the current through R_s which in turn increases the voltage drop across R_s, offsetting the increase in the output voltage. The output voltage decreases and remains constant. Similarly, when the output voltage decreases, a reverse action is followed, resulting in an increase in the output voltage.

Since the voltage drop across the collector-base is small, we can take $V_o' = V_o$. From the circuit of Fig. 9.59,

$$V_o' = A_V(V_z - \beta V_o) = V_o$$

where $\quad \beta = \dfrac{R_2}{R_1 + R_2}$ \hfill (9.96)

Hence, $\quad V_o = \dfrac{A_V}{1 + \beta A_V}$

Since A_V is very large,

$$V_o = \dfrac{1}{\beta} = \dfrac{R_1 + R_2}{R_2} V_z \tag{9.97}$$

The output current is given by

$$I_o = \dfrac{V_o}{R_L} \tag{9.98}$$

and the input current,

$$I_i = \dfrac{V_i - V_o}{R_s} \tag{9.99}$$

Hence, the collector current of the transistor,

$$I_C = I_i - I_o \tag{9.100}$$

The shunt regulator has the advantage of a simple circuitry. The second advantage of the shunt regulator is its inbuilt short circuit protection. When the output terminals are shorted, the output voltage V_o becomes zero. Hence, the input current I_i reduces to V_i/R_s. This current is not large enough to damage any component. However, the disadvantage of this regulator is its low efficiency due to power losses in R_s and transistor T.

9.5.2 Series Voltage Regulator

In the shunt voltage regulator of Fig. 9.59, the variation of output voltage controls the conduction of the transistor T to maintain the output voltage constant. The transistor is shunting the load. In series voltage regulator the controlling transistor is connected in series with the load. The circuit of series voltage regulator is shown in Fig. 9.60. In this series voltage regulator, again, the operational amplifier compares the Zener voltage with the feedback voltage. The constant d.c. voltage V_z of the Zener diode is applied to the noninverting terminal and the feedback voltage obtained from voltage divider R_1 and R_2 is fed to the inverting terminal of the operational amplifier. The feedback gain β is given by Eq. (9.96) and the output is given by Eq. (9.97).

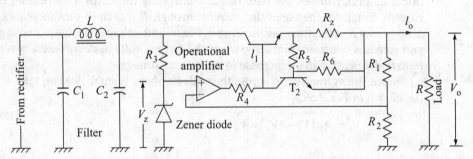

FIGURE 9.60 A series voltage regulator.

When the output voltage increases, the feedback voltage also increases, whereas the output voltage of operational amplifier decreases. The output of the operational amplifier controls the conduction of transistor T_1. A lesser base voltage offsets the increase in V_o. A reverse action maintains the output voltage when it tends to decrease.

The transistor T_2 and resistances R_4, R_5, R_6, and R_7 are used to provide overcurrent and short circuit protections. The voltage divider network of R_5 and R_6 senses the voltage drop across R_7. When the load current increases to its maximum value, the voltage across R_7 becomes large; therefore, the voltage to the base of T_2 is large enough to drive T_2 on. The collector current of T_2 flows through R_4, which in turn reduces the base voltage of T_1. This decreases the output voltage and load current. In case load terminals are shorted, transistor T_2 conducts heavily and reduces the base voltage of transistor T_1 to such a low value that reduces the short circuit current to the limiting value.

The efficiency of the series voltage regulator is higher than the shunt regulator and is commonly used for power level of about 10 W. The circuit is also simple.

9.5.3 Switching Voltage Regulator

Buck Type of Switching Regulator

The circuit of a buck regulator is shown in Fig. 9.61. The output voltage of this regulator is less than the input voltage. The switching device used is a power transistor. The PWM switches on the power transistor for a time interval T_{on} and switches it off for a time interval T_{off}. The total time $(T_{on} + T_{off})$ is called switch period. The switching frequency, therefore, is

$$f_s = \frac{1}{T_{on} + T_{off}} \qquad (9.101)$$

and the duty cycle is

$$\delta = \frac{T_{on}}{T_{on} + T_{off}} \qquad (9.102)$$

The operational amplifier acts as a voltage comparator. When the output voltage increases, the feedback signal V_{FB} increases, and hence the comparator output decreases. The switch on period decreases decreasing the duty cycle. Finally, the output voltage is maintained. A reverse action takes place when the output voltage decreases.

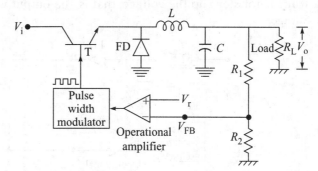

FIGURE 9.61 Buck type of switching regulator.

When pulse from the PWM is high, the transistor conducts and the input voltage feeds the load through the inductor L. The freewheeling diode (FD) is reverse biased, and hence does not conduct. The circuit during the interval T_{on} is represented by the circuit of Fig. 9.62(a). When the pulse of PWM is low, the transistor is turned off. The current through the inductor starts decreasing and a reverse voltage, as shown in Fig. 9.62(b), is induced in the inductor. The freewheeling diode becomes forward biased, and hence conducts. The energy stored in the inductor is dissipated in the load. The load current, now, flows through the diode. The capacitor also discharges through the load. The current, thus, continues to flow through the load. The turn-off period as mentioned above is T_{off}.

Due to very high gain of the operational amplifier, there is a virtual short circuit across the input terminals, and hence,

$$V_r = V_{FB} = \frac{R_2}{R_1 + R_2} V_o$$

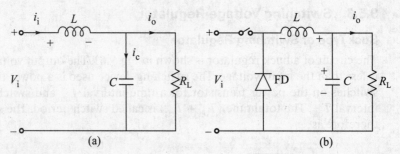

FIGURE 9.62 Equivalent circuit during (a) turn-on and (b) turn-off operations.

and
$$V_o = \frac{R_1 + R_2}{R_2} V_r \qquad (9.103)$$

Thus, the output voltage is a constant.

Boost Type of Switching Regulator

The circuit of boost type of switching regulator is shown in Fig. 9.63. Unlike the circuit of buck type, the circuit of boost type uses a MOSFET in place of the transistor. A boost regulator steps up the voltage, that is, the output voltage is higher than the input voltage.

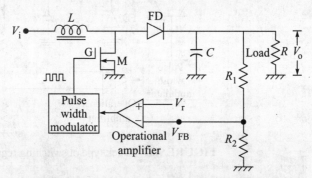

FIGURE 9.63 A boost type of switching regulator.

During the high pulse the MOSFET is on and the input is short circuited through the inductor. The energy is stored in the magnetic field of the inductor. The freewheeling diode is reverse biased by the capacitor voltage. The capacitor-stored energy is dissipated in the load. The current through load is fed by the capacitor. The equivalent circuit during this mode of operation is shown in Fig. 9.64(a).

During low pulse the MOSFET is open. The capacitor is discharged, and hence the freewheeling diode is forward biased. The stored magnetic energy in the inductor decays and a current flows through the load via the inductor. The capacitor is also charged. The equivalent circuit during this mode of operation is shown in Fig. 9.64(b).

The output voltage is again given by Eq. (9.103). The duty cycle δ increases when the output voltage tends to decrease, increasing the output voltage. When V_o tends to increase, δ decreases decreasing the output voltage.

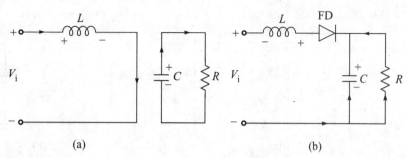

FIGURE 9.64 An equivalent circuit during (a) turn-on and (b) turn-off operations.

Buck-Boost Type of Switching Regulator

The circuit is similar to that of the buck regulator except that the feedback signal is fed to the comparator after inverting it by an inverting operational amplifier, as shown in Fig. 9.65. The inductor is connected across the transistor, whereas in the buck regulator it is connected in series. It gives both higher and lower output than the input, depending on whether the duty cycle is high or low. Moreover, the polarity of the output voltage is opposite to that of the input voltage. Hence, this regulator is also known as inverting regulator.

When the pulse is high, the transistor is on and the inductor is fed from the source. The freewheeling diode is reverse biased. The voltage across the inductor is equal to V_i. The capacitor feeds the load during this period. The equivalent circuit is as shown in Fig. 9.66(a).

When the pulse is low, the transistor is turned off. The stored energy in the inductor decays and the polarity of induced voltage in the inductor is reversed, as shown in Fig. 9.66(b). The inductor feeds the current through the load as well charges the capacitor.

The negative feedback helps in keeping the output voltage constant during both the modes of operation.

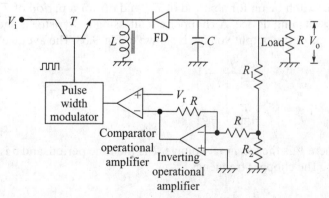

FIGURE 9.65 A buck-boost type of switching regulator.

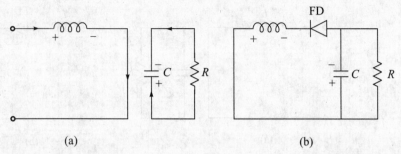

FIGURE 9.66 An equivalent circuit during (a) turn-on and (b) turn-off operations.

9.6 CHOPPERS

Many industrial applications require to be fed from variable d.c. voltage sources. Examples of such d.c. systems are subway cars, trolley buses, battery-operated vehicles, battery charging, etc. As discussed, a variable d.c. voltage is obtained by using phase-controlled rectifiers from an a.c. voltage source. A variable d.c. voltage from a constant d.c. voltage is obtained by using choppers. A chopper converts a constant d.c. voltage to an adjustable d.c. voltage.

One way of obtaining an adjustable d.c. voltage from a constant d.c. voltage is to first convert the constant d.c. voltage to an a.c. voltage by using an inverter. The a.c. voltage output of the inverter is stepped up or stepped down by a transformer, which is then converted again to d.c. voltage by use of a diode rectifier. This method is called an a.c. link chopper. This type of chopper operates in two stages. First from d.c. to a.c. and then from a.c. to d.c. Hence, an a.c. link chopper is costly, bulky, and less efficient. A d.c. chopper converts a constant d.c. voltage to an adjustable d.c. voltage. Since d.c. choppers convert a constant voltage to an adjustable voltage in one stage, this type of chopper is very popular. Thus, a chopper is a d.c. to d.c. converter.

9.6.1 Working Principle

Basically, a chopper is a switching device. When the switch is on, the supply voltage is applied to the load, and when the switch is off, the voltage applied to the load is zero. Let the switch be on for a period of T_{on} and off for a period of T_{off}. A thyristor can be used as switching device. A chopper circuit using a thyristor is shown in Fig. 9.67. The wave shape of the output voltage is shown in Fig. 9.68. The average load voltage is given by

$$V_{av} = \frac{T_{on}}{T_{on}+T_{off}} V$$

$$= \frac{T_{on}}{T} V = \delta V \qquad (9.104)$$

where V is the d.c. input voltage, T is the time period, and δ is the duty cycle.

The chopper frequency,

$$f = \frac{1}{T} \qquad (9.105)$$

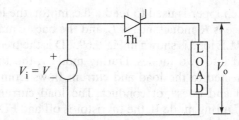

FIGURE 9.67 A chopper circuit using a thyristor.

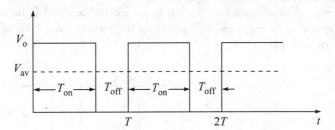

FIGURE 9.68 Wave shape of voltage output.

Thus, the output voltage depends on the value of duty cycle, and it can be controlled by controlling the duty cycle δ. Hence, it is evident that there may be two techniques of controlling the output voltage. The first is the constant-frequency technique and the second is the variable-frequency technique. In constant-frequency technique, the frequency and, hence, the time period T are constant. The duty cycle is varied by varying the period of conduction T_{on}. This technique is also known as time ratio control or PWM technique. In the second technique, the frequency and, hence, the time period are varied for varying the duty cycle. The period of conduction T_{on} is kept constant, whereas the period of blocking T_{off} is varied. The second technique is also known as frequency modulation (FM) technique.

9.6.2 Analysis of Chopper

As the output voltage of the chopper circuit shown in Fig. 9.67 is always less than the input voltage, the chopper is called step down chopper. When the load connected to the chopper of Fig. 9.67 is a resistance, the load current follows the load voltage, and hence the wave shape of load current is similar to the wave shape of the load voltage. The average load voltage,

$$V_{r.m.s.} = \sqrt{\frac{1}{T} \int_0^{\delta T} V^2 dt} = \sqrt{\delta} V \qquad (9.106)$$

and the output power,

$$P_o = \frac{V_{r.m.s.}^2}{R} = \frac{\delta V^2}{R} \qquad (9.107)$$

If losses are neglected, the input power is same as the output power.

When the chopper is used to feed a d.c. motor, the load is designated as RLE. The motor resistance is R, inductance is L, and the back e.m.f. of the motor is E. A chopper circuit with a RLE load is shown in Fig. 9.69. FD is the freewheeling diode. The operation can be divided into two modes. During mode I, the thyristor is conducting and the source is connected to the load and current flows from the source to the load. FD is reverse biased and does not conduct. The load current increases from I_{min} to I_{max}, exponentially. During mode II, the thyristor is off and FD is forward biased. The current flows through the load via FD. The current decreases from I_{max} to I_{min}, exponentially. Depending on the circuit conditions the load current may be continuous or discontinuous. However, discontinuous operation is not desirable. An LC filter, as shown in Fig. 9.70, can be added in Fig. 9.69. This converts the chopper d.c. output to a steady d.c. output with some ripple.

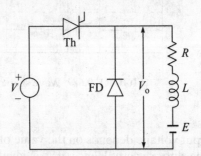

FIGURE 9.69 A chopper with RLE load.

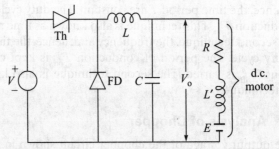

FIGURE 9.70 A chopper with LC filter and RLE load.

9.6.3 Classification of Choppers

The choppers are classified on the basis of the direction of voltage and current. In a two-coordinate system representing current and voltage as X and Y coordinates, respectively, the choppers are classified in the following five categories.

Class A Chopper

Choppers discussed are classified under this category. The load current flows from the source to the load. Both the load voltage and current are positive, as shown in Fig. 9.71(a). It is a one-quadrant chopper.

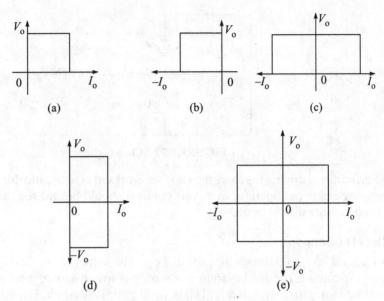

FIGURE 9.71 Chopper quadrants of operations.

Class B Chopper

It is also a one-quadrant chopper, but it operates in second coordinate, as shown in Fig. 9.71(b). The current is negative, that is, it flows from the load to the source. The circuit of a class B chopper is shown in Fig. 9.72. When the thyristor is on, the load e.m.f. E feeds current through the resistance and inductance and the load voltage are zero. When the thyristor is off, the energy stored in the inductor is returned to supply through the diode, and the current decreases.

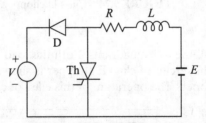

FIGURE 9.72 Class B chopper.

Class C Chopper

It is a two-quadrant chopper and operates in the first and second quadrants. The circuit of a class C chopper is shown in Fig. 9.73. It uses two thyristors and two diodes. Basically, it is a combination of class A and class B choppers. When Th_2 and D_1 are off, the circuit looks like Fig. 9.69 and works as class A chopper and the load current is positive. When Th_1 and D_2 are off, the circuit looks like Fig. 9.72 and operates as class B chopper and the current is negative. Thus, in this chopper power can flow both from source to load and from load to source. It is used in operation and regenerative braking of a d.c. motor.

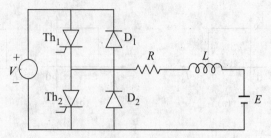

FIGURE 9.73 Class C chopper.

When motor is driving the load, the chopper works as class A, and during braking of the motor, the chopper operates in class B mode. It should be ensured that in no condition both thyristors should be on.

Class D Chopper

It is also a two-quadrant chopper. But it operates either in the first quadrant or in the fourth quadrant. The load current is always positive, whereas the load voltage can be positive or negative. The circuit of class D chopper is shown in Fig. 9.74. If both Th_1 and Th_2 are on, the circuit looks like Fig. 9.69 and both the load current and voltage are positive. When Th_1 and Th_2 are off and D_1 and D_2 are on, the load voltage is negative. In both the cases the load current is positive.

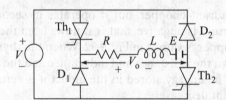

FIGURE 9.74 Class D chopper.

Class E Chopper

It can operate in all the four quadrants. Both the load voltage and current can be positive or negative. The circuit of class E chopper is shown in Fig. 9.75. It is a combination of two class C choppers. The operation of this class of choppers is explained as follows.

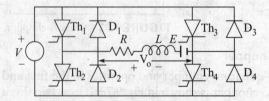

FIGURE 9.75 Class E chopper.

Quadrant-I Thyristor Th_4 is on and thyristors Th_2 and Th_3 are off. When Th_1 is switched on, current flows from source to load. When Th_1 is switched off, the current flows through Th_4 and D_2. Both the load voltage and current are positive.

Quadrant-II Thyristors Th_1, Th_3, and Th_4 are off. When Th_2 is switched on, negative current flows through L, R, Th_2, D_4, and E. Energy is stored in L. When Th_2 is off, current flows back to source through D_1 and D_4. For this operation $V < E +$ voltage drop across the inductor.

Quadrant-III Thyristors Th_1 and Th_4 are off, and Th_2 is on. When Th_3 is switched on, the current flows to the source through load, Th_2 and Th_3. Both the load voltage and current are negative. When Th_3 is turned off, negative current flows through Th_2 and D_4. The load voltage is still negative.

Quadrant-IV For this operation the polarity of E should be reversed. When Th_4 is switched on, positive current flows through E, Th_4, D_2, and L. When Th_4 is turned off, energy stored in L is dissipated and positive current fed back to source through D_2 and D_3. The load voltage is still negative.

9.6.4 Commutation Methods for Choppers

As the chopper is a d.c. to d.c. converter, line commutation is not possible. Hence, a chopper must be associated with a commutation circuit. The following types of commutation methods are discussed here.

Auxiliary Commutation

The chopper circuit of Fig. 9.69 with a commutation circuit is shown in Fig. 9.76. An auxiliary thyristor Th_A is used to commutate the main thyristor Th_1. Hence, the commutation method is called auxiliary commutation. It is also known as voltage commutation as the voltage across a capacitor is applied across the thyristor to reverse bias it. It is also called classical commutation. The diode D_1, L, and C together with Th_A form the commutation circuit. The inductor L is used to maintain the applied voltage to be constant.

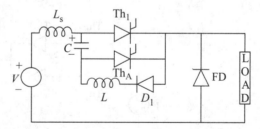

FIGURE 9.76 An auxiliary circuit for a chopper.

The capacitor is assumed to be charged to supply voltage with polarity, as shown in Fig. 9.76. At the beginning of the operation, when the main thyristor is switched on by a gate pulse, the capacitor starts discharging through Th_1, D_1, and L. The load current is supplied from the source through Th_1. Total energy stored in the capacitor is discharged and stored in the inductor. But the current continues to flow and the capacitor is charged with reverse polarity, and the auxiliary thyristor is forward biased. The diode D_1 is also nonconducting. When Th_A is switched on, the main thyristor is reverse biased by the capacitor voltage, and the load current flows through C, Th_A. The thyristor Th_1 is turned off. The capacitor is again charged with the supply voltage as in the beginning. Thus, the capacitor is charged with upper plate positive and its current decays to zero. At this stage, the load current continues to flow in the same direction but through FD by the inductor-stored energy. The load current continues to flow through FD until Th_1 is switched on again.

This commutation circuit is simple. However, it has the following disadvantages:

1. The main thyristor requires to have higher peak rating as load current as well as commutation current flows through it.
2. The charging and discharging time of the capacitor depends on the load current. Hence, when the load current is low, the high frequency operation is not possible.
3. Testing of the circuit without the load is not possible.

Load Commutation

The load commutated chopper circuit is shown in Fig. 9.77. This circuit consists of four thyristors, a capacitor, and a freewheeling diode. Assume that at the beginning of commutation cycle, the capacitor C is charged to a negative voltage equal to the supply voltage. The upper plate of the capacitor is negative. At the beginning, thyristors Th_1 and Th_2 are switched on by gate pulses. The load current flows from the source to the load via Th_1, C, and Th_2. The load voltage at this time is twice the supply voltage, that is, $2V$. The capacitor reverses the voltage across it. When the capacitor voltage becomes V, the current through Th_1, C, and Th_2 decreases to zero. The load current, in the same direction, flows through FD. At the end of the cycle, Th_3 and Th_4 are switched on by gate pulses, and the load current again flows from the source to the load through Th_3, C, and Th_4. Again the capacitor voltage starts changing from V to $-V$. When the capacitor voltage become $-V$, that is, negative with upper plate negative, the current through Th_3, C, and Th_4 becomes zero. The load current, then, flows through FD. The load current flows through FD until Th_1 and Th_2 are switched on at time $2T$. The operation is repeated.

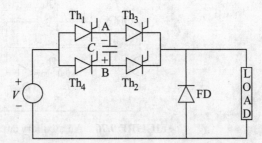

FIGURE 9.77 A load commutated chopper circuit.

9.6.5 Step Up Chopper

All the chopper circuits discussed above are for step down choppers, that is, the output or the load voltage is always less than the supply voltage. In the chopper circuit of Fig. 9.78, the load voltage is higher than the supply voltage. When the thyristor Th is switched on, the current builds up (approximately linearly) in inductance L. Energy is stored in L. The voltage drop across L is as shown in Fig. 9.78. Let Th conducts for a

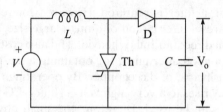

FIGURE 9.78 A step up chopper.

period T_{on} and remains off for a period of T_{off}. During the on period, current rises from an initial value I_1 to I_2. When Th is turned off, the polarity of voltage drop across L reverses as current decays. Now diode is forward biased and the current flows through diode D to the load capacitor. During the off period, the energy stored in L (during the on period) is absorbed in the load. The current through L decreases from I_2 to I_1 during the off period. The load voltage is equal to the supply voltage plus the voltage drop across L. During the on period energy is stored in L,

$$E_{on} = VI_{av} \times T_{on}$$

$$= V\left(\frac{I_1 + I_2}{2}\right)T_{on} \tag{9.108}$$

During the off period, the energy released by L to the load,

$$E_{off} = (V_o - V)I_{av} \times T_{off} = (V_o - V)\left(\frac{I_1 + I_2}{2}\right)T_{off} \tag{9.109}$$

Neglecting the losses,

$$E_{on} = E_{off}$$

$$V\left(\frac{I_1 + I_2}{2}\right)T_{on} = (V_o - V)\left(\frac{I_1 + I_2}{2}\right)T_{off}$$

or

$$V_o = V\frac{T_{on} + T_{off}}{T_{off}} = V\frac{T}{T_{off}} \tag{9.110}$$

$$= V\frac{T}{T - T_{on}} = \frac{V}{1 - \delta} \tag{9.111}$$

Thus, the output voltage is greater than the supply voltage.

9.7 SWITCHED MODE POWER SUPPLY

As discussed, a controlled d.c. supply can be obtained from a phase-controlled rectifier. This rectifier operates at a supply frequency of 50 Hz. To remove the ripples from the rectified output, a filter is required. However, for removing the ripple at power frequency, it requires a filter circuit of quite a large size. This makes the d.c. power supply using a controlled rectifier bulky, heavier, and also costlier. A switched mode power supply (SMPS) is a good alternative for obtaining a d.c. power from an a.c. source. SMPS works on the principle of d.c. chopper. By operating the on/off switch very rapidly, a.c. ripple frequency rises to a very high value. Hence, LC filters, which are small in size and light in weight, can be used. Therefore, the small physical size and light weight of an SMPS made it very popular. As SMPS operates on the chopper principle, the d.c. output voltage is controlled by varying the duty cycle of the chopper. For this either PWM or FM technique is used. In this section, however, PWM technique is used.

An SMPS is a multistage power supply. An a.c. supply is first converted into a d.c. supply. This d.c. voltage is converted into a.c. by using the chopper principle and a transformer. The transformer isolates the output from the input. The isolation means the output connections can be reversed to change the polarity. The transformer also can be used to change the output magnitude by changing the turns ratio. The a.c. output at this stage is converted into a d.c. output by a diode rectifier. The chopper operates at very high frequency and a very small-size filter is used to filter out the ripples. In case the switching device is a power transistor, the chopping frequency is limited to about 40 kHz, whereas it is of the order of 200 kHz in case a power MOSFET is used as the switching device. For simplicity, the controlling circuits for thyristors are omitted in the following circuits of the SMPS. The category of the SMPS is based on the type of d.c. regulator (converter) being used. On that basis broad categories of SMPS are as given below:

1. Flyback SMPS
2. Push-pull SMPS
3. Half-bridge SMPS
4. Full-bridge SMPS

9.7.1 Flyback SMPS

This category of SMPS uses a d.c. to d.c. converter known as flyback converter. The function of the flyback converter is identical to the buck-boost regulator, discussed in Section 9.11. In the buck-boost regulator, the input source forces the current into the inductor and raises its stored energy level when the input side switching transistor conducts. When the input switch is off and the output switch (the diode) is on, the stored energy is allowed to flow from the inductor into the load. The total energy in the inductor during this period is dissipated in the load. If the inductor in the buck-boost regulator is replaced by a transformer, it forms a flyback converter. As a flyback converter converts a d.c. voltage to a d.c. voltage at different level, it is also called d.c. transformer. The circuit of a flyback SMPS with a flyback converter is shown in Fig. 9.79. A transistor T is used in the circuit as a switching device. However, for increasing the chopping frequency a MOSFET can be used. As far as the function of the flyback converter is concerned it is identical to the buck-boost regulator (converter). The energy stored in the magnetic circuit of the transformer when the transistor T is on is dissipated into the load through the secondary winding due to mutual inductance when T is off. The operation of the flyback SMPS circuit, as shown in Fig. 9.79, can be explain as follows.

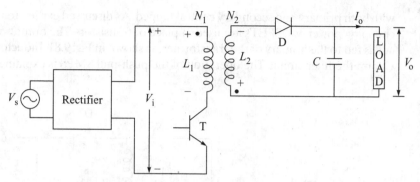

FIGURE 9.79 The circuit of a flyback SMPS.

The a.c. voltage of the mains is converted to d.c. by using a rectifier. The d.c. output is fed to flyback converter. When transistor is on, the voltage is fed to the transformer primary. Voltages are induced in the primary and secondary of the transformer. Primary voltage $v_1 = V_i$ and secondary voltage $v_2 = (N_2/N_1)V_i$. The circuit of Fig. 9.79 reduces as shown in Fig. 9.80(a). The polarities of the induced voltages in the two windings are as shown in Fig. 9.80(a). The output voltage obtained is of the same polarity as that of the input voltage. The output voltage can be step down or step up. The secondary voltage reverse biases the diode D. The filter capacitor has a capacitance large enough to make the output voltage constant. The capacitor voltage is equal to the load voltage.

When T is turned off, voltages of polarities opposite to those shown in Fig. 9.80(a) are induced in the primary and secondary windings. The diode D is, now, forward biased and starts conducting a current. The circuit under this condition reduces as shown in Fig. 9.80(b). As a result the energy stored in the transformer is delivered partially to the load and partially to charge the capacitor.

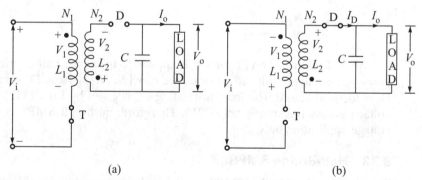

FIGURE 9.80 A flyback SMPS equivalent circuit during (a) T is on and (b) T is off.

9.7.2 Push-Pull SMPS

The circuit of a push-pull SMPS that uses a push-pull converter (or regulator) is shown in Fig. 9.81. The converter uses two power transistors as switching devices and a transformer with both primary and secondary central tapped. As discussed earlier, for higher frequency chopping power MOSFETs are used in place of transistors. The output of the rectifier circuit is fed to the primary of the transformer, as shown in Fig. 9.81. Inductor L and capacitor C form the filter circuit. The operation of the push-pull SMPS is explained as follows.

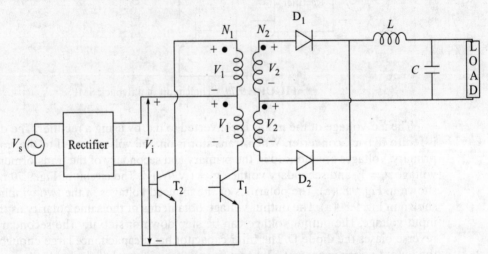

FIGURE 9.81 A circuit of push-pull SMPS.

When T_1 is turned on V_i is applied to the lower half of the primary winding. Hence, $v_1 = V_i$. Voltage $v_2 = (N_2/N_1)V_i$ is induced in both of the secondary windings. The voltage in the upper secondary winding forward biases diode D_1, while that in the lower winding reverse biases diode D_2. The load voltage, therefore, is given by

$$V_o = \frac{N_2}{N_1} V_i \tag{9.112}$$

When T_2 is turned on (T_1 off) V_i is applied to the upper half of primary winding and $v_1 = -V_i$. Secondary induced voltages are $v_2 = (N_2/N_1)V_i$. Now, D_1 is reverse biased and D_2 is forward biased. The load voltage, again, is given by Eq. (9.112). The open circuit voltage across the transistors is $2V_i$. Therefore, push-pull SMPS is suitable for low-voltage application only.

9.7.3 Half-Bridge SMPS

The circuit of half-bridge SMPS using half-bridge regulator is shown in Fig. 9.82. The circuit differs from that of the push-pull regulator in that the input to the transformer primary is fed through a bridge. The bridge is formed by using two transistors and two capacitors. There is single winding on the primary side. MOSFET replaces the transistor if operation is at higher frequency. The capacitances of the two capacitors are equal, and hence the voltage drop across each capacitor is $V_i/2$. The operation of the half-bridge SMPS is explained as follows:

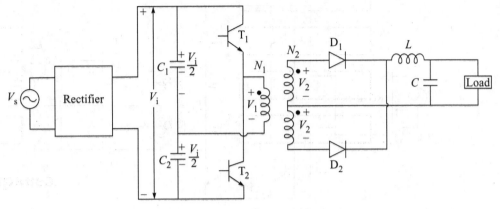

FIGURE 9.82 A circuit of half-bridge SMPS.

When T_1 is turned on, the voltage of C_1 is applied to the primary of the transformer. Hence, $v_1 = (V_i/2)$ and voltage induced in each secondary is $v_2 = (N_2 V_i/2N_1)$. The diode D_1 is forward biased and so it conducts. The output voltage, therefore, is given by

$$V_o = \frac{N_2}{2N_1} V_i = 0.5 K V_i \tag{9.113}$$

where K is the transformation ratio of the transformer. It is inverse of the turns ratio.

When T_2 is turned on (T_1 off), the voltage across the capacitor C_2 appears across the primary of the transformer. Hence, $v_1 = -(V_i/2)$, the voltage induced in each secondary is $v_2 = -(N_2 V_i/2N_1)$. The diode D_2 is forward biased and so it conducts. The output voltage is again given by Eq. (9.113). The open circuit voltage across the transistors is V_i. Therefore, a half-bridge SMPS is preferred over a push-pull SMPS for high-voltage application.

9.7.4 Full-Bridge SMPS

When two capacitors of half-bridge SMPS circuit are replaced by two transistors, it results in the circuit of full-bridge SMPS. The circuit of a full-bridge regulator is shown in Fig. 9.83. The operation of the full-bridge SMPS is explained as follows:

When transistors T_1 and T_2 are turned on together, voltage V_i appears across the primary. Hence, $v_1 = V_i$. Secondary induced voltage $v_2 = (N_2 V_i/N_1)$. The diode D_1 is forward biased and so it conducts. The output voltage, therefore, is given by

$$V_o = \frac{N_2}{N_1} V_i = K V_i \tag{9.114}$$

where K is the transformation ratio of the transformer.

When transistors T_3 and T_4 are turned on together (T_1 and T_2 off), the voltage across the primary, $v_1 = -V_i$. Secondary induced voltage $v_2 = -(N_2 V_i/N_1)$. The diode D_2 is forward biased and so it conducts. The output voltage, again, is given by Eq. (9.114). The open circuit voltage across each transistor is V_i.

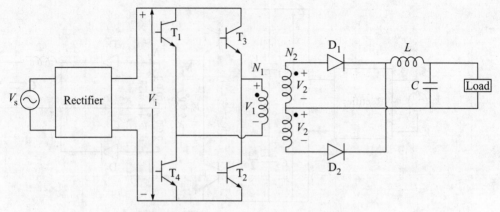

FIGURE 9.83 A circuit of full-bridge SMPS.

The main advantages of SMPS over conventional power supply are as follows:

1. For the same power rating, SMPS is smaller in size, lighter in weight, and has higher frequency. This is because of its high-frequency operation.
2. SMPS is less sensitive to input voltage variation.

SMPS has the following disadvantages also:

1. SMPS has higher output ripple and its regulation is worse.
2. SMPS is a source of electromagnetic and radio interference due to high-frequency switching.
3. Control of radio frequency noise requires filter on both input and output sides of an SMPS.

However, the advantages of SMPS far outweigh its disadvantages, and therefore SMPS has widespread popularity and growth.

9.8 UNINTERRUPTED POWER SUPPLY

There are several applications in which continuous power supply is needed. Even a short-time power failure can cause great problems. Examples of such applications are major computer installations, process control in chemical plants, safety monitors, general communication systems, hospital intensive care units, etc. A temporary failure of power system may cause great public inconvenience due to even short-time closure of computer installation in railway booking counters and general communication system. A short-time failure of power in hospital intensive care units may cause loss of a life. It may cause large economic losses in chemical plants. For such critical loads, it is very important to have a UPS system. Nowadays, static UPS systems are used for this purpose. There are two types of UPS systems, namely, offline (or short-break) UPS and online (or no-break) UPS. However, the main elements of a UPS system of both types are rectifier, filter, battery, inverter, automatic voltage regulator, and static switches. Static switches have no moving part. Power semiconductor devices, such as thyristors, triac, and power transistors, are used as switching devices. These switching devices can be turned on/off within a few microseconds. On time of a static switch is of the order of 3 μs.

The standby battery in UPS is either nickel-cadmium or lead-acid type. The nickel-cadmium batteries are two to three times costlier than lead-acid batteries. However, nickel-cadmium batteries have the following advantages:

1. Their electrolyte is noncorrosive.
2. The electrolyte does not emit any explosive gas during charging.
3. These batteries have longer life, as these do not get damaged due to overcharging or discharging.

9.8.1 Offline UPS

The offline UPS is used where short break in mains supply can be tolerated. An offline UPS circuit is shown in Fig. 9.84. In normal condition, the load is supplied from the mains supply. At the time of power failure UPS is turned on by the static switch. The main a.c. is rectified to d.c. by a rectifier. The d.c. output of the rectifier charges the battery. At the time of power failure the battery feeds the inverter, which in turn converts the d.c. voltage to the a.c. voltage. The filter removes the harmonics. The a.c. voltage thus obtained is fed to the load through the static switch, which turns on at the time of power failure. Under normal circumstances, the on switch is closed, whereas the off switch is open. The load is supplied from the mains; however, the rectifier continuously charges the battery. At the time of power failure, the normally off switch is closed and the normally on switch is open. The battery supplies the load through inverter and filter. A momentary interruption of about 5 ms in supply to the load can be noticed in case when incandescent lamps and fluorescent tubes are parts of the load. When the mains supply comes back the load is connected to the mains supply through the normally on switch as it is turned on. The normally off switch is turned off. Again, an interruption in illumination of lamps can be noticed. The advantage of this UPS is that it is cheaper.

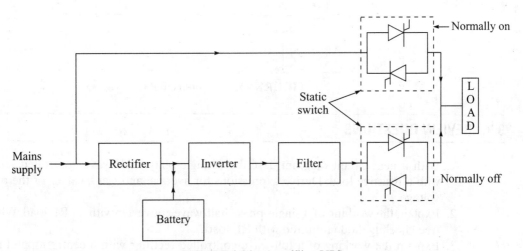

FIGURE 9.84 An offline UPS.

9.8.2 Online UPS

In offline or short-break UPS the load is momentarily disconnected from the supply. In some cases even momentarily break of the supply is not desirable. In such cases, online or no-break UPS is used. If static switches in Fig. 9.84 is interchanged, the circuit forms an online UPS. The circuit of an online UPS is shown in Fig. 9.85. In this type of UPS the rectifier continuously charges the battery as well as supplies the inverter. Inverter continuously supplies to the load through the filter and the normally on switch. In case of failure of the mains, battery instantaneously starts feeding to load through the inverter, filter, and normally on switch. Thus, no discontinuity in illumination is observed in online UPS. The output of the inverter is always conditioned. Conditioning means constant voltage, constant frequency inverter output. However, if required, a voltage regulator can be used. If the inverter circuit develops some fault, the load is directly supplied by the mains by closing the normally off switch. An online UPS has the following additional advantages:

1. The supply to the load can be conditioned by using a voltage regulator before normally on switch.
2. As load is not supplied directly from the mains, it is protected from any transient in the a.c. supply.
3. The inverter output frequency can be maintained at a desired value.

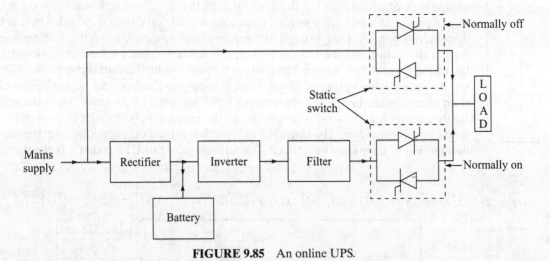

FIGURE 9.85 An online UPS.

9.9 REVIEW QUESTIONS

1. With a neat circuit diagram describe a single-phase half-wave controlled rectifier with a resistive load. Derive expressions for average and r.m.s. values of the output voltage.
2. Explain the working of a single-phase half-wave converter with an RL load. Why is a freewheeling diode required with RL load?
3. Explain the working of a full-wave controlled rectifier with a centre tapped transformer when the rectifier feeds (a) resistive load and (b) RL load.

4. In what respect is the operation of a fully controlled rectifier different from R load, RL load, and RLE load? Draw waveforms of input voltage, triggering pulses, output voltage, output current, and input current for RLE load.
5. What is a single-phase semiconverter? How does the operation of a semiconverter differ from a fully controlled converter?
6. Discuss the operation of a three-phase fully controlled converter with an RLE load. In what way does it differ from a three-phase semiconverter?
7. Draw the circuit diagram and explain in detail the operation of a series inverter. How can the drawbacks of a series inverter be removed?
8. With a circuit diagram explain in detail the different modes of operation of a parallel inverter. What are its advantages and disadvantages over the series inverter?
9. Draw the circuit diagram and explain in detail the operation of a single-phase half-bridge inverter. What are its advantages and disadvantages?
10. Explain the working of a single-phase full-bridge inverter for different types of loads. How does it differ from single-phase half-bridge inverter?
11. With a circuit diagram, explain in detail the different modes of operation of a three-phase bridge inverter. What are its advantages and disadvantages?
12. Discuss the working of a single-phase and three-phase current source inverters.
13. How do integral cycle-control a.c. voltage controllers differ from phase-control a.c. voltage controllers?
14. Derive expressions for r.m.s. value of output voltage of an integral cycle-control a.c. voltage controller (regulator). Define duty cycle.
15. How does a single-phase half-wave regulator differ in operation from a single-phase full-wave regulator? Why are half-wave regulators not popular?
16. Derive expressions for average and r.m.s. values of output voltage of a single-phase half-wave regulator. Why is the average value of output voltage negative, but its r.m.s. value positive?
17. With a circuit diagram explain the operation of a single-phase full-wave regulator with R load and RL load.
18. Draw waveforms of input voltage, output voltage, and load current of a single-phase full-wave regulator. Derive the average and r.m.s. values of a output voltage when the load is resistive.
19. Draw the circuit configuration and explain the working of a three-phase a.c. regulator.
20. Draw the circuit diagram and explain the operation of d.c. shunt regulator. What are its advantages and disadvantages?
21. Draw the circuit diagram and explain operation of d.c. series regulator. What are its advantages and disadvantages?
22. How does a switching regulator differ from shunt and series regulators? How do you classify the switching regulator? Discuss the function of each class.
23. Draw the circuit diagram and explain the different modes of operation of buck regulator. What are its advantages and disadvantages?
24. Draw the circuit diagram and explain the modes of operation of boost regulator. What are its advantages and disadvantages?
25. Draw the circuit diagram and explain the modes of operation of buck-boost regulator. What are its advantages and disadvantages?
26. What is chopper? Explain its working principle. Derive expression for the average value of output voltage. Discuss different techniques of controlling a chopper.

27. Discuss about different classes of choppers.
28. Draw the circuit diagram and explain the operation of a class A chopper feeding a resistive load. Derive expressions for the average and r.m.s. values of output voltage and output power.
29. Draw the circuit diagram and explain the operation of a class A chopper feeding an RLE load.
30. Why is natural commutation not possible in a chopper? Discuss the operation of an auxiliary commutated chopper. Discuss different modes of operation.
31. Discuss the operation of a load commutated chopper. Discuss different modes of operation.
32. Explain the working of a step up chopper.
33. What are different categories of SMPS? Discuss the function of the SMPS in general. What advantages does an SMPS have over other power supply?
34. Draw the circuit diagram and explain the modes of operation of a flyback SMPS. What are its advantages and disadvantages?
35. Draw the circuit diagram and explain the modes of operation of a push-pull SMPS. What are its advantages and disadvantages?
36. Draw the circuit diagram and explain the operation of a half-bridge SMPS. What are its advantages and disadvantages?
37. Draw the circuit diagram and explain the operation of a full-bridge SMPS. What are its advantages and disadvantages?
38. Draw the circuit diagram and explain the working of offline and online UPS. What are their advantages and disadvantages?

9.10 SOLVED PROBLEMS

1. A single-phase half-wave converter supplies a 230-V, 1-kW heater. Find the average load voltage, r.m.s. load voltage, average load current, and r.m.s. load current for firing angle of 45° and 90°. Calculate the power absorbed in the heater.

    Solution

    Heater resistance,

    $$R = \frac{(230)^2}{1000} = 52.9 \, \Omega$$

    From Eq. (9.1), the average voltage,

    $$V_{d.c.} = \frac{V_m}{2\pi}(1+\cos\alpha)$$

    $$= \frac{230\sqrt{2}}{2\pi}(1+\cos 45) = \textbf{88.37 V} \qquad \text{(For } \alpha = 45°\text{)}$$

    $$= \frac{230\sqrt{2}}{2\pi}(1+\cos 90) = \textbf{51.77 V} \qquad \text{(For } \alpha = 90°\text{)}$$

From Eq. (9.3), the r.m.s. voltage,

$$V_{r.m.s.} = \frac{V_m}{2}\left(\frac{\pi-\alpha}{\pi}+\frac{\sin 2\alpha}{2\pi}\right)^{0.5}$$

$$= \frac{230\sqrt{2}}{2}\left(\frac{\pi-\pi/4}{\pi}+\frac{\sin 2\times 45}{2\pi}\right)^{0.5} = \mathbf{155.07\ V} \qquad \text{(For } \alpha = 45°\text{)}$$

$$= \frac{230\sqrt{2}}{2}\left(\frac{\pi-\pi/2}{\pi}+\frac{\sin 2\times 90}{2\pi}\right)^{0.5} = \mathbf{115\ V} \qquad \text{(For } \alpha = 90°\text{)}$$

The average current,

$$I_{d.c.} = \frac{V_{d.c.}}{R} = \frac{88.37}{52.9} = \mathbf{1.67\ A} \qquad \text{(For } \alpha = 45°\text{)}$$

$$I_{d.c.} = \frac{V_{d.c.}}{R} = \frac{51.77}{52.9} = \mathbf{0.98\ A} \qquad \text{(For } \alpha = 90°\text{)}$$

The r.m.s. value of current,

$$I_{r.m.s.} = \frac{V_{r.m.s.}}{R} = \frac{155.07}{52.9} = \mathbf{2.93\ A} \qquad \text{(For } \alpha = 45°\text{)}$$

$$I_{r.m.s.} = \frac{V_{r.m.s.}}{R} = \frac{115.0}{52.9} = \mathbf{2.17\ A} \qquad \text{(For } \alpha = 90°\text{)}$$

The power absorbed by the heater,

$$P = V_{r.m.s.}\ I_{r.m.s.} = 155.07 \times 2.93 = \mathbf{454.36\ W} \qquad \text{(For } \alpha = 45°\text{)}$$

$$P = V_{r.m.s.}\ I_{r.m.s.} = 115.0 \times 2.17 = \mathbf{249.55\ W} \qquad \text{(For } \alpha = 90°\text{)}$$

2. A single-phase 230-V, 50-Hz supply is rectified by a continuously fired half-wave converter and is fed to a d.c. motor. The armature resistance of the motor is 8 Ω and its armature inductance is negligible. Assume that the back e.m.f. of the motor is 150 V. Find the power supplied to the motor and supply power factor.

Solution

The thyristor conducts when the instantaneous value of source e.m.f. is more than the back e.m.f. of the motor. Hence, the instantaneous value of source current,

$$i_o = \frac{V_m \sin\omega t - E}{R} \qquad (\theta_1 < \omega t < \theta_2)$$

$$= 0 \qquad \qquad \text{in rest of the cycle}$$

θ_1 and θ_2 are given by

$$V_m \sin\omega t = E$$

or $\theta_1 = \sin^{-1}\left(\dfrac{150}{230\sqrt{2}}\right) = 27.46° = 0.479 \text{ rad}$

$\theta_2 = 180 - 27.46 = 152.54° = 2.662 \text{ rad}$

The average value of current,

$$I_{d.c.} = \dfrac{1}{2\pi R}\int_{0.479}^{2.662}(V_m \sin\omega t - E)d\omega t$$

$$= \dfrac{1}{2\times 8\pi}[-V_m \cos\omega t - E\omega t]_{0.4796}^{2.662}$$

$$= \dfrac{1}{2\times 8\pi}\left[230\sqrt{2}(\cos 0.479 - \cos 2.662) - 150(2.662 - 0.479)\right]$$

$$= 4.969 \text{ A}$$

The r.m.s value of current,

$$I_{r.m.s.} = \dfrac{1}{\sqrt{2\pi}R}\left[\int_{0.479}^{2.662}(V_m^2 \sin^2\omega t - 2V_m E \sin\omega t + E^2)\right]^{0.5}$$

$$= \dfrac{1}{\sqrt{2\pi}\times 8}\left[\left\{\omega t\left(\dfrac{V_m^2}{2} + E^2\right) - \dfrac{V_m^2 \sin 2\omega t}{2\times 2} + 2V_m E \cos\omega t\right\}_{0.479}^{2.662}\right]^{0.5}$$

$$= 9.291 \text{ A}$$

Mechanical power developed in the motor,

$$P_M = EI_{d.c.} = 150\times 4.969 = 745.35 \text{ W}$$

Copper loss in the motor

$$P_{cu} = I_{r.m.s.}^2 R = (9.291)^2 \times 8 = 690.58 \text{ W}$$

Total power supplied to the motor,

$$P = 745.35 + 690.58 = \mathbf{1435.93 \text{ W}}$$

Supply power factor,

$$\text{pf} = \dfrac{1435.93}{230\times 9.291} = \mathbf{0.672}$$

3. A single-phase half-wave converter is supplied by a transformer whose secondary voltage is 230 V, 50 Hz. The converter is used to supply an inductive load of resistance of 5 Ω and inductance of 2 mH. The firing angle is 45° and the extinction angle is 200°. Find the average and r.m.s values of load voltage, and the average value of load current. Find the circuit turn-off time.

Solution

From Eq. (9.19), the average value of load voltage,

$$V_{d.c.} = \frac{V_m}{2\pi}(\cos\alpha - \cos\beta)$$

$$= \frac{230\sqrt{2}}{2\pi}(\cos 45 - \cos 200) = \mathbf{85.25\ V}$$

From Eq. (9.21), the r.m.s. value of load voltage,

$$V_{r.m.s.} = \frac{V_m}{2\sqrt{\pi}}\sqrt{(\beta - \alpha) + \frac{\sin 2\alpha - \sin 2\beta}{2}}$$

$$= \frac{230\sqrt{2}}{2\sqrt{\pi}}\left[\left(\frac{200\pi}{180} - \frac{45\pi}{180}\right) + \frac{\sin 2\times 45 - \sin 2\times 200}{2}\right]^{0.5}$$

$$= \mathbf{155.82\ V}$$

The average value of load current,

$$I_{d.c.} = \frac{V_{d.c.}}{R} = \frac{85.25}{5} = \mathbf{17.05\ A}$$

From Fig. 9.5, the turn-off time

$$T_{off} = \frac{2\pi - \beta}{\omega} = \frac{2\pi - (200\pi/180)}{2\pi \times 50} = \mathbf{8.889\ ms}$$

4. The converter of Solved Problem-3 is used to charge a battery of voltage 150 V through a resistance of 5 Ω in series with an inductance of 2 mH. Find the average values of load voltage and current. Find the circuit turn-off time.

Solution

From Eq. (9.32), the average charging current,

$$I_{d.c.} = \frac{1}{2\pi R}(V_m(\cos\alpha - \cos\beta) - E(\beta - \alpha))$$

$$= \frac{1}{2\pi \times 5}\left(230\sqrt{2}(\cos 45 - \cos 200) - 150(200 - 45)\times \frac{\pi}{180}\right)$$

$$= \mathbf{4.134\ A}$$

The average load voltage,

$$V_{d.c.} = I_{d.c.}R + E = 4.134 \times 5 + 150 = \mathbf{170.67\ V}$$

The minimum value of firing angle,

$$\alpha_{min} = \sin^{-1}(E/V_m) = \sin^{-1}(150/230\sqrt{2})$$

$$= \mathbf{27.46°}$$

The turn-off time,

$$T_{off} = \frac{2\pi + \alpha_{min} - \beta}{\omega} = \frac{360 + 27.46 - 200}{100\pi} \times \frac{\pi}{180}$$

$$= 10.41 \text{ ms}$$

5. (a) A single-phase full-wave converter uses a centre tapped transformer. The thyristor peak forward voltage rating is 1050 V. It feeds a load of resistance of 30 Ω. If the firing angle is 40°, find the average and r.m.s. values of load voltage and current. Use factor of safety of 2.5. (b) Use same thyristor in a single-phase bridge converter for feeding the same load. Find the average and r.m.s. values of load voltage and current.

Solution

(a) The peak inverse voltage in this case is $2V_m$. Hence, the converter can be designed for maximum value,

$$V_m = \frac{1050}{2 \times 2.5} = 210 \text{ V}$$

From Eq. (9.41), the average value of voltage,

$$V_{d.c.} = \frac{V_m}{\pi}(1+\cos\alpha) = \frac{210}{\pi}(1+\cos 40)$$

$$= 118.05 \text{ V}$$

The average value of load current,

$$I_{d.c.} = \frac{V_{d.c.}}{R} = \frac{118.05}{30} = 3.93 \text{ A}$$

From Eq. (9.43), the r.m.s. value of voltage,

$$V_{r.m.s.} = V_m \sqrt{\frac{\pi - \alpha}{2\pi} + \frac{\sin 2\alpha}{4\pi}}$$

$$= 210 \sqrt{\frac{\pi - 40\pi/180}{2\pi} + \frac{\sin 80}{4\pi}} = 143.5 \text{ V}$$

The r.m.s. value of load current,

$$I_{d.c.} = \frac{V_{r.m.s.}}{R} = \frac{143.5}{30} = \textbf{4.783 A}$$

(b) The peak inverse voltage in this case is V_m. Hence, the converter can be designed for maximum value,

$$V_m = \frac{1050}{2.5} = 420 \text{ V}$$

The average value of voltage,

$$V_{d.c.} = \frac{V_m}{\pi}(1+\cos\alpha) = \frac{420}{\pi}(1+\cos 40)$$

$$= 236.1 \text{ V}$$

The average value of load current,

$$I_{d.c.} = \frac{V_{d.c.}}{R} = \frac{236.1}{30} = 7.86 \text{ A}$$

The r.m.s. value of voltage,

$$V_{r.m.s.} = V_m \sqrt{\left(\frac{\pi-\alpha}{2\pi} + \frac{\sin 2\alpha}{4\pi}\right)}$$

$$= 420\sqrt{\frac{\pi - 40\pi/180}{2\pi} + \frac{\sin 80}{4\pi}} = 287.0 \text{ V}$$

The r.m.s. value of load current,

$$I_{d.c.} = \frac{V_{r.m.s.}}{R} = \frac{287.0}{30} = 9.5667 \text{ A}$$

6. Thyristors having a peak voltage rating of 1000 V and average forward current of 40 A are available for building converter circuits. Determine the average power output for single-phase full-wave (a) midpoint converter and (b) bridge converter. The factor of safety of 2.0 may be taken for both current and voltage ratings.

Solution

(a) The peak inverse voltage in this case is $2V_m$. Hence, the converter can be designed for maximum value,

$$V_m = \frac{1000}{2 \times 2.0} = 250 \text{ V}$$

From Eq. (9.41), the maximum average value of voltage is for $\alpha = 0$,

$$V_{d.c.} = \frac{V_m}{\pi}(1+\cos 0) = \frac{250 \times 2}{\pi}$$

$$= 159.15 \text{ V}$$

The average value of load current,

$$I_{d.c.} = \frac{\text{Current rating}}{\text{Factor of safety}} = \frac{40}{2} = 20 \text{ A}$$

The average output power,

$$P_{d.c.} = V_{d.c.} I_{d.c.} = 159.15 \times 20 = \textbf{3.183 kW}$$

(b) The peak inverse voltage in this case is V_m. Hence, the converter can be designed for maximum value,

$$V_m = \frac{1000}{2.0} = 500 \text{ V}$$

From Eq. (9.41), the maximum average value of voltage,

$$V_{d.c.} = \frac{V_m}{\pi}(1+\cos 0) = \frac{500 \times 2}{\pi}$$

$$= 318.3 \text{ V}$$

The average value of load current,

$$I_{d.c.} = \frac{\text{Current rating}}{\text{Factor of safety}} = \frac{40}{2} = 20 \text{ A}$$

The average output power,

$$P_{d.c.} = V_{d.c.} I_{d.c.} = 318.3 \times 20 = \mathbf{6.366 \text{ kW}}$$

7. A single-phase full-controlled bridge converter is used to charge a battery of 120 V through an inductive load of $R = 0.5\ \Omega$ and $L = 2$ mH. The input supply is 220 V, 50 Hz, and the average current of 12 A is constant over the working range. Calculate the firing angle (a) when the battery is connected in charging mode and (b) when the polarity of the battery is reversed. Also calculate the input power factor in each case.

Solution

The average output voltage for RL load,

$$V_{d.c.} = \frac{2V_m}{\pi}\cos\alpha = \frac{2\sqrt{2}\times 220}{\pi}\cos\alpha = 198.07\cos\alpha$$

(a) KVL equation of the load circuit,
$$198.07\cos\alpha = E + I_{d.c.}R = -120 + 12 \times 0.5 = 126.0$$

$$\therefore \quad \alpha = \cos^{-1}\left(\frac{126.0}{198.07}\right) = 50.5°$$

(b) KVL equation of the load circuit,
$$198.07\cos\alpha = -E + I_{d.c.}R = -120 + 12 \times 0.5 = -114.0$$

$$\therefore \quad \alpha = \cos^{-1}\left(\frac{-114.0}{198.07}\right) = 125.14°$$

The power factor is given by

$$pf = \frac{\text{Output power}}{\text{Input VA}}$$

Input VA $= 220 \times 12 = 2640$

(a) Output power = $120 \times 12 + 12^2 \times 0.5 = 1512.0$

$$pf = \cos\phi = \frac{1512}{2640} = \mathbf{0.573}$$

(b) Output power = $120 \times 12 - 12^2 \times 0.5 = 1368.0$

$$pf = \cos\phi = \frac{1368}{2640} = \mathbf{0.518}$$

In case (a) the power is being supplied to the battery, and hence the circuit is operating as a converter. In case (b) the battery is supplying power to the supply, and hence the circuit is operating as inverter.

8. A highly inductive load is supplied by a full-controlled bridge converter. The converter is supplied by a transformer. The primary is fed by a voltage source of $400 \sin \omega t$. The average value of load voltage and current are 120 V and 15 A, respectively. The firing angle is 45°. Find (a) transformer turns ratio, (b) transformer VA rating, (c) PIV of thyristor, and (d) peak and r.m.s. currents of thyristors. Assume that the forward voltage drop across the thyristor is 1.5 V.

Solution

Average voltage,

$$V_{d.c.} = \frac{2V_m}{\pi}\cos\alpha - 2 \times 1.5$$

$$120 = \frac{2V_m}{\pi}\cos 45 - 3.0$$

$$V_m = \frac{(120+3)\pi}{2\cos 45} = 273.24$$

Secondary r.m.s. voltage,

$$V_s = \frac{V_m}{\sqrt{2}} = \frac{273.24}{\sqrt{2}} = 193.2 \text{ V}$$

Primary r.m.s. voltage,

$$V_s = \frac{V_m}{\sqrt{2}} = \frac{400}{\sqrt{2}} = 282.84 \text{ V}$$

(a) Turns ratio of transformer,

$$N = \frac{282.84}{193.2} = \mathbf{1.464}$$

(b) The secondary current is constant for the cycle. Hence,

$$I_{r.m.s.} = 15 \text{ A}$$

and $VA = 193.2 \times 15 = \mathbf{2898 \text{ VA}}$

(c) **PIV = 273.24 V**

(d) The current through the thyristor flows for half cycle. Hence, the peak and r.m.s. currents,

$I_m = 15$ **A**

$$I_{r.m.s.} = \sqrt{\frac{15^2}{2}} = \frac{15}{\sqrt{2}} = \textbf{10.607 A}$$

9. A three-phase full-controlled bridge converter feeds a resistive load of $R = 80\ \Omega$. The three-phase input voltage is 400 V, 50 Hz. The power supplied to the load is 450 W. Find (a) firing angle, (b) r.m.s. value of input current, (c) input apparent power, and (d) input power factor. Assume ripple-free load current due to perfect filtering.

Solution

Peak phase voltage,

$$V_m = \frac{400\sqrt{2}}{\sqrt{3}} = 326.6\ \text{V}$$

The r.m.s. output voltage is given by

$$\frac{V_{r.m.s.}^2}{80} = 450$$

Hence, $V_{r.m.s.} = 189.74$ V

(a) The r.m.s. output voltage,

$$V_{r.m.s.} = \sqrt{3} V_m \sqrt{\frac{1}{2} + \frac{3\sqrt{3}}{4\pi}\cos 2\alpha}$$

$$\left(\frac{1}{2} + \frac{3\sqrt{3}}{4\pi}\cos 2\alpha\right) = \frac{V_{r.m.s.}^2}{3V_m^2} = \frac{189.74^2}{3 \times 326.6^2} = 0.1125$$

$$\cos 2\alpha = -0.16023$$

$$\alpha = \textbf{49.61°}$$

(b) The load current,

$$I_{d.c.} = \sqrt{\frac{450}{80}} = 2.372\ \text{A}$$

The supply current flows for $2\pi/3$ in π; hence, for continuous load current, the r.m.s. value of supply current,

$$I_{r.m.s.}^s = \sqrt{I_{d.c.}^2 \times \frac{2\pi/3}{\pi}} = 2.372\sqrt{\frac{2}{3}} = \textbf{1.937 A}$$

(c) The input apparent power,

$$\sqrt{3}VI = \sqrt{3} \times 400 \times 1.937 = \textbf{1342 VA}$$

(d) Assuming no loss the actual power supplied by the source,

$P_s = 450$ W

Hence, the input power factor,

$$\text{pf} = \frac{450}{1342} = \mathbf{0.335}$$

10. A three-phase semiconverter is supplied by a 440-V, 50-Hz, three-phase supply. The load resistance is 15 Ω. The average output voltage is 50% of maximum possible output voltage. Find (a) firing angle, (b) average output current, (c) r.m.s. output current and voltage, (d) average and r.m.s. thyristor current.

Solution

The peak value of phase voltage,

$$V_m = \frac{440\sqrt{2}}{\sqrt{3}} = 359.26 \text{ V}$$

(a) The maximum possible output is for $\alpha = 0$. Hence, the maximum average output voltage,

$$V_d^m = \frac{3\sqrt{3}V_m}{2\pi}(1+\cos 0) = \frac{3\sqrt{3} \times 359.26}{\pi}$$

$$= 594.21 \text{ V}$$

The required average output voltage,

$$V_{d.c.} = 0.5 \times 594.21 = 297.1 \text{ V}$$

Hence, $297.1 = \dfrac{3\sqrt{3} \times 359.26}{2\pi}(1+\cos\alpha)$

or $\quad \alpha = \pi/2$

The firing angle is more than $\pi/3$; therefore, the output voltage waveform is discontinuous.

(b) For discontinuous output voltage waveform, the average output voltage is same as the continuous output voltage waveform. Hence,

$$V_{d.c.} = 297.1 \text{ V}$$

and $\quad I_{d.c.} = \dfrac{V_{d.c.}}{R} = \dfrac{297.1}{15} = \mathbf{19.806 \text{ A}}$

(c) For discontinuous output voltage, the r.m.s. output voltage, from Eq. (9.63),

$$V_{r.m.s.} = \sqrt{3}V_m\left[\frac{3}{4\pi}\left(\pi - \alpha + \frac{\sin 2\alpha}{2}\right)\right]^{0.5}$$

$$= \sqrt{3} \times 359.26\left[\frac{3}{4\pi}\left(\pi - \frac{\pi}{2} + \frac{\sin \pi}{2}\right)\right]^{0.5} = \mathbf{381.05 \text{ V}}$$

and $I_{r.m.s.} = \dfrac{V_{r.m.s.}}{R} = \dfrac{381.05}{15} = 25.403 \text{ A}$

(d) Average thyristor current,

$$I_{d.c.}^{Th} = \dfrac{I_{d.c.}}{3} = \dfrac{19.806}{3} = 6.602 \text{ A}$$

The r.m.s. value of thyristor current,

$$I_{r.m.s.}^{Th} = \dfrac{I_{r.m.s.}}{\sqrt{3}} = \dfrac{25.403}{\sqrt{3}} = 14.67 \text{ A}$$

11. In a series inverter the load resistance $R = 25\,\Omega$ is connected in series with commutating elements of $L = 7.5$ mH and $C = 1.5\,\mu$F. Check whether the circuit works as a series inverter. Take $E_C = 0$. Determine the voltage across the capacitor and inductor at the time of commutation. The input voltage is 110 V.

Solution

Here

$$R^2 = 25 \times 25 = 625 \text{ and } \dfrac{4L}{C} = \dfrac{4 \times 7.5 \times 10^{-3}}{1.5 \times 10^{-6}} = 20000$$

Since, $R^2 < \dfrac{4L}{C}$, the circuit is underdamped and hence it works as a series inverter. Also,

$$\xi = \dfrac{R}{2L} = \dfrac{25}{2 \times 7.5 \times 10^{-3}} = 1667$$

$$\omega_r = \sqrt{\dfrac{1}{LC} - \dfrac{R^2}{4L^2}} = \sqrt{\dfrac{1}{7.5 \times 10^{-3} \times 1.5 \times 10^{-6}} - \dfrac{25^2}{4 \times 7.5^2 \times 10^{-6}}}$$

$$= 9.28 \times 10^3 \text{ rad/s}$$

and $\omega_o = \dfrac{1}{\sqrt{LC}} = \dfrac{1}{\sqrt{7.5 \times 10^{-3} \times 1.5 \times 10^{-6}}} = 9.43 \times 10^3 \text{ rad/s}$

From Eq. (9.71), the voltage across the capacitor at $t = (\pi/\omega_r)$ and for $E_C = 0$,

$$V_C = V_i \left(1 + e^{-\pi\xi/\omega_r}\right) = 110\left(1 + e^{-\pi \times 1.667/9.28}\right)$$

$$= 172.56 \text{ V}$$

The voltage across the inductor, as the voltage across the resistance is zero, is given by

$$V_L = V_i - V_C = 110 - 172.56 = -62.56 \text{ V}$$

12. For a series inverter $R = 80\,\Omega$, $L = 5$ mH, and $C = 1.5\,\mu$F. Find the output frequency. If the load resistance is varied from 50 to 110 Ω, find the range of the output frequency. Take $T_{off} = 0.2$ ms.

Solution

Here

$$R^2 = 80 \times 80 = 6400 \text{ and } \frac{4L}{C} = \frac{4 \times 5 \times 10^{-3}}{1.5 \times 10^{-6}} = 13333$$

Since, $R^2 < (4L/C)$, the circuit is underdamped, and hence it works as a series inverter. The damped frequency of the RLC circuit,

$$\omega_r = \sqrt{\frac{1}{LC} - \frac{R^2}{4L^2}} = \sqrt{\frac{1}{5 \times 10^{-3} \times 1.5 \times 10^{-6}} - \frac{80^2}{4 \times 5^2 \times 10^{-6}}}$$

$$= 8.33 \times 10^3 \text{ rad/s}$$

Hence, time 0a in Fig. 9.32,

$$t = \frac{\pi}{\omega_r} = \frac{\pi}{8.33} = 0.377 \text{ ms}$$

and the output frequency,

$$f_o = \frac{1}{2(t + T_{off})} = \frac{10^3}{2(0.377 + 0.2)} = 866.55 \text{ Hz}$$

For $R = 50 \, \Omega$,

$$\omega_r = \sqrt{\frac{1}{LC} - \frac{R^2}{4L^2}} = \sqrt{\frac{1}{5 \times 10^{-3} \times 1.5 \times 10^{-6}} - \frac{50^2}{4 \times 5^2 \times 10^{-6}}}$$

$$= 10.41 \times 10^3 \text{ rad/s}$$

Hence, time 0a in Fig. 9.32,

$$t = \frac{\pi}{\omega_r} = \frac{\pi}{10.41} = 0.302 \text{ ms}$$

and the output frequency,

$$f_o = \frac{1}{2(t + T_{off})} = \frac{10^3}{2(0.302 + 0.2)} = \mathbf{996.02 \text{ Hz}}$$

For $R = 110 \, \Omega$,

$$\omega_r = \sqrt{\frac{1}{LC} - \frac{R^2}{4L^2}} = \sqrt{\frac{1}{5 \times 10^{-3} \times 1.5 \times 10^{-6}} - \frac{110^2}{4 \times 5^2 \times 10^{-6}}}$$

$$= 3.51 \times 10^3 \text{ rad/s}$$

Hence, time 0a in Fig. 9.32,

$$t = \frac{\pi}{\omega_r} = \frac{\pi}{3.51} = 0.895 \text{ ms}$$

and the output frequency,

$$f_o = \frac{1}{2(t+T_{off})} = \frac{10^3}{2(0.895+0.2)} = 456.62 \text{ Hz}$$

Range of output frequency = **456.62 Hz to 996.02 Hz**.

13. The circuit parameters of a series inverter are $R = 2\,\Omega$, $L = 40\,\mu H$, and $C = 5\,\mu F$. The output frequency is 5 kHz. (a) Find the circuit turn-off time. (b) For a factor of safety of 2, determine the maximum possible output frequency. The thyristor turn-off time is 25 μs.

Solution

(a) The output angular frequency,

$$\omega = 2\pi \times 5000 = 31.42 \times 10^3 \text{ rad/s}$$

$$\xi = \frac{R}{2L} = \frac{2}{2 \times 40 \times 10^{-6}} = 25 \times 10^3$$

$$\omega_o = \frac{1}{\sqrt{LC}} = \frac{1}{\sqrt{40 \times 10^{-6} \times 5 \times 10^{-6}}} = 70.71 \times 10^3 \text{ rad/s}$$

and $\omega_r = \sqrt{\omega_o^2 - \xi^2} = \sqrt{70.71^2 - 25.0^2} \times 10^3 = 66.14 \times 10^3 \text{ rad/s}$

The circuit turn-off time,

$$T_{off}^c = \frac{\pi}{\omega} - \frac{\pi}{\omega_r} = \pi\left(\frac{10^{-3}}{31.42} - \frac{10^{-3}}{66.14}\right) = \mathbf{52.5\,\mu s}$$

The circuit turn-off time is more than the thyristor turn-off time; hence, the inverter operates satisfactorily.

When the circuit turn-off time (or dead zone) is just equal to the thyristor turn-off time × factor of safety (= 25 × 2 = 50 μs), the maximum possible output frequency is given by

$$50 \times 10^{-6} = \left(\frac{\pi}{2\pi f_{max}} - \frac{\pi \times 10^{-6}}{0.06614}\right)$$

$$f_{max} = \frac{10^6}{2 \times 97.5} = \mathbf{5.128\text{ kHz}}$$

14. A single-phase half-bridge inverter has input voltage $V_i = 110$ V and resistive load $R = 5\,\Omega$. Determine the (a) r.m.s. value of fundamental component of the output voltage, (b) output power, (c) peak current in each thyristor, (d) average current of each thyristor, and (e) peak reverse blocking voltage.

Solution

(a) The output voltage is a square wave with amplitude of $V_i/2 = 110/2 = 55$ V. Since the wave is an odd periodic function, it has only sine terms. Also, as it is antisymmetrical about the vertical axis, it contains only odd harmonics. Hence, the amplitude of the fundamental is given by

$$V_{1m} = \frac{2V_i}{T}\int_0^{T/2} \sin\omega t\, dt$$

$$= \frac{2V_i}{\pi} = \frac{2\times 110}{\pi} = 70 \text{ V}$$

Hence, the r.m.s. value of fundamental output voltage,

$$V_{r.m.s.} = \frac{V_{1m}}{\sqrt{2}} = \frac{70}{\sqrt{2}} = 49.5 \text{ V}$$

(b) The r.m.s value of square wave,
$$V_o = 55 \text{ V}$$
Hence, the output power
$$P_o = \frac{V_o^2}{R} = \frac{55^2}{5} = 605 \text{ W}$$

(c) Peak current in each thyristor = Load current
$$I_{Th}^m = \frac{V_o}{R} = \frac{55}{5} = 11 \text{ A}$$

(d) Each thyristor conducts for 50% of time; hence, the average value of current,
$$I_{av} = \frac{11}{2} = 5.5 \text{ A}$$

(e) Peak reverse blocking voltage $= 2\times V_o = 2\times 55 = 110 \text{ V}$

15. Repeat Solved Problem-14 for a single-phase full-bridge inverter.

Solution

(a) The output voltage is a square wave with amplitude of $V_i = 110$ V. Since the wave is an odd periodic function, it has only sine terms. Also, as it is antisymmetrical about the vertical axis, it contains only odd harmonics. Hence, the amplitude of the fundamental is given by

$$V_{1m} = \frac{4V_i}{T}\int_0^{T/2} \sin\omega t\, dt$$

$$= \frac{4V_i}{\pi} = \frac{4\times 110}{\pi} = 140 \text{ V}$$

Hence, the r.m.s. value of fundamental output voltage,

$$V_{r.m.s.} = \frac{V_{1m}}{\sqrt{2}} = \frac{140}{\sqrt{2}} = 99.0 \text{ V}$$

(b) The r.m.s. value of square wave,
$$V_o = 110 \text{ V}$$

Hence, the output power,

$$P_o = \frac{V_o^2}{R} = \frac{110^2}{5} = \textbf{2420 W}$$

(c) Peak current in each thyristor = Load current

$$I_{Th}^m = \frac{V_o}{R} = \frac{110}{5} = \textbf{22 A}$$

(d) Each thyristor conducts for 50% of time; hence, the average value of current

$$I_{av} = \frac{22}{2} = \textbf{11.0 A}$$

(e) Peak reverse blocking voltage = V_o = **110 V**

16. An integral cycle-control a.c. voltage controller is fed by 220-V, 50-Hz supply and feeds a resistive load of 14 Ω. It conducts for 10 cycles and remains off for 30 cycles. Find (a) duty cycle, (b) r.m.s. output voltage, (c) input power factor, and (d) average and r.m.s. values of thyristor currents.

Solution

(a) Duty cycle,

$$\delta = \frac{n}{n+m} = \frac{10}{10+30} = \textbf{0.25}$$

(b) The r.m.s. value of output voltage,

$$V_o = 220 \times \sqrt{0.25} = \textbf{110 V}$$

The power output,

$$P_o = \frac{V_o^2}{R} = \frac{110^2}{14} = 864.3 \text{ W}$$

Since losses are neglected, the input power,

$$P_i = 864.3 \text{ W}$$

The r.m.s. value of load current,

$$I_o = \frac{110}{14} = 7.86 \text{ A}$$

$$I_i = 7.86 \text{ A}$$

and input VA = 220 × 7.86 = 1728.57 VA

(c) The input power factor,

$$\text{pf} = \frac{P}{VA} = \frac{864.3}{1728.57} = \textbf{0.5}$$

(d) Peak value of thyristor current,

$$I_{Th}^m = \frac{\sqrt{2} \times 220}{14} = \mathbf{22.22 \text{ A}}$$

The thyristor current flows for only one half of a cycle, that is, during the positive half, upper thyristor conducts, whereas during the negative half, current flows through the lower thyristor. Hence, the average value of current through each thyristor,

$$I_{av} = \frac{\delta}{T}\int_0^{T/2} 22.22 \sin \omega t \, dt$$

$$= \frac{22.22 \times 0.25}{\pi} = \mathbf{1.77 \text{ A}}$$

and the r.m.s. value of thyristor current,

$$I_{Th} = \sqrt{\frac{\delta}{T}\int_0^{T/2} 22.22^2 \sin^2 \omega t \, dt}$$

$$= \frac{22.22\sqrt{0.25}}{2} = \mathbf{5.555 \text{ A}}$$

17. A single-phase half-wave a.c. regulator is supplied by a 220-V, 50-Hz supply and feeds a resistive load of 10 Ω. The firing angle of thyristor is 45° in each positive half cycle. Determine the (a) average output voltage, (b) r.m.s. output voltage, (c) output power, (d) input power, (e) input power factor, and (f) average input current over one cycle.

Solution

(a) From Eq. (9.76), the average value of output voltage,

$$V_{av} = \frac{V_m}{2\pi}(\cos\alpha - 1) = \frac{\sqrt{2} \times 220}{2\pi}(\cos 45° - 1)$$

$$= \mathbf{-14.5 \text{ V}}$$

The average output voltage is negative, as only a part of positive half of the cycle appears at the output, whereas the whole negative half of the cycle appears at the output.

(b) From Eq. (9.77), the r.m.s. value of output voltage,

$$V = \frac{V_m}{2\sqrt{\pi}}\sqrt{\left(2\pi - \alpha + \frac{\sin 2\alpha}{2}\right)}$$

$$= \frac{\sqrt{2} \times 220}{2\sqrt{\pi}}\sqrt{\left(2\pi - \frac{45\pi}{180} + \frac{\sin 90°}{2}\right)} = \mathbf{214.95 \text{ V}}$$

(c) Output power,

$$P_o = \frac{V^2}{R} = \frac{214.95^2}{10} = \mathbf{4620.35 \text{ W}}$$

(d) Neglecting the losses, input power = output power; hence,

$P_i = 4620.35$ W

(e) As input current is same as the output current; hence,

$$I_i = \frac{214.95}{10} = 21.5 \text{ A}$$

Input VA = 220 × 21.5 = 4730 VA

∴ Input pf = $\frac{4620.35}{4730}$ = **0.977**

(f) As input current is same as the output current; hence,

$$I_{av} = \frac{-14.5}{10} = -1.45 \text{ A}$$

18. A single-phase full-wave a.c. voltage controller is fed from 230-V, 50-Hz supply. The load connected is resistive load of 12 Ω. The firing angle of thyristors is 55°. Determine the (a) average output voltage over half a cycle, (b) r.m.s. output voltage, (c) output power, (d) input power factor, and (e) average and r.m.s. values of thyristor currents.

Solution

(a) From Eq. (9.78), the average value of output voltage over half cycle,

$$V_{av} = \frac{V_m}{\pi}(1+\cos\alpha) = \frac{230\sqrt{2}}{\pi}(1+\cos 55°)$$

= **163 V**

(b) From Eq. (9.79), the r.m.s. value of output voltage over half cycle,

$$V = \frac{V_m}{\sqrt{2\pi}}\sqrt{\left(\pi-\alpha+\frac{\sin 2\alpha}{2}\right)}$$

$$= \frac{230\sqrt{2}}{\sqrt{2\pi}}\sqrt{\left(\pi-\frac{55\pi}{180}+\frac{\sin 110°}{2}\right)} = \textbf{211.3 V}$$

(c) Output power,

$$P_o = \frac{V^2}{R} = \frac{211.3^2}{12} = \textbf{3720.6 W}$$

(d) Neglecting the losses, input power = output power; hence,

$P_i = 3720.6$ W

As the input current is same as the output current; hence,

$$I_i = \frac{211.3}{12} = 17.6 \text{ A}$$

Input $VA = 230 \times 17.6 = 4050$ VA

$\therefore \quad$ Input pf $= \dfrac{3720.6}{4050} = 0.92$

(e) As the current through each thyristor flows only for half of a cycle, the average value of thyristor current,

$$I_{av} = \dfrac{163}{2 \times 12} = -6.79 \text{ A}$$

and the r.m.s. value of thyristor current,

$$I_{r.m.s.} = \dfrac{211.3}{\sqrt{2} \times 12} = -12.45 \text{ A}$$

19. A single-phase a.c. voltage controller using a triac is fed from a single-phase supply and feeds a resistive load. Find the firing angle if the power delivered to the load is (a) 75% and (b) 25% of the maximum power.

Solution

(a) The power delivered is maximum when $\alpha = 0$. When the power delivered is $X\%$ of the maximum power, then the voltage output is $\sqrt{0.01X}\, V$, where V is the supply voltage. Hence, from Eq. (9.79),

$$\sqrt{0.01X}\, V = \dfrac{V}{\sqrt{\pi}} \sqrt{\pi - \alpha + \dfrac{\sin 2\alpha}{2}}$$

or $0.01X\pi = \pi - \alpha + \dfrac{\sin 2\alpha}{2}$

When $X = 75$,

$\sin\theta - \theta = -1.57$ $\hfill (\theta = 2\alpha)$

$\therefore \quad \alpha = \mathbf{66.2°}$

(b) When $X = 25$,

$\sin\theta - \theta = -4.712$ $\hfill (\theta = 2\alpha)$

$\therefore \quad \alpha = \mathbf{114°}$

20. A single-phase full-wave a.c. voltage controller is fed from a single-phase supply of 220 V, 50 Hz and feeds an inductive load having $R = 4\,\Omega$ and $L = 18$ mH. Determine the (a) control range of firing angle, (b) conduction period of each thyristor if $\alpha = \phi$, and (c) maximum possible r.m.s. load current and corresponding output power and input power factor.

Solution

From Eq. (9.82), the power factor angle of the load,

$$\phi = \tan^{-1}\left(\dfrac{\omega L}{R}\right) = \tan^{-1}\left(\dfrac{2\pi \times 50 \times 18 \times 10^{-3}}{4}\right)$$

$= \mathbf{54.73°}$

(a) Hence, the minimum value $\alpha = \phi = 54.73°$ and the range of firing angle is given by

$$54.73° < \alpha < 180°$$

(b) For $\alpha = \phi$, the conduction period of each thyristor is 180°.

(c) The maximum r.m.s. value of load current is for $\alpha = \phi$ and for this value of firing angle, conduction angle $\beta = 180° + \phi$. From Eq. (9.89), the maximum value of r.m.s. load current for this value of α and β,

$$V = \frac{V_m}{\sqrt{2}} = \frac{220\sqrt{2}}{\sqrt{2}} = 220 \text{ V}$$

and the maximum value of load current,

$$I_m = \frac{V}{Z} = \frac{220}{\sqrt{4^2 + (2\pi \times 18 \times 10^{-3})^2}} = \textbf{31.76 A}$$

21. A three-phase a.c. voltage controller is fed from a 440-V, 50-Hz supply and feeds a balance resistive load of 18 Ω in each phase. The active device is two thyristors in antiparallel in each phase. The firing angle is 30°. Determine the (a) r.m.s. value of output voltage per phase, (b) output power, (c) line current, and (d) input power factor.

Solution

(a) Input voltage per phase

$$V_{ph} = \frac{440}{\sqrt{3}} = 254 \text{ V}$$

Since $0 \leq \alpha \leq 60°$, from Eq.(9.93), the r.m.s. value of output voltage per phase,

$$V_o = \sqrt{3} V_m \sqrt{\frac{1}{\pi}\left(\frac{\pi}{6} - \frac{\alpha}{4} + \frac{\sin 2\alpha}{8}\right)}$$

$$= \sqrt{3}\sqrt{2} \times 254 \sqrt{\frac{1}{\pi}\left(\frac{\pi}{6} - \frac{\pi}{4 \times 6} + \frac{\sin 60}{8}\right)}$$

$$= \textbf{248.45 V}$$

(b) Output power

$$P_o = 3 \times \frac{248.45^2}{18} = 10288 \text{ W} = \textbf{10.288 kW}$$

(c) Line current is same as phase current; hence,

$$I_L = \frac{254}{18} = \textbf{14.11 A}$$

(d) Input VA = $3 \times 254 \times 14.11 = 10752.7$ VA = 10.7527 kVA

$$\text{Input pf} = \frac{10288}{10752.7} = \textbf{0.957}$$

22. In a d.c. voltage regulator, the output voltage at full load is 25 V. When the load is removed the output voltage rises to 25.5 V. (a) Calculate the percentage voltage regulation. (b) If percentage regulation is 1.5%, find the no-load voltage if full-load voltage remains same.

Solution

(a) Percentage regulation $= \dfrac{\text{No-load voltage} - \text{full-load voltage}}{\text{Full-load voltage}}$

$= \dfrac{25.5 - 25.0}{25.0} \times 100 = \mathbf{2\%}$

(b) $1.5 = \dfrac{V_{NL} - 25.0}{25.0} \times 100$

or $V_{NL} = \dfrac{1.5 \times 25.0}{100} + 25.0 = \mathbf{25.375\ V}$

23. The input and output voltages of a d.c. chopper are 110 V and 80 V. It operates on constant frequency principle with a frequency of 1 kHz. Find the period of conduction and blocking in each cycle.

Solution

The time period,

$$T = \dfrac{1}{f} = \dfrac{1}{10^3} = 1\ \text{ms}$$

The average value of output voltage,

$$V_o = V_i \times \dfrac{T_{on}}{T}$$

The period of conduction,

$$T_{on} = \dfrac{V_o T}{V_i} = \dfrac{80 \times 10^{-3}}{110} = \mathbf{0.727\ ms}$$

The period of blocking,

$$T_{off} = T - T_{on} = 1 - 0.727 = \mathbf{0.273\ ms}$$

24. A chopper with the input voltage of 110 V is used to supply a d.c. series motor. The armature and field resistances of the motor are 0.05 Ω and 0.04 Ω, respectively. The average current of the motor is 12 A, while its back e.m.f. is 65 V. The chopper frequency is 1 kH. Find the period of conduction and blocking.

Solution

The average output of the chopper is the input voltage to the motor. Hence, the average value of the chopper,

$$V_{av} = 65 + 12(0.05 + 0.04) = 66.08\ V$$

But $V_{av} = \dfrac{T_{on}}{T} V_i = T_{on} f\, V_i$

The period of conduction, therefore, is given by

$$T_{on} = \dfrac{V_{av}}{V_i f} = \dfrac{66.08}{110 \times 10^3} = \textbf{0.6 ms}$$

The period of blocking,

$$T_{off} = T - T_{on}$$

$$= \dfrac{1}{10^3} - 0.6 \times 10^{-3} = \textbf{0.4 ms}$$

25. A separately excited d.c. motor rated at 220 V, 20 A has a rated speed of 800 rpm. The armature resistance is 1.0 Ω. The speed is controlled by adding a chopper between the supply of 220 V and the motor. It is desired to run the motor at 600 rpm. (a) Find the duty cycle if the torque developed is the rated torque. (b) Find the duty cycle if the torque developed is half the rated torque.

Solution

Back e.m.f. at 800 rpm $E = 220 - 20 \times 1 = 200$ V.

Since back e.m.f. is proportional to speed, back e.m.f. at 600 rpm is given by

$$E' = \dfrac{600}{800} \times 200 = \textbf{150 V}$$

(a) Torque $T = \Phi I_a$. Since Φ in a separately excited motor is constant, the torque is proportional to I_a. As the torque is same, the current even at 600 rpm is 20 A. The required input voltage to the motor or the output voltage of the chopper,

$$V_{av} = 150 + 20 \times 1 = 170 \text{ V}$$

Hence, the duty cycle $\delta = \dfrac{170}{220} = \textbf{0.773.}$

(b) The torque developed is half of the rated torque; hence, the motor current is also half of the rated current, that is, $Ia = 10$ A. Hence,

$$V_{av} = 150 + 10 \times 1 = 160 \text{ V}$$

and duty cycle $\delta = \dfrac{160}{220} = \textbf{0.727}$

26. A step up chopper has an input and output voltages of 220 V and 265 V, respectively. The blocking period in each cycle is 0.5 ms. Find the period of conduction in each cycle.

Solution

From Eq. (9.110),

$$V_o = \frac{T}{T_{off}} V$$

or $\quad T = \dfrac{V_o}{V} T_{off} = \dfrac{265}{220} \times 0.5 = 0.6$ ms

The period of conduction,

$$T_{on} = 0.6 - 0.5 = \mathbf{0.1 \text{ ms}}$$

9.11 EXERCISES

1. A single-phase 110-V, 50-Hz supply is used to charge a 60-V battery through a continuously fired half-wave converter and a 11-Ω resistance. Find the average and r.m.s. values of currents, the power supplied to the battery, the power dissipated in the resistance, the total power supplied by the supply, and the supply power factor.

2. A single-phase half-wave converter is fed by a transformer whose secondary voltage is 400 sin ωt. The firing angle is 60°. The load is resistive with $R = 15\ \Omega$. (a) Find the average values of voltage and current, and the r.m.s. values of voltage and current. (b) Determine $P_{d.c.}$, $P_{a.c.}$, and rectification efficiency. (c) Calculate form factor, ripple factor, and peak inverse voltage across the thyristor. (d) Find VA rating of the transformer and the transformer utilization factor.

3. A single-phase half-wave converter is supplied by a transformer whose secondary voltage is 400 sin 100πt. The converter is used to supply an inductive load with $R = 4\ \Omega$ and $L = 3$ mH. The firing angle is 40° and the extinction angle is 220°. Find the average and r.m.s. values of load voltage, and the average value of load current. Find the circuit turn-off time.

4. The converter of question-3 is used to charge a battery of voltage 110 V through a resistance of 4 Ω in series with an inductance of 3 mH. Find the average values of load voltage and load current. Find the circuit turn-off time.

5. A 220-V, 1000-W heater is connected to a single-phase full-controlled bridge converter. If the heater is delivering a power of 800 W, find the r.m.s. output voltage, firing angle, and r.m.s. value of load current. For a safety factor of 2, calculate the peak inverse voltage rating of the thyristor.

6. A single-phase full-controlled bridge converter is used to charge a battery of 150 V through an inductive load of $R = 0.4\ \Omega$ and $L = 3$ mH. The input supply is 230 V, 50 Hz, and the average current of 10 A is constant over the working range. Calculate the firing angle (a) when battery is connected in charging mode and (b) when the polarity of the battery is reversed. Also calculate the input power factor in each case.

7. A full-controlled bridge converter is fed by a transformer the primary of which is fed by a voltage source of 240 V, 50 Hz. The converter feeds a highly inductive load. The average value of load voltage and current are 150 V and 12 A, respectively. Firing angle is 40°. Find (a) transformer turns ratio, (b) transformer VA rating, (c) PIV of thyristor and (d) peak and r.m.s. currents of thyristors. Assume that the forward voltage drop across thyristor is 1.6 V.

8. A three-phase full converter is supplied by a 440-V, 50-Hz, three-phase supply. The load is highly inductive and the average value of load current is 100 A. The firing angle is 45°. Find (a) output power $P_{d.c.}$; (b) average, r.m.s., and peak currents through thyristors; and (c) peak inverse voltage of thyristors.

9. A three-phase full converter is supplied by a 440-V, 50-Hz, three-phase supply. The load resistance is 12 Ω. The average output voltage is 50% of the maximum possible output voltage. Find (a) firing angle, (b) average output current, (c) r.m.s. output current and voltage, (d) average and r.m.s. thyristor current.

10. A three-phase semiconverter is supplied by a 400-V, 50-Hz, three-phase supply. The load resistance is 10 Ω. Find (a) average output voltage for $\alpha = \pi/4$, (b) average output current, (c) r.m.s. output voltage and current, (d) average and r.m.s. thyristor currents.

11. In a series inverter $R = 20$ Ω, $L = 6$ mH, and $C = 1.5$ μF. Check whether the circuit is self-commutable. Take $E_C = 0$. Determine the voltage across the capacitor and inductor at the time of commutation. The input voltage is 220 V.

12. For a series inverter $R = 90$ Ω, $L = 6$ mH, and $C = 1.5$ μF. Find the output frequency. If the load resistance is varied from 45 to 125 Ω, find the range of the output frequency. Take $T_{off} = 0.2$ ms.

13. The circuit parameters of a series inverter are $R = 3$ Ω, $L = 50$ μH, and $C = 6$ μF. The output frequency is 7.5 kHz. (a) Find the circuit turn-off time. (b) For a factor of safety of 1.5, determine the maximum possible output frequency. Thyristor turn-off time is 15 μs.

14. A single-phase half-bridge inverter has input voltage $V_i = 50$ V and resistive load $R = 3$ Ω. Determine (a) r.m.s. value of fundamental component of the output voltage, (b) output power, (c) peak current in each thyristor, (d) average current of each thyristor, and (e) peak reverse blocking voltage. Repeat when the half-bridge inverter is replaced by a full-bridge inverter.

15. An integral cycle-control a.c. voltage controller conducts for 30 cycles and remains off for 70 cycles. It is fed by 200-V, 50-Hz supply and feeds a resistive load of 12 Ω. Determine (a) r.m.s. value of output voltage, (b) power output, power input, (c) input power factor, (d) average and r.m.s. values of thyristor current. Neglect losses.

16. A single-phase half-wave a.c. regulator is fired at a firing angle of thyristor of 45° in each positive half cycle. It is supplied by a 200-V, 50-Hz supply and feeds a resistive load of 15 Ω. Determine (a) average output voltage, (b) r.m.s. output voltage, (c) output power, (d) input VA, (e) input power factor, and (f) average input current over one cycle.

17. A single-phase full-wave a.c. voltage controller is fed from 220-V, 50-Hz supply and feeds a resistive load of 11 Ω. The delay angle of thyristors is 35°. Determine (a) average output voltage over half a cycle, (b) r.m.s. output voltage, (c) output power, (d) input power factor, and (e) average and r.m.s. values of thyristor currents.

18. A single-phase a.c. voltage controller using a triac is fed from 230-V, 50-Hz supply and feeds a resistive load of 15 Ω. The triac is in phase angle control mode. The firing angle is 30°. Determine (a) average output voltage over half a cycle, (b) r.m.s. output voltage, (c) output power, and (d) input power factor.

19. A single-phase full-wave a.c. voltage controller is fed from 220-V, 50-Hz supply and feeds an inductive load of $R = 4$ Ω and $L = 22$ mH. The delay angle of thyristors is 60°. Find the conduction angle of thyristors and the r.m.s. value of output voltage.

20. A three-phase a.c. voltage controller is fed from a 440-V, 50-Hz supply and feeds a three-phase heating load of 25 kW. The active device is (a) triacs in each phase and (b) two thyristors in antiparallel in each phase. Find current and voltage ratings of active devices in each case.

21. If the duty cycle of a d.c. chopper is 0.5 and the frequency of chopping is 1.5 kHz, find the output voltage. The input voltage is 120 V. Also, find on and off periods.

22. A chopper with input voltage of 120 V and frequency of 500 Hz is used to supply a d.c. shunt motor. The armature and field resistances of the motor are 0.1 Ω and 200 Ω, respectively. The average current of the motor is 15 A, whereas its back e.m.f. is 75 V. Find the period of conduction and blocking.

23. A separately excited d.c. motor, rated at 250 V, 25 A has rated speed of 750 rpm. The armature resistance is 0.5 Ω. The speed is controlled by adding a chopper between the supply and the motor. It is desired to run the motor at 500 rpm. (a) Find the duty cycle if the torque developed is the rated torque. (b) Find the duty cycle if the torque developed is half the rated torque.

24. A step up chopper has an input and output voltages of 220 V and 280 V, respectively. Find the duty cycle.

10
DIGITAL SYSTEMS

Outline

10.1 Introduction 492
10.2 Binary Logic and Logic Gates 493
10.3 Number Systems 501
10.4 Boolean Algebra 513
10.5 Simplification of Logical Functions 518
10.6 Review Questions 527
10.7 Solved Problems 528
10.8 Exercises 536

10.1 INTRODUCTION

The information in digital systems is represented by binary (two) digits. A digit in digital computer is known as a bit. A bit may assume one of two values. The two values are 1 and 0. A mathematical system that uses only two digits is called a binary system. The binary mathematical system was first investigated by the British mathematician George Boole in 1854. The binary system first found applications when Shannon developed a switching theory in 1930.

Modern digital computers, communication systems, and automatic control systems are either totally or partially digital systems. Digital systems process the discrete data. While analogue systems deal with continuously varying quantities in time, digital systems deal with discrete signals sampled at discrete time. When the sampling time is very small a digital system behaves like an analogue system.

Most of electronic switches are binary in nature and are known as binary switches or logic gates. Logic gates are employed in digital computers, communication networks, and digital control systems. The logic gates are of two types. They are either without memory or with memory. They are known as combinational logic gates or sequential logic gates, respectively.

A digital computer consists of a large number of elementary logic gates in a complex fashion so as to perform the various operations in a computer. These computers form the heart of most of the digital systems and fulfill the complex functions of information processing.

Because of the various advantages of the digital systems, every field of engineering, nowadays, is working in digital mode. Various advantages of digital systems are listed below:

1. *Accurate representation and manipulation of data.* Data can be represented and manipulated to any desired accuracy by digital techniques. Digital systems can also acquire data from the physical systems with greater accuracy and resolution.
2. *Easy manipulation of multiple information.* It is easy to manipulate information coming from different sources of a large complex system. This is possible because all the information, irrespective of its nature, can be converted into numbers with desired accuracy. These numbers can be manipulated by an arithmetic logic unit.
3. *Greatly reduced errors.* The chances of errors are greatly reduced in digital systems. The information is represented by sequences of 1's and 0's. During the transmission of information the sequences of 1's and 0's representing the information does not change. Hence, the original information remains unaffected.
4. *Easy storage of the data.* In digital systems the data can be easily stored. Unlike analogue method of data storage, the information stored in digital mode does not deteriorate with time.
5. *High speed of operation.* Switching circuits in digital systems work with high speed. Therefore, a very large number of operations can be carried out in a specified sequence within a fraction of a second. Digital systems, therefore, offer vast opportunities for solving complex problems. The solutions of these complex problems were otherwise almost impossible.

As discussed in the beginning, a computer works on binary digits 0 and 1. It is because of the fact that a digital computer uses integrated circuits with thousands of transistors. Due to parameter variations the behaviour of a transistor may be very erratic and the quiescent point may shift from one position to another. However, the cutoff and saturation points are fixed. Thus, a transistor is a very reliable two-state device. One state represents digit 0 and another 1. All input voltages are recognized as either 0 or 1. A magnetic core of ferrite is also a two-state device. It has a rectangular hysteresis loop, and hence remains either in nonmagnetized or magnetized state. Nonmagnetized state represents 0 and magnetized state 1.

A group of many two-state devices is known as a register. Each two-state device can be either in state 0 or in state 1. Thus, a register represents a sequence of 0's and 1's.

10.2 BINARY LOGIC AND LOGIC GATES

A digital computer is essentially a logic machine, which performs a large number of logical operations at a very high speed. Hence, symbolic logic is of prime importance in digital computers. A digital computer performs complicated operations by interconnecting a large number of logic gates. The different interconnections are designed to implement the laws of logic. The mathematical technique for logical analysis of systems is called *Boolean algebra*. Boolean algebra is named after George Boole who developed it for describing logical statements. Boolean algebra is also known as symbolic logic. It is possible to make systematic manipulation of a large number of statements and their relationship into simpler statements with the help of symbolic logic. Boolean algebra is based on only two possible states. The two states are ON and OFF states of logic gates. ON state is represented by 1 and OFF by 0. The information is reduced to one of these two states.

The binary logics or logical functions involved in the design of digital systems are OR, AND, NOT, NOR, NAND, EXCLUSIVE OR, INHIBIT (ENABLE). An electronic circuit can be designed to perform each logic stated above. The electronic circuit is known as a gate. Each gate is named after its function such as AND gate, OR gate, etc.

10.2.1 OR Gate

An OR gate has two or more inputs but only one output. In case all the inputs are low (i.e., 0), the output is low (i.e., 0). In case any of the inputs is high (i.e., 1), the output is high (i.e., 1). Symbols of OR gate for 2, 3, and 4 inputs are shown in Fig. 10.1. Circuits for OR gates having 2, 3, and 4 inputs are shown in Fig. 10.2. If all the inputs are low (i.e., 0), all diodes are off (i.e., nonconducting), and so the output is low (i.e., 0). If any of the

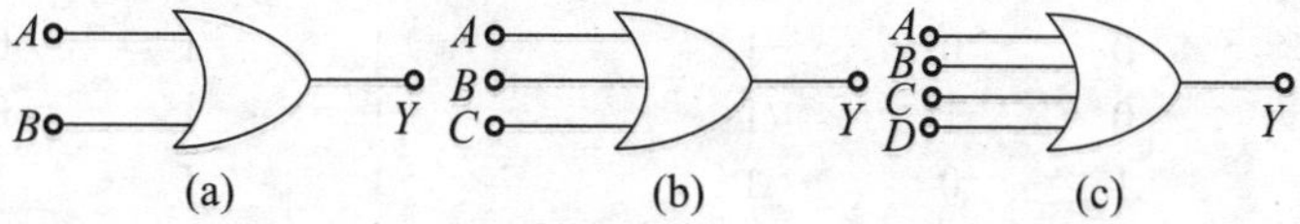

FIGURE 10.1 The symbols of OR gate for (a) 2 inputs, (b) 3 inputs, and (c) 4 inputs.

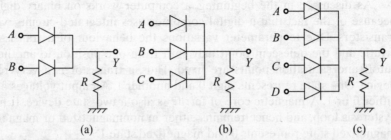

FIGURE 10.2 The circuits of OR gate for (a) 2 inputs, (b) 3 inputs, and (c) 4 inputs.

inputs is high, the corresponding diode is on (i.e., conducting). Hence, the output is high (i.e., 1). An OR gate (for 2 inputs) is mathematically expressed as

$$Y = A \text{ OR } B = A + B \tag{10.1}$$

Equation (10.1) means that, in a 2-input OR gate, when $A = 1$ or $B = 1$ or $A = B = 1$, $Y = 1$. '+' sign indicates OR in logic. The output Y is zero when all inputs to the OR gate are zero. Even if only one of the inputs is 1, the output $Y = 1$.

A summary of options for a logic function can be expressed by a table called *truth table*. Truth tables for 2 inputs, 3 inputs, and 4 inputs of OR gates of Fig. 10.1 are given in Tables 10.1 to 10.3.

In an equivalent electrical system, an OR gate can be represented by switches in parallel. The number of switches in parallel is equal to the number of inputs in the OR

Table-10.1 Truth table for Fig. 10.1(a).

A	B	Y
0	0	0
0	1	1
1	0	1
1	1	1

Table-10.2 Truth table for Fig. 10.1(b).

A	B	C	Y
0	0	0	0
0	0	1	1
0	1	0	1
0	1	1	1
1	0	0	1
1	0	1	1
1	1	0	1
1	1	1	1

Table-10.3 Truth table for Fig. 10.1(c).

A	B	C	D	Y
0	0	0	0	0
0	0	0	1	1
0	0	1	0	1
0	0	1	1	1
0	1	0	0	1
0	1	0	1	1
0	1	1	0	1
0	1	1	1	1
1	0	0	0	1
1	0	0	1	1
1	0	1	0	1
1	0	1	1	1
1	1	0	0	1
1	1	0	1	1
1	1	1	0	1
1	1	1	1	1

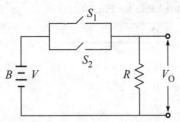

FIGURE 10.3 An electrical equivalent of OR gate.

gate. For example, a 2-input OR gate is equivalent to two electrical switches in parallel, as shown in Fig. 10.3.

10.2.2 AND Gate

Like an OR gate an AND gate can have two or more inputs and one output. For a high (i.e., 1) output all the inputs must be high (i.e., 1). Figure 10.4 shows the symbols of AND gates for 2 inputs, 3 inputs, and 4 inputs. Circuits for AND gates having 2 inputs, 3 inputs, and 4 inputs are shown in Fig. 10.5. If any of the inputs of the AND gate is low, the corresponding diode conducts, and hence the output is low. An AND gate (for 2 inputs) is mathematically expressed as

$$Y = A \text{ AND } B = A \cdot B \tag{10.2}$$

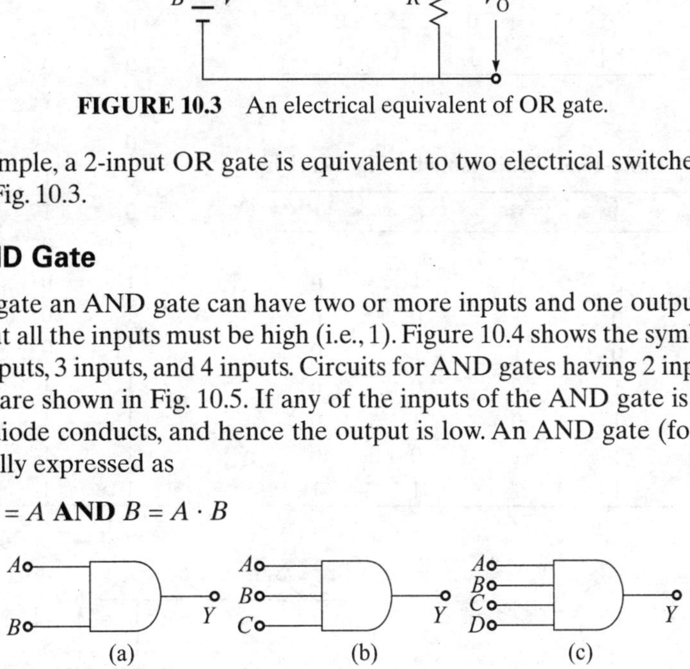

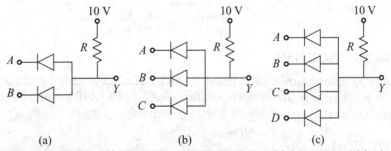

FIGURE 10.4 The symbols of AND gate for (a) 2 inputs, (b) 3 inputs, and (c) 4 inputs.

FIGURE 10.5 The circuits of AND gate for (a) 2 inputs, (b) 3 inputs, and (c) 4 inputs.

Equation (10.2) means that only when $A = B = 1$, $Y = 1$. '·' sign indicates AND in logic.

Truth tables for 2 inputs, 3 inputs, and 4 inputs of AND gate of Fig. 10.4 are given in Tables 10.4 to 10.6.

An AND gate can be represented by switches in series in an equivalent electrical system. The number of switches in series is equal to the number of inputs in the AND gate. For example, a 2-input AND gate is equivalent to two electrical switches in series, as shown in Fig. 10.6.

Table-10.4 Truth table for Fig. 10.3(a).

A	B	Y
0	0	0
0	1	0
1	0	0
1	1	1

Table-10.6 Truth table for Fig. 10.3(c).

A	B	C	D	Y
0	0	0	0	0
0	0	0	1	0
0	0	1	0	0
0	0	1	1	0
0	1	0	0	0
0	1	0	1	0
0	1	1	0	0
0	1	1	1	0
1	0	0	0	0
1	0	0	1	0
1	0	1	0	0
1	0	1	1	0
1	1	0	0	0
1	1	0	1	0
1	1	1	0	0
1	1	1	1	1

Table-10.5 Truth table for Fig. 10.3(b).

A	B	C	Y
0	0	0	0
0	0	1	0
0	1	0	0
0	1	1	0
1	0	0	0
1	0	1	0
1	1	0	0
1	1	1	1

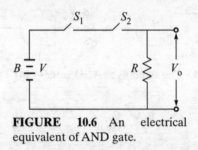

FIGURE 10.6 An electrical equivalent of AND gate.

10.2.3 NOT Gate

A NOT gate is also called an inverter. This gate has only one input and one output. The output is always negative of the input. Two symbols of a NOT gate are shown in Fig. 10.7. The truth table is given in Table-10.7. The logic equation of the NOT gate is given by Eq. (10.3).

$$Y = \overline{A} \text{ or } Y = \textbf{NOT } A \tag{10.3}$$

Figure 10.8(a) uses a simple switch. Figure 10.8(b) shows a circuit using a transistor for realization of a NOT gate. When the switch is closed (i.e., 1), the output is 0 V (i.e., 0). When the switch is open (i.e., 0), the output is 10 V (i.e., 1). When 10 V is applied to the

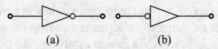

FIGURE 10.7 The symbols for NOT gate.

Table-10.7 Truth table of Fig. 10.5.

A	Y
1	0
0	1

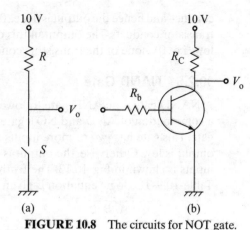

FIGURE 10.8 The circuits for NOT gate.

input of the gate, the transistor is saturated; therefore, the output is 0 V (i.e., 0). When the input applied is 0, the transistor is cut off; hence, the output is 10 V (i.e., 1).

10.2.4 NOR Gate

A NOR gate is an OR gate followed by a NOT gate (Fig. 10.9). The output of the OR gate is fed to the NOT gate. This gate has two or more inputs and only one output. For a high output, all inputs must be low. The symbol for a NOR gate with 3 inputs is shown in Fig. 10.10. The truth table of the NOR gate of Fig. 10.10 is given in Table-10.8. The logic equation of NOR gate is given by

$$Y = \overline{A + B + C} \tag{10.4}$$

The electronic circuit using transistors for realizing a NOR gate of 3 inputs is shown in Fig. 10.11. When all the three inputs are high (e.g., 10 V, i.e., 1), all three transistors

FIGURE 10.9 A combination of OR and NOT gates.

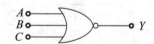

FIGURE 10.10 The symbol for NOR gate.

Table-10.8 Truth table of Fig. 10.9.

A	B	C	Y
0	0	0	1
0	0	1	0
0	1	0	0
0	1	1	0
1	0	0	0
1	0	1	0
1	1	0	0
1	1	1	0

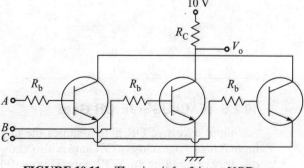

FIGURE 10.11 The circuit for 3-input NOR gate.

conduct, and hence the output is 0 (i.e., 0). When any of the inputs is high, the corresponding transistor conducts. The output in this case is also 0 (i.e., 0). When all the three inputs are low (i.e., 0), none of the transistors conducts, and hence the output is high (i.e., 1).

10.2.5 NAND Gate

A NAND gate is an AND gate followed by a NOT gate (Fig. 10.12). This means that it is a combination of AND and NOT gates. The output of the AND gate is fed to the NOT gate. This gate has two or more inputs and only one output. When all inputs are high the output is low. Otherwise, the output is always high. The symbol for a NAND gate with 3 inputs is shown in Fig. 10.13. The truth table of the NAND gate of Fig. 10.13 is given in Table-10.9. The logic equation is given by

$$Y = \overline{A \cdot B \cdot C} \tag{10.5}$$

FIGURE 10.12 A combination of AND and NOT gates.

FIGURE 10.13 The symbol for NAND gate.

The electronic circuit for the realization of a NAND gate using transistors for 2 inputs is shown in Fig. 10.14. When both inputs are high (e.g., 10 V, i.e., logic 1), both transistors conduct, and hence the output is 0 V (i.e., 0). When any of the inputs is low, the corresponding transistor does not conduct. There is no current through the load resistance. Hence, the output is high (e.g., 10 V i.e., 1).

Table-10.9 Truth table of Fig. 10.13.

A	B	C	Y
0	0	0	1
0	0	1	1
0	1	0	1
0	1	1	1
1	0	0	1
1	0	1	1
1	1	0	1
1	1	1	0

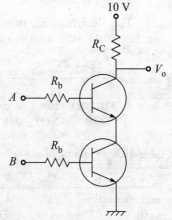

FIGURE 10.14 The circuit for NAND gate.

10.2.6 EXCLUSSIVE OR Gate

A 2-input exclusive OR gate assumes the high value (i.e., 1 state) if one and only one input is high. An exclusive OR gate is also known as XOR gate. Figure 10.15(a) illustrates one way of building an XOR gate. Figure 10.15(b) shows the symbol for the exclusive

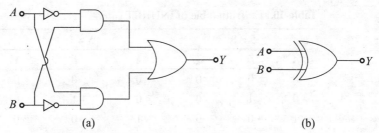

FIGURE 10.15 An XOR gate (a) building blocks and (b) symbol.

Table-10.10 Truth table of XOR.

A	B	Y
0	0	0
0	1	1
1	0	1
1	1	0

OR gate. The truth table of a 2-input exclusive gate is given in Table-10.10. In Fig. 10.15(a), the output of the upper AND of XOR is $\overline{A} \cdot B$ and that of the lower AND is $A \cdot \overline{B}$. Hence, the output of the XOR gate of Fig. 10.15(a) is given by the following logical expression:

$$Y = \overline{A} \cdot B + A \cdot \overline{B} \qquad (10.6)$$

From the truth table of XOR, $Y = 1$, if $A \neq B$. This property of an XOR gate is used to test the inequality of two bits. Also, $Y = 0$, if $A = B$. This property of an XOR gate is used to test the matching of two bits.

10.2.7 INHIBIT (ENABLE) Gate

An INHIBIT gate is actually an AND gate with m inputs ($m = 2, 3, \ldots$). There is an additional input S which is fed to the AND gate via a NOT gate, as shown in Fig. 10.16. The output is high when all inputs are high and the enable input is low. In case the enable input is high, the output is low for all combinations of inputs. Building blocks of this gate are shown in Fig. 10.16 and symbol in Fig. 10.17. The truth table for $m = 2$ is given in Table-10.11.

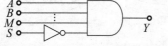

FIGURE 10.16 The building blocks of an INHIBIT gate.

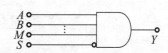

FIGURE 10.17 The symbol for an INHIBIT gate.

Table-10.11 Truth table of INHIBIT gate.

S = 0			S = 1		
A	B	Y	A	B	Y
0	0	0	0	0	0
0	1	0	0	1	0
1	0	0	1	0	0
1	1	1	1	1	0

10.2.8 Universal Gates

As NAND and NOR gates can be used to realize OR, AND, and NOT gates, the NAND and NOR gates are known as universal gates. All the logic systems can be implemented by using either NAND or NOR gate. The realization of these gates is easier. These gates also consume less power than other gates.

Realizations of Logic Gates Using NOR Gate

1. The NOT gate can be realized by joining two inputs of the NOR gate together, as shown in Fig. 10.18(a).
2. To realize OR gate, two NOR gates are used in cascade, as shown in Fig. 10.18(b). The output of the first NOR gate is fed to the tied inputs of the second NOR gate.
3. The AND gate can be realized by using three NOR gates, as shown in Fig. 10.18(c). Two NOR gates are in parallel. One of the two inputs of OR gate is fed to the tied inputs of the first NOR gate. Another input is fed to the tied inputs of the second NOR gate. The outputs of these two NOR gates are fed to the inputs of the third NOR gate.

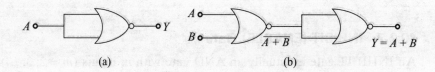

(a) (b)

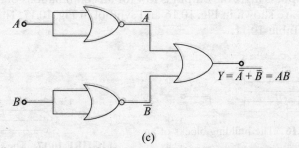

(c)

FIGURE 10.18 Realization, using NOR gates, of (a) NOT gate, (b) OR gate, and (c) AND gate.

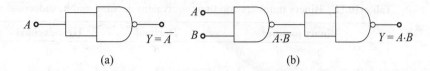

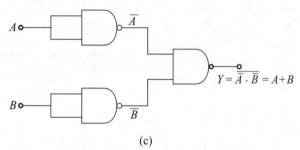

FIGURE 10.19 Realization using NAND gates of (a) NOT gate, (b) AND gate, and (c) OR gate.

Realizations of Logic Gates Using NAND Gate

1. The NOT gate can be realized by joining the two inputs of the NAND gate together, as shown in Fig. 10.19(a).
2. To realize AND gate, two NAND gates are used in cascade, as shown in Fig. 10.19(b). The output of the first NAND gate is fed to the tied inputs of the second NAND gate.
3. The OR gate can be realized by using three NAND gates, as shown in Fig. 10.19(c). Two NAND gates are in parallel, and the one input is fed to the tied inputs of each NAND gate. Two outputs of these NAND gates are fed to the inputs of the third NAND gate.

10.3 NUMBER SYSTEMS

In conventional arithmetic we use decimal number system. This system uses 10 basic symbols: 0, 1, 2, 3, 4, 5, 6, 7, 8, and 9. The symbols used in a number system are known as *digits* of the number system. As discussed earlier, a digital computer system does not use this decimal number system. A digital computer uses a binary number system. Binary number system uses only first two digits (i.e., 0 and 1) of the decimal system. The number of basic symbols used in a number system is known as the *radix* or the *base*. Hence, the radix or the base for the decimal system is 10, whereas that for the binary is 2. A hexadecimal number system uses 16 symbols. First 10 symbols are same as those used for the decimal number system and remaining six are A, B, C, D, E, and F. Hence, the base of the hexadecimal number system is 16. Hereinafter simply the word computer will be used in place of digital computers.

10.3.1 Binary Number System

As the binary number system uses digits 0 and 1, a binary number is a string of 0's and 1's. The digit in the binary number is called a bit. A string of eight bits is called a byte.

Table-10.12 Binary numbers and their equivalent decimal and hexadecimal numbers.

Binary number				Decimal	Hexadecimal
0	0	0	0	0	0
0	0	0	1	1	1
0	0	1	0	2	2
0	0	1	1	3	3
0	1	0	0	4	4
0	1	0	1	5	5
0	1	1	0	6	6
0	1	1	1	7	7
1	0	0	0	8	8
1	0	0	1	9	9
1	0	1	0	10	A
1	0	1	1	11	B
1	1	0	0	12	C
1	1	0	1	13	D
1	1	1	0	14	E
1	1	1	1	15	F

A byte is a basic unit of data in computers. In computers, data are processed in strings of 8 bits or multiples of 8 bits (i.e., 16, 32, etc.). Computer memories also store data in strings of 8 bits or multiples of 8 bits. Total numbers represented by N bits binary is given by Nth power of 2, that is, 2^N. Hence, a 4-bit binary number can form 16 combinations of 4-bits strings. The 16 combinations of 4-bit binary number (called 4-bit binary word) and their equivalent decimal numbers and hexadecimal numbers are given in Table-10.12.

A number in any system has two parts: integer part and fractional part. First we discuss the conversion of integer part from binary system to decimal system and from decimal system to binary system. Later, conversions of fractional part are considered.

A decimal number
$$2543 = 2 \times 10^3 + 5 \times 10^2 + 4 \times 10^1 + 3 \times 10^0$$
Similarly, a binary number
$$1101 = 1 \times 2^3 + 1 \times 2^2 + 0 \times 2^1 + 1 \times 2^0 = 13$$

The procedure to convert a binary number into an equivalent decimal number is as follows:

1. Multiply the bits of the binary number from right to left by their corresponding weights as $2^0, 2^1, 2^2, 2^3, ...$, respectively.
2. Add all the weighted values of bits.
3. Result is the decimal number.

Example-10.1

Convert the following binary numbers into their equivalent decimal numbers:

(1) 10101, (2) 1101, (3) 101110, (4) 100111, (5) 10111

Solution

Following the above procedure, the equivalent decimal numbers are as follows:

(1) $1\times2^4 + 0\times2^3 + 1\times2^2 + 0\times2^1 + 1\times2^0 = 21$

(2) $1\times2^3 + 1\times2^2 + 0\times2^1 + 1\times2^0 = 13$

(3) $1\times2^5 + 0\times2^4 + 1\times2^3 + 1\times2^2 + 1\times2^1 + 0\times2^0 = 46$

(4) $1\times2^5 + 0\times2^4 + 0\times2^3 + 1\times2^2 + 1\times2^1 + 1\times2^0 = 39$

(5) $1\times2^4 + 0\times2^3 + 1\times2^2 + 1\times2^1 + 1\times2^0 = 23$

A decimal number can be converted to an equivalent binary number by repetitive division of the decimal number by 2. Let the decimal number be N. Then, this decimal number can be expressed by an equivalent m-bit binary number as follows:

$$N = R_m 2^{m-1} + R_{m-1} 2^{m-2} + \ldots + R_2 2^1 + R_1 2^0 \tag{10.7}$$

The bits $R_1, R_2, \ldots, R_m$ of the equivalent binary number can be obtained by following the steps given in Table-10.13.

Table-10.13 Procedure for decimal to binary conversion.

Division	Quotient	Remainder
$N \div 2$	$N_1 = R_m 2^{m-2} + \ldots + R_3 2^1 + R_2$	R_1 (LSB)
$N_1 \div 2$	$N_2 = R_m 2^{m-3} + \ldots + R_4 2^1 + R_3$	R_2
$N_2 \div 2$	$N_3 = R_m 2^{m-4} + \ldots + R_4 2^1 + R_4$	R_3
.....	.....	.....
$N_{m-3} \div 2$	$N_{m-2} = R_m 2^1 + R_{m-1}$	R_{m-2}
$N_{m-2} \div 2$	$N_{m-1} = R_m$	R_{m-1}
$N_{m-1} \div 2$	0	R_m (MSB)

Example-10.2

Find the binary equivalent of the decimal numbers: (1) 26, (2) 12, (3) 45, (4) 121, (5) 37.

Solution

The quotient and remainder can be written in a simpler form as follows:

(1)

2	26	Remainder
2	13	0(LSB)
2	6	1
2	3	0
2	1	1
	0	1(MSB)

The equivalent binary number is **11010**.

(2)

2	12	Remainder
2	6	0(LSB)
2	3	0
2	1	1
	0	1(MSB)

The equivalent binary number is **1100**.

(3)

2	45	Remainder
2	22	1(LSB)
2	11	0
2	5	1
2	2	1
2	1	0
	0	1(MSB)

The equivalent binary number is **101101**.

(4)

2	121	Remainder
2	60	1(LSB)
2	30	0
2	15	0
2	7	1
2	3	1
2	1	1
	0	1(MSB)

The equivalent binary number is **1111001**.

(5)

2	37	Remainder
2	18	1(LSB)
2	9	0
2	4	1
2	2	0
2	1	0
	0	1(MSB)

The equivalent binary number is **100101**.

Binary Addition and Subtraction

Rules for binary arithmetic are given in Table-10.14.

In both addition and subtraction, we start from the LSB and proceed towards MSB, like the decimal addition and subtraction.

Table-10.14 Rules for addition and subtraction.

Addition	Subtraction
0 + 0 = 0	0 − 0 = 0
1 + 0 = 0 + 1 = 1	1 − 0 = 1
1 + 1 = 10	1 − 1 = 0
1 + 1 + 1 = 11	10 − 1 = 1

Example-10.3

(a) Convert decimal numbers 45 and 37 into their equivalent binary numbers. (b) Add the binary numbers and convert the result into decimal number.

Solution

(a) The decimal numbers 45 and 37 have already been converted into binary numbers in Example-10.2. Hence, the equivalent binary numbers are

$$45 = \mathbf{101101} \text{ and } 37 = \mathbf{100101}$$

(b) Binary addition

$$
\begin{array}{r}
1\ 0\ 1\ 1\ 0\ 1 \\
(+)\ 1\ 0\ 0\ 1\ 0\ 1 \\
\hline
1\ 0\ 1\ 0\ 0\ 1\ 0
\end{array}
\qquad
\begin{array}{r}
45 \\
(+)\ 37 \\
\hline
82
\end{array}
$$

The resulted binary number is **1010010**.

the equivalent decimal number = $1 \times 2^6 + 0 \times 2^5 + 1 \times 2^4 + 0 \times 2^3 + 0 \times 2^2 + 1 \times 2^1 + 0 \times 2^0$
$= 82$

Example-10.4

Subtract the binary number 01101 from 11011.

Solution
Binary subtraction

```
      1 1 0 1 1        27
(–)   0 1 1 0 1   (–)  13
      ─────────        ──
      0 1 1 1 0        14
```

The 3rd bit in the first number is 0. Hence, 1 is borrowed from the 4th bit of the first number and 1 is subtracted to give 1. Because of borrowing, 4th bit of the first number becomes 0.

Example-10.5

Subtract the binary number 10101 from 01101.

Solution
In this case the number to be subtracted is greater; hence, the result is negative.
Binary subtraction

```
      0 1 1 0 1        13
(–)   1 0 1 0 1   (–)  21
      ─────────        ──
    – 0 1 0 0 0       –08
```

Signed Binary Number

To mention the sign of the binary numbers, digit 0 is used for + sign in the positive number and 1 for – sign of the negative number. The MSB is the sign bit is followed by the magnitude bits. The numbers expressed in this manner are called signed binary numbers. In the 16-bit system most significant bit is used for the sign of the number and the remaining 15 bits for the number. For example +121 and –121 are

+121 = 0000 0000 0111 1001
–121 = 1000 0000 0111 1001

The signed binary numbers require too much electronic circuitry for performing addition and subtraction. Therefore, positive decimal numbers are expressed in sign-magnitude form, but the negative numbers are expressed in 2's compliment form and not in sign-magnitude form of binary numbers. To understand the 2's complement form for binary system, it is helpful to discuss 10's compliment form of the decimal system.

10's Complement

The 10's complement of a decimal number is equal to the 9's complement plus one. The 9's complement of a decimal number is obtained by replacing each of its digits by a digit equal to 9 minus the digit to be replaced. For example, 9's complement of 324 is 675 (= 999 − 324). Hence, 10's complement of 324 is 676. Now, let us add 324 and its 10's complement and examine the result.

$$
\begin{array}{rr}
& 324 \\
(+) & 676 \\
\hline
& 000
\end{array}
$$

There is a carry 1. The number considered is of three digits and if the sum of the number and its 10's complement is considered only up to three digits and the carry is neglected, the result is zero. Thus, it can be concluded that the 10's complement of a number gives its negative value.

2's Complement

The 2's complement of a binary number is equal to the 1's complement plus one. The 1's complement of a binary number is obtained by replacing 1's of the number by 0's and 0's of the number by 1's. For example 1's complement of 101 is 010. Hence, the 2's complement of 101 is 011 (= 010 + 1). Now, let us add 101 and its 2's complement and examine the result.

$$
\begin{array}{rr}
& 101 \\
(+) & 011 \\
\hline
& 000
\end{array}
$$

There is a carry 1. The number considered is of three bits and if the sum of the number and its 2's complement is considered only up to three bits and the carry is neglected, the result is zero. Thus, it can be concluded that the 2's complement of a number gives its negative value.

From the above discussion, it is concluded that the method of 2's complement can be used for subtracting a binary number from another binary number. When one binary number is to be subtracted from the other, the 2's complement of the number to be subtracted is added to the other number. If the resulting addition contains a carry, the result is positive. The carry is discarded and the result gives the magnitude. When the addition does not contain a carry the result is negative. To get the result, the 2's complement of the addition gives the magnitude. The following examples can illustrate the method.

Example-10.6

Using 2's complement, subtract 01101 from 11011.

Solution

2's complement of 01101 = 10010 + 1 = 10011. Hence,

$$
\begin{array}{r}
1\,1\,0\,1\,1 \\
(+)\ 1\,0\,0\,1\,1 \\
\hline
0\,1\,1\,1\,0
\end{array}
$$

The carry is 1 and is discarded. Hence, the result is positive and the magnitude is 01110 (=14). The same result has been obtained in Example-10.4.

Example-10.7

Using 2's complement, subtract 10101 from 01101.

Solution

2's complement of 10101 = 01010 + 1 = 01011. Hence,

$$
\begin{array}{r}
0\,1\,1\,0\,1 \\
(+)\,0\,1\,0\,1\,1 \\
\hline
1\,1\,0\,0\,0
\end{array}
$$

There is no carry. Hence, the result is negative. For the magnitude, the 2's complement of the result is taken.

2's complement of 11000 = 00111 + 1 = 01000.

The same result has been obtained in Example-10.5.

10.3.2 Hexadecimal Number System

Hexadecimal number system, as shown in Table-10.12, has 16 digits. Hence, the base (radix) of the hexadecimal system is 16. As given in Table-10.12, the digits of hexadecimal system are 0, 1, 2, 3, 4, 5, 6, 7, 8, 9, A, B, C, D, E, F. Table-10.12 gives binary and decimal equivalents of the hexadecimal digits. The hexadecimal number system is very convenient and is extensively used because the numbers in this system are very short compared to binary numbers. For convenience of the programmer, codes for mnemonics and data are entered into a microprocessor kit in hexadecimal system. Internally they are converted into binary system by the microprocessor.

Conversion of Decimal Fraction to Binary Fraction

So far only whole numbers have been discussed. A decimal fraction can be converted to an equivalent binary fraction by successive multiplication of the decimal fraction by 2.

Let the decimal fraction be M. The decimal fraction can be expressed by an equivalent m-bit binary fraction as follows:

$$M = I_1 2^{-1} + I_2 2^{-2} + \cdots + I_{m-1} 2^{-m+1} + I_m 2^{-m} \tag{10.8}$$

where I's are non-negative integers less than 2 (i.e., 0 or 1).

The bits $I_1, I_2, \ldots, I_m$ of the equivalent binary fraction can be obtained by the steps given in Table-10.15.

Table-10.15 Procedure for decimal fraction to binary fraction conversion.

Fraction × 2	Remainder fraction	Integer
$M \times 2$	$M_1 = I_2 2^{-1} + I_3 2^{-2} + \cdots + I_m 2^{-m+1}$	I_1 (MSB)
$M_1 \times 2$	$M_2 = I_3 2^{-1} + I_4 2^{-2} + \cdots + I_m 2^{-m+2}$	I_2
$M_2 \times 2$	$M_3 = I_4 2^{-1} + I_5 2^{-2} + \cdots + I_m 2^{-m+3}$	I_3
...	....	...
$M_{m-3} \times 2$	$M_{m-2} = I_{m-1} 2^{-1} + I_m 2^{-2}$	I_{m-2}
$N'_{m-2} \times 2$	$N_{m-1} = I_m 2^{-1}$	I_{m-1}
$N'_{m-1} \times 2$	0.00	I_m (LSB)

In some cases it is possible that the remainder fraction does not become zero even for large value of m. In such a case, an approximation is made and the result is taken up to a certain number of bits after the binary point (equivalent to decimal point). Similar procedure is followed for a number having both integer and fraction. Binary fraction can be added, subtracted, etc., like the decimal numbers.

Example-10.8

Express the decimal number 0.375 into its binary equivalent.

Solution

The table below shows the procedure:

Fraction	Fraction × 2	Remainder fraction	Integer
0.375	0.75	0.75	0 (MSB)
0.75	1.50	0.50	1
0.50	1.00	0.00	1 (LSB)

The binary equivalent is **0.011**.

Example-10.9

Express the decimal number 0.475 into its binary equivalent.

Solution

The table below shows the procedure:

Fraction	Fraction × 2	Remainder fraction	Integer
0.475	0.95	0.95	0 (MSB)
0.95	1.90	0.90	1
0.90	1.80	0.80	1
0.80	1.60	0.60	1
0.60	1.20	0.20	1

From the above table, it is clear that it is not possible to get a zero. Hence, the process is terminated after five stages. The binary equivalent is **0.01111**. For further accuracy, process may be continued after five stages till the error is within the permissible limit.

Example-10.10

Find binary equivalent of the decimal number 12.75.

Solution

First, the integer part is considered.

2	12	Remainder
2	6	0 (LSB)
2	3	0
2	1	1
	0	1 (MSB)

The equivalent binary number is **1100**.
Now, consider the fraction part.

Fraction	Fraction × 2	Remainder fraction	Integer
0.75	1.50	0.50	1 (MSB)
0.50	1.00	0.00	1 (LSB)

The equivalent binary number of fraction is 11. Hence, the binary equivalent of 12.75 = **1100.11**.

Example-10.11

Add the binary numbers 1011.011 and 0010.11.

Solution

$$
\begin{array}{rr}
1011.011 & 11.375 \\
(+)\ 0010.110 & (+)\ 2.75 \\
\hline
1110.001 & 14.125
\end{array}
$$

The result = **1110.001**.

Hexadecimal to Binary Conversion

Each digit of a hexadecimal number is converted to its 4-bit binary equivalence using Table-10.12. For example, converting 9C3 to its binary equivalence,

9	C	3	Hexadecimal number
1 0 0 1	1 1 0 0	0 0 1 1	Binary number

Binary to Hexadecimal Conversion

We know that each digit of a hexadecimal number in its binary equivalence is made of 4 bits. Hence, groups of 4 bits are made from right to left. Each 4-bit binary group is replaced by its equivalent hexadecimal digit. For example,

0 1 1 1	1 1 1 1	1 1 0 0
7	F	C

Hexadecimal to Decimal Conversion

The base of the hexadecimal number system is 16. Hence, the weights of different digits of the hexadecimal number from right to left are $16^0, 16^1, 16^2, 16^3, 16^4, 16^5, \ldots$. The equivalent decimal number is equal to the sum of all digits multiplied by their weights. For example,

$$
\begin{array}{cccc}
A & E & 5 & C
\end{array}
$$
$$A \times 16^3 + E \times 16^2 + 5 \times 16^1 + C \times 16^0$$
$$10 \times 16^3 + 14 \times 16^2 + 5 \times 16^1 + 12 \times 16^0 = \mathbf{44636}$$

In indirect method, the hexadecimal number is converted first to its binary equivalence, and then the binary number is converted to its decimal equivalence. Example-10.12 illustrates the statement.

Example-10.12

Convert the hexadecimal number 3B8 into its decimal equivalence: (a) by first converting into binary and (b) by direct method.

Solution

(a) The binary equivalence

$$\begin{array}{cccc} 3 & B & 8 & \text{Hexadecimal} \\ 0011 & 1011 & 1000 & \text{Binary} \end{array}$$

$$0 \times 2^{11} + 0 \times 2^{10} + 1 \times 2^9 + 1 \times 2^8 + 1 \times 2^7 + 0 \times 2^6 + 1 \times 2^5 + 1 \times 2^4$$
$$+ 1 \times 2^3 + 0 \times 2^2 + 0 \times 2^1 + 0 \times 2^0$$
$$= 2^9 + 2^8 + 2^7 + 2^5 + 2^4 + 2^3 = \mathbf{952} \qquad \text{Decimal}$$

(b) Direct method

$$3 \times 16^2 + B \times 16^1 + 8 \times 16^0 = 3 \times 256 + 11 \times 16 + 8 = \mathbf{952} \qquad \text{Decimal}$$

Decimal to Hexadecimal Conversion

In indirect method, the decimal number is converted to its binary equivalence, and then the binary number to its hexadecimal equivalence. In the direct method, the successive division by 16 (like decimal to binary conversion) is followed. The remainders are the digits of the hexadecimal number, first remainder as LSB and last one as MSB.

Example-10.13

Convert decimal number 275 into its hexadecimal equivalence by (a) indirect method and (b) direct method.

Solution

(a) The binary equivalence is given by

2	275	Remainder
2	137	1 (LSB)
2	68	1
2	34	0
2	17	0
2	8	1
2	4	0
2	2	0
2	1	0
	0	1 (MSB)

Hence, $275 = 100010011 = 0001 \quad 0001 \quad 0011$
In hexadecimal = **113**

(b) Direct method:

16	275	Remainder
16	17	3 (LSB)
16	1	1
16	0	1 (MSB)

Hexadecimal number is **113**.

10.4 BOOLEAN ALGEBRA

In the conventional algebra, symbols representing numerical quantities are related by arithmetic processes; for example, $+, -, \times, \div, =$. In Boolean algebra, however, symbols represent statements or propositions. These propositions are related by logic. Each proposition has either of the two values. It can be either *true* or *false*. For example, consider the following statements or propositions. Their truth values are shown against the propositions in Table-10.16.

Table-10.16 Truth values of propositions.

Propositions	Truth values
1. Himalaya is a river.	False
2. Moon shines at night.	True
3. Sun moves round the earth.	False
4. Earth moves round the sun.	True
5. Sun rises in the east.	True

Boolean algebra offers an attractive method of designing switching circuits. A switch is either open or closed, only two states. The logic and memory elements in a digital computer also have only two possible states, that is, 0 or 1.

Propositions can be combined or manipulated by AND, OR, and NOT operations. The AND combination of two propositions is true if and only if both propositions are true. If either one or both of the two propositions are false, the combination is false. For example, 'Moon shines at night AND earth moves round the sun' is true because both these propositions are true. The AND combination of two propositions is often called *logical product* or *conjunction*. This combination is denoted by dot (·) between two symbols. Sometimes dot is not placed between the two symbols and still AND combination implied.

The OR combination is false if and only if both propositions are false. If either one or both of the two propositions are true, the OR combination is true. For example, 'Moon shines at night OR Himalaya is a river' is true, as one of the two propositions is true. The OR combination is called logical sum and is denoted by plus (+) sign between the two symbols.

The NOT operation of a proposition is true if and only if the proposition is false. For example NOT of 'Himalaya is a river' is true, as the proposition is false. The NOT operation is denoted by (−) or (') placed over the original symbol.

10.4.1 Boolean Functions

Many propositions, say $A, B, C, ...$, when combined by AND, OR, and NOT operations, form Boolean functions. A function $Y = f(A, B, C, ...)$ is also called a *switching function*. The propositions $A, B, C, ...$ are also called *switching variables*. For example $Y = f(A, B) = A\overline{B} + \overline{A}B$ is a Boolean function. In this, if A is true and B is false, $\overline{A}$ is false and $\overline{B}$ is true. Hence, $A\overline{B}$ is true and $\overline{A}B$ is false. The function Y, therefore, is true. The function is also true when A is false and B is true. The function is known as EXCLUSIVE OR, since the function is true if and only if only one of the two propositions is true.

10.4.2 Truth Table

Let $Y = f(A_1, A_2, ..., A_n)$ is a Boolean or switching function of n switching variables or propositions $A_1, A_2, ..., A_n$. Since any of the propositions can assume independently any one of the two values: True (T) or False (F), there are 2^n combinations of $A_1, A_2, ..., A_n$ to be considered for determining the value of the function Y. A table can be prepared for the values of the function for 2^n possible combinations. The table is known as the truth table of the Boolean function.

For example consider a Boolean function of two propositions A_1 and A_2 as

$$Y = f(A_1, A_2) = A_1 \cdot \overline{A_2} + \overline{A_1} \cdot A_2 \tag{10.9}$$

The truth table of the Boolean function can be prepared as in Table-10.17.

From the Table-10.17, it is clear that Y is true when either A_1 or A_2 is true. Generally, false is represented by 0 and true by 1 in Boolean algebra. If F is replaced by 0 and T by 1, then the truth table is given by Table-10.18, which represents an XOR and the truth table is same as given in Table-10.10.

It is also possible to write the switching or logical function from a given truth table. To illustrate the statement, let us consider the truth table in Table-10.18. The function $Y = 1$, when $A_1 = 0$ AND $A_2 = 1$ OR $A_1 = 1$ AND $A_2 = 0$. Hence, the function can be written as

$$Y = f(A_1, A_2) = \overline{A_1}A_2 + A_1\overline{A_2} \tag{10.10}$$

Table-10.17 Truth table.

A_1	A_2	Y
F	F	F
F	T	T
T	F	T
T	T	F

Table-10.18 Truth table with 0,1.

A_1	A_2	Y
0	0	0
0	1	1
1	0	1
1	1	0

The procedure for writing the switching function can easily be extended to more number of propositions. The following procedure may be followed:

1. Consider the first row for which $Y = 1$.
2. Complement the variables in the row with 0.
3. A product term of the function is obtained by ANDing the various variables or complements of the variables with 0 in the row.
4. Repeat step 2 with other rows of the truth table.
5. All the product terms obtained above are connected by '+'.

Example-10.14

For the truth table (Table-10.19), find the switching or logic function.

Solution

From the truth table:

$Y = 1$, when $\overline{A_1} \cdot \overline{A_2} \cdot A_3 = 1$ or $\overline{A_1} A_2 A_3 = 1$
OR $Y = 1$, when $\overline{A_1} \cdot A_2 \cdot \overline{A_3} = 1$ or $A_1 A_2 \overline{A_3} = 1$
OR $Y = 1$, when $A_1 \cdot A_2 \cdot \overline{A_3} = 1$ or $A_1 A_2 \overline{A_3} = 1$

Hence, $Y = f(A_1, A_2, A_3) = \overline{A_1}\overline{A_2}A_3 + \overline{A_1}A_2\overline{A_3} + A_1 A_2 \overline{A_3}$

Table-10.19 Truth table for Example 10.14.

A_1	A_2	A_3	Y
0	0	0	0
0	0	1	1
0	1	0	1
1	0	0	0
0	1	1	0
1	0	1	0
1	1	0	1
1	1	1	0

Example-10.15

Make a truth table for the function

$$Y = f(A_1, A_2, A_3) = A_1 A_2 \overline{A_3} + A_1 \overline{A_2} A_3 + \overline{A_1} A_2 \overline{A_3} + \overline{A_1} \overline{A_2} \overline{A_3}$$

Solution

The truth table is given in Table-10.20.

Table-10.20 Truth table for Example-10.15.

A_1	A_2	A_3	Y	
0	0	0	1	(Fourth term is 1)
0	0	1	0	(No term is 1)
0	1	0	1	(Third term is 1)
1	0	0	0	(No term is 1)
0	1	1	0	(No term is 1)
1	0	1	1	(Second term is 1)
1	1	0	1	(First term is 1)
1	1	1	0	(No term is 1)

When all the three variables are 0, the last term is 1. Hence, 1 is entered in Y column for this combination of variables. Similarly, entries for other combinations of variables are completed.

10.4.3 Postulates of Boolean Algebra

A mathematical system consisting of a set of binary elements X together with two operations (+) and (·) is called a Boolean algebra if and only if the following postulates hold good:

1. *Commutative property*: The operations (+) and (·) are commutative. This means that for each pair of elements A_1 and A_2 in the set of A,

   (i) $A_1 + A_2 = A_2 + A_1$ (10.11)
   (ii) $A_1 \cdot A_2 = A_2 \cdot A_1$ (10.12)

2. *Associative property*: The operations (+) and (·) are associative. This means that the order of association is of no consequence in (+) and (·). This means that for any three elements $A_1, A_2,$ and A_3 in the set A,

   (i) $A_1 + (A_2 + A_3) = (A_1 + A_2) + A_3$ (10.13)
   (ii) $A_1 \cdot (A_2 \cdot A_3) = (A_1 \cdot A_2) \cdot A_3$ (10.14)

3. *Distributive property:* Each operation of (+) and (·) is distributive over the other. This means that for any three elements $A_1, A_2,$ and A_3 in the set A,

   (i) $A_1 + (A_2 \cdot A_3) = (A_1 + A_2) \cdot (A_1 + A_3)$ (10.15)
   (ii) $A_1 \cdot (A_2 + A_3) = (A_1 \cdot A_2) + (A_1 \cdot A_3)$ (10.16)

4. (i) There exist an element 0 in A such that for every element (A_1) in A,

   $A_1 + 0 = A_1$ (10.17)

(ii) There exist an element 1 in A such that for every element (A_1) in A,

$$A_1 \cdot 1 = A_1 \tag{10.18}$$

5. For every element (A_1) in A, there exists a complementary element $\overline{A}_1$ such that

(i) $A_1 \cdot \overline{A}_1 = 0$ (10.19)

(ii) $A_1 + \overline{A}_1 = 1$ (10.20)

where 0 and 1 are the elements defined in postulates 4(i) and 4(ii). The element 0 is called additive identity element and 1 is called multiplicative identity element.

The above postulates were suggested by Huntington in 1904. These postulates exhibit a duality between logical operations AND and OR and between 1 and 0. If '·' is replaced by '+', and '+' by '·', in any of the postulates, the result is the dual of original postulates. For example, postulate 3(ii) is dual of that of 3(i) or vice versa. Because of the duality in the fundamental postulates, if a new theorem is deduced using these postulates, a dual theorem always exists corresponding to the original theorem. The principle of duality can be stated as follows:

Every proved theorem can be transformed into a second valid theorem if the operations AND and OR and identity elements 0 and 1 are interchanged throughout.

There are two important theorems by De Morgan and are known as De Morgan's theorems. Theorems are:

De Morgan's first theorem: $\overline{A_1 + A_2} = \overline{A}_1 \cdot \overline{A}_2$ (10.21)

De Morgan's second theorem: $\overline{A_1 \cdot A_2} = \overline{A}_1 + \overline{A}_2$ (10.22)

Proof. Let us assume that $\overline{A}_1 \cdot \overline{A}_2 = \overline{A_1 + A_2}$ and $X = A_1 + A_2$, then $\overline{X} = \overline{A}_1 \cdot \overline{A}_2$. Now to prove the De Morgan first theorem, it is sufficient to prove that $X \cdot \overline{X} = 0$ and $X + \overline{X} = 1$

$$X \cdot \overline{X} = (A_1 + A_2) \cdot (\overline{A}_1 \cdot \overline{A}_2) = (\overline{A}_2 \cdot \overline{A}_1) \cdot A_1 + (\overline{A}_1 \cdot \overline{A}_2) \cdot A_2$$

$$= \overline{A}_2 \cdot (\overline{A}_1 \cdot A_1) + \overline{A}_1 \cdot (\overline{A}_2 \cdot A_2) = \overline{A}_2 \cdot 0 + \overline{A}_1 \cdot 0$$

$$= 0 + 0 = 0$$

and

$$X + \overline{X} = (A_1 + A_2) + (\overline{A}_1 \cdot \overline{A}_2) = [(A_1 + A_2) + \overline{A}_1] \cdot [(A_1 + A_2) + \overline{A}_2]$$

$$= [A_2 + (A_1 + \overline{A}_1)] \cdot [A_1 + (A_2 + \overline{A}_2)] = [A_2 + 1] \cdot [A_1 + 1]$$

$$= 1 \cdot 1 = 1$$

Thus, the first theorem is proved. The second theorem is the dual of the first theorem. Hence, the principle of duality applies and the second statement is also true. These theorems can be extended for any number of variables. In addition to above postulates and De Morgan's theorems, some commonly used Boolean theorems are given in Table-10.21.

Above-stated postulates and theorems are very useful tools for manipulating and simplifying complex switching functions. The following example illustrates the use of Boolean algebra theorems to simplify a logical function.

Example-10.16

Simplify the following logical function:

$$f(A_1, A_2, A_3, A_4) = A_1 + A_1 A_2 A_3 + \overline{A_1} A_2 A_3 + A_1 A_4 + A_1 \overline{A_4} + \overline{A_1} A_2$$

Solution

$$\begin{aligned}
f &= A_1 + (A_1 + \overline{A_1}) A_2 A_3 + A_1 (A_4 + \overline{A_4}) + \overline{A_1} A_2 \\
&= A_1 + A_2 A_3 + A_1 + \overline{A_1} A_2 \qquad (\because \quad A + \overline{A} = 1) \\
&= A_1 + A_2 A_3 + A_1 + A_2 \qquad \text{(from Theorem-9)} \\
&= A_1 + A_2 (1 + A_3) = A_1 + A_2 \qquad \text{[from Theorem-1 and Postulate 4(ii)]}
\end{aligned}$$

Table-10.21 Some commonly used Boolean theorems.

S.N.	Theorem	Proof
1.	$A_1 + 1 = 1$	A_1 can take either 0 or 1. Hence, in both case, the result is 1. Dual is also true.
2.	$A_1 \cdot 0 = 0$	
3.	$\overline{0} = 1$	As $A_1 + 0 = A_1$, if $A_1 = \overline{0}, \overline{0} + 0 = \overline{0}$. Also, as $A_1 + \overline{A_1} = 1$, if $A_1 = \overline{0}, \overline{0} + 0 = 1$. Therefore, $\overline{0} = 1$. Hence, dual is also true.
4.	$\overline{1} = 0$	
5.	$\overline{\overline{A}} = A_1$	If $A_1 = 0$, $\overline{A_3} \overline{A_4} = \overline{0} = 1$, and $\overline{\overline{A_1}} = \overline{1} = 0 = A_1$. Similarly, taking $A_1 = 1$, same result comes.
6.	$A_1 + A_1 \cdot A_2 = A_1$	$= A_1 \cdot 1 + A_1 \cdot A_2 = A_1 (1 + A_2) = A_1 \cdot 1 = A_1$. Dual is also true.
7.	$A_1 \cdot A_1 + A_2 = A_1$	
8.	$A_1 \cdot (\overline{A_1} + A_2) = A_1 A_2$	$= A_1 \overline{A_1} + A_1 \cdot A_2 = 0 + A_1 \cdot A_2 = A_1 \cdot A_2$. Dual is also true.
9.	$A_1 + \overline{A_1} \cdot A_2 = A_1 + A_2$	

10.5 SIMPLIFICATION OF LOGICAL FUNCTIONS

A logic or switching circuit is required to produce an output as a combination of its inputs such that the conditions given in the truth table are satisfied. A mathematical model representing the logic circuit is developed from the truth table. It is, now, possible to realize the mathematical model by means of logic gates. However, the design is not unique and it is possible to get many circuits performing the same function. Now, the designer task is to select an optimum circuit configuration. The general idea is to find a least costly circuit satisfying a given set of input-output relations. In some cases, however, due to some other considerations, a different design is considered.

As mentioned earlier, it is possible to simplify the logic functions using Boolean algebra. However, this is not convenient if the expressions are long and variables involved are many. In this section, various forms of representing a Boolean function are discussed first. Simplifying techniques are discussed next.

10.5.1 Forms of Boolean Functions

Let us consider the following two forms of Boolean functions:

$$Y = f(A_1, A_2, A_3) = A_1 + A_1 A_2 A_3 + \overline{A}_1 A_2 \overline{A}_3 \tag{10.23}$$

$$Y = f(A_1, A_2, A_3) = (A_1 + \overline{A}_2 + A_3)(A_2 + A_3)(\overline{A}_1 + A_2 + \overline{A}_3) \tag{10.24}$$

Before further discussion, let us define some terms here.

Literal: A literal is defined as a variable or its complement in a function. The function of Eq. (10.23) has seven literals, while the function of Eq. (10.24) has eight literals.

Conjunction: A product term in a function is defined as a conjunction. A product term may contain only one literal or a product of literals. In Eq. (10.23), there are three conjunctions.

Disjunction: A sum term in a function is defined as a disjunction. A sum term may contain only one literal or a sum of literals. In Eq. (10.24), there are three disjunctions.

10.5.2 Disjunctive Normal Form of Function

A logic function that contains a single conjunction or a sum of conjunctions is said to be in disjunctive normal form. It is also known as *sum of product (S of P) form*. In a special case, if each product term in a function consists of all the variables either in complemented form or in uncomplemented form, the function is called as *disjunctive canonical form or minterm canonical form or standard sum of product form*. The product terms in this case is known as standard products or minterms. Example of minterm canonical form of function is given in Eq. (10.25).

$$Y = f(A_1, A_2, A_3) = A_1 \overline{A}_2 \overline{A}_3 + A_1 A_2 A_3 + \overline{A}_1 A_2 \overline{A}_3 \tag{10.25}$$

10.5.3 Conjunctive Normal Form of Function

A logic function that contains a single disjunction or a product of disjunctions is said to be in conjunctive normal form. It is also known as *product of sum (P of S) form*. In a special case, if each sum term in a function consists of all the variables either in complemented form or in uncomplemented form, the function is called as *conjunctive canonical form or maxterm canonical form or standard product of sum form*. The sum terms in this case are known as standard sums or maxterms. Example of maxterm canonical form of function is given in Eq. (10.26).

$$Y = f(A_1, A_2, A_3) = (A_1 + \overline{A}_2 + A_3) \cdot (\overline{A}_1 + A_2 + A_3) \cdot (\overline{A}_1 + A_2 + \overline{A}_3) \quad (10.26)$$

If 0 is assigned to a complemented variable and 1 to an uncomplemented variable, a minterm (or product term) is represented by a binary number. The corresponding decimal number d is determined and the minterm is denoted by the symbol m_d. Hence, minterms of the function of Eq. (10.25) are represented by m_4, m_7, and m_2 and the function as $\Sigma m(4, 7, 2)$. For representing maxterm (sum term), 0 is assigned to an uncomplemented variable and 1 to a complemented variable. The maxterm is represented by M_d, where d is the decimal equivalent of binary number representing the maxterm. For example maxterms of the function of Eq. (10.26) are represented by M_2, M_4, and M_5 and the function by $\Pi M(2, 4, 5)$.

Example-10.17

Find the standard sum of product form for the following logical function:

$$Y = f(A_1, A_2, A_3, A_4) = A_1 \overline{A}_2 \overline{A}_3 + A_1 A_2 A_4 + A_1 A_3 A_4$$

Solution
Using Postulate-5(ii),

$$\begin{aligned} Y = f(A_1, A_2, A_3) &= A_1 \overline{A}_2 \overline{A}_3 (A_4 + \overline{A}_4) + A_1 A_2 A_4 (A_3 + \overline{A}_3) \\ &\quad + A_1 A_3 A_4 (A_2 + \overline{A}_2) \\ &= A_1 \overline{A}_2 \overline{A}_3 A_4 + A_1 \overline{A}_2 \overline{A}_3 \overline{A}_4 + A_1 A_2 A_3 A_4 + A_1 A_2 \overline{A}_3 A_4 \\ &\quad + A_1 A_2 A_3 A_4 + A_1 \overline{A}_2 A_3 A_4 \\ &= A_1 \overline{A}_2 \overline{A}_3 A_4 + A_1 \overline{A}_2 \overline{A}_3 \overline{A}_4 + A_1 A_2 A_3 A_4 (1 + 1) \\ &\quad + A_1 A_2 \overline{A}_3 A_4 + A_1 \overline{A}_2 A_3 A_4 \quad\quad (\because\ 1 + 1 = 1) \\ &= A_1 \overline{A}_2 \overline{A}_3 A_4 + A_1 \overline{A}_2 \overline{A}_3 \overline{A}_4 + A_1 A_2 A_3 A_4 + A_1 A_2 \overline{A}_3 A_4 + A_1 \overline{A}_2 A_3 A_4 \end{aligned}$$

Example-10.18

Find the standard product of sum form for the following logical function:

$$Y = f(A_1, A_2, A_3, A_4) = A_1 + A_2 + \overline{A}_3 \overline{A}_4$$

Solution

Using Postulate-3(i),

$$Y = f(A_1, A_2, A_3, A_4) = (A_1 + A_2 + \overline{A}_3)\cdot(A_1 + A_2 + \overline{A}_4)$$

$$= (A_1 + A_2 + \overline{A}_3 + A_4\overline{A}_4)\cdot(A_1 + A_2 + \overline{A}_4 + A_3\overline{A}_3) \qquad (\because \ A\cdot\overline{A} = 0)$$

$$= (A_1 + A_2 + \overline{A}_3 + A_4)\cdot(A_1 + A_2 + \overline{A}_3 + \overline{A}_4)\cdot$$
$$(A_1 + A_2 + A_3 + \overline{A}_4)\cdot(A_1 + A_2 + \overline{A}_3 + \overline{A}_4) \qquad \text{[from Postulate 3(i)]}$$

$$= (A_1 + A_2 + \overline{A}_3 + A_4)\cdot(A_1 + A_2 + \overline{A}_3 + \overline{A}_4)\cdot\ (A_1 + A_2 + A_3 + \overline{A}_4) \qquad (\because \ A\cdot A = A)$$

10.5.4 Simplification by K-Map Method

Karnaugh map (K-map) is a very useful tool in representing and simplifying logical functions. K-map is also known as Veitch representation of a logical function. It may be regarded as a pictorial representation of the truth table for a function.

For a 2-variable function, K-map is shown in Fig. 10.20. As there are four possible combinations of 2 variables, the K-map is made of four squares, each representing one combination. In each square, a product term of the sum of product form function is written in its respective square. Each square is designated by a decimal number corresponding to its binary number for the product term. For example, a product term (minterm) $A_1\overline{A}_2$ is represented by 10 (decimal number 2). The number of the square is written on the right-hand upper corner of the square. The uncomplemented variable is given 1 and the complemented variable is given 0. The two rows correspond to $\overline{A}_2(0)$, and $A_2(1)$, whereas two columns correspond to $\overline{A}_1(0)$ and $A_1(1)$.

Figure 10.21 illustrates the K-map for a logical function of 3 variables A_1, A_2, A_3. Rows correspond to $\overline{A}_3(0)$ and $A_3(1)$, whereas four columns correspond to four combinations $\overline{A}_1\overline{A}_2(00)$, $\overline{A}_1A_2(01)$, $A_1A_2(11)$, and $A_1\overline{A}_2(10)$.

Figures 10.22 to 10.24 show K-maps for 4-, 5-, and 6-variable functions, respectively.

A_2 \ A_1	$\overline{A}_1$ 0	A_1 1
$\overline{A}_2$ 0	0 $\overline{A}_1\overline{A}_2$	2 $A_1\overline{A}_2$
$\overline{A}_2$ 1	1 $\overline{A}_1A_2$	3 A_1A_2

FIGURE 10.20 The K-map for a 2-variable function.

A_3 \ A_1A_2	$\overline{A}_1\overline{A}_2$ 00	$\overline{A}_1A_2$ 01	A_1A_2 11	$A_1\overline{A}_2$ 10
$\overline{A}_3$ 0	0 $\overline{A}_1\overline{A}_2\overline{A}_3$	2 $\overline{A}_1A_2\overline{A}_3$	6 $A_1A_2\overline{A}_3$	4 $A_1\overline{A}_2\overline{A}_3$
A_3 1	1 $\overline{A}_1\overline{A}_2A_3$	3 $\overline{A}_1A_2A_3$	7 $A_1A_2A_3$	5 $A_1\overline{A}_2A_3$

FIGURE 10.21 The K-map for a 3-variable function.

A_3A_4 \ A_1A_2	$\bar{A}_1\bar{A}_2$ 0 0	$\bar{A}_1A_2$ 0 1	A_1A_2 1 1	$A_1\bar{A}_2$ 1 0
$\bar{A}_3\bar{A}_4$ 00	0	4	12	8
$\bar{A}_3A_4$ 01	1	5	13	9
A_3A_4 11	3	7	15	11
$A_3\bar{A}_4$ 10	2	6	14	10

FIGURE 10.22 The K-map for a 4-variable function.

A_4A_5 \ $A_1A_2A_3$	$\bar{A}_1$ 0				A_1 1			
	$\bar{A}_2\bar{A}_3$ 0 0	$\bar{A}_2A_3$ 0 1	A_2A_3 1 1	$A_2\bar{A}_3$ 1 0	$\bar{A}_2\bar{A}_3$ 0 0	$\bar{A}_2A_3$ 0 1	A_2A_3 1 1	$A_2\bar{A}_3$ 1 0
$\bar{A}_4\bar{A}_5$ 00	0	4	12	8	16	20	28	24
$\bar{A}_4A_5$ 01	1	5	13	9	17	21	29	25
A_4A_5 11	3	7	15	11	19	23	31	27
$A_4\bar{A}_5$ 10	2	6	14	10	18	22	30	26

FIGURE 10.23 The K-map for a 5-variable function.

$A_1A_5A_6$ \ $A_2A_3A_4$		$\bar{A}_2$ 0				A_2 1			
		$\bar{A}_3\bar{A}_4$ 0 0	$\bar{A}_3A_4$ 0 1	A_3A_4 1 1	$A_3\bar{A}_4$ 1 0	$\bar{A}_3\bar{A}_4$ 0 0	$\bar{A}_3A_4$ 0 1	A_3A_4 1 1	$A_3\bar{A}_4$ 1 0
$\bar{A}_1$ 0	$\bar{A}_5\bar{A}_6$ 00	0	4	12	8	16	20	28	24
	$\bar{A}_5A_6$ 01	1	5	13	9	17	21	29	25
	A_5A_6 11	3	7	15	11	19	23	31	27
	$A_5\bar{A}_6$ 10	2	6	14	10	18	22	30	26
A_1 0	$\bar{A}_5\bar{A}_6$ 00	32	36	44	40	48	52	60	56
	$\bar{A}_5A_6$ 01	33	37	45	41	49	53	61	57
	A_5A_6 11	35	39	47	43	51	55	63	59
	$A_5\bar{A}_6$ 10	34	38	46	42	50	54	62	58

FIGURE 10.24 The K-map for a 6-variable function.

K-map for a logical function expressed in minterm representation can be drawn by putting 1's in the squares, which correspond to the minterms of the function. In remaining squares 0's are written. In case the function is expressed in maxterm form, 0's are written in the squares corresponding to the maxterms and 1's are written in the rest of the squares. Following examples illustrate the above statement.

Example-10.19

Draw K-maps for the following logical functions:

(a) $f_1(A_1, A_2, A_3, A_4) = \sum m(0, 3, 5, 7, 12, 14, 15)$

(b) $f_2(A_1, A_2, A_3, A_4) = \prod M(1, 2, 4, 7, 10, 15)$

Solution

(a) The function is the sum of product (or minterms) form. Hence, K-map is drawn by writing 1's in the squares designated by 0, 3, 5, 7, 12, 14, 15, and 0's in rest of the squares. The K-map is drawn in Fig. 10.25.

$A_3A_4 \diagdown A_1A_2$	$\bar{A}_1\bar{A}_2$ 00	$\bar{A}_1A_2$ 01	A_1A_2 11	$A_1\bar{A}_2$ 10
$\bar{A}_3\bar{A}_4$ 00	0 / 1	4 / 0	12 / 1	8 / 0
$\bar{A}_3A_4$ 01	1 / 0	5 / 1	13 / 0	9 / 0
A_3A_4 11	3 / 1	7 / 1	15 / 1	11 / 0
$A_3\bar{A}_4$ 10	2 / 0	6 / 0	14 / 1	10 / 0

FIGURE 10.25 The K-maps for (a) f_1.

(b) Contrary to (a), the K-map is drawn by writing 0's in the squares designated by 1, 2, 4, 7, 10, 15, and 1's in rest of the squares, as the function is the product of sum (or maxterms) form. The K-map is drawn in Fig. 10.26.

$A_3A_4 \diagdown A_1A_2$	$\bar{A}_1\bar{A}_2$ 00	$\bar{A}_1A_2$ 01	A_1A_2 11	$A_1\bar{A}_2$ 10
$\bar{A}_3\bar{A}_4$ 00	0 / 1	4 / 0	12 / 1	8 / 1
$\bar{A}_3A_4$ 01	1 / 0	5 / 1	13 / 1	9 / 1
A_3A_4 11	3 / 1	7 / 0	15 / 0	11 / 1
$A_3\bar{A}_4$ 10	2 / 0	6 / 1	14 / 1	10 / 0

FIGURE 10.26 The K-maps for (b) f_2.

The following identity is useful in simplifying the logical function:

$$Af + \overline{A}f = (A + \overline{A}) \cdot f = f \qquad (10.27)$$

Let us assume that two terms in a function are identical except for only one variable, which appears in uncomplemented form in one term and in complemented form in another term. If both the terms are combined, it gives a term that contains only the common factor. For example, if two terms of the function of Example-10.19(a) taken are $A_1A_2A_3A_4$ and $A_1A_2A_3\overline{A}_4$, then after the combination of the two terms, the result is the common factor $A_1A_2A_3(A_4 + \overline{A}_4) = A_1A_2A_3$. The common factor contains only three of the four variables of the function. One of the variables is missing from the resulting product term. It can be observed from Fig. 10.25 that such two terms appear as adjacent 1's in the K-map. Thus, two 1's makes a loop, as shown in Fig. 10.25. If four 1's forms a loop, then two of the variables would be missing from the resulted product term. In general, the number of missing variables from the resulting product term would be j (= 1, 2, 3, ..), if adjacent 1's makes a loop containing 2^j squares. It is to be noted that the first and last rows and the first and last columns are also treated as adjacent. The following examples illustrate the above statement.

Example-10.20

Draw the K-maps for the following logical functions:

(a) $f_1(A_1, A_2, A_3, A_4) = A_1 + A_2A_4 + A_1A_2\overline{A}_3 + A_1\overline{A}_2A_3 + A_1A_2A_3A_4$

(b) $f_2(A_1, A_2, A_3, A_4) = (A_1 + \overline{A}_3) \cdot (\overline{A}_1 + \overline{A}_3 + A_4) \cdot (\overline{A}_1 + A_2 + A_3 + \overline{A}_4)$

(c) $f_3(A_1, A_2, A_3, A_4) = \sum m(0, 2, 4, 7, 11, 14, 15)$

Solution

(a) To explain the procedure clearly, first we draw the K-map for each term separately, and then the resultant K-map for the function is obtained by combining the individual maps. The maps are drawn in Fig. 10.27.

A_1

A_3A_4 \ A_1A_2	00	01	11	10
00			1	1
01			1	1
11			1	1
10			1	1

(a)

A_2A_4

A_3A_4 \ A_1A_2	00	01	11	10
00				
01		1	1	
11		1	1	
10				

(b)

(c) $A_1 A_2 \bar{A}_3$

A_1A_2 \ A_3A_4	00	01	11	10
00			1	
01			1	
11				
10				

(d) $A_1 \bar{A}_2 A_3$

A_1A_2 \ A_3A_4	00	01	11	10
00				
01				
11				1
10				1

(e) $A_1 A_2 A_3 A_4$

A_1A_2 \ A_3A_4	00	01	11	10
00				
01				
11		1		
10				

(f) $f(A_1 A_2 A_3 A_4)$

A_1A_2 \ A_3A_4	00	01	11	10
00	0	0	1	1
01	0	1	1	1
11	0	1	1	1
10	0	0	1	1

FIGURE 10.27 The K-maps for (a) A_1, (b) $A_2 A_4$, (c) $A_1 A_2 \bar{A}_3$, (d) $A_1 \bar{A}_2 A_3$, (e) $A_1 A_2 A_3 A_4$, and (f) the function of Example 10.20(a).

As discussed above, the number of adjacent 1's represent the number of missing variables in a minterm. Hence, for drawing the K-map for a minterm, the following steps are taken:

1. Depending on the missing variables in the considered minterm, calculate the number of adjacent squares containing 1's from 2^j, where j is the number of missing variables.
2. Choose the squares. The squares represent the variables (literals) present as well as missing variables (both uncomplemented and complemented).
3. Write 1 in all the chosen squares.
4. Repeat the procedure for minterms of the function.
5. Write 0 in the rest of the squares.

(b) The complemented function is given by

$$\bar{f}_2(A_1,A_2,A_3,A_4) = \bar{A}_1 \cdot A_3 + A_1 \cdot A_3 \cdot \bar{A}_4 + A_1 \cdot \bar{A}_2 \cdot \bar{A}_3 \cdot A_4$$

The K-map of the complemented function is drawn as in (a), except 0's are written in place of 1's. The rest of the squares are written with 1's. The K-map is shown in Fig. 10.28.

(c) The function is given in minterm form. Hence, 1's are written in the squares representing the minterm numbers and 0's are written in the rest of the squares. The complete K-map is shown in Fig. 10.29.

Now, we discuss the procedure for simplifying of logical functions by using K-map. The following steps are followed:

1. Draw the K-map of the given function.
2. Draw the different loops consisting adjacent 1's in the map.
3. Find the common factor of the terms enclosed by the loop. Start with the biggest loop. Replace all terms with their common factors.

A_3A_4 \ A_1A_2	00	01	11	10
00	1	1	1	1
01	1	1	1	0
11	0	0	1	1
10	0	0	0	0

FIGURE 10.28 The K-map for Example 10.20(b).

A_3A_4 \ A_1A_2	00	01	11	10
00	1	1	0	0
01	0	0	0	0
11	0	1	1	1
10	1	0	1	0

FIGURE 10.29 The K-map for Example 10.20(c).

The following example illustrates the procedure for simplifying the function.

Example-10.21

Determine a minimum sum of product expression for the following logical function:

$$f(A_1, A_2, A_3, A_4) = \sum m(0, 2, 4, 6, 10, 14, 15)$$

Solution

The K-map for this function is shown in Fig. 10.30.
From the K-map, the minimum sum of product expression for the function of Example-10.21 is given below:

$$f(A_1, A_2, A_3, A_4) = A_3\overline{A}_4 + A_1A_2A_3 + \overline{A}_1\overline{A}_3\overline{A}_4$$

A_3A_4 \ A_1A_2	$\overline{A}_1\overline{A}_2$ 0 0	$\overline{A}_1A_2$ 0 1	A_1A_2 1 1	$A_1\overline{A}_2$ 1 0
$\overline{A}_3\overline{A}_4$ 0 0	1 (0)	1 (4)	0 (12)	0 (8)
$\overline{A}_3A_4$ 0 1	0 (1)	0 (5)	0 (13)	0 (9)
A_3A_4 1 1	0 (3)	0 (7)	1 (15)	0 (11)
$A_3\overline{A}_4$ 1 0	1 (2)	1 (6)	1 (14)	1 (10)

FIGURE 10.30 The K-map of Example 10.21.

10.6 REVIEW QUESTIONS

1. Name different logic gates. What are their functions? What do you understand by truth table of a logic gate?
2. Draw the symbols and circuits for 2-input and 3-input AND gates. Write their Boolean expressions and truth tables.
3. What is the difference between the OR gate and Exclusive OR gate? Draw the building blocks and symbol for a 2-input XOR gate. Write the truth tables for OR and XOR gates, if there are three inputs.
4. Why are NAND and NOR gates called universal gates? Realize OR, AND, NOT gates by using NAND and NOR gates.
5. What do you know about the binary number? Why is binary system used in a digital computer?
6. Discuss the procedures for converting (a) a decimal number to a binary number and (b) a binary number to a decimal number.

7. Discuss the procedures of binary addition and subtraction.
8. Discuss about (a) signed binary numbers, (b) 1's complement, and (c) 2's complement. How is subtraction carried out in 2's complement representation?
9. Give the procedure of converting the fractional part of a decimal number to its binary equivalence.
10. What is a hexadecimal system of numbers? Discuss the procedure for the following conversions:
    (a) Hexadecimal to binary.
    (b) Binary to hexadecimal.
    (c) Hexadecimal to decimal.
    (d) Decimal to hexadecimal.
11. Explain the method of simplifying a logical function by using Karnaugh maps. Explain by taking a 2-variable logical function.

10.7 SOLVED PROBLEMS

1. Find the equivalent decimal numbers for the following binary numbers:
   (a) 11101,   (b) 1011,   (c) 100101,   (d) 101010,   (e) 111010

   Solution

   The equivalent decimal numbers can be calculated as follows:
   (a) $1\times 2^4 + 1\times 2^3 + 1\times 2^2 + 0\times 2^1 + 1\times 2^0 = \mathbf{29}$
   (b) $1\times 2^3 + 0\times 2^2 + 1\times 2^1 + 1\times 2^0 = \mathbf{11}$
   (c) $1\times 2^5 + 0\times 2^4 + 0\times 2^3 + 1\times 2^2 + 0\times 2^1 + 1\times 2^0 = \mathbf{37}$
   (d) $1\times 2^5 + 0\times 2^4 + 1\times 2^3 + 0\times 2^2 + 1\times 2^1 + 0\times 2^0 = \mathbf{42}$
   (e) $1\times 2^5 + 1\times 2^4 + 1\times 2^3 + 0\times 2^2 + 1\times 2^1 + 0\times 2^0 = \mathbf{58}$

2. Convert the following decimal numbers into their equivalent binary numbers.
   (a) 105,   (b) 112.67,   (c) 12.6875

   Solution

   The decimal number can be converted into binary numbers as follows:

   (a)
2	105	Remainder
2	52	1 (LSB)
2	26	0
2	13	0
2	6	1
2	3	0
2	1	1
	0	1 (MSB)

   The equivalent binary number is **1101001**.

(b) The number contains both integer and fractional parts. Let us first convert the integer into its equivalent binary number and then the fractional part. In fractional part conversion, the process is terminated after the seventh term.

Integer conversion

2	112	Remainder
2	56	0 (LSB)
2	28	0
2	14	0
2	7	0
2	3	1
2	1	1
	0	1 (MSB)

The integer binary number is 1110000.

Fractional conversion

Fraction	Fraction × 2	Integer
0.67	1.34	1 (MSB)
0.34	0.68	0
0.68	1.36	1
0.36	0.72	0
0.72	1.44	1
0.44	0.88	0
0.88	1.76	1 (LSB)

The fractional binary number is 1010101.

Hence, the equivalent binary number is **1110000. 1010101**.

(c) Integer conversion

2	12	Remainder
2	6	0 (LSB)
2	3	0
2	1	1
	0	1 (MSB)

The integer binary number is 1100.

Fractional conversion

Fraction	Fraction × 2	Integer
0.6875	1.375	1 (MSB)
0.375	0.75	0
0.75	1.5	1
0.5	1.00	1 (LSB)

The fractional binary number is 1011.

Hence, the equivalent binary number is **1100. 1011**.

3. (a) Convert the decimal numbers 71 and 67 into binary numbers. (b) Add the binary numbers and convert the result into decimal number.

Solution

(a) The decimal numbers are converted into binary as follows:

Decimal number 71

2	71	Remainder
2	35	1 (LSB)
2	17	1
2	8	1
2	4	0
2	2	0
2	1	0
	0	1 (MSB)

The binary number is **1000111**.

Decimal number 67

2	67	Remainder
2	33	1 (LSB)
2	16	1
2	8	0
2	4	0
2	2	0
2	1	0
	0	1 (MSB)

The binary number is **1000011**.

(b) Binary addition

$$
\begin{array}{r}
1\,0\,0\,0\,1\,1\,1 \\
(+)\,1\,0\,0\,0\,0\,1\,1 \\
\hline
1\,0\,0\,0\,1\,0\,1\,0
\end{array}
$$

Equivalent decimal number

$= 1\times 2^7 + 0\times 2^6 + 0\times 2^5 + 0\times 2^4 + 1\times 2^3 + 0\times 2^2 + 1\times 2^1 + 0\times 2^0$

$= \mathbf{138}$

4. (a) Convert the decimal numbers 67 and 71 into binary numbers. (b) Subtract binary equivalent of 67 from binary equivalent of 71 and convert the result into decimal number.

Solution

(a) The decimal numbers have already been converted into their equivalent binary numbers in Solved Problem-3.

(b) Binary subtraction

$$
\begin{array}{r}
1\,0\,0\,0\,1\,1\,1 \\
(-)\,1\,0\,0\,0\,0\,1\,1 \\
\hline
0\,0\,0\,0\,1\,0\,0
\end{array}
$$

Equivalent decimal number = **4**

5. Solve Problem-4(b) using 2's complement method of subtraction.

Solution

2's complement of 1000011 = 0111100 + 1 = 0111101

$$
\begin{array}{r}
1\,0\,0\,0\,1\,1\,1 \\
(+)\,\,0\,1\,1\,1\,1\,0\,1 \\
\hline
1\,0\,0\,0\,0\,1\,0\,0
\end{array}
$$

The carry is 1 and is discarded. Hence, the result is positive and the magnitude is 0000100 = **4**. The result is same as obtained in Solved Problem-4.

6. Using 2's complement method of subtraction, subtract 1110011 from 1011010.

Solution

2's complement of 1110011 = 0001100 + 1 = 0001101

$$
\begin{array}{r}
1\,0\,1\,1\,0\,1\,0 \\
(+)\,0\,0\,0\,1\,1\,0\,1 \\
\hline
1\,1\,0\,0\,1\,1\,1
\end{array}
$$

There is no carry. Hence, the result is negative. The magnitude is obtained by taking 2's complement of the result. 2's complement of 1100111 = 0011001 = $1 \times 2^4 + 1 \times 2^3 + 1 \times 2^0 =$ **25**.

7. Add the binary numbers 10110.111 and 0111.01.

Solution

The addition is performed as follows:

$$
\begin{array}{r}
10110.111 \\
(+)\ 00111.010 \\
\hline
11110.001
\end{array}
$$

8. Convert the hexadecimal number 4A9 into its decimal equivalent (a) by first converting into binary and (b) by direct method.

Solution

(a) The binary equivalence

4	A	9	Hexadecimal
0 1 0 0	1 0 1 0	1 0 0 1	Binary

$0 \times 2^{11} + 1 \times 2^{10} + 0 \times 2^9 + 0 \times 2^8 + 1 \times 2^7 + 0 \times 2^6 + 1 \times 2^5 + 0 \times 2^4$
$\quad + 1 \times 2^3 + 0 \times 2^2 + 0 \times 2^1 + 1 \times 2^0$
$= 2^{10} + 2^7 + 2^5 + 2^3 + 2^0 =$ **1193**   Decimal

(b) Direct method

$4 \times 16^2 + A \times 16^1 + 9 \times 16^0 = 4 \times 256 + 10 \times 16 + 9 =$ **1193**   Decimal

9. Convert decimal number 371 into its equivalent hexadecimal number by (a) indirect method and (b) direct method.

Solution

(a) The binary equivalence is given by

2	371	Remainder
2	185	1 (LSB)
2	92	1
2	46	0
2	23	0
2	11	1
2	5	1
2	2	1
2	1	0
	0	1 (MSB)

Hence, $371 = 101110011 = 000101110011$
In hexadecimal $= \mathbf{173}$

(b) Direct method:

16	371	Remainder
16	23	3 (LSB)
16	1	7
16	0	1 (MSB)

Hexadecimal number is **173**.

10. The truth table of a logic circuit is given in Table-10.22. (a) Find the logic function and (b) draw the logic circuit satisfying the truth table.

Table-10.22 Truth table for Solved Problem-10.

A_1	A_2	A_3	Y
0	0	0	0
0	0	1	1
0	1	0	0
0	1	1	0
1	0	0	1
1	0	1	1
1	1	0	1
1	1	1	1

Solution

(a) From the truth table $Y = 1$ for the second row. Hence, the first minterm is $\overline{A_1}\overline{A_2}A_3$. Similarly, other minterms are $A_1\overline{A_2}\overline{A_3}$, $A_1\overline{A_2}A_3$, $A_1A_2\overline{A_3}$, and $A_1A_2A_3$, which corresponds to row numbers 5, 6, 7, and 8. The logic function in canonical form, therefore, is given by

$$Y = \overline{A_1}\overline{A_2}A_3 + A_1\overline{A_2}\overline{A_3} + A_1\overline{A_2}A_3 + A_1A_2\overline{A_3} + A_1A_2A_3$$

The function can be simplified by using Boolean theorems and postulates.

$$Y = \overline{A_1}\overline{A_2}A_3 + A_1\overline{A_2}\overline{A_3} + A_1\overline{A_2}A_3 + A_1A_2\overline{A_3} + A_1A_2A_3$$
$$= \overline{A_1}\overline{A_2}A_3 + A_1\overline{A_2}(\overline{A_3} + A_3) + A_1A_2(\overline{A_3} + A_3)$$
$$= \overline{A_1}\overline{A_2}A_3 + A_1\overline{A_2} + A_1A_2 = \overline{A_1}\overline{A_2}A_3 + A_1(\overline{A_2} + A_2) = A_1 + \overline{A_1}\overline{A_2}A_3$$
$$= (A_1 + \overline{A_1}) \cdot (A_1 + \overline{A_2}A_3) \qquad \text{(from associative property)}$$
$$= A_1 + \overline{A_2}A_3$$

(b) The logic circuit is as given in Fig. 10.31.

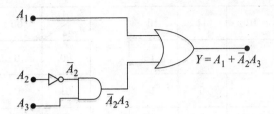

FIGURE 10.31 A logic circuit for Solved Problem 10.

11. (a) Draw the switching circuit for the logic function
$$Y = \overline{\overline{A_1 \cdot A_2} + \overline{A_1 \cdot A_3}}$$

(b) Simplify the logic function and draw the logic circuit for the simplified function.

Solution

(a) The logic or switching circuit for $Y = \overline{\overline{A_1 \cdot A_2} + \overline{A_1 \cdot A_3}}$ is given in Fig. 10.32(a).

(b) $Y = \overline{\overline{A_1 \cdot A_2} + \overline{A_1 \cdot A_3}} \quad = \overline{\overline{A_1} + \overline{A_2} + \overline{A_1} + \overline{A_3}}$ (by De Morgan's theorem)

$= \overline{\overline{A_1} + \overline{A_2} + \overline{A_3}} = \overline{\overline{A_1}} \cdot \overline{\overline{A_2}} \cdot \overline{\overline{A_3}} = A_1 \cdot A_2 \cdot A_3$ (by De Morgan's theorem)

The logic circuit for the simplified function is shown in Fig. 10.32(b).

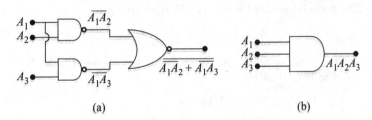

(a)   (b)

FIGURE 10.32 A logic circuit for Solved Problem 11.

12. Draw the K-maps for the following switching functions:
$$Y_1 = \overline{A}_2 + A_1 A_3 + A_2 A_3 \overline{A}_4 + A_1 \overline{A}_2 A_3 + A_1 \overline{A}_2 A_3 A_4$$
$$Y_2 = \sum m(2, 5, 6, 9, 10, 13, 14)$$

Solution

The K-map for the following switching function is drawn in Fig. 10.33(a):
$$Y_1 = \overline{A}_2 + A_1 A_3 + A_2 A_3 \overline{A}_4 + A_1 \overline{A}_2 A_3 + A_1 \overline{A}_2 A_3 A_4$$

The K-map for the following switching function is shown in Fig. 10.33(b):
$$Y_2 = \sum m(2, 5, 6, 9, 10, 13, 14)$$

Figure 10.33

(a)

$A_3A_4 \backslash A_1A_2$	00	01	11	10
00	1	0	0	1
01	1	0	0	1
11	1	0	1	1
10	1	1	1	1

(b)

$A_3A_4 \backslash A_1A_2$	00	01	11	10
00	0	0	0	0
01	0	1	1	1
11	0	0	0	0
10	1	1	1	1

FIGURE 10.33 The K-maps for the switching functions of Solved Problem 12.

13. Draw the K-maps for the following switching functions:

$$Y_1 = (\overline{A}_1 + A_3) \cdot (A_1 + A_3 + \overline{A}_4) \cdot (A_1 + \overline{A}_2 + \overline{A}_3 + A_4)$$

$$Y_2 = \prod M(0, 3, 5, 7, 11, 13)$$

Solution

The complemented function of Y_1 is given by

$$\overline{Y}_1 = A_1 \cdot \overline{A}_3 + \overline{A}_1 \cdot \overline{A}_3 \cdot A_4 + \overline{A}_1 \cdot A_2 \cdot A_3 \cdot \overline{A}_4$$

and its K-map is shown in Fig. 10.34(a). The K-map for the following logic function is given in Fig. 10.34(b):

$$Y_2 = \prod M(0, 3, 5, 7, 11, 13)$$

(a)

$A_3A_4 \backslash A_1A_2$	00	01	11	10
00	1	1	0	0
01	0	0	0	0
11	1	1	1	1
10	1	0	1	1

(b)

$A_3A_4 \backslash A_1A_2$	00	01	11	10
00	0	1	1	1
01	1	0	0	1
11	0	0	1	0
10	1	1	1	1

FIGURE 10.34 The K-maps logic function for Solved Problem 13.

14. A logic function in canonical form is given by
$$Y = \overline{A}_1\overline{A}_2 A_3 + A_1\overline{A}_2\overline{A}_3 + A_1\overline{A}_2 A_3 + A_1 A_2 \overline{A}_3 + A_1 A_2 A_3$$

Use K-map method to simplify the logic function.

Solution

The K-map is drawn in Fig. 10.35.

The four adjacent squares that contain 1 are 6, 4, 5, 7. The common variable in these squares is A_1. Hence, one term of the function is A_1. The two adjacent squares that contain 1 are 1 and 5. The common factor in these squares is $\overline{A}_2 A_3$. Hence, the second term of the function is $\overline{A}_2 A_3$. The simplified function, therefore, is
$$Y = A_1 + \overline{A}_2 A_3$$

The result is same as obtained in Solved Problem 10.

A_3 \ $A_1 A_2$	$\overline{A}_1\overline{A}_2$ 0 0	$\overline{A}_1 A_2$ 0 1	$A_1 A_2$ 1 1	$A_1 \overline{A}_2$ 1 0
$\overline{A}_3$ 0	0 (0)	2 (0)	6 (1)	4 (1)
A_3 1	1 (1)	3 (0)	7 (1)	5 (1)

FIGURE 10.35 The K-maps logic function for Solved Problem 14.

15. Determine a minimum sum of product expression for the following logical function:
$$Y = \sum m(0, 2, 5, 7, 10, 12, 15)$$

Solution

The K-map for the function is shown in Fig. 10.36.

The logic function is given as follows:
$$Y = \overline{A}_1 A_3 \overline{A}_4 + A_2 A_3 A_4 + \overline{A}_1 A_2 A_4 + \overline{A}_1 \overline{A}_2 \overline{A}_4 + A_1 A_2 \overline{A}_3 \overline{A}_4$$

$A_3 A_4$ \ $A_1 A_2$	$\overline{A}_1\overline{A}_2$ 0 0	$\overline{A}_1 A_2$ 0 1	$A_1 A_2$ 1 1	$A_1 \overline{A}_2$ 1 0
$\overline{A}_3\overline{A}_4$ 0 0	0 (1)	4 (0)	12 (1)	8 (0)
$\overline{A}_3 A_4$ 0 1	1 (0)	5 (1)	13 (0)	9 (0)
$A_3 A_4$ 1 1	3 (0)	7 (1)	15 (1)	11 (0)
$A_3 \overline{A}_4$ 1 0	2 (1)	6 (0)	14 (0)	10 (1)

FIGURE 10.36 The K-maps logic function for Solved Problem 15.

10.8 EXERCISES

1. Find the equivalent decimal numbers for the binary numbers given below:
   (a) 1101, (b) 10111, (c) 100111, (d) 10110, (e) 111011

2. Find binary equivalent of the following decimal numbers:
   (a) 13.75, (b) 123.37, (c) 38, (d) 120.39, (e) 15.625

3. (a) Convert the decimal numbers 87 and 76 into binary numbers. (b) Add the binary numbers and convert the result into decimal number.

4. (a) Convert the decimal numbers 95 and 125 into binary numbers. (b) Subtract the binary equivalent of 95 from the binary equivalent of 125 and convert the result into decimal number.

5. Use 2's complement method of subtraction to subtract 0110101 from 1101011.

6. Using 2's complement method of subtraction subtract 1101011 from 0110101.

7. Add the binary numbers 1101.011 and 1110.101.

8. Convert the hexadecimal number 4C3 into its decimal equivalent (a) by first converting into binary and (b) by direct method.

9. Find decimal equivalent of the hexadecimal number 9B4 by (a) indirect method and (b) direct method.

10. Convert decimal number 295 into its hexadecimal equivalence by (a) indirect method and (b) direct method.

11. The truth table of a logic circuit is given in Table-10.23. (a) Find the logic function and (b) draw the logic circuit satisfying the truth table.

   Table-10.23 Truth table for question 11.

A_1	A_2	A_3	Y
0	0	0	1
0	0	1	1
0	1	0	1
0	1	1	1
1	0	0	0
1	0	1	0
1	1	0	1
1	1	1	0

12. (a) Draw the logic circuit for the logic function
    $$Y = \overline{\overline{A_1} \cdot A_2 + \overline{A_1}} + A_1 A_2$$
    (b) Simplify the logic function and draw the logic circuit for the simplified function.

13. Draw the K-maps for the following switching functions:

$$Y_1 = A_2 + \overline{A}_1\overline{A}_3 + A_2\overline{A}_3\overline{A}_4 + A_1\overline{A}_2A_3 + A_1\overline{A}_2\overline{A}_3A_4$$

$$Y_2 = \sum m(3, 5, 7, 8, 11, 14, 15)$$

$$Y_3 = (A_1 + A_2) \cdot (\overline{A}_2 + A_3 + A_4) \cdot (A_1 + \overline{A}_2 + \overline{A}_3 + \overline{A}_4)$$

$$Y_4 = \prod M(0, 3, 6, 9, 12, 15)$$

14. A logic function in canonical form is given by

$$Y = \overline{A}_1 A_2 \overline{A}_3 + A_1 \overline{A}_2 \overline{A}_3 + \overline{A}_1 \overline{A}_2 A_3 + \overline{A}_1 \overline{A}_2 \overline{A}_3 + A_1 A_2 A_3$$

Use K-map method to simplify the logic function.

15. Determine a minimum sum of product expression for the following logical function:

$$Y = \sum m(0, 3, 5, 8, 10, 11, 14)$$

11
COMPONENTS OF DIGITAL SYSTEMS

Outline

11.1 Introduction 540
11.2 Encoders 543
11.3 Adders 544
11.4 Subtractors 550
11.5 Decoders and Demultiplexers 552
11.6 Data Selectors/Multiplexers 556
11.7 ROM Unit 558
11.8 Flip Flops 561
11.9 Registers 568
11.10 Counters 572
11.11 RAM Unit 581
11.12 D/A Converter 584
11.13 A/D Converter 590
11.14 Review Questions 593
11.15 Solved Problems 595
11.16 Exercises 621

11.1 INTRODUCTION

A digital system is constructed from a few basic network configurations. These basic blocks are used frequently in digital systems in various combinations. As discussed in the previous chapter, it is possible to perform all logic operations by using a single logic gate, namely, NAND or NOR gate. In addition, to perform logic operations, a digital system requires storing binary numbers. For this purpose a memory cell, called flip flop, is used. Therefore, it can be said that any digital system can be constructed entirely from NAND gates and flip flops. To perform functions in a digital system different combinations of gates and/or flip flops are required. A combination of gates and/or flip flops, performing a particular function, available on a single chip, is called an integrated circuit. These integrated circuits form the building components for a digital system. These integrated circuits may perform the following functions:

1. Binary addition.
2. Decoding (demultiplexing).
3. Data selection (multiplexing).
4. Counting.
5. Digital-to-analogue (D/A) conversion.
6. Analogue-to-digital (A/D) conversion.
7. Storage of binary information (memories and registers).

In addition, there are a few other related operations.

The building components are divided into the following three categories:

1. Combinational logic circuits.
2. Sequential logic circuits.
3. D/A and A/D converters.

11.1.1 Combinational Logic Circuits

A combinational logic circuit is composed of gate structures and is driven by a set of input signals and gives a set of output signals. The input signals are completely independent of the state of the output signals. At any instant of time, the binary outputs are functions of binary combinations of input signals. A block diagram of a combinational logic circuit is shown in Fig. 11.1. The n binary input variables come from an external source and the m output variables go to an external destination. A combinational logic circuit transforms binary information from the given input signals to the required output signals. Combinational logic circuits are employed in digital computers for the generation of binary control decisions and work as digital components required for data processing.

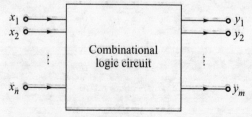

FIGURE 11.1 A combinational logic circuit.

A combinational logic circuit satisfies the instructions derived from a truth table or a given logic function. For n binary input variables, the truth table lists the corresponding output binary values for each of 2^n input combinations. A combinational logic circuit can also be specified with m Boolean functions, one for each output variable. Each output function is expressed in terms of the n input variables.

11.1.2 Sequential Logic Circuits

A sequential logic circuit differs from a combinational logic circuit in two important ways:

1. Sequential logic circuits possess memory to store the output in the form of binary bits. The stored output remains in its state till it is not instructed to do otherwise.
2. Sequential logic circuits have a feedback loop that serve to apply the stored outputs to logic gates at an appropriate point in the circuit.

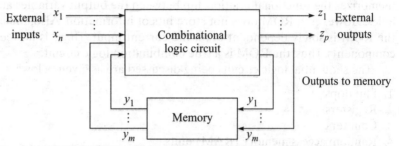

FIGURE 11.2 A block diagram of a sequential logic circuit.

The block diagram of a sequential logic circuit is shown in Fig. 11.2. The memory elements are called flip flops. From the block diagram it is clear that a sequential logic circuit consists of a combinational logic circuit and a number of clocked flip flops. The combinational logic block receives binary signals from external inputs and from the outputs of flip flops. The outputs of the combinational logic circuit go to external destination and to inputs of the flip flops. The combinational logic circuit gates determine the binary value to be stored in the flip flops after each clock transition. The outputs of the flip flops, in turn, are fed back to the combinational logic circuit. Thus, the outputs of the flip flops determine the behaviour of the sequential logic circuit. Hence, the external outputs of a sequential logic circuit are functions of both external inputs and the present state of the flip flops. The next state of the flip flops is also a function of their present state and external inputs. Thus, a sequential logic circuit is specified by a time sequence of external inputs, external outputs, and internal flip flops binary states.

From above discussion we conclude that the outputs of a combinational logic circuit at a given instant of time is solely dependent upon the external inputs at that instant of time, whereas the current state of the outputs of a sequential logic circuit is dependent not only upon the current external inputs but also upon a time sequence of external inputs, external outputs, and internal flip flop binary states.

Sequential logic circuits are of two types, namely, synchronous and asynchronous sequential logic circuits. The output states of synchronous sequential logic circuits change only at discrete clocked instants of time. In asynchronous sequential logic circuits, output states depend on the order in which the input signals change. However, we discuss only about synchronous sequential logic circuits.

The following combinational logic circuits are discussed in the succeeding sections:

1. Encoders.
2. Adders.
3. Subtractors.
4. Decoders and demultiplexers.
5. Data selectors/multiplexers.
6. Read only memory (ROM) units.

Classifying the ROM under the combinational logic circuits requires some clarifications. The ROM does not possess memory in the sense as required by a sequential logic circuit. It is important to understand that the memory of a ROM is a memory that memorizes the functional relationship between the output variables and the input variables. However, a ROM does not store bits of information. Also, the current output of the ROM is solely dependent upon the current input. There is no feedback of output components. Thus, the ROM is a true combination logic circuit.

The sequential logic circuits being discussed are as given below:

1. Flip flops.
2. Registers.
3. Counters.
4. Random access memory (RAM) units.

11.1.3 D/A and A/D Converters

A digital system is used as a digital controller in a control system for controlling a physical system. The block diagram of a control system using a digital controller is shown in Fig. 11.3. Signals used in physical systems are, in general, analogue signals. The input signals to the digital controller must be digital signals. Therefore, a device that converts an analogue signal into a digital signal is required. This device is known as an A/D converter or ADC. Outputs of the digital controller are digital signals, whereas the controlling signals to the physical system need to be analogue signals. Hence, we need a device that converts digital signals into analogue signals. This device is called a digital-to-analogue (D/A) converter or DAC.

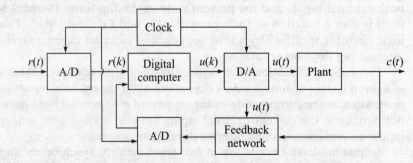

FIGURE 11.3 A digital feedback control system.

11.2 ENCODERS

An encoder converts alphanumeric characters to binary code. The decimal numbers and alphabet together are called alphanumeric characters. An encoder may be a decimal to binary encoder, hexadecimal to binary encoder, octal to binary-coded decimal (BCD) encoder, etc. If the coded binary number is to be expressed with n bits, the number of input lines (i.e., decimal number to be encoded) is given by 2^n. For example, for coded binary number of 3 bits, the number of lines that can be encoded is 8. Accordingly, as the output is coded with 3 bits, the number of OR gates required is three. It is also important to know that at any one time only one of the input lines can be activated (i.e., closed or at logic '1') while encoding takes place. A decimal to binary encoder for 10 inputs is shown in Fig. 11.4. As 10 inputs can be coded in binary by 4 bits, the number of OR gates required is four.

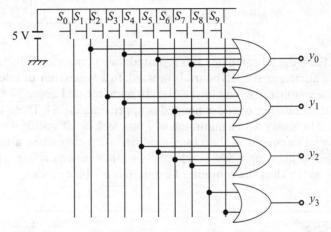

FIGURE 11.4 A decimal to binary encoder with 10 inputs (0–9).

From the circuit diagram, it is clear that when switch S_0 is pressed for encoding 0, none of the OR gates is activated and all the gate outputs are low. So, the encoded output word is

$$Y_3Y_2Y_1Y_0 = 0\,0\,0\,0$$

When decimal number 4 is to be encoded, switch S_4 is pressed activating only OR gate Y_2. The encoded output word is

$$Y_3Y_2Y_1Y_0 = 0\,1\,0\,0$$

When decimal number 7 is to be encoded, switch S_7 is pressed activating OR gates Y_2, Y_1, and Y_0. The encoded output word is

$$Y_3Y_2Y_1Y_0 = 0\,1\,1\,1$$

Hence, the encoder of Fig. 11.4 converts the decimal numbers 0 to 9 into their corresponding binary words. The truth table for decimal to binary encoder for 10 inputs is given in Table-11.1.

Table-11.1 Truth table for decimal to binary encoder.

S_0	S_1	S_2	S_3	S_4	S_5	S_6	S_7	S_8	S_9	Y_3	Y_2	Y_1	Y_0
1	0	0	0	0	0	0	0	0	0	0	0	0	0
0	1	0	0	0	0	0	0	0	0	0	0	0	1
0	0	1	0	0	0	0	0	0	0	0	0	1	0
0	0	0	1	0	0	0	0	0	0	0	0	1	1
0	0	0	0	1	0	0	0	0	0	0	1	0	0
0	0	0	0	0	1	0	0	0	0	0	1	0	1
0	0	0	0	0	0	1	0	0	0	0	1	1	0
0	0	0	0	0	0	0	1	0	0	0	1	1	1
0	0	0	0	0	0	0	0	1	0	1	0	0	0
0	0	0	0	0	0	0	0	0	1	1	0	0	1

For a digital computer, it is required that a binary code be transmitted when any key of an alphanumeric keyboard is pressed. In a typewriter or teletype, there are 26 letters for each upper case and lower case, 10 numeral, and about 22 special characters. Hence, the total number of codes needed is approximately 84. These 84 alphanumerics can be coded in binary with a minimum of 7 bits, as $2^7 = 128$ and $2^6 = 64$. A simple keyboard for a digital computer is, therefore, modified such that when a key is pressed, a switch is closed, connecting a 5 V (say) supply to the corresponding input line, making the line state as '1'. Thus, the computer keyboard is a 7-bit encoder.

11.3 ADDERS

Like decimal addition, the addition of two binary numbers is carried out column by column starting from the least significant bit (LSB) and moving to the most significant bit (MSB). The addition of LSB is the simplest operation of addition. This is because the operation does not need to handle a carry bit. The combinational logic circuit that performs binary addition without carry is called *half adder*. This circuit is called half adder, as two such circuits are used to take care of the carry from the preceding bit.

11.3.1 Half Adder

The symbol for a half adder is shown in Fig. 11.5(a) and the truth table is given in Table-11.2. It is called a half adder as two half adders make a full adder. The inputs A and B to the half adder are the LSBs of the two binary numbers to be added. There are two output variables S and C for two inputs. The column S gives the sum of inputs A and B, and the column C represents the carry. A carry occurs when $A = 1$ and $B = 1$. For the binary system the addition 1 + 1 yields 10, the binary equivalent of decimal number 2. The column S of the truth table of Table-11.2 represents the logical operation XOR (Exclusive OR). The column C, obviously, represents the logical operation AND. Hence, the truth table of Table-11.2 represents the logical circuit of Fig. 11.5(b).

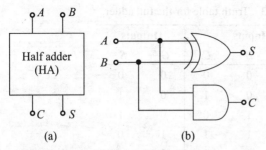

Table-11.2 Truth table of half adder.

Inputs		Outputs	
A	B	C	S
0	0	0	0
0	1	0	1
1	0	0	1
1	1	1	0

FIGURE 11.5 (a) Symbol and (b) implementation of a half adder.

Two alternative implementation of the half adder can be obtained by expressing the logical function of the half adder in sum of product form and product of sum form. The logical function of sum (or S) column of the half adder in sum of product form is

$$S = \overline{A}B + A\overline{B} \tag{11.1}$$

The carry column is output of an AND gate. Hence, one of the alternative implementations of the half adder is shown in Fig. 11.6(a).

The logical function of sum (or S) column of the half adder in product of sum form is

$$S = (A+B) \cdot (\overline{A}+\overline{B}) \tag{11.2}$$

The carry column is output of an AND gate. Hence, another alternative implementation of the half adder is shown in Fig. 11.6(b).

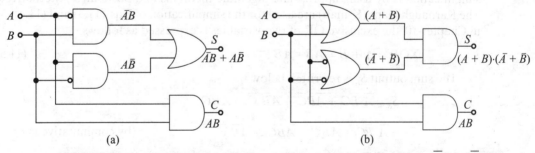

FIGURE 11.6 An alternative logic diagram of the half adder when (a) $S = \overline{A}B + A\overline{B}$ and (b) $S = (A+B) \cdot (\overline{A}+\overline{B})$.

11.3.2 Full Adder

This logic circuit is called a full adder as it accommodates the carry from the preceding bit addition as well as two present bits of the two binary numbers to be added. Thus, the full adder has three inputs. The truth table for the full adder is given in Table-11.3.

To design the logic circuit for the implementation of the full binary addition we follow the truth table of Table-11.3. The three inputs are A, B, and C_i. A and B are concerned bits of the two binary numbers to be added, whereas the C_i is the carry bit from the preceding bits addition. The outputs are sum S_o and carry C_o. The carry output C_o is carried as one of the three inputs to subsequent addition of the next significant bits of the two binary numbers to be added.

Table-11.3 Truth table for the full adder.

Inputs			Outputs	
A	B	C_i	C_o	S_o
0	0	0	0	0
0	0	1	0	1
0	1	0	0	1
0	1	1	1	0
1	0	0	0	1
1	0	1	1	0
1	1	0	1	0
1	1	1	1	1

Like half adder one design for the logic circuit that performs full addition of two binary numbers is obtained by expressing the outputs S_o and C_o in sum of products form. From Table-11.3,

$$S_o = \overline{A}\,\overline{B}C_i + \overline{A}B\overline{C_i} + A\overline{B}\,\overline{C_i} + ABC_i \tag{11.3}$$

and $$C_o = \overline{A}BC_i + A\overline{B}C_i + AB\overline{C_i} + ABC_i \tag{11.4}$$

The full adder can be implemented by simplifying Eqs. (11.3) and (11.4). One way of simplification is by using appropriate algebraic theorems and postulates. Alternatively, The Karnaugh maps (K-map) are used for the simplification of Eqs. (11.3) and (11.4). As in Chapter-10, the exclusive OR of two variables is expressed as follows:

$$\text{XOR} = A \oplus B = \overline{A}B + A\overline{B} \tag{11.5}$$

The sum output S_o is rewritten below

$$S_o = \overline{A}\,\overline{B}C_i + \overline{A}B\overline{C_i} + A\overline{B}\,\overline{C_i} + ABC_i$$

$$= \overline{A}\,\overline{B}C_i + ABC_i + \overline{A}B\overline{C_i} + A\overline{B}\,\overline{C_i} \quad \text{(by commutative law)}$$

$$= (\overline{A}\,\overline{B} + AB)C_i + (\overline{A}B + A\overline{B})\overline{C_i} \quad \text{(by associative law)}$$

$$= (\overline{A}A + \overline{A}\,\overline{B} + AB + B\overline{B})C_i + (\overline{A}B + A\overline{B})\overline{C_i} \quad (\because \overline{A}A = 0, B\overline{B} = 0)$$

$$= (\overline{A}(A+\overline{B}) + B(A+\overline{B}))C_i + (\overline{A}B + A\overline{B})\overline{C_i}$$

$$= (A+\overline{B})(\overline{A}+B)C_i + (\overline{A}B + A\overline{B})\overline{C_i} \quad \text{(by factorization)}$$

$$= (\overline{A\overline{B}} + \overline{\overline{A}B})C_i + (\overline{A}B + A\overline{B})\overline{C_i} \quad \text{(by De Morgan theorem)}$$

$$= (\overline{A \oplus B})C_i + (A \oplus B)\overline{C_i}$$

$$= (A \oplus B) \oplus C_i \tag{11.6}$$

From the above expression it is clear that two XORs are required to implement the full adder. The inputs to the first XOR are A and B, whereas those for the second XOR are output of the first XOR and C_i.

Let us apply similar procedure to the carry output of Eq. (11.4).

$$C_o = \overline{A}BC_i + A\overline{B}C_i + AB\overline{C}_i + ABC_i$$

$$= (\overline{A}B + A\overline{B})C_i + AB(\overline{C}_i + C_i) \quad \text{(by associative law)}$$

$$= (A \oplus B)C_i + AB \quad [\because (\overline{C}_i + C_i) = 1] \quad (11.7)$$

To implement the above logic we need two AND gates and one OR gate. The inputs for the first AND gate are A and B and those for the second are output of the first XOR and the carry from preceding bits addition C_i. The logic circuit for the full adder is shown in Fig. 11.7.

The statement made earlier that two half adders make a full adder is very clear from the logic circuit of Fig. 11.7.

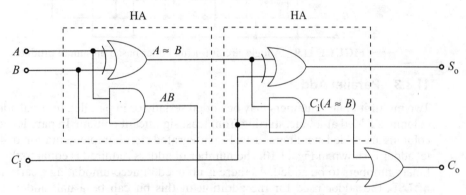

FIGURE 11.7 A logic circuit for the full adder.

Alternatively, for simplifying the above functions of full adder the K-maps are used. The K-maps for the two functions are given in Fig. 11.8.

Since there are no adjacent 1s in K-map of S_o, the function of Eq. (11.3) is in its simplest form. On the contrary, there are adjacent 1s in the K-map of C_o and the function of Eq. (11.4) is simplified as follows:

$$C_o = AB + AC_i + BC_i \quad (11.8)$$

C_i\\AB	$\bar{A}\bar{B}$	$\bar{A}B$	AB	$A\bar{B}$
$\bar{C}_i$	0	1	0	1
C_i	1	0	1	0

(a)

C_i\\AB	$\bar{A}\bar{B}$	$\bar{A}B$	AB	$A\bar{B}$
$\bar{C}_i$	0	0	1	0
C_i	0	1	1	1

(b)

FIGURE 11.8 The K-maps for (a) output sum S_o and (b) output carry C_o.

For implementing logic function of Eq. (11.3), four AND gates and one OR gate are needed. For implementing Eq. (11.8), three AND gates and one OR gate are needed. The logic circuit is shown in Fig. 11.9.

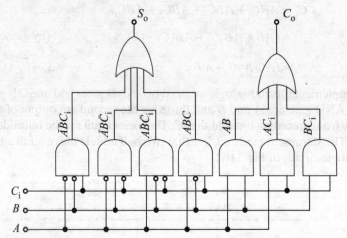

FIGURE 11.9 A logic circuit for full adder based on sum of products.

11.3.3 Parallel Adder

Two multibit binary numbers may be added serially or in parallel. In serial addition, one column is added at a time, starting from least significant column. In parallel addition, all columns are added simultaneously. The integrated circuit implementation of 4-bit parallel adder is shown in Fig. 11.10. The number of adders required is equal to the bits in the binary numbers to be added. As there is no need of accommodating a carry in addition of LSB, the adder used for the addition of this bit can be a half adder. The adders required for remaining bits, however, must be full adders. In parallel operation, it is required that all bits are available simultaneously and that they are applied to each adder at the same time. Therefore, the binary numbers to be added are frequently stored in storage registers before being applied to the adders.

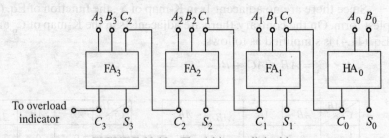

FIGURE 11.10 The 4-bit parallel adder.

The advantage of parallel addition is its fast operation. However, because of the time required to propagate the carry information through the various levels of the logic gates, the delay in operation is encountered. This is called *carry ripple*. In practical version of the adder, the ripple effect is minimized by modifying the design.

11.3.4 Serial Adder

As discussed earlier, in a serial adder one column is added at a time. In computers, the input A and B are fed to two lines in the form of two synchronous pulse trains. The adder takes as inputs the two trains of pulses. The output is also a train of pulses, representing the sum of two numbers. Figure 11.11(a) and (b) show typical trains of pulses, representing decimal numbers 14 and 11. The pulse train representing the sum (25) is shown in Fig. 11.11(c).

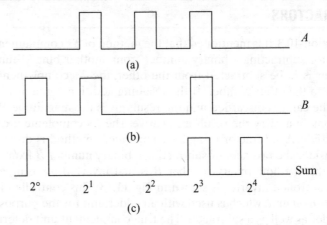

FIGURE 11.11 Pulse trains representing numbers (a) $A = 14$, (b) $B = 11$, and (c) sum = 25.

As we know, the sum of two multibit numbers is obtained by adding the carry to the sum of the bits of like significance. The carry is from the addition of just lower significant bit. The logic circuit of the serial adder is shown in Fig. 11.12. This circuit includes a time delay unit TD. The time delay unit is a *D flip flop*. Time delay is equal to time T between pulses. The carry is moved to the D flip flop. The serial numbers A_n, B_n, and S_n are stored in *shift registers*. D flip flop and shift register are discussed in succeeding sections of this chapter.

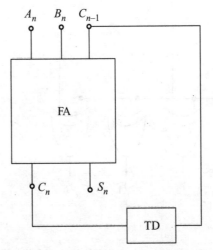

FIGURE 11.12 A serial binary full adder.

For example, consider pulse trains of Fig. 11.11(a) and (b). Let us suppose that we are adding the bit corresponding to $n = 2$. Hence, $A_n = 1$ and $B_n = 0$. At this time the carry, that is, C_{n-1}, in the D flip flop is 1. The resulted outputs are $S_n = 0$ and $C_n = 1$. Now, in shift registers $A_n = 1$, $B_n = 1$, and $S_n = 0$, and D flip flop output is C_{n-1} is again 1. The process of addition of next bit is repeated.

11.4 SUBTRACTORS

In Section-10.3, it is mentioned that the method of 2's complement is used in digital computer for subtracting a binary number from another binary number. When one binary number is to be subtracted from the other, the 2's complement of the first number is added to the other number. If the resulting addition contains a carry, the result is positive. The carry is discarded and the result gives the magnitude. When the addition does not contain a carry the result is negative. The 2's complement of the addition gives the magnitude. A subtractor using 2's complement method is discussed in this section. Let us consider the problem of subtracting a binary number B from another binary number A. In this method, binary number B is first inverted by using a controlled inverter. A 4-bit controlled inverter is shown in Fig. 11.13. This controlled inverter is really a *true/complement* unit, which is used with an adder unit for the purpose of using the device as an adder as well as a subtractor. The true/complement unit determines whether addition or subtraction is performed. The mode input M controls the operation. If $M = 1$, the true/complement unit inverts the input binary number, whereas if $M = 0$, the unit returns the input binary number without inversion.

A 4-bit adder with controlled inverter is shown in Fig. 11.14. This unit is called 2's complement adder/subtractor. When M is low (i.e., $M = 0$), the unit performs addition and when M is high (*i.e.*, $M = 1$), the unit performs subtraction. When binary number B

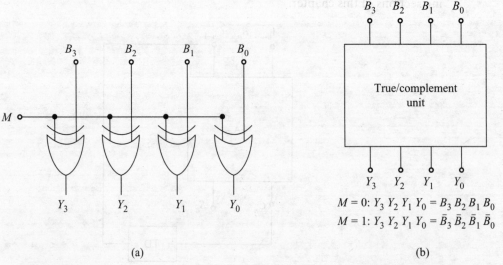

FIGURE 11.13 A 4-bit controlled inverter (a) detailed circuit, and (b) block diagram.

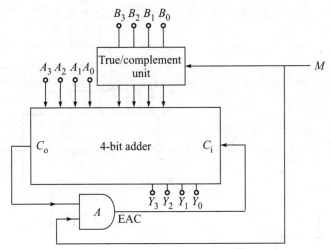

FIGURE 11.14 A block diagram of a 4-bit adder/subtractor.

is to be subtracted from the binary number A, M is made high. The binary number B is inverted before being fed to the full adder. The carry $C_o = 1$ from the 4-bit adder may be used to supply the 1, which must be added to $\overline{B}$. This bit is called the *end around carry* (EAC) because this carry out is fed back to the carry input to the least significant adder. It is to be noted that the first adder is a full adder and not a half adder, as in Fig. 11.10. Thus, in subtraction mode, the adder/subtractor of Fig. 11.14 adds A and 2's complement of B. As discussed in Section-10.3, if $A > B$, the carry is 1, the result is positive and discarding the carry the result gives the magnitude (A minus B). If $A < B$, the carry is zero; hence, EAC is zero. The output is sum of A and $\overline{B}$. The circuit of Fig. 11.14, therefore, needs to be modified.

Now, we demonstrate that if no carry results in the system of Fig. 11.14, the correct result of A minus B is a negative value, and the magnitude can be obtained by complementing the output bits. The NOT function changes a 1 to a 0, and vice versa. To illustrate let us consider 4-bit binary numbers. Hence,

B plus $\overline{B}$ = 1111

B = 1111 minus $\overline{B}$

A minus B = A minus (1111 minus $\overline{B}$)

$\quad\quad\quad\quad\quad$ = (A plus $\overline{B}$) minus 1111

$\quad\quad\quad\quad\quad$ = minus [1111 minus (A plus $\overline{B}$)]

As 1111 minus a 4-bit binary number is the complement of the number,

$$A \text{ minus } B = \text{minus } (\overline{A \text{ plus } \overline{B}}) \quad\quad\quad (11.9)$$

Hence, it proves the statement that the magnitude of A minus B (when $A < B$) is given by the complement of A plus $\overline{B}$. The modified circuit of adder/subtractor of Fig. 11.14 is shown in Fig. 11.15.

In modified circuit of Fig. 11.15, the mode input N of true/complement unit-2 is 1 only when $C_o = 0$ and $M = 1$. Thus, the true/complement unit-2 results the output equal

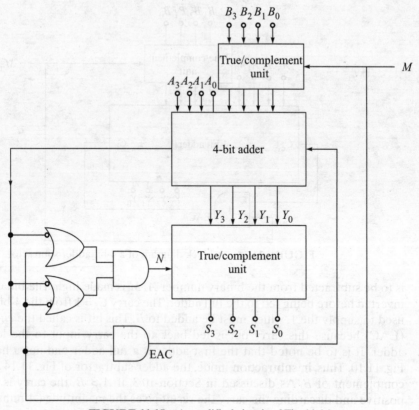

FIGURE 11.15 A modified circuit of Fig. 11.14.

to $(\overline{A \text{ plus } \overline{B}})$, which is the magnitude of A minus B when $A < B$. When $M = 0$, the output is A plus B. When $M = 1$ and $C_o = 1$, the output is A plus 2's complement of B, and with carry discarded, the output is the magnitude of A minus B.

11.5 DECODERS AND DEMULTIPLEXERS

11.5.1 Decoders

In a digital system, the numbers or instructions are conveyed by binary levels or pulse trains. For a 4-bit data, there are 16 combinations, and hence 16 instructions can be coded. At one time, only one of these 16-coded instructions can be conveyed. Accordingly, one and only one line corresponding to that instruction is excited by a multiposition switch. This process of identifying the code of the instruction is known as decoding. Decoding is necessary in many applications such as digital display, digital-to-analogue conversion, and memory addressing. The device used for decoding is called decoder.

Decoders can be of several types depending on the code used for representing the instruction. For example, a decoder may be used to decode the BCD-coded decimal number. The BCD code of a decimal number, say 6, is 0110. The operation of decoding may be carried out with a 4-input AND gate, as shown in Fig. 11.16. The output of an

Sec. 11.5 / Decoders and Demultiplexers

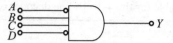

FIGURE 11.16 A 4-input AND gate.

AND gate is 1 if and only if the inputs are $A = 0, B = 1, C = 1$, and $D = 0$. Since this code represents the decimal number 6, the output is level as line 6.

A BCD to decimal decoder is shown in Fig. 11.17. The decoder has four inputs A, B, C, and D, and 10 outputs. For the time being, the dashed line is ignored. In Fig. 11.17, NAND gates are used in place of AND gates in Fig. 11.16. Hence, an output is 0 (low) for the correct BCD code and 1 (high) for other codes. Since a 4-bit input code selects 1 of the 10 output lines, the system of Fig. 11.17 is referred to as a 4- to 10-line *decoder*. The decoder, thus, operates as a 10-position switch, which responds to a BCD input instruction.

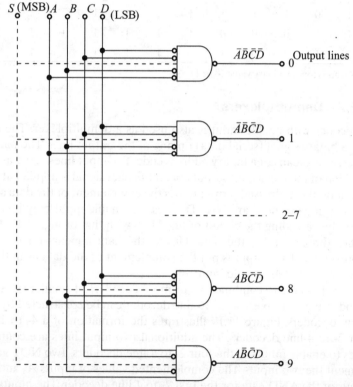

FIGURE 11.17 A BCD to decimal decoder (it may be used as a demultiplexer).

It is sometimes desired to decode only during certain intervals of time. In that case, an additional input S is added to each gate as shown by dashed line in Fig. 11.17. This input is called a *strobe*. The strobe inputs to all 10 NAND gates are same. If $S = 1$, gates are *enabled* and decoding takes place. If $S = 0$, decoding is inhibited. Hence, the strobe input is also called enable input. The enable input can be used with a decoder of any number of inputs and outputs. The truth table for a 4- to 10-line decoder is given in Table-11.4.

Table-11.4 Truth table for the 4- to 10-line decoder of Fig. 11.17.

Inputs*					Outputs									
S	A	B	C	D	Y_0	Y_1	Y_2	Y_3	Y_4	Y_5	Y_6	Y_7	Y_8	Y_9
0	X	X	X	X	1	1	1	1	1	1	1	1	1	1
1	0	0	0	0	0	1	1	1	1	1	1	1	1	1
1	0	0	0	1	1	0	1	1	1	1	1	1	1	1
1	0	0	1	0	1	1	0	1	1	1	1	1	1	1
1	0	0	1	1	1	1	1	0	1	1	1	1	1	1
1	0	1	0	0	1	1	1	1	0	1	1	1	1	1
1	0	1	0	1	1	1	1	1	1	0	1	1	1	1
1	0	1	1	0	1	1	1	1	1	1	0	1	1	1
1	0	1	1	1	1	1	1	1	1	1	1	0	1	1
1	1	0	0	0	1	1	1	1	1	1	1	1	0	1
1	1	0	0	1	1	1	1	1	1	1	1	1	1	0

* X stands for any binary value (0 or 1).

11.5.2 Demultiplexers

A decoder with enable input is also used as a demultiplexer. The demultiplexer transmits a binary signal (serial data) on one of the output lines. The particular output line is selected by means of a binary address code. The input lines serve as address code for the selection of the particular output line. If the data signal is applied at the enable input, the output on the addressed output line is the complement of the data signal. This is because the output gates used are NAND gates. An enable input may be applied to a demultiplexer by cascading the circuit of Fig. 11.18 with that of Fig. 11.17. If enable input is 0, S is the complement of the data. Hence, the data appears on the line selected by the address code. The output is not the complement of the data as in the case of the demultiplexer without an enable input.

It is possible to combine two or more decoders of smaller sizes to form a larger size decoder. For example, a 4- to 16-line decoder can be constructed by combining four 2- to 4-line decoders. Figure 11.19 illustrates the formation of a 4- to 16-line decoder using four 2- to 4-line decoders. The additional two input lines are connected through AND gates to enable inputs of the four 2- to 4-line decoders. Two NOT gates are used to complement the two inputs. The complemented values of inputs A_2 and A_3 are connected to inputs of the AND gate for the first 2- to 4-line decoder. The inputs to other AND gates are A_2 and $\overline{A_3}$ to the second gate, $\overline{A_2}$ and A_3 to the third gate, and A_2 and A_3 to the fourth gate. The working of the resulted 4- to 16-line decoder is given in Table-11.5.

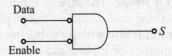

FIGURE 11.18 Data with enable input to a demultiplexer.

Sec. 11.5 / Decoders and Demultiplexers   555

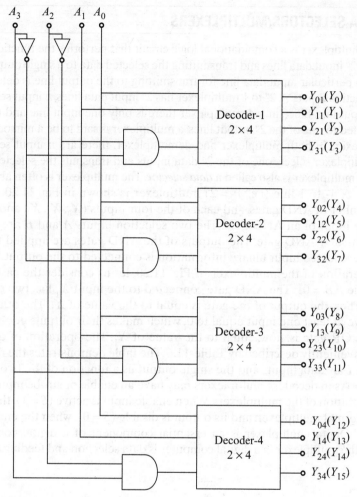

FIGURE 11.19 A 4- to 16-line decoder forms from four 2- to 4-line decoders.

Table-11.5 Truth table of 4- to 16-line decoder of Fig. 11.19.

Input	A_3	A_2	A_1	A_0	Y	Input	A_3	A_2	A_1	A_0	Y
0	0	0	0	0	Y_{01}	8	1	0	0	0	Y_{03}
1	0	0	0	1	Y_{11}	9	1	0	0	1	Y_{13}
2	0	0	1	0	Y_{21}	10	1	0	1	0	Y_{23}
3	0	0	1	1	Y_{31}	11	1	0	1	1	Y_{33}
4	0	1	0	0	Y_{02}	12	1	1	0	0	Y_{04}
5	0	1	0	1	Y_{12}	13	1	1	0	1	Y_{14}
6	0	1	1	0	Y_{22}	14	1	1	1	0	Y_{24}
7	0	1	1	1	Y_{32}	15	1	1	1	1	Y_{34}

11.6 DATA SELECTORS/MULTIPLEXERS

A multiplexer is a combinational logic circuit that performs the function of selecting one of the 2^n input data lines and transmitting the selected data to a single output line. The selection of a particular input data line for transmitting to the output line is determined by a set of n selection inputs. A 2^n to 1 multiplexer has 2^n input data lines, n input selection lines, and one output line. Since in a demultiplexer there is only one input line and this input line is connected to one of the 2^n output lines, a multiplexer is said to be a mirror image of the demultiplexer. In both multiplexer and demultiplexer, there are n input selection lines. Since a multiplexer selects one of the 2^n data inputs and transmits the selected data to the output, the multiplexer is also called a *data selector*. The multiplexer is often abbreviated as MUX.

A 4- to 1-line (i.e., $n = 2$) multiplexer is shown in Fig. 11.20. There are four (2^n) number of AND gates, and each of the four inputs ($X_0, X_1, X_2,$ and X_3) is connected to one input of an AND gate. The two selection inputs A and B are decoded to select a particular AND gate. The outputs of the AND gates are applied to a single OR gate. The selected input binary information is connected to the output. To demonstrate the operation of the multiplexer of Fig. 11.20, let us consider the case when the selected code $AB = 01$. The AND gate connected to the input X_1 has two of its inputs equal to 1. Thus, the output of the gate is equal to the value of X_1. The other three AND gates have at least one input equal to 0, which makes their outputs equal to 0. The output of the OR gate is, now, equal to the value of X_1. The operation of a multiplexer can be conveniently described by Table-11.6. The table demonstrates the relationship between the four data inputs and the single output as a function of the two selected inputs.

As in decoders, multiplexers may have an enable or strobe input, which controls the operation of the multiplexers. When enable input is active ($S = 1$), the circuit functions as a normal multiplexer, and its output is disable ($S = 0$), when the enable input is inactive.

Digital multiplexer is an essential component of a digital computer. The purpose of the multiplexer in a digital computer is data selection and feeding it to the output regis-

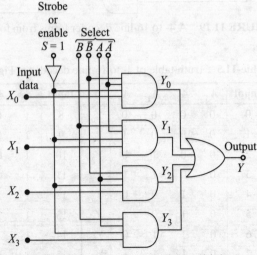

FIGURE 11.20 A 4- to 1-line multiplexer.

Table 11.6 Functional table for 4- to 1-line multiplexer.

Selection inputs		Outputs
A	B	Y
0	0	X_0
0	1	X_1
1	0	X_2
1	1	X_3

ter. There are occasions when there is a need of certain size of multiplexer, but only smaller size of multiplexers are available. In this situation, it is possible to make a stack or tree of the smaller-size multiplexers to form the multiplexer of the required size. Figure 11.21 illustrates the formation of 16- to 1-line multiplexer using five 4- to 1-line multiplexers.

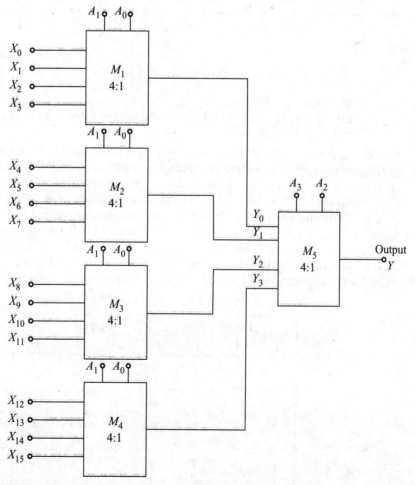

FIGURE 11.21 A stack of 4- to 1-line multiplexer forming a 16- to 1-line multiplexer.

11.7 ROM UNIT

Basically, a ROM consists of a decoder connected in a permanent and prespecified fashion to an encoder as shown in Fig. 11.22 (a). Thus, both the input and output of the ROM are in coded form. Hence, a ROM may be said to be a code conversion system having m inputs and n outputs. The m bits of inputs may be greater than, equal to, or less than the n bits of the outputs. The decoder first decodes m bit codes into 2^m word lines. The encoder section encodes these word lines into desired n-bit coded outputs. If the inputs to the encoder section are all 2^m combinations of 1s and 0s, then 2^m n-bit words are read at the output. The ROM can be shown by a block diagram as in Fig. 11.22(b).

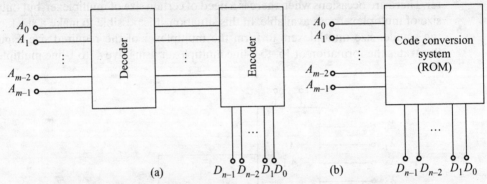

FIGURE 11.22 (a) A ROM as a combination of a decoder and an encoder and (b) its block diagram.

As the purpose of a ROM is to provide a specific output for a specific input, the ROM generates an input-output relation specified by a truth table. As such it can implement any combination with m input and n output. To illustrate, let us design a ROM for a particular application specified by a truth table given in Table-11.7. For a given input code of 3 bits, the ROM is to provide a 4-bit output as specified in the truth table of Table-11.7. From Table-11.7, consider the input logic of 010. For this, the desired output to be read is 1010. The 4-bit out word requires an encoder with four OR gate. The decoder section of the ROM have eight (2^3) NAND gates. Thus, in order to design the ROM for the application of Table-11.7, it is required to provide interconnections between

Table-11.7 Truth table for a 3 × 4 ROM.

Inputs			Outputs			
A_2	A_1	A_0	D_3	D_2	D_1	D_0
0	0	0	0	0	1	1
0	0	1	0	1	0	1
0	1	0	1	0	1	0
0	1	1	1	1	0	0
1	0	0	0	0	1	0
1	0	1	0	1	1	0
1	1	0	0	0	0	1
1	1	1	1	1	1	0

eight AND gates of the decoder section and four OR gates of the encoder section. Each OR is responsible to generate a bit of the output word. For the 4 bits (D_3, D_2, D_1, D_0) of the output, the logic function representing the truth table are as given below:

$$D_0 = \overline{A_2}\ \overline{A_1}\ \overline{A_0} + \overline{A_2}\ \overline{A_1}\ A_0 + A_2 A_1 \overline{A_0} \tag{11.10}$$

$$D_1 = \overline{A_2}\ \overline{A_1}\ \overline{A_0} + \overline{A_2} A_1 \overline{A_0} + A_2 \overline{A_1}\ \overline{A_0} + A_2 \overline{A_1} A_0 + A_2 A_1 A_0 \tag{11.11}$$

$$D_2 = \overline{A_2}\ \overline{A_1} A_0 + \overline{A_2} A_1 A_0 + A_2 \overline{A_1} A_0 + A_2 A_1 A_0 \tag{11.12}$$

$$D_3 = \overline{A_2} A_1 \overline{A_0} + \overline{A_2} A_1 A_0 + A_2 A_1 A_0 \tag{11.13}$$

The ROM designed as per the truth table and satisfying the above logic functions is shown in Fig. 11.23.

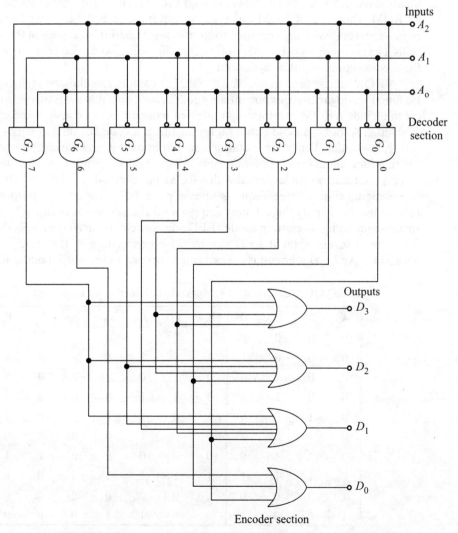

FIGURE 11.23 An 8×4 ROM simulating truth table of Table-11.7.

The three input variables A_2, A_1, A_0 essentially serve the purpose of a coded address to select a specific 4-bit output word. The ROM capacity is indicated by $2^m \times n$, where m is the number of input variables and n is the number of bits in the output words. Thus, the capacity of the ROM of Fig. 11.23 is $8 \times 4 = 32$ bits. For example, a ROM with 8-bit address and 8-bit output word has a capacity of $2^8 \times 8 = 256 \times 8 = 2048$ bits. This ROM is commonly called as a 2K ROM, where 2048 is rounded of as 2000 and K stands for thousand.

The interconnections of the decoder and the encoder sections in a ROM is permanent, and so a particular code of input signals generates one and only one output word. That is why this digital device is called ROM. In some special ROM units, the interconnections can be programmed by the user. The interconnections of the decoder section to the encoder section, which are fusible links are preserved or broken by the programming. However, once connections are made, the pattern of output corresponding to the inputs is fixed. Such type of ROM is called **PROM**. The third type of ROM is known as **EPROM**. This type of ROM incorporates an erasable programming feature. For this purpose, gates (which are sensitive to short-wave radiation) are used in the interconnections. The radiation may be ultraviolet light. The radiation can be used to make or break these connections as often as desired.

A ROM can be used as a *look-up table*. For routine calculations such as trigonometric functions, logarithms, exponentials, square roots, etc., it is sometimes more economical to include a ROM as a look-up table. For example, let us consider the calculation of $Y = \sin X$. A look-up table (truth table) for this is a code conversion between the input code representing X in binary notation and output code giving the corresponding value of the sine function in binary notation. Similarly, any calculation for which a truth table can be prepared may be implemented with a ROM. For each table a different ROM is used.

A digital system often requires several pulse trains for control purposes. A ROM may be used to supply these binary sequences if the address is changed by a counter. A seven-segment light-emitting diode (LED) display can be driven by a ROM. To display 0 to 9 by a seven-segment LED can be driven according to the truth table given in Table-11.8. An LED segment is excited when corresponding output of the ROM is 0 and

Table 11.8 Conversion of a 4-bit-coded decimal to an LED display code.

A	B	C	D	Y_6	Y_5	Y_4	Y_3	Y_2	Y_1	Y_0
0	0	0	0	1	0	0	0	0	0	0
0	0	0	1	1	1	1	1	0	0	1
0	0	1	0	0	1	0	0	1	0	0
0	0	1	1	0	1	1	0	0	0	0
0	1	0	0	0	0	1	1	0	0	1
0	1	0	1	0	0	1	0	0	1	0
0	1	1	0	0	0	0	0	0	1	1
0	1	1	1	1	1	1	1	0	0	0
1	0	0	0	0	0	0	0	0	0	0
1	0	0	1	0	0	1	1	0	0	0

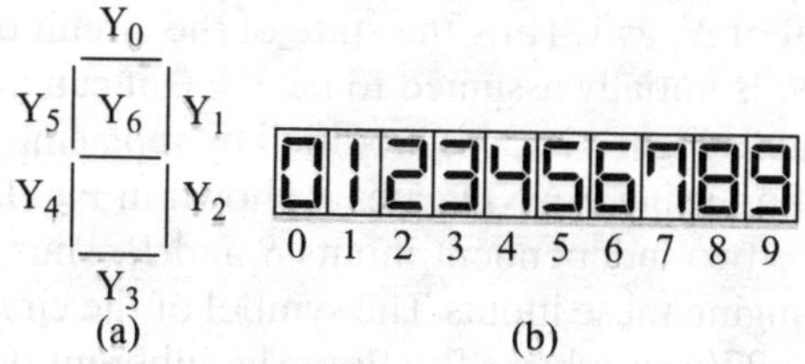

FIGURE 11.24 (a) A seven-segment LED and (b) display for 0 to 9.

dark when 1. The seven-segment LED is shown in Fig. 11.24(a), and its display for 4-bit-coded decimal numbers from 0 to 9 is shown in Fig. 11.24(b). Alphanumeric characters may be displayed by replacing LED with 5×7 dot matrix display and taking a ROM implementing the truth table for the dot matrix display unit.

Combinational logic can be implemented with a ROM. If n logical functions of m variables are expressed in the sum of product form, these logical equations may be implemented with an m input and n output ROM. This is economical if m and n are large.

11.8 FLIP FLOPS

Memory devices have the property to remain in a certain state even after the removal of the input that caused this state to appear. In sequential circuit, memory elements used are of bistable type and have two states 1 and 0 to convey the information. A block diagram representation of a memory is shown in Fig. 11.25. There are two inputs X_1 and X_2, and two outputs Y and $\overline{Y}$. The two outputs are complement of each other. The memory element is said to be of 1 state if $Y = 1$ and $\overline{Y} = 0$, and 0 state if $Y = 0$ and $\overline{Y} = 1$. The state of memory can be changed only by applying suitable inputs X_1 and X_2. The basic memory element is obtained by cross coupling two NOT gates as shown in Fig. 11.26. As discussed in Chapter-10, a NOT gate can be obtained by a NAND gate. For this, two inputs of the NAND gate are shorted and a single input terminal is taken out. This circuit is called a flip flop. The flip flop, shown in Fig. 11.26, can exist in one of the two states. Being a bistable device, the flip flop is also called a binary or bistable multivibrator. Since flip flop can store 1 bit of information, $Y = 1$ or $Y = 0$, it is a 1-bit memory unit. Also, the flip flop is known as a latch, as its present state is locked or latched.

To illustrate that two stable states exist for the circuit of Fig. 11.26, let us assume that after the removal of the inputs, gate N_1 is in state $Y = 1$. Thus, the input to gate N_2 is 1, and hence, the output of N_2 is $\overline{Y} = 0$. But this output of N_2 serves as the input to N_1,

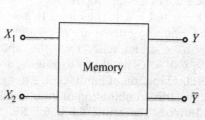

FIGURE 11.25 A memory element.

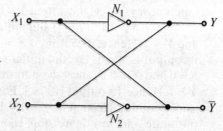

FIGURE 11.26 A simple flip flop or latch.

resulting output of N_1 as 1. Thus, the state of the circuit of Fig. 11.26 is stable. Similarly, if the output of N_1 is initially assumed to be $Y = 0$, it can be shown that the state of the circuit remains stable. Figure 11.26 is modified by replacing NOT gates by two input NAND gates. The flip flop using NAND gates is shown in Fig. 11.27(a). In addition to feedback inputs, there are two independent inputs S and R. Thus, the state of the flip flop can be changed by changing these inputs. The symbol of the circuit is shown in Fig. 11.27(b). The circuit of Fig. 11.27(a) is a basic flip flop. The inherent delays in the gates play an important role in working of the flip flop. Because of the feedback paths, the circuit works as an asynchronous sequential circuit, and the current state of the circuit depends on the history of the circuit. The circuit is sometimes called *direct coupled SR flip flop* or *SR latch*.

The operation of the circuit of Fig. 11.27(a) is analyzed by examining the state of the output terminal for the four combinations of the inputs S and R. To illustrate the operation, the NAND gate circuit is considered.

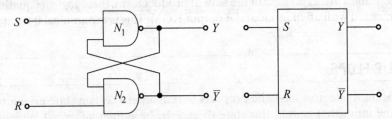

FIGURE 11.27 (a) An SR flip flop using NAND gates and (b) its symbol.

When $S = 1$ and $R = 1$, the circuit behaves like the latch of Fig. 11.26. If initially $Y = 0$, then being complement $\overline{Y} = 1$. The inputs to gate N_1 are $S = 1$ and $\overline{Y} = 1$. Hence, its output is $Y = 0$. For the initial value of $Y = 1$, and hence $\overline{Y} = 0$, the inputs of N_1 are $S = 1$ and $\overline{Y} = 0$. Hence, the output of N_1 is $Y = 1$. This agrees with the assumption. Thus, the previous output state remains unchanged.

When $S = 0$ and $R = 1$, assume that the initial state is $Y = 1$, and hence $\overline{Y} = 0$. The inputs of N_2 are $R = 1$ and $Y = 1$, which makes the output of N_2, $\overline{Y} = 0$, and hence $Y = 1$. For the initial value of $Y = 0$ and $\overline{Y} = 1$, the inputs of N_1 are $S = 0$ and $\overline{Y} = 1$. Hence, the output of N_1 is $Y = 1$. Thus, we conclude that for this combination of inputs the flip flop is in state 1. The state of the flip flop is called *set* or *preset*.

When $S = 1$ and $R = 0$, let us assume that the initial state is $Y = 1$ and then being complement $\overline{Y} = 0$. The inputs of N_2 are $R = 0$ and $Y = 1$, which makes the output of N_2, $\overline{Y} = 1$, and hence $Y = 0$. For the initial value of $Y = 0$ and $\overline{Y} = 1$, the inputs of N_1 are $S = 1$ and $\overline{Y} = 1$. Hence, the output of N_1 is $Y = 0$, and hence $\overline{Y} = 1$. In summary, this combination of inputs always resets the output $Y = 0$. The state of the flip flop is called *reset* or *clear*.

When $S = 0$ and $R = 0$ and initially $Y = 0$ and $\overline{Y} = 1$. The inputs to gate N_1 are $S = 0$ and $\overline{Y} = 1$. Hence, its output is $Y = 1$. The inputs to gate N_2 are $R = 0$ and $Y = 0$. Hence, its output is $\overline{Y} = 1$. For the initial value of $Y = 1$, and hence $\overline{Y} = 0$, the inputs of N_1 are $S = 0$ and $\overline{Y} = 0$. Hence, the output of N_1 is $Y = 1$. The inputs to gate N_2 are $R = 0$ and $Y = 1$. Hence, its output is $\overline{Y} = 1$. For this combination of inputs both gates generate the same output logic, that is, $Y = \overline{Y} = 1$. Hence, this combination of inputs makes the output state indeterminate and is not allowed. The truth table for SR flip flop of Fig. 11.27(a) is given in Table-11.9.

Table-11.9 Truth table for NAND SR flip flop.

S	R	Y	$\overline{Y}$	Comment
0	0	1	1	Indeterminate
0	1	1	0	Set or preset
1	0	0	1	Reset or clear
1	1	NC		No change

11.8.1 Clocked SR Flip Flop

The SR flip flop circuits discussed above are essentially asynchronous sequential circuits. These circuits can be modified to operate in synchronism with a clock. These circuits are, therefore, called synchronous sequential flip flips. The clock is a train of pulses of period T as shown in Fig. 11.28. The pulse width t_p is small in comparison with T. The asynchronous SR flip flop of Fig. 11.27(a) is modified as in Fig. 11.29 to make it a clocked SR flip flop. From the circuit of Fig. 11.29, it is clear that between the clock pulses (Ck = 0), the output of Fig. 11.29 remains unchanged, independent of the values of inputs S and R. Hence, if $Y = 1$, it remains 1, whereas if $Y = 0$, it remains 0. Hence, the flip flop of Fig. 11.29 does not change state between clock pulses. Now, consider the operation of the circuit when clock pulse is present (Ck = 1).

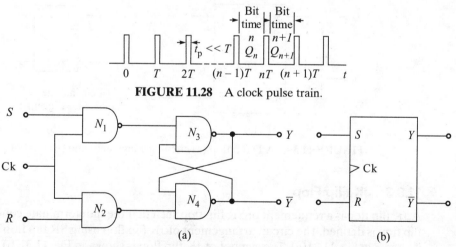

FIGURE 11.28 A clock pulse train.

FIGURE 11.29 A clocked SR flip flop (a) circuit and (b) symbol.

For the four combinations of S and R, the output states of the clocked SR flip flop are as given in the truth table of Table-11.10.

In addition to SR flip flop, there are various types of flip flops, which are available commercially. They are JK flip flop, T flip flop, and D flip flop. The JK flip flop removes the ambiguity (indeterminate condition) of SR flip flop. The T flip flop acts as a toggle switch and changes the output state with each clock pulse. The D flip flop acts as a delay unit, which causes the output to follow the input D, but delayed by 1-bit time, that is, $Y_{n+1} = D_n$.

Table-11.10 Truth table for clocked SR flip flop.

S	R	Y	$\overline{Y}$	Comment
0	0	NC		No change
0	1	0	1	Reset or clear
1	0	1	0	Set or preset
1	1	1	1	Indeterminate

11.8.2 D Flip Flop

A D flip flop is a modification of the SR flip flop. An inverter is inserted between S and R inputs of the SR flip flop and a single input is applied to the D input terminal. The circuit arrangement is shown in Fig. 11.30(a) and the symbol in Fig. 11.30(b). When Ck = 1 and the data input signal $D = 0$, it follows that $S = 0$ and $R = 1$. Under this condition, therefore, the output follows the truth table of the clocked SR flip flop. Hence, the output $Y = 0$, which is the value of D. For Ck = 1 and $D = 1$, $S = 1$, and $R = 0$, the output $Y = 1$, which is again equal to D. Therefore, we can conclude that the next state of output is equal to the current state of input. That means

$$Y_{i+1} = D_i$$

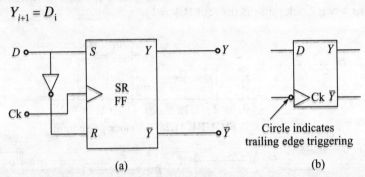

FIGURE 11.30 A D flip flop (a) circuit arrangement and (b) symbol.

11.8.3 JK Flip Flop

A JK flip flop is a refinement of SR flip flop such that the indeterminate condition of SR flip flop is defined. The circuit arrangement of JK flip flop using SR flip flop of Fig. 11.29 is shown in Fig. 11.31(a). The symbol of JK flip flop is shown in Fig. 11.31(b). The input J in combination with $\overline{Y}$ provides the input to the S terminal of the SR flip flop through the AND gate A_1. Similarly, Y in combination with input K, through AND gate A_2, provides input to the R terminal. Thus, due to gates A_1 and A_2, the inputs to the terminals S and R of the SR flip flop depend not only on the input signals but also on the output of the flip flop. For Ck = 1, the operation of the JK flip flop can be explained as follows:

When $J = 0$ and $K = 0$ and initially $Y = 0$, then $\overline{Y} = 1$, $S = 0$, and $R = 0$. The output follows the first row of Table-11.10. Hence, $Y = 0$. When initially $Y = 1$, $S = 0$, and $R = 0$, the output again follows the first row of Table-11.10 and $Y = 1$. Hence, for this combination of inputs, there is no change in the output.

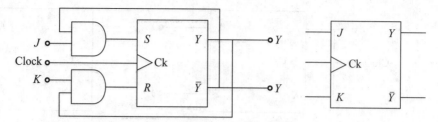

FIGURE 11.31 A JK flip flop (a) circuit and (b) symbol.

When $J = 0$ and $K = 1$ and initially $Y = 0$, $S = 0$, and $R = 0$. The output follows the first row of Table-11.10; hence, $Y = 0$. When initially $Y = 1$, $S = 0$, and $R = 1$, the output follows the second row of Table-11.10. Again, $Y = 0$. Hence, for this combination of inputs, the output reset to 0.

When $J = 1$ and $K = 0$ and if initially $Y = 0$, then $\bar{Y} = 1$, $S = 1$, and $R = 0$. The output follows the third row of Table-11.10; hence, $Y = 1$. When initially $Y = 1$, $S = 0$, and $R = 0$, the output follows the first row of Table-11.10. Again, $Y = 1$. Hence for this combination of inputs, the output sets to 1.

When $J = 1$ and $K = 1$ and initially $Y = 0$, $S = 1$ and $R = 0$. The output follows the third row of Table-11.10. Hence, $Y = 1$ (complement of initial output). When initially $Y = 1$, $S = 0$, and $R = 1$. The output follows the second row of Table-11.10. Hence, $Y = 0$ (complement of the initial output). Hence, for this combination of inputs, the output is complement of the initial output.

The result is given in the truth table of Table-11.11.

Table 11.11 Truth table of JK flip flop.

J	K	Y	$\bar{Y}$	Comment
0	0	NC		No change
0	1	0	1	Reset or clear
1	0	1	0	Set or preset
1	1	$\bar{Y}$	Y	Complemented value

11.8.4 T Flip Flop

A T flip flop is obtained by joining the J and K terminals of the JK flip flop and taking out a single terminal. The T stands for toggle. The circuit of T flip flop is shown in Fig. 11.32(a) and its symbol in Fig. 11.32(b). Thus, a single input T is applied to both J and K terminals. When $T = 0$ ($J = 0$ and $K = 0$) and the output follows the first row of Table-11.11, there is no change in the output. When $T = 1$ ($J = 1$ and $K = 1$) and the output follows the fourth row of Table-11.11, the output is complement of the initial output.

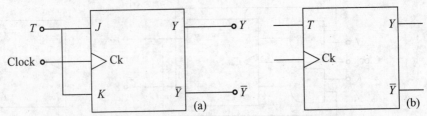

FIGURE 11.32 A T flip flop (a) circuit arrangement (b) symbol.

11.8.5 Edge Triggered Flip Flop

In clocked flip flop the output can change state when Ck is high (1) and remains in the same state when Ck is low (0). Thus, the output can change during the entire half cycle when Ck is high. This may be a disadvantage in many situations. To explain the situation, let us consider the clocked JK flip flop of Fig. 11.31. For example, let the inputs be $J = 1$ and $K = 1$ and output $Y = 0$. When Ck is made high, the output becomes $Y = 1$ (according to the fourth row of Table-11.11). However, this change in Y from 0 to 1 takes place after a time interval Δt. The time interval Δt is equal to the propagation delay of NAND gates. The delay time Δt is very small compared to the pulse width. Now, $J = 1$ and $K = 1$ and output $Y = 1$, and as Ck = 1 still; hence, according to the fourth row of Table-11.11, Y changes from 1 to 0. Thus, we must conclude that for the duration Ck = 1, the output Y oscillates back and forth between 0 and 1. At the end of the clock pulse (Ck = 0), the value of output Y is ambiguous. This situation is called *race around* condition. This situation can be avoided by making the output to change its state only for very short interval of time (smaller than the pulse width) in the positive half cycle of the clock. This is known as *edge triggering* and the resulting flip flop as *edge triggered flip flop*. Edge triggering is possible by using an RC circuit. The capacitor charges fully when Ck = 1. The exponential charging produces a narrow positive voltage spike across the resistor. Thus, the Ck input is a positive voltage spike, as shown in Fig. 11.33(a).

In edge triggered flip flop, output transitions (changes of output state) occur at specific level of clock pulse. When the pulse input level exceeds the threshold level, the inputs are locked out so that the flip flop is irresponsive to further change in the input until the clock pulse returns to 0 and another pulse occurs. Some edge triggered flip flops cause a transition on the rising edge of the clock pulse and other cause a transition on the falling edge of clock pulse. The former is called *positive edge transition*, whereas the later is called *negative edge transition*.

Figure 11.33(b) and (c) shows the clock pulses used in the edge triggered flip flops. For positive edge transition, the inputs of the flip flop are transferred to the output Y when the clock pulse makes a positive transition (changes from 0 to 1 level) as shown in Fig. 11.33(b). The output cannot change when the clock is in 1 level, in 0 level, or in a transition from 1 to 0 level. For the negative edge transition, the inputs of the flip flop are transferred to the output Y when the clock pulse makes a negative transition (changes from 1 to 0 level) as shown in Fig. 11.33(c). The output cannot change when the clock is in 1 level, in 0 level, or in a transition from 0 to 1 level.

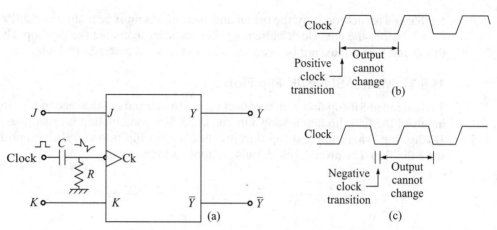

FIGURE 11.33 (a) An edge triggered JK flip flop and (b and c) edge triggering clock pulse.

The effective transition period is usually a very small fraction of the total period of the clock pulse.

11.8.6 Flip Flop with Preset and Clear

The output of flip flops discussed earlier is random before the clock pulse is applied. Therefore, it is necessary to reset or clear the output before clock pulse is applied. It may also be required to preset the flip flop into state 1. For this purpose flip flops available in integrated circuit packages are sometimes provided with special input terminals. These inputs are usually called *preset* and *clear*. An SR flip flop with preset and clear terminals is shown in Fig. 11.34(a) and its symbol in Fig. 11.34(b).

The clear operation may be accomplished by setting clear input to 0 and preset input to 1. Hence, Ck = 0, Cr = 0, and Pr = 1. Since Cr = 0, the output of gate N_2 is $\overline{Y} = 1$. As Ck = 0, the output of N_3 is 1, and as Pr = 1, all inputs to N_1 are 1 and Y = 0. This is the desired state of the flip flop. Similarly, for presetting the state of the flip flop into state 1, it is required to put Cr = 1, Pr = 0, and Ck = 0. The preset and clear data are called direct or asynchronous input as they are not in synchronism with the clock. However, these inputs are applied at any time between clock pulses. Once the state of the flip flop is

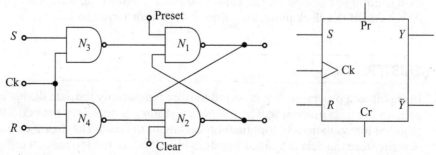

FIGURE 11.34 An SR flip flop with preset and clear terminals (a) circuit and (b) symbol.

established asynchronously, the preset and clear inputs must be maintained at Pr = 1 and Cr = 1, before the next clock pulse is applied in order to enable the flip flop. The inputs Pr = 0 and Cr = 0 must not be used since they lead to an ambiguous state.

11.8.7 Master-Slave JK Flip Flop

Two cascaded SR flip flop with feedbacks from the outputs of the second flip flop to the inputs of the first flip flop is shown in Fig. 11.35. The combination is called master-slave JK flip flop, where the first flip flop is called master flip flop and the second is called slave flip flop. The inverted clock pulse is applied to the slave.

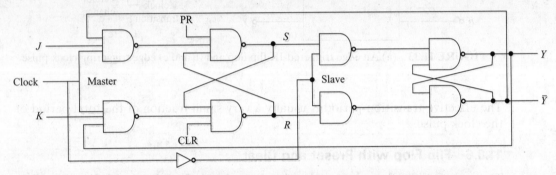

FIGURE 11.35 A master-slave JK flip flop (a) circuit and (b) symbol.

The operation of the master-slave JK flip flop is explained as follows:

For Pr = 1, Cr = 1, and Ck = 1, the master is enabled and its operation follows the JK truth table of Table 11.11. Since, inverted clock $\overline{Ck} = 0$ is applied to the slave flip flop, the slave is inhibited. Hence, there is no change in the slave output during the pulse duration. Clearly, the race around difficulty is absent in the master-slave JK flip flop. As soon as Ck = 0, the master flip flop is inhibited and $\overline{Ck} = 1$, which enables the slave flip flop. The slave is an SR flip flop, hence, follows the truth table of Table 11.10.

When the slave is enabled, that is, Ck = 0, and if $S = Y_M = 1$ and $R = \overline{Y}_M = 0$, then from Table-11.10, $Y = 1$ and $\overline{Y} = 0$. Similarly, if $S = Y_M = 0$, and $R = \overline{Y}_M = 1$, then $Y = 0$ and $\overline{Y} = 1$. Hence, we can conclude that in the interval between the clock pulses, the output of the master (i.e., Y_M) is transferred to the slave output Y. The output Y does not change during a clock pulse. The master output Y_M follows the truth table of JK flip flop. At the end of the clock pulse, the value of Y_M is transferred to Y.

11.9 REGISTERS

In digital data processing it is required to store temporarily the data during intermediate point in the data processing. The temporary storage is necessary for performing certain required manipulations before final processing of the data. The digital devices that temporarily store the data are called registers. Flip flops are used to make a register. As each flip flop stores 1 bit of information, therefore, n flip flops are required for making an n-bit register for storing an *n*-bit word information. For example, a computer employing 16-bit word length requires a register made of 16 flip flops to hold the number before it is manipulated. The simplest form of register consists of flip flops only. However, a register may have combinational gates, in addition to flip flops. The combinational gates perform certain data manipulation. Therefore, in a broadest definition, a register is a digital device consisting of a group of flip flops and gates. The flip flops hold the binary information and the gates control when and how new information is transferred into the register.

Basically, registers are classified into two types. One type is called *parallel register* and the second type is called serial or *shift register*. The transfer of new data into a resister is known as loading of the register. In parallel register, all the bits are loaded simultaneously with a common clock pulse transition and the loading is referred to as parallel loading. This makes the operation of the register very fast. The shift register processes each bit of the data in succession. Hence, the shift register is much slower than the parallel register.

11.9.1 Parallel Register

A master clock generator supplies a continuous train of clock pulses. The clock pulses are applied to all flip flops of the register. A 4-bit register made of four D flip flops is shown in Fig. 11.36. A load control input is directed to the D inputs through control

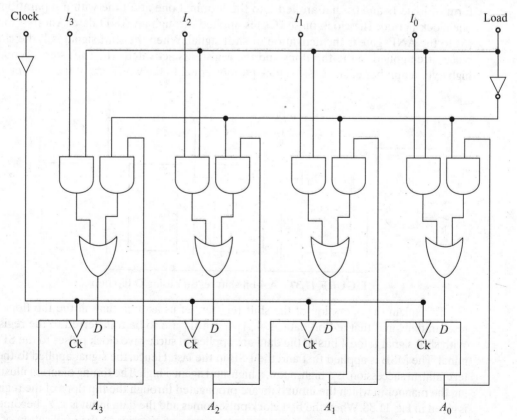

FIGURE 11.36 A 4-bit register with parallel load.

gates. The load input in the register controls the transfer of data with the clock pulse. When the load input is 1, the data in the four inputs are transferred into the register with the next clock pulse. When load input is 0, the data inputs are inhibited and the D inputs of the flip flops are connected to their respective outputs. The feedback connections from outputs to inputs are necessary because the D flip flops do not have no change condition. With each clock pulse, the D input determines the next state of the output. To leave the output unchanged, it is necessary to make the D input equal to the present value of the output. It is to be noted that the clock pulses are applied to the clock inputs all the time. It is the load input that determines whether the next pulse transfers the data to the register or not. The transfer of data from the inputs into the register is done simultaneously with all bits during a single pulse transition.

11.9.2 Shift Register

A shift register differs from the parallel register in the sense that the data are shifted from one flip flop to the next in line by application of clock pulses. An n-bit shift register consists of n flip flops connected in cascade. The output of one flip flop is connected to the input of the following flip flop. The number of flip flops to be used depends on the bits to be held in the register. A 4-bit shift register using D flip flops is shown in Fig. 11.37. The data to be written into the shift register appear at the serial input (SI) terminal in the form of logic 1s and 0s that are fed into the terminal one at a time with a separation of one clock period. The clock pulse (Ck) is applied through an AND date. Another input (S) to the AND gate is the activation or shift signal. When the shift signal is 1, the clock pulses are applied to the flip flops and the register is activated. The shift signal remains high over a number of successive clock periods equal to the word size of the register.

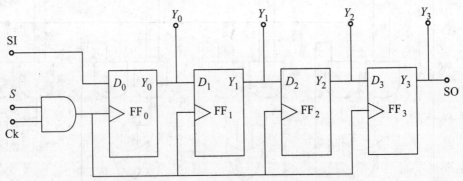

FIGURE 11.37 A 4-bit shift register using D flip flops.

To explain the operation of the shift register, let us assume that all the flip flops are reset initially and that a 4-bit data $A_3 A_2 A_1 A_0 = 1011$ are to be transferred to the register. With shift signal 1, the 4 bits of the data are applied in successive clock pulses to the SI terminal. The MSB is applied first and the LSB in the last. Hence, the signal applied to the SI terminal, in succession, is a high, a low, a high, and again a high. The timing diagram illustrating the manner in which the input data are propagated through the flip flops of the register is given in Fig. 11.38. When the first clock pulse comes and the data input is 1, Y_0 becomes 1. At the same time, the state of Y_1 is same as the state of Y_0 prior to the application of the

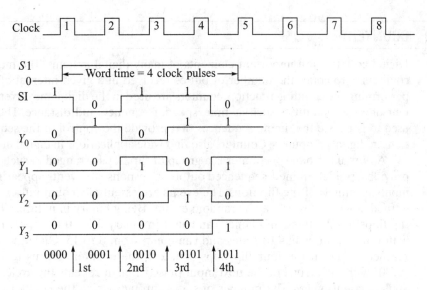

FIGURE 11.38 Timing diagram of the shift register with input data 1011.

clock pulse. Similarly, state of Y_2 and Y_3 are determined by the states of Y_1 and Y_2 prior to the application of clock pulse. Hence, Y_1, Y_2, and Y_3 remain in 0 state. With the second clock pulse when data input is 0, Y_0 becomes 0 and Y_1 goes to 1 and the register reads $Y_3 Y_2 Y_1 Y_0$ = 0010. With the end of the third clock pulse the register reads $Y_3 Y_2 Y_1 Y_0$ = 0101. Similarly, with the end of the fourth clock pulse the register reads $Y_3 Y_2 Y_1 Y_0$ = 1011, which is exactly the input word applied to the input terminal SI, The return of the shift signal to 0 prevents further shifting from occurring in the flip flops. Thus, the register has read the serial input and the data remain recorded in the memory of the flip flops.

The shifting action associated with the circuit of Fig. 11.37 has given the name of the register as *shift register*. The register, as shown in Fig. 11.37, is connected to provide shifting to the right. But a left shift is equally easy to arrange by simply entering the input data on the right and reversing the input/output connection between the flip flops.

The shifting of the data illustrated by the timing diagram is conveniently summarized in the Table-11.12.

We see the shifting of data by 1 bit in the right direction with the application of a new clock pulse. At the end of the nth clock pulse (in above example n = 4), the last row of the table represents the data applied in serial fashion at the SI terminal in parallel form.

Table 11.12 Shifting of data.

Flip flop state	Y_3	Y_2	Y_1	Y_0	Serial input data
Initial	0	0	0	0	1
First clock	0	0	0	1	0
Second clock	0	0	1	0	1
Third clock	0	1	0	1	1
Fourth clock	1	0	1	1	

11.10 COUNTERS

Digital counters are important elements of many digital systems. The main function of counters is to count the number of pulses received in a given interval of time. Besides performing the counting function, counters are used in the digital measurement of important physical quantities such as time, speed, frequency, and distance. They can also be used to generate the timing sequences that provide the control of the sequential operations in digital computers. Counters also find wide applications in digital instrumentation.

A digital (or binary) counter is a group of flip flops arranged in such a way so as to provide a predetermined, sequenced output in response to events appearing at the clock inputs terminals of the flip flops. The events are essentially count pulses, which may be normal clock pulses or may be random events. Being binary in nature, a counter with n flip flops has 2^n states, and hence can count in binary from 0 to $2^n - 1$. Thus a counter with four flip flops has 16 states and can count from 0 to 15. Once the last number is reached (1111, in the four flip flops counter), the counter returns to its initial state (0000) with the arrival of the next input pulse. Then, it repeats the cycle in response to applied count pulses. The number of states through which the counter cycles is called *modulo* of the counter. Thus, modulo M of an n state counter is given by $M = 2^n$. Like other sequential logic circuits, counters can be asynchronous or synchronous.

11.10.1 Asynchronous Counter

In asynchronous counters, different flip flops change their states at different times, as the flip flops are not under the command of a single clock pulse. In an asynchronous counter, the input pulse effectively ripples through the chain of flip flops. Hence, asynchronous counters are also known as *ripple counters*.

Ripple Counter

The ripple counter is a basic and simple counter. Because of its simplicity, it is a very commonly used counter. A 16-state ripple counter using four master-slave JK flip flops is shown in Fig. 11.39. The output of each flip flop is connected to the clock input of the following flip flop. The pulses to be counted are applied to the clock input of the first flip flop. For all flip flops, J and K inputs are tied to the supply voltage. Hence, $J = K = 1$ and the flip flops are converted to T flip flops (Fig. 11.32) with $T = 1$. For a T flip flop with $T = 1$, the output of master changes its state every time the clock input changes from 0 to 1. The new state of the master is transferred to the slave when clock changes from 1 to 0. Thus, the output Y_0 of the first flip flop changes state at the falling edge of each pulse. All the other output Y's change their states when and only when the output of the preceding flip flop changes from 1 to 0. The negative transition ripples through the counter from LSB to MSB.

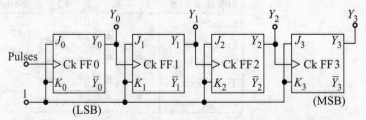

FIGURE 11.39 A ripple counter using master-slave JK flip flops.

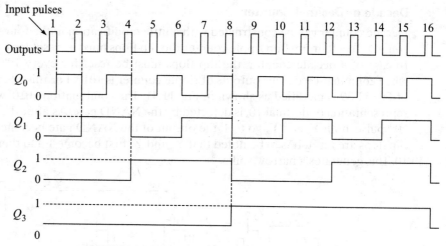

FIGURE 11.40 Timing diagram of the ripple counter.

The counter is first cleared by applying a '0' to the rest input. The timing diagram of the counter is given in Fig. 11.40. States of the flip flops as the incoming pulses arrive are given in Table-11.13. It is to be noted that the outputs of the flip flops in Table-11.13 have been ordered in the reverse direction from the order of the flip flops in Fig. 11.39. The order of array of 0s and 1s in any row in Table-11.13 is precisely the binary representation of the number of input pulses. Thus, the chain of flip flops counts the number of pulses in the binary system.

Table 11.13 States of the flip flops of ripple counter.

Number of input pulses	Flip flop outputs			
	Y_3	Y_2	Y_1	Y_0
0	0	0	0	0
1	0	0	0	1
2	0	0	1	0
3	0	0	1	1
4	0	1	0	0
5	0	1	0	1
6	0	1	1	0
7	0	1	1	1
8	1	0	0	0
9	1	0	0	1
10	1	0	1	0
11	1	0	1	1
12	1	1	0	0
13	1	1	0	1
14	1	1	1	0
15	1	1	1	1

Decade or Decimal Counter

Decade counter can be constructed with a little modification in the four flip flop ripple counter. In a four flip flop ripple counter the flip flops must be reset for every 16th pulse. In case of a decade counter, the flip flops must be reset for every 10th pulse, that is, when combination of four outputs of ripple counter is 1010. For this, the ripple counter of Fig. 11.39 is modified as shown in Fig. 11.41. The combination 1010, which is binary representation of decimal 10, is detected by the NAND gate. As after the application of 10th pulse, both Y_1 and Y_3 go to 1, the output of the NAND gate becomes 0, and all the flip flops are reset. It is to be noted that Y_1 and Y_3 first become 1 and then 0, after pulse 10. This generates a narrow spike.

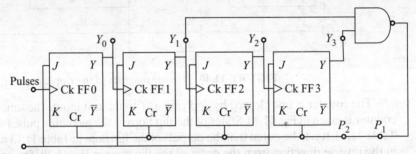

FIGURE 11.41 A decade counter circuit with resetting NAND gate.

If the propagation delay from the clear input to the flip flop output is different for each flip flop, the clear operation may not be reliable. In Fig. 11.41, if the first flip flop takes appreciably longer time to reset than the third flip flop, when Y_3 goes to 0, output of NAND gate goes to 1. Hence, Cr = 1 and Y_1 does not reset. To eliminate the difficulty with resetting, the decade counter of Fig. 11.41 is modified as shown in Fig. 11.42. A latch is used between the NAND gate output and clear input terminal.

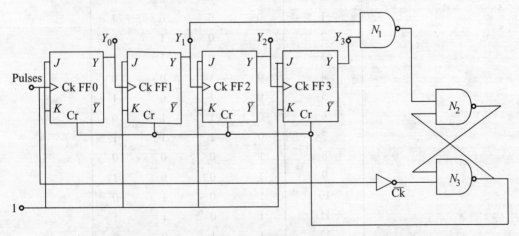

FIGURE 11.42 A decade counter circuit with resetting latch.

States of the flip flops of the decade counter as the incoming pulses arrive are given in Table-11.14.

Table 11.14 States of the flip flops of decade counter.

Number of input pulses	Flip flop outputs			
	Y_3	Y_2	Y_1	Y_0
0	0	0	0	0
1	0	0	0	1
2	0	0	1	0
3	0	0	1	1
4	0	1	0	0
5	0	1	0	1
6	0	1	1	0
7	0	1	1	1
8	1	0	0	0
9	1	0	0	1
10	0	0	0	0

The operation of the latch can be explained as follows. As flip flops used are of master-slave JK type. The outputs Y_1 and Y_2 are 1 after the end of the 10th pulse, that is, when Ck becomes zero. Now, as both Y_1 and Y_2 are 1 and so the output of N_1 is 0, which is an input of N_2. Hence, the output of N_2 is 1, whatever be the output of N_3. Until clock pulse is 0, both inputs of N_3 are 1; hence, its output is 0. Thus, a 0 input is applied to the Cr input terminals for a period long enough to reset the flip flops of the counter.

Similarly, a counter of any modulus can be constructed from ripple counter by adding a logic gate to detect the maximum desired sequence and forcibly resetting the counter.

Up-Down Counter

An up-down counter counts in either forward or reverse direction. It is also called a reversible or forward-backward counter. From the earlier discussions, we know that for forward counting the trigger input of a succeeding flip flop is the output Y of its preceding flip flop. The counting proceeds in the reverse direction if the trigger input is connected to the complemented output $\overline{Y}$, instead of the output Y. If the output Y of a flip flop transits from 0 to 1, its output $\overline{Y}$ transits from 1 to 0. Thus, a negative going transition in $\overline{Y}$ induces a change in state in the succeeding flip flop. Hence, for the reverse connection the succeeding flip flop changes its state when and only when the preceding flip flop goes from state 0 to state 1. The first flip flop makes a transition as usual at each externally applied pulse. If this rule is applied to any number in Table-11.13, the next small number in the table results. For example, let us consider the number 10, which is 1010 in binary representation. At the next pulse, the rightmost zero (which is output Y_0) becomes 1. This change of state of Y_0 causes Y_1 to change state from 1 to 0, as the input of flip flop1 (FF1) ($\overline{Y_0}$) changes from 1 to 0. The change of state of Y_1 from 1 to 0 does not change the state of Y_2. As there is no change of state of Y_2, the state of Y_3 does not change. The net result is that the counter reads 1001. As we started with 10 and ended with 9, a reverse count has taken place.

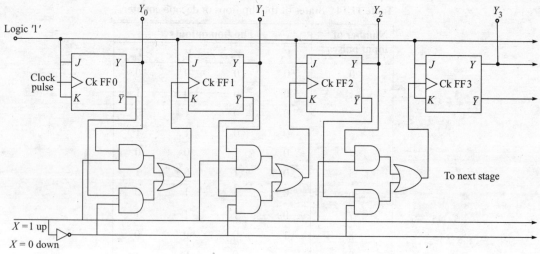

FIGURE 11.43 An up-down ripple counter.

A logic block circuit of an up-down counter is shown in Fig. 11.43. The Fig. 11.43 is obtained by adding three 2-level AND-OR gates between the stages. These gates control the direction of the counter. If the input $X = 1$, the output Y of the preceding flip flop is connected to the input of the succeeding flip flop. When $X = 0$, the input of the succeeding flip flop is connected to the complemented output $\overline{Y}$ of the preceding flip flop. In other words, $X = 1$ converts system to an up counter and $X = 0$ to a down counter.

11.10.2 Synchronous Counter

In the asynchronous counters discussed so far, the change of state occurs when a transition occurs at the output of preceding state. Hence, counting in an asynchronous counter takes finite time. The time required for a counter to complete its response to an input pulse is known as *carry propagation delay*. The carry time of an asynchronous counter is maximum when each state is to change from 1 to 0. In this situation, the next pulse must cause all previous states of flip flops to change. No flip flop responds until the previous state has completed its transition. Hence, a carry propagation delay of a counter is of the order of magnitude of the sum of the carry times of all flip flops of the chain. If there are many stages of flip flops, the carry time may be longer than the interval between the two input pulses. In such case, the counter may fail to read.

In a synchronous counter, all the flip flops are clocked simultaneously (synchronously) by the input pulse; hence, all flip flops change their states simultaneously. The propagation delay, therefore, may be reduced considerably in a synchronous counter. Synchronous counters, therefore, are capable of operating at higher frequency. However, synchronous counters are generally more complicated and costlier as they require more logical gates.

A circuit configuration of a 4-bit synchronous counter is shown in Fig. 11.44. Each flip flop is a T type, obtained by tying the J terminal to the K terminal of a master-slave JK flip flop. As we know, if $T = 0$, there is no change of state when the flip flop is clocked. The output of the flip flop is complemented with each clock pulse when $T = 1$. Unlike ripple counter, each flip flop of the synchronous counter is clocked by the input pulse simultaneously. In addition, AND gates are used to provide the logic 1 at the appropriate T terminal at the proper time. The AND gate A_1 can be removed as the output of this gate is always equal to

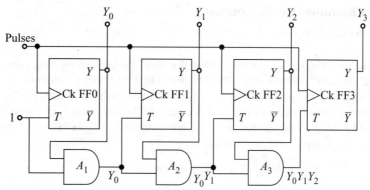

FIGURE 11.44 A 4-bit synchronous counter with serial carry.

Y_0. Hence, Y_0 can directly be connected to T terminal of FF1. When count enable signal is 0, all T inputs are maintained at 0 and the output of the counter does not change. The first stage output Y_0 is complemented at the trailing edge of the clock pulse when the enable input is 1. Each of the other flip flops is complemented when and only when all previous least significant flip flops are at state 1. The operation of the counter can be explained as follows. Before counting starts, the flip flops are reset to 0. The counter reads 0000. As the first clock pulse goes from 1 to 0, the FF0 output is complemented and $Y_0 = 1$ as $T = 1$. The remaining succeeding flip flops do not change their states as their T inputs are 0. The counter reads 0001. During the second clock period the inputs of AND gate A_1 are $T = 1$ and $Y_0 = 1$. The output of A_1 gate is 1; hence, $T = 1$ for FF1. The output Y_1 becomes 1 at the trailing edge of the second clock pulse. Also, Y_0 goes to 0 as $T = 1$. The other flip flop does change their states as $Y_0 = 0$. The counter reads 0010. At the trailing edge of third clock pulse, only Y_0 changes its state from 0 to 1. Hence, the counter reads 0011. At the trailing edge of the fourth clock pulse, flip flops FF0, FF1, and FF2 change their states as their $T = 1$. Hence, $Y_2 = 1$, $Y_1 = 0$, and $Y_0 = 0$. The counter reads 0100. Similarly, for other clock pulses the counter reads as given in Table-11.13.

In the above synchronous counter, the carry (i.e., the enable input) passes through all the control gates in series; hence, the synchronous counter is with *series carry*. The maximum operation frequency can be improved by using *parallel carry*. For this, the enable input to each flip flop comes from a multi-input AND gate excited by the outputs from all preceding flip flops as shown in Fig. 11.45.

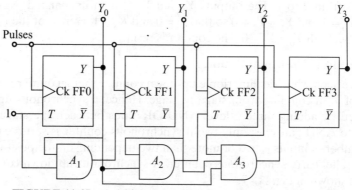

FIGURE 11.45 A 4-bit synchronous counter with parallel carry.

Synchronous Decade Counter

For a decade counter, the circuit of Fig. 11.45 is slightly modified, as shown in Fig. 11.46. From Fig. 11.45, it is clear that FF1 toggles only if $Y_0 = 1$. However, to prevent Y_1 from going to state 1 after the 10th pulse, it is inhibited by connecting one of inputs of A_1 gate to $\overline{Y}_3$ as modified in Fig. 11.46. It is also required that FF3 changes state from 0 to 1 after the eighth pulse and returns to 0 after the tenth pulse. For this, the tie between the J_3 and K_3 terminals of FF3 in Fig. 11.45 is opened. The terminal J_3 is connected to the output of AND gate A_3, whereas terminal K_3 is connected to Y_0 as in Fig. 11.46. Now the logics for T, J, and K inputs are as follows:

$$T_0 = J_0 = K_0 = 1 \qquad T_1 = J_1 = K_1 = Y_0 \overline{Y}_3$$

$$T_2 = J_2 = K_2 = Y_0 Y_1, \qquad J_3 = Y_0 Y_1 Y_2, \qquad \text{and} \qquad K_3 = Y_0 \qquad (11.14)$$

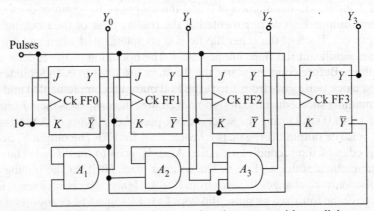

FIGURE 11.46 A 4-bit synchronous decade counter with parallel carry.

Till the seventh pulse, the gate A_1 output is Y_0 as $\overline{Y}_3 = 1$. Hence, the decade counter of Fig. 11.46 works as the counter of Fig. 11.45. Before the eighth pulse, $Y_0 = Y_1 = Y_2 = 1$ so that $J_3 = 1$ and $K_3 = 1$. Hence, FF3 toggles and at the trailing edge of the eighth pulse, making $Y_3 = 1$. The counter reads 1000. Before the ninth pulse, $Y_0 = Y_1 = Y_2 = 0$ so that $J_3 = 0$ and $K_3 = 0$. Hence, FF3 is inhibited and its state remains unchanged, whereas Y_0 changes from 0 to 1. The outputs Y_1 and Y_2 remain unchanged. Now, before the 10th pulse, $Y_0 = 1$ and $Y_1 = Y_2 = 0$ so that $J_3 = 0$ and $K_3 = 1$. Hence, at the trailing edge of the 10th pulse, both $Y_0 = Y_3 = 0$. The counter is reset.

Synchronous Up-Down Counter

We have seen while dealing ripple up-down counter that a counter is reversed if $\overline{Y}$ is in place of Y in coupling from stage to stage. Hence, a synchronous up-down counter is obtained by adding three 2-level AND-OR gates between the stages. The number of inputs to AND gates are only two if synchronous counter is series carry type, whereas the number of inputs is two or more than two depending on the position of the flip flops in a parallel carry synchronous counter. A synchronous up-down counter with parallel carry is shown in Fig. 11.47.

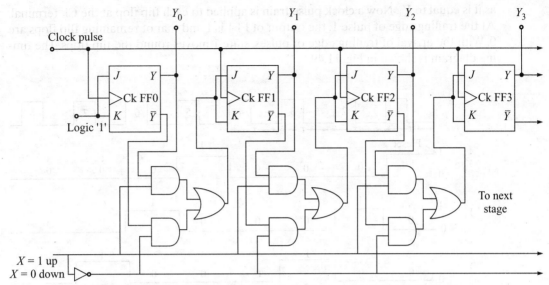

FIGURE 11.47 A 4-bit synchronous up-down counter with parallel carry.

Ring Counter

In a ring counter D flip flops are used in place of JK flip flops. A 4-bit ring counter is shown in Fig. 11.48. From the circuit it is clear that the connection between the Y output terminal of a flip flop and the D input terminal of the succeeding flip flop forms a ring. Hence, the counter is given its name as ring counter. Although in structure this device is a counter, its use in digital systems is for the purposes other than counting. Chiefly, it is used for generating timing sequences for the purpose of controlling the operation of the digital system.

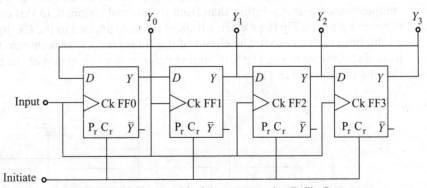

FIGURE 11.48 A 4-bit ring counter using D flip flops.

The operation of the ring counter is explained as follows. As we know the output state of a D flip flop at a next count pulse is equal to the current value of the state at the D input. Hence $Y_{i+1} = D_i$. At starting a preset, command is applied to the FF0 to set its output state at 1 and simultaneously resetting the remaining flip flops. The input D_0 is 0, as it is equal to Y_3. Now a clock pulse train is applied to each flip flop at the Ck terminal. At the trailing edge of pulse 1, the output of FF1 is 1 and that of remaining flip flops are 0. With the arrival of trailing edge of pulses, state 1 moves round the flip flops. The timing diagram is shown in Fig. 11.49.

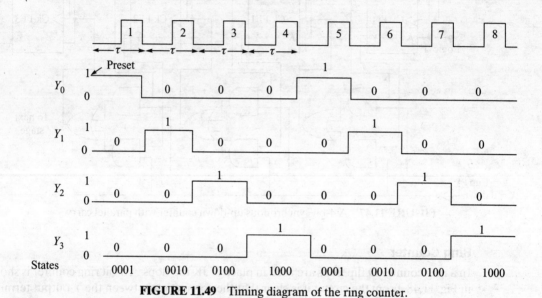

FIGURE 11.49 Timing diagram of the ring counter.

Switchtail (Johnson) Counter

A slight modification in the ring counter results in an interesting change. Unlike the ring counter, the connection from the last flip flop to the first flip flop is made from the complemented terminal rather than from the normal terminal. In this counter, the state 0 moves round the flip flops when a train of pulses is applied to the Ck inputs. At starting all flip flops are reset to 0. The circuit of the Johnson counter is shown in Fig. 11.50 and its timing diagram in Fig. 11.51. Its timing diagram is an inversion of the time diagram of the ring counter of Fig. 11.48.

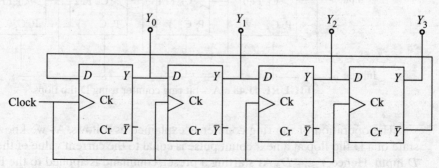

FIGURE 11.50 A 4-bit Johnson counter using D flip flops.

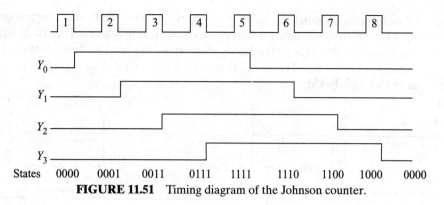

FIGURE 11.51 Timing diagram of the Johnson counter.

11.11 RAM UNIT

The RAM is an array of storage cells. These cells memorize information in binary form. Unlike ROM, information can be randomly written into or read out of each storage cell as required. That is why this memory device is given the name random access or read/write memory. The basic storage cell is the latch or basic flip flop discussed earlier. To understand the operation of the RAM, let us first discuss about a 1-bit read/write device. A 1-bit read/write memory using a simple SR flip flop with control gates is shown in Fig. 11.52(a). Its symbolic block showing different external terminals is shown in Fig. 11.52(b). The terminal I is the data input terminal, X is the address terminal, E is read/write enable terminal, and O is the data read terminal. To read data out of or write data into the cell, it is necessary to make the address X high ($X = 1$). Now, the stored information is available at the output of the flip flop. If E is made low (i.e., $E = 0$), one can read the data at the output terminal O. To write the data into the memory cell, it is required to make the read/write enable input E high. As $E = 1, X = 1$, then $S = I$ and $R = \bar{I}$. Hence, $Y = I$. Thus, the input 1(0) is written in the elementary memory cell (latch or basic flip flop).

Suppose we wish to have a read/write memory for one word of 8-bit. For this it is required to have eight elementary memory cells connected in parallel. This combination

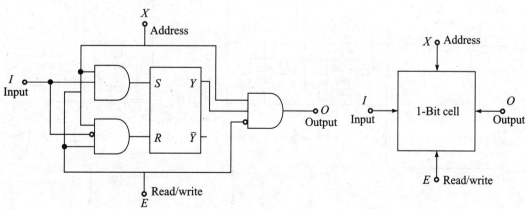

FIGURE 11.52 A 1-bit read/write memory (a) circuit and (b) symbol.

of eight memory cells has eight data input lines and eight data output lines. All eight cells have a common address line and excited by one address. It is made write enable by a common enable signal. The block diagram of the read/write memory for one word of 8-bit data is shown in Fig. 11.53. Thus a 1-bit memory cell of Fig. 11.52 is the basic building block of a RAM.

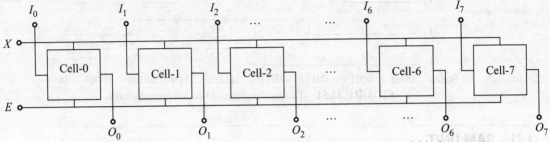

FIGURE 11.53 Block diagram of a read/write memory for a 8-bit data.

Practically, a RAM having capacity to store more than one word of data of 8-bit each is a requirement. Hence, a RAM of capacity to store 16 words of 8-bit each must have 16 blocks of Fig. 11.53 connected in parallel and is shown in Fig. 11.54. This system, therefore, also has eight data input lines and eight data output lines. The total number of storage cells required is $16 \times 8 = 128$. A group of eight cells arranged in a

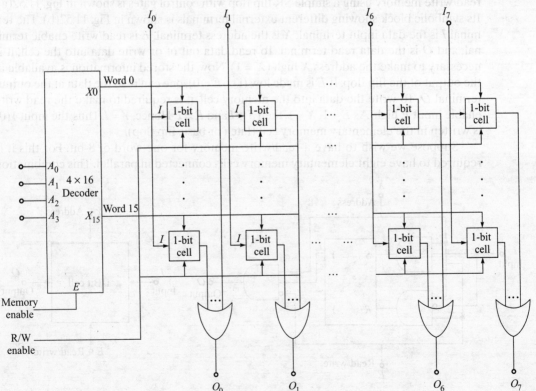

FIGURE 11.54 A block diagram of a read/write memory of 16×8 cells.

horizontal line as in Fig. 11.53 is excited by the same address line. Hence, there are 16 such lines, each line excited by its own address line. In other words, addressing is provided by exciting one of the 16 lines at a line. To address one of the 16 lines, a 4×16 decoder is required. In addition to address line, each memory cell must have three connections as shown in Fig. 11.54. These are input terminal I, output terminal O, and read/write enable terminal E. The input bits appear at the terminal I of the cell and the same bit is applied to all cells along a column. The output line connection from each cell along a given column forms one of the 16 inputs to an OR gate as shown in Fig. 11.54. Similarly, all eight input and output terminals are formed. This type of addressing is called **linear selection**.

To illustrate the operation of the RAM of Fig. 11.54, let us consider an example in which an input word $A_7A_6A_5A_4A_3A_2A_1A_0 = 11010011$ is to be written into the memory register of the RAM represented by the input word 7 (i.e., row 8) and in binary by 0111. Thus, the address code of data line is 0111. To write data into the cells, read/write input E is made high to make cells write enable. Although the write enable signal is applied to all the cells, the input signal appears only for the cells on the eighth row (row 7), which have been addressed.

In linear addressing RAM of Fig. 11.54 of sixteen 8-bit words there are 16 lines with eight cells per line. Here, we use a 4×16 decoder for addressing. However, there is a more commonly used addressing technique, called *coincident selection*. In coincident selection, a matrix topology, in which 16 elementary memory cells are arranged in a rectangular 4×4 array, is used. Each cell of the array stores 1 bit for one word. For the 8-bit word eight matrix planes of storage cells are used. Thus, one such matrix plane is used for each bit of the 8-bit word. The eight cells, one each from the eight matrix planes, are connected together. One plane of the above matrix is symbolically shown in Fig. 11.55. Each bit (storage cell) is located by addressing an X address line and a Y address line. Thus, a cell is located by X-Y coordinate addresses, as shown in Fig. 11.55. Two 2×4 decoders are used. One is used for X addressing and another for Y addressing. Basic elements of which a RAM is constructed are shown in Fig. 11.56. The elements used are the rectangular array of cells, and the X and Y decoders.

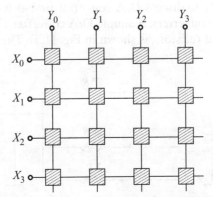

FIGURE 11.55 A 4×4 array of storage cells.

584 Chapter 11 / Components of Digital Systems

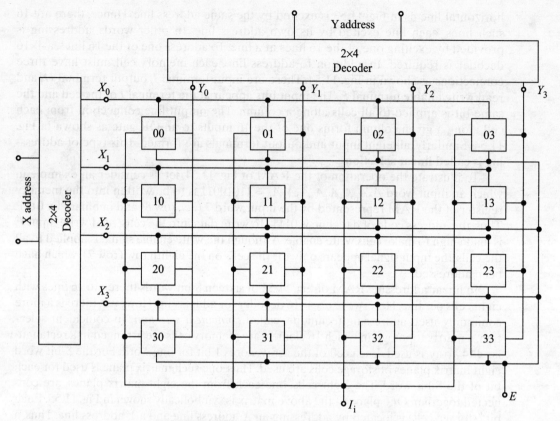

FIGURE 11.56 RAM with coincident selection.

11.12 D/A CONVERTER

As mentioned earlier, in many cases we need a device that converts a digital signal into an analogue signal. The device is called a D/A converter or a DAC. A D/A converter, therefore, accepts a digital word as its input and converts it to an analogue voltage or current. We prefer to discuss D/A converter first as it is used as a constituent block in some of the A/D converters. A simple D/A converter circuit, which is a summing amplifier with weighted resistors, is shown in Fig. 11.57. The resistors are weighted in binary

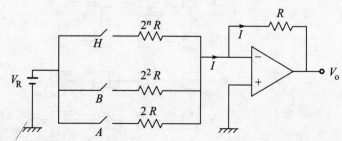

FIGURE 11.57 A simple D/A converter with binary weighted resistances.

progression. The switches $S_0, S_1, S_2, \ldots S_{n-1}$ are controlled by digital input word. Logic '1' indicates a closed switch, whereas logic '0' indicates an open switch.

From Fig. 11.57

$$I = I_0 + I_1 + \ldots + I_{n-1} \tag{11.15}$$

or
$$I = r_0 \frac{V_R}{2^n R} + r_1 \frac{V_R}{2^{n-1}} + \cdots + r_{n-1} \frac{V_R}{2R}$$

or
$$V_0 = IR = \left[r_0 2^{-n} + r_1 2^{-(n-1)} + \cdots + r_{n-1} 2^{-1} \right] V_R \tag{11.16}$$

where V_R is the reference voltage and the coefficients r_i ($i = 0, 1, 2, 3, \ldots, n-1$) represent the bits of the binary word. Hence, coefficients $r_i s$ take values either 0 or 1. The voltage V_R is a stable reference voltage used in the circuit. The MSB corresponds to r_{n-1} and its weight is $V_R/2$. The LSB is r_0 and its weight is $V_R/2^n$.

For example, consider a 4-bit word, then Eq. (11.16) can be written as

$$V_o = \left(2^3 r_3 + 2^2 r_2 + 2 r_1 + r_o \right) \times \frac{V_R}{16} \tag{11.17}$$

For simplicity, if V_R is assumed to be 16 V, then the 16 binary words of 8-bits and their respective analogue outputs are as given in Table-11.15.

From Table-11.15 it is clear that the analogue output of the D/A converter is proportional to the digital input.

Table-11.15 Inputs and outputs of a 4-bit D/A converter.

Digital inputs				Analogue output (V)
r_3	r_2	r_1	r_0	
0	0	0	1	1
0	0	1	0	2
0	0	1	1	3
0	1	0	0	4
0	1	0	1	5
0	1	1	0	6
0	1	1	1	7
1	0	0	0	8
1	0	0	1	9
1	0	1	0	10
1	0	1	1	11
1	1	0	0	12
1	1	0	1	13
1	1	1	0	14
1	1	1	1	15

The switches in the D/A converter can be implemented using transistor, FET, or MOFET. The accuracy of the D/A converter depends on the absolute accuracy of the resistor values and change of their values with temperature. The resistances used are of different values, and the largest value is equal to $2^n R$, $2R$ being the smallest value of resistance. With the increase of the number of bits, the value of resistance $2^n R$ may become excessively large. The D/A converter suggested in Fig. 11.57 is, therefore, not very convenient and the accuracy is likely to be poor. The problems faced while designing the D/A converter of Fig. 11.57 are listed as below:

1. Wide range of resistors is needed, especially when the number of bits increases. For example, we need resistors of resistance values $R, 2R, 4R, \ldots, 128R$ for an 8-bit D/A converter. Here, the largest resistor is 128 times the smallest one. For a 12-bit D/A converter, the largest resistor needs to be 2048 times the smallest one. Hence, the mass production of weighted resistors is impractical.
2. It is difficult to maintain the accurate resistance ratio.
3. The switches are in series with the resistors; hence, their 'on resistance' needs to be very low. They should have zero off-set voltage. The designer meets these requirements by using good metal oxide semiconductor field-effect transistors.

An alternative circuit of a D/A converter is the ladder-type D/A converter. This type of converter uses only two values of resistors, R and $2R$. A 3-bit ladder-type of D/A converter is shown in Fig. 11.58. The R-2R ladder avoids the problem of a wide range of resistor values. Since this circuit requires only two resistor values, it is well suited to integrate circuit realization. This type of D/A converter, however, requires twice the number of resistors than that required in weighted resistor type D/A converter for the same number of bits.

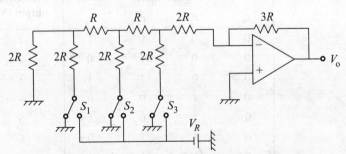

FIGURE 11.58 An R-2R ladder-type D/A converter.

From the circuit of Fig. 11.58, it is observed that at any of the ladder nodes the resistance is $2R$ looking to the left or right or toward the switch. This is true as the input terminal is at the virtual ground. Hence, the current at any node of the circuit splits equally toward the left or the right. For digital number 100 (MSB bit), switch S_3 is closed, and hence the circuit of Fig. 11.58 reduces as shown in Fig. 11.59. The current from the source V_R,

$$I = \frac{V_R}{3R}$$

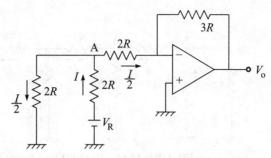

FIGURE 11.59 For MSB bit (digital number 100).

Hence, $I_o = \dfrac{I}{2} = \dfrac{V_R}{6R}$

and $V_o = \dfrac{V_R}{6R} \times 3R$

$= \dfrac{V_R}{2}$ V

For digital number 010, switch S_2 is closed, and hence the circuit of Fig. 11.58 reduces as shown in Fig. 11.60. As said above the currents at nodes A and B equally splits toward the left and right. The current distribution is as shown in Fig. 11.60. The current from the source V_R,

$$I = \dfrac{V_R}{3R}$$

Hence $I_o = \dfrac{I}{4} = \dfrac{V_R}{12R}$

and $V_o = \dfrac{V_R}{12R} \times 3R$

$= \dfrac{V_R}{4} = \dfrac{V_R}{2^2}$ V

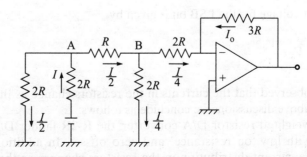

FIGURE 11.60 For next to MSB bit (digital number 010).

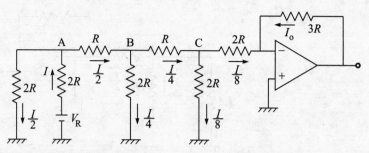

FIGURE 11.61 For LSB bit (digital number 001).

For digital number 001 (LSB bit), switch S_1 is closed, and hence the circuit of Fig. 11.58 reduces as shown in Fig. 11.61. As said above the currents at nodes A, B, and C equally splits toward the left and right. The current distribution is as shown in Fig. 11.61. The current from the source V_R,

$$I = \frac{V_R}{3R}$$

Hence $I_o = \frac{I}{8} = \frac{V_R}{24R}$

and $V_o = \frac{V_R}{24R} \times 3R$

$$= \frac{V_R}{8} = \frac{V_R}{2^3} \text{ V}$$

From the above discussion, we conclude that the output voltage reduces by 2 when we proceed from MSB bit to LSB bit. Hence for an n bit DAC the output voltage for the MSB bit is given by

$$V_o = \frac{V_R}{2} \tag{11.18}$$

The output voltage for the LSB bit is given by

$$V_o = \frac{V_R}{2^n} \tag{11.19}$$

It can be observed that the currents in the resistors change as the input word changes. From the above discussions we conclude as follows:

Like weighted resistor D/A converter, the R-2R ladder D/A converter requires switches with low 'on resistance' and zero off-set. In addition, in both the above circuits, the current distribution in the resistors changes as the input word changes.

Hence, the power dissipation and heating in the network change, introducing nonlinearity in the D/A converter. Also, the load on the reference source depends on the binary input. Therefore, most of integrated D/A converters, employing ladder network, use the configuration of Fig. 11.62. This circuit is known as *inverted R-2R ladder circuit*.

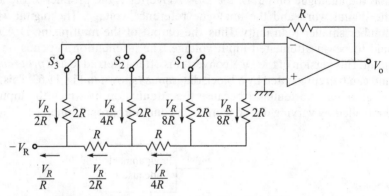

FIGURE 11.62 An inverted R-2R ladder D/A converter.

In the circuit of Fig. 11.62, the currents flowing in the ladder resistors and in the reference voltage source are independent of input word. The input data cause the ladder currents to be switched to either the real ground or the input (virtual ground) of the operational amplifier. Since both the input terminals (inverted or noninverted) of the operational amplifier are at the ground potential, the ladder currents are independent of the switch position.

As both the contacts of the switches are at the same potential, the current distributions in the ladder are same for any input word and any switch position, as is clear from Fig. 11.62. Hence, for any input word the current drawn from the reference voltage source V_R is given by

$$I = \frac{V_R}{R}$$

When all the switches are to the right, as shown in the Fig. 11.62, the currents through $2R$ resistors flow to the ground. Hence, under this condition, the output current I_o is zero. When all the switches are to the left, the current through $2R$ resistors sink to the virtual ground. Hence, the output current

$$I_o = \frac{V_R}{R}\left[\frac{1}{2}+\frac{1}{4}+\frac{1}{8}\right] = \frac{7V_R}{8R} \qquad (11.20)$$

Since the currents throughout the ladder resistors remain constant, the voltages across them are constant. The stray capacitance of the resistors, therefore, has little effect. The inverted R-2R ladder is a popular D/A converter configuration. It is often implemented in CMOS technology. MOS switches are excellent.

11.12.1 Multiplying D/A Converter

A D/A converter that employs a variable analogue signal in place of a fixed reference voltage is called a *multiplying D/A converter*. From Eqs. (11.16) and (11.17), it is clear that the analogue output of the D/A converter is the product of the analogue value of the digital word and the analogue reference voltage. The digital word represents a number smaller than unity. Thus, the output of the multiplying D/A converter can be said to be an attenuated input voltage. The attenuation depends on the digital word input. This version of the D/A converter is often referred to as a *programmable attenuator* and can be represented by a block diagram as shown in Fig. 11.63. This is referred to as a programmable attenuator because the attenuation by which the input is attenuated is controlled by varying digital word by a computer programme.

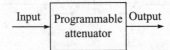

FIGURE 11.63 A programmable attenuator.

11.13 A/D CONVERTER

As mentioned earlier, an A/D converter converts an unknown analogue input signal into an *n*-bit binary or digital number. An A/D converter is also called in short as ADC. The *n*-bit number is a binary fraction representing the ratio of input signal to a fixed reference signal. A comparator circuit is the heart of all A/D converters. The comparator circuit is shown in Fig. 11.64. The comparator compares an unknown analogue input voltage V_i with a reference voltage V_R. The comparator is basically a multistage high-gain differential amplifier. The output of the comparator is determined by relative polarity of the two inputs V_i and V_R. When V_R is less than V_i, the output is maximum and the comparator is ON (logic '1'). Because of high gain of the amplifier, it is either saturated (at high difference of inputs) or cut off (at low difference of inputs). Hence, it acts as a binary device. To perform analogue to digital conversion, the analogue reference is varied to determine which of the 2^n possible digital word is closest to the unknown analogue voltage.

There are a number of A/D converter circuits available. However, we will discuss only two of them. The first A/D converter uses counter method, whereas another uses successive approximation method.

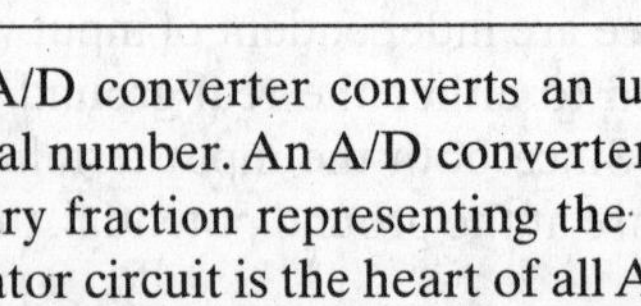

FIGURE 11.64 A comparator.

11.13.1 A/D Converter Using Counter

An 8-bit A/D converter using a counter is shown in Fig. 11.65(a). It consists of a counter, a D/A converter, a comparator, and logic gates. The logic gates control the clock input to the counter. An analogue to digital conversion is initiated by making the start of conversion (SOC) signal high. When SOC is low, the counter is clear. When SOC is made high the clock pulses pass through the counter and it advances. A voltage (V_o) proportional to the digital count is generated by the D/A converter. The voltage V_o is of the form of a staircase ramp. The A/D converter, therefore, is also called a staircase A/D converter.

The analogue input voltage is V_i, and B is the 8-bit digital output, and as discussed above, the output of the D/A converter is an analogue output V_o. The operation of the A/D converter is explained as follows: First, the SOC signal goes low, clearing the counter and the voltage V_o is zero. When SOC signal is made high, the counter is ready to count. As $V_o = 0$, the comparator output V_{PS} is high, and hence the counter's COUNT is high. The counter starts counting upwards from zero and gives a voltage staircase, as shown in Fig. 11.65(b), as the output of the D/A converter. As long as the analogue input V_i is greater than V_o, the comparator has a positive output and COUNT remains high. At some point of staircase voltage, the next step makes V_o greater than V_i. This forces COUNT to go low and the counter stops. Now, the digital output B is the digital equivalent of the analogue input V_i. The negative going edge of COUNT signal is used as the end of conversion (EOC) signal to tell that the conversion is complete. If V_i is changed, an external circuit sends an SOC SIGNAL. This clears the counter and a new cycle begins.

The main disadvantage of the A/D converter using counter is its slow speed. It takes the maximum time when the counter has to reach the maximum count before the staircase is greater than V_i. For an 8-bit D/A converter, the worse conversion time is 2^8 (=256) clock periods.

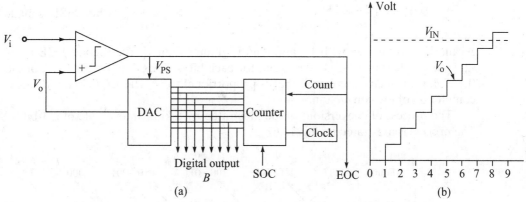

FIGURE 11.65 (a) An A/D converter using a counter and (b) staircase output.

11.13.2 Successive Approximation Type A/D Converter

Because of its high resolution and high speed, the successive approximation type of A/D converter is most widely used in practice. The circuit of the successive approximation

A/D converter is organized as shown in Fig. 11.66. Its principal components are a D/A converter, a comparator, a successive approximation register (SAR), a clock pulse generator, and control and status logics. In the circuit of Fig. 11.66, the role of a counter in Fig. 11.65 is played by the SAR. The operation of the A/D converter is explained as follows:

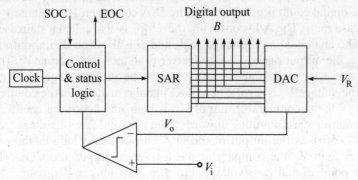

FIGURE 11.66 A successive approximation type A/D converter.

At the SOS, the SAR is cleared and its MSB is set to 1. This results in $V_o = V_R/2$. The analogue input V_i is compared with the output V_o. If V_i is greater than V_o, the MSB is left unchanged; otherwise the MSB is made 0. Next, MSB is made 1. At this stage there are two possible values of V_o. These are

$$V_o = \frac{V_R}{2} + \frac{V_R}{4} = \frac{3V_R}{4} \qquad \text{(when MSB is left 1)}$$

$$V_o = 0 + \frac{V_R}{4} = \frac{V_R}{4} \qquad \text{(when MSB is made 0)}$$

Again, V_i is compared with V_o, and if V_i is greater than V_o, the second MSB is left 1, otherwise 0. This process is carried out for each bit of SAR. Finally, the SAR output is the digital or binary number B that is proportional to V_i. The EOC line indicates the completion of the conversion.

The process of conversion can be illustrated by taking an example of a 3-bit SAR. The process is illustrated by Fig. 11.67.

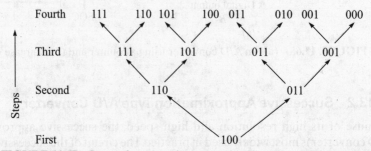

FIGURE 11.67 Steps of a successive approximation type A/D converter.

Advantage of a successive approximation type A/D converter is its high speed. There are many other A/D converters commercially available from 6-bit to 16-bit, in the market.

11.14 REVIEW QUESTIONS

1. What do you understand by an encoder? What are the design considerations? How many inputs are there for an n-bit encoder?
2. Draw a logic circuit for a decimal to binary encoder and explain its operation.
3. Find the truth table for the half adder. Draw the logic circuit for the half adder.
4. Write the logic functions for the sum and the carry of the half adder in both sum of minterms and product of maxterms. Draw the logic circuits for both of the logic functions.
5. Find the truth table for the full adder. Write the Boolean functions for the sum and the carry in both sum of product form and product of sum form.
6. Draw the logic circuit for the full adder. Explain its operation.
7. Draw the system of a 4-bit parallel binary adder using 1-bit full adders. Explain the operation of the system.
8. Explain the operation of a serial binary full adder.
9. If A and B are 4-bit binary numbers with $B > A$, show that to subtract A from B it is only required to add B, $\overline{A}$, and 1.
10. Draw the block diagram of a 4-bit subtractor using full adders.
11. Draw the schematic block diagram of 4-bit adder/subtractor. Describe in short each block of the circuit. Also explain the operation of the circuit when used as (a) an adder and (b) a subtractor.
12. Suggest a circuit for a 4-bit adder/subtractor. Describe in brief each block of the circuit. Also explain its operation when binary number A is subtracted from another binary number B when $A < B$ and when $A > B$.
13. What do you know about a decoder? How is the decoder different from the encoder?
14. A decoder has four input lines. How many output lines does the decoder have? How many AND gates are needed? How many minterms are there? What will happen if AND gates are replaced by NAND gates?
15. What is a demultiplexer? Show how to convert a decoder into a demultiplexer. How can a strobe be added to this system?
16. What is a multiplexer? Draw a logic block diagram of a 4- to 1-line multiplexer. Explain its operation. Why is the multiplexer called as data selector as well?
17. Distinguish among the multiplexer, the demultiplexer, and the decoder regarding their operation, application, and construction.
18. Show how a multiplexer may be used as (a) parallel to serial converter and (b) a sequential data selector.
19. What is a ROM? Name different building elements of the ROM. Distinguish among the ROM, the PROM, and the EPROM.
20. Explain how a truth table can be implemented by a ROM. What does it mean by the phrase 2 K ROM?
21. What is a sequential logic circuit? How does it differ from the combinational logic circuit? How does a synchronous sequential circuit differ from an asynchronous sequential circuit?

22. List three applications of the ROM and explain them briefly. Show the following two lines of the conversion table from BCD to seven-segment indicator code: 0111 and 1000.
23. What is a latch? Why is it called so? Explain how a latch is constructed using NOT gates. Show that the latch so obtained has two stable states.
24. Draw the basic NAND flip flop. Describe completely the operation of NAND flip flop for all four combinations of the two input logical variables S and R. Find the appropriate truth table. What are its shortcomings?
25. What is the usefulness of operating a flip flop in synchronous manner? Describe the operation of the clocked SR flip flop.
26. What is the D flip flop? How is it different from the SR flip flop. How is the uncomplemented output related to the inputs?
27. Draw the JK flip flop using the SR flip flop and two AND gates. Explain its operation. What are its advantages over the SR flip flop?
28. Draw a schematic block diagram of the T flip flop. What is its distinctive feature?
29. Explain the race around condition in connection with the JK flip flop.
30. Draw the clocked JK flip flop with preset and clear inputs. Explain the function of the preset and clear inputs.
31. Draw a schematic block diagram of the master-slave JK flip flop. Explain its operation. Show that the race around condition is eliminated.
32. What is a register? What purpose does the register serve in digital systems? How are registers classified?
33. Draw the schematic block diagram of the parallel register and explain its operation.
34. Construct a shift register from SR flip flops. Explain its operation. How does it differ from the parallel register?
35. Describe the constructional details of a digital counter. Name some of its applications in digital systems.
36. Define modulo of the counter. Illustrate how the number of states of the counter is determined.
37. With block diagram explain the operation of the ripple counter. Why is it so called? What are its drawbacks?
38. Explain with a neat block diagram the operation of the decimal ripple counter. What is the effect of propagation delay in the counter? How is this effect eliminated in the decimal counter with resetting latch?
39. Explain with neat schematic block diagram the operation of an up-down ripple counter.
40. What is the synchronous counter? How does it differ in construction and circuit arrangement from the ripple counter? What are its advantages and disadvantages?
41. Describe the operation of the synchronous counter with serial carry and parallel carry.
42. Describe how a decade synchronous counter is constructed. Explain its operation.
43. Describe the construction of an up-down synchronous counter and explain its operation.
44. Describe the operation of the Johnson counter. How does it differ from the ripple counter?
45. How does a RAM differ from a ROM? Draw the schematic diagram of the basic cell of the RAM. Identify the input and output signals to each cell. Describe the function of the basic cell.
46. Explain linear selection and coincident selection in a RAM. How are registers related to the RAM?

47. Explain the operation of a binary weighted resistor D/A converter. What are its advantages and disadvantages?
48. What is a ladder-type D/A converter? What are its advantages over those of weighted register type D/A converter? What are its drawbacks?
49. How are drawbacks of ladder D/A converter eliminated? Illustrate your points.
50. Describe the operation of an A/D converter using a counter. List its merits and demerits.
51. Describe the operation of a successive approximation A/D converter. List its merits and demerits.

11.15 SOLVED PROBLEMS

1. Design an encoder that converts hexadecimal to binary form. Write the logical expression for each output.

   Solution

   For converting hexadecimal to binary, the number of input lines is 16. Hence, the number of outputs lines required is four. The number of OR gates needed, therefore, is four. For the different inputs the outputs are given in Table-11.16.

   Table-11.16 Truth table for hexadecimal to binary encoder.

S_0	S_1	S_2	S_3	S_4	S_5	S_6	S_7	S_8	S_9	S_A	S_B	S_C	S_D	S_E	S_F	Y_3	Y_2	Y_1	Y_0
1	0	0	0	0	0	0	0	0	0	0	0	0	0	0	0	0	0	0	0
0	1	0	0	0	0	0	0	0	0	0	0	0	0	0	0	0	0	0	1
0	0	1	0	0	0	0	0	0	0	0	0	0	0	0	0	0	0	1	0
0	0	0	1	0	0	0	0	0	0	0	0	0	0	0	0	0	0	1	1
0	0	0	0	1	0	0	0	0	0	0	0	0	0	0	0	0	1	0	0
0	0	0	0	0	1	0	0	0	0	0	0	0	0	0	0	0	1	0	1
0	0	0	0	0	0	1	0	0	0	0	0	0	0	0	0	0	1	1	0
0	0	0	0	0	0	0	1	0	0	0	0	0	0	0	0	0	1	1	1
0	0	0	0	0	0	0	0	1	0	0	0	0	0	0	0	1	0	0	0
0	0	0	0	0	0	0	0	0	1	0	0	0	0	0	0	1	0	0	1
0	0	0	0	0	0	0	0	0	0	1	0	0	0	0	0	1	0	1	0
0	0	0	0	0	0	0	0	0	0	0	1	0	0	0	0	1	0	1	1
0	0	0	0	0	0	0	0	0	0	0	0	1	0	0	0	1	1	0	0
0	0	0	0	0	0	0	0	0	0	0	0	0	1	0	0	1	1	0	1
0	0	0	0	0	0	0	0	0	0	0	0	0	0	1	0	1	1	1	0
0	0	0	0	0	0	0	0	0	0	0	0	0	0	0	1	1	1	1	1

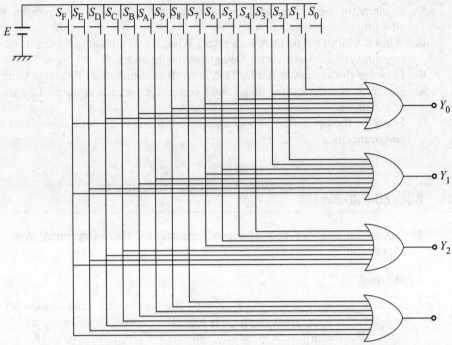

FIGURE 11.68 A hexadecimal to binary encoder with 16 inputs (0–F).

Table-11.17 Logical expression for output bits.

Output	Logical expressions
Y_0	$S_1 + S_3 + S_5 + S_7 + S_9 + S_B + S_D + S_F$
Y_1	$S_2 + S_3 + S_6 + S_7 + S_A + S_B + S_E + S_F$
Y_2	$S_4 + S_5 + S_6 + S_7 + S_C + S_D + S_E + S_F$
Y_3	$S_8 + S_9 + S_A + S_B + S_C + S_D + S_E + S_F$

The logic circuit of the encoder is similar to Fig. 11.4, except that there are 16 switches, named as $S_0, S_1, S_2, ..., S_D, S_E, S_F$. The output OR gates are still four. The numbers of inputs to the gates are changed as shown in Fig. 11.68. The logical expressions for output bits are as given in Table-11.17.

2. Design an encoder satisfying the truth table given in Table-11.18. Write the logical expression for each output.

Solution

From the truth table for the encoder, the logical expression for each output bit are $Y_0 = S_0 + S_1$, $Y_1 = S_2 + S_3$, $Y_2 = S_0 + S_1 + S_3$, and $Y_3 = S_1 + S_2 + S_3$. The logical circuit of the encoder is as shown in Fig. 11.69.

Table-11.18 Truth table for Solved Problem 2.

Inputs				Outputs			
S_3	S_2	S_1	S_0	Y_3	Y_2	Y_1	Y_0
0	0	0	1	0	1	0	1
0	0	1	0	1	1	0	1
0	1	0	0	1	0	1	0
1	0	0	0	1	1	1	0

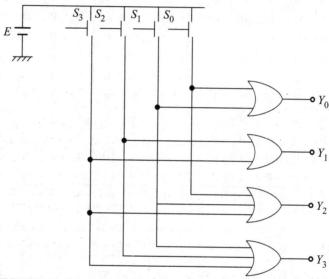

FIGURE 11.69 A logical circuit of encoder of Solved Problem 2.

3. Write the truth table of an encoder to display decimal numbers (0–9) by the seven-segment LED display with positive logic. The seven-segment LED assembly is shown in Fig. 11.24(a). Write logical expression for each segment. Implement the logical expressions by using diodes instead of OR gates.

Solution

The truth table for the encoder is shown in Table-11.19.

The logical expressions for output bits are given below:

$$Y_0 = S_0 + S_2 + S_3 + S_5 + S_7 + S_8 + S_9$$

$$Y_1 = S_0 + S_1 + S_2 + S_3 + S_4 + S_7 + S_8 + S_9$$

$$Y_2 = S_0 + S_1 + S_3 + S_4 + S_5 + S_6 + S_7 + S_8 + S_9$$

$$Y_3 = S_0 + S_2 + S_3 + S_5 + S_6 + S_8$$

$$Y_4 = S_0 + S_2 + S_6 + S_8$$

$$Y_5 = S_0 + S_4 + S_5 + S_6 + S_8 + S_9$$

$$Y_6 = S_2 + S_3 + S_4 + S_5 + S_6 + S_8 + S_9$$

Table-11.19 Truth table of decimal to a LED display encoder.

S_0	S_1	S_2	S_3	S_4	S_5	S_6	S_7	S_8	S_9	Y_6	Y_5	Y_4	Y_3	Y_2	Y_1	Y_0
1	0	0	0	0	0	0	0	0	0	0	1	1	1	1	1	1
0	1	0	0	0	0	0	0	0	0	0	0	0	0	1	1	0
0	0	1	0	0	0	0	0	0	0	1	0	1	1	0	1	1
0	0	0	1	0	0	0	0	0	0	1	0	0	1	1	1	1
0	0	0	0	1	0	0	0	0	0	1	1	0	0	1	1	0
0	0	0	0	0	1	0	0	0	0	1	1	0	1	1	0	1
0	0	0	0	0	0	1	0	0	0	1	1	1	1	1	0	0
0	0	0	0	0	0	0	1	0	0	0	0	0	0	1	1	1
0	0	0	0	0	0	0	0	1	0	1	1	1	1	1	1	1
0	0	0	0	0	0	0	0	0	1	1	1	0	0	1	1	1

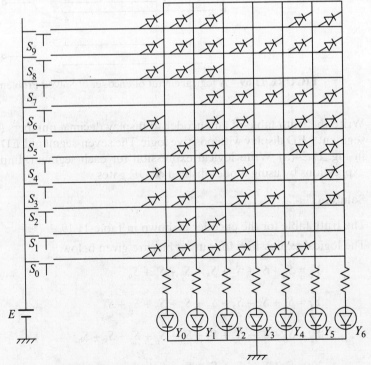

FIGURE 11.70 An encoder for Solved Problem 3.

The encoders have been implemented by using multi-inputs OR gates. The encoder can also be implemented by diodes. In this problem, the encoder, therefore, is implemented by using diodes instead of OR gates. For this a rectangular array (or matrix) of wires, as shown in Fig. 11.70, is used. Diodes are used to interconnect these wires so as to generate the binary codes of encoder outputs as given in Table-11.19.

To explain the operation, let us consider that the switch 6 is closed. From Fig. 11.68 it is clear that except the LED segments Y_0 and Y_1, all the segments of the LED are activated and digit 6 is displayed. Similarly, any of the decimal digits can be enlighted by closing the corresponding switch.

4. Show how the half adder logic circuit of Fig. 11.6(a) can be implemented with the NOR gates.

Solution

We have discussed in Section-10.2 that AND and OR gates can be realized by NOR gates. Utilizing Fig. 10.18(b) and (c), Fig. 11.6(a) can be implemented by using only NOR gates. The half adder using NOR gates is shown in Fig. 11.71.

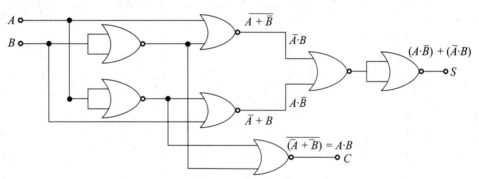

FIGURE 11.71 A half adder using NOR gates.

5. A half subtractor is a combinational logic circuit that subtracts 1-bit binary number B from A and produces the difference. Develop the truth table and write the logical expressions in sum of minterms for the difference and borrow. Implement the logical expressions into a logical circuit.

Solution

The inputs of the subtractor are A and B and outputs are D (difference) and P (borrow). The truth table is given as follows.

Inputs		Outputs	
A	B	D	P
0	0	0	0
0	1	0	1
1	0	1	0
1	1	0	0

The logical expressions for difference and borrow in sum of minterms are given as

$$D = A\bar{B} \text{ and } P = \bar{A}B$$

The above logical expressions are implemented by the logical circuit shown in Fig. 11.72.

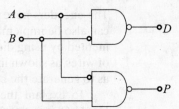

FIGURE 11.72 A logic circuit of a half subtractor.

6. Design the logical circuit for a 3×8 decoder. Draw the truth table for inputs A, B, C and outputs Y_7, Y_6, Y_5, Y_4, Y_3, Y_2, Y_1, Y_0. Show the logical expression for each minterm.

Solution

The truth table is given below.

Inputs				Outputs							
S	A	B	C	Y_7	Y_6	Y_5	Y_4	Y_3	Y_2	Y_1	Y_0
0	X	X	X	X	X	X	X	X	X	X	X
1	0	0	0	0	0	0	0	0	0	0	1
1	0	0	1	0	0	0	0	0	0	1	0
1	0	1	0	0	0	0	0	0	1	0	0
1	0	1	1	0	0	0	0	1	0	0	0
1	1	0	0	0	0	0	1	0	0	0	0
1	1	0	1	0	0	1	0	0	0	0	0
1	1	1	0	0	1	0	0	0	0	0	0
1	1	1	1	1	0	0	0	0	0	0	0

The 3×8 decoder using AND gates is shown in Fig. 11.73, with expressions for output minterms.

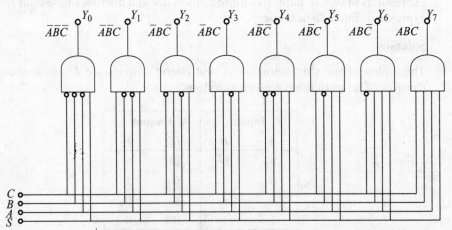

FIGURE 11.73 A 3×8 decoder for Solved Problem 6.

7. Use a 3×8 decoder to make a full adder. Why is the 3×8 decoder especially suitable for this purpose?

Solution

From Eqs. (11.3) and (11.4), the sum and the carry of a full adder are logically expressed in sum of product forms as follows:

$$S_o = \overline{A}\,\overline{B}C_i + \overline{A}B\overline{C}_i + A\overline{B}\,\overline{C}_i + ABC_i$$
$$C_o = \overline{A}BC_i + A\overline{B}C_i + AB\overline{C}_i + ABC_i$$

The minterms for both sum and carry are of three variable and the minterms of 3×8 decoder are also of three variables. Hence, the 3×8 decoder is suitable and economical for making a full adder. The 3×8 decoder with two 4-input OR gates, one each for sum and carry, is used for the full adder. The logical circuit is shown in Fig. 11.74.

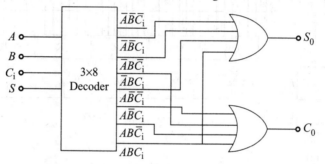

FIGURE 11.74 A full adder using a 3×8 decoder.

8. A seven-segment LED display is activated from a 4×7 decoder to display decimal numerals. The input lines are A, B, C, D and the output lines are a, b, c, d, e, f, g denoting seven segments of the LED display, respectively. Write the truth table and draw the block diagram representation using the decoder and other gates. Determine the logical expression for g segment as a sum of minterms.

Solution

The truth table is as follows:

A	B	C	D	g	f	e	d	c	b	a
0	0	0	0	0	1	1	1	1	1	1
0	0	0	1	0	0	0	0	1	1	0
0	0	1	0	1	0	1	1	0	1	1
0	0	1	1	1	0	0	1	1	1	1
0	1	0	0	1	1	0	0	1	1	0
0	1	0	1	1	1	0	1	1	0	1
0	1	1	0	1	1	1	1	1	0	0
0	1	1	1	0	0	0	0	1	1	1
1	0	0	0	1	1	1	1	1	1	1
1	0	0	1	1	1	0	0	1	1	1

The logical expression for g segment,

$$G = \bar{A}\,\bar{B}\,C\,\bar{D} + \bar{A}\,\bar{B}\,C\,D + \bar{A}\,B\,C\,\bar{D} + \bar{A}\,B\,\bar{C}\,D$$
$$+ \bar{A}\,B\,C\,\bar{D} + A\,\bar{B}\,\bar{C}\,\bar{D} + A\,\bar{B}\,\bar{C}\,D$$

The block diagram for the display system is shown in Fig. 11.75.

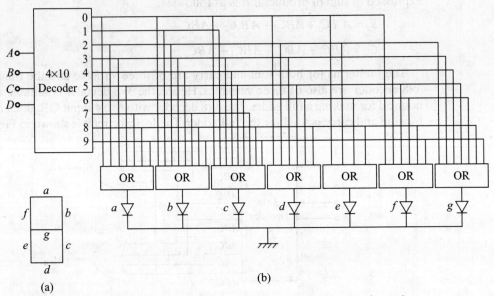

FIGURE 11.75 (a) A seven-segment LED and (b) a 4×7 decoder to activate the seven-segment LED display.

9. A process is expressed by three input logical variables A, B, C, and the following Boolean functions for the output variables X, Y, Z:

$$X = A\bar{B} + \bar{B}C$$
$$Y = \bar{C}$$
$$Z = \bar{A} + \bar{B}C$$

Design the required combinational circuit using an appropriate decoder and gates.

Solution

$$X = A\bar{B} + \bar{B}C = A\bar{B}(C + \bar{C}) + (A + \bar{A})\bar{B}C$$

$$= A\bar{B}C + A\bar{B}\,\bar{C} + A\bar{B}C + \bar{A}\,\bar{B}C = A\bar{B}C + A\bar{B}\,\bar{C} + \bar{A}\,\bar{B}C \qquad (1)$$

$$Y = (A + \bar{A})(B + \bar{B})\bar{C} = AB\bar{C} + \bar{A}B\bar{C} + A\bar{B}\,\bar{C} + \bar{A}\,\bar{B}\,\bar{C} \qquad (2)$$

$$Z = \bar{A}(B + \bar{B})(C + \bar{C}) + (A + \bar{A})\bar{B}C$$

$$= \bar{A}BC + \bar{A}B\,\bar{C} + \bar{A}\,\bar{B}C + A\bar{B}C + \bar{A}\,\bar{B}\,\bar{C} \qquad (3)$$

Thus, the logical expressions of X, Y, and Z in forms sum of output minterms of the encoder. Hence, these logical expressions can be written in sums of encoder outputs as follows:

$$X = Y_5 + Y_4 + Y_1 \tag{1}$$

$$Y = Y_6 + Y_2 + Y_4 + Y_0 \tag{2}$$

$$Z = Y_3 + Y_2 + Y_1 + Y_5 + Y_0 \tag{3}$$

The logical circuit is shown in Fig. 11.76.

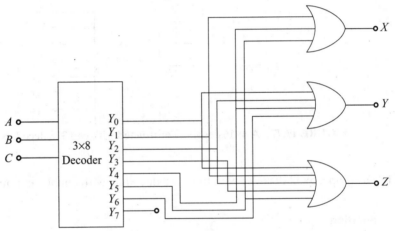

FIGURE 11.76 A logical circuit of Solved Problem 9.

10. Draw a block diagram representation of a 4×16 decoder using appropriate number of 3×8 decoders and gates. Explain the operation.

Solution

Figure 11.77 shows a 4×16 decoder made from two 3×8 decoders and one NOT gate. The MSB bit of the 4×16 decoder is connected to the enable input of one of the 3×8 decoders and through an inverter to the enable input of the other. The three LSBs of the four inputs of 4×16 decoder are connected to inputs of both 3×8 decoders. It is assumed that each 3×8 decoder is enabled when $E = 1$. When $E = 0$, the 3×8 decoders are disabled and all its outputs are in the 0 level. When $A_3 = 0$, the upper 3×8 decoder is enabled and the lower is disabled. The lower 3×8 decoder outputs are inactive with outputs at 0. The output of upper 3×8 decoder are active and generate outputs Y_0 through Y_7, depending on the values of A_0, A_1, A_2, whereas $A_3 = 0$. When $A_3 = 1$, the lower 3×8 decoder is enabled and the upper is disabled. The lower 3×8 decoder generates outputs Y_8 through Y_{15}, equivalent of the binary numbers having $A_3 = 1$.

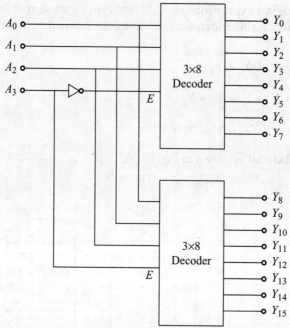

FIGURE 11.77 A 4×16 decoder constructed with two 3×8 decoders.

11. Develop a 5×32 decoder with four 3×8 decoders with enable and one 2×4 decoder.

   Solution

   The block diagram of the 5×32 decoder is shown in Fig. 11.78. The two MSBs inputs A_3 and A_4 of the 5×32 decoder are connected to the two inputs of the 2×4 decoder. The four outputs of the 2×4 decoder are connected to enable inputs of the four 3×8 decoders as shown in Fig. 11.76. When A_3 and A_4 are 0, the leftmost 3×8 decoder is enabled. Remaining 3×8 decoders are disabled. For $A_3 = 1$ and $A_4 = 0$, only the second leftmost 3×8 decoder is enabled. When $A_3 = 0$ and $A_4 = 1$, only the third 3×8 decoder is enabled and the rightmost 3×8 decoder is enabled only when $A_3 = 1$ and $A_4 = 1$.

12. Develop a logic circuit for a 1×4 demultiplexer. How many address control lines are there? Where is such a circuit useful?

   Solution

   A decoder with enable input is also used as a demultiplexer. The demultiplexer transmits a binary signal (serial data) on one of the output lines. The particular output line is selected by means of a binary address code. The input lines of the decoder serve as the address code. If the decoder is constructed with NAND gates, the data and the enable input are applied through inverters to two inputs of an AND gate and the

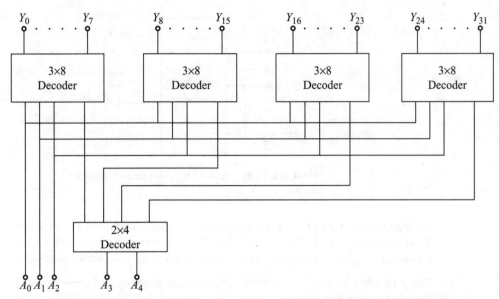

FIGURE 11.78 The 5×32 decoder with four 3×8 decoder and one 2×4 decoder.

output of the AND gate is applied to the enable terminal of the demultiplexer as shown in Fig. 11.79(a). When $E = 0$, S is equal to the complement of the data. Hence, the data appears at one of the outputs. The data is directly applied to the enable terminal of the AND gate demultiplexer as shown in Fig. 11.79(b).

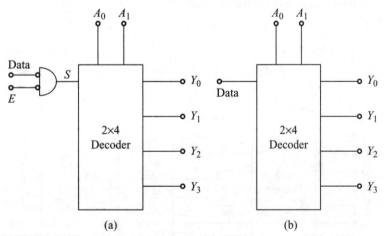

FIGURE 11.79 A demultiplexer with (a) NAND gates and (b) AND gates.

13. Explain how to convert a 4×10 decoder unit into a 3×8 decoder.

Solution

The block diagram of a 4×10 decoder used as a 3×8 decoder is shown in Fig. 11.80. When decoder is enabled, E is assumed to be 1. As MSB of 4×10 decoder, A_3, is connected

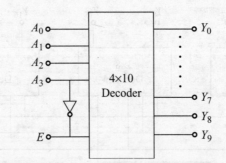

FIGURE 11.80 A 4×10 decoder as 3×8 decoder.

through an inverter to E, A_3 is always 0 till the decoder is enabled. Hence, outputs Y_8 and Y_9 are always 0 (for AND gate decoder). Decoder output Y_0 through Y_7 depends only on values of A_0, A_1 and A_2 only. The 4×10 decoder, therefore, works as a 3×8 decoder.

14. Design an 8×1 multiplexer using an appropriate decoder and gates. What is the number of selection variables.

Solution

As it is an 8×1 multiplexer, the decoder used to make the multiplexer has eight AND gates, three selection variables, and one enable input. There are eight data input terminals. The logical circuit is shown in Fig. 11.81. The function of the multiplexer is illustrated in the following table.

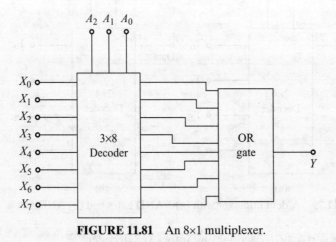

A	B	C	Y
0	0	0	X_0
0	0	1	X_1
0	1	0	X_2
0	1	1	X_3
1	0	0	X_4
1	0	1	X_5
1	1	0	X_6
1	1	1	X_7

FIGURE 11.81 An 8×1 multiplexer.

15. Draw the block diagram of a 32×1 multiplexer using (a) two 16×1 multiplexers and appropriate gates and (b) two 16×1 multiplexers and one 2×1 multiplexer. Explain the operations of both the options.

Solution

(a) There are five address inputs in a 32×1 multiplexer. MSB bit (A_4) of the address inputs is connected through an inverter to the enable input of the upper 16×1 multiplexers and directly to the enable input of the lower 16×1 multiplexer. When $A_4 = 0$, the enable input of the upper 16×1 multiplexer is high and so it is enabled. The lower 16×1 multiplexer is disabled. When $A_4 = 1$, the enable input of the lower 16 × 1 multiplexer is high and so it is enabled. The upper one is disabled. Thus, when $A_4 = 0$, one of the data X_0 through X_{15} is selected as the upper multiplexer is active, while the output of the lower multiplexer is at 0 level. During the selection of one of the data from A_{16} to A_{31} ($A_4 = 1$) the lower multiplexer is active, while the output of the upper multiplexer is at 0 level. The outputs of the two 16×1 multiplexers are connected to the two inputs of an OR gate. The output of the OR gate gives the selected input data. The block diagram is shown in Fig. 11.82(a).

(b) The logical block diagram of the 32×1 multiplexer is shown in Fig. 11.82(b). The MSB bit of address inputs of the multiplexer is connected to the address input of the 2×1 multiplexer and remaining 4 bits are connected to the four address inputs of the two 8×1 multiplexer. The 8×1 multiplexers operate as usual. The data X_0 through X_{15} are selected by the address input, and the selected data appear at the output of the upper 8×1 multiplexer. Similarly, the data X_{16} through X_{31} are selected by the address input, and the selected data appear at the output of the lower 8×1 multiplexer. When MSB is low, the output of the upper 8×1 multiplexer is selected

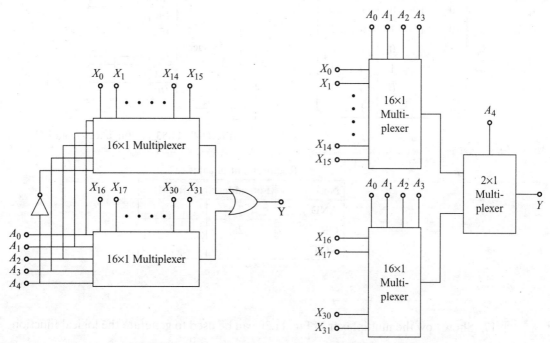

FIGURE 11.82 Two methods of constructing 32 × 1 multiplexer.

by the 2×1 multiplexer, and the selected data from X_0 through X_{15} appears as the final output at the output of 2×1 multiplexer. When MSB is high, the output of the lower 8×1 multiplexer is selected by the 2×1 multiplexer, and the selected data from X_{16} through X_{31} appear at as the final output.

16. How many AND, OR, and NOT gates are required if a 3-input adder is implemented with a ROM? Compare these numbers with those used in a full adder.

Solution

The truth table of the required ROM is given below. From the truth table the sum and the carry in the sum of product of three variables are given as follows:

$$S_o = \overline{A}\,\overline{B}C_i + \overline{A}B\overline{C}_i + A\overline{B}\,\overline{C}_i + ABC_i$$
$$C_o = \overline{A}BC_i + A\overline{B}C_i + AB\overline{C}_i + ABC_i$$

The logic diagram of the adder is shown in Fig. 11.83. The comparison of the number of gates is given in the table below.

Truth table

A	B	C_i	S_o	C_o
0	0	0	0	0
0	0	1	1	0
0	1	0	1	0
0	1	1	0	1
1	0	0	1	0
1	0	1	0	1
1	1	0	0	1
1	1	1	1	1

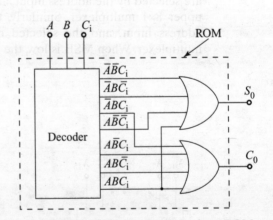

FIGURE 11.83 A full adder from ROM.

Requirement of gates

No.	Full adder	ROM
AND	7	7
OR	2	2
NOT	3	7

17. Show how the multiplexer of Fig. 11.20 can be used to generate the logical function $Y = \overline{B}$

Solution

The logical function can also be written as
$$Y = (\overline{A} + A)\overline{B} = \overline{A}\,\overline{B} + A\overline{B}$$

Refer Fig. 11.20. For generating $\overline{B}$, it is clear that the outputs Y_0 and Y_2 of AND gates should be 1, and outputs Y_1 and Y_3 should be 0. From that the data inputs $X_0 = 1$, $X_1 = 0$, $X_2 = 1$, and $X_3 = 0$.

18. A ROM is to be used to generate the following functions:

$$Y_0 = A\overline{C} + B\overline{C}$$
$$Y_1 = A + B\overline{C}$$

where A, B, and C are the input variables. Design the ROM. Show all interconnections between the decoder and the encoder.

Solution

The logical functions can be written in canonical forms as follows:
$$Y_0 = A(B + \overline{B})\,\overline{C} + (A + \overline{A})B\overline{C}$$
$$= AB\overline{C} + A\overline{B}\,\overline{C} + AB\overline{C} + \overline{A}B\overline{C}$$
$$= \overline{A}B\overline{C} + A\overline{B}\,\overline{C} + AB\overline{C}$$

Similarly, $Y_1 = A + B\overline{C} = \overline{A}B\overline{C} + A\overline{B}\,\overline{C} + A\overline{B}C + AB\overline{C} + ABC$

The interconnection between the decoder and the encoder is shown in Fig. 11.84.

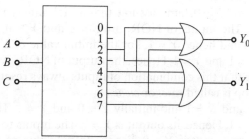

FIGURE 11.84 A ROM to generate Y_0 and Y_1.

19. Draw the logic diagram for an SR flip flop. Use NOR gates instead of NAND gates. Explain the operation for all combinations of inputs.

Solution

The logic diagram of the SR flip flop is shown in Fig. 11.85.

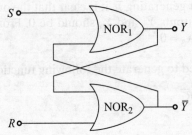

FIGURE 11.85 An SR flip flop using NOR gates.

The operation of basic SR flip flop may be analyzed by examining the output state for the four combinations of the two inputs. The state of output for four combinations of input variables are analyzed as follows:

(a) When $S = 0$ and $R = 0$, initially $Y = 0$ and $\overline{Y} = 1$. The inputs to gate NOR_1 are $S = 0$ and $\overline{Y} = 1$. Hence, its output is $Y = 0$. The inputs to NOR_2 are $R = 0$ and $Y = 0$. Hence, $\overline{Y} = 1$. For the initial value of $Y = 1$, and hence $\overline{Y} = 0$, the inputs of NOR_1 are $S = 0$ and $\overline{Y} = 0$. Hence, the output of NOR_1 is $Y = 1$. The inputs of NOR_2 are $R = 0$ and $Y = 1$. Hence, the output of NOR_2 is $\overline{Y} = 0$. Thus, the previous output state remains unchanged.

(b) When $S = 0$ and $R = 1$, assuming initial state $Y = 1$ and hence $\overline{Y} = 0$. The inputs of NOR_2 are $R = 1$ and $Y = 1$, which makes the output of NOR_2, $\overline{Y} = 0$ or $Y = 1$. It is to be remembered that the cross feedback connections ensure the output states are complementary. For the initial value of $Y = 0$ and $\overline{Y} = 1$, the inputs of NOR_2 are $R = 1$ and $Y = 0$. Hence, the output of NOR_2 is $\overline{Y} = 0$ or $Y = 1$. In summary we conclude that for this combination of inputs the flip flop is in state 1. The state of the flip flop is called *set* or *preset*.

(c) When $S = 1$ and $R = 0$, let us assume that initial state is $Y = 1$ and then being complement $\overline{Y} = 0$. The inputs of NOR_1 are $S = 1$ and $\overline{Y} = 0$, which makes the output of NOR_1, $Y = 0$ and hence $\overline{Y} = 1$. For the initial value of $Y = 0$ and $\overline{Y} = 1$, the inputs of NOR_1 are $S = 1$ and $\overline{Y} = 1$. Hence, the output of NOR_1 is $Y = 0$ and hence $\overline{Y} = 1$. Thus, we conclude that this combination of inputs always resets the output $Y = 0$. The state of the flip flop is called *reset* or *clear*.

(d) When $S = 1$ and $R = 1$ and initially $Y = 0$ and $\overline{Y} = 1$. The inputs to gate NOR_1 are $S = 1$ and $\overline{Y} = 1$. Hence, its output is $Y = 0$. The inputs to gate NOR_2 are $R = 1$ and $Y = 0$. Hence, its output is $\overline{Y} = 0$. For the initial value of $Y = 1$, and hence $\overline{Y} = 0$, the inputs of NOR_1 are $S = 1$ and $\overline{Y} = 0$. Hence, the output of NOR_1 is $Y = 0$. The inputs to gate NOR_2 are $R = 1$ and $Y = 1$. Hence, its output is $\overline{Y} = 0$. For this combination of inputs both gates generate the same output logic. Hence, this combination of inputs makes the output state indeterminate and is not allowed. The truth table for SR flip flop of Fig. 11.83 is given in the following table.

Truth table

S	R	Y	$\overline{Y}$	Comment
0	0	NC		No change
0	1	1	0	Set or preset
1	0	0	1	Reset or clear
1	1	0	0	Indeterminate

20. The excitation table for a flip flop is given in the following table. Derive the truth table and identify the flip flop.

Excitation table

Y_i	Y_{i+1}	J_i	K_i
0	0	0	X
0	1	1	X
1	0	X	1
1	1	X	0

Truth table

J_i	K_i	Y_i	Y_{i+1}	Conclusion
0	0	0 (1)	0 (1)	No change
0	1	0 (1)	0	Clear
1	0	0 (1)	1	Set
1	1	0 (1)	1 (0)	Complemented

Solution

The truth tables of flip flops specify the next state when the inputs and the present state are known. During the design, however, the required transition from the present state to the next state is usually known. It is wished to find the flip flop input conditions that cause the required transition. For this reason a table is needed that lists the required input conditions for a given transition of state. Such a table is called excitation table for a flip flop. The truth table satisfying the excitation table is given above. From analyzing the excitation table, we arrived at the following conclusion.

For transition of the output from 0 to 0, there are two combinations of inputs. They are $J = 0$, $K = 0$, and $J = 0$ and $K = 1$. These combinations are shown as the first and the second rows of the truth table. For transition of the output from 0 to 1, again there are two combinations of inputs. These combinations are shown as the third and the fourth rows of the truth table. The transition from 1 to 0 is satisfied by the second and fourth rows. The transition from 1 to 1 is satisfied by the first and third rows of the truth table.

From the truth table it is clear that the flip flop is a JK flip flop.

21. The waveforms applied to J, K, and Ck inputs of a JK flip flop are as shown in Fig. 11.86. Plot the output waveforms for Y and $\overline{Y}$ lined up with respect to the clock pulses.

Solution

In the interval between the clock pulses, there is no change in the output state. In Fig. 11.86(a) the inputs J and K change their states just before the end of the clock pulses. Taking this into consideration, the output waveforms for Y and $\overline{Y}$ along with the clock pulses are plotted in Fig. 11.86(b).

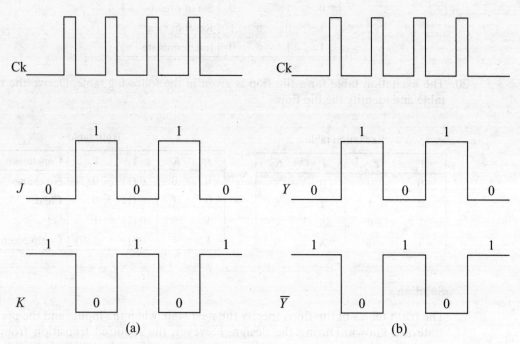

FIGURE 11.86 (a) The waveforms applied to JK flip flop and (b) output waveforms.

22. Design a JK flip flop by adding two AND gates and an OR gate to a flip flop. Write the truth table and compare it with truth table of JK flip flop.

Solution

The block diagram is shown in Fig. 11.87.

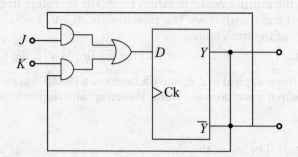

FIGURE 11.87 A JK flip flop using D flip flop and gates.

The operation of the circuit is illustrated in the following table.

J_i	K_i	D_i	Y_i	Y_{i+1}	Comment
0	0	1	1	1	No change
0	0	0	0	0	
0	1	0	1	0	Reset or clear
0	1	0	0	0	
1	0	1	1	1	Set or preset
1	0	1	0	1	
1	1	0	1	0	Complemented
1	1	1	0	1	

Hence, the truth table is given as follows:

Truth table of JK flip flop of Fig. 11.87

J	K	Y	$\overline{Y}$	Comment
0	0	NC		No change
0	1	0	1	Reset or clear
1	0	1	0	Set or preset
1	1	$\overline{Y}$	Y	Complemented value

The truth table of the JK flip flop of Fig. 11.87 is same as the truth table of JK flip flop of Table-11.11. Hence, the flip flop of Fig. 11.87 is a JK flip flop.

23. The inputs of JK flip flop are connected and logic 1 is applied to *J* input as shown in Fig. 11.88(a). Clock input applied is as shown in Fig. 11.88(b). Sketch the corresponding waveform of *Y*.

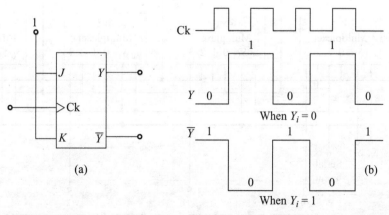

FIGURE 11.88 (a) The flip flop and (b) the output waveforms with clocks.

Solution

The flip flop of Fig. 11.88(a) is a T flip flop. Hence, the output of the flip flop changes its state when there is a clock pulse. The output waveforms corresponding to two initial states of the output are shown in Fig. 11.88(b).

24. Use multiplexer to make a 4-bit shift register bidirectional with parallel load. Explain the operation.

Solution

A 4-bit bidirectional shift register with parallel load is shown in Fig. 11.89. Each stage of the shift register consists of D flip flop and a 4×1 multiplexer. Two selection inputs S_0 and S_1 of the multiplexer select one of the data inputs of the multiplexer for the D flip flop. The selection lines control the mode of operation of the shift register as follows. When $S_0 S_1 = 00$, data input 0 of each multiplexer is selected. This condition connects the output of each D flip flop to its own input. Hence, there in no change in the output state. When $S_0 S_1 = 01$, the terminal marked 1 in each multiplexer is connected to the D input of the corresponding flip flop. This makes the register right shift. The serial input data is transferred into flip flop FF_0. The output of each flip flop is transferred into the succeeding flip flop. When $S_0 S_1 = 10$, the register becomes left shift register. The other serial input data is transferred into the flip flop FF_3. The output of each flip flop is transferred into the next significant flip flop. When $S_0 S_1 = 11$, the binary information from each input I_0 through I_3 is transferred into the corresponding flip flop. This results in a parallel load operation.

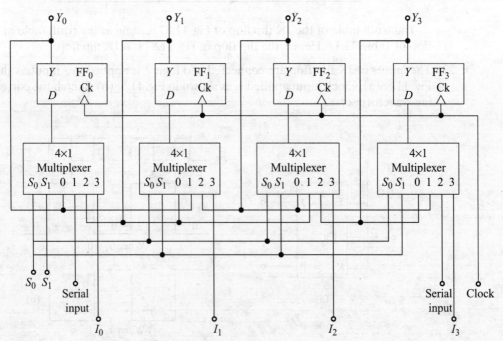

FIGURE 11.89 The bidirectional shift register with parallel load.

25. The content of a 4-bit register is initially 1101. The register is shifted six times to the left with the serial input being 101101. What is the content of the register after shift?

Solution

The shifting of the data is illustrated by the timing diagram in Fig. 11.90.

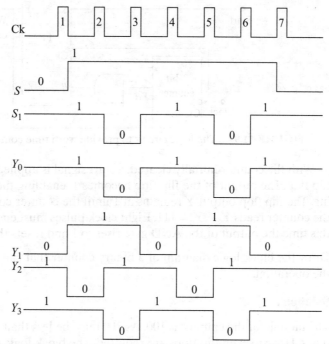

FIGURE 11.90 A shifting diagram of data for Solved Problem 25.

The shifting of the data is conveniently summarized in the following table.

Shifting of data

Flip flop state	Y_3	Y_2	Y_1	Y_0	Serial input data
Initial	1	1	0	1	1
First clock	1	0	1	1	0
Second clock	0	1	1	0	1
Third clock	1	1	0	1	1
Fourth clock	1	0	1	1	0
Fifth clock	0	1	1	0	1
Sixth clock	1	1	0	1	

26. Design a logic circuit that provides a digital system with word time control for 8-bit data words. Explain the operation.

Solution

A counter with 8 modulo is needed. Hence, a 3-bit counter is required. This counter is used with an SR flip flop together with an AND gate to provide the specified control. The block diagram of the logic circuit is shown in Fig. 11.91.

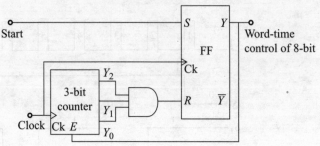

FIGURE 11.91 The logic circuit to provide word time control for 8-bit data words.

With the counter initially cleared, a start signal is applied at the S input of the SR flip flop. The output of the flip flop becomes 1, enabling the counter to begin counting. The flip flop output Y remains at 1 until the counter completes the cycle, that is, the counter reads $Y_2 Y_1 Y_0 = 111$. Eight clock pulses must elapse to reach this state. At this time the output of the AND gate rises to 1 and resets the flip flop.

27. Draw the block logic diagram of a binary counter that can count up to 100. Explain the operation.

Solution

The modulo of the counter is 100. As 100 must be less than 2^n, the minimum value of n is 7. Hence, seven flip flops are required. The block logic diagram of the counter is shown in Fig. 11.92.

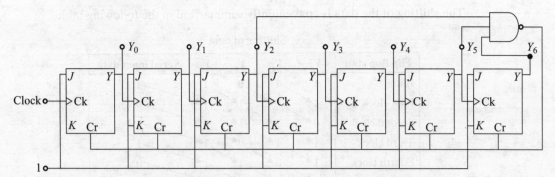

FIGURE 11.92 A binary counter counting up to 100.

Without AND gate and with $Cr = 1$, the counter behaves like a ripple (or binary) counter that can count from 0 to 127. For 100 clock pulses, $Y_6 = Y_5 = Y_2 = 1$, and $Y_4 = Y_3 = Y_1 = Y_0 = 0$. To make the counter to count from 0 to 99 and to reset at 100, a 3-input NAND gate is used. The outputs Y_6, Y_5, and Y_2 are connected to input terminals of the NAND gate. The output of the NAND gate is connected to Cr of all the flip flops. At clock pulse 100, when $Y_6 = Y_5 = Y_2 = 1$, the output of the NAND gate is 0; hence, all the flip flops are reset to 0. For other clock pulses from 1 to 99, the output of the NAND gate is 1, and hence the counter is active. The cycle is repeated after 100.

Alternatively, two decade counters of Fig. 11.41 can be used for making a 100 modulo counter. The block logic diagram is shown in Fig. 11.93. The unit counter counts up to 9 and at 10th clock pulse, it resets. The $NAND_3$ gives a clock pulse to 10's counter and it counts 1. The counter displays the number in BCD form.

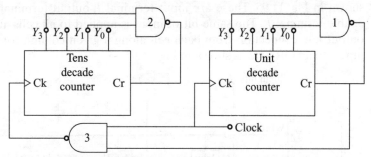

FIGURE 11.93 An alternative form of Fig. 11.92.

28. For the logic diagram of the synchronous counter shown in Fig. 11.94, write the truth table for Y_0, Y_1, and Y_2 after each pulse and verify that this is a 5:1 counter.

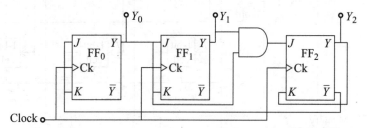

FIGURE 11.94 A synchronous counter for Solved Problem 28.

Solution

The truth table of the counter is given in the following table.

Pulse	Y_0	Y_1	Y_2
Initially	0	0	0
1st	1	0	0
2nd	0	1	0
3rd	1	1	0
4th	0	0	1
5th	0	0	0

From the truth table it is clear that the counter is 5:1 counter.

29. Draw the complete circuit diagram of a 4-bit four-word RAM. Use the basic 1-bit cell and show all inter connections. Describe how a word is written into the memory and how it is read out of memory.

Solution

The RAM of capacity to store four words of data of 4-bit requires $4 \times 4 = 16$ basic 1-bit cells. The connections of the cells are shown in Fig. 11.95. A group of four cells arranged in a horizontal line are excited by the same address line. Hence, a 2×4 decoder is required. In addition to address line, each cell has three connections as shown in Fig. 11.95. These are input terminal I, output terminal O, and read/write enable terminal E. The same bit of input is applied to all cells along a column. The output line connection from each cell along a given column forms one of the four inputs to an OR gate as shown in Fig. 11.95.

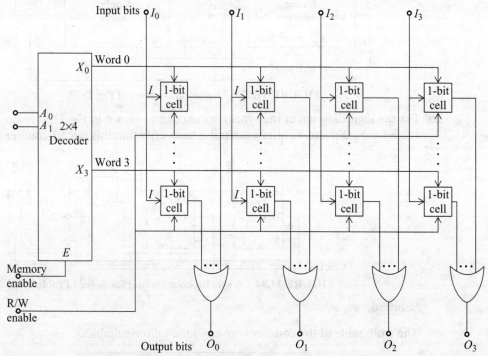

FIGURE 11.95 A block diagram of a 4-bit four words RAM.

To illustrate the operation of the RAM, let us consider an input word $A_3A_2A_1A_0$ = 1101. We wish to write this data into the memory register of the RAM represented by input word 3. So, the address code of data line is 11. To write data into the cells, read/write enable input E is made 1. The data is written (or stored) in the memory cells. To read the data, the read/write enable input is made 0. The stored data appear on the four output terminals in binary form.

30. A 64-word 1-bit RAM with on-chip decoding is available. The memory accepts a 6-bit address word. Using the above memory unit as a building block, construct a 64-word by 4-bit memory.

Solution

For constructing 4-bit 64-word RAM using 64-word 1-bit RAM chip, we require four number of chips. As decoding is available on the chip, six address input terminals are available in each chip. The connection diagram of the 64-word 4-bit memory RAM is shown in Fig. 11.96.

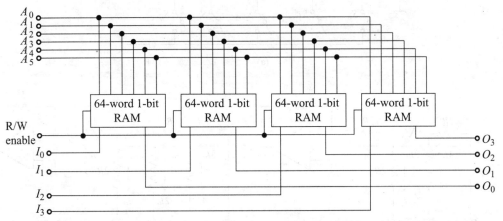

FIGURE 11.96 Connection of a 4-bit 64-word RAM using 64-word 1-bit RAM chip.

31. It is to design a RAM that has 256 words of 8 bits per word. If 16×8 RAM chips are available, find

    (a) The number of 16×8 RAM chips that are needed to built the memory.

    (b) The number of flip flops needed in the memory address register (MAR).

Solution

(a) The number of 16×8 RAM chip needed to build a 256 words of 8-bit RAM is **16**.

(b) A word of m bit is called storage or memory register. Each memory register is assigned a specific address to facilitate its location To address each of the memory registers, there is another register called MAR. To address M memory register, the MAR has n flip flop, where n is given by $2^n = M$. Hence, for $M = 256$, the MAR must have **8** flip flops.

32. (a) For the R-2R ladder D/A converter, show that when the second MSB is 1 and all other bits are 0, the output is $V_o = V_R/4$. (b) Find V_o if only the third MSB is 1. (c) Find V_o if only the third LSB is 1.

Solution

Consider an n-bit D/A converter as shown in Fig. 11.97(a).

(a) When only the second MSB is 1, the circuit of Fig. 11.97(a) reduces as Fig. 11.97(b). The output current is given by

$$I_o = -\frac{V_R}{12R}$$

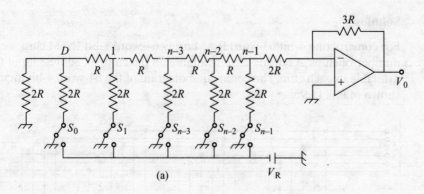

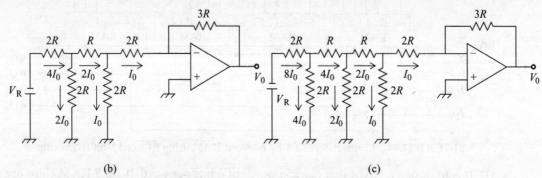

FIGURE 11.97 A D/A converter of n bits.

and hence, the output voltage,

$$V_o = -I_o \times 3R = \frac{V_R}{12R} \times 3R = \frac{V_R}{2^2} = \frac{V_R}{4}$$

(b) When only third the MSB is 1, the circuit of Fig. 11.97(a) reduces as Fig. 11.97(c). The output current,

$$I_o = -\frac{V_R}{24R}$$

and hence, the output voltage,

$$V_o = -I_o \times 3R = \frac{V_R}{24R} \times 3R = \frac{V_R}{2^3} = \frac{V_R}{8}$$

(c) Similarly, when only the ith MSB is 1, the output voltage can be given by

$$V_o = \frac{V_R}{2^i}$$

The LSB can be said as the nth MSB, and hence when only the LSB is 1, the output voltage is given by

$$V_o = \frac{V_R}{2^n}$$

The output voltage, when only the third LSB is 1, hence, is given by

$$V_o = \frac{V_R}{2^{n-3}}$$

For a special case, when $n = 8$, the output voltage, when only the third LSB (which is also the fifth MSB) is 1, is given by

$$V_o = \frac{V_R}{2^5} = \frac{V_R}{32}$$

11.16 EXERCISES

1. Design an encoder that converts octal to binary form. Write the logical expression for each output.
2. Design an encoder satisfying the following truth table. Write the logical expression for each output.

Inputs				Outputs		
S_3	S_2	S_1	S_0	Y_2	Y_1	Y_0
0	0	0	1	1	0	0
0	0	1	0	0	1	1
0	1	0	0	1	0	1
1	0	0	0	1	1	0

3. Design an encoder satisfying the following truth table, using diode matrix.

Inputs				Outputs			
S_3	S_2	S_1	S_0	Y_3	Y_2	Y_1	Y_0
0	0	0	1	1	1	1	0
0	0	1	0	0	1	1	1
0	1	0	0	1	0	1	1
1	0	0	0	0	1	0	1

4. Show how the half adder circuit of Fig. 11.6(b) can be implemented with the NAND gates.
5. A half subtractor is a combinational logic circuit that subtracts 1-bit binary number B from A and produces the difference. Develop the truth table and write the logical expressions in product of maxterms for the difference and borrow. Implement the logical expressions into a logical circuit.
6. Design the logical circuit for a 2×4 decoder. Draw the truth table for inputs A, B, and outputs Y_3, Y_2, Y_1, Y_0. Show the logical expression for each minterms.
7. A combinational circuit is defined by the following logical expressions:

   $Y = \overline{A}B$

   $Z = AB + A\overline{C}$

   Design the logical circuit using 3×8 decoder and any other gates.
8. Draw a block diagram representation of a 3×8 decoder using appropriate number of 2×4 decoders and gates. Explain the operation.
9. Convert a 4×10 decoder using NAND gates with enable input into a 1×10 demultiplexer. How many address control lines are there?
10. Draw a logic diagram of a 4×10 decoder using OR gates instead of AND gates. Explain its operation.

11. Design a 4×1 multiplexer using OR gates. How many inputs, outputs, and address variables are there? Draw the logic circuit and write the truth table.
12. There is a need of 8×1 multiplexer. However, only 4×1 and 2×1 multiplexers are available in the store. Is it possible to design the 8×1 multiplexer? If yes, draw the schematic block diagram and explain the operation.
13. A process involves three logical variables. The description of the desired logical function is given as

$$Y = A\overline{C} + \overline{B}C$$

Design a multiplexer consistent with the logical function.

14. A process with two input variables (A and B) and 4-bit outputs are described by the following truth table.

Inputs		Outputs			
A	B	Y_3	Y_2	Y_1	Y_0
0	0	1	0	1	1
0	1	0	1	0	1
1	0	0	0	1	1
1	1	1	1	0	0

Design a ROM satisfying the truth table. Specify the size of the decoder section and the number of gates in the encoder section as well the inter connection.

15. From the truth table of the JK flip flop, derive its excitation table.
16. Show that a JK flip flop can be converted to a D flip flop with an inverter in the J and K inputs.
17. Verify that an SR flip flop can be converted to a T flip flop if S is connected to $\overline{Y}$ and R to Y.
18. Waveforms applied to S, R, and Ck inputs of an SR flip flop of Fig. 11.98(a) are shown in Fig. 11.98(b). Plot the output waveforms for Y lined up with respect to the clock pulses.
19. The logic circuit shown in Fig. 11.99 employs the universal NAND gates. Which type of flip flop is this?
20. The content of a 4-bit register is initially 1010. A serial input of 110101 is applied to the register by shifting right six times. What is the content of the register after each shift?

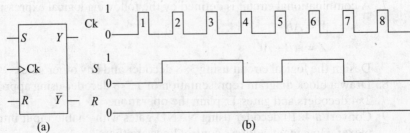

FIGURE 11.98 (a) An SR flip flop and (b) waveforms applied to S, R, and Ck.

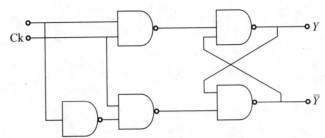

FIGURE 11.99 A logic circuit for question 19.

21. In the 4-bit register of Fig. 11.37 assume that $Y_3Y_2Y_1Y_0 = 1100$. Draw the waveform of each flip flop output if the input sequence of 01011 is applied to D_0. Assume that the shift command is at logic 1 for the required length of time.
22. Use multiplexer to make a 4-bit shift register bidirectional. Explain the operation.
23. Draw the block diagram of a binary counter that can count up to 50.
24. Augment the shift register of Fig. 11.37 with a 4-input NOR gate. The output of the NOR gate is connected to the serial input terminal. The NOR gate inputs are $Y_3Y_2Y_1Y_0$.
    (a) Verify that regardless of the initial state of each flip flop, the register assumes correct operation as a ring counter after three clock pulses.
    (b) If initially $Y_3 = 1$, $Y_2 = 0$, $Y_1 = 0$, and $Y_0 = 1$, sketch the waveform at Y_0 for the first eight pulses.
25. Draw the logic diagram of a 5-bit up-down synchronous counter with serial carry.
26. Use linear selection to draw a 16-word 4-bit RAM matrix using 1-bit basic RAM.
27. A binary weighted register D/A converter is shown in Fig. 11.100.
    (a) Show that the output resistance is independent of the digital word and that
    $$R_o = \frac{8}{15}R$$
    (b) Show that the analogue voltage for the MSB is
    $$V_o = \frac{8}{15}V_R$$
    (c) Show that the analogue voltage for the LSB is
    $$V_o = \frac{1}{15}V_R$$

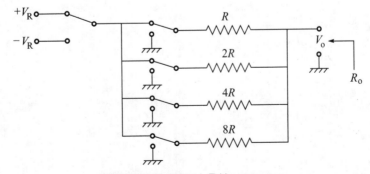

FIGURE 11.100 A D/A converter.

12

DIGITAL COMPUTER

Outline

- 12.1 Introduction 626
- 12.2 Central Processing Unit 628
- 12.3 Computer Registers 633
- 12.4 Control Unit 636
- 12.5 Memory 637
- 12.6 Input-Output Devices 642
- 12.7 Computer Instructions 648
- 12.8 Instruction Cycle 652
- 12.9 Computer Programming 653
- 12.10 Microprocessor 661
- 12.11 Microcomputer 671
- 12.12 Microcontroller 673
- 12.13 Review Questions 677
- 12.14 Solved Problems 678
- 12.15 Exercises 687

12.1 INTRODUCTION

A digital computer is a machine that manipulates the data in digital form according to a predefined set of instructions. The machine is called hardware and the set of instructions is called a program (software). Hence, the digital computer is a programmable machine. It has the ability to store and manipulate data. It receives input and provides output in a useful format. The first electronic computers were developed in the late 1940s. These electronic computers were used primarily for numerical computations. These computers were of the size of a large room and consumed as much power as several hundred modern personal computers.

Since the late 1940s the digital computers have evolved through many generations to a wide variety of forms available today. The earlier large computers used electronic valves and were slow. Today's computers use modern integrated circuits (ICs) and can be of the size of a normal book. There are mainly three types of digital computers, namely,

1. Large-scale general-purpose computers.
2. Mini digital computers.
3. Microdigital computers.

12.1.1 Large-Scale General-Purpose Computer

Large-scale general-purpose computers have been available since 1950s. However, the power of this type of computers has increased thousand times today. It is mainly used for data processing and scientific work on large scale; for example, in meteorology, astronomy, space vehicle simulation, etc. Examples are CRAY1 and CDC Cyber 205.

12.1.2 Mini Digital Computer

This type of computer first came into existence in the 1960s as small pieces of equipment used to control processes and other real-time systems. The features were short word lengths, small storages, few peripherals, and low cost. Recently mini computer is regarded as a mainframe device and is used in online commercial work. Examples are DEC PDP11, Data General Eclipse, DEC VAX 780, and PRIME 9950.

12.1.3 Microdigital Computer

In the 1970s, it became possible to have a processor on a single chip using IC technology, and the microprocessor was developed. A microprocessor is a central processing unit (CPU) on a single chip. It has its word length from 4 to 32 bits. It rapidly became popular in applications such as real-time control and instrumentation. This enabled the development of a microcomputer on a board. Its applications are in commercial and scientific data processing and real-time control. Examples are Intel 8008, 8080, 8086; iAPX386; Zilog Z80, Z8000, Z80000; Motorola 6800, 68000, 68020; NatSem 32032.

A computer system can be divided into two functional entities, namely, *hardware* and *software*. The hardware consists of the electronic and electromechanical devices. This forms the physical entity of the computer. The software consists of the instructions and data. On the set of instructions the hardware manipulates the data to perform various data processing tasks. The sequence of instructions for the computer is called a *program*, whereas the data manipulated by the program is called a *data base*.

The programs included in the computer software package are referred to as the operating system. These programs are different from the programs written by the user for the purpose of solving a particular problem. These programs are referred to as high-level language programs. For example, the high-level language program written by the user is an application program, and the compiler translates the application program written in high-level language to program in machine language. The machine language program is a system program; therefore, for the effective operation of the computer, the software is also needed in addition to the hardware. The function of the software is to enable the user to solve a problem by utilizing the capability of the hardware.

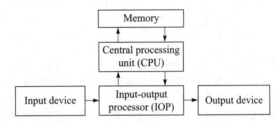

FIGURE 12.1 A block diagram of a digital computer.

The computer hardware is divided into three major parts, namely, CPU, memory, and input-output device. These components are shown in Fig. 12.1. Main functions of the CPU are to (i) point to the current instruction in a program; (ii) fetch, decode, and execute the instructions; and (iii) supervise the activity of the input-output devices. The CPU contains an arithmetic and logic unit (ALU), registers, and control unit. The ALU manipulates the data. It contains different logic gates and performs arithmetic and logical operations. There are several registers in the CPU. Registers are small memories within the CPU. These registers are used for temporary storage and manipulation of data and instructions. Data remain in the registers till they are not sent to the memory or input-output devices. The control unit controls the entire operations of the CPU. It controls flow of data and instruction within the computer. The memory stores instructions and data. There are two types of memories, namely, semiconductor memories and magnetic memories. The semiconductor memory is available for users and is called random access memory (RAM). This is called RAM because the CPU can access any location in the memory at random and retrieve the binary information (instruction or data) in a fixed interval of time. These memories are volatile, and hence the contents of the memories are lost when power is interrupted. Some examples of magnetic memories are cassettes and floppy disks. These are used to store programs and data permanently and can be erased when desired. The input-output devices are also known as peripheral devices. The peripheral devices are connected to the CPU through input-output processor (IOP). The IOP contains electronic circuits for communicating and controlling the transfer of information between the computer and the input and output devices. The input and output devices are keyboards, printers, magnetic disk drives, and other communication devices. Memory and input-output devices are connected to the CPU through a group of connecting wires. The group of wires is called a bus. The buses are to carry information from one element to another. There are three types of buses, namely, address bus, data bus, and control bus. An address bus carries the address of a memory location or an input-output device that the CPU wants to access. It is unidirectional bus. The data and control buses are bidirectional, as the data can flow either from the CPU to memory (or I/O devices) or from memory (or I/O devices) to the CPU.

A general-purpose digital computer executes various operations. In addition, it can be instructed to perform some specific sequence of instructions. The user can control the process of a computer by means of a program. A program is a set of instructions specifying the operations, operands, and the sequence of processing. A computer instruction is a binary code that specifies a sequence of operation for the computer. Instruction codes together with data are stored in the memory. The computer reads each instruction from the memory and places it in a control register. The control interprets the instruction code and executes the operations in a sequence. Every computer has its own unique instruction set. As instruction set for a computer is a binary code, it is difficult for a user to work with and understand the binary-coded instructions. Also, the user is required to write a different set of instructions if he changes the computer. It is, therefore, preferable to write programs with more familiar symbols of the alphanumeric character set. Since a computer recognizes its own instruction code, there is need of a translator that translates user-written programs into binary programs recognized by the computer.

The ALU performs arithmetic and logical operations in response to commands obtained from the instruction register (IR) and the results are placed in the accumulator. The IR is the register that holds the instruction (coded in binary) to be executed until it is decoded. The accumulator is the primary working register. Data to be operated are obtained from memory under the control of a memory address register (MAR). The MAR stores the address of memory location from where data are to be read. The program counter is a register that stores the address of the instruction to be read next. The whole process is controlled by a set of predetermined instructions (the program) decoded by the control decoder.

12.2 CENTRAL PROCESSING UNIT

The CPU is that part of the computer system that carries out the instructions of a computer program. It is the primary element of the computer system carrying out the function of the computer. It is also referred to as the processor. The term CPU has been in use in the computer industry since the early 1960s. Its form, design, and implementation have changed dramatically since the 1960s, but its fundamental operation remains much the same.

Early CPUs were, as a part of a computer, for a specific application. However, this was a costly method of designing a specific CPU. This has led to the development of mass-produced CPUs that are made for one or many purposes. This has become possible with the development of transistor and IC. This was the era of miniaturization and standardization. In the 1970s, it became possible to have a CPU on a chip. It was the microprocessor. A microprocessor is a complete CPU on a single chip.

As discussed earlier, the CPU consists of three major parts, namely, an ALU, a number of registers, and a control unit. The basic structure of a CPU is illustrated in Fig. 12.2. This figure also includes the associated memory and IOP. The ALU is discussed in this section, whereas registers and control units are discussed in Sections 12.3 and 12.4, respectively.

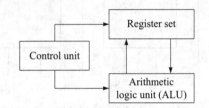

FIGURE 12.2 A block diagram of a CPU.

12.2.1 Arithmetic Logic Unit

As discussed earlier, ALU performs the following arithmetic and logic operations:

1. Addition and subtraction.
2. Multiplication and division.
3. Logic AND and OR.
4. Logic EXCLUSIVE OR (XOR).
5. Complement (logic NOT).
6. Increment (addition of 1).
7. Decrement (subtraction of 1).
8. Left or right shift.
9. Clear.

The left or right shift means that the content of the accumulator is shifted left or right by 1 bit. The clear makes the content of the accumulator or carry flag zero. However, the mathematical operations such as exponent, logarithm, and trigonometric operations are not performed by the ALU. The operations are performed either by software or by an auxiliary special-purpose arithmetic processor called coprocessor. The coprocessor is on a separate IC chip.

The binary adder and subtractor have already been discussed in Sections 11.3 and 11.4. The increment operation adds 1 to a binary number in a register. For example, a 4-bit binary number 1010 in a register becomes 1011 after increment. The increment operation is easily implemented with a binary counter (see Section 11.10). Whenever the count enable is active, the clock pulse transition increments the content of the register by 1. When the increment operation is required to be done with a combinational circuit independent of a particular register, it can be implemented by means of half adders connected in series, as shown in Fig. 12.3.

Figure 12.3 illustrates a 4-bit increment circuit using four half adders. It can be extended to an n bit by using n half adders. One input of the least significant half adder is connected to logic 1 and other input is the least significant bit (LSB) of the number to be

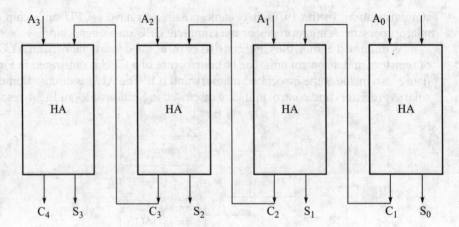

FIGURE 12.3 A 4-bit binary increment.

incremented. The carry output of a half adder is connected to one of the inputs of next significant half adder. The circuit gets 4 bits from A_0 to A_3, adds 1 to it, and gives incremented output in S_0 to S_3. The output carry C_4 is 1 only when the number to be incremented is 1111. This makes the outputs S_0 through S_3 to go 0.

12.2.2 Arithmetic Operations

An arithmetic circuit performs addition, 1's complement, 2's complement, subtraction, increment, and decrement. The circuit performing the above arithmetic operations is a parallel adder, and a 4-bit circuit using four full adders is shown in Fig. 12.4. The 4-bit inputs are A and B and the 4-bit output is D. The 4 bits of A go directly to the X inputs of full adders. The

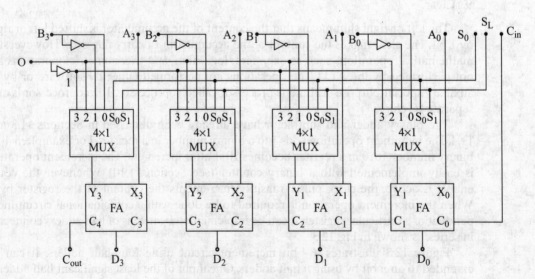

FIGURE 12.4 A 4-bit arithmetic circuit.

4 bits of B and their complements are connected to data inputs of four 4×1 multiplexers (MUXs), respectively. The other two data inputs of MUXs are connected to logic-0 and logic-1. The MUXs are controlled by two selection inputs S_1 and S_0. The input carry C_{in} goes to the carry input of the least significant full adder. The C_{in} can be 0 or 1. The other carry inputs are connected from one stage to the next. The output of the circuit is given by

$$D = A + Y + C_{in}$$

The + sign denotes the arithmetic plus. By controlling the value of Y with the selection of S_1 and S_0 and making C_{in} equal to 0 or 1, it is possible to generate the eight arithmetic operations given in Table-12.1.

Table-12.1 Arithmetic operation of the circuit of Fig. 12.4.

Select			Input	Output	Operation
S_1	S_0	C_{in}	Y	$D = A + Y + C_{in}$	
0	0	0	B	$D = A + B$	Addition
0	0	1	B	$D = A + B + 1$	Addition with carry
0	1	0	$\overline{B}$	$D = A + \overline{B}$	Subtraction with borrow
0	1	1	$\overline{B}$	$D = A + \overline{B} + 1$	Subtraction
1	0	0	0	$D = A$	Transfer A
1	0	1	0	$D = A + 1$	Increment A
1	1	0	1	$D = A - 1$	Decrement A
1	1	1	1	$D = A$	Transfer A

12.2.3 Logic Operations

Logic operations are binary operations for strings of bits stored in registers. In logic operations, each bit of the register is considered separately and is treated as binary variable. For hardware implementation of the logic operations, computers use mostly four logic operations, namely, AND, OR, XOR, and NOT. All other logic operations are derived from the above four logic operations. Figure 12.5 shows a circuit that generates the four logic operations. The circuit

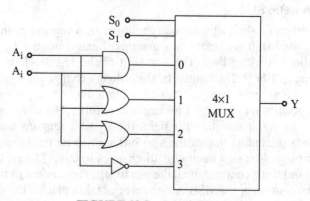

FIGURE 12.5 A logic circuit.

consists of four gates (AND, OR, XOR, and NOT) and a 4 × 1 MUX. The two selection inputs S_1 and S_0 select one of the data inputs of the MUX and direct its value to the output of the MUX. The circuit shows one typical stage with ith bit of the n-bit registers. Hence, for a logic circuit of n bits, the circuit must be repeated n times for $i = 0, 1, 2, 3, …, n−1$. The selection inputs are applied to all stages. The list of operations is given in Table-12.2.

Table-12.2 Logic operations.

S_1	S_0	Output	Operation
0	0	$E = A \cdot B$	AND
0	1	$E = A + B$	OR
1	0	$E = A \oplus B$	XOR
1	1	$E = \overline{A}$	NOT

12.2.4 Shift Operations

The shift operation is used for serial transfer of data. There are three types of shifts, namely, logic shift, circular shift, and arithmetic shift.

Logical Shift

The shift may be left shift or right shift. In logical shift, left of register R (denoted by shl R) the content of the register is shifted left by 1 bit. The rightmost bit becomes 0 after shifting. The content of the register is shifted right by 1 bit in logical shift right of register R (denoted by shr R). The leftmost bit becomes 0 after shifting. Thus, in logical shift, 0 is transfer from the serial input.

Circular Shift

This shift is also known as rotate operation. It circulates the bits of the register around the two ends without loss of any bit content. This is accomplished by connecting the serial output to the serial input of the register. The circular shift left of the register R is denoted by cil R and the circular shift right of the register by cir R. In cil R, the most significant bit (MSB) moves to LSB and other bits shift to their left. In cir R, the LSB shifts to MSB and other bits move to their right.

Arithmetic Shift

In arithmetic shift, signed binary number in a register is shifted either left or right. An arithmetic shift left multiplies a signed binary number by 2. In shift right, the binary number is divided by 2. The first shift of the register R is denoted by *ashl* R, whereas second as *ashr* R. The arithmetic shifts do not change the sign bit. As we know, the leftmost bit in a register is the sign bit and the remaining bits represent the number. The sign bit is 0 for positive number and 1 for negative number. Negative numbers are in 2's complement form. In the arithmetic shift right, the sign bit remains unchanged and contents of all the bits (including that of the sign bit) shift to right. The content of the LSB is lost. For example, a 4-bit sign number 0101 changes to 0010. The arithmetic shift left inserts a 0 in LSB and shifts contents of all the bits to left. The content of the sign bit is lost and filled up by the content of the MSB. A sign reversal takes place if the sign bit changes its value after the shift. This happens when overflow occurs. An overflow occurs in ashl if before the shift the contents of sign bit and MSB is not equal. An overflow flip flop is used to detect this overflow. For n bit register,

$$V_S = R_{n-1} \oplus R_{n-2}$$

The output V_s of the XOR gate is transferred into the overflow flip flop with the same clock pulse that shifts the register. If $V_s = 1$, there is overflow, and if $V_s = 0$, there is no overflow. The overflow, thus, can be indicated.

A combinational circuit shifter using MUXs is shown in Fig. 12.6. The circuit uses four MUXs, 4-bit data inputs A_0 through A_3 and output H_0 through H_3. There are two serial inputs, one for shift left (I_L) and another for shift right (I_R). When selection input $S = 0$, the input data are shifted right, and when $S = 1$, the input data are shifted left. The function of shifter circuit is given in Table 12.3. The two serial inputs can be controlled by another MUX to provide the three types of shifts.

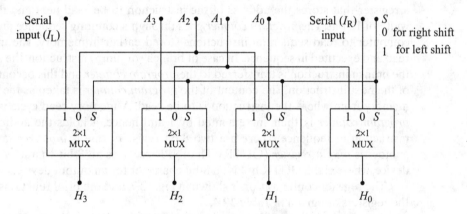

FIGURE 12.6 A 4-bit combinational circuit shifter.

Table-12.3 Function of shifter circuit of Fig. 12.6.

Selection	Output			
S	H_0	H_1	H_2	H_3
0	I_R	A_0	A_1	A_2
1	A_1	A_2	A_3	I_L

The arithmetic, logic, and shift circuits discussed earlier in this section are combined using a MUX to form one ALU. Thus, the ALU performs arithmetic, logic, and shift operations.

12.3 COMPUTER REGISTERS

The computer instruction codes are normally stored in consecutive memory locations. A memory unit has word length of 16 bits. Twelve-bit word length specifies the address of the data. Three bits store the instruction to be operated. The last bit specifies whether the address is direct or indirect address. As said earlier, the CPU of a digital computer consists of several registers. For executing an instruction the control reads the instruction from the memory location that has a specific address. The instruction code read by the control unit is stored in a register. The computer also has a register where data is placed during processing. For holding a memory address the computer needs another register.

A register that holds the data read from the memory is called *data register* (DR). It is also called memory buffer register. The *accumulator* (AC) is a general-purpose processing register. It holds one of the operands before the execution of an instruction. It also receives the result of most of the arithmetic and logical operations. The instruction read from the memory is stored in the IR. It holds the instruction until it is decoded. A register that stores temporary data during the processing is called *temporary register* (TR). A 12-bit register is used to store the address of the memory and is called *address register* (AR), since the width of the memory address is of 12 bits. As said earlier, the *program counter* is a register that stores the address of the instruction to be read next and therefore has a width of 12 bits. The *program counter* goes through a counting sequence and instructs the computer to read sequential instructions stored earlier in memory. The instructions are read and executed in sequence. In case of branch (or jump) instruction, the address part of the branch instruction is transferred to the *program counter* and this becomes the address of the next instruction. The content of the *program counter* is taken as the address of the memory from where the instruction is to be read. A memory read cycle is initiated. The *program counter* is, then, incremented by 1 and hence, it holds the address of the next instruction in sequence. There are two 8-bit registers, one is *input register* (INPR) and another is *output register* (OUTR). The INPR receives an 8-bit character from an input device, whereas the OUTR holds an 8-bit character for an output device.

The register configuration is shown in Fig. 12.7, and the brief functions and widths of the registers are given in Table-12.4.

As mentioned above, the basic computer has eight registers. The information is transferred from one register to another register and between memory and registers through buses. An efficient method of transferring information is to use a common bus. A bus is formed by a set of lines, one for each bit of a register. Through these lines binary

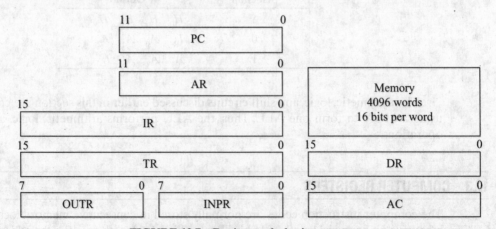

FIGURE 12.7 Registers of a basic computer.

Table-12.4 List of registers for a basic computer.

Register symbol	No. of bits	Name of register	Function
DR	16	Data register	Holds memory operand
AC	16	Accumulator	Processor register
IR	16	Instruction register	Holds instruction code
TR	16	Temporary register	Holds temporary data
AR	12	Address register	Holds address for memory
PC	12	Program counter	Holds address of instruction
INPR	8	Input register	Holds input character
OUTR	8	Output register	Holds output character

information is transferred one at a time. Control signals decide which register is selected by the bus during a particular register transfer.

A method of constructing a common bus system is with MUXs. The MUXs select the source register whose information is then placed on the bus. To illustrate the method, a common bus system for four registers is shown in Fig. 12.8. It uses four 4×1 MUXs. For the sake of clarity, labels are used to show the connections from the outputs of the registers to the inputs of the MUXs. The block diagram of Fig. 12.8 shows that the bits in the same significant position in each register are connected to the data inputs of one MUX to form one line of the bus. Thus MUX0 multiplexes the 0 bits of four registers, MUX1 multiplexes the 1 bits of four registers, and so on. The selection inputs S_1 and S_0 of the MUXs select the 4 bits of one register and transfer them into the 4-line common bus. $S_1 S_0 = 00$ select register A, $S_1 S_0 = 01$ select register B, $S_1 S_0 = 10$ select register C, and $S_1 S_0 = 11$ select register D.

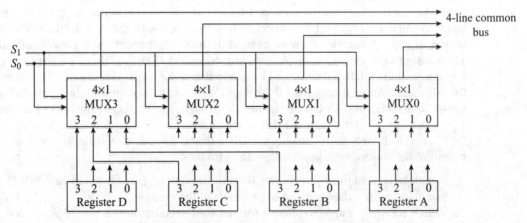

FIGURE 12.8 A bus system for four registers.

For the registers listed in Table-12.4, a common bus having 16 lines is used, the maximum number of bits to be transferred into the bus is 16. The registers DR, AC, IR, and TR are of 16 bits; hence, their contents are applied to the 16 lines of the common bus. In case of the registers AR and PC, the contents of the 12 bits of the registers are transferred into the 12 least significant lines of the bus. The four MSBs are set to 0. When AR and PC receive information from the bus, only the 12 LSBs are transferred into the registers.

12.4 CONTROL UNIT

The control unit is the brain of the computer and controls the entire operations of the CPU. It also controls the peripheral devices, such as memory, input, and output devices connected to the CPU. It controls the data flow between the CPU and the memory, and input-output devices. It generates timing and control signals for all operations. It provides status, control, and timing signals that the peripheral devices require. Hence, the control unit is also called *timing and control unit*.

A master clock generator controls the timing for all registers in the computer. The clock pulses are applied to all flip flops and registers in the system. The system includes the control unit as well. The control unit generates control signals that provide control inputs for MUXs in the common bus and for processor register. It also controls operations in the accumulator.

There are two types of control organizations, namely, hardware control and software control. The hardware control is implemented with gates, flip flops, decoders, and other digital circuits. It has the advantage of being fast in operation. However, its disadvantage is that it requires changes in the wiring among the various components if the design has to be modified or changed. In software control, the control information is stored in the control memory. The control memory is programmed to initiate the required sequence of operations. It has the advantage that any change or modification can be done by upgrading the program in control memory.

The block diagram of the hardware control unit is shown in Fig. 12.9. The circuit consists of two decoders (3×8 and 4×16 decoders), a sequence counter (SC), and a number of control logic gates. An instruction read from the memory is placed in IR. The IR, as shown in Fig. 12.9, is divided into three parts: I bit (bit 15), the operation code (bits 12–14), and bits 0 through 11. The 3-bit operation code is decoded with the 3×8 decoder. The eight output of the 3×8 decoder represents eight decimal numbers by their binary equivalent. These outputs correspond to eight operation codes. Bit 15 of the IR is transferred to a flip flop designated by I. Bits 0 through 11 are applied to the control logic gates. The output of the 4-bit SC counts in binary from 0 through 15. The outputs of the counter are decoded by the 4×16 decoder into 16 timing signals T_0 through T_{15}. The other inputs to the control logic gates are AC (bits 0 through 15) to check if AC = 0 and to detect the sign bit in AC (15); DR (bits 0 through 15) to check if DR = 0; and the values of other six flip flops. The output of the control logic gates are given as follows:

1. Signals to control the inputs of the SC and the eight registers listed in Table-12.4.
2. Signals to control the read and write inputs of the memory.
3. Signals for $S_2, S_1,$ and S_0 to select a register for the common bus.
4. Signals to control the AC adder and logic circuit.

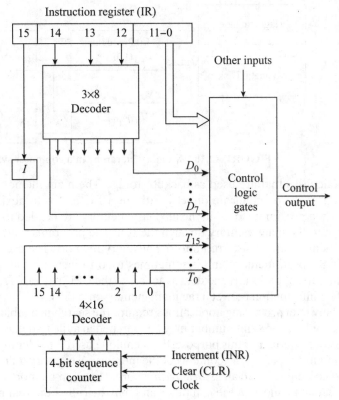

FIGURE 12.9 A control unit of basic computer.

12.5 MEMORY

The memory is an essential component of a digital computer. It is used to store programs and data. There are two types of memories. The memory that communicates directly with the CPU is called *main* (or *primary*) memory. The storage devices used as backup storage are called *auxiliary* (or *secondary*) memory. The auxiliary memory devices commonly used in computer systems are magnetic disks and tapes. These memories are used to store system programs, large data files, and other backup information. The main memory is used to store only programs and data that are currently needed by the processor. All the other information is stored in the auxiliary memory and is transferred to main memory when needed for processing by the CPU.

The auxiliary memory is slow, but a high-capacity device, whereas the main memory is relatively fast, but smaller in capacity. Another is the cache memory that is even smaller in capacity, but faster in speed. The cache memory is accessible to high-speed processing logic. The above three memories and their hierarchy is shown in Fig. 12.10. The speed of the magnetic tape is relatively slow and is at the bottom in hierarchy of the computer systems. The magnetic tapes are used to store removable files. Next in hierarchy are the

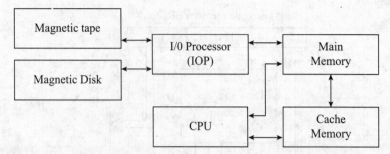

FIGURE 12.10 Memory hierarchy in a computer system.

magnetic disks that are used as backup storage. The main memory is in central position and is able to communicate directly with the CPU and with auxiliary memory through IOP. When programs stored in the auxiliary memory are needed to be processed, they are brought to the main memory through the IOP. The programs not currently needed in the main memory are transferred into the auxiliary memory. The cache memory is a special very high speed memory and is sometimes used to increase the speed of processing by making current programs and data available to the CPU at a rapid rate. The cache memory transfers information between the main memory and the CPU.

Many data processing applications require the search in a table stored in the memory. For example, an account number may be searched in a file to determine the holder's name and account status. For this purpose, the technique used is to identify the data for access by the content (or part of the content) of the memory. A memory unit accessed by content or part of content is called an *associative memory*. The main memory available in a computer is the actual memory of the computer and may be called the real memory. In some large computers, the user is permitted to construct programs as if a large memory is available. For this a special technique is used to execute this program. Thus, we have a feeling of having a memory larger than the real memory available with the computer. This memory that is not a real memory but is only a concept is called a virtual memory.

The most important property of a memory device is its access time. The access time is the time needed to transfer information from the memory to the output circuit. The access time for the main memory is 80 to 100 ns (nanosecond), whereas for the auxiliary memory the access time is about 20 ms (millisecond). The access time for the cache memory is 15 to 25 ns (nanosecond).

The reading of data in some memories may destroy the information stored in the memory. This property of the memory is called destructive readout (DRO). The example of a memory having this property is the dynamic RAM. In some memories the reading of information does not destroy the information. This property is called nondestructive readout (NDRO). The static RAM, ROM, magnetic disks, and magnetic tapes are examples of memories having this type of property.

The other properties of memories that help in selecting a memory are per bit cost, power requirement, physical characteristics, organization required, and the capacity.

12.5.1 Main Memory

The main memory is the central storage device in the computer. It directly communicates with the CPU. It is relatively a large and fast memory used to store programs and data during the operation of the computer. The main memory consists of RAM chips. There are two possible modes of operations of RAM chips: *static mode* and *dynamic mode*. The static RAM essentially consists of flip flops that store binary information. The information remains stored until power is on. In the dynamic RAM, binary information is stored in the form of electric charges. For storing the electric charges, capacitors are used in the chip by metal-oxide semiconductor transistors. The stored capacitor charge tends to discharge with time. Hence, the capacitors are recharged periodically for refreshing the dynamic memory. The refreshing of memory is done by cycling through the words every few milliseconds to restore the decaying charge. The dynamic RAM consumes less power and provides larger storage capacity in a single chip. The static RAM is easier to use and has shorter read and write cycles.

In a general-purpose computer the main memory is mostly made up of RAM IC chips. However, a portion of the main memory may be constructed with ROM chips. We know RAM is a read/write memory, whereas ROM is read only memory; however, both are random access. RAM is used to store the program and data that are subject to change. ROM stores that programs that are to remain permanently in the computer. Since RAM is volatile, its contents are destroyed when power goes off, whereas the contents of ROM remain unchanged even after power is removed. RAM and ROM chips are available in various sizes. If the memory needed is larger than the capacity of a single chip, a number of chips are combined to fulfill the required memory size, as discussed in Chapter 11.

12.5.2 Cache Memory

As discussed earlier, cache memory is a fast memory, faster than the main memory, in the computer system. In memory hierarchy, cache memory is placed between the CPU and the main memory. The most frequently accessed instruction and data are placed in the cache memory. It reduces the average memory access time. Because of the high cost, the capacity of the cache is only a small fraction of that of the main memory. The technique to access the cache memory is different from that of the main memory. When the CPU needs to access an address in the memory, it first searches the cache. If it succeeds, the data is read from the cache. If the CPU fails to access, the main memory is accessed for the data. A block of data that contains the accessed data is then transferred from the main memory to the cache memory. The data, thus transferred to the cache, can be used in future.

The performance of the cache is measured in terms of the quantity called hit ratio. When access in the cache is a success, it is called a *hit,* and in case of failure, it is called a *miss*. The hit ratio is defined as follows:

$$\text{Hit ratio} = \frac{\text{Number of hits}}{\text{Number of hits} + \text{Number of miss}}$$

The hit ratio of 0.9 and higher has been found.

The basic property of the cache memory is fast access time. Hence, very little time must be wasted when searching the data in the cache memory. The transfer of data from the main memory to the cache memory is known as *mapping*.

12.5.3 Auxiliary Memory

It is also called secondary or external memory. It differs from the main (or primary) memory in that it is not directly accessible by the CPU. The computer usually uses its input-output devices to access the auxiliary memory and transfers the desired data using intermediate area in the main memory. The auxiliary memory is nonvolatile and retains its data even after power is withdrawn. Although it is slower than the main memory, modern computer systems have larger capacity of auxiliary memory than that of the main memory because of the fact that the auxiliary memory is less expensive. The data are kept for a longer time in the auxiliary memory. The auxiliary memories that are very frequently used are magnetic disks and tapes. Some other examples of secondary memories are flash memory (e.g., USB flash or keys) and rotating optical devices, such as CD and DVD drives. Other memories of this group that are not used frequently include magnetic drums, floppy drives, paper tapes, and punch cards.

12.5.4 Magnetic Disks

Magnetic disks are made of metal or plastic circular plates coated with magnetic material. Both sides of the disk are used and a number of disks may be stacked on a common spindle. Read/write heads are available on both sides of the disks. All disks rotate together at a high speed. The disks are accessed while rotating. There are concentric circles along which bits are stored in spots on the magnetic surface. The concentric circles are called tracks. The tracks are divided in sections called sectors. In most computer systems, the minimum quantity of information that can be transferred is a sector. A magnetic disk with tracks and sectors is shown in Fig. 12.11.

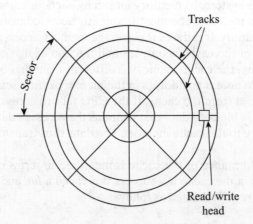

FIGURE 12.11 A magnetic disk with tracks and sectors.

In some units there is a single read/write head for each disk surface. In this type of units, track address bits are used by a mechanical assembly to place the head at the specified track position before reading or writing. In other type of units there are separate read/write heads for each track on each surface. In this type, a particular track can be selected electronically by the address bits through a decoder. The second type of disks are expensive.

The address bits in the disk system specify the disk number, the disk surface, the sector number, and the track within the sector. After the read/write head is positioned at the specific track, the system has to wait. The reading or writing process begins when the specified sector reaches under the read/write head. Data transfer is very fast once the beginning of the specified sector has reached. In some systems, disks have multiple heads, making simultaneous transfer of bits possible from several tracks at the same time. In a given sector, the track near the circumference is longer than the track near the centre of the disk. If density of bits recorded is equal, some tracks record more bits than others. To make all tracks in a sector to record, some disks uses variable density with highest density on the track near the centre and lowest density on the track near the circumference.

The magnetic disk that is permanently attached to the computer system and cannot be removed by the user is known as a *hard disk*. A removable disk used with a disk drive is known as a *floppy disk*. It is a single disk made up of plastic coated with magnetic material. Floppy disks are thin disks and are available in two sizes; one with diameter of 5.25 inch and other of 3.5 inch. They were used extensively in old personal computers. They were used as backup storage devices.

12.5.5 Magnetic Tape

The magnetic tape is a strip of plastic coated with magnetic material. Bits are recorded as magnetic spots on the tape along several tracks. A typical width of the tape is 0.5 in. and it has seven or nine tracks. It is a bulk-storage device and is used for backup. Each track has its own read/write head so that data can be recorded and read as a sequence of characters.

12.5.6 CD-ROM

CD-ROM is read only optical memcry. In this device, laser beam is used for reading the data. The manufacture of the disk writes data on it. The disk is made of a resin, such as polycarbonate. The material coated on the disk changes its state when a high-intensity laser beam is focused on it. The high-intensity laser beam makes tiny pits along the track. For reading data, laser beam of lower intensity is required, while for writing data, high-intensity laser beams are needed. The reflected beam is sensed by a photodiode. A pit spreads the light, so the light reaching at the photodiode is less. The surface without pit reflects sufficient light to the photodiode. The change in reflected light to the photodiode is converted into electrical signals that represent data to be read.

12.5.7 CD-RAM

It is also an optical disk. It is read/write memory. Information can be written and read from this type of CD. The content of the CD can be erased and new information can be written on it. It can, therefore, be used as a secondary memory of a computer system. It has the following advantages over the hard disk:

1. It has a very high storage capacity.
2. It can be removed from the drive.
3. It has long life.
4. It is more reliable.

12.5.8 Associative Memory

A memory unit accessed by the content and not by the memory address is called associative memory. It is also known as *content addressable memory*. When data are written in an associative memory, no address is written. When the data are read, the content or its part is specified. The memory locates the complete information that has a match. This type of memory is very costly compared to RAM. This type of memory is required in the storage and retrieval of rapidly changing database, radar signal tracking, image processing, computer vision, and artificial intelligence.

12.5.9 Virtual Memory

The programs and data are first stored in secondary memory. The main memory receives portions of a program or data as required by the CPU. Thus, a programmer can write a program requiring memory space that can be at the most equal to the memory space of the secondary memory. However, the main memory space is much less than that. Virtual memory is a concept used in some large computers. In this concept, portions of program or data are transferred from the secondary memory to the main memory as needed by the CPU. The programmers, thus, have the feeling that they have a very large memory at their disposal, even though the computer actually has a relatively small main memory.

An address used by the programmer is called a virtual address, and set of such addresses is the address space. An address in the main memory is called a physical or real address. The set of physical addresses is called memory space. Thus, the address space is the set of addresses generated by the programmer, whereas the memory space is the set of locations in the main memory that are directly addressable for the CPU. The address space may be larger than the memory space in computer. The address space is called virtual memory.

A virtual memory system provides a mechanism that translates program-generated addresses into physical addresses. This is done dynamically while the CPU is executing the program. The translation called mapping is handled automatically by the hardware.

12.6 INPUT-OUTPUT DEVICES

The input-output unit of a computer provides an efficient mode of communication between the control unit and the outside world. A computer does not serve useful purpose without the ability to receive information from the outside source and to transmit its results in a meaningful form. The input-output devises and secondary memory are called peripheral devices.

The most familiar device to enter information into a computer is a typewriter-like keyboard. The keyboard allows a person to enter alphanumeric information directly into the computer. As the entering of information through the keyboard is very slow with respect to the speed of the processor, programs and data are prepared in advance and transmitted into storage media such as magnetic disks. Before execution, the program and data are transferred into the computer main memory at a rapid rate. The CPU access the information from the main memory while processing. The results of the program are also transferred into high-speed storages before transferring them to the printer. Commonly used output device with general-purpose computers are cathode-ray tube (CRT) screen and printer. The CRT gives visual display of the output, whereas the printer provides a printed output of the results.

An input device transfers data and instructions into binary form acceptable by the computer. As said above, a keyboard is a commonly used input device. Nowadays a number of input devices are being used. These devices do not need typing of information for entering into the secondary memory like a keyboard. These input devices include mouse, light pen, graphic tablet, joystick, trackball, and touch screen. These devices are called pointing devices, as the user entering the information has to select something on CRT screen by pointing to it. A microphone is also used as an input device.

The CRT is a visual display unit and is used with a keyboard. The keyboard is used for input and the CRT for output. Printers give hard copy of the results, programs, and data. Other types of display units are also available such as LED (light emitting diode) and LCD (liquid crystal display). The secondary memory is also used for storing the results of the computer for future use. In this condition, the secondary memory is an output device as well.

12.6.1 Input Devices

The input devices mentioned above are described in brief here.

Keyboard: A keyboard is similar to a typewriter keyboard. It contains alphanumeric characters, some special characters, and some control keys. When a character is typed on the keyboard, an electrical signal is generated. The electrical signal is encoded into a binary-coded signal (corresponding to the typed character) by an encoder. The binary-coded signal is stored in the memory.

Light pen: It is a light-sensitive pen-like device. It is used to choose a menu option displayed on the CRT screen. When the tip of the pen is moved over the screen surface, its photocell detects the light coming out of the screen and the corresponding signal is sent to the CPU. The menu is a set of programmed choices offered to the user. The user gives his choice by touching the light pen against a menu. When the signal is sent by the light pen to the CPU, it identifies the menu option.

Mouse: It is the most commonly used input devices. It is called a mouse because its shape is like a mouse. It is moved across a flat surface. In the older version, there are two wheels on underside of the mouse. The wheels are connected to a shaft encoder. The movement of the wheels is converted into electrical signals, which indicate the position of a cursor on the screen of the CRT. The cursor points out the selected menu. In modern mouse, the wheels are replaced by light-sensing device.

Joystick: Its function is same as that of the mouse. A joystick has a stick with balls on its two ends. The lower end ball moves in a socket. The upper end ball is used to handle it. The joystick can be moved left or right, forwards or backwards. The electronic circuit inside the stick detects and measures the position of the lower end ball in the socket from the central position.

Trackball: Like mouse and joystick, trackball is also a pointing device. It contains a ball with electronic circuit in it. The cursor can be moved on the CRT screen by spinning the ball in different directions. The electronic circuitry detects the direction and speed by which the ball spins. The information in electrical signal is sent to the CPU.

Optical scanner: The scanner is also an input device. It is capable of entering information directly into the computer. It employs a light source and a light-sensing device. The information to be entered is typed. The scanner scans the information and enters it into the computer by encoding in binary code.

12.6.2 Output Devices

The output devices receive information (results after processing of a program) from the computer and make the information available in a form acceptable to the user. The computer sends information in binary-coded form to the output device. The output device translates the information into a form needed by the user. The CRT gives a visual display of the output, while the printer provides a printed output. Sometimes the computer's output is converted by the output device in a form that can be used as an input to other devices. The signal may be used as a control input to a process. A digital-to-analogue converter (DAC) is used to convert the digital output into analogue output.

Printer

The printer is a very commonly used output device. The printer provides printed outputs of results, programs, and data. The printers are classified as follows:

1. Character printer.
2. Line printer.
3. Page printer.

The character printer prints one character at a time. This is a low-speed printer. This type of printers have printing speed of 30 to 600 characters per second. The line printer prints a line of the text at a time. Thus, this type of printers is relatively faster. Its printing speed is 300 to 3000 lines per minute. The page printer prints a page at a time. Its speed is very high relative to the speed of character and line printers.

Printers are also classified on the basis of technology used in their construction. On the basis of technology the printers are classified as follows:

1. Impact printer:
    (a) Dot matrix printer.
    (b) Letter quality printer.
2. Nonimpact printer:
    (a) Thermal printer.
    (b) Electrostatic printer.
    (c) Inkjet printer.
    (d) Laser printer.

Impact printers are electromechanical devices. A printer of this type causes hammers or pins to strike against a paper through an ink-soaked ribbon for printing the text. Nonimpact printers are not electromechanical devices, and hence do not use hammers or pins as printing heads to strike against a ribbon. They use thermal, electrostatic, inkjet, or laser technology for printing. A nonimpact printer, usually, is faster than an impact printer. The advantage of the impact printer is that it can produce multiple copies of the print by using carbon papers between the layers of the paper. The nonimpact printers produce only a single copy of the text at a time. The developed version of the nonimpact printers has an offline device to produce additional copies of the computer output. For this purpose the printer accepts data from the magnetic tape to print more copies of the output.

Dot Matrix Impact Printer: A character is printed by printing the selected number of dots (pins) from a 5×7 matrix of dot (pins). The printing head consists of 7 rows and 5 columns of pins per character. This pattern of pins is called 5×7 dot matrix. For example, a 5×7 dot matrix printer printing 80 characters per line has seven horizontal lines, each line casting of $5 \times 80 = 400$ dots. The printing of a character is done by taking one column of the dot matrix at a time. The selection of dots in the column depends on the specific character that is to be printed. The selected dots of the column are printed by the printing head at a time as the head moves across a line. These printers operate at two or three speeds. Lower speed gives better quality printing. The dot matrix printers are very flexible, as they do not have fixed character fonts. Hence, any shape of character can be printed by the software.

Letter Quality Impact Character Printer: The impact type of letter quality printer is like a typewriter without a keyboard. It consists of a daisywheel with character arms placed along the circumference. To print a character, the wheel rotates to the proper position and an energized magnet then presses the character against the ribbon. The quality of this type of printer is better than that of the dot matrix character printer. However, its speed is much slower compared to that of the dot matrix printer. Its speed is in the range of 10 to 90 characters per second. It is also costlier than the dot matrix printer. It cannot print graphics whereas the dot matrix printer can.

Dot Matrix Thermal Character Printer: This is nonimpact type of printer. It is similar in principle to the dot matrix character printer of impact type. The differences are that it uses a special paper coated with heat-sensitive material, and its printing head consists of 5×7 matrix of tiny heating elements. The specific heating elements selected for a character are heated by electric current. To print a character the head is moved to position of printing on the paper. Thereafter the heating elements for desired character are turned on. After a short time the elements are turned off. Then the print head is moved to next character position.

A new version of this type of printer uses special heat-sensitive ribbon that holds ink in a wax binder. When the hot pins of the print head heats the ribbon against the paper, the wax melts and the ink, in the dots, is transferred to the paper. Its print approaches embossed character quality. The speed is about 200 characters per second.

Electrostatic Printer: This printer is also a nonimpact dot matrix-type printer. The pins of the print head put electric charges on the paper and not ink. The paper is then passed through a bath of oppositely charged toner particles. Thus, the paper picks up the oppositely charged toner particles on the electrically charged spots on the paper. When the paper is passed through the fussing process, the toner particles on the paper melt forming the character impression. It can be used for graphics.

Inkjet Character Printer: This is again a nonimpact dot matrix printer; hence, it can be used for printing both the texts and graphics. One of the inkjet printers uses an ink cartridge that contains columns of tiny heaters. When a set of heater elements is activated, drops of inks are applied to the paper. The cartridge moves from left to right while the paper rolls upwards. The print quality is very near to the letter quality. The speed of such printers is in the same range as that of the dot matrix printers.

Line Printers: As discussed above, a line printer prints a line of the text at a time. The speed of printing of this type of printers is in the range of 300 to 3000 lines per minute. These are used where a large volume of printing is required. The more commonly used line printer is drum printer. It consists of a rotating drum. The drum contains a large number of identical full character sets, arranged in rings round the periphery of the drum. Each ring contains a character in the same position along the axis of the drum. The number of rings of characters is same as the number of characters in the line to be printed. Appropriate characters are selected by rotating the drum at high speed and hammering the paper and ink ribbon against the printer drum with solenoid hammer at the right instant. A complete line is printed after a full revolution of the drum. Then, paper moves to the next line.

The printer contains sufficient memory to store all characters of one line. For example if there are 64 characters, a 6-bit binary code is required. Hence, for a text of 160 characters per line, the storage capacity required is $160 \times 6 = 960$ bits. It can be used online or offline.

Page Printers: The page printers are of nonimpact type and can print one page of the text at a time. The page printer uses laser or other light beam to produce an image of the page on a photosensitive drum. The laser beam is controlled by the computer. The laser-exposed areas attract toner (ink powder). The drum transformed the toner to the paper. The paper is then moved to the fusing section where the toner is permanently fused to the paper with heat or pressure. The drum is discharged and cleaned. Now, the drum is ready to print the next page.

The page printers are expensive and require a lot of maintenance. The speed of these printers is of the range of 10 to 100 pages per minute. The laser printers are popular for large printing work.

12.6.3 CRT Terminal

A CRT terminal consists of a CRT display unit, a keyboard, CRT refresh RAM and controller, and USART or UART. One or more microprocessors are used with modern CRT terminal for controlling and coordinating the keyboard, CRT display unit, and data transmission from the terminal to the computer and vice versa. The data entered by the keyboard is displayed on the CRT screen so that the user can verify whether he/she has entered the data correctly. There is a buffer memory within the CRT terminal that keeps the data entered till the enter key is not pressed. There pointer is called cursor, which appears on the screen indicating the position of the next character appearing on the screen after entering the character through the keyboard.

CRT display unit is a very commonly used output device. It displays the data or information received from the computer. It can display both characters and graphics. The CRT is very similar to a TV screen. The CRT controller uses IC circuits. ICs are used to provide the interface between the computer and the CRT display unit. Such circuits that transfer asynchronous serial data are called *universal asynchronous receiver transmitter* (UART). The circuits capable of transferring data either in synchronous or asynchronous mode are called *universal synchronous-asynchronous receiver transmitter* (USART).

12.6.4 Input-Output Interface

For transferring information between internal storage and input-output devices, there is need of a communication link. The input-output devices include keyboard and display terminal, printer, secondary memory, and magnetic tape. The external devices connected to the computer system are also called peripherals. The method of providing this link is known as input-output interface. The reasons for the need of an interface are as follows:

1. The input-output devices are electromechanical devices and their manner of operation is different from that of CPU, which is an electronic device.
2. The data transfer rate of the input-output device is slower than that of the CPU.
3. Data codes and formats in input-output devices differ from the word format in the CPU.
4. The operating modes of input-output devices are different from each other. Each of the input-output devices must be controlled.

The interface includes special hardware components between the CPU and peripherals to supervise and synchronize all input and output transfers. In addition, each input-output device may have its own controller that supervises the internal mechanism in the device. A typical communication link between the CPU and peripherals is shown in Fig. 12.12. The peripherals include keyboard and display terminal, printer, secondary memory, and magnetic tape. The link lines are known as *input-output bus*. The input-output bus has data lines, address lines, and control lines. Each peripheral device has its own interface unit. Each interface decodes the address and control received from the bus and interprets them. When the interface detects its own address,

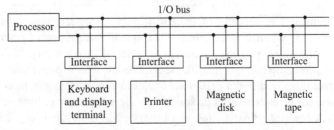

FIGURE 12.12 Connection of input-output bus to peripheral devices.

it activates the path between the bus and its own device otherwise disables it. It also provides signals for the device.

12.6.5 Input-Output Processor

To avoid individual interface for each peripheral device, some computer systems have one or more external processors. Each processor is assigned to communicate directly with all input-output devices. This processor is called IOP. Thus, the IOP relieves the CPU from the task of input-output transfers. A processor that communicates with remote terminals over telephone lines is called *data communication processor* (DCP).

The IOP is similar to a CPU. However, it is designed to handle the details of input-output processing. It can fetch and execute its own instructions. Its instructions are specially designed for facilitating input-output transfers. The IOP can also perform other tasks, such as arithmetic, logic, branching, and code translation. The block diagram of a computer with two processors (CPU and IOP) is shown in Fig. 12.13. The IOP provides a data transfer path between various peripherals and the memory unit. However, the memory unit can communicate CPU and IOP by means of direct memory access. The CPU task is to initiate the input-output program.

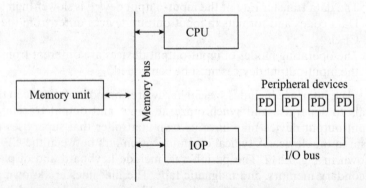

FIGURE 12.13 The block diagram of a computer with an IOP.

12.7 COMPUTER INSTRUCTIONS

A computer is capable of executing various operations. However, it can execute the operations when it is instructed to a specific sequence of operations. It is the user who can control the process by means of a set of instructions. These instructions specify the operations, operands, and the sequence by which processing has to occur. The set of instructions is called a *program*. The data processing task may be changed by specifying a new set of instructions or specifying the same instructions with different data.

A computer instruction is a binary code that specifies a sequence of operations for the computer. Instruction codes with data are stored in a memory. Computer reads the instruction from the memory and places it in a control register. The control interprets the binary code of the instruction and starts executing the instruction by issuing a sequence of operations. A computer has its own instruction set.

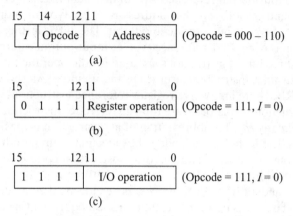

FIGURE 12.14 A computer instruction formats: (a) memory-reference, (b) register reference, and (c) Input-output reference instructions.

There are three formats for instruction codes, and the formats depend on the type of instructions. The instruction types are memory-reference instruction (MRI), register-reference instruction, and input-output reference instruction, as shown in Fig. 12.14. Each format has 16 bits of which 3 bits (12–14) contain instruction code (opcode). The remaining 13 bits depend on the type of formats. For MRI, 12 bits (0–11) specify an address and 1 bit (15) specify the mode of addressing (I). $I = 0$ specify direct address and $I = 1$ indirect address. When the address bits of instruction code (0–11) is not used as address bits, rather specify an operand, the instruction is said to have an immediate operand. The register-reference and input-output reference instructions belong to this type of instructions. When bits 0-11 specify the address of an operand, the instruction is called a direct address instruction. In an MRI of direct address ($I = 0$), the opcode specifies an instruction (say ADD) and the address part specifies the address of an operand. When the address bits (0–11) do not specify the address of an operand, the instruction is said to have an indirect address. In an indirect address MRI ($I = 1$), the opcode specifies an instruction (say ADD) and the address part does not specify the address of an operand, rather it specifies the address of the memory location where the address of the operand is placed.

The register-reference and input-output reference instructions are immediate operand type of instructions. A register-reference instruction is recognized by the operation code 111 with $I = 0$. It specifies an operation on an accumulator (an AC register). An operand from the memory is not needed. Hence, its address bits (0–11) are used to specify the operation to be executed. Similarly, an input-output reference instruction does not need a reference to the memory. It is recognized by opcode 111 with $I = 1$. Hence, its address bits (0–11) are used to specify the type of input-output operation to be performed.

The instruction type is recognized by the computer control from the last 4 bits (12–15) of the instruction. If opcode is not 111, the instruction is an MRI. If $I = 0$, it is direct address type of MRI, and if $I = 1$, it is indirect type. If the opcode is 111, then control inspects the mode bit I (15). If $I = 0$, the instruction is register-reference type and if $I = 1$, it is input-output reference type of instruction. It is a point to be noted that bit I is mode bit and decides whether address is direct type or indirect type. However, when opcode is 111, it is not used as mode bit. The bit I, in this case, decides whether the instruction is of register-reference type or input-output reference type.

The three bits of instructions are used for the operation code in MRIs, the maximum of eight distinct instructions can be in the memory. As shown in Table-12.5, there are seven MRIs. There are twelve register-reference instructions, and six input-output reference instructions. Thus, there are total 25 reference instructions. *The instruction symbol is a three-letter word.* This abbreviation of an instruction is for the use of the programmers and users. The hexadecimal code is hexadecimal number equivalent to the binary code used for the instruction. Thus, a 16-bit instruction code is reduced to a four-digit code. The 12-bit address of a MRI is denoted by three X's (Table-12.5). Digit X stands for a hexadecimal digit. The XXX represents hexadecimal equivalent of 12-bit address. When $I = 0$, the last 4 bits of an instruction have an equivalent hexadecimal digit from 0 to 6. When $I = 1$, the last 4 bits of an instruction have an equivalent hexadecimal digit from 8 to E, since last bit is 1. A register-reference instruction uses 16 bits to specify an operation. The leftmost 4 bits are always 0111 with hexadecimal equivalent 7. The other three hexadecimal digits give the binary equivalent of the remaining 12 bits. An input-output reference instruction uses 16 bits to specify an input-output operation. The leftmost 4 bits are always 1111 equivalent to hexadecimal F. The other three hexadecimal digits represent the binary equivalent of the remaining 12 bits.

The set of instructions are said to be complete if a computer has sufficient number of following categories of instructions:

1. Arithmetic, logic, and shift instructions.
2. Instructions for transferring information between memory and processor.
3. Program control instructions and status condition check instructions.
4. Input and output instruction.

Arithmetic, logic, and shift instructions provide computational capabilities for processing unit of the computer. As computations on the data are done in processor register, the data stored in the memory must be transferred to the processor register before computations start. Hence, it is required to have an instruction that moves data between two units. It is required to compare two numbers and proceed differently if the first is greater than the second and if the first is not greater than the second. The branch instructions are used to change the sequence of executing the program. Input-output instructions are needed for communication between the computer and the peripherals.

The instructions listed in Table-12.5 form a minimum set of instructions that provide all the processing capabilities mentioned above. The instructions ADD, CMA (complement AC), and INC (increment AC) can be used to add and subtract binary numbers. The circulate instructions, CIR and CIL, can be used for any shift including arithmetic shift. Multiplication and division can be performed using addition and subtraction and shift. By using logic operations AND, CMA, and CLA (clear AC), it is possible to implement any logic operation. Moving information from memory to AC can be performed by using LDA (load AC). Transferring information from AC to memory is done with

Table-12.5 Set of instructions for a computer.

Symbol	Hexadecimal code		Description
	$I = 0$	$I = 1$	
AND	0XXX	8XXX	AND memory word to AC
ADD	1XXX	9XXX	Add memory word to AC
LDA	2XXX	AXXX	Load memory word to AC
STA	3XXX	BXXX	Store content of AC to memory
BUN	4XXX	CXXX	Branch unconditionally
BSA	5XXX	DXXX	Branch and save return address
ISZ	6XXX	EXXX	Increment and skip if zero
CLA	7800		Clear AC
CLE	7400		Clear E
CMA	7200		Complement AC
CME	7100		Complement E
CIR	7080		Circulate right AC and E
CIL	7040		Circulate left AC and E
INC	7020		Increment AC
SPA	7010		Skip next instruction if AC positive
SNA	7008		Skip next instruction if AC negative
SZA	7004		Skip next instruction if AC zero
SZE	7002		Skip next instruction if E zero
HLT	7001		Halt computer
INP	F800		Input character to AC
OUT	F400		Output character from AC
SKI	F200		Skip on input flag
SKO	F100		Skip on output flag
ION	F080		Interrupt on
IOF	F040		Interrupt off
ORG N	These pseudoinstructions give information to the assembler that		Origin of the program is at location N
END			Denotes the end of symbolic program
DEC N			Signed decimal N to be converted to binary
HEX N			Hexadecimal N to be converted to binary

ATA (store AC). Three branch instructions and four skip instructions together can be used for program control and status conditions checkup. The INT (input) and OUT (output) instructions can transfer information between the computer and peripherals.

12.8 INSTRUCTION CYCLE

A program stored in the memory unit is a sequence of instructions. The program is executed in the computer by going through a cycle for each instruction. The instruction cycle is subdivided into four subcycles or phases as follows:

1. Fetch an instruction from the memory.
2. Decode the instruction.
3. Read the effective address from the memory if the instruction has an indirect address.
4. Execute the instruction.

After completion of the fourth step, the control goes to the first step and fetch, decode, and execute the next instruction in the sequence. This process continues unless HALT instruction is met.

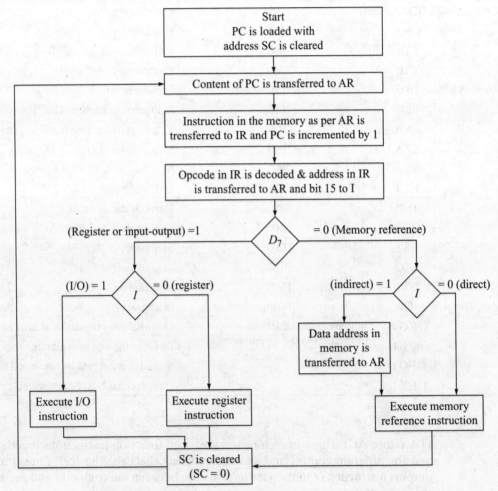

FIGURE 12.15 Flowchart for execution of an instruction.

To begin the program counter is loaded with the address of the first instruction in the program. The SC is cleared to 0 providing a decoded timing signal T_0. After each clock pulse, SC is incremented by 1 and the timing signals go through a sequence $T_0, T_1, T_2, \ldots$. As AR (address register) is connected to the address inputs of the memory where the program is stored, the address is transferred to the AR from the PC during the clock transition associated with timing signal T_0. The instruction in the memory is read and placed in IR with the clock transition associated with timing signal T_1. At the same time, the PC is incremented by 1 to make it ready for the address of the next instruction in the program. At T_2, the opcode in IR is decoded (see Fig. 12.9). The indirect bit is transferred to flip flop I. The address part is transferred to AR. If D_7 (in Fig. 12.9) is 0, it is an MRI. In this case, if $I = 1$, it is an indirect address and the data address from the memory is transferred to AR. If $I = 0$, the data address is already in AR. During time sequence T_3, the MRI is executed on the data and SC is cleared (SC = 0). If D_7 is 1, the instruction is a register or input-output reference instruction and it is executed during time sequence T_3. The SC is cleared. The process of execution of the next instruction in program is fetch in step 1. The flowchart is shown in Fig. 12.15.

12.9 COMPUTER PROGRAMMING

A computer program is a set of instructions for directing the computer to perform a required data processing task. There are various types of programming languages, but the computer can execute only those programs that are in binary form. Programs written in any other language must, therefore, be translated into binary forms before they are executed by the computer. Programs may be one of the following categories:

1. *Binary code.* This is a sequence of instructions and operands in binary and placed in the computer memory.
2. *Hexadecimal code.* This is a hexadecimal equivalent translation of the binary code.
3. *Symbolic code.* This is a user written sequence of instructions in symbols. The symbols consist of alphanumeric or special characters. Each symbolic instruction can be translated into one binary-coded instruction. This translation is done by a special program called an **assembler.** As assembler translates the symbolic program in binary-coded program. The symbolic program is called *assembly language program.*
4. *High-level programming languages.* These are special languages developed for users to write the procedures for solving a problem. No way, computer hardware behaviour is concerned. A high-level programming language is user-friendly. FORTRAN is an example of a high-level programming language. The translation program that translates the high-level language program into a sequence of binary-coded instructions is called a **compiler.**

12.9.1 Machine Language

A computer understands only binary-coded instructions, and therefore a binary program is referred to as a machine language program of category 1. Since there is a simple equivalency between binary and hexadecimal representations, the program written in hexadecimal is referred to as a machine language program of category 2. There is one-to-one relationship between symbolic and binary instructions. Hence, an assembly language is considered to be a machine level language.

A program is a sequence of instructions, and these instructions are executed one by one in a sequence in which they are stored in the memory. Thus, each instruction has a location in the memory. These locations are identified by serial numbers in binary. These numbers are called *memory address*. The execution of a program starts from the instruction having 0 (in binary) address and proceeds executing the program in serial order of the memory address. Each instruction has 16 digits of binary number. Writing 16 bits for each instruction is tedious. However, the number of digits in an instruction can be reduced by using hexadecimal digits. The 16-digit binary instructions can be represented by four digits in hexadecimal. The hexadecimal decimal representation is easy to use. However, the program written in hexadecimal is converted into binary code by replacing each digit of the hexadecimal instruction by its binary equivalent. The instructions with symbolic operation code, as listed in Table-12.5, uses the symbolic names of instructions instead of their binary or hexadecimal equivalent. The memory addresses and operand addresses, however, are written in hexadecimal.

12.9.2 Assembly Language and Assembler

Machine language is not convenient both for programmers and for users, and it was used during early developmental stages of the computer. Later, symbolic names of instructions (listed in Table-12.5) were used for writing a program. The symbols are known as mnemonics. A program written in mnemonics is called an assembly language program. Almost every commercial computer has its own assembly language. Like any programming language, assembly language also is defined by a set of rules. The rules for writing assembly language programs are documented and published by the computer manufacturer for the user.

Each line of an assembly language program is arranged in three columns. The columns are called fields. The fields give the following information:

1. The first column is the *label field*. It may be empty or may specify a symbolic address.
2. The second column is the *instruction field*. It specifies a machine instruction or a pseudo instruction.
3. The third column is the *comment field*. It may be empty or may have a comment.

A symbolic address consists of one, two, or three alphanumeric characters. It should not have more than three characters. The first character of the address must be a letter followed by letters or numerals. The address symbols can be chosen arbitrarily by the programmer. Its symbol must be terminated by a comma. It indicates that the symbol is a label and differentiates from the instruction.

The instruction field specifies one of the three forms of instructions, namely, an MRI, a register-reference, an input-output reference instruction, and a pseudo instruction with or without an operand. An MRI consists of two or three symbols. The first is a three-letter symbol defining an instruction code, the second is a symbolic address, and the third, if it is there, is *I*, indicating an indirect address instruction. If *I* is missing, it is a direct instruction. A register-reference or input-output reference (non-MRI) instruction consists of only one symbol (listed in Table-12.5). The following symbols illustrate the above statement:

    CLE   non-MRI
    ADD PDR   direct address MRI
    ADD MTR I   indirect address MRI

The first three-letter symbols in instruction field of each line of a program must be one of the computer instructions listed in Table-12.5. A symbolic address in the instruction field specifies the memory location of the operand. This symbolic address must be defined somewhere in the program as a label in the first column. It is absolutely necessary for translating the assembly language program to a binary program.

As a computer recognizes only machine language program in binary, the assembly language program is to be translated into a machine language program in binary before the program is executed by the computer. An **assembler** translates the assembly language program into a machine language program in binary. The assembler is a program that accepts an assembly language program and translates it into its equivalent machine language program in binary. To illustrate how an assembly language program is translated, let us consider assembly language program to add two numbers given in Table-12.6.

Table-12.6 Assembly language program to add two numbers.

	ORG 100	/Origin of program is at location 100
	LDA FIR	/Load operand from location FIR
	ADD SEC	/Add operand from location SEC
	STA RES	/Store sum in location RES
	HLT	/Halt computer
FIR,	DEC 73	/Decimal operand
SEC,	DEC -13	/Decimal operand
RES,	HEX 0	/Location RES hexadecimal address
	END	/End of assembly language program

The translation of the assembly language program of Table-12.6 into an equivalent binary code may be done as follows. At the first line, a pseudoinstruction ORG is encountered. It says that the binary program should start from hexadecimal location 100 (Hex). The second line's first symbol is a direct address MRI LDA and $I = 0$. Hence, the binary code in hexadecimal of LDA from Table-12.5 is 2. The binary value of the address part must be obtained from label FIR, which is the second symbol of line 2. The line 2 contains an instruction for location 100 (Hex), and every other line specifies a machine instruction or an operand for sequential memory locations. Thus, label FIR is line 6, which corresponds to memory location 104 (Hex). By combining two parts of the instruction LDA, the hexadecimal instruction code is 2104. The other lines of machine instructions are translated in a similar manner. The translated program is given in Table-12.7.

There are three pseudoinstructions. The first is ORG and other two are at label FIR and SEC, specifying decimal operands with pseudoinstruction DEC. The 4th pseudoinstruction HEX specifies hexadecimal. The pseudoinstruction DEC could have been

Table-12.7 Translated program of Table-12.6.

Hexadecimal code		Assembly program	
Location	Content		
			ORG 100
100	2104		LDA FIR
101	1105		ADD SEC
102	3106		STA RES
103	7001		HLT
104	0049	FIR,	DEC 73
105	FFF3	SEC,	DEC -13
106	0000	RES,	HEX 0
			END

used in place of HEX as well. Decimal 73 is converted to binary and placed in location 104 in its hexadecimal equivalent. Decimal 13 is a negative number and must be converted into binary in signed 2's complement form. The hexadecimal equivalent of this binary number (in 2's complement form) is placed in location 105. Hex 0 binary code in 12 bits is hexadecimal 000 and is placed in location 106. The END symbol tells the translator that the program has ended and there are no more lines to be translated.

12.9.3 Assembler

An assembler, as discussed earlier, translates the assembly language program into an equivalent machine language program in binary code. It is a software program that accepts an assembly language program and produces an equivalent binary coded machine language program. The input program is called the *source program* and the output program the *object program*. The assembler operates on character strings and produces an equivalent binary interpretation.

An assembly program is first stored in the memory. The user enters the assembly program in the memory through the keyboard. A loader program inputs characters of the assembly program into the memory. As the characters of the assembly program can only be entered into the memory in binary code, we need a binary code for each characters used in the program. In a basic computer, each character is represented by an 8-bit binary code with MSB always 0. The other 7 bits are as specified by ASCII. The ASCII code stands for American Standard Code for Information Interchange. It is a 7-bit code and is extensively used in basic computer, peripherals, instrumentation, and communication devices. The ASCII code in hexadecimal form is given in Table-12.8. The last code 0D is for CR (carriage return) and is produced when the return key on the keyboard is pressed. It does not print a character, but is associated with the physical movement of the cursor on the screen. When this code is encountered the assembler recognizes a CR code as the end of a line of code.

The coded form of assembly program is stored in consecutive memory location. As a character is represented in binary code by 8 bits, two characters can be stored in one-memory word, which has a capacity of 16 bits. For example, the following line of program in Table-12.6 is stored in six memory locations as given in Table-12.9:

Table-12.8 SCII code in hexadecimal form.

Character	Code	Character	Code	Character	Code	Character	Code
0	30	C	43	O	4F	Space	20
1	31	D	44	P	50	(	28
2	32	E	45	Q	51	)	29
3	33	F	46	R	52	*	2A
4	34	G	47	S	53	+	2B
5	35	H	48	T	54	,	2C
6	36	I	49	U	55	-	2D
7	37	J	4A	V	56	.	2E
8	38	K	4B	W	57	/	2F
9	39	L	4C	X	58	=	3D
A	41	M	4D	Y	59	CR	0D
B	42	N	4E	Z	5A		

Table-12.9 Memory storage of the line of code FIR, DEC 73.

Memory location	Symbol	Hexadecimal code	Binary representation
1	F I	46 49	0100 0110 0100 1001
2	R ,	52 2C	0101 0010 0010 1100
3	D E	44 45	0100 0100 0100 0101
4	C space	43 20	0100 0011 0010 0000
5	7 3	37 33	0011 0111 0011 0011
6	CR space	0D	0000 1101 0010 0000

Last location stores only code for carry return which occupies 8 bits only. Remaining 8 bits are blank. The assembler neglects all characters in the comment field.

FIR, DEC 73

The input for the assembler is the assembly language program in ASCII. This input is scanned by the assembler twice and generates equivalent binary program. An assembler that scans the assembly program twice is called a *two-pass assembler*. There is *one-pass assembler* as well. The one-pass assembler scans the entire program only once. During the first pass, the two-pass assembler collects all labels and assigns addresses to all labels counting their position from the starting address. For assigning addresses to all labels, one-pass assembler must be equipped with some means. During the second pass it produces machine code and assigns address for each instruction in the program. The major tasks the two-pass assembler performs during translation process are briefly described now.

First, the assembler uses a memory word called location counter (LC). The LC stores the memory location assigned to the instruction or operand presently being processed. The ORG pseudoinstruction initializes the LC to the value of the first location. Since instructions are stored in sequential locations, the content of LC is incremented by 1 after processing each line of code. In case ORG is missing in the program, the assembler set LC to 0 initially.

During the first pass, the assembler LC is initially set to 0. If the first line is ORG, LC is set to the number that follows ORG. The assembler scans the next line of code and

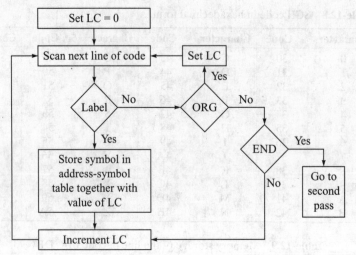

FIGURE 12.16 The flowchart for first-pass processing of the assembler.

checks whether it is a label. If it is a label, it is stored in the address symbol table. The binary equivalent number of the address as specified by the content of LC is also stored. Nothing is stored in the table if no label is encountered. LC is incremented by 1 and a new line of code is processed. A flowchart is shown in Fig. 12.16.

For the program of Table-12.6, the assembler-generated address symbol table is given in Table-12.10.

During the second pass, machine instructions are translated by means of lookup table procedures. In a lookup table procedure the specific instruction is matched with the entries of the lookup tables. Any symbol that is encountered in the program must be there in the one of the lookup table. The assembler uses four lookup tables, namely, pseudoinstruction (PI) table, MRI table, register and input-output reference (non-MRI) table, and address symbol table. As in Table-12.5, there are four symbols in PI table, seven in MRI table, and 18 in non-MRI table. The address symbol table (as in Table-12.9) is generated during the first pass. The assembler searches these four tables to find the symbol that is currently processing in order to determine its binary code.

Table-12.10 Address symbol table for the program in Table-12.6.

Memory location	Symbol or (LC)*	Hexadecimal code	Binary representation
1	F I	46 49	0100 0110 0100 1001
2	R,	52 2C	0101 0010 0010 1100
3	(LC)	01 04	0000 0001 0000 0100
4	SE	53 45	0101 0011 0100 0101
5	C,	43 2C	0100 0011 0010 1100
6	(LC)	01 05	0000 0001 0000 0101
7	RE	52 45	0101 0010 0100 0101
8	S,	53 2C	0101 0011 0010 1100
9	(LC)	01 06	0000 0001 0000 0110

* (LC) designates content of location counter.

Sec. 12.9 / Computer Programming   **659**

The flowchart shown in Fig. 12.17 describes the tasks performed by the assembler during the second pass (scan). Labels are neglected this time. LC is set to be 0 initially. The assembler goes directly to the instruction field and checks the first symbol. It first checks PI table. If it matches with ORG, the assembler goes to a subroutine that set LC to the number that follows ORG. A match with END terminates the translation process. An operand PI converts the operand into binary and placed this binary code in the memory location specified the content of LC in the first pass. LC is incremented by 1 and the assembler continues to analyze the next line of code.

If the search in PI table fails, the assembler goes to MRI table. When a symbol is matched, the assembler extracts its 3 bit code and inserts it in the instruction bits with of the memory. The second symbol in MRI is a symbolic address which is converted to

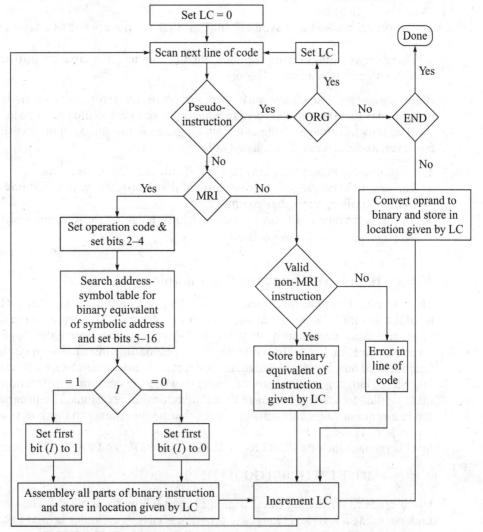

FIGURE 12.17 Flowchart for the second pass processing of the assembler.

binary by searching the address symbol table. The I bit is set to 0 if I, as the third symbol in MRI is absent, and it is set to 1 if the third symbol is I. Thus, all 16 bits of the memory location specified by LC are filled. LC is incremented by 1 and the assembler continues to analyze the next line of code.

If no match is found in the MRI table, the assembler searches in non-MRI table. If a symbol matches in this table, the assembler stores 16-bit instruction code into the memory location specified by LC. LC is incremented by 1 and the assembler continues to analyze a new line.

Before the assembler starts processing of translation of a program, it checks possible errors in the symbolic program. This is an important task of an assembler. This is called error diagnostics. The possible errors may be listed as follows:

1. An invalid machine code symbol which is detected by its being absent in the MRI and non-MRI table.
2. The program also has a symbolic address that did not appear as a label in the program.
3. There may be other errors such as a missing comma (,) in label, an instruction code without address symbol.

If the assembler found any such error, it prints an error message to inform the programmer that the symbolic program has an error at a specific line of code.

The only advantage of the assembly language is that its computation time is less. However, its disadvantages are listed as follows:

1. Programming in assembly language is difficult and time-consuming.
2. The assembly language is machine oriented and program written for one computer cannot be run on any other computer.
3. The programmer must have a detailed knowledge of the computer structure for which he/she is writing a program.

12.9.4 High-Level Language Programming

The assembly language discussed earlier is known as low-level language. The demerits of the assembly language do not make it user-friendly. To overcome the difficulties in programming in assembly language, high-level languages were developed. In low-level language, each machine code has its corresponding instruction in symbols called mnemonic. Thus, low-level language instructions have one-to-one correspondence. High-level languages allow programmers to write the computer instructions in statement similar to English language or mathematical expressions. The program is written in a sequence of statements in a form that people prefer to think in when solving a problem. Hence, a high-level language is procedure oriented. Examples of high-level languages include FORTRAN, BASIC, COBOL, and PASCAL. For example,

If (SETPT.GT.0.0) GO TO 215

which reads as "if the set point is greater than 0.0 go to line 215". Thus, writing and understanding a high-level language program is easy. A compiler is used to convert the high-level language source program statement into machine language instructions. However, the compiler converts a statement into many machine language instructions. Thus, one-to-one correspondence like that in low-level language is lost. A high-level language is independent of a computer. A program written in high-level language can run on any computer. However, each computer has its own compiler. Thus, the compiler is computer dependent. The advantages of high-level language programs are listed below:

1. As compared to assembly language, high-level language program is easier to write and understand.
2. The programmer need not know about the computer hardware orientation and behaviour.
3. A high-level language program can run on any computer. The only requirement is that the computer must have a compiler of the language in which the program is written.
4. It provides better documentation.

The compiler for a high-level language is a more complicated program than an assembler. It requires knowledge of system to fully understand its operation. A compiler may use an assembly language as an intermediate step in the translation or may translate the program directly to binary.

12.10 MICROPROCESSOR

As discussed earlier, microprocessor is a CPU on a single chip. When a microprocessor is combined with a RAM, it forms a microcomputer. The block diagram of a microprocessor with a RAM is shown in Fig. 12.18. The microprocessor contains an ALU, a timing and control unit, and several registers. The ALU is responsible for

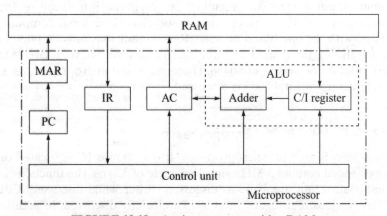

FIGURE 12.18 A microprocessor with a RAM.

performing addition and subtraction. It is also capable of performing logic operations AND, OR, and XOR. It includes a full adder of the required number of bits as well as a register called accumulator (AC) and a complement/increment register (C/I). The purpose of C/I register is to perform subtraction by addition of the 2's complement of the number to be subtracted. The 2's complement of the number is obtained by complementing the binary form and incrementing by 1. The PC register is a program counter. It provides the memory address of RAM to the MAR. Starting with the memory address of the first instruction of the program, the PC provides successive instruction memory address to the MAR. The RAM stores instructions and data (operand). The IR has double tasks. Its first task is to inform the control unit about the kind of operation that is to be performed. The operation may be addition or subtraction. Its second task is to provide the address of the operand. The operand is copied from the RAM to the C/I register of the ALU. The time and control unit serves as the director of all operations. It generates timing and control signals. It controls the peripheral devices, such as memory, input and output devices. It controls the data flow between the microprocessor, and the memory and input-output devices. It provides status, control, and timing signals that the peripheral devices required.

The practical form of the microprocessors discussed above contains many thousands of electronic components on a single chip. The electronic components include hundreds of AND and OR gates, and flip flops in order to form a very sophisticated control unit. In addition, there are many special and general-purpose registers, MUXs, decoders, and other necessary digital components. The first microprocessor was introduced by Intel Corporation in 1972. It was a 4-bit microprocessor named as Intel 4004. However, the first commercially successful unit was 8-bit microprocessor Intel 8080. It was introduced in 1973 and has 64 KB memory, 40 pins, and 2 MHz clock frequency. The more popular 8-bit microprocessor Intel 8085 was introduced in 1976. It is faster than Intel 8080, as it has clock frequency of 3 to 6 MHz. Some other 8-bit microprocessors are Motorola 6800 and Zilog Z80 microprocessor. Intel Corporation also developed 16-bit microprocessors Intel 8088 and Intel 8086. Intel 8088 internal architecture is of 16 bits, but its external data bus is only 8-bit wide. Hence, 8-bit input-output devices and memory can easily be interfaced to Intel 8088. As 8-bit input-output devices are cheaper, a 8088 based personal computer is cheaper compared with the one based on 8086. Recent additions of 8086 family include the third-generation units named as Intel 80186 and Intel 80286. More recent developments are Intel 80386 and Intel 80486. These are 32-bit microprocessors. In this section, the architecture of Intel 8085 is discussed.

12.10.1 Intel 8085 Microprocessor

Intel 8085 is an 8-bit microprocessor. It is a 40 pins IC fabricated on a single chip. Its clock speed is about 3 MHz and clock cycle of 320 ns. The functions of ALU and timing and control unit are same as discussed earlier while discussing CPU in Section 12.2. Here, various registers of Intel 8085 are briefly discussed. In addition to accumulator, Intel 8085 contains general-purpose registers and special-purpose registers. The block diagram of Intel 8085 is shown in Fig. 12.19.

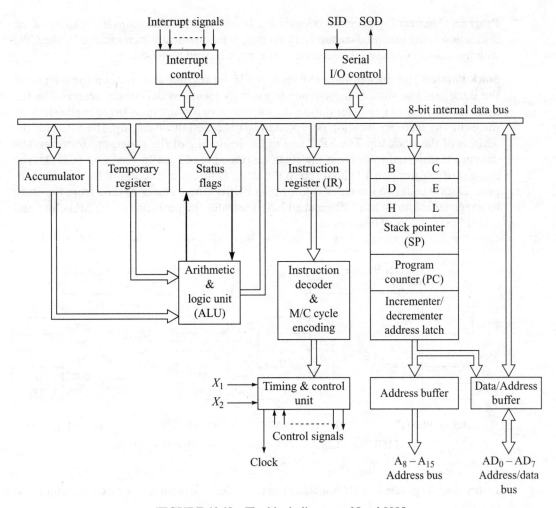

FIGURE 12.19 The block diagram of Intel 8085.

Accumulator

It the key register of the microprocessor. One input of the ALU is held in the accumulator (A). The other input may be the memory or one of the registers. The results of arithmetic and logic operations are first stored in A.

General-Purpose Registers

These registers are used for temporary storage of data or addresses. In case of data being of 16 bit, two 8-bit registers are combined. The two registers together are called register pair.

Special-Purpose Registers

There are six 8-bit registers named as B, C, D, E, H, and L. One of the operand is stored in these registers. Other registers of this type are program counter, stack pointer, IR, and status register.

Program Counter: The program counter is a 16-bit register. It contains the address of the instruction being executed currently. Its content is automatically incremented by the CPU with the address of the next instruction. At start, it is fixed by REST.

Stack Pointer: The stack pointer (SP) is also a 16-bit register and contains the address of the stack top. The stack is a sequence of memory locations define and reserved by the user. The content of a register during the execution of a program is saved in the stack by the user. The memory location occupied in last is called the stack top. The SP holds the address of the stack top. The SP is rest in the beginning of the program execution. The data are transferred to or from the stack on principle of Last In First Out (LIFO). The content of register pair BC is stored to the stack as this pair is required for another purpose. Data transfer to the stack is known as PUSH. The transfer of data from the stack to a register is known as POP. Figure 12.20 illustrates the positions of the SP before and after PUSH and POP operations.

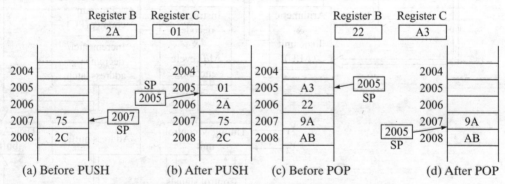

FIGURE 12.20 Stack before and after PUSH and POP.

Instruction Register: The IR holds the binary-coded instruction to be executed until it is decoded.

Status Register: The status register (SR) is an 8-bit register. The 5 bits of the register, as shown in Fig. 12.21, are used for five special status indications. These 5 bits are known as

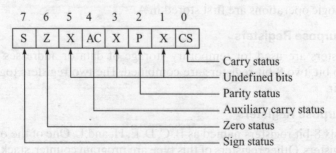

FIGURE 12.21 Position of various status flags.

status flags. These flags are carry flag, zero flag, sign flag, parity flag, and auxiliary flag. Three bits are undefined.

Carry (CS): The CS flag holds carry out of MSB resulting from the execution of an arithmetic operation. If the carry is from addition or borrow from subtraction or comparison, the carry flag is set to 1, otherwise 0.

Zero (Z): The Z flag is set to 1 if the result of an arithmetic or logic operation is zero, otherwise 0.

Sign (SS): The SS flag is set to 1, if MSB of the result of an arithmetic or logic operation is 1, otherwise 0.

Parity (PS): The PS flag is set to 1, if the result of an arithmetic or logic operation contains even number of 1s; if it is odd, it is 0.

Auxiliary carry (AC): The AC flag is set to 1, if MSB of the execution of an arithmetic or logic operation results a carry out of bit-3 (bits are numbered from 0 to 7) to bit-4, otherwise 0.

Bus Structure

There are three types of buses. These are address bus, data bus, and control bus. The width of a bus is defined as the number of communication lines contained by the bus.

Address Bus: Each memory location and input-output devices are associated with it. The microprocessor sends the address (in binary) of a device over the address bus. The address is received by all devices connected to the microprocessor, but only that device which has been addressed responds. Address bus is unidirectional and is of 16 bit wide.

Data Bus: The data bus is 8-bit wide and is bidirectional. Hence, data are sent to and received from different devices by the microprocessor.

Control Bus: This bus carries the control signals between the microprocessor and peripheral devices. Hence, the control bus is also bidirectional and has a width of 8 bits. Control signals are generated by timing and control units.

Instruction Cycle

The necessary steps carried out by the CPU to fetch an instruction and necessary data from the memory and to execute the instruction constitute an instruction cycle. Thus, an instruction cycle is a combination of fetch cycle and execution cycle. The total time of execution of an instruction includes the time required in fetching and executing the instruction. Thus, instruction cycle is fetch cycle plus execution cycle.

Fetch Cycle: It is the time required for fetching the machine code (opcode) of the instruction from the memory. The entire operation of fetching an opcode takes three clock cycles. Sending memory address from the PC to the memory takes one clock cycle. The second clock cycle is taken in reading the opcode. During the third clock cycle opcode is sent to the processor.

Execution Cycle: When opcode is received by the processor, it is decoded by the decoder. The execution of the instruction begins. The process of decoding and execution takes one clock cycle. If execution requires an operand, the processor fetches the operand from the memory before starting the execution. In this case, execution cycle requires more than clock cycles. If instruction is only 1 byte long, its execution cycle requires only one clock cycle. However, if instruction contains data or address of data,

Table-12.10 Descriptions of pins of Intel 8085.

Pin No.	Pin name	Descriptions of pins	Pin type
1–2	x_1–x_2	To be connected to external clock generator.	Input
3	RESETOUT	It is high when processor is resetting.	Output
4	SOD	It is data line for serial output.	Output
5	SID	It is data line for serial input.	Input
6	TRAP*	It is highest priority nonmaskable interrupt.	Input
7–9	RESET 7.5, 6.5, 5.5*	These are restart interrupts and they have programmable mask.	Input
10	INTR*	It is lowest priority interrupt request signal.	Input
11	INTA	It is an active low interrupt acknowledgement signal sent by microprocessor after INTR is received.	Output
12–19	AD0–AD7	The 8 LSB bits of address of memory or I/O are sent through these lines.	Output
20	V_{ss}	Ground terminal.	
21–28	A8–A15	The 8 MSB bits of address of memory or I/O are sent through these lines.	Output
29,33	S0, S1	They are status signals sent by microprocessor to distinguish various types of operations.	Output
30	ALE	It is address latch enable signal.	Output
31	WR	It is active low signal to control WRITE operation	Output
32	RD	It is active low signal to control READ operation	Output
34	IO/M	It is control signal distinguishing between memory address and I/O address.	Output
35	READY	It is active high signal to sense whether peripheral is ready or not.	Output
36	RESETIN	It reset the program counter to zero.	Output
37	CLOCKOUT	It is clock output for user.	Output
38	HLDA	It is hold signal saying that hold request has been received.	Output
39	HOLD	This signal high indicates that another device is requisitioning the use of address and data bus.	Output
40	V_{cc}	+5 V supply	

* Intel 8085 has five interrupts, namely, TRAP, RESET 7.5, 6.5, 5.5, and INTR. Among these, TRAP has the highest priority followed by REST 7.5, RESET 6.5, and RESET 5.5. The INTR has the lowest priority.

one or two extra fetch cycles are required. In this case, the execution cycle may require several clock cycles.

Pin Configuration

Intel 8085 has 40 pins and its pin configuration is shown in Fig. 12.22. Pin descriptions are given in Table-12.10.

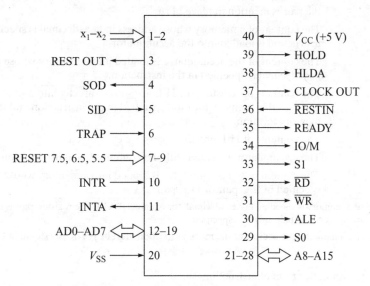

FIGURE 12.22 Pin configuration of Intel 8085.

Instruction Set of Intel 8085

We have discussed the instruction set of a basic computer with a minimum set of 25 instructions, as given in Table-12.6 in Section-12.7. Similarly, Intel 8085 has a set of 80 instructions. As in Section-12.7, the instructions are divided into five groups, namely, data transfer group, arithmetic group, logical group, branch group, and control group. Data transfer group instructions move data between registers, between memory and register, and data directly to register. Arithmetic group instructions are addition, subtraction, increment, and decrement of data. Logical group instructions perform logical operation AND, OR, and XOR. Branch group instructions include jump instruction, and subroutine call and return instruction. These instructions may be conditional or unconditional. The conditional instructions are executed when a particular condition is satisfied. The unconditional instructions operate unconditionally. Control group of instructions perform data transfer between the microprocessor and input-output devices, manipulate the stack, and alter internal control flags. They also make it possible for the programmer to halt the microprocessor, put it in a 'no operation state', enable/disable its interrupt system. The five groups of instructions are given in Tables-12.11, 12.12, 12.13, 12.14, and 12.15, respectively.

Table-12.11 Data transfer group of instructions* §.

Instruction	Description
MOV x, y	Moves the content of register (memory) y into register (memory) x.
MVI x, data	The data are immediately moved to register (memory) x. The first byte of the instruction is the opcode and the second byte is data to be moved.
LXI rp, data 16	16-bit data are immediately loaded to a register pair. Name of the first register of the pair is mention in place of rp.
LDA addr	The content of a memory whose address in hexadecimal is specified in the instruction is loaded into the accumulator.
STA addr	The content of the accumulator is loaded into a memory whose address in hexadecimal is specified in the instruction.
SHLD addr	The contents of register pair H-L are stored directly into memories. The content of L is stored in memory location specified in the instruction and that of H in the next memory location.
XCHG	The contents of H-L and D-E pairs are exchanged.
LDAX rp	The accumulator is loaded with the content of memory location specified in rp.
STAX rp	The content of the accumulator is loaded into a memory whose location is specified in B-C pair or D-E pair.

* The location of a memory is specified in hexadecimal, for example, 2A10. The register pair is specified by its first register; for example, for B-C pair only B is specified.

§ In case x (or y) is a memory, the address of the memory location must be previously stored in H-L pair.

Table-12.12 Arithmetic group of instructions* §.

Instruction	Description
ADD x	Content of register (memory) x is added to the content of the accumulator. The result is stored in the accumulator.
ADC x	Content of register (memory) x and carry status are added to the content of the accumulator. The result is stored in the accumulator.
ADI data	Data are immediately added to the content of the accumulator and the result is stored in the accumulator itself.
ADC data	Data and carry status are added to the content of the accumulator and the result is stored in the accumulator itself.
SUB x	The content of register (memory) x is subtracted from the content of the accumulator and the result is stored in the accumulator itself. In case x is a memory its location must be previously stored in H-L pair.
SBB x	The content of register (memory) x and borrow are subtracted from the content of the accumulator and the result is stored in the accumulator itself.
SUI data	Data are subtracted from the content of the accumulator and the result is stored in the accumulator itself.
SBI data	Data and borrow are subtracted from the content of the accumulator and the result is stored in the accumulator itself.
INR x	The content of register (memory) x is incremented by 1. In case x is a memory its location must be previously stored in H-L pair.
DCR x	The content of register (memory) x is decremented by 1.

Table-12.12. Arithmetic group of instructions* §. (*Continued*)

Instruction	Description
INX rp	The content of the register pair is incremented by 1.
DCX rp	The content of the register pair is decremented by 1.
DAA	Converts the hexadecimal result of addition (stored in the accumulator) into decimal number. The instruction is used after ADD or ADC instruction.
DAD	The content of the register pair is added to the content of the H-L pair.

* The location of a memory is specified in hexadecimal, for example, 2A10. The register pair is specified by its first register; for example, B-C pair only B is specified.

§ In case x is a memory its location must be previously stored in H-L pair.

Table-12.13 Logic group of instructions* #§.

Instruction	Description
ANA x	AND the content of the register (memory) x with that of the accumulator and the result is stored in the accumulator.
ANI data	AND the immediate data with that of the accumulator and the result is stored in the accumulator.
ORA x	OR the content of the register (memory) x with that of the accumulator and the result is stored in the accumulator.
ORI data	OR the immediate data with that of the accumulator and the result is stored in the accumulator.
XORA x	Exclusive OR the content of the register (memory) x with that of the accumulator and the result is stored in the accumulator.
XORI data	Exclusive OR the immediate data with that of the accumulator and the result is stored in the accumulator.
CMA	Complements (0 to 1 and 1 to 0) the content of the accumulator.
CMC	Complements the carry status flag.
STC	Sets carry status (CS) to 1.
CMP x	Compares the content of register (memory) x with that of the accumulator.
CPI data	Compares the immediate data with the content of the accumulator.
RLC	Rotates left the content of the accumulator by 1 bit. The 7th bit is moved both to CS and 0th bit of the accumulator.
RRC	Rotates right the content of the accumulator by 1 bit. The 0th bit is moved both to CS and 7th bit of the accumulator.
RAL	Rotates left the content of the accumulator by 1 bit through carry. The 7th bit is moved to CS and CS to 0th bit of the accumulator.
RAR	Rotates right the content of the accumulator by 1 bit through carry. The 0th bit is moved to CS and CS to 7th bit of the accumulator.

* The location of a memory is specified in hexadecimal, for example, 2A10. The register pair is specified by its first register; for example, B-C pair only B is specified.

In all AND, OR, and XOR operations, the carry flag (CS) and auxiliary carry (AC) are cleared, that is, CS and AC are set to 0. For comparing the content of the register (memory) or the data are subtracted from the content of the accumulator, the status flags are set according to the result. But the result of the subtraction is discarded and the content of the accumulator remains unchanged.

§ In case x is a memory its location must be previously stored in H-L pair.

Table-12.14 Branch group of instructions.

Instruction	Description
JMP addr (label)	Specifies the address of the label where the program jumps unconditionally.
JZ addr (label)	Program jumps to the specified address of label when the result is zero.
JNZ addr (label)	Program jumps to the specified address of label when the result is non-zero.
JC addr (label)	Program jumps to the specified address of label if there is a carry.
JNC addr (label)	Program jumps to the specified address of label if there is no carry.
JP addr (label)	Program jumps to the specified address of label if the result is plus.
JM addr (label)	Program jumps to the specified address of label if the result is minus.
JPE addr (label)	Program jumps to the specified address of label if the result contains even number of 1s.
JPO addr (label)	Program jumps to the specified address of label if the result contains odd number of 1s.
CALL(RET) addr (label)	Call (return) a subroutine unconditionally. The program jumps to (back from) the subroutine that is at specified address of label.
CC (RC) addr (label)	Call (return) a subroutine that is at specified address of label, if CS = 1.
CNC (RNC) addr (label)	Call (return) a subroutine that is at a specified address of label, if CS = 0.
CZ (RZ) addr (label)	Call (return) a subroutine that is at specified address of label, if Z = 1.
CNZ (RNZ) addr (label)	Call (return) a subroutine which is at specified address of label, if Z = 0.
CP (RP) addr (label)	Call (return) a subroutine which is at specified address of label, if SS = 0.
CM (RM) addr (label)	Call (return) a subroutine which is at specified address of label, if SS = 1.
CPE (RPE) addr (label)	Call (return) a subroutine which is at specified address of label, if P = 1.
CNO (RNO) addr (label)	Call (return) a subroutine which is at specified address of label, if P = 0.
RST n	The content of the PC is saved in the stack and then program jumps unconditionally to the address which is eight times 'n' where n may be 0 to 7.
PCHL	The content of HL pair is transferred to the PC.

Table-12.15 Control group of instructions.

Instruction	Description
PUSH rp	Push the content of specified register pair to the stack.
PUSH psw	The content of the accumulator and status flags are pushed into stack.
POP rp	The last saved content of the stack is moved to the register pair.
POP psw	The content of the stack (top) is moved to the status flags (psw) and the accumulator.
HLT	Stops the program execution until a RESET or interrupt signal occurs.
XTHL	Exchange the stack with H-L pair.
SPHL	Moves the content of H-L pair to the stack pointer.
EI	Enables all maskable interrupts.
DI	Disables all maskable interrupts.
SIM	Enables or disables maskable interrupts according to the bit pattern of the accumulator.

Table-12.15 Control group of instructions. (*Continued*)

Instruction	Description
RIM	The accumulator is loaded with pending interrupts.
NOP	No operation is performed when this instruction is executed.
IN port-addr	The content of the port specified by the port address is read by the processor.
OUT port-addr	The content of the accumulator is moved to the port specified by its address.

12.11 MICROCOMPUTER

When a microprocessor is combined with a RAM, it forms a computing system. However, to make it communicate with people, it is necessary to add a third section, called input-output section. The input-output section consists of input-output interface, which connects the computing system to appropriate input and output devices. The input and output devices may include keyboard, CRT terminals, printer, and others. This combination forms a microcomputer. The block diagram of a microcomputer is shown in Fig. 12.23.

The input-output devices have already been discussed earlier. The input-output interface is discussed here.

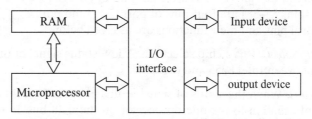

FIGURE 12.23 The block diagram of a microcomputer.

12.11.1 Input-Output Interface

The memories and input-output interfaces are connected to the microprocessor to form a microcomputer. Connecting microprocessor with the input-output devices is known as *interfacing*. The memory and input-output devices are known as peripherals. For successful transformation of information between the microprocessor and the peripherals, the peripherals must be compatible. In case a particular peripheral is not compatible, some electronic circuits have to be designed through which the peripherals may be interfaced with the CPU. Mostly, memories and input-output devices are interfaced with CPU by the manufacturers. The electronic circuits used for interfacing the peripherals are known as interfacing devices. A number of general-purpose and special-purpose single-chip interfacing devices are developed. One of them is programmable peripheral interface (PPI). The PPI developed by Intel is discussed below.

The PPI is a multiport device. The ports may be programmed according to requirement. Intel 8255 is a PPI and is very useful for interfacing peripherals. It has two versions, Intel 8255A and Intel 8255A.5. The general description of the device is same. However, their electrical characteristics are different. Only their common functions are discussed here and for further discussion it is referred as Intel 8255. It contains three ports, namely, port A, port B, and port C. Each of the port is 8-bit wide. Port C is further divided into two 4-bit ports, namely, port C_{upper} and port C_{lower}. Each port can be programmed either as input port or as output port. Architecture of Intel 8255 is shown in Fig. 12.24.

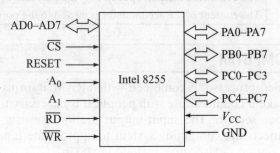

FIGURE 12.24 Pins of Intel 8255.

There are 40 pins in Intel 8255 IC. It operates by a single 5-V supply. AD0 to AD7 (pins 1–8) are connected to data bus. PA0 to PA7 (pins 15–22) are eight pins of port A and PB0 to PB7 (pins 23–30) are eight pins of port B. PC0 to PC3 (pins 31–34) and PC4 to PC7 (pins 35–38) are four pins of port C_{lower} and port C_{upper}, respectively. Other pins are control pins and their functions are as given below.

$\overline{CS}$ *(Chip select):* It is a chip select pin. A low-status signal at this pin enables communication between Intel 8255 and the CPU.

$\overline{RD}$ *(Read):* When the status of signal at this pin goes low, Intel 8255 sends out data or status information to the microprocessor on the data bus. In other words, a low signal allows the microprocessor to read data from the input port of Intel 8255.

$\overline{WR}$ *(Write):* When signal at this pin goes low, the microprocessor sends data or control word to Intel 8255 at output port. The control word is 8-bit information written by the programmer, and it defines the ports as either input or output port.

Pin A_0 and A_1: In conjunction with $\overline{CS}\ \overline{RD}\ \overline{WR}$, pins A_0 and A_1 are used to select input port and control word register. These pins are normally connected to the LSBs of the address bus. The address of ports and control word register of the two version of Intel 8255 are different and are given in Table-12.16.

Table-12.16 Address of ports and control word register.

Port/control word register	Addresses	
	Intel 8255A	Intel 8255A.5
Port A	00	08
Port B	01	09
Port C	02	0A
Control word register	03	0B

RESET   This pin resets Intel 8255.

V_{CC} A 5-V d.c. supply is connected to this pin.

GND   This pin is to be connected to ground.

12.12 MICROCONTROLLER

A microcontroller is a small computer on a single chip. It contains a processor, programmable memory, and input-output devices. The programmable memory may include programmable read only memory (PROM)/erasable programmable read only memory (EPROM) and a typically small amount of RAM. In contrast to the microprocessor in a personal computer and other general-purpose applications, microcontrollers are designed for embedded applications. Because of reduced size and cost, microcontrollers make it economical to digitally control other devices and processes. Microcontrollers are used in automatically controlled products and devices. These devices may be automobile engine control systems, implantable medical devices, remote controls, office machineries, appliances, power tools, and toys.

A self-contained microcontroller with a processor, memory, and peripherals can be used as an embedded system. The majority of microcontrollers in use today are embedded in other systems. These may include automobiles, telephones, appliances, and peripherals for computer system. Typical input-output devices of embedded microcontrollers include switches, relays, LED and LCD displays, and sensors for data such as temperature, humidity, light level, etc. Embedded microcontrollers usually have no keyboard, screen, disks, or other recognized input-output devices of a personal computer. Microcontrollers may lack human interaction devices of any kind.

Microcontrollers must provide real-time response to events in the system they are controlling. In case of occurrence of a certain event, an interrupt signal is sent to the processor to suspend processing the current sequence of instructions and to begin an interrupt source routine. The interrupt source routine performs a required processing based on the source of interrupt before returning to the original sequence of instructions. Possible source of interrupts are device dependent, such as internal timer overflow, completing analogue to digital conversion, and a logic level change on an input such as from a button being pressed. In battery operated devices, an interrupt may give a caution signal when the battery goes low.

Microcontroller programs must fit in the available on-chip programmable memory. An external expandable memory increases the cost of the microcontrollers. A compact machine code program is, therefore, stored in the microcontroller memory. Depending on the device, the programmable memory may be permanent ROM that can only be programmed at the factory. A programmable memory may be field alterable flash or erasable ROM.

Microcontrollers usually contain a number of general-purpose input/output (GPIO) pins. GPIO pins are software configurable to either as an input or as an output. When configured as an input, pins are often used to read sensors or external signals. The pins can drive external devices such as LEDs or motors when they are configured as an output. Many embedded systems are required to read sensors that produce analogue signals. In this case, an analogue-to-digital converter (ADC) is needed to convert an analogue signal into digital signal. Sometimes, a DAC is required with the microcontroller for giving outputs in the form of analogue signals. In addition to ADC and DAC, many embedded microcontrollers include various types of timers as well. One of the most common type of timers is programmable interval timer (PIT). A PIT just counts down from some value to zero. When it reaches zero, it sends an interrupt to the processor, indicating that it has finished its job. This timer is useful for devices such as thermostats, which periodically test the temperature around them to see if they need to turn on the air conditioner, heater, etc. Another timer may be the time processing unit (TPU). This is a sophisticated timer and, in addition to counting, it detects input events, generates output events, and performs other useful operations. A detects input, generates output events, and perform other operations. A dedicated pulse width modulation (PWM) block makes it possible for the processor to control power converters, resistive loads, motors, etc., without using lots of processor resources in tight timer loop. UART block facilitates to receive and transmit data over a serial line with very little load on the processor. Dedicated on-chip hardware also often includes capabilities to communicate with other devices (chips) in digital format such as serial peripheral interface (SPI).

12.12.1 Intel Microcontroller

Intel introduced 8048 family of microcontrollers in 1976 and is popularly known as MCS-48. This family includes 8048, 8049, 8748, 8749, 8035, and 8039. Intel 8049 is 8048 except it has on-chip ROM of double size and higher execution speed. The ROM size is 1K for 8048 and 2K for 8049 and their speeds are 2.5 µs and 5 µs, respectively. There is no other difference between the two chips. Both 8035 and 8039 have no on-chip memory. The difference between the two is that the execution speed of 8035 is same as that of

Table-12.17 Summary of Intel families of microcontroller.

Name	I/O ports	Timer/ counter	External interrupt	ADC	Memory in byte	RAM in byte
8021	3×8 bit	1, 8 bit	0	No	ROM 1024	64
8022	3×8 bit	1, 8 bit	1	Yes	ROM 2048	64
8031	4×8 bit	2, 8 bit	5	No	No	128
8035	3×8 bit	1, 8 bit	1	No	No	64
8039	3×8 bit	1, 8 bit	1	No	No	64
8048	3×8 bit	1, 8 bit	1	No	ROM 1024	64
8049	3×8 bit	1, 8 bit	1	No	ROM 2048	64
8748	3×8 bit	1, 8 bit	1	No	EPROM 1024	64
8749	3×8 bit	1, 8 bit	1	No	EPROM 2048	64
8051	4×8 bit	2, 16 bit	5	No	ROM 4096	128
8751	4×8 bit	2, 16 bit	5	No	EPROM 4096	128
8096	5×8 bit	4, 16 bit	8	No	No	256
8097	5×8 bit	4, 16 bit	8	Yes	No	256

8048 and the execution speed of 8039 equals that of 8049. The 8748 and 8749 are new versions of 8048 and 8049, respectively, except that their ROMs have been replaced by EPROMs. All these chips operate from a single +5 V d.c. supply. However, 8748 and 4849 require additional +25 V supply for EPROM programming.

More powerful microcontrollers of MCS51 series were introduced by Intel in 1980. These have 8-bit CPU, 4K/8K ROM/EPROM, a 128/256-byte RAM, timer/counter, parallel and series input-output ports. Also, these have expandable EPROM and RAM.

The MCS-96 series of microcontrollers were introduced by Intel in 1983. These contain 16-bit CPU, 8K ROM, 252 byte RAM, timer/counter, parallel and series input-output ports, 10-bit ADC, high-speed input-output, pulse width modulated output, watch dog timer. Also, these have expandable EPROM and RAM.

The summary of Intel microcontroller discussed above is given in Table-12.17. All microcontrollers have expandable memory. Intel 8748 is very popular and is used for industrial control. Hence, it is discussed in detail in this section.

Intel 8748

It is a microcomputer on single chip. It consists of a processor, 1024-byte EPROM, 64-byte RAM, and three 8-bit ports. It has an 8-bit accumulator, 12-bit program counter. There is an 8-bit program status word (PSW). Three LSBs of PSW act as stack pointer. Flags C and AC are carry and auxiliary carry flags, respectively. The bit FO is a software flag. The block diagram of Intel 8748 is shown in Fig. 12.25. Intel 8243 is input-output expander, and 8155, 8355, and 8755 are multifunction devices. These can be used with Intel 8048 family of microcontrollers.

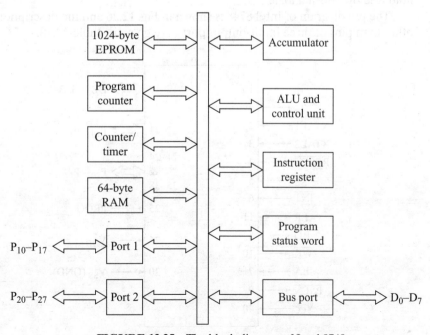

FIGURE 12.25 The block diagram of Intel 8748.

The RAM can be used as either read/write memory or general-purpose registers. Two sets of 8 bytes of RAM are used as general-purpose registers. In each set, 2 bytes, designated as R_0 and R_1, are used as data counter to address the memory and external memory. The stack pointer is maintained in the PSW. 16 bytes of RAM are used as stack. Remainder 16 bytes are used as general-purpose registers. The system memory is divided into two groups, namely, program memory and data memory. 1024 bytes of EPROM are used as program memory and 64 bytes of RAM as data memory. External program and data memories can also be used.

Intel 8748 can operate in a variety of modes, internal execution mode, external execution mode, debug mode, programming mode, and verify mode. In the internal execution mode of operation, program is executed without accessing the external program and data. The processor has to issue additional control signals when external program and/or data memory are accessed. The control signals distinguish whether the processor want to access the external or internal program and/or data memory. The debug mode of operation is used to address the external program memory. In this mode of operation, the processor is disconnected from the internal program memory. In the programming mode of operation, EROM is programmed. In verify mode of operation, the contents of EROM are verified.

Intel 8748 has three 8-bit input-output ports. One of the three ports is a bus port. It is a true bidirectional bus with input and output strobes. Outputs are statically latched, whereas inputs are not latched. Hence, the external logic must hold input data stable at bus port till the data has been read. All 8 pins of the bus must be assigned either as input or as output. Other two ports named port 1 and port 2 are secondary input-output ports. Output parallel data are latch on port 1 or port 2 till data have been read. Inputs are hold free by external logic.

The pin diagram of Intel 8748 is shown in Fig. 12.26 and the descriptions of the pins, other than pins of three input-output ports, are given in Table-12.18.

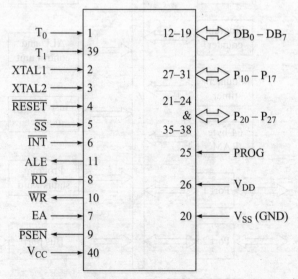

FIGURE 12.26 The pin diagram of Intel 8748.

Table-12.18 Description of pins other than that of input-output ports.

Pin	Description
$\overline{WR}$	Write control for data memory.
$\overline{RD}$	Read control for data memory.
EA	External program memory access.
$\overline{PSEN}$	Read control for external program memory.
$\overline{SS}$	Single step control.
ALE	External clock signal and address latch enable. It becomes high at beginning of every instruction machine cycle. It acts as address latch enable for memory access. It is also used as clock for external circuit.
V_{CC}	+5 V d.c. supply.
V_{SS}	Ground.
V_{DD}	+25-V d.c. supply to program 8748, +5 V d.c. supply for RAM.
PROG	+25-V d.c. supply to program 8748, control output for 4-bit I/O.
T0	Test input, Program/Verify mode select and optional clock output.
T1	Test input, optional event counter input.
XTAL1, XTAL2	External crystal connections.

12.13 REVIEW QUESTIONS

1. What is a digital computer? Name different types of digital computers and discuss briefly its development from a computing machine to the today's microprocessor.
2. What are the main components of a digital computer? Describe each component.
3. With a neat block diagram discuss the operation of a 3-bit arithmetic circuit.
4. Suggest a circuit for binary increment using half adders and explain the operation of the circuit.
5. Design a logic circuit using a MUX for logic operation AND, OR, XOR, and NOT. Explain working of the circuit.
6. What are the different types of shift operations? Design a shifter circuit using MUXs. Explain the operation of the circuit.
7. What are the different types of registers used in a basic computer? What are their functions? Design a common bus system for transferring information of registers using MUX.
8. What do you know about a control unit of a computer? With a block diagram explain the operations of the control unit.
9. What are the different types of memories used with a computer system? How are these memories interfaced with the computer?
10. Write briefly about different input-output devices used in a computer system.
11. Name different types of printers. Explain briefly about each of them.

12. What do you understand by input-output interface? What are its functions?
13. What is a computer instruction? Name and describe about different types of computer instructions.
14. With a flowchart diagram explain how an instruction is executed in a computer.
15. What are different categories of programming? What are their advantages and disadvantages?
16. With a flowchart diagram explain the first pass processing of an assembler.
17. With a flowchart diagram explain the second pass processing of an assembler.
18. What are the main components of a microprocessor? Describe each component.
19. What are the different types of registers employed in a microprocessor? Discuss their functions.
20. What are the different types of instructions used in a microprocessor Intel 8085?
21. Draw a block diagram of a microcomputer. Discuss each block.
22. Discuss different families of microcontroller. Give architecture of Intel 8748. Explain each of its blocks.

12.14 SOLVED PROBLEMS

1. If the common bus system, using MUXs, has 8 registers of 16 bits each, determine (a) How many selection inputs are there in each MUX? (b) What size of MUXs are needed? (c) How many MUXs are there in the bus?

Solution

The common bus system is of 8 lines.

(a) The number of selection inputs is equal to the number of binary digit needed to express maximum decimal number 8. Hence, the number of selection inputs = **3**.

(b) The size of each MUX = **8 × 1**.

(c) The number of MUXs = The number of registers = **8**.

2. Design a 4-bit combination circuit decrementer using four full adders (FA).

Solution

The circuit of the decrementer is shown in Fig. 12.27. The operation is self-explanatory. The circuit adds 2's complement of 0001 (decimal 1) to the number to be decremented. Thus, decimal number 1 is subtracted from the number.

3. The 4-bit arithmetic circuit shown in Fig. 12.4 is embedded in an IC. Show the connections among two such ICs to make an 8-bit arithmetic circuit.

Solution

The connection of two ICs is shown in Fig. 12.28. The pins of the ICs are two sets of 4 bits for connecting the two numbers (A_3, A_2, A_1, A_0 and B_3, B_2, B_1, B_0), S_0 and S_1 are

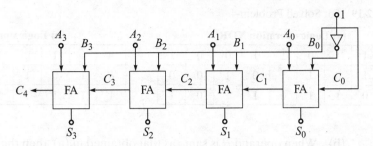

FIGURE 12.27 A 4-bit decrementer circuit using full adders.

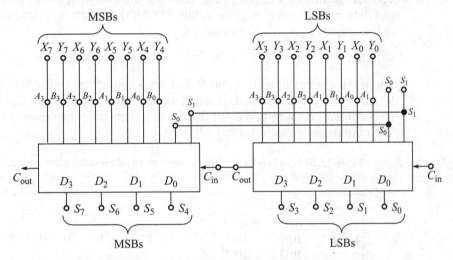

FIGURE 12.28 Connections of two arithmetic operation ICs.

selection inputs and C_{in} and C_{out} are input and output carries, respectively. X_0 through X_7, and Y_0 through Y_7 are 8-bit input numbers. The output number is shown as S_0 through S_7.

4. Register A holds an 8-bit binary number 11011001. Determine the operand B and the logic operation to be performed in order to change the value of A to (a) 01101101 and (b) 11111101.

Solution

(a) The inputs of the gate are A and B. The LSBs of number A (input) is 1 and the output of the gate is 1; hence, the logic gate cannot be a NOT gate. The third LSB of input A is 0, whereas the output required is 1; hence, the logic gate cannot be an AND gate. The gate cannot be an OR gate, as one input is 1 and the output is 0. Hence, the operation to be performed must be XOR. The inputs A and B and required outputs after XOR operation is given in Table-12.19(a).

Table-12.19 For Solved Problem-4.

(a) Logic operation XOR

A	1	1	0	1	1	0	0	1
B	1	0	1	1	0	1	0	0
A'	0	1	1	0	1	1	0	1

(b) Logic operation OR

A	1	1	0	1	1	0	0	1
B	1	0	1	1	0	1	0	0
A'	1	1	1	1	1	1	0	1

(b) When operand B is same as that obtained in (a), then the logic operation can be OR. The result is shown in Table-12.19(b).

5. The adder subtractor circuit of Fig. 11.14 has the following values for input mode M and data input A and B as given in Table-12.20(a). In each case, determine the values of the outputs S_3, S_2, S_1, S_0, and carry C_4.

Solution

For $M = 0$, the circuit adds A and B, and the output gives A plus B, C_4 may be 0 or 1. When $M = 1$, the circuit adds A and 2's complement of B. When C_4 is 1, the output gives A minus B, and when $C_4 = 0$, the 2's complement of the output gives A minus B. The outputs are given in Table-12.20(b).

Table-12.20 (a) Inputs for and (b) outputs of adder-subtractor circuit.

(a)

	M	A	B
a	0	0101	0110
b	0	1010	1011
c	1	1011	1001
d	1	0110	1101
e	1	0101	0011

(b)

	S_3	S_2	S_1	S_0	C_4
a	1	0	1	1	0
b	0	1	0	1	1
c	0	0	1	0	1
d	1	0	0	1	0
e	0	0	1	0	1

6. A register contains the value 01011011. Perform the six operations (shl R, shr R, cil R, cir R, ashl R, ashr R). Each time the shift starts from the value given above.

Solution

The shifted values are given in Table-12.21.

Table-12.21 For Solved Problem-6.

Initial value	After shl R	After shr R	After cil R	After cir R	After ashl R	After ashr R
01011011	10110110	00101101	10110110	10101101	001101101 overflow	00101101

7. A computer uses a memory unit with 256K words. The word width is of 32 bits. A binary instruction code is stored in one word of memory. The instruction has four

parts: an indirect bit, an opcode, a register code part to specify one of 64 registers, and an address part.

(a) Determine the number of bits specified for the opcode, register code, and address part.

(b) Draw the instruction word format and indicate the number of bits in each part.

Solution

(a) Beside the MSB (31) for the indirect bit, the bits specified are as follows:

(1) For opcode—7 bits (24 through 30).

(2) For register code part—The bits required to address 64 registers are 6 (18 through 23).

(3) For address part—The bits required to address 256K = 256 × 1024 words of memory locations are 18 (0 through 17).

(b) The instruction word format is shown in Fig. 12.29.

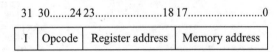

FIGURE 12.29 Instruction word format.

8. A computer uses a memory of 65536 words. The word is 8 bit wide. The registers used with the computer are PC, AR, TR (each of 16 bits) and AC, DR, IR (each of 8 bits). An MRI consists of three words: an 8-bit opcode (one word) and 16-bit address (in next two words). All operands are of 8 bits. There is no indirect bit.

(a) Draw the block diagram showing the memory and registers of the computer as in Fig. 12.7.

(b) Draw the block diagram showing the placement of a three-word instruction and the corresponding 8-bit operand in the memory.

Solution

(a) The block diagram showing the memory and registers is shown in Fig. 12.30.

(b) The block diagram showing the memory locations of instruction and operand in the memory is shown in Fig. 12.31.

9. Consider the instruction formats of the basic computer for which the list of instructions given in Table-12.5. For each of the following 16-bit instructions, give the equivalent 4-bit hexadecimal code and explain in own words what it is that the instruction is going to perform.

1. 0001 0000 0010 0100
2. 1011 0001 0010 0100
3. 0111 0000 0010 0000

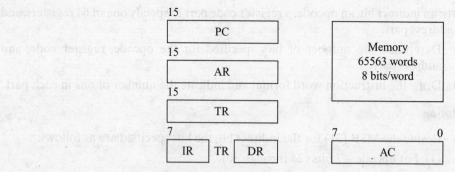

FIGURE 12.30 The memory and registers of the computer.

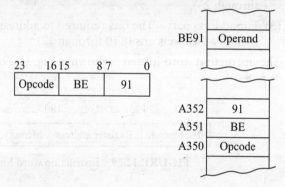

FIGURE 12.31 Locations of instruction and operand.

Solution

The equivalent 4-bit hexadecimal codes and the operation performed by the instructions are given in Table-12.22.

Table-12.22 Equivalent 4-bit hexadecimal codes and operation.

	Hexadecimal code	Operation
1	1024	It is a direct MRI. The address of operand stored in the memory is 024 (Hex). The symbol of opcode is ADD. It adds the operand in the memory to the content of the accumulator (AC). The result is stored in the AC.
2	B124	It is an indirect MRI. The symbol of opcode is STA. It stores the content of the AC to the memory location 124 (Hex).
3	7020	It is a register-reference instruction. The symbol of opcode is INC. It increments the content of the AC by 1.

10. A line of code in an assembly language program is as follows:

   STA RES

   (a) Show that four memory words are required to store the line of code and give their binary content.

(b) Show that one memory word stores the binary translated code and give its binary content.

Solution

The memory location of the above line of assembly program is given in Table-12.23. (a) The table shows that four memory locations are needed to store the line of code. (b) The binary code of each memory word is given in the table.

Table-12.23 Memory storage of the line of code FIR, DEC 73.

Memory location	Symbol	Hexadecimal code	Binary representation
1	S T	53 54	0101 0011 0101 0100
2	A space	41 20	0100 0001 0010 0000
3	R E	52 45	0101 0010 0100 0101
4	S CR	53 0D	0101 0011 0000 1101

11. A program is stored in the memory unit in hexadecimal code as given below. Show the contents of the AC, PC, and IR in hexadecimal, at the end of execution of each instruction.

    Location   010 011 012 013 014 015 016 017
    Instruction 7800 1016 4014 7001 0017 4013 C145 93C6

Solution

The contents of the above-mentioned registers are given in Table-12.24.

Table-12.24 Contents of AC, PC, and IR after execution of each instruction.

	AC	PC	IR	Comment
0	Initial value	010	7800	It clears the AC.
1	0000	011	1016	The operand at 016 is added to AC.
2	C145	012	4014	Control moves to location 014.
3	8144	014	0017	ANDs AC with operand at 017.
4	8144	015	4013	Control moves to location 013.
5	8144	013	7001	Program halt.

12. What happens during the first pass of the assembler (Fig. 12.16) if the line of code of a pseudoinstruction ORG or END has a label? Modify the flowchart to include an error message if this occurs.

Solution

It is wrong to have a label with pseudoinstruction ORG or END. The flowchart is shown in Fig. 12.32.

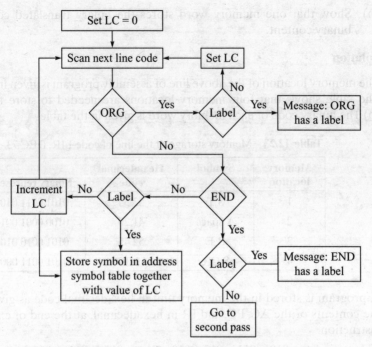

FIGURE 12.32 The flowchart of first pass of assembler.

13. Write an assembly language program to subtract a number from another number.

Solution

The assembly language program to subtract a number from another number is listed below:

	ORG 101	/The program starts from location 101
	LDA SEC	/Load second number to AC
	CMA	/Complement AC
	INC	/Increment AC
	ADD FIR	/Add first number to AC
	STA THR	/Store difference at THR
	HLT	/Halt computer
FIR,	DEC 75	/First number
SEC,	DEC 54	/Second number
THR,	HEX 0	/Difference number
	END	/End of assembly program

14. Translate the assembly program of Solved Problem 13 into hexadecimal code.

Solution

The translated program is given in Table-12.25.

Table-12.25 Translated program of Solved Problem 13.

Hexadecimal code		Assembly program
Location	Content	
		ORG 101
101	2108	LDA SEC
102	7200	CMA
103	7020	INC
104	1107	ADD FIR
105	3109	STA THR
106	7001	HLT
107	004B	FIR, DEC 75
108	0036	SEC, DEC 54
109	0000	THR HEX 0
		END

15. Write an assembly program to output one character.

Solution

Before writing an assembly program for output, it is helpful to discuss how an input terminal sends information to computer and an output terminal receives information from computer. The input-output configuration is shown in Fig. 12.33.

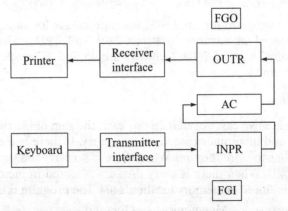

FIGURE 12.33 The block diagram of input-output configuration.

As shown in the block diagram, the AC receives information from the keyboard through a transmitter interface and input register (INPR). There is an input flag (flip flop) (FGI). When information is stored in INPR, FGI goes 1 from 0. INPR and FGI are needed as there is difference of speed between input-output devices and computer. The computer receives information when FGI is 1. Similarly, the printer receives information through receiver interface and output register (OUTR). The output flag (flip flop) (FGO) goes 1 to indicate readiness to receive information from the computer. Then computer sends the information.

The assembly program to output one character (say A) is listed as follows:

	ORG 100	
	LDA CHR	/Load character into AC
CHE,	SKO	/Check output flag
	BUN CHE	/FGO = 0, branch to check again
	OUT	/FGO = 1, output character
	HLT	/Halt computer
CHR	HEX 0041	/Character is 'A'

16. Write a program for Intel 8085 microprocessor to find 2's complement of an 8-bit number. The decimal number is 49.

Solution

The equivalent binary is 0011 0001 and its hexadecimal is 31. The number is stored in the memory location 2401 (Hex) and the result is stored in 2402 (Hex). The program is listed as follows:

Address	Mnemonics	Operand	Comment
2001	LDA	2401	Data is transfer from memory to A
2002	CMA		Content of A is complemented
2003	INR	A	Content of A is incremented
2004	STA	2402	Stores 2's complement in 2402

17. Write a program for Intel 8085 microprocessor for summing two 8-bit binary numbers stored in memory locations 2401 and 2402 and store the sum in memory locations 2403 and 2404. Numbers are 45 and 2C, respectively. All numbers are in hexadecimal.

Solution

There are two cases of sum. In one case, the sum of the two numbers does not have a carry, and in another case, there is a carry. If there is a carry, it needs two memory locations to store the sum. When there is no carry the sum is stored in memory location 2403. When there is carry the sum is stored in memory location 2403, and the carry is stored in memory location 2404. The program is listed below:

Address	Mnemonics	Operand	Comment
2001	LXI	H, 2401	1st number addr is stored in HL pair
2002	MVI	C, 00	Register C is set to 00
2003	MOV	A, M	First number is moved to A
2004	INX	H	Increment content of HL pair
2005	ADD	M	Adds two numbers
2006	JNC	NC	Jumps to NC (label) in which there is no carry

2007	INR	C	If carry, C is incremented
2008 NC	STA	2403	The sum is stored in 2403
2009	MOV	A, C	The carry is stored in A
200A	STA	2404	The carry is stored in 2404

12.15 EXERCISES

1. In constructing a common bus system of 16 registers with 32 bits for a computer using MUXs, (a) how many selection inputs are there in each MUX? (b) what size of MUXs is needed? (c) how many MUXs are there in the bus?

2. Design a digital circuit that performs the four logic operations of XOR, XNOR NOR, and NAND. Use two selection variables.
   [**Hint:** see Fig. 12.5]

3. Register A holds an 8-bit binary number 11101011. Find the operand B and the arithmetic operation to be performed in order to change the value of A to (i) 01101101 and (ii) 11111001.

4. The adder subtractor circuit of Fig. 11.14 has the following values for input mode M and data input A and B as given in Table-12.26. In each case, determine the values of the outputs S_3, S_2, S_1, S_0, and C_4.

Table-12.26 Inputs for adder-subtractor circuit of Fig. 11.14.

	M	A	B
a	0	1101	1110
b	0	1110	1010
c	1	1011	0101
d	1	1110	1001
e	1	0101	0011

5. What is the value of the output H of the shifter circuit of Fig. 12.6 if A $(A_0 A_1 A_2 A_3)$ is 1011, IR = 1 and IL = 0, if $S = 0$ and $S = 1$.

6. An instruction at address 021 (Hex) in the basic computer has $I = 0$, and an operation code of the AND instruction. The address part is 083 (Hex). The memory word at address 083 contains the operand A8F2 and the content of AC is B936. Determine the contents of registers PC, AR, DR, AC, and IR after the execution of the instruction.

7. For the memory registers shown in Fig. 12.7, draw a block diagram showing the placement of instruction with $I = 0$ in the memory.

8. An 8-bit register contains the binary number 10101100. What is the register value after an arithmetic shift right? Starting from the initial value 10101100, determine the register value after an arithmetic shift left, and state whether there is an overflow.

Table-12.27 Instruction for question 9.

Location	Instruction
101	BSA 104
102	CMA
103	HLT
104	0000
105	CLA
106	INC
107	BUN 103

9. The program given in Table-12.27 is stored in the memory unit of the computer. The computer executes the instructions starting from address 101(Hex). What are the contents of AC and the memory word at address 104 when computer halts?

10. A line of code in an assembly language program is as follows:

    PL3, LDA A5B I

    (a) Show that seven memory words are required to store the line of code and give their binary content.

    (b) Show that one memory word stores the binary-translated code and give its binary content.

11. Modify the flowchart of Fig. 12.16 to include an error message when a symbolic address is not defined by a label.

12. Write an assembly program to input one character.

13. Write an assembly program to add two numbers 23 and 5A stored at memory locations 2501 and 2502 and store the result in location 2503.

14. Write a program for microprocessor Intel 8085 to transfer the content of the memory location 2401 to register C. The content of memory location 2401 is 14. All numbers are in hexadecimal.

15. Write a program for microprocessor Intel 8085 to subtract a number B from another number A.

13
COMMUNICATION SYSTEMS

Outline

13.1 Introduction   690
13.2 Modulation and Demodulation   697
13.3 Transmitters   703
13.4 Receivers   705
13.5 Digital Communication   707
13.6 Quantization   711
13.7 Digital Modulation   715
13.8 Multiplexing and Multiple Accessing   719
13.9 Optical Fibre and Communication Systems   722
13.10 Review Questions   732
13.11 Solved Problems   733
13.12 Exercises   738

13.1 INTRODUCTION

In a broad sense, communication means to establish a link between two points. Both points may be situated on the earth or one on the earth and another in the space or both in the space. The modes of communication may be roads, rails, aeroplanes, telegraphs, telephones, radios, and televisions. However, our concern is the electrical communication. In electrical communication, electrical signals establish a link between the two points. The first successful transmission of message telegraphically was done by Samuel F.B. Morse in 1938. Today, electrical communication is an integral part in every sphere of our lives. In modern age, communication plays an important role by providing exact locations of ships, aircrafts, rockets, and satellites in space. Moreover, these days the communication system helps in linking computers situated at distant locations.

The field of communication has seen a tremendous development in the methods and ways of transmitting information from one place to another during a century. The high-speed digital communication has gradually replaced analogue communication during the past several years. Starting from the age of telegraphy it has entered an age of optical communication. The tremendous development, however, has become possible only because of the development in the area of electronic devices, circuits, and systems. The year-wise development of the communication system is listed in Table 13.1.

In the communication process, information is transmitted from one point called source to the other point called destination. In the electrical communication system, information is converted in the electrical form by a transducer. The electrical form of the information is called a signal in communication system. The signals are classified on the following bases.

Periodic and nonperiodic signals. The periodic signal is the one that repeats itself after every time interval T, called the time period. Mathematically, a periodic signal $S_p(t)$ is given by

$$S_p(t) = S_p(t + T)$$

The signal that does not satisfy the above condition is the nonperiodic or aperiodic signal.

Analogue and digital signals. The source or information signal can be analogue or digital. Common examples of analogue signals are audio and video signals. Digital signals are derived from analogue signals or are data, for example, alphanumeric characters. An analogue signal is a signal defined over a continuous range of time. Its amplitude assumes a continuous range of values. An analogue signal is shown in Fig. 13.1(a). A digital signal is a discrete-time signal with quantized amplitude. In quantization a variable is set to a distinct value. The resulting distinct values are called quantized values. Quantized values of the signal in Fig. 13.1(a) are shown in Fig. 13.1(b). The digital form of the signal of Fig. 13.1(a) is shown in Fig. 13.1(c).

Deterministic and random (stochastic) signals. A signal completely specified by a function of time is called a deterministic signal. There is no uncertainty regarding its value at any time. On the contrary, a random signal has a certain amount of uncertainty associated with its value at any time. A random signal is described in terms of its statistical properties. It cannot be predicted precisely beforehand.

Sec. 13.1 / Introduction

Table-13.1 Developments of modern communication systems.

Year	Developments
1838	Birth of telegraphy
1876	Birth of telephony
1887	Wireless telegraphy
1904	Electronic communication (radio and telephone)
1923	Birth of television
1931	Teletypewriter
1936	FM radio
1937	Pulse code modulation
1938	Radar and microwave system
1950	Time division multiplexing in telephony
1955	Satellite communication
1956	Transoceanic telephone cable (36 voice channel)
1958	Long-distance data transmission
1962	Satellite communication with Telstar I
1962	Birth of high-speed digital communication
1964	Electronic telephone switching system
1966	Cable TV systems, picture phone
1977	First-generation telephone system
1979	Intercity optical link (with repeaters at 10 km)
1988	Under-sea fibre-optic link

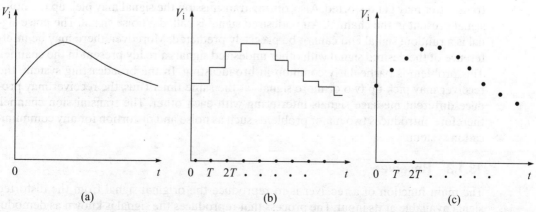

FIGURE 13.1 A signal (a) continuous, (b) quantized, and (c) digital form.

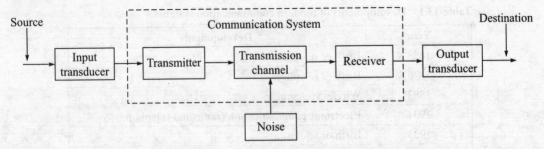

FIGURE 13.2 A typical communication system.

The elements of a communication system can be represented by a block diagram shown in Fig. 13.2. Each block represents a functional element of the communication system. The essential elements of a communication system are the *transmitter*, *transmission channel*, and *receiver*. In addition, an input transducer is required to convert the information (or message) signal into an electrical signal. The electrical signal is the input to the transmitter. The output transducer is used to convert the received electrical signal into the desired message.

13.1.1 Transmitter

The purpose of a transmitter is to modify the electrical signal of the message into a suitable form that can be transmitted through the transmission channel. This can be achieved through a process known as modulation. The modulation processes are discussed in Section 13.2.

13.1.2 Transmission Channel

It is basically a medium that electrically connects the transmitter to the receiver. The medium may be a pair of wires, coaxial cable, free space, or optical fibre. The properties of the channel can strongly influence the performance of the communication system. Because of nonlinearities and imperfection in the channel, the signal launched by the transmitter may get distorted. Also, during transmission the signal may pick up the noise signal present in the channel. An undesired signal is called a noise signal. The noise signal is a random signal and cannot be precisely predicted. Moreover, there may be interference of the desired signal with other undesired signals already present in the channel. This problem is particularly common in broadcasting. In the broadcasting system, the receiver may pick up two or more signals at the same time. Thus, the receiver may produce different message signals interfering with each other. The transmission channel, therefore, introduces two major problems such as noise and distortion for any communication system.

13.1.3 Receiver

The main function of a receiver is to reproduce the original signal from the distorted signal available at its input. The process that reproduces the signal is known as demodulation. It is the reverse of modulation. The process of demodulation is discussed in Section 13.2. Whatever be the modulation-demodulation scheme, the receiver cannot exactly reproduce the original signal because of the degradation of the signal caused by the transmission channel. However, some modulation-demodulation schemes offer better performance compared to others.

Need for Modulation

Electrical communication, with which we are concerned, is transmission of message through the space. However, the message signal being a low-frequency signal cannot be directly transmitted through the space, whereas radio frequency (RF) (above 20 kHz) signals can be transmitted through the space both efficiently and economically. In communication through space (which is called radio communication), the message is transmitted in the form of electromagnetic waves from the transmitting antenna. A radio communication system is shown in Fig. 13.3. There are two antennas, one is the transmitting antenna and the other is the receiving antenna. However, for efficient radiation from an antenna, it is necessary that the size of the antenna should be of the order of one-tenth of the wavelength (λ) of the signal to be radiated. Unfortunately, many signals, such as audio frequency signals, have frequency components down to 100 Hz or even less. Efficient radiation, say, at frequency of 100 Hz, therefore, requires an antenna of length of 300 km. This size is quite impractical. The modulation translates this low-frequency signal to high-frequency range and this translated signal can, subsequently, be radiated efficiently from a practically possible size of antenna.

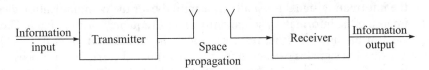

FIGURE 13.3 A radio communication system.

Modulation makes it possible to translate the signals of similar frequency ranges to different regions in the frequency spectrum. This allows a user to tune his radio or television set to a particular broadcasting station. In the absence of a modulation scheme, the signals from various broadcasting stations would be a cluster of interfering signals. Sometimes, it is necessary to transmit a number of signals through the same channel simultaneously. Modulation schemes make it possible to multiplex a number of signals at the same time through the same channel without any interference among the signals. The multiplexing scheme is utilized in long-distance telephony, data telemetry, etc. Modulation schemes can suppress the noise and interference to some extent. As discussed earlier, the noise and interference are two major limitations of any communication system.

Modulation schemes alter some characteristics of a high-frequency signal called carrier in accordance with the low-frequency message signal called the modulating signal. The characteristics of the carrier may be its amplitude, frequency, phase angle, etc. The frequency of the carrier is, sometimes, used to categorize the communication system. The various ranges of frequencies, in radio spectrum, used for electrical communication are given in Table 13.2.

Table-13.2 An RF spectrum.

Frequency ranges	Designation	Abbreviation
30–300 Hz	Extremely low frequency	ELF
300–3000 Hz	Voice frequency	VF
3–30 kHz	Very low frequency	VLF
30–300 kHz	Low frequency	LF
0.3–3 MHz	Medium frequency	MF
3–30 MHz	High frequency	HF
30–300 MHz	Very high frequency	VHF
0.3–3 GHz	Ultra high frequency	UHF
3–30 GHz	Super high frequency	SHF
30–300 GHz	Extra high frequency	EHF

The RF spectrum given in Table 13.2 is a portion of the electromagnetic spectrum (shown in Fig. 13.4). In recent years, an increasingly larger portion of the electromagnetic spectrum is being used in the communication system. This has happened because the information signal is usually transmitted over the communication channel by superimposing the information signal onto a high-frequency carrier signal. The carrier signals used in modulation schemes are usually sinusoidal in nature. At the destination, the information signal is extracted from the modulated carrier by a process called demodulation. The amount of information that can be transmitted is directly related to the frequency of the carrier. An increase in frequency of the carrier increases the available transmission bandwidth. Hence, it provides a larger information capacity. For increasing information capacity, carriers of higher frequencies were progressively employed. This has led to the development of various means of communication such as television, radar, microwave, and optical links.

Figure 13.4 shows the portion of the electromagnetic spectrum used for various communication media. The communication media used in various portions of the spectrum include twin wire lines, coaxial cables, waveguides, and optical fibres. Various communication systems operating in different ranges of the spectrum include telephone, AM and FM broadcasting, television, satellite communication, etc. These applications cover a range of the spectrum ranging from 300 Hz to 90 GHz. The optical communication operates is the region of the electromagnetic spectrum that spreads from 100 μm (far infrared) to 10 nm (ultraviolet) region of the wavelength. This portion of the electromagnetic spectrum is popularly known as optical region. The optical communication can be of two types: one is through optical fibre (guided) and another through atmospheric channel (unguided).

Fundamental limitations of analogue communication system are noise and bandwidth. The noise is undesired signals that tend to disturb the transmission and processing of the desired signals in communication systems. The noise signals cannot be controlled totally. However, it is possible to minimize the effect of the noise on the desired signals.

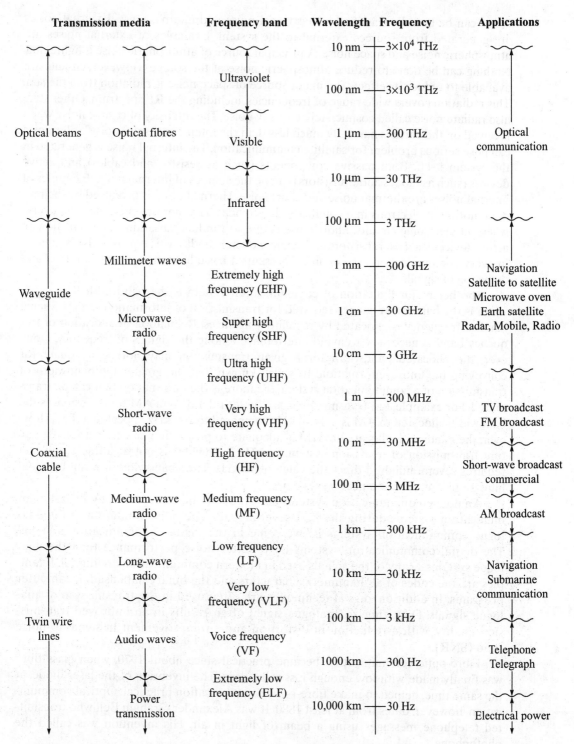

FIGURE 13.4 The electromagnetic spectrum and various communication systems operating in different regions.

Noise can be broadly classified into external noise and internal noise. External noises are generated from sources external to the system. Examples of external noises are atmospheric noise and space noise. A principal source of atmospheric noise is lightening. Nothing can be done to reduce atmospheric noise at the source. However, circuits are available to reduce its effect. The major source of space noise is radiation from the sun. This radiation covers wide range of frequencies, including the RF spectrum. Other stars also radiate noise called cosmic, stellar, or sky noise. The intensity of cosmic noise when received on the earth is naturally much less than the intensity of solar noise. Solar noise can pose serious problem for satellite communication. The internal noise is generated in the system itself. Both passive components (such as resistor and cables) and active devices (such as diodes and transistors) can be the sources of internal noise. Examples of internal noises are thermal noise and short noise. Thermal noise is produced by the random motion of electrons in a conductor due to heat. The thermal noise is found everywhere in electronic circuits. Short noise is due to random variation in current flow in active devices such as transistors and semiconductor diodes. This type of noise is present in every communication system and represents a basic limitation on transmission and detection of signals.

Another major limitation of communication system is the bandwidth. The bandwidth is the band of frequency required for transmission of information signals. Such a band of frequency is allocated by regulatory agencies. Regulation in allocation of frequency band is necessary to avoid interference among the signals of frequency. However, the allocated bandwidth for a given transmission may not be adequate for conveying the entire information. It is well known that the greater the bandwidth of transmission of a communication system, the more is the information that can be transmitted. For example, FM transmission has a better fidelity than AM transmission, as the bandwidth allocated to AM is only 10 kHz compared to 200 kHz allocated to FM. However, the bandwidth allocated to AM is adequate to please human ear. But for the picture transmission of television system, 10-kHz bandwidth is not at all acceptable. It requires several hundred times the value of 10 kHz. Thus, bandwidth is a fundamental limitation of any communication system.

Analogue communication systems are gradually being replaced by digital communication systems during the past several years. The noise limitation in the analogue communication systems is overcome in the digital communication systems. The digital communication systems have better noise performance than the analogue systems. Many of the signals used in modern communication are digital. Examples are the codes of alphanumeric characters and the binary data used in computer programs. In addition, digital techniques are often used in the transmission of analogue signals. Digitizing an analogue signal often results in an improved transmission quality, with a reduction in distortion and an improvement in signal-to-noise ratio (SNR).

Fibre-optic communication became practical since about 1970, when glass fibre was finally made with low enough loss to be used. The invention of the laser diode, at the same time, helped to make fibre-optic communication practical. Optical communication, however, dates from at least 1880. It was Alexander Graham Bell who transmitted telephone messages using a beam of light in air. His invention was called the photophone.

13.2 MODULATION AND DEMODULATION

The audio frequency (20 Hz to 20 kHz) cannot be transmitted directly through the space, whereas RF (above 20 kHz) signals can be transmitted through the space, both efficiently and economically. Therefore, a common method of transmitting audio or RF signals through space is to transmit an RF signal, called a *carrier*, which is modified in some way by the information signal. This modification of the carrier signal is called *modulation*. After the modulated signal is received, the original information is extracted from the carrier signal by a process known as *demodulation*.

A sinusoidal carrier signal is described as

$$v_c(t) = V_c \sin(\omega_c + \theta) \tag{13.1}$$

where

V_c = amplitude of carrier signal.

ω_c = angular frequency.

θ = relative phase.

There are three parameters in the carrier signal that can be modified by the process of modulation. These parameters are amplitude, frequency, and relative phase. On the basis of variation in these three parameters, the modulation can be classified as

1. Amplitude modulation.
2. Frequency modulation.
3. Phase modulation.

13.2.1 Amplitude Modulation

In amplitude modulation, the amplitude of the carrier signal is altered in response to the information signal. This type of modulation is simple, but still useful. Let us consider a carrier signal of high-frequency sinusoid given by

$$v_c(t) = V_c \cos \omega_c t \tag{13.2}$$

For the sake of mathematical convenience, suppose that the information signal is also a sinusoid given by

$$v_s(t) = V_s \cos \omega_s t \tag{13.3}$$

The carrier and modulating signals are shown in Fig. 13.5. The carrier frequency ω_c must be much higher than the information (or modulating) signal frequency ω_s, that is, $\omega_c \gg \omega_s$. Now, let the amplitude of the modulated signal be

$$V_m = V_c + k V_s \cos \omega_s t \tag{13.4}$$

where k is a positive constant such that $V_c \geq kV_s$. Hence, amplitude-modulated signal can be expressed as

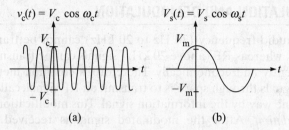

FIGURE 13.5 (a) Carrier signal and (b) information signal.

$$v_m(t) = (V_c + kV_s \cos \omega_s t) \cos \omega_c t$$
$$= V_c \left(1 + \frac{kV_s}{V_c} \cos \omega_s t\right) \cos \omega_c t$$
$$= V_c (1 + m \cos \omega_s t) \cos \omega_c t \tag{13.5}$$

where $m = kV_s/V_c$ and is called *modulation index*. Since $V_c \geq kV_s$, $m \leq 1$. When $m = 1$, the modulation is referred as 100%. For the 50% modulation, the modulated signal is shown in Fig. 13.6(a). The upper boundary for the amplitude modulation is described by $+V_c (1 + m \cos \omega_s t)$ and is referred to as *upper envelope*. Similarly, $-V_c (1 + m \cos \omega_s t)$ is known as *lower envelope*.

The modulated signal (Eq. 13.5) can also be expressed as

$$v_m(t) = V_c \cos \omega_c t + V_c m \cos \omega_c t \cos \omega_s t$$
$$= V_c \cos \omega_c + \frac{mV_c}{2}[\cos(\omega_c + \omega_s) + \cos(\omega_c - \omega_s)] \tag{13.6}$$

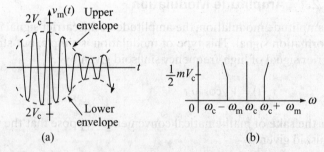

FIGURE 13.6 (a) An amplitude-modulated signal and (b) its frequency spectrum.

Hence, from Eq. (13.6), it is clear that the amplitude-modulated signal is actually a sum of three sinusoidal signals. One signal is just the carrier, $v_c(t) = V_c \cos \omega_c t$. The other two signals have amplitudes equal to $(mV_c/2)$, one having frequency of $\omega_c + \omega_s$ and another having frequency of $\omega_c - \omega_s$. A amplitude versus frequency plot of $v_m(t)$ is shown in Fig. 13.6(b). The plot is referred to as *frequency spectrum* or simply *spectrum*. The sinusoid whose frequency is $\omega_c + \omega_s$ is called upper sideband and the sinusoid whose frequency is $\omega_c - \omega_s$ is known as lower sideband. The bandwidth of the amplitude-modulated signal is equal to twice the bandwidth of the information signal.

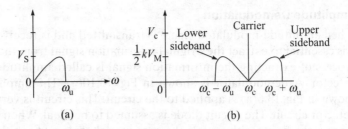

FIGURE 13.7 Spectra of (a) nonsinusoidal modulating and (b) modulated signal.

In above discussion, modulating signal considered is a sinusoid, that is, $v_s(t) = V_s \cos \omega_s t$. Now, suppose that the modulating signal is nonsinusoidal having a spectrum from 0 to frequency ω_u as shown in Fig. 13.7(a). The maximum amplitude of the frequency component between 0 and ω_u is V_s. The spectrum of the amplitude-modulated signal by this modulating signal is shown in Fig. 13.7(b).

A transistor amplitude modulator is shown in Fig. 13.8. This circuit is primarily an amplifier with relatively high-frequency carrier $v_c(t)$ as the input and the gain of the amplifier depends on relatively low-frequency modulating signal $v_s(t)$. The coupling capacitors C_1 and C_2 and the bypass capacitor C_E are chosen so that they behave as small impedances (usually zero) for $v_c(t)$ and as large impedances (ideally infinite) for $v_s(t)$. Practically ω_c should be at least 100 times ω_s. The modulating signal in series with the emitter varies with the effective value of the emitter resistance r_e; hence, the gain of the amplifier as $v_s(t)$ varies. The result is an output voltage $v_m(t)$, which is a sinusoid with carrier frequency ω_c and its amplitude varying with $v_s(t)$. Thus, the output is an amplitude-modulated signal.

To eliminate the problem due to coupling and bypass capacitors, a d.c. amplifier modulator, as shown in Fig. 13.9, can be used. Again, the gain of the differential pair varies according to the modulating signal $v_s(t)$.

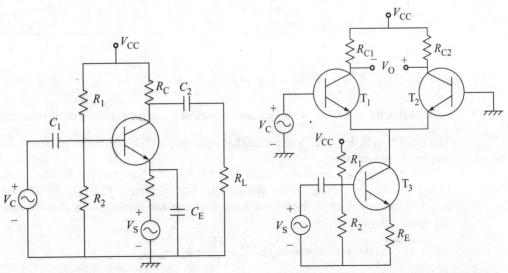

FIGURE 13.8 A transistor modulator.   **FIGURE 13.9** A differential modulator.

Amplitude Demodulation

When amplitude-modulated signal is transmitted and is received at the receiving end, it is essential to extract the original information signal from the modulated signal. The process of extracting the information signal is called amplitude demodulation or AM detector. A demodulator is shown in Fig. 13.10(a). The amplitude-modulated signal shown in Fig. 13.6(a) is applied to the circuit. This circuit is very well known as a peak detector circuit. The circuit diode is assumed to be ideal. When the applied signal $v_m(t)$ reaches the first positive peak, the capacitor charges up to this value. During the interval between the first and second positive peaks, the capacitor discharges with a time constant of $\tau = RC$. The time constant τ should be large compared to time period T_c of the carrier signal. In other words, $RC \gg T_c = 1/f_c = 2\pi/\omega_c$, f_c being the frequency of the carrier signal. However, τ should not be too large, otherwise, the capacitor voltage will fail to follow the successive peak of the amplitude modulated signal. In particular, the circuit time constant τ should be small compared to the time period T_s of the information signal. In other words, $RC \ll T_s = 1/f_s = 2\pi/\omega_s$, f_s being the frequency of the information signal. In general, when the information signal $v_s(t)$ is nonsinusoidal, the time constant of the circuit (RC) should be small compared to the period corresponding to highest frequency component. If the highest frequency is f_s^h, then $RC \ll T_s = 1/f_s^h = 2\pi/\omega_s^h$ So, the time constant of the circuit of Fig. 13.10(a) must satisfy the following inequalities:

$$\frac{2\pi}{\omega_c} = \frac{1}{f_c} \ll RC \ll \frac{1}{f_s^h} = \frac{2\pi}{\omega_s^h} \tag{13.7}$$

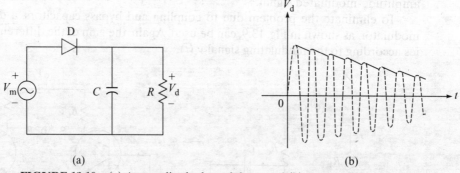

FIGURE 13.10 (a) An amplitude demodulator and (b) output of the demodulator.

The condition in Eq. (13.7) requires that $\omega_c \gg \omega_s$, which was an original condition for amplitude modulation.

When time constant of the circuit of Fig. 6.42(a) is chosen as per the condition of Eq. (6.67), the output voltage of the circuit $v_o(t)$ is as shown in Fig. 6.42(b). This output is equal to the upper envelope of the applied amplitude-modulated signal. The output of the demodulator is given by

$$v_o(t) = V_c(1 + m\cos\omega_s t) = V_c + \frac{kV_s}{V_c} \times V_c \cos\omega_s t$$
$$= V_c + kV_s \cos\omega_s t \tag{13.8}$$

Thus, the output signal $v_o(t)$ contains the original information signal $V_s \cos \omega_s t$ and a d.c. component of magnitude V_c. To filter out the d.c. component, $v_o(t)$ is passed through a simple RC high-pass filter.

13.2.2 Frequency Modulation

In frequency modulation, carrier frequency is varied in response to the information signal without affecting the amplitude and relative phase angle of the carrier signal. The carrier, modulating, and modulated waves are shown in Fig. 13.11(a), (b), and (c), respectively. The frequency-modulated wave is expressed as

$$V_{fm} = V_c \sin (\omega_c t + M_I \sin \omega_s t) \tag{13.9}$$

where M_I is the modulation index.

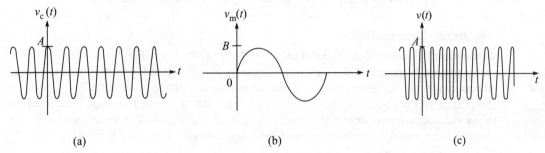

FIGURE 13.11 (a) Carrier, (b) modulating, and (c) modulated waves.

The frequency components of the frequency-modulated signal extend to infinity on both sides of the carrier frequency. But, the amplitudes of the components fall off rapidly outside a certain range as shown in Fig. 13.12. When frequency outside a certain bandwidth is neglected, the signal after demodulation is distorted. For selecting a bandwidth, Carson's rule may be followed. According to Carson's rule, bandwidth is selected as follows:

$$\text{BW} = 2(D + f_h) \tag{13.10}$$

where D is the maximum deviation in the carrier frequency from its unmodulated frequency and f_h is the highest frequency in the information signal.

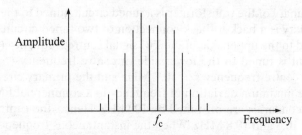

FIGURE 13.12 Bandwidth of modulated signal.

The frequency modulation can be produced by varying capacitance or inductance of the tank circuit of an oscillator according to the information signal. A schematic diagram of the frequency modulator is shown in Fig. 13.13. Frequency modulation requires greater bandwidth than amplitude modulation.

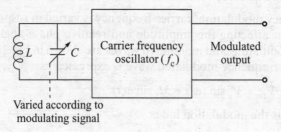

FIGURE 13.13 A schematic diagram of a frequency modulator.

Frequency Demodulation

A simplest frequency demodulator or FM detector is shown in Fig. 13.14. This FM detector is known as *balanced slope detector*. It uses a parallel-tuned circuit. If its input signal frequency is varying over a range within the bandwidth of the tuned circuit response, the amplitude of the output of the tuned circuit varies according to the frequency deviation from the centre frequency in the input signal. If the output signal is rectified and filtered, it gives a d.c. voltage representative of the instantaneous frequency deviation.

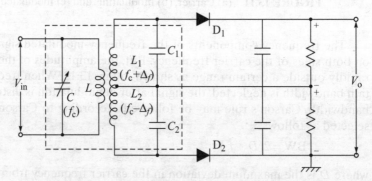

FIGURE 13.14 A circuit of a balanced slope detector.

The primary of the transformer is a tuned circuit, tuned to the central frequency (f_c). The secondary is a back-to-back connection of two tuned circuits. The upper-tuned circuit is tuned to the upper side of the FM signal centre frequency $(f_c + \Delta f)$ and the lower-tuned circuit is tuned to the lower side of centre frequency $(f_c - \Delta f)$. In commercial circuit, FM centre frequency $f_c = 10.7$ MHz and the primary circuit is tuned to this frequency. The minimum deviation of frequency in a commercial FM broadcast is 75 kHz. The two secondaries are off tuned by 100 kHz. Hence, the centre frequencies of these circuits are 10.6 and 10.8 MHz. When the instantaneous frequency is $f_c = 10.7$ MHz, the two tuned circuits produce same amplitude outputs. Thus, they cancel each other when

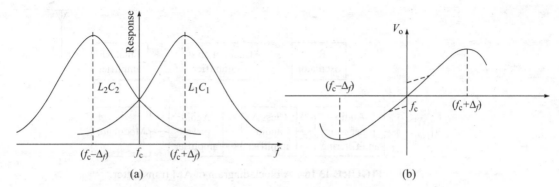

FIGURE 13.15 (a) Responses of secondaries and (b) output of the detector.

combined to produce a zero output. When the instantaneous frequency is $(f_c + \Delta f)$ = 10.8 MHz, the output is positive maximum. The output is negative maximum when the instantaneous frequency is $(f_c - \Delta f)$ = 10.6 MHz. For other frequencies between 10.6 and 10.8 MHz, the output amplitude lies between negative and positive maximum values. The responses of the two secondaries with frequencies are shown in Fig. 13.15(a) and the output of FM detector or demodulator is shown in Fig. 13.15(b).

13.2.3 Phase Modulation

In phase modulation the relative phase angle θ is varied about unmodulated value according to the information signal. Its analysis is similar to that of the frequency modulation.

13.3 TRANSMITTERS

A transmitter converts a message signal in a convenient signal that can be sent through an antenna or an aerial in the surrounding atmosphere. An RF transmitter receives an electrical signal from the information source and transmits a high-power RF carrier signal modulated with the information signal. The transmitter generates the modulated signal with sufficient power and reasonable efficiency. An AM transmitter is shown in the block diagram of Fig. 13.16. RF oscillator is a stable source of RF carrier signal. The RF signal is amplified in a chain of class C amplifiers until it reaches the final RF amplifier. The audio signal from the microphone is fed to an audio processing and filtering unit. In this block, the audio signal is filtered to restrict the bandwidth within the specified frequency range of transmission. The audio signal is compressed to reduce fluctuations in its amplitude by the audio amplifier. This is done by automatically controlling the gain of the audio amplifier as per the amplitude of the audio signal. The gain is reduced for stronger signals and increased for weaker signals. This is done to improve the overall percentage of modulation. The processed signal is amplified to raise its power to a level adequate for achieving modulation. The RF signal from the class C amplifiers and the audio signal from the modulator are coupled to the final RF amplifier. In RF amplifier, the RF carrier signal is amplitude modulated by the audio signal. The modulated signal is fed to the transmitting antenna.

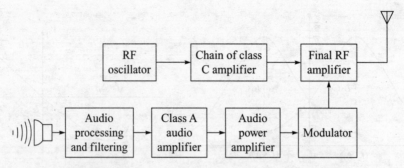

FIGURE 13.16 A block diagram of AM transmitter.

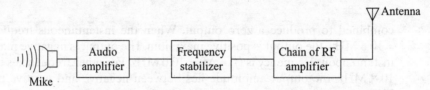

FIGURE 13.17 A basic block diagram of FM transmitter.

A basic FM transmitter is shown in the block diagram of Fig. 13.17. FM oscillator is the heart of the transmitter. The FM oscillator block generates the FM signal. The frequency of the FM oscillator is made to vary linearly in accordance with the modulating signal. In addition to linear variation of frequency, another design requirement is the stability of centre frequency of the FM oscillator. As the information signal is carried by the frequency deviation from the centre frequency, any drift in the centre frequency would seriously affect the information signal received by the receiver. The chain of RF amplifiers is used to raise the power of the modulated signal to a required level. Class B or class C amplifiers can be used to achieve higher efficiency. The RF amplifiers should, however, have adequate bandwidth to handle the FM signal. The de-emphasis circuit restores the amplitude frequency relationship of the modulating signal that existed at the transmitter prior to pre-emphasis. *Although pre-emphasis is done prior to modulation, de-emphasis is done after demodulation.*

One of the various schemes available for stabilizing the centre frequency of the FM oscillator is illustrated by the block diagram shown in Fig. 13.18. The block diagram within the dotted lines provides the frequency stabilization. The centre frequency has the stability of a crystal oscillator that is being used as a reference. Divide-by-M and divide-by-N are frequency counters. 'M' and 'N' are so chosen that the FM frequency (f_o) to be stabilized is given by

$$f_o = \frac{M}{N} f_c \qquad (13.11)$$

where f_c is the frequency of crystal oscillator.

The outputs of 'M' and 'N' are fed to a phase detector. The phase detector produces a d.c. voltage in accordance with the frequency difference between the two inputs. If FM

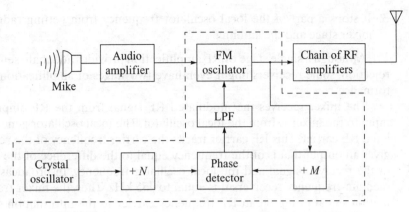

FIGURE 13.18 An FM transmitter with frequency stabilization.

oscillator circuit is varactor based, the d.c. voltage is used to correct the oscillator output frequency, whenever it tends to drift. The varactor is a variable capacitance diode. The value of the capacitance changes with the reverse-biased voltage. The varactor is used in the filter circuit of the FM oscillator. The other blocks of Fig. 13.18 shown outside the dotted lines are the blocks of Fig. 13.17.

13.4 RECEIVERS

Receivers perform the inverse operation to transmitters. They have to amplify a low-level signal as received from an antenna and separate it, as much as possible, from noise and interference. The receiver also demodulates the signal and amplifies the information signal to a power level sufficient for the intended application.

13.4.1 AM Receiver

Figure 13.19 shows the block diagram depicting the essential building blocks of an AM broadcast communication receiver. The RF stage is a tuned amplifier stage. The resonant frequency can be tuned over the entire carrier frequency range of interest. The RF stage performs the following three major functions:

1. It increases the SNR of the received signal.
2. It provides selectivity and image rejection.

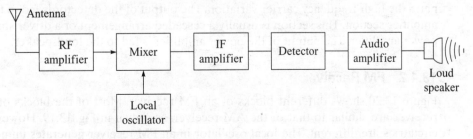

FIGURE 13.19 A block diagram of an AM receiver.

3. It stops a part of the local oscillator frequency from getting radiated through the mixer stage and the antenna.

Although a single stage of RF amplification provides adequate sensitivity and image rejection, some receivers might even have two stages of amplification for superior performance.

The mixer receives the modulated RF signal from the RF amplifier. The second input to the mixer is from the local oscillator. The local oscillator generates an unmodulated RF carrier. This RF carrier tracks the carrier frequency of the received signal and gives an output that is of the frequency equal to the difference of the carrier frequency of the received signal and the local oscillator frequency. In broadcast band and other medium-frequency receivers it is equal to 455 kHz. Thus, the mixer translates the information signal riding on an RF frequency carrier to the information signal riding on a relatively lower-frequency carrier. The modulated signal present at the output of the mixer stage has a fixed carrier frequency, that is, intermediate frequency (IF). The IF is independent of the carrier frequency being tuned in. The mixer usually generates the following frequencies:

1. The individual frequencies participating in the mixing operation. These are received carrier frequency and local oscillator frequency.
2. Sum of the two frequencies.
3. Difference of the two frequencies.

The difference of the two frequencies is the desired frequency, and others are undesired frequencies. The undesired frequencies are eliminated at the output of the mixer. It is done by tuning to the difference frequency by a double-tuned circuit with the primary forming the mixer output and secondary feeding the input of the first IF stage.

The major voltage gain and sensitivity are provided by the IF-stage amplifiers. The chain of the IF amplifiers are tuned to a fixed frequency. A higher IF provides better image rejection, whereas lower IF provides higher voltage gain and sensitivity for a given number of IF amplifier stages. For high-frequency carrier applications, such as FM broadcast, television, radar, etc., the IF chosen is also higher because it is easier to design broadband IF amplifier stages with higher frequency. For example, for FM broadcast band of 88 to 108 MHz, the IF is 10.7 MHz, whereas for television receiver applications it is between 30 and 40 MHz.

The job of the detector is to recover the information signal from the amplitude-modulated signal. It is similar to a half-wave rectifier circuit. The rectifier portion usually eliminates the negative half of the modulated signal, whereas the filter section eliminates the high-frequency carrier variation. The output of the detector is fed to the audio amplifier section. This section is usually a cascaded arrangement of a driver stage and a power amplifier. The output of the power amplifier is fed to the loudspeaker.

13.4.2 FM Receiver

Figure 13.20 shows different blocks of an FM receiver. Most of the blocks of the FM receiver are similar to that of the AM receiver, as shown in Fig. 13.19. However, their circuitries are different. The local oscillator in an FM receiver generates unmodulated RF carrier in VHF/UHF range. For FM broadcast receiver the local oscillator works

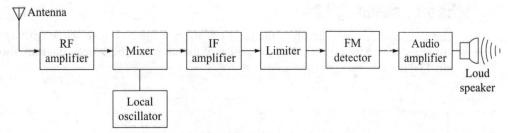

FIGURE 13.20 An FM broadcast receiver.

over a frequency range of 98.7 to 118.7 MHz, in which 10.7 MHz is the IF. The RF and IF stages of the FM and AM receivers differ only in their bandwidths. In comparison with a bandwidth of 10 kHz in an AM receiver, the bandwidth in the FM receiver is 100 kHz.

The limiter circuit shown immediately after the IF stage is used to remove any amplitude variation. The amplitude variation might have crept into the FM signal due to other interference signals. Although the information in the FM signal is unaffected by the amplitude variations, yet it is essential to remove it in case the FM detector used is sensitive to the amplitude variation as well. However, the limiter circuit can be eliminated if the FM detector used is insensitive to the amplitude variations. The FM detector extracts information signal from the FM signal. An FM detector has been discussed in Section 13.2.

The audio amplifier does the usual work of amplifying the information signal. The audio amplifier in an FM receiver, however, has a wider bandwidth. It is 15 kHz against 5 kHz in the case of that of an AM receiver.

13.5 DIGITAL COMMUNICATION

As mentioned earlier, digital communication has the following advantages over analogue communication:

1. Primary advantage of the digital communication is noise immunity. The analogue signals are more sensitive to undesired variations in amplitude, frequency, and phase than digital pulses. In digital transmission, threshold detection makes their reception simple.
2. It is easier to multiplex digital pulses than analogue signals. They can be stored easily. The transmission rate of digital communication system can easily be varied to suit the different environment. It is also easy to interface with different types of equipment.
3. Error performance, that is, error detection and correction can easily be computed.

To transmit an analogue signal using digital technique, it is must to convert the analogue signal to a digital signal. This is accomplished by using an analogue-to-digital converter. It is known as A/D converter or ADC. Before converting an analogue signal into a digital signal, the analogue signal (called baseband signal) is converted to a discrete signal by a process called sampling. The device used for sampling is called a sampler. Thus, after sampling, the analogue signal, which was continuous, is converted into a discontinuous signal defined at discrete intervals of time.

13.5.1 Sampling

The sampling basically involves the multiplication of the analogue signal with a train of impulses with a period T_s. The process is illustrated by the block diagram shown in Fig. 13.21.

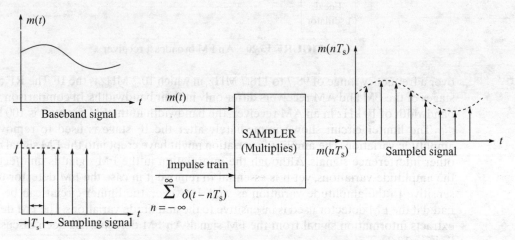

FIGURE 13.21 Sampler with input and output signals.

The sampled signal at each sampling instant is equal to the baseband signal at that instant, that is, the instant at which the impulse occurs. An impulse signal is defined as a pulse signal with width $\Delta T \to 0$ and height $h \to \infty$. The sampling process with a train of true impulses is called instantaneous sampling.

In 1928, Harry Nyquist showed mathematically that a band-limited analogue signal can be completely specified by its periodic samples, as long as the sampling rate is at least twice the frequency of the highest-frequency component of the signal. An analogue signal with its highest-frequency component of f_m is said to be band limited to f_m. In practice, however, the sampling frequency considered is considerably greater than twice the maximum frequency to be transmitted. For example, in telephony, a sample rate of 8 kHz is used for a maximum audio frequency of 3.4 kHz, and compact disc system has a 44.1 kHz sampling rate for a maximum audio frequency of 20 kHz.

In case of instantaneously sampled signal (i.e., a train of impulses is used for sampling), each sample is of very small width ΔT. So, the strength of each sample is weak. When the instantaneously sampled signal is transmitted through the channel, these samples may be badly affected by channel noise. To reduce the effect of the noise, on the sampled signal, the strength of each sample must be increased by increasing the width ΔT. For this, sampling signal used is a train of pulses of width τ and amplitude of 1 V. This sampling is called natural sampling. The naturally sampled signal is shown in Fig. 13.22. The tops of the pulses follow the analogue signal. Thus, the amplitudes of samples at the leading and trailing edges are not the same. In some situations, a *sample-and-hold circuit* is used to keep the pulse amplitude constant for the duration of the pulse. This type of sampling is shown in Fig. 13.23 and is called *flat-topped* sampling.

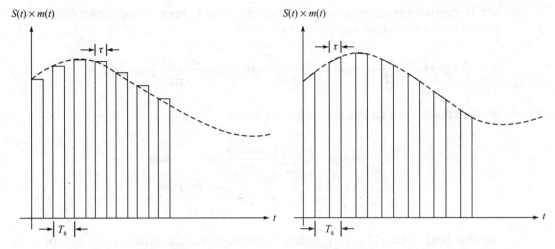

FIGURE 13.22 A naturally sampled signal.   **FIGURE 13.23** A flat-topped sampled signal.

If sampling rate is too low, the original signal cannot be recovered from the sampled signal. This distortion due to low sampling rate is called *aliasing* or *foldover distortion*. Due to this form of distortion, the frequencies are translated downwards. Figure 13.24 demonstrates how aliasing develops. In Fig. 13.24(a), the rate of sampling is adequate and the signal can be reconstructed. In Fig. 13.24(b), the rate of sampling is too low and the attempt to reconstruct the original signal resulted in a lower-frequency signal. This is due to aliasing distortion. Once aliasing is present, it cannot be eliminated.

The effect of aliasing can be easily demonstrated as follows. For simplicity, let us assume a sinusoidal signal as the modulating signal,

$$e_m = E_m \sin \omega_m t \tag{13.12}$$

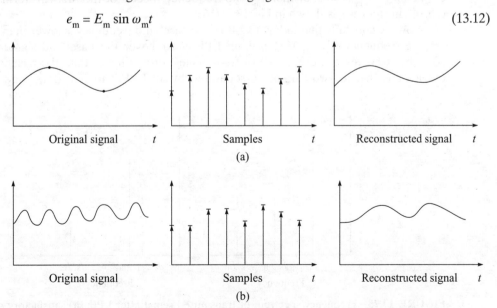

FIGURE 13.24 Aliasing effect: (a) satisfactory sampling rate and (b) sampling rate too low.

The carrier signal is a pulse train of width τ, amplitude unity and period T. The spectrum (Fourier series) for pulse train is given by

$$e_m = \frac{\tau}{T} + 2\frac{\tau}{T}\left(\frac{\sin \pi\tau/T}{\pi\tau/T}\cos\omega_s t + \frac{\sin 2\pi\tau/T}{2\pi\tau/T}\cos 2\omega_s t + \frac{\sin 3\pi\tau/T}{3\pi\tau/T}\cos 3\omega_s t + \ldots\right) \quad (13.13)$$

Multiplying the two signals, the output is given by

$$v_O(t) = E_m\frac{\tau}{T}\sin\omega_m t + 2E_m\frac{\tau}{T}\left(\frac{\sin \pi\tau/T}{\pi\tau/T}\sin\omega_m t\cos\omega_s t\right.$$
$$\left. + \frac{\sin 2\pi\tau/T}{2\pi\tau/T}\sin\omega_m t\cos 2\omega_s t + \frac{\sin 3\pi\tau/T}{3\pi\tau/T}\sin\omega_m t\cos 3\omega_s t + \ldots\right) \quad (13.14)$$

The first term of Eq. (13.14) is just the modulating signal (analogue signal) multiplied by a constant. From this, it can be concluded that when the output of the multiplier is passed through a low-pass filter (LPF), we would get the original signal, provided that there is no component of the signal occupying the baseband frequency range. Next, the second term contains $\sin\omega_m t\cos\omega_s t$, which represents the sum and difference of frequencies f_s and f_m, that is, $(f_s - f_m)$ and $(f_s + f_m)$, where $f_s = \omega_s/2\pi$ and $f_m = \omega_m/2\pi$. The other terms of Eq. (13.14) produce still higher frequency components, and they are not of our concern. When the sampling frequency f_s is greater than $2f_m$, the difference frequency $f_s - f_m$ is higher than f_m, and the frequency spectrum of the first two terms of Eq. (13.14) is shown in Fig. 13.25(a). When $f_s = 2f_m$, $f_s - f_m = f_m$; hence, the difference frequency coincides with the signal frequency. When the sampling rate is too low, that is, f_s is lower than $2f_m$, then $f_s - f_m$ is smaller than f_m. The frequency spectra of the first two terms of Eq. (13.14), in this case, is shown in Fig. 13.25(b).

Now we conclude that in Fig. 13.24(a), the baseband frequency is lower than the difference frequency $(f_s - f_m)$. Hence, an LPF can recover the baseband signal. In Fig. 13.24(b), however, the difference of frequencies term is lower than the baseband frequency and, hence, two cannot be separated by an LPF. This is a frequency domain

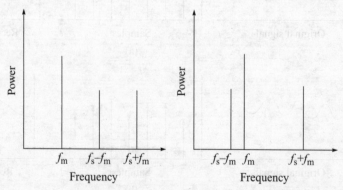

FIGURE 13.25 Frequency spectrum of transmitted signal after LPF: (a) satisfactory sampling rate and (b) sampling rate too low.

representation of the aliasing that has been seen earlier in the time domain in Fig. 13.23(b). When $f_s = 2f_m$, the difference frequency coincides with the baseband frequency. This is just on the edge of aliasing. If the aliasing does take place, the interfering component is at a frequency

$$f_a = f_s - f_m \tag{13.15}$$

The frequency f_a is the frequency of aliasing distortion.

13.5.2 Analogue Pulse Modulation Techniques

Sampling alone is not a digital technique. The immediate result of sampling is a pulse amplitude modulation (PAM) by the analogue signal. The PAM signal is shown in Fig. 13.23. PAM is an analogue scheme in which the amplitude of each pulse is proportional to the amplitude of the analogue signal at the instant at which it is sampled. The main use of PAM is the intermediate step, before being transmitted. In the next step, the PAM signal has to be quantized. Similarly, at the receiver end, in the first step, the digital signal must be converted back to PAM signal. In the next step, the baseband signal is recovered using an LPF. The other two pulse modulation techniques are pulse duration modulation (PDM) and pulse position modulation (PPM). In PDM, all pulses have the same amplitude, but the duration of each pulse depends on the amplitude of the signal at the instant of sampling. The PDM signal is shown in Fig. 13.26. The PDM has its communication use in the high-power audio amplifier used in AM transmitters. It is also used for telemetry systems. Although still an analogue mode, PDM is more robust than PAM because it is insensitive to amplitude variations due to noise and distortion. Like PAM, PDM can be demodulated by an LPF. Sometimes the PDM is also called pulse width modulation (PWM).

Pulse position modulation (PPM) is closely related to PDM. All pulses have the same amplitude and the same duration. However, the timing of the pulses varies with the amplitude of the original signal. PPM signal is shown in Fig. 13.27.

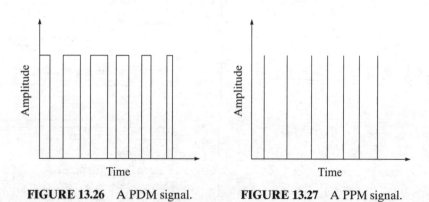

FIGURE 13.26 A PDM signal.   **FIGURE 13.27** A PPM signal.

13.6 QUANTIZATION

The operation of an A/D converter is based on converting the continuous analogue signal into a number of discrete levels. This operation is called quantization. These levels are allocated a digital code (binary code). However, the quantization of an analogue signal distorts the signal information. Figure 13.28 provides the typical input-output characteristic of a 3-bit A/D converter. The A/D converter may be called a quantizer. Figure 13.28 illustrates the round off quantization relation between the analogue input signal and the quantized output signal. In the round off operation, for example, the decimal number 3.55 is rounded off to 3.6, and the decimal number −3.55 to −3.6, if the round off is to one decimal place. The parameter q, which is equal to least significant bit, is known as the quantization level or resolution. For the 3-bit word, as illustrated in Fig. 13.28, the least significant bit (or q) is (1/8) FS. FS is equal to the maximum variation of the analogue signal. If V_H is the highest value and V_L is the lowest value of the analogue signal, then FS = $V_H - V_L$. The dotted straight line in Fig. 13.28 represents the ideal output, if there is no quantization. The quantizer gives the quantization error. The difference between the dotted straight line and the quantization characteristic is the quantization error. This error is not a deviation from an ideal characteristic, but an inherent feature of the quantizer. As can be seen from Fig. 13.28, the error oscillates between +0.5q and −0.5q.

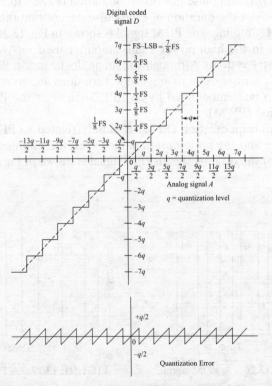

FIGURE 13.28 The input-output characteristic of the quantizer.

The output signal of a quantizer can be considered to be the input analogue signal with quantization error added to it. For example, if the quantizer output is mq, then the analogue equivalent of the quantizer output is $mQ + 0.5q$. The noise, as shown in Fig. 13.28, has an r.m.s. value of $0.5q/\sqrt{3}$. The SNR is given by

$$\text{SNR} = \frac{\text{r.m.s. value of the signal}}{\text{r.m.s. value of the noise}} \tag{13.16}$$

Hence, for a sinusoidal input, the SNR that is also known as *dynamic range* (DR) is given by

$$\text{SNR} = \text{DR} = 6n + 1.8 \text{ dB} \tag{13.17}$$

Example 13.1

Find the maximum dynamic range for a quantizer using 16-bit quantizing.

Solution

The maximum dynamic range is given by

$$\begin{aligned}\text{DR} &= 6n + 1.8 \\ &= 6 \times 16 + 1.8 = \mathbf{97.8 \text{ dB}}\end{aligned}$$

The q for an n-bit quantizer is $1/2^n$ FS. Although the quantization error (noise) can be reduced by increasing the resolution of the quantizer, there is always a quantization uncertainty of $\pm 0.5q$. The SNR, for different bits, is given in Table 13.3.

Increasing the number of bits per sample increases the data rate, which is given by

$$D = f_s\, n \text{ bits/s} \tag{13.18}$$

where f_s is the sampling frequency and n is the number of bits.

Table-13.3 Quantizer SNR*.

Number of bits (including sign bit)	SNR (dB)
6	38
8	50
10	62
12	74
14	86

*SNR increases at 6 dB/bit.

Example 13.2

Find the minimum data rate required to transmit audio signal with a sampling rate of 40 kHz and 14 bits per sample.

Solution

Data rate is given by

$$D = f_s n$$
$$= 40 \times 10^{14} \times 14$$
$$= 560 \text{ kbit/s}$$

The operation of a quantizer involves a finite amount of time called conversion time. The conversion time again introduces a time delay into the signal processing equal to half of its period. There is also a minimum bound in the conversion process called aperture time (t_a), where there is an uncertainty of reading the analogue signal. Aperture time determines the highest frequency that can be converted. For sinusoidal signals this can be found by taking the point of maximum rate of change of value, that is, at a zero crossing. The maximum change for sine wave at zero is ωV, where sine wave is given by $v(t) = V \sin \omega t$. The voltage change during aperture time t_a, hence, is $\omega V t_a$. Therefore, the relation between q and T_a can be given by

$$\omega V t_a = q$$

But for number of bits equal to n,

$$q = \frac{2V}{2^n - 1}$$

Hence, $$t_a = \frac{1}{\pi f (2^n - 1)} \qquad (\omega = 2\pi f) \qquad (13.19)$$

Example 13.3

What would the aperture time be for a 1-kHz signal using a 10-bit quantizer. The input sine wave amplitude equals the maximum input for the quanter.

Solution

The aperture time is given by

$$t_a = \frac{1}{\pi f (2^n - 1)}$$

$$= \frac{1}{2 \times 1000 \times (2^{10} - 1)}$$

$$= 156 \text{ ns}$$

13.6.1 Companding

The quantization error depends on the step size. The SNR for the lower-amplitude signals is poorer compared to the larger amplitude signals, when uniform quantization level is employed. So, for a uniform SNR throughout the signal, the resolution or quantization level is to be adjusted according to the amplitude of the signal components. This means that the step size Q should be small for small amplitude components and large for large amplitude components. This process is known as companding. To perform this operation, the signal is passed through a nonlinear network. The response of the network is shown in Fig. 13.29. The response curve is known as mu (μ)-law curve.

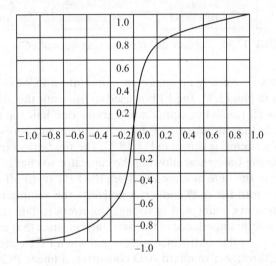

FIGURE 13.29 Response of companding network.

The network is known as compressor. It reduces the quantization error (i.e., quantization level Q) for small signals and increases it for large signals. The compression on the signal is done at the transmitting end and the reverse effect of compression (known as expansion) is performed at the receiving end.

13.7 DIGITAL MODULATION

The following are the commonly used digital modulation techniques:

1. Pulse code modulation (PCM)
2. Differential pulse code modulation
3. Delta modulation
4. Adaptive delta modulation

13.7.1 Pulse Code Modulation

PCM is the most commonly used digital modulation scheme. In PCM, the available range of signal voltages is divided into levels. This is discussed while quantizing the signal. Each level in quantization is assigned a binary number. Each sample is then represented by the binary number representing the level closest to its amplitude. This binary number is transmitted in serial form.

Coding and Decoding

The process of converting an analogue signal into a PCM signal is called coding. The inverse operation of coding is called decoding. Both processes are often performed by a single integrated circuit device called *codec*.

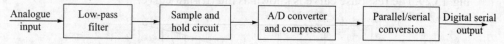

FIGURE 13.30 A block diagram showing steps of PCM coding.

The steps for converting an analogue signal into a PCM code is given in the block diagram shown in Fig. 13.30. The LPF is used to filter out the aliasing. As shown earlier, the filter blocks all frequency components above one half the sampling rate. The next step is to sample the output of LPF using a sample and zero-order hold. One of the sample-and-hold circuits is shown in Fig. 13.31. The field-effect transistor (FET) turns on during the sampling time; thus, allowing the capacitor to charge to the amplitude of the incoming signal at the time of sampling. Then the FET turns off and the capacitor stores the signal value until the A/D converter converts the sample to digital form. The two operational amplifiers, connected as voltage followers, isolate the circuit from the other stages. The low output impedance of the first stage ensures that capacitor quickly charges or discharges to the value of the incoming signal when FET conducts. Next, the sampled signal is passed through a standard A/D converter, if linear PCM is used. Compression can be applied to the analogue signal. However, it is more common to use compression into the coding process.

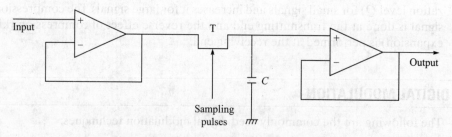

FIGURE 13.31 A sample-and-hold circuit.

The codes used in telephony generally employs compression by using a piecewise linear approximation to the μ-law curve shown in Fig. 13.29. Each of the positive and negative going curves is divided into eight segments; thus, giving total 16 segments. The segmented curve is shown in Fig. 13.32. Segments 0 and 1 have same slope and do not compress the segment. For each higher-numbered segment, the step size is double to that of the previous segment. Each segment has 16 steps. The result is a close approximation to the actual curve.

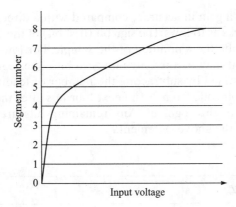

FIGURE 13.32 Positive half of segmented μ-law curve.

The binary number produced by the codec in a telephone system has eight bits. The most significant bit (i.e., first bit) is a sign bit, which is 1 for positive voltage and 0 for negative voltage. The next three bits (i.e., bits 2, 3, 4) represent the segment number, from 0 to 7. The last four bits determine the step within the segment. If the signal is normalized, that is, the maximum input level is set 1 V, the step sizes can easily be calculated as follows. Let the step size for the 0 and 1 segments be x mV. Then segment 2 has a step size of $2x$ mV, segment 3 a step size of $4x$ mV, and so on. Since each segment has 16 steps, the value of x can be obtained as follows:

$$16(x + x + 2x + 4x + 8x + 16x + 32x + 64x) = 1000 \text{ mV}$$
$$x = 0.488 \text{ mV}$$

The relationship between input voltage and segment is shown in Table 13.4.

Many modem codes achieve compression by encoding the signal using a 12-bit linear PCM code first. Then the 12-bit linear code is converted into an 8-bit compressed code by discarding some of the bits. This is an example of digital signal processing. However, some precision is lost in the conversion for large amplitude samples. But the data rate needed to transmit the information is much less than that for a 12-bit PCM. Since most of the samples in an audio signal have amplitudes much less than the maximum

Table-13.4 μ-Law curve compressed PCM coding.

Segment	Voltage range (mV)	Step size (mV)
0	0–7.8	0.488
1	7.8–15.6	0.488
2	15.6–31.25	0.9772
3	31.25–62.5	1.953
4	62.5–125	3.906
5	125–250	7.813
6	250–500	15.625
7	500–1000	31.25

amplitude, there is a gain in accuracy compared with a direct 8-bit linear PCM. Briefly, the conversion works as follows. The sign bit (first bit) of the 12-bit PCM is retained. The other 11 bits describe the amplitude of the sample. For low-level samples, the last few bits and the sign bit may be the only non-zero bits. The segment number for the 8-bit code can be determined by subtracting the number of leading zeros (not counting the sign bit) in the 12-bit code from 7. The next four bits after the first 1 are counted as the level number within the segment. Any remaining bits are discarded. The following example illustrates the above statement.

Example 13.4

Convert the 12-bit sample 100110100100 into an 8-bit compressed code.

Solution

The sign bit is 1. So, sign bit of 8-bit code is 1. The number of leading zeros is 2. So, 7 − 2 = 5 (101 in binary). Thus, the first four bits of the 8-bit code are 1101. The next four bits after the first 1 are 1010. Discarding the remaining bits, the 8-bit code is 11011010.

The decoding process is the reverse of coding and is illustrated in the block diagram in Fig. 13.33. The expansion process follows an algorithm analogous to that used in the compression. The LPF at the output removes the high-frequency components in PAM signal that exist from the digital-to-analogue converter.

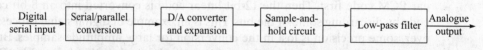

FIGURE 13.33 A block diagram showing the steps of PCM decoding.

13.7.2 Differential PCM

In the differential coding, only the difference between the amplitude of the current sample and that of the previous sample is coded and transmitted, and not the entire sample amplitude of the current sample. Since successive samples often have similar amplitudes, it is possible to use fewer bits to encode the changes.

13.7.3 Delta Modulation

In delta modulation, only one bit per sample is transmitted. The bit is 1 if the current sample is more positive than the previous sample. It is zero if the current sample is more negative than the previous sample. Since only a small amount of information about each sample is transmitted, this modulation requires a much higher rate of sampling than PCM for equal amount of reproduction.

13.7.4 Adaptive Delta Modulation

In this type of modulation, the step size varies according to the previous values. This method is more efficient. After a number of steps in the same direction, the step size increases. A well-designed adaptive delta modulation scheme can transmit voice at about half the bit rate of PCM system with equivalent quality.

13.8 MULTIPLEXING

Channel sharing is a common phenomenon in communication systems. Examples include a cable television system, telephone trunks, and radio transmission. There are two closely related terms describing the combining of signals. When all signals passing through a channel originate from the same source, the technique is called *multiplexing*. When signals originating from several sources are combined on a single channel, the technique is called *multiple accessing*. Following are the two major types of multiplexing or multiple accessing:

1. Frequency division multiplexing (FDM)
2. Time division multiplexing (TDM)

A frequency division multiplexing or multiple accessing can simultaneously transmit several inputs on a single communication channel by modulating them with different carrier frequencies. These carrier frequencies are nonoverlapping portions of the frequency spectrum. In time division multiplexing or multiple accessing different inputs are sequentially connected, one after another, to the single communication channel.

13.8.1 Frequency Division Multiplexing

The block diagram of a four-channel FDM is shown in Fig. 13.34. Four input signals are first applied to their channel modulators, which modulate their respective inputs

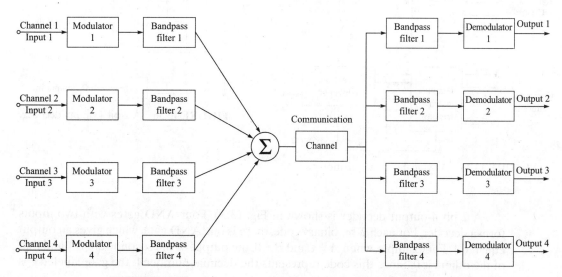

FIGURE 13.34 A frequency division multiplexing or multiple accessing.

with different carrier frequencies. The carrier oscillator frequencies are so chosen that they avoid the overlapping of frequency spectra between each other. A band-pass filter for each local channel is used so that only the working frequencies are allowed to pass. The harmonics and other spurious frequencies are blocked. At the receiving ends, the signals are separated by selective filters and demodulators, as shown in Fig. 13.34.

13.8.2 Time Division Multiplexing

There are cases where analogue signals are multiplexed with the communication channel, and signals are transmitted in analogue form. In other cases, the analogue signals are multiplexed with a digital computer for analysis and/or control. In these cases, an A/D converter is used after the multiplexer. A sample-and-hold circuit is used before multiplexing, as shown in Fig. 13.31, when simultaneous samples of input signals are required. Before discussing different types of TDMs, an address decoder is described first.

Address Decoder

An address decoder receives an input from a digital computer via address lines that serve to select a particular analogue channel to be sampled. The address decoder associates the particular channel a computer address codes. A binary code is sent to the computer through special input/output devices to select an analogue channel and to input the data on that channel. A data acquisition system may be as shown in Fig. 13.35.

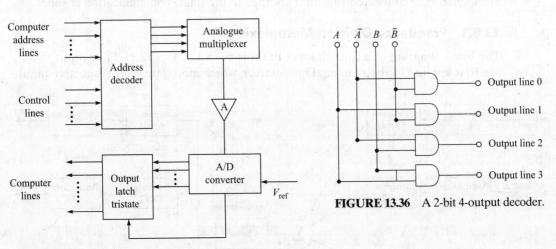

FIGURE 13.35 A typical data acquisition system.

FIGURE 13.36 A 2-bit 4-output decoder.

A 2-bit 4-output decoder is shown in Fig. 13.36. Four AND gates with two inputs form a decoder. For each 2-bit binary code, there is one AND gate, which gives an output equal to 1. For example, when $A = 1$ and $B = 0$, the output of the second gate is 1 and that of the others is 0. Since this code represents the decimal number 1, the gate labeled 1 is on and others are off.

Analogue Multiplexing This is essentially a solid-state switch that works according to the decoded addressed signal and selects the data on the selected channel by closing the switch of the channel. A four-channel multiplexing circuit is shown in Fig. 13.37. The multiplexer receives an input from the address decoder and uses this to close the appropriate switch. For example, an address code **10** selects channel 2, whereas code **00** selects channel 0, **01** channel 1, and **11** channel 3. Thus, the address decoder must convert the computer address line to one of these four possibilities. The actual switch element is usually an FET. An FET has 'on' resistance of a few hundred ohms and 'off' resistance of hundreds of thousands of megaohms.

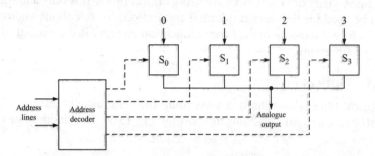

FIGURE 13.37 A four-channel analogue multiplexing circuit.

Digital Multiplexing The logic circuit for a 4-line to 1-line multiplexing is shown in Fig. 13.38. Ignore the dash lines for the moment. If the select code is **01**, then X_1 appears at the output Y; if the address is 11, then the output is X_3. Similarly, other channels are multiplexed.

Sometimes, it is desired to decode only during certain intervals of time. In such cases, an additional input, known as a *strobe* or an *enable* input, is added through an

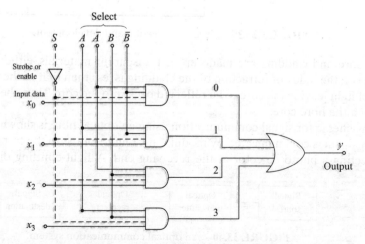

FIGURE 13.38 A four-channel analogue multiplexing circuit.

inverter to each AND gate. Enable input is excited by a binary signal S after inversion. If $S = 0$, gates are enabled, whereas if $S = 1$, none of the gates is enabled and the decoding is inhibited.

13.9 OPTICAL FIBRE AND COMMUNICATION SYSTEMS

Fibre-optic communication systems are becoming very important today. Most of the new telephone cables of any great length, whether on land or under water, use fibre optics in place of coaxial cables. Cable television systems employ fibre optics for new trunk lines. Many data networks are using optical fibres. It is only a matter of time, fibre would be used for different residential connections for telephone and/or television services. Before discussing optical communication systems, it is essential to discuss optical fibre and fibre optics.

13.9.1 Optical Fibre

An optical fibre is essentially a waveguide for light, usually infrared. An optical fibre consists of a core and a cladding as shown in Fig. 13.39. The cladding surrounds the core.

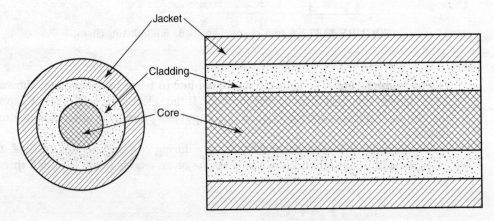

FIGURE 13.39 Fibre cross section and development.

Both core and cladding are made up of transparent materials, either glass or plastic. However, the index of refraction of the cladding is less than that of the core. This causes rays of light leaving the core to be reflected back into the core. Thus, the light propagates through the fibre core.

A scheme for signal communication using an optical fibre is shown in Fig. 13.40.

A signal is communicated by modulating a light source to the transmitting end and connecting a photo detector at the receiving end. A light-emitting diode (LED) or a

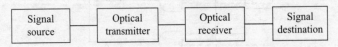

FIGURE 13.40 An optical communication system.

laser diode can be used for light source. Any of modulation schemes discussed earlier can be used. However, digital transmission is more common than analogue. The light source can be turned on and off or switched between two power levels for digital transmission. Analogue signals are usually transmitted using PCM. Amplitude modulation and frequency modulation schemes are also used sometimes.

In practice, optical fibre is used as substitute for copper cable. Its other uses include point-to-point microwave radio link, telephone cables, point-to-point transmission of television signals, and computer networks. Optical fibre has many advantages over electrical cable for communications. It handles greater data rates than coaxial cable, as it has greater bandwidth. The higher bandwidth allows more signals to be multiplexed. The optical fibre has loss lower than that of copper cable. Hence, it helps in increasing the distance between two repeaters. However, it is more expensive than copper cable. Since optical cables do not carry electric currents, they are immune to crosstalk between cables. They are also not influenced by the electromagnetic interference from other sources. Since there is no electrical connection, optical cables are useful in explosive environments where a spark hazard may exist. They can also be used to isolate electrically one circuit from another. In medical electronics, fibre optics is often used to isolate the patient from circuit connected to power line. Thus, it avoids the shock hazards.

One application where optical fibres cannot substitute the copper cable or waveguide is the transmission of power. For example, they cannot be used to connect a transmitter to an antenna. Optical fibres are used with power levels in milliwatt range. Hence, they are strictly for transmission of signals, not energy.

In size and weight, optical fibres have advantages over copper cables. On cost factor, optical fibre can be less expensive than coaxial cable. However, it is more expensive than twisted pair. For wide bandwidth long-distance applications optical fibre is economical, whereas for low bandwidth short-distance applications copper is still cheaper. The cost of converting an electrical signal to and from optical form makes use of optical fibres costlier.

The fibre used in communication systems is a thin strand of glass or plastic. Since the fibre has very little mechanical strength, it is enclosed in a protective jacket, as shown in Fig. 13.39. The jacket is made up of plastic. Generally, two or more fibres are included in one cable. This increases the bandwidth and redundancy in case of fibre breaks. It is easier to build duplex system using two fibres, one for transmission in each direction.

Working of Optical Fibres

Optical fibres work on the *principle of total reflection*. When the ray of light enters from one medium into another, it does not travel in the same direction. It bends as shown in Fig. 13.41(a), if the indexes of refraction are not same. The index of refraction is defined as the ratio of velocity of light in free space to the velocity of light in the medium. In Fig. 13.41(a), the angles θ_1 and θ_2 are *angle of incidence* and *angle of refraction*, respectively. The two angles are related by Snell's law:

$$n_1 \sin \theta_1 = n_2 \sin \theta_2 \qquad (13.20)$$

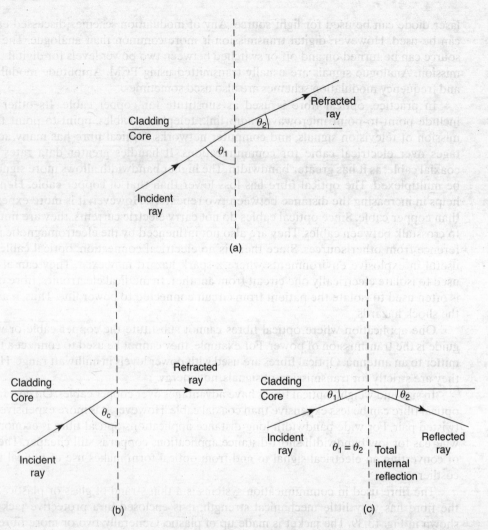

FIGURE 13.41 (a) Refraction, (b) angle of incidence is critical angle, and (c) total internal reflection.

where n_1 and n_2 are refraction indexes of the two media, respectively. In each case, the angle is measured between the ray of light and the *normal* (line perpendicular to the interface between the media). The angle of refraction is given by

$$\sin\theta_2 = \frac{n_1}{n_2}\sin\theta_1 \qquad (13.21)$$

For an optical fibre, the first medium is the core, whereas the second medium is the cladding. The refraction index of the core is greater than that of the cladding; that is, $n_1 > n_2$. There is a value of $\theta_1 < 90°$, for which $\theta_2 = 90°$. The refracted ray, in this case, lies along the interface between the media, as shown in Fig. 13.41(b). This value of the incident angle θ_1 is called critical angle θ_c.

Total Internal Reflection When $\theta_1 > \theta_c$, the ray is reflected rather than refracted. Snell's law fails, and the angle of reflection is equal to angle of incidence, as shown in Fig. 13.41(c). For $\theta_2 = 90°$, $\sin \theta_2 = 1$ and $\theta_1 = \theta_c$. Hence, from Eq. (13.20)

$$\sin \theta_c = \frac{n_2}{n_1} \sin 90° = \frac{n_2}{n_1}$$

Hence, $\quad \theta_c = \sin^{-1}\left(\frac{n_2}{n_1}\right) \quad$ (13.22)

Example 13.5

A fibre has indexes of refraction for the core and the cladding equal to 1.6 and 1.4, respectively. Calculate (a) the critical angle, (b) θ_2 for $\theta_1 = 30°$, and (c) θ_2 for $\theta_1 = 70°$.

Solution

(a) From Eq. (13.22)

$$\theta_c = \sin^{-1}\left(\frac{n_2}{n_1}\right)$$

$$= \sin^{-1}\left(\frac{1.4}{1.6}\right) = 61°$$

(b) Since 30° is less than the critical value; hence, Snell's law applies. Therefore, from Eq. (13.21)

$$\sin \theta_2 = \frac{n_1}{n_2} \sin \theta_1$$

$$= \frac{1.6}{1.4} \sin 30° = 0.571$$

$$\theta_2 = \sin^{-1} 0.571 = \mathbf{34.8°}$$

(c) Since 70° is greater than the critical value; hence, Snell's law does not apply. The law of reflection applies in this case. Hence,

$$\theta_2 = \theta_1 = \mathbf{70°}$$

Numerical Aperture The numerical aperture (NA) is one of the specifications often used for optical fibre and the components that work with it. It is closely related to the critical angle of the optical fibre. It is defined as the sine of the maximum angle which a ray entering the fibre can have with the axis of the fibre and still propagate by total internal reflection. It is assumed that the light enters the core from the air. Air can be considered equivalent to the free space. As illustrated in Fig. 13.42(a), the total *angle of acceptance* is twice of $\sin^{-1}(\text{NA})$. Any rays of light entering the cone of acceptance are reflected internally and propagate within the fibre. The rays of light outside the cone of acceptance are refracted into the cladding and do not propagate. The NAs for optical fibres vary about 0.1 to 0.5. To derive expression for the numerical aperture, the projection of Fig. 13.42(a) is shown in Fig. 13.42(b). Applying Snell's law at interface of air and core, the natural aperture is given by

$$\begin{aligned} NA = \sin \phi &= n_1 \sin (90° - \theta_c) \\ &= n_1 \cos \theta_c \\ &= n_1 \sqrt{1 - \sin^2 \theta_c} \\ &= n_1 \sqrt{1 - \frac{n_2^2}{n_1^2}} \\ &= \sqrt{n_1^2 - n_2^2} \end{aligned} \qquad (13.23)$$

where θ_c is the critical angle for the interface of core and cladding.

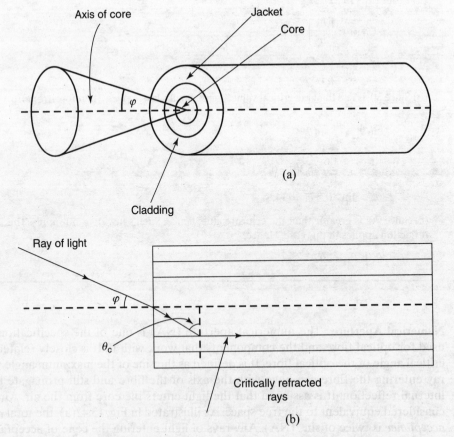

FIGURE 13.42 Cone (angle) of acceptance and numerical angle.

Example 13.6

Calculate the NA and the maximum angle of acceptance for the fibre of Example 13.5.

Solution

From Eq. (13.23),

$$\text{NA} = \sqrt{n_1^2 - n_2^2}$$
$$= \sqrt{1.6^2 - 1.4^2}$$
$$= \mathbf{0.775}$$

The maximum angle of acceptance is given by

$$\theta_{ac} = 2\sin^{-1}(\text{NA})$$
$$= 2\sin^{-1}(0.775)$$
$$= \mathbf{101.6}$$

Mode and Materials The propagation of light in a fibre can be in different modes. For a fibre of relatively larger diameter, the light entering at different angles excites different modes. On the contrary, a fibre with a sufficiently small diameter may support only one mode. Multimode propagation causes *dispersion*. Dispersion limits the usable bandwidth of the fibre. Single-mode fibre has much less dispersion. However, single-mode fibre is more expensive to manufacture and its small diameter gives less NA, which makes it more difficult to couple light into and out of the fibre. The maximum allowable diameter (d_{max}), in metres, for a single-mode fibre varies with the wavelength (λ) of light and is given by the following equation:

$$d_{max} = \frac{0.766\lambda}{\text{NA}} \qquad (13.24)$$

Example 13.7

Calculate the maximum diameter of a single-mode fibre for use with infrared light with a wavelength of 820 nm. Find the maximum allowable diameter if the NA of the fibre is 0.15.

Solution

The maximum diameter of the single-mode fibre is given by

$$d_{max} = \frac{0.766\lambda}{NA}$$

$$= \frac{0.766 \times 820 \times 10^{-9}}{0.15}$$

$$= 4.2 \times 10^{-6} \text{ m}$$

$$= \mathbf{4.2\ \mu m}$$

The single-mode and multimode fibres are known as *step-index fibres* because, in both types, the index of refraction changes sharply between core and cladding. Another type of fibre is known as *graded-index fibre*. In graded-index fibre the index of refraction gradually decreases away from the centre of the core. The graded-index fibre is a multimode fibre. It has less dispersion than the multimode step-index fibre. However, single-mode fibres are still preferred for most demanding applications.

Most optical fibres are made up of high-quality glass because of its very great transparency. It reduces the losses. Some low-cost multimode fibres designed for short-distance applications are made up of acrylic plastic. The short-distance applications include optical links in computer electronics and control signal lines in automobiles. The high losses in these low-cost fibres are very much higher than in glass fibre. But this is of no importance when distances involved are a few metres or less.

13.9.2 Fibre-Optic Cables

Optical fibres are very thin, and so they need protection from external forces, both while being installed and during use. Damage from corrosive environments must also be avoided. It is often desirable to run several fibres together. For these reasons, individual fibre is made in the form of a cable. The cable, in addition to the optical fibre, includes jackets and strength members.

There are two basic types of fibre cables, namely, loose-tube cable and tight-buffer cable. The difference is in whether the fibre is free to move inside a tube with a diameter much larger than the fibre or it is in a relatively tight-fitting jacket. The Kevlar, a plastic material, and steel are used as strength members in tight-fitting cables. Kevlar makes the cable completely nonmetallic. Both types of cables are shown in Fig. 13.43. The advantages of the two types of cables are given in Table 13.5.

Due to the greater strength of the loose-tube cables, they are used as telephone cables that have to be pulled for long distances through ducts. Kevlar tight-buffer cables are nonmetallic, and so are used in the communication systems, as they have no electromagnetic interference. Fibre cables are available in a wide variety of types. There are flat cables to install under carpets, armoured and waterproof cables for underwater use, and fibre cables with copper pairs included for communication during installation or the provision of power to repeaters.

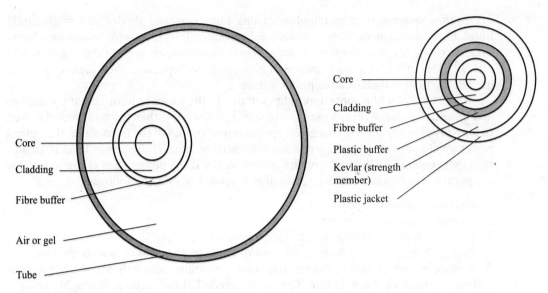

FIGURE 13.43 Fibre-optic cables (a) loose-tube construction and (b) tight-buffer construction

Table-13.5 Advantages of fibre-optic cables.

Loose-tube cable	Tight-buffer cable
All the stress of cable pulling is taken by strength members.	Smaller, cheaper, and generally easier to use.
The fibre is free to expand and contract with temperature.	Kevlar makes the cable flexible.

Splices and Connectors

The splices are used to join two lengths of cables permanently, and connectors are used to couple the fibres to sources and detectors. For practical reasons, the length of a spool of cable is limited to about 10 km. Hence, longer spans between repeaters need splices to be made in the field. Generally, the loss in a properly made splice is less than that in a connector. Splices can have losses of 0.02 dB or less, whereas connector losses are about 0.2 dB. The ends of the fibres touch in a well-made splice, giving small losses. In a connector, there is a small air gap left to avoid damage of polished surfaces of fibres. Hence, the losses in connectors are more.

Optical Communication System

In optical communication, the transmitter can be either an LED or a laser diode and the receiver can be either a PIN diode or an avalanche photodiodes. PIN diode is the P-Intrinsic-N diode. This diode is used as an electronic switch and attenuator. When reverse biased, it represents a large resistance in parallel with small capacitance. In forward bias it behaves as a variable resistance. The resistance is zero when the diode is fully conducting. In general, short-range systems, such as communication system in offices, factories, or automobiles, LED emitters and multimode fibres are used.

Long-range systems, such as telephone trunk lines, use laser diodes and single-mode fibres. Early single-mode fibre systems used wavelength of 1.3 μm for lowest loss. However, currently the zero-dispersion single-mode fibre systems use a wavelength of 1.55 μm. Bit rates of 10 Gb/s are common in high-speed systems, and even higher rates are used in the newest undersea telephone cables.

The length of a fibre-optic link is limited due to the losses in fibre, and at connectors and splices. Because of these losses, the received power is less than the transmitted power. In long-distance fibre links, for a given optical power output of the transmitter, the optical power level at the receiver may not be of acceptable level. If the power level is greater than that required by the receiver, the system works. If the power level at the receiver is not per the requirement of the receiver, it is necessary to take the following steps:

1. Increase transmitter power.
2. Increase receiver sensitivity.
3. Reduce length of cable; hence, reducing losses.
4. Reduce losses by using better cable and/or connectors and splices with less loss.
5. Reduce power needed at the receiver. One possibility, although not a very practical one, is to reduce the data rate. The power needed at the receiver is roughly proportional to the data rate.

For designing a communication system it is needed to estimate the transmitter power in dBm. This is obtained by adding all the losses to the power in dBm required at the receiver. The dBm expresses the power in decibels, when power in milliwatts is transferred in decibels. That is, if power in milliwatts is P, the power in dBm is given by $10 \log P$. If there is excess of power at the receiver, it is called the *system margin*. A margin of 5 to 10 dBm is required to allow for the deterioration of components over time. It may require additional splices in future.

As discussed earlier, the data rate that can be used with a length of fibre is limited by dispersion. Dispersion in step-index multimode fibres is greatest, whereas it is least in single-mode fibres. Particularly, dispersion is greatest when used at a wavelength of about 1.55 μm. The effects of dispersion increase with the length of the fibre. This is because the difference in time reaching the end of the fibre by two signals with different velocity increases with the length of the fibre. The effect of dispersion is also proportional to the bandwidth of the information signal. In addition to fibre, the finite rise times of transmitters and receivers also limit the bandwidth. The effects of the rise times must, therefore, also be included while calculating the maximum data rate.

Repeaters and Optical Amplifiers

Because of loss and dispersion, there is a limit to the length of a single span of fibre-optic cable. Therefore, when distances are large, some form of gain must be provided. Following are the two methods to increase the gain:

1. Change the signal to electrical form, amplify it, regenerate it, if it is in digital form, and then convert it back to optical form.
2. Amplify the optical signal directly, without converting the optical signal into electrical form. This is a newer technique.

The first method uses regenerative repeaters, and the second method uses optical amplifiers. Both are briefly discussed below.

Regenerative Repeaters A most common repeater converts the optical signal to electrical energy, amplifies it, and then converts it back to optical form. Figure 13.44 shows the block diagram of a typical repeater.

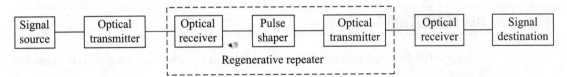

FIGURE 13.44 Optical communication system with regenerative repeater.

In digital communication, the receiver at the repeater converts optical signal into a digital signal. The signal may be distorted and noisy and can be decoded into ones and zeros. From the decoded signal, a new perfect pulse train can be reconstructed in pulse shaper. For closely located repeaters the noise accumulation may be minimum. Thus, the digital systems can avoid the accumulation of noise and distortion that affect the analogue system badly. Although most fibre-optic systems are digital, analogue signals are still used for some applications. One example that uses analogue fibre trunks is the television system. The use of analogue technique in television systems allows easier interconnection between the fibre and coaxial cable portions of the system.

Repeaters need both electrical power and maintenance. For these reasons, the trend in optical system design is to minimize the number of repeaters.

Optical Amplifier In case only the loss, and not the dispersion, is the limiting factor on the length of a fibre span, it is possible to amplify the optical signal directly. It does not need to convert the optical signal into electrical form and back. Furthermore, an optical amplifier can be used with any type of signal, analogue or digital, whether multiplexed or not.

The construction of optical amplifiers is based on principles similar to those that govern the operation of lasers. There are three possible uses of optical amplifiers. In Fig. 13.45(a), amplifier is placed after the transmitter and used as a power amplifier. The

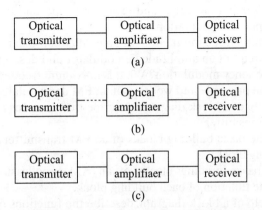

FIGURE 13.45 Uses of optical amplifiers: (a) as power amplifier, (b) as a preamplifier, and (c) as a repeater.

amplifier is installed just before the receiver in Fig. 13.45(b). It is used as a preamplifier. Figure 13.45(c) shows the amplifier in the middle of the transmitter and receiver.

13.10　REVIEW QUESTIONS

1. What are the classifications of the communication signals? Discuss different types of signals.
2. Show an analogue communication system in block diagram. Discuss in brief the different blocks of the communication system.
3. What are the different types of noises? Give their names and describe the ways in which they are generated.
4. What is meant by the signal-to-noise ratio (SNR)? Why is it important in communication systems?
5. Why is modulation needed in communication systems? What are the different types of modulation? What do you understand by carrier signal and modulating signal?
6. A voice signal of 400 Hz is transmitted on an AM radio station operating at 1020 kHz. Which of these frequencies is
   (a) The information frequency?
   (b) Carrier frequency?
   (c) The baseband frequency?
   (d) The modulating frequency?
7. Identify the band of each of the following frequencies:
   (a) 10 MHz (used for shortwave commercial broadcast)
   (b) 3.45 GHz (used for microwave oven)
   (c) 100 kHz (used for navigation systems)
   (d) 4 GHz (used for satellite television)
   (e) 900 MHz (used for mobile)
8. What is amplitude modulation? What does it mean by percentage of modulation? What is modulation index?
9. What do you mean by upper and lower envelope, and frequency spectrum in reference to amplitude modulation?
10. Discuss an amplitude modulator with a neat circuit diagram.
11. Draw the circuit of an amplitude demodulator and describe its working.
12. What is frequency modulation? What is maximum frequency deviation?
13. Describe the principle and working of an FM demodulator.
14. With the help of a block diagram, describe the functions of different building blocks of an AM transmitter.
15. What are the basic building blocks of an FM transmitter? Describe the function of each building block.
16. What are basic building blocks of an AM broadcast communication receiver? Describe the function of each building block.
17. With the help of a block diagram, describe the functions of different building blocks of an FM broadcast communication receiver.

18. Discuss sampling of analogue signals for digital communication. Describe a sample and zero hold circuit. What are natural sampled signal and flat-topped sampled signal?
19. What happens when a signal is sampled at less than the Nyquist rate? What is aliasing effect?
20. Discuss the three types of analogue pulse modulation.
21. What is quantization? What is quantization error? For a sinusoidal input, find the expression for SNR.
22. What is linear pulse code modulation (PCM)? What advantage does companded PCM have over linear PCM?
23. Discuss frequency division multiplexing.
24. Discuss time division multiplexing.
25. Discuss digital multiplexing.
26. List four possible advantages of optical fibre over wire cables, and explain each of them.
27. What are single-mode, multimode, step-index, and graded-index fibres?
28. Derive the expression for numerical angle. What is angle of acceptance?
29. Name different types of fibre-optic cables. What are their advantages and disadvantages?
30. Describe briefly the optical communication system.
31. What is system margin in connection with optical communication? Why is it necessary?
32. What advantage does a regenerative repeater have over an amplifier as a means of extending the length of a fibre-optics system? Does a fibre amplifier have any advantage over a regenerative repeater? If so, state them.

13.11 SOLVED PROBLEMS

1. An information signal having a bandwidth of 1 kHz is to be transmitted using (a) amplitude modulation, (b) frequency modulation with a maximum deviation of 1.5 kHz in the carrier frequency, and (c) pulse code modulation with a five-digit code. Determine the minimum carrier channel bandwidth in each case.

   Solution

   (a) For an AM, the minimum channel bandwidth is twice the bandwidth of the information signal. Therefore, the channel minimum bandwidth

   $$BW = 2 \times 1 = \mathbf{2\ kHz}$$

   (b) For frequency modulation, the channel minimum bandwidth is twice the bandwidth of the information signal plus maximum deviation and is given in Eq. (13.10). Therefore, the channel minimum bandwidth

   $$BW = 2(D + f_h)$$
   $$= 2(1.5 + 1) = \mathbf{5\ kHz}$$

(c) The channel bandwidth for PCM signal with a five-digit code is given by

$$BW = 2nf_h = 2 \times 5 \times 1 = \mathbf{10\ kHz}$$

2. The peak-to-peak amplitude of the modulating signal is 80 V and the peak-to-peak amplitude of the carrier is 120 V. Find the percentage of modulation.

Solution

The percentage modulation is given by the ratio

$$\frac{V_{max} - V_{min}}{V_{max} + V_{min}} \times 100$$

where V_{max} is the maximum peak-to-peak amplitude and V_{min} is the minimum peak-to-peak amplitude of the modulated signal. Hence,

$$V_{max} = 120 + 80 = 200\ V$$
$$V_{min} = 120 - 80 = 40\ V$$

Hence, the percentage modulation

$$= \frac{200 - 40}{200 + 40} \times 100$$

$$= \frac{160}{240} \times 100 = \mathbf{66.67\%}$$

3. An FM signal is represented by

$$v(t) = 10 \sin(5 \times 10^9 + 3.5 \sin 1500t)$$

Determine the carrier frequency, the information signal frequency, and the modulation index.

Solution

The carrier frequency,

$$f_c = \frac{5 \times 10^9}{2\pi} = \mathbf{795.8\ MHz}$$

The information signal frequency,

$$f_m = \frac{1500}{2\pi} = \mathbf{238.7\ Hz}$$

The modulation index

$$M_I = \mathbf{3.5}$$

4. An FM signal is represented by

$$v(t) = 15 \cos(2\pi \times 10^8 t + 150 \cos 2\pi \times 10^3 t)$$

Determine the bandwidth of the signal. Also, write the expression for the instantaneous frequency of the signal.

Solution

From the modulated signal, the modulation index
$$M_I = 150$$
and modulating frequency
$$f_m = \frac{2\pi}{2\pi} = 1 \text{ kHz}$$

Hence, the frequency deviation
$$D = 150 \times 1 = 150 \text{ kHz}$$
and the bandwidth
$$BW = 2(D + f_m)$$
$$= 2(150 + 1) = \mathbf{302 \text{ kHz}}$$

The expression for the instantaneous frequency is the derivative of the instantaneous phase. The instantaneous phase
$$\theta(t) = 2\pi \times 10^8 t + 150 \cos 2\pi \times 10^3 t$$
Hence, the instantaneous angular frequency
$$\omega(t) = 2\pi \times 10^8 - 150 \times 2\pi \times 10^3 \sin 2\pi \times 10^3 t$$
and the instantaneous frequency
$$f(t) = \frac{\omega(t)}{2\pi} = 10^8 - 150 \times 10^3 \sin 2\pi \times 10^3 t$$

5. A 75-MHz carrier is frequency modulated by such a sinusoidal signal that the maximum frequency deviation achieved is 40 kHz. Determine the modulation index and bandwidth of the modulated signal if the modulating frequency is (a) 1 kHz and (b) 100 kHz.

Solution

(a) The modulation index is given by
$$M_I = \frac{\text{Maximum frequency deviation}}{\text{Modulating signal frequency}} = \frac{D}{f_m}$$
$$= \frac{50}{1} = 50$$

The bandwidth
$$BW = 2(D + f_m) = 2(50 + 1) = \mathbf{102 \text{ kHz}}$$

(b) $M_I = \dfrac{50}{100} = 0.5$

$$BW = 2(50 + 100) = \mathbf{300 \text{ kHz}}$$

6. A signal given by
$$v(t) = 3.5 \cos 5\pi \times 10^3 t + 0.5 \cos \pi \times 10^4 t$$
is instantaneously sampled. The sampling interval is T_s. Find the maximum value of T_s.

Solution

The signal can be written as
$$v(t) = 3.5 \cos 2\pi(5/2) \times 10^3 t + 0.5 \cos 2\pi \times 5 \times 10^3 t$$
Hence, the two component frequencies of the signal are
$$f_{m1} = \frac{5}{2} = 2.5 \text{ kHz}$$
and $f_{m2} = 5$ kHz

The band limiting frequency, maximum of the two,
$$f_m = 5 \text{ kHz}$$
Hence, the minimum sampling rate (Nyquist rate),
$$f_{s\,min} = 2 f_m = 2 \times 5 = 10 \text{ kHz}$$
and $$T_s = \frac{1}{f_{s\,min}} = \frac{1}{10} \text{ ms}$$
$$= 0.1 \text{ ms}$$

7. A signal given by
$$v(t) = 2.5 \cos 150 \pi t + 05.5 \cos 180\pi t$$
is instantaneously sampled at sampling frequency of 200 Hz. If the sampled signal is passed through an ideal LPF with a cut-off frequency of 170 Hz, what frequency components appear at the output of the LPF?

Solution

The component frequencies of the signal are
$$f_{m1} = \frac{150}{2} = 75 \text{ Hz}$$
and $$f_{m2} = \frac{180}{2} = 90 \text{ Hz}$$

The sampling frequency
$$f_s = 200 \text{ Hz}$$
The different frequency components after the sampling are
$f_{m1} = 75$ Hz, $f_{m2} = 90$ Hz, $f_{m3} = f_s + f_{m1} = 200 + 75 = 275$ Hz,
$f_{m4} = f_s - f_{m1} = 200 - 75 = 125$ Hz, $f_{m5} = f_s + f_{m2} = 200 + 90 = 290$ Hz,
$f_{m3} = f_s - f_{m2} = 200 - 90 = 110$ Hz

Hence, the frequencies that appear at the output of the LPF are

75 Hz, 90 Hz, 110 Hz, and 125 Hz

8. Calculate the number of levels if the number of bits per sample (n) is (a) 12 bits and (b) 16 bits.

Solution

The number of levels is given by $N = 2^n$

(a) $N = 2^{12} = $ **4096**

(b) $N = 2^{16} = $ **65536**

9. Find the maximum dynamic range (or SNR) for a linear PCM system using 32-bit quantization.

Solution

From Eq. (13.17), the maximum dynamic range

$$DR = 6n + 1.8 \text{ dB}$$
$$= 6 \times 32 + 1.8$$
$$= \mathbf{193.8 \text{ dB}}$$

10. A composite video signal with a baseband frequency range from d.c. to 4 MHz is transmitted by linear PCM, using 8 bits per sample and a sampling rate of 10 MHz.

(a) How many quantization levels are there?

(b) Calculate the bit rate.

(c) Calculate SNR.

Solution

(a) The number of quantization levels

$$N = 2^n = 2^8$$
$$= \mathbf{256}$$

(b) The bit rate or data rate

$$D = f_s n = 10 \times 8$$
$$= \mathbf{80 \text{ MHz}}$$

(c) From Eq. (13.17), the SNR

$$SNR = 6n + 1.8$$
$$= 6 \times 8 + 1.8$$
$$= \mathbf{49.8 \text{ dB}}$$

11. A fibre has an index of refraction for core and cladding as 1.8 and 1.5, respectively. Calculate (a) the critical angle, (b) angle of refraction (θ_r) for angle of incidence (θ_i) equal to 45°, and (c) θ_r for $\theta_i = 65°$.

Solution

(a) From Eq. (13.22),

$$\theta_c = \sin^{-1}\left(\frac{n_2}{n_1}\right)$$

$$= \sin^{-1}\left(\frac{1.5}{1.8}\right) = 56.44°$$

(b) Since 45° is less than the critical value, Snell's law applies. Therefore, from Eq. (13.21).

$$\sin\theta_r = \frac{n_1}{n_2}\sin\theta_1$$

$$= \frac{1.8}{1.5}\sin 45° = 0.849$$

$$\theta_r = \sin^{-1} 0.849 = 58°$$

(c) Since 65° is greater than the critical value, Snell's law does not apply. The law of reflection applies in this case. Hence,

$$\theta_r = \theta_i = 65°$$

12. Calculate the NA and the maximum angle of acceptance for the fibre of question 11.

Solution

From Eq. (13.23), the NA

$$NA = \sqrt{n_1^2 - n_2^2}$$

$$= \sqrt{1.8^2 - 1.5^2}$$

$$= \mathbf{0.995}$$

The maximum angle of acceptance is given by

$$\theta_{a.c.} = 2\sin^{-1}(NA)$$

$$= 2\sin^{-1}(0.995)$$

$$= \mathbf{168.5}$$

13.12 EXERCISES

1. An information signal having bandwidth of 750 Hz is to be transmitted using:
   (a) amplitude modulation
   (b) frequency modulation with a maximum deviation of 1 kHz in the carrier frequency, and
   (c) pulse code modulation with a four-digit code.
   Determine the minimum carrier channel bandwidth in each case.

2. The peak-to-peak amplitude of the modulating signal is 50 V and the peak-to-peak amplitude of the carrier is 110 V. Find the percentage of modulation.

3. An FM signal is represented by
$$V = 10 \sin (2 \times 10^9 + 0.5 \sin 1 \times 10^3 t)$$
Determine the carrier frequency, the information signal frequency and the modulation index.

4. An FM signal is represented by
$$v(t) = 5 \cos (2\pi \times 10^8 t + 120 \cos 2\pi \times 10^3 t)$$
Determine the bandwidth of the signal. Also, write the expression for the instantaneous frequency of the signal.

5. A 75-MHz carrier is frequency-modulated by a sinusoidal signal. The maximum frequency deviation achieved is 45 kHz. Determine the modulation index and bandwidth of the modulated signal if the modulating frequency is (a) 2.5 kHz and (b) 50 kHz.

6. A signal given by
$$v(t) = 5 \cos 3\pi \times 10^3 t + \cos \pi \times 10^4 t$$
is instantaneously sampled. The sampling interval is T_s. Find the maximum value of T_s.

7. A signal given by
$$v(t) = 2.5 \cos 120\pi t + 05.5 \cos 280\pi t$$
is instantaneously sampled at sampling frequency of 250 Hz. If the sampled signal is passed through an ideal LPF with a cut-off frequency of 175 Hz, what frequency components appear at the output of the LPF?

8. Calculate the number of levels if the number of bits per sample is (a) 8 bits and (b) 32 bits.

9. Find the SNR for a linear PCM system using a 12-bit quantization.

10. A composite video signal with a baseband frequency range from d.c. to 4 MHz is transmitted by linear PCM, using 16 bits per sample and a sampling rate of 5 MHz. Calculate (a) the number of quantization levels, (b) bit rate, and (c) SNR.

11. A fibre has an index of refraction for core and cladding as 1.5 and 1.2, respectively. Calculate (a) the critical angle, (b) angle of refraction (θ_r) for angle of incidence (θ_i) equal to 45°, and (c) θ_r for $\theta_i = 65°$. Also, determine the NA and angle of acceptance.

ANSWERS TO EXERCISES

Chapter 1

1. 2.57 Ω
2. (a) 2.074 mA, (b) 193 Ω, (c) 41.46 Ω
3. 7.6 Ω
4. 3.2
5. (a) 62.637 µA, (b) (i) 358 Ω, (ii) 773.5 Ω
6. 0.1 µA, 34.7 mV, −14.9053 V
7. $v_o = 0.5\,v_i$

Chapter 2

1. (a) −19.3 mA, −1.175 V,
   (b) 0.2 V, 4.825 V, (c) −7.5 mA
2. (a) −15.5 mA, −2.25 V, −2.25 V, 1.21 V,
   (b) 9 V
3. 4.5 kΩ
4. (a) 9.7 V, 4.1 mA
5. (a) 0.15 kΩ, 18.6 V, (b) 0.159 Ω, 37.5 V
6. (a) 18 V, (b) 25.65 V, (c) 1.4 V.

Chapter 3

1. $i(t) = 9.66 \sin\theta - 0.58$ mA
9. (a) 12.47 ns, (b) 0.5 ns
11. (a) 0.1614 A, (b) 0.05135, (c) 0.0807 A,
    (d) −102.71 V, 325.22 V, (e) 5.276 W,
    (f) 0.75%
12. (a) 121.2 mA, (b) 60.6 mA,
    (c) 276.15 V, (d) 3.672 W, (e) 5%
13. 22.47 kΩ

Chapter 4

1. −38.46, 1088.46 Ω, −5.97, −424.01, −358.19,
   24.04 µA/V, 41.6 kΩ
2. (a) −40, (b) (i) 1 kΩ, (ii) 0.976 kΩ,
   (c) 1229.5, (d) (i) 36.73 kΩ, (ii) 50 kΩ
3. (a) 2.838 mA, 7.787 V, (b) 3.54 mA, 6.06 V
4. (a) 1.765 kΩ, 83 kΩ, (b) 4.5 mA, 6.927 V
5. 83.59 kΩ
6. (a) 0.417 kΩ, 0.409 kΩ, 38.77 kΩ,
   (b) 13.47 mA, 3.78 V
7. (a) 1.3 mA, 1.54 V, (b) − 8.68 V
8. 1.28 kΩ, 0.164 kΩ
9. 2.5 kΩ, 7.5 kΩ
10. −49.5, −39.15, −96.77, 3788.5, 0.87 kΩ,
    2.38 kΩ
11. −60.0, −20.0, −60.0, 1200.0, 2.95 µF
12. 17.84 Hz, 19.42 kHz, from 17.84 Hz to
    29.42 kHz; 41 Hz, 12.8 kHz, from 41 Hz to
    12.8 kHz
13. −32.35, −23.8, 769.79
14. (a) −10.17, −7.7, (b) −0.982, −0.957,
    (c) 39.24 kΩ

15. −0.672 mA/V, −1.01, 44.48 kΩ, 5 kΩ, 1.154 kΩ

16. (a) −82.66 V/mA, −82.66, 1 kΩ, 16.67 kΩ, (b) −2.98 V/mA, −2.98, 1 kΩ, 8.85 kΩ

Chapter 5

1. 18750, 85.46 dB
2. (a) 0.025%, (b) 0.0025%
3. 246.86, 47.71 dB
4. (a) −15 V, 0.15 mV, 1 mA, −1 mA, (b) 1 mW, 1.5 mW, −1 mW, −10 mW
5. $A/[R_2 + (1 + A)R_1]$
7. 13.75 V
8. (a) 120 Ω, (b) 300 kΩ
9. $RR_L/(R + R_L)$, $A(R + R_L)/RR_L$
10. 40 Ω, 30 Ω
11. 70.72%
12. 19.3 mA, 386 mA, 10.35 V, 24.09%
13. 40 mA, 10.87 V, 26.67%
14. (a) 2 W, (b) 4 V, (c) 12 V, (d) 0.5 A, 166.67 mA, (e) 10%
15. 26.5 W, 16.71 W, 63.06%

Chapter 6

1. (a) 650 kHz, 290 kΩ, (b) 6.9 kΩ, 188.3 kHz
2. (a) 2.9 kΩ, 65 kΩ, (b) 5.8 mΩ^{-1}, 0.650 μF
3. (a) 895 Ω, 310 Ω, (b) 0.0376 μF, 715 Ω
4. (a) 3.18 kΩ, 10 kΩ, (b) 0.00159 μF, 2.5 kΩ
5. $(2\pi RC)^{-1}$, $R_2 = 2R_1$
6. (a) 433 kHz, 1.25 kΩ^{-1}, (b) 0.00166 μF, 0.25 kΩ
7. (a) 712 kHz, 0.267 mΩ^{-1}, (b) 187 μF, 0.16 kΩ

8. (a) 2.35 MHz, 2.362 MHz, (b) 7071000
9. (a) 207 Hz, (b) 737 kΩ, (c) 1.11 kΩ
10. (a) 7.35 mA, (b) 5.29 + 312.5t, (c) 0.335 − 312.5t, (d) 1.07 ms
11. 8.57 V
12. (a) 3 kHz, (b) 6.5 kHz

Chapter 7

1. 0.106, −19.52 dB, 9.46, 19.52 dB, 11.8 μF
2. 51.7 Ω, 10.065 dB
3. 408.1 Ω, 101.0 Ω
4. Three 20 dB and one 10 dB attenuators. For 20 dB: $R_1 = 450.0$ Ω, $R_2 = 111.1$ Ω and for 10 dB: $R_1 = 2855.7$ Ω, $R_2 = 386.5$ Ω
5. (a) 338 Ω, 11.92, (b) 247.5 Ω, 61.1 Ω
6. 250 Ω, 27.78 Ω, 2.25 kΩ
7. (a) 1, 35.37 kHz, (b) 0.833, 42.45 kHz, (c) 0.769, 30.67 kHz
8. 0.74, 7.873 kHz
9. 0.99, 1.06 Hz, 63.66 kHz
10. 0.177 μF, 0.707 μF
11. −1.67, 637 Hz
12. −1.67, 942 Hz
13. Assuming $C = 0.1$ μF, $L = 112.6$ mH, $R = 10.61$ Ω
14. Taking $C_1 = C_2 = 0.1$ μF, $R_1 = 2122$ Ω, $R_3 = 212$ Ω, $R' = 5.31$ Ω, $R_2 = 5.32$ Ω
15. $\omega_o = \dfrac{1}{C}\sqrt{\dfrac{1+R_1/R_2}{R_1 R_3}}$,

    $\text{BW} = \dfrac{2}{2\pi R_3 C}$ and

    $Q = \dfrac{1}{2}\sqrt{\dfrac{R_3}{R_1}\left(1+\dfrac{R_1}{R_2}\right)}$

Chapter 8

1. 3.0 cm
2. 2.965×10^7 m/s, 0.3 mm/V
3. 8.75 cm, 250 V, 0.35 mm/V, 2.86 V/mm
4. 0.994 mWb/m^2
5. (i) 48.6°, (ii) 131.4°

Chapter 9

1. 2.11 A, 7.71 A, 126.6 W, 653.89 W, 780.49 W, 0.92
2. (a) 95.5 V, 6.367 A, 160.9 V, 10.73 A, (b) 608.05 W, 1726.46 W, 35.2%, 1.685, 1.356, 400 V, 3035 VA, 0.2
3. 97.53 V, 200 V, 23.88 A, 7.78 ms
4. 10.634 A, 152.54 V, 8.66 ms
5. 196.77 V, 61.5°, 622.25 V
6. (a) 41.95°, 0.67, (b) 134.84°, 0.635
7. (a) 1.08, (b) 2665.56 V, (c) 314.14 V, (d) 12 A, 8.49 A
8. (a) 42017 W, (b) 33.33 A, 57.74 A, 100 A, (c) 359.26 V
9. (a) 60°, (b) 24.76 A, (c) 28.081 A, 336.97 V, (d) 8.253 A, 16.21 A
10. (a) 461.08 V, (b) 46.108 A, (c) 475.56 V, 47.556 A, (d) 15.37 A, 27.456 A
11. 353 V, –133 V
12. 801.28 Hz, from 233 Hz to 961.5 Hz
13. (a) 47.44 µs, (b) 5.8 kHz
14. For half bridge inverter: (a) 22.5 V, (b) 208.33 W, (c) 8.33 A, (d) 4.167 A, (e) 50 V; for full bridge inverter: (a) 45.0 V, (b) 833.33 W, (c) 16.67 A, (d) 8.33 A, (e) 50 V
15. (a) 109.5 V, (b) 1000 V, (c) 1000 W, (d) 0.548, (e) 2.25 A, 6.45 A
16. (a) –6.03 V, (b) 198.55 V, (c) 2628.22 W, (d) 2647 VA, (e) 0.993, (f) –0.402 A
17. (a) 180 V, (b) 215 V, (c) 4202.5 W, (d) 4202.5 W, (e) 0.98, (f) 8.2 A, 13.82 A
18. (a) 193.2 V, (b) 226.66 V, (c) 3425 W, (d) 3425 W, (e) 0.985
19. 240°, 220 V
20. (a) 149.2 A, 660 V, (b) 34.8 A, 660 V
21. 72 V, 0.402 ms, 0.268 ms
22. 1.5 ms, 0.5 ms
23. (a) 0.683, (b) 0.626, 24.214

Chapter 10

1. (a) 13, (b) 23, (c) 39, (d) 22, (e) 59
2. (a) 1101.0110, (b) 11110011.0101110, (c) 100110, (d) 1111000.0110001, (e) 1111.1010
3. (a) 1010111, 1001100, (b) 10100011, 163
4. (a) 1011111, 1111101, (b) 0011110, 30
5. 0110110, 54
6. 0110110, 54
7. 11100.000
8. (a) 1219, (b) 1219
9. 2484, 2484
10. (a) 127, (b) 127
11. (a) $\overline{A}_1 + A_2 \overline{A}_3$
14. $\overline{A}_1 \overline{A}_2 + \overline{A}_1 \overline{A}_3 + A_1 A_2 A_3$
15. $\overline{A}_1 \overline{A}_3 \overline{A}_4 + \overline{A}_1 A_3 A_4 + A_2 A_3 \overline{A}_4 + \overline{A}_1 A_2 A_3 + \overline{A}_1 A_2 \overline{A}_3 A_4$

Chapter 11

1. $Y_1 = S_1 + S_3 + S_5 + S_7$, $Y_2 = S_2 + S_3 + S_6 + S_7$, $Y_3 = S_4 + S_5 + S_6 + S_7$
2. $Y_0 = S_1 + S_2$, $Y_1 = S_1 + S_3$, $Y_2 = S_0 + S_2 + S_3$
5. Difference $D = A\overline{B}$, borrow $B' = \overline{A}B$
9. Control lines = 4
11. Inputs = 4, output = 1, address variables = 2
14. Decoder size is 2 × 4, AND gates = 4, outputs = 4, OR gates = 4
19. D flip flop

Chapter 12

1. (a) 4, (b) 16, (c) 16
2. (i) XOR, (ii) OR
8. Asr R: 11010110, Asl R: 01011000 (overflow)

Chapter 13

1. (a) 1.5 kHz, (b) 1.75 kHz, (c) 6 kHz
2. 45.45%
3. Carrier: 318.3 MHz, information signal: 159.15 Hz, modulation index: 0.5
4. BW = 242 kHz, $f(t) = 10^8 - 120 \times 10^3 \sin 2\pi \times 10^3 t$
5. (a) 18, 95 kHz, (b) 0.9, 190 kHz
6. 333.33 μs
7. 60 Hz, 140 Hz, and 110 Hz
8. (a) 256, (b) 4294967296
9. 73.8 dB
10. (a) 65536, (b) 80 MHz, (c) 97.8 dB
11. (a) 53.13°, (b) 62.1°, (c) 65°, 0.9, 128.3

INDEX

2's Complement 507
10's Complement 507
π attenuator 303

accumulator 634, 663
active filters 319
active RC bandpass filter 327
active resonant bandpass filter 325
a.c. voltage controllers 433
ADC 707
A/D converter 590, 707
A/D converter using counter 591
adders 542
address bus 665
address decoder 720
address register 634
aliasing 709
ALU 629
amplification 122
amplifiers 120
amplitude modulation 697
AM receiver 705
analogue and digital signals 690
analogue communication 690
analogue multiplexing 721
analogue storage CRO 374
analogue technique 374
analogue-to-digital (A/D) conversion 540
analysis of chopper 451
AND gate 495
angle of acceptance 725
angle of incidence 723
angle of refraction 723
applications of cathode-ray oscilloscopes 364
arithmetic, logic, and shift instructions 650
arithmetic logic unit 629
arithmetic operations 630
arithmetic shift 632
ASCII code 656
assembler 653
assembly language and assembler 654
assembly language program 653
associative memory 642

astable blocking oscillator 269
astable multivibrator 261
asynchronous counter 572
attenuator 279, 296
audio signal generators 280
auxiliary commutation 455
auxiliary memory 640
auxiliary secondary memory 637
avalanche breakdown 19
avalanche or breakdown diodes 18

bandpass filter 310, 324
bandpass LC filter 318
bandpass RC filter 313
bandstop filter 310, 324
bandstop RC filter 314
bandwidth 701
bandwidth of passband 327
base 34
base clipping circuit 80
biasing means 125
biasing of transistor amplifier 125
binary addition 540
binary addition and subtraction 505
binary code 653, 712
binary logic 493
binary number system 501
binary to hexadecimal conversion 511
bipolar junction transistor 34
blanking circuit 363
blocking oscillators 265
Boolean algebra 513
Boolean functions 514
boost type of switching regulator 448
bridge converter 387
bridged T attenuators 305
buck-boost type of switching regulator 449
buck type of switching regulator 447
bus structure 665
Butterworth filter 320

cache memory 639
capacitance 15

746 Index

carrier 697
cascading T attenuators 302
cascading π attenuators 304
cathode-ray tube 344, 345
CD-RAM 641
CD-ROM 641
central processing unit 628
character printer 644
choppers 450
circular shift 632
cladding 722
class A amplifiers 133
class AB amplifiers 133
class A chopper 452
class A power amplifier 222
class B amplifiers 133
class B amplifier without output transformer 231
class B chopper 453
class B power amplifier 227
class B push-pull amplifier 228
class C amplifier 133
class C chopper 453
class D chopper 454
class E chopper 454
classification of amplifiers 132
classification of choppers 452
clipper circuits 79
clocked SR flip flop 563
codec 716
coding 716
collector 34
Colpitts oscillator 255
combinational logic circuits 540
common-base configuration 37
common-collector configuration 37
common-emitter configuration 37
common-mode rejection ratio (CMRR) 203
common-mode signal 202
communication system 690
commutating diode 395
commutation methods for choppers 455
comparator 81
compensated attenuator 297
compiler 653
complementary MOSFET 53
computer instructions 648
computer programming 653
computer registers 633
conduction 11
conjunctive normal form of function 519
connectors 729
constant-current sweep generator 360
control bus 665
controlled rectifiers 387
control unit 636
conversion efficiency 223
conversion of decimal fraction to binary fraction 508
counters 572

counting 540
CPU 628
CRO 344
CRO block diagram 362
CRT terminal 646
crystal oscillator 258
current 11
current amplifiers 132
current gain 120
current series feedback 150
current shunt feedback 148
current source 208
current source inverters 431
current-to-voltage converter 219

D/A and A/D converters 540
D/A converter 584
data bus 665
data register 634
data selection (multiplexing) 540
data selectors/multiplexers 542, 556
D.C. restorer 95
D.C. voltage regulators 444
decade or decimal counter 574
decimal to hexadecimal conversion 512
decoders 552
decoders and demultiplexers 542
decoding 540
deflecting forces 349
deflection sensitivity 350
deflection systems 344
delay line 344, 363
delta modulation 718
demodulation 692
demultiplexers 554
demultiplexing 540
depletion 15
depletion capacitance 15
depletion MOSFET 51
determination of B-H loops for magnetic materials 364
deterministic and random (stochastic) signals 690
D flip flop 564
differential amplifier 202
differential coding 718
differential-mode signal 202
differential output 203
diffusion capacitance 18
diffusion current 5
digital code 712
digital communication 690
digital computer 626
digital multiplexing 721
digital storage CRO 376
digital technique 374
digital-to-analogue (D/A) conversion 540
diode 13, 14
diode clamper 94
diode-compensation circuit 129

Index

disjunctive normal form of function 519
distortion factor 221
distortions 133
dot matrix impact printer 645
dot matrix printer 644
dot matrix thermal character printer 645
double-ended clipping circuit 79
drift 11
drift current 5
dual-beam CRO 372
dual-trace CRO 371
dynamic characteristic 75

edge triggered flip flop 566
effect of feedback on bandwidth 145
effect of feedback on frequency distortion 145
effect of feedback on noise 146
effect of feedback on sensitivity 145
effect of feedback on stability 145
efficiency of rectification 88, 91
electromagnetic deflection 351
electromagnetic focusing 348
electron gun 345
electrostatic deflection 349
electrostatic focusing 347
electrostatic printer 644, 645
emitter 34
emitter follower 153
encoders 542
enhancement MOSFET 51
EPROM 560
EXCLUSIVE OR gate 498
execution cycle 666

feedback amplifier 141
feedback gain 212
FET biasing 130
fetch cycle 665
field-effect transistor (FET) 45
filters 310
first pass 657
flat-topped sampling 708
flip flops 561
flip flop with preset and clear 567
fluorescent screen 346
flyback SMPS 458
FM receiver 706
focusing devices 347
foldover distortion 709
forms of Boolean functions 519
forward-bias voltage 10
frequency demodulation 702
frequency division multiplexing 719
frequency modulation 697
frequency response of an amplifier 134
frequency spectrum 698
full adder 545
full-bridge SMPS 458

full-wave phase-control A.C. voltage controller 437
full-wave rectifier 89
function generators 280

gate triggering 57
general-purpose registers 663

half adder 544
half-bridge SMPS 458
half-wave phase-control a.c. voltage controller 436
half-wave rectifier 84
Hartley oscillator 255
hexadecimal number system 508
hexadecimal to binary conversion 511
hexadecimal to decimal conversion 511
high-frequency gain 135
high-level programming languages 653
high-pass filter 310, 322
high-pass LC filter 317
high-pass RC filter 313
high-temperature triggering 57
horizontal amplifier 358
horizontal deflection system 344, 355
hybrid model 124

impedance matching 224
information signal 697
INHIBIT (ENABLE) gate 499
inkjet character printer 646
inkjet printer 644
input and output instruction 650
input attenuator 353
input devices 643
input impedance 121
input-output devices 642
input-output interface 647, 671
input-output processor 648
input register 634
input selector 353
input static characteristics 38
instantaneous sampling 708
instruction cycle 652, 665
instruction register 664
instruction set of Intel 8085 667
instructions for transferring information 650
integral cycle-control a.c. voltage controllers 433
Intel 8085 microprocessor 662
Intel 8748 675
Intel microcontroller 674
inverted R-2R ladder D/A converter 589
inverters 417
inverting operational amplifier 212

JK flip flop 564
joystick 643
junction field-effect transistor 45

keyboard 643

748 Index

ladder-type D/A converter 586
large-scale general-purpose computers 626
laser printer 644
LC filters 316
letter quality impact character printer 645
letter quality printer 644
light-emitting diode (LED) 22, 722
light pen 643
light triggering 57
linear applications of operational amplifiers 214
linear-circuit model for the transistor 43
linear selection 583
line printers 644, 646
Lissajous figures 364
load commutation 456
logical shift 632
logic gates 493
logic operations 631
lower envelope 698
low-frequency response 134
low-frequency response of an RC-coupled 140
low-pass filter 310, 320
low-pass LC filter 316
low-pass RC filter 312

machine language 653
magnetic disks 640
magnetic tape 641
main memory 638
main primary memory 637
master-slave JK flip flop 568
measurement of dielectric loss 364
measurements of frequency and phase 364
measurements of sinusoidal voltages and currents 364
memories 540
mesh storage CRT 374
metal oxide semiconductor field-effect transistor (MOSFET) 51
microcomputer 671
microcontroller 673
microdigital computers 626
microprocessor 661
midpoint converter 387
mini digital computers 626
modified series inverters 421
modulation 692
modulation index 698
monostable blocking oscillator 266
monostable multivibrator 264
MOSFET biasing 131
MOSFET gate protection 54
mouse 643
multiple accessing 719
multiplexing 719
multiplying D/A converter 590
multistage amplifiers 132
multivibrators 260
NAND gate 498

n-channel MOSFET 51
noise 696
noise signal 692
noninverting operational amplifier 213
NOR gate 497
NOT gate 496
n-type semiconductor 5
number systems 501
numerical aperture 725

object program 656
offline UPS 463
online UPS 464
OP AMPs 209
operating point 74
operational amplifiers 209
optical amplifiers 730
optical communication 690, 729
optical fibre 722
optical scanner 644
OR gate 493
oscillators 250
output admittance 123
output devices 644
output register 634
output static characteristics 38

padding sources and loads 308
page printers 644, 646
parallel adder 548
parallel register 569
passive filters 310
p-channel MOSFET 51
PCM 716
peak clippers 80
peak-clipping circuit 80
peak inverse diode voltage 86
peak rectifier 93
periodic and nonperiodic signals 690
phase modulation 697
phase shifter 215
phase-shift oscillator 251
phosphor storage CRT 375
photo detector 722
photodiode 21
piecewise linear characteristic of a diode 76
pin configuration 667
p-n junction 7
postulates of Boolean algebra 516
power amplifiers 132, 220
power efficiency 223, 225
power supply 344, 363
printer 644
program control instructions 650
program counter 634, 664
PROM 560
p-type semiconductor 6
pulse amplitude modulation (PAM) 711

Index

pulse and pulse wave generators 273
pulse duration modulation (PDM) 711
pulse position modulation (PPM) 711
pulse width modulation (PWM) 711
push-pull amplifiers 225
push-pull SMPS 458

quantization 712
quiescent point 74

RAM unit 581
RC-coupled amplifier 139
RC filters 312
read only memory (ROM) units 542
receiver 692
rectifier 84
registers 540, 568
regulation 86
repeaters 730
resonant-circuit oscillator 253
reverse-bias voltage 10
ring counter 579
ripple counter 572
ripple factor 89
ROM 560

sample-and-hold circuit 708
sampling 708
sampling CRO 376
sampling gate 81
scale changer 215
Schmitt trigger 271
second pass 659
semiconductors 2
sequential logic circuits 540
serial adder 549
series voltage regulator 446
shift instructions 650
shift operations 632
shift register 570
shunt voltage regulator 444
signal generators 277
signal-to-noise ratio 696
sign changer 214
signed binary number 506
silicon-controlled rectifier (SCR) 54
simplification by K-map method 521
simplification of logical functions 518
single-ended output 203
single-phase current source inverter 431
single-phase full-bridge inverter 427
single-phase full-controlled bridge converter 404
single-phase full-wave converters 400
single-phase half-bridge inverter 425
single-phase half-wave converters 387
single-phase parallel inverter 424
single-phase phase-control a.c. voltage controller 436
single-phase semiconverter 408

single-phase series inverters 418
slicer circuit 80
SMPS 458
source follower 156
source program 656
special purpose CROs 371
special-purpose registers 663
spectrum 698
splices 729
square wave generators 270
stack pointer 664
static characteristics of transistors 37
static resistance 14
status condition check instructions 650
status register 664
step up chopper 456
storage CRO 373
subtractors 542, 550
successive approximation type A/D converter 591
summer 215
sweep-frequency generator 281
switched mode power supply 457
switching voltage regulator 447
switchtail (Johnson) counter 580
symbolic code 653
symmetrical attenuators 300
symmetrical T attenuator 301
symmetrical π attenuators 303
synchronization of sweep 358
synchronous counter 576
synchronous decade counter 578
synchronous up-down counter 578

T attenuators 301
temporary register 634
T flip flop 565
thermal printer 644
three-phase a.c. voltage controller 442
three-phase bridge inverter 430
three-phase current source inverter 432
three-phase full-controlled bridge converters 411
three-phase semiconverter 413
three-phase series inverter 423
thyristor 56
time-base generator 344, 355
time division multiplexing 720
total reflection 723
trackball 644
transconductance amplifiers 132
transfer characteristic 48, 75
transfer gain 143
transistor amplifier 120
transistor clipping circuit 271
transmission channel 692
transmitter 692
transresistance amplifiers 132
triac 59
triac (bidirectional SCR) 59

triggered sweep   359
truth table   514
two-pass assembler   657
types of feedbacks   146

UART   646
unijunction transistor (UJT)   60
uninterrupted power supply   462
universal gates   500
up-down counter   575
upper envelope   698
UPS   462
USART   646

variable attenuators   307
vertical amplifier   354
vertical deflection system   344, 352

V-I characteristic   13, 21
virtual memory   642
voltage   122
voltage amplifiers   132
voltage doubler   96
voltage-rate triggering   57
voltage series feedback   146
voltage shunt feedback   151
voltage-to-current converter   217
voltage triggering   57

Wien-bridge oscillator   257
working principle   450

Zener breakdown   19
Zener diodes   18